INVERTEBRATE STRUCTURE AND FUNCTION

INVERTEBRATE

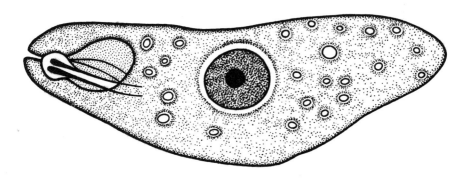

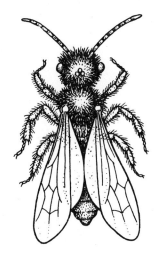

E. J. W. **BARRINGTON** FRS
formerly
Professor of Zoology, Nottingham University

STRUCTURE AND FUNCTION

SECOND EDITION

NELSON

Thomas Nelson and Sons Ltd
Lincoln Way Windmill Road
Sunbury-on-Thames
Middlesex TW16 7HP
P.O. Box 73146 Nairobi Kenya
P.O. Box 943 95 Church Street Kingston 5 Jamaica

Thomas Nelson (Nigeria) Ltd
8 Ilupeju Bypass PMB 1303 Ikeja Lagos

First published in Great Britain 1967
Reprinted 1968, 1969, 1970, 1972, 1976, 1979
Second edition first published in Great Britain 1979
ELBS edition first published 1970
Reprinted 1971 and 1974
ELBS edition of second edition 1979

Copyright © E. J. W. Barrington 1967, 1979

All Rights Reserved. No part of this
publication may be reproduced, stored in
a retrieval system, or transmitted, in
any form or by any means, electronic,
mechanical, photocopying, recording or
otherwise, without the prior permission
of the publishers

ISBN 0 17 771051 9

*Photoset and printed in Malta
by Interprint (Malta) Ltd*

Preface

The line traditionally drawn in zoological teaching between the vertebrates and the invertebrates is an unfortunate one: it obscures the fundamental unity that underlies the organization of living material. Of course it is possible to use the vertebrates for studying the working-out of biological principles within a compact system of closely interrelated groups. To this extent these animals are a convenient demonstration of unity in diversity; in consequence, they appeal to some students as an 'easy' group.

But the vertebrates are part of the Phylum Chordata, which contains truly invertebrate members. These in their turn are closely related to other invertebrate groups. Vertebrate studies by themselves, therefore, tell us little, if anything, of the origin of vertebrates, or of the origin of the principles of biological organization that have determined the course of their adaptive evolution. Indeed, the appeal that the vertebrates make to our anthropocentric tendencies can be dangerously deceptive. It can easily lead to over-optimistic generalization from limited data, obtained from some laboratory mammal that has nothing to recommend it for the purpose other than its convenience and its compliant behaviour.

If, therefore, we are to evaluate and exploit the dramatic advances of contemporary biology (those that are being made, for example, at the growing points of molecular and ultrastructural biology) we need as one essential condition the widest possible extension of our understanding of the principles of animal organization. This must come in large measure from invertebrate studies, and it is the purpose of this book to provide some help to students who wish to widen their viewpoint in this way. It is only too easy for them to become discouraged by the readiness with which the rich diversity of the animal kingdom can fragment into what Aldous Huxley has called, in another context, 'a quantity of mutually irrelevant happenings dotted, like so many unexplored and fantastic islands, on the face of a vast ocean of incomprehension'. I hope that this book may provide them with some protection against this occupational hazard of zoologists, and that it may make the invertebrates a little less 'difficult'.

My underlying theme is a self-evident one: that the business of animals is to stay alive until they have reproduced themselves, and that the business of zoologists is to try to understand how they do it. This understanding demands the recognition that structure and function are two indissociable aspects of animal organization, linked in patterns that have been determined by the course of events in the remote past. Those events, which can to some extent be reconstructed by deduction, and which are always worth speculating about, are reflected to the best of our ability in our schemes of animal classification.

These considerations have moulded the form of this book. In selecting its subject matter I have assumed that readers will already be using textbooks of descriptive and experimental zoology, of physiology, and of biochemistry. I have assumed also that they will be studying invertebrate structure and function at first hand in the laboratory and in the field. Detail has therefore been restricted to what seemed needed to set the scene, as it were, for a particular line of thought. What I have tried to do, in short, is to open up discussion of some of the problems and questions that force themselves on the attention of anyone who wishes to know why an animal is built in a particular way, and how it manages to survive in its characteristic habitat and community. As Wordsworth put it to the Leech Gatherer: 'How is it that you live, and what is it you do?'

It may seem illogical that a book taking unity as its theme does not include the vertebrates with the invertebrates. They are not, in fact, left without mention. But to have dealt with them in any detail would have made the book unwieldy, and I should in any case have deemed it presumptuous to suppose myself capable of writing such an account. For the same reasons the book is not a complete survey of invertebrate biology. But I hope that I have managed to select subjects that will at least serve as a guide and a stimulus to the achievement of some of that synthesis of knowledge that ought to flow from a biological education. I hope, in fact, that the book may serve as an accompaniment to the more general and reflective parts of a university zoology course; as a companion for the seminars, the tutorials, and the essay writing. Perhaps, too, it will illuminate the profound truth of Edith Sitwell's belief: that it is odious to regard oneself as 'superior' to any living being, human or animal.

I am grateful to Sir Gavin de Beer, F.R.S., who invited me to write this book; to Professor E. W. Knight-Jones, Professor O. E. Lowenstein, F.R.S., Dr Sidnie M. Manton, F.R.S., and Professor J. D. Robertson, who gave advice and encouragement during its preparation; and to my publishers, who have brought to its production more skill and patience than any author could resonably have expected.

I am indebted also to Blackwell Scientific Publications Ltd for permission to quote two passages from *Larval Forms* by Walter Garstang; to Mrs George Bambridge, Macmillan & Co. Ltd, and Doubleday & Co. Inc for permission to quote from *Kim* by Rudyard Kipling; to the Clarendon Press for permission to quote from the *Oxford Translation of Aristotle*, 'Historia Animalium'; and to the American Association for the Advancement of Science for permission to quote from 'Are there any "Acellular Animals"?' by A. Boyden in *Science*, **125** (3239), 155–156 (1957), and from 'Concerning the "Cellularity" or Acellularity of the Protozoa' by S. H. Hutner and L. Provasoli in *Science*, **125** (3255), 989 (1957).

All illustrations in this book which have been taken from any other sources, with or without amendment, are used by permission of the publishers and, as far as possible, of the authors, to all of whom my grateful thanks are due. Full details of the sources are given underneath each illustration.

E.J.W.B.

Preface
to second edition

The first edition of this book was planned as a selective treatment of aspects of invertebrate biology, designed to be a companion for students during the more general parts of their zoology courses. But the plan was frustrated from the beginning by the book itself, which took charge of me, as books tend to do of their author, and grew to a size that I had not foreseen. Perhaps because of this, it immediately took its place as a primary text, and one that has given much enjoyment—or so I am led to believe. I was well aware, however, that it retained some unevenness in treatment which was inherent in the selective basis of the original plan.

Revision has given me a welcome opportunity to modify the balance. Movement, for example, is treated more comprehensively, greater attention being given in this regard to medusae, ctenophores, nemertines, nematodes and molluscs, while the echinoderms are handled in a way more commensurate with the important advances that have been made in our understanding of their biology and phylogeny. The discussion of chemical communication has been extended to take account of recent developments in invertebrate endocrinology, and there is a more extended analysis of types of animal associations.

More generally, and no less important, I have scrutinized the book rigorously throughout, and have reworked many passages. New findings have been presented where these seemed to strengthen the argument, and I have provided more explicit accounts of the experimental analysis of function, where in the earlier edition I had assumed prior reading and understanding.

These additions and modifications are accompanied by many new illustrations. Altogether some 170 text figures and tables have been added, thanks to the generosity of my publishers and of the many colleagues who have helped us by providing the necessary material. Mostly these illustrations are taken direct from original publications, by courtesy of the editors and publishers concerned, to whom also I am greatly indebted. I hope that this will stimulate students to explore the remarkable range of recent biological research.

PREFACE

It is, of course, invertebrate biologists who make a book like this possible, by constantly stimulating the interest and extending the knowledge of those who write about their work. I have tried to do them justice within the limited space available, and I apologise in advance to all those specialists who would have wished (as indeed would I) to have seen their own studies more extensively represented. There is a continuing temptation during any revision to include new material just because it is so fascinating, but I have tried to control myself. Of course, the book has become longer, but I hope that it will be found to retain in the new format the emphasis upon principles which I believe to have been a welcome characteristic of the first edition. Tone, expression, choice of subject, and knowing where to stop, were taken by Anne Elliot as indices of a sensible, discerning mind, and I hope that I have not fallen too far short of the searching standard which Jane Austen thus commended to her.

It is a pleasure to thank Dr S. L. Sutton, who, initially disguised as an anonymous referee, studied the not too legible typescript of this edition with resolution and independence, and came up with many wise suggestions, upon which I very gladly drew. The responsibility for what remains is my own. It is a pleasure, too, to thank my publishers, and especially Dr Dominic Recaldin, for the care and understanding with which they have handled this revision. And I must thank, in conclusion, the many friends, both teachers and students, who have told me from time to time of the pleasure that the book has given them, and especially the lecturer who is reliably said to have selected it as his favourite bedside reading. The revision could never have been completed without that background of support.

E. J. W. B.

Contents

Part 1: The Foundations of Animal Life

1 LIVING SYSTEMS
1–1 Approach to animal life, 1; 1–2 Origin of living systems, 5; 1–3 Evolution of energy relationships, 13; 1–4 Organisms and ecosystems, 20

2 ORGANIZATION AND LIFE
2–1 Homology and analogy in organization, 27; 2–2 Eucellular plan of structure, 29; 2–3 Cell theory, 31; 2–4 Membranes and organization, 34

Part 2: Movement

3 MOVEMENT AND FIBRILS
3–1 Muscle filaments and myonemes, 42; 3–2 Flagella and cilia, 48; 3–3 Amoeboid movement, 58

4 MOVEMENT AND HYDROSTATICS Read
4–1 Principle of the hydrostatic skeleton, 65; 4–2 Coelenterate organization and the hydrostatic skeleton, 68; 4–3 Ctenophore organization and the evasion of hydrostatics, 84; 4–4 Some limitations of sponges, 88; 4–5 Triploblastic structure: Platyhelminthes, 92; 4–6 Triploblastic structure: Nemertinea, 95; 4–7 Triploblastic structure: Nematoda, 101

5 MOVEMENT, HYDROSTATICS, AND THE COELOM
5–1 Significance of the coelom, 108; 5–2 Sipuncula, Entoprocta, Polyzoa (Ectoprocta), 112; 5–3 Echinoderms, 117; 5–4 Molluscs, 133; 5–5 Molluscs: Cephalopoda, 145

6 MOVEMENT AND METAMERISM
6–Significance of metamerism, 153; 6–2 Locomotion of oligochaetes, 156; 6–3 Locomotion of polychaetes, 160; 6–4 Locomotion of leeches, 170

7 MOVEMENT AND ANTHROPODIZATION
7–1 Skeleton of arthropods, 172; 7–2 Tagmata and the head, 182; 7–3 Locomotion and the crustacean limb, 192; 7–4 Locomotion in terrestrial arthropods, 197; 7–5 Flight in insects, 210

Part 3: Aspects of Metabolism

8 NUTRITION OF PROTOZOA
8–1 Feeding, 217; 8–2 Digestion, 226

9 NUTRITION OF SOME LOWER METAZOA
9–1 Food resources, 229; 9–2 Microphagy and macrophagy, 235; 9–3 Filter feeding and digestion in sponges, 237; 9–4 Feeding and digestion in coelenterates, 243; 9–5 Intracellular and extracellular digestion: coelenterates, 249; 9–6 Feeding and digestion in Platyhelminthes and Nemertinea, 253

10 FILTER FEEDING
10–1 Filter feeding in polychaetes, 256; 10–2 Filter feeding and digestion in molluscs, 264; 10–3 Filter feeding and digestion in Deuterostomia, 275; 10–4 Filter feeding in Crustacea, 281

11 RESPIRATION
11–1 Some physical factors, 293; 11–2 Respiratory pigments, 296; 11–3 Gills and lophophores, 304; 11–4 Gills and lungs in molluscs, 307; 11–5 Gills and tracheae in arthropods, 318

12 EXCRETION
12–1 Excretory organs, 328; 12–2 Excretion of nitrogen, 342

13 OSMOTIC AND IONIC REGULATION
13–1 Ionic regulation in marine animals, 351; 13–2 The mammalian nephron as a model, 355; 13–3 Some principles of invertebrate renal function, 357; 13–4 Protozoa and fresh-water life, 364; 13–5 Metazoa and fresh-water life, 368; 13–6 From water to dry land, 379

Part 4: Information and Control

14 SOURCES OF INFORMATION
14–1 Coded signals, 388; 14–2 Properties of receptors, 391; 14–3 Mechanoreception, 395; 14–4 Chemoreception, 399; 14–5 Photoreception, 401

15 PRIMITIVE NERVOUS SYSTEMS
15–1 Components of behaviour, 413; 15–2 Coelenterates and the nerve net, 421; 15–3 Echinoderms, 431; 15–4 Hemichordates, 439

16 ADVANCED NERVOUS SYSTEMS
16–1 Trends in neural evolution, 444; 16–2 Metameric nervous system and locomotion in annelids, 445; 16–3 Aspects of arthropod behaviour, 460; 16–4 Giant nerve fibres, 474; 16–5 Molluscan nervous system, 480; 16–6 Nervous system and learning in cephalopods, 482

17 CHEMICAL COORDINATION
17–1 Neurohumours, 492; 17–2 Hormones and neurohormones, 494; 17–3 Endocrine regulation in crustaceans, 497; 17–4 Endocrine regulation in insects, 508; 17–5 Endocrine regulation in annelids, 516; 17–6 Endocrine regulation in molluscs, 528; 17–7 Evolution of endocrine systems, 533; 17–8 Pheromones and Allelochemicals, 536

Part 5: Reproduction

18 PATTERNS OF REPRODUCTION
18–1 Protozoan life cycles, 542; 18–2 Asexual reproduction and polychaete life cycles, 550; 18–3 Unity in the early development of Metazoa, 555; 18–4 Embryology and phylogeny; the origin of Metazoa, 559; 18–5 Protostomia and Deuterostomia, 575

19 LARVAL FORMS
19-1 Modes of larval development, 586; 19-2 Some protostome larvae, 588; 19-3 Crustacean larvae, 597; 19-4 Some deuterostome larvae: the origin of vertebrates, 603; 19-5 Insect larvae, 617

20 LARVAL LIVES
20-1 Competition and cooperation, 620; 20-2 Marine larvae and habitat selection, 623; 20-3 Larval life in fresh water, 634

Part 6: Associations

21 COLONIAL AND SOCIAL LIFE
21-1 Protozoan colonies, 639; 21-2 Sponge, coelenterate, and ectoproctan colonies, 642; 21-3 Asexual reproduction and colonial life in urochordates, 647; 21-4 Social life in insects, 654

22 INTERSPECIFIC ASSOCIATIONS
22-1 Types of association, 669; 22-2 Commensalism and mutualism, 671; 22-3 Behavioural aspects, 679

23 FURTHER INTERSPECIFIC ASSOCIATIONS
23-1 Mutualism and nutrition, 684; 23-2 Symbiotic algae, 691; 23-3 Corals and symbiosis, 698; 23-4 Host–parasite relationships, 706

Classification   730

Selected Bibliography   736

Index   753

1
Living Systems

1-1 APPROACH TO ANIMAL LIFE

They say of Scandinavian furniture that 'good design is timeless—it is the product of evolution'. The student of other aspects of animal life would agree with this claim; adding that good design expresses aptness for function. Upon this point of view our survey of invertebrate biology is based.

It recognizes that animals are constructed upon patterns of organization that have been tested and proved through immense periods of competition and differential survival. It presupposes, therefore, that the way in which animals function can only be understood in the light of their past history. Further, it recognizes that the animals that share our life today are not imperfect creations that would fit better into their environment if they had some of our own advantages. The fact that they have survived at all (and mostly for very much longer than we have yet succeeded in doing) is a tribute to the fitness of their organization.

This organization is an expression of the properties of systems of carbon compounds, but this does not necessarily mean that life is no more than a fortuitous association of molecules, nor does it necessarily follow that the humanist is correct in supposing that 'man must rely only upon himself'. But it is at least certain that the activities of living organisms depend upon the operation of physical and chemical principles no different from those that govern the properties of non-living systems.

A fundamental characteristic of living systems is that they carry on a continuous exchange of energy and materials with their environment; we say that they are open systems, involved in exchanges that are the driving force of the complex systems of chemical reactions that we call metabolism. One result of their metabolic activity is that they are able to build up some of the products of metabolism into the substance of their bodies, thereby providing for the replacement of worn-out material and for growth. Indeed, no part of a living body escapes the consequences of this continuous flux. Studies with radioactive tracers have shown that even the molecules of apparently permanent, inert material, such as supporting skeletal structures, are steadily replaced by corresponding molecules taken into the body from outside. A further result of metabolic activity is the capacity for irritability and for adaptive response to stimulation, so that by movement of part or of the whole of the body the organism behaves in a way that makes possible a further consequence: the reproduction of the individual and hence the perpetuation of its species.

Reproduction depends upon the capacity of living systems for making copies of themselves—the process that we call replication. The perpetuation of the species, however, depends in the long run upon occasional imperfections in the replication, and as a result of these the copy may differ from the parental form in certain respects. These differences, which we call mutations, are likely either to aid or to impede the adjustments of a particular organism to its environment. But the resources of the environment are not limitless, so that the maintenance and growth of organisms involves competition between them for limited supplies of materials. Organisms tend by their own activities to extend the range of their distribution and thus to exploit their environment to the limits of their capacities. This tendency, as Hardy has argued, has probably been of immense evolutionary importance. Populations which develop mutations that aid such extension will probably be more successful in this competitive exploitation. They will tend to survive and reproduce at the expense of other populations, a consequence that is the basis of the process that we call natural selection. Thus we conceive the relations between living material and its environment to have been continuously moulded, with the resulting production of organisms that are ever more complex and ever more efficient in the exploitation of the environment. This is what we call evolution, which we see as a continuous sequence of change leading from the simplest forms of life to the most complex.

This concept of levels of complexity may seem self-evident to even the most superficial observer of animal life, yet it deserves some attention here, for it is not easy to translate it into more concrete terms. An early expression of it, and one that has powerfully influenced man's approach to other animals, is seen in Aristotle's *Scala Naturae* (Fig. 1–1), or Ladder of Nature. According to his interpretation:

> Nature proceeds little by little from things lifeless to animal life in such a way that it is impossible to determine the exact line of demarcation, nor on which side thereof an intermediate form should lie. Thus, next after lifeless things in the upward scale comes the plant, and of plants one will differ from another as to its amount of apparent vitality; and, in a word, the whole genus of plants, whilst it is devoid of life as compared with an animal, is endowed with life as compared with other corporeal entities. Indeed, as we have just remarked, there is observed in plants a continuous scale of ascent towards the animal. . . . In regard to sensibility, some animals give no indication whatsoever of it, whilst others indicate it but indistinctly. Further, the substance of some of these intermediate creatures is fleshlike, as is the case with the so-called tethya [ascidians] and the acalephae [coelenterates]; but the sponge is in every respect like a vegetable. And so throughout the entire animal scale there is a graduated differentiation in amount of vitality and in capacity for motion.

Aristotle's interpretation was not an evolutionary one in our modern use of the term, but it does carry a clear implication of relative status. We have been accustomed to place at the top of the ladder the evil, flesh-eating beast that Sartre finds in us. Once

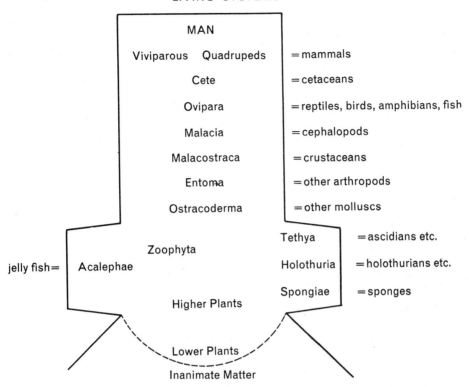

Fig. 1-1. The *Scala Naturae* or Ladder of Nature, according to the descriptions of Aristotle. From Singer, 1931. *A Short History of Biology.* Clarendon Press, Oxford.

we accept this position, however, there remains an implied corollary that other animals are in some sense 'lower', and that the 'lowest' are those at the bottom of the ladder. We do, in fact, regularly speak of 'lower' and 'higher' animals and because of this it is necessary to consider exactly what we mean by these terms.

We shall see that animals must be organized so as to function and behave in a manner best calculated to ensure survival and reproduction. From this point of view some environments are more 'difficult' than others. The littoral zone of the sea, particularly that part of it below the tide marks, is easier to occupy than either dry land or the air, for example, and we shall find some reasons later why this is so. Exploitation of these more difficult environments has required the development of new devices that are not needed by animals living in the easier habitats: waterproofing for land, and wings for the air, are obvious examples. From this point of view animals living in more difficult environments may be regarded as higher animals; the possession by them of new and specialized devices is an objective criterion by which their rightful place on the ladder may be defined.

But this analysis is not sufficient. Animals may inhabit very difficult environments, yet we may still feel that they are truly lower organisms. For example, life within the alimentary canal of another animal presents many problems. Few of us would expect to survive the experience of Jonah, but intestinal parasites regularly do so; yet this seems an inadequate reason for calling them higher animals. The important consideration here is that the possibilities of life on this planet may be exploited in many ways. One species may survive because it possesses a narrow and inflexible range of responses, allowing it to sample only a small fraction of the potential resources by which it is surrounded. Such an animal is *Peripatus*, which has reacted to the danger of desiccation on land by restricting itself to damp and concealed niches. Like the city financier in the garden, 'he looks importantly about him, while all the spring goes on without him'. Other animals may exploit their environment much more fully; perhaps because they possess devices that enable them to resist a wider range of stresses, or perhaps because they can sample a wider choice of food. These animals may be regarded as higher than those that lead more restricted lives.

Here we have another objective criterion, and an approach to an explanation of the biological significance of more complex organization. We have, too, an objective justification of the dominant status of man in the *Scala Naturae*. It can be justified by his ability to manipulate his environment to his own purpose. It can be justified also by the flexibility of his behaviour, and by the unique capacity of his nervous system, which results, among many other things, in making him the only animal that can scrutinize the rest of the animal kingdom in sufficient depth to be able to write books about it.

Two other concepts may conveniently be mentioned here, since they are closely associated with this matter of status. In our comparisons of animals we customarily refer to them, or to the groups to which they belong, as being either 'primitive' or 'specialized'. By specialized we mean that they possess characters that tend to debar them from further evolutionary change. Primitive groups or primitive animals, by contrast, possess many characters that are theoretically capable of further change. For example, we shall speak of the nerve net as a primitive type of nervous system, because we can conceive it as the forerunner of the polarized and centralized type of nervous system of higher animals.

Finally, it is our common habit to speak of animals or groups as being 'successful' or 'unsuccessful'. These terms, like 'higher' and 'lower', are relative, and can only be usefully employed if we provide ourselves with some objective standard. Since life is always a struggle, and the environment fundamentally hostile to its maintenance, it is fair to say that any group of animals that has survived at all is a successful one. But we may reasonably go further, and say that the more successful ones are those that have not merely survived, but have made the fullest use of the potentialities of the environment. In this sense the successful animals are what we have just defined as higher ones. But this is not all, for at any particular level of evolution there will be some groups that may be judged more successful than others. A useful objective criterion here is to consider relative abundance. Groups that have exploited most successfully a particular level of organization will tend to be more abundant than the less successful ones. This abundance will be reflected in the number of individuals representing the group at any given moment, and in the gross mass of their material (or biomass, as it is called).

Success will be reflected also in the number of species within the group, for diversification usually results when natural selection acts on a particular plan of organization to adapt it more closely to the environment. A number of subsidiary groups become established, each adapted to some particular mode of life. Within each of these the process continues until it has produced a range of species, each of which, by virtue of its own unique combination of characters, avoids competition with related species. This is the process that we call adaptive radiation. The more 'successful' the initial plan of organization, the greater the resultant diversification, and hence the greater the number of species to which it gives rise. The effect can be seen in Fig. 1–2, which shows the relative abundance of species in the major groups of animals. From this we can see, among other things, that the insects can be regarded as a highly successful group, in so far as they contain more species than all the rest of the animal kingdom put together. It is part of our present purpose to examine some of the reasons for these differing degrees of success. In doing so, we shall then see how success consolidated at one evolutionary level has provided a basis for fresh progress. Thus we shall come to appreciate the essential unity of living organisms, and our own dependence upon events that occurred early in the history of our planet.

1–2 — ORIGIN OF LIVING SYSTEMS

The continuity of evolution is a fundamental element in the biologist's interpretation of the history of the earth. So much so, that he finds it logical to extend the concept to include also the origin of life from non-living material. At first sight it may seem formidably difficult to justify this extension. Living organisms are poised in such delicately balanced relationships with their environment that they are often said to present a highly improbable state of matter; a state of which it is therefore very difficult to conceive the origin. Until recently, indeed, the problem of the origin of life seemed to be beyond human understanding; but this was an over-pessimistic view, based, perhaps, upon the feeling that the facts of the situation were for ever beyond our reach.

Facts, however, are not the only tools of the scientist; a powerful element of creative thought is also involved in his activity. It is true that a scientist faced with a particular problem will need to deploy with the utmost efficiency his training and experience, and to bring into his consideration all relevant information. But given these, he will devise a hypothesis on which to base his further study of the problem. This is the creative aspect of his work. In everyday speech, he might claim to have 'had an idea', and he may not always find it easy to account for its emergence.

If the hypothesis is to be a useful one, however, it must be one that can be put to the test; it must, in fact, be tested to destruction. We cannot expect to be able to prove that our hypotheses are 'true'. Our sense organs, for example, and the instruments that extend their scope, can at best give us only a limited range of information. Moreover, in biological research we are dealing with such highly variable material that not until we have studied a wide range of species can we begin to feel confidence in the generalizations that we base upon the evidence available to us. In any case these generalizations have at best a limited validity. Their formulation is an exercise of the inductive method, in which universal statements are inferred from particular ones, and we should not forget Popper's fundamental criticism of this procedure. As he succinctly remarks, 'no matter how many instances of white swans we may have observed, this does not justify the conclusion that *all* swans are white'.

What we can do, however, is to apply the deductive method, and state certain

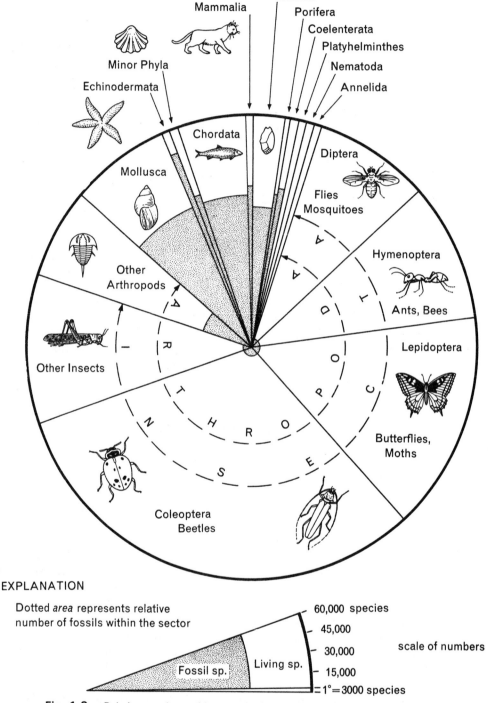

Fig. 1-2. Relative numbers of known species, living and fossil, of various animal phyla. From Muller and Campbell, 1954. *Syst. Zool.*, **3**, 168–170.

consequences, or make certain predictions, that would be expected to follow from a particular hypothesis. By observation and experiment we can then determine whether or not these consequences materialize. If we find that they do not, then we say that our predictions have been falsified, and we begin to suspect that the hypothesis is ill founded. Continued falsification may then make it necessary to put the hypothesis away, and to make a fresh start. This is no disaster. As Popper has emphasized, an essential criterion for an empirical scientific system is that it must be possible to refute it by experience. However, we may be fortunate and find that our deductions are justified; we then say that we have verified them. Within these limits we have corroborated the hypothesis, in Popper's terminology. It may therefore be well founded, and at least we can go on to use it as a basis for further investigation and speculation. Additional corroboration will strengthen our confidence in it, but we can never afford to forget that one day it may need modification. Even the most cherished hypothesis is always open to eventual falsification; to forget this is to forget a fundamental principle of scientific procedure.

The problem of the origin of life can certainly be attacked in the way outlined above. We start with a belief in the existence of unifying principles in the organization of the universe. It is a belief derived from the work of Newton and Darwin, whose demonstration that unity can be found as well in the depths of the universe as in the life of this planet has permanently influenced our approach to the study of natural phenomena. At this stage in the history of human thought, therefore, we may reasonably accept, as a working hypothesis, that life may have originated during the early history of the earth as part of a continuous process of cosmic evolution. It is equally reasonable to suppose that its emergence was determined by the working-out of principles that are still demonstrable today, even though they operate in a very different context. In all of this we may be wrong. But this does not weaken the scientific validity of our procedure. What we are saying, in effect, is that living organisms of today are the products of evolution, and that therefore life itself may have originated out of non-living systems through an evolutionary process. Indeed, so confident are we of the validity of the principle of unity of plan in the cosmos that the possibility of life existing elsewhere in the solar system, and in other and remoter parts of the universe, has now become a matter for serious discussion and investigation.

If we are correct in this approach it follows that the history of animals and plants as we find it recorded in fossil form in the rocks must have been preceded by a much longer phase of evolution. This phase, in its earliest stages, would have been essentially chemical in character. During it there would have been laid down, under the influence of natural selection, the ground plan of the organization of living systems. The events of that remote period must therefore have had a profound influence upon the subsequent history of animals and plants, providing them with a common inheritance, which must have gone far towards determining the patterns of organization that they were later to achieve. It is because of this, and not only because of the intrinsic interest of the problem, that it becomes essential to attempt some interpretation of the possible course of this primeval phase of evolutionary history.

Our analysis may arbitrarily begin with the origin of the solar system. It seems likely that our sun, an average star about 5,000 million years old, was formed, like other stars, by condensation in a cloud of gas composed largely of hydrogen. Other kinds of molecules, including water, may have been present as well, and there may have been heavy elements formed elsewhere in the galaxy by the violent explosions of stars that constitute supernovas. At a certain stage of condensation the temperature would have risen to a level of many millions of degrees, permitting thermonuclear reactions and their associated explosions, with the release of heavy elements formed by thermonuclear fusion. The sun then entered into a relatively stable condition; stable, that is, as astronomers understand the term, with thermonuclear fusion transforming 564 million tons of hydrogen into 560 million tons of helium for every second of time. The earth and its fellow planets would then have arisen by further condensation in the remains of the cloud. So also around other suns, for planetary systems are probably common features of the universe. The early atmosphere of the earth must have contained large amounts of hydrogen. Dust particles would also have been present, including particles of silicates and of iron, and it is possible that simple, carbon compounds (hydrocarbons, for example), already present in interstellar space, would have condensed on them. There they could have reacted with each other, and have given rise to other and larger molecules, providing the raw material of future terrestrial life.

In this vast evolutionary perspective we find an extension and strengthening of the views of Darwin, who in summarizing the arguments of *The Origin of Species*, remarked that 'there is a grandeur in this view of life, with its several powers, having been gradually breathed by the creator into a few forms or into one'. The grandeur has been powerfully and ironically enhanced by the extension of the principle of evolution into the field of cosmical physics. Thermonuclear reactions provide the elements out of which terrestrial life has been fashioned, and the radiant solar energy that drives it, while, in another context, they may yet provide the means by which it will eventually destroy itself.

The earth may then, from the beginning, have contained primeval (non-living) molecules, borne on the interstellar dust particles that we have mentioned, but even if this were not so, the prevailing high temperature would have promoted the combination of some of the available elements. In this way there could have arisen ammonia, methane, and water vapour, which are believed to have been the first constituents of the earth's atmosphere. This belief is supported by the identification of these same substances in the atmospheres of the larger and more distant planets, where conditions are believed to have changed less rapidly than on the earth. In the course of time the earth would have cooled sufficiently for water to condense on its surface. This would initially have been fresh water, but material swept from the land would have slowly accumulated in it: thus the salt-water oceans would have formed. According to one view, it is in these that the earliest forms of life may have arisen, their origin dependent upon the solvent properties of water, and its consequent facilitation of chemical reactions.

The periods of time that we are discussing ('before the birth of consciousness, When all went well') are almost too vast for our minds to grasp. Chemical evolution may perhaps have passed into biological evolution, characterized by growth, reproduction, and natural selection, some 3,000 million years ago, for molecular palaeontology (the study of the organic molecules of rocks) has given at least a suggestion of the existence of a mixture of abiogenic and biogenic hydrocarbons in shale of about that age. Fossils described from much later Pre-Cambrian sediments include algal and fungal remains that are perhaps 1,600,000,000 years old, a segmented worm about

650 million years old, and blue-green algae not less than 800 million years old. And it is in any case certain, from the abundance of Cambrian fossils, that there must have been a long period of biological evolution during Pre-Cambrian time.

What can be more easily grasped than these estimates, however, is an hypothesis that was first clearly formulated by Oparin, and that is in line with this analysis of the sequence of chemical events. This hypothesis proposes that life in that inconceivably remote period must have originated in reducing conditions: the abundant supply of oxygen, on which it now depends, could not at that time have been available. In accordance with the methodology outlined above, this deduction has been tested by laboratory experiments, and the results of these tests are found to support this general analysis. They have shown that organic material can actually be formed in such a reducing atmosphere, provided that an adequate supply of energy is available. In 1953–54 it was shown by Miller, in what have now become classical experiments, that the passage of electrical discharges through a mixture of hydrogen, ammonia, methane, and water vapour could lead to the formation of the fundamental substrates required by living organisms (e.g. formic acid, acetic acid, succinic acid) and also to amino acids (Fig. 1–3). Moreover, amino acids have been polymerized to form peptide-like structures, under conditions comparable with those that might have existed during the early history of the earth. The appearance of these substances, which are the essential structural units of living material, may, therefore, have been inevitable and predictable during those remote times. Electrical energy was probably available, resulting from lightning displays such as are recorded as taking place today on Jupiter. Ultraviolet light, however, would probably have been a more important energy source; it would not at that time have been reduced in intensity by the ozone layer that is formed now in our oxygen-rich atmosphere, and it would have been continuously available. Experiments similar to those of Miller, but using ultraviolet light as the energy source, regularly produce amino acids, provided that sufficient hydrogen is present to make the environment a reducing one.

Of course, the production of amino acids and peptide chains is a very long

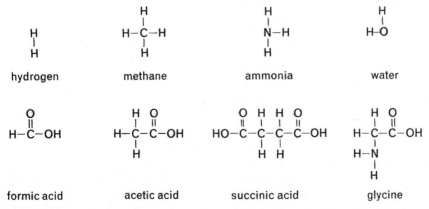

Fig. 1-3. The earliest molecules believed to have been present on the surface of the earth (*top row*), and molecules derived from them by electrical discharge or ultraviolet radiation (*bottom row*). After Urey.

way indeed from the establishment, maintenance, and replication of the organized patterns of living systems; further assumptions are clearly needed to develop this interpretation. We must assume that subsequently there was a building-up of increasingly complex molecular chains and of the molecular associations known as coacervates. This might have taken place in ancient seas, perhaps by the adsorption of the molecules onto mineral particles. We must further assume that these molecular aggregations developed the power of self-replication.

This replication must have been achieved through the capacity of complex organic molecules to undergo polymerization, for the reproduction of organisms today depends upon the properties of the polymeric molecules of deoxyribonucleic acid (DNA), organized within the chromosomes as a double helix, in which two molecular strands are coiled around a common axis (Fig. 1–4). Each DNA strand is composed of repeating units called nucleotides, which are formed of three constituents, a sugar (deoxyribose), a phosphate, and a nitrogenous base. The sugar and phosphate are always the same, but the base may be any one of four compounds, cytosine and thymine, which are pyrimidines, and adenine and guanine which are purines. The association of the two chains into a double helix is interpreted as a result of hydrogen bonding between complementary base pairs, adenine combining with thymine, and cytosine with guanine.

These four bases constitute a four-letter alphabet which gives a coded specification of all the proteins that an individual needs to synthesize. Each of the 20 common amino acids is coded as a sequence of three bases, called a codon. Several different codons, differing in their third base, may act as synonyms, specifying the same amino acid, for which reason the code is said to be degenerate. This feature, it has been thought, may be an evolutionary relic of a time when the third base had some

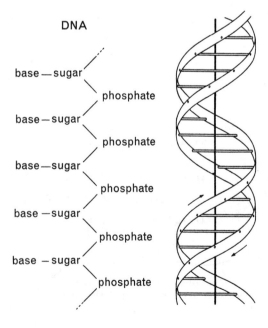

Fig. 1-4. Left, chemical formula of a single chain of deoxyribonucleic acid. Right, a purely diagrammatic figure, in which the two ribbons symbolize the two phosphate sugar chains, and the horizontal rods the pairs of bases holding the chains together. The vertical line marks the fibre axis. Adapted from Watson and Crick, 1953. Nature, Lond., **171**, 737-738.

non-specific function. Despite this, however, the order of the codons in the DNA molecule determines the order in which amino acids are built up to form particular proteins. Moreover, this code can be replicated and transmitted, for the double helix can synthesize another helix like itself, each of its two strands producing a complementary strand with the aid of an enzyme, DNA polymerase.

The translation of the code into biological organization involves the synthesis of a strand of ribonucleic acid (RNA) under the influence of another enzyme, RNA polymerase. This enzyme moves along the DNA fibre and effects a transcription of one of the DNA strands, which thus acts as a template for the newly formed RNA strands (Fig. 1–5). The codons of the latter are complementary to those of the template, but with uracil always replacing thymine. RNA strands formed in this way pass into the cytoplasm as messenger RNA (m-RNA). Here they are found attached to the ribosome granules, which are, indeed, responsible for removing them from their DNA templates.

Also in the cytoplasm is another form of RNA called transfer RNA (t-RNA). This is present as a number of different t-RNA molecules about 80 bases long, there being one type (or at most a few types) of molecule for each amino acid. Each t-RNA molecule becomes attached to its specific amino acid, and then conveys it to the m-RNA. The insertion of the amino acids into their correct positions in developing polypeptide strands depends upon each kind of t-RNA molecule having its own specific sequence of three nucleotides called an anticodon. This recognizes its complementary codon on an m-RNA strand, and attaches to it, with the cooperation of a ribosome. Its amino acid is now added to the developing polypeptide chain, while the t-RNA molecule, its function discharged, is freed to move away (Fig. 1–5). So the process continues until the information coded on the m-RNA strand has

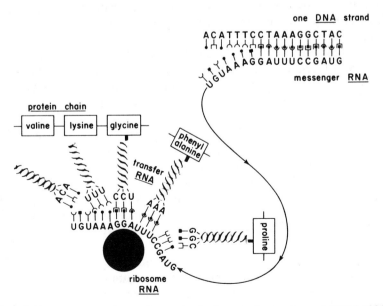

Fig. 1-5. Diagram of protein synthesis in cytoplasm. From Leblond, 1965. *Symp. Int. Soc. cell Biol.*, **4**, 321–339.

been translated into the appropriate protein molecule. The end of the molecule is coded by the codon UAA or UAG, while another signal indicates the start of a new one.

We cannot, on any evolutionary interpretation, suppose that this astonishing sequence of events arose ready-made: it must have been established by stages. Its ultimate source eludes our grasp, and may be for ever inaccessible to our scientific method. But we can at least suggest some possible stages in its evolution—using the same type of analysis as we should apply to the interpretation of morphological and physiological evolution. In effect, we can compare patterns of organization and, when we find degrees of resemblance that seem too close to have arisen independently by chance, we can accept them as evidence of a common ancestry.

The appearance of nucleic acids was clearly crucial for the elaboration of proteins. Conceivably, RNA may have appeared before DNA, for the synthesis of proteins might have been facilitated at first by the random collision of amino acids with a primordial transfer RNA. This might later have come to catalyse their bonding, although there could have been no coding of protein structure at this early stage. The organization of precise replication, and the establishment of the genetic code, would have required further advances in nucleoprotein evolution, including the formation of DNA, and also the diversification of t-RNA into molecular variants which became associated with specific amino acids.

This latter assumption follows from the fact that existing t-RNA molecules have certain structural features in common, as well as each carrying their own specific anticodon. We are justified in interpreting their similarities as evidence that they evolved from a common ancestral molecule, the diversification of which could have resulted from reduplication and other forms of genetic mutation in the DNA which had come to direct its synthesis. The alternative view, that different forms of t-RNA might have arisen independently within the earliest organisms, and yet with essentially identical properties as regards participation in protein synthesis, is wholly improbable.

Nor is it possible easily to conceive that the genetic code could have arisen more than once. Deciphered initially through studies of the bacterium *Escherichia coli*, it is known to be substantially similar in principle in the toad *Xenopus laevis* and also in the guinea-pig. Moreover, it can be used to provide a plausible interpretation of the course of molecular variation in, for example, abnormal human haemoglobins. Repeated independent origin of such a complex system is too remote a possibility to be credible in terms of evolutionary theory. We conclude, then, that the ability to replicate must have arisen in some primordial form of life, from which all present-day organisms have descended, and that the genetic code—and the associated mechanism of protein synthesis—is also part of their common inheritance. Recognition of this fundamental unity in the organic world must always colour our interpretation of the diversity of present-day organisms, and create a sense of kinship with them.

The assumption that the conditions obtaining during the phase of chemical evolution could have led to the establishment of substances with such remarkable properties as those of the nucleic acids is an immense one. Yet we are helped to accept it by the knowledge that ribose, deoxyribose, adenine, and guanine have been produced in the laboratory in experiments similar in principle to those of Miller, while nucleotides have been polymerized to yield nucleic acids containing at least 200 residues. This, too, can be said in favour of it: in the reducing conditions then prevailing it would have been theoretically possible for organic molecules to accumulate and interact, whereas a similar accumulation could not occur today simply because the molecules would be oxidized by the atmosphere or broken down by living organisms. Moreover, the evolution of living material need not have been dependent upon entirely random processes. Calvin has suggested that simple inorganic compounds, or heavy metals, may have acted from an early stage as catalysts; they may thereby have served as driving forces that could have been favoured and canalized by natural selection.

1–3 EVOLUTION OF ENERGY RELATIONSHIPS

Whatever the means by which this organic complex evolved, its growth and replication would have required the supply of materials and energy that we have seen to be the foundation of living systems. We must assume, therefore, that in the primeval oceans there were other complex and energy-rich molecules that could be taken up into these systems, and that the latter could release and make use of the energy so obtained. This would have constituted the first appearance of metabolic processes.

The metabolism of organisms as we know them today depends upon a very peculiar way of storing and transferring energy, and of releasing it in a form that is immediately available for use in biological processes. In principle, a large proportion of the energy released by the metabolic breakdown of organic compounds is taken up by adenosine diphosphate (ADP), which is thus transformed by oxidative phosphorylation into adenosine triphosphate (ATP). In due course this is broken down again by hydrolysis into ADP, the energy so released becoming available for some form or other of biological activity:

$ATP \rightarrow ADP + P_1$ free energy change $= -31.0$ kJ mol^{-1} ($= -7.4$ kcal mol^{-1})

The energy released by this exergonic reaction has often been regarded as associated with a high-energy phosphate bond, represented as $\sim$P. In fact, however, ATP is one of a number of high-energy compounds, the properties of which in this respect are an expression of the structure of the molecule as a whole. But however regarded, the reaction is one of the unique features of living material, permitting as it does the ready storage and release of energy under biological conditions and in a form readily available for the organism.

Adenosine triphosphate is so central and uniform a feature of the organization of living systems that we may reasonably suppose it to have evolved at an early stage of chemical evolution and so to have become inseparable from life. There is good evidence to support this supposition. We have seen that adenine can be formed in the laboratory in conditions analogous to those that might have existed during the earliest stages of chemical evolution. It is thus all the more significant that high-energy phosphates are generated when ferrous iron is oxidized by hydrogen peroxide in the presence of orthophosphate. These conditions could probably have existed during those early stages, and may well have promoted the incorporation of these phosphates into living systems. Thus we may think of the earliest forms of life, according to this analysis, as precariously evolving in a reducing atmosphere, and dependent upon energy that was already stored in the complex molecules of their environment.

Organisms that now obtain their energy by breaking down complex and energy-rich carbon compounds taken in from their environment are known as heterotrophs. The earliest forms of anaerobic life that we have been postulating can therefore be termed primitive heterotrophs. Their emergence was a major achievement of chemical evolution, yet their future was not assured, for the reserves of energy stored in the molecules around them could not have lasted indefinitely. The molecules could not have been unlimited in abundance, and the supplies of them must sooner or later have been exhausted. It is supposed that this barrier to the maintenance and further evolution of living material was overcome by the emergence of the capacity for photosynthesis, the process in which (as we see it today) the electromagnetic energy of solar radiation is trapped by chlorophyll pigments (Fig. 1–6) and transformed into the potential energy of carbohydrates. This process, in the highly evolved form found today in green plants and algae, takes place in two stages: a light-dependent 'light phase' and a light-independent 'dark phase'. The first of these involves the splitting of water molecules, with the consequent production, by electron transfer, of the strongly reducing compound $NADPH_2$ (reduced nicotinamide adenine dinucleotide phosphate); ATP is also formed, while oxygen is released as a by-product. In the dark phase, ATP drives the reduction of CO_2 to sugar or other carbohydrates through the energy of $NADPH_2$. The main events may be summarized thus:

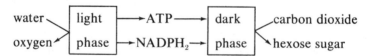

They result in a gain to the plant of 48×10^4 joules of energy for the fixation of each mole of CO_2 as glucose.

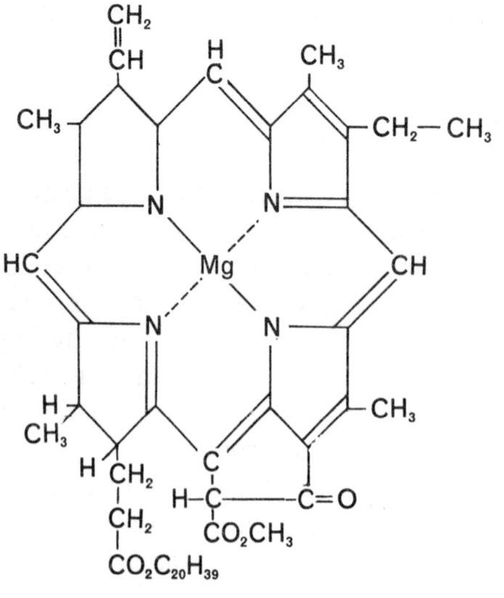

Fig. 1-6. Chlorophyll *a*. After Calvin.

Chlorophyll is a magnesium-porphyrin complex. We shall see later (sec. 11–2) that the production of porphyrins is so widespread in living organisms that we must suppose these substances to have appeared at a very early stage of evolution. They have, in fact, been produced in electric discharge experiments like those mentioned earlier. Their use in photosynthesis would have provided a continuous supply of energy-rich carbon compounds, so that living organisms needed no longer to depend upon ready-made sources of these in the environment. No less important was the release of oxygen into the atmosphere as an end result of photosynthetic reactions. Because of this, and probably also because of some further release of the element through the dissociation of water vapour by solar radiation, the atmosphere changed from reducing to oxidizing conditions. This made possible a change in the metabolism of living systems from the primitive anaerobic type to the aerobic type that is so characteristic of living organisms today.

Organisms that obtain their energy through photosynthesis are termed phototrophs. On the general hypothesis outlined above, the earliest phototrophs are regarded as having evolved from primitive heterotrophs, with a mode of metabolism exemplified today in the fermenting anaerobic bacteria. The descendants of those early heterotrophs are the green plants, which can build their organic molecules out of carbon dioxide and simple salts, so that their carbon needs can be met entirely from inorganic sources. For this reason they are said to be autotrophic as regards their carbon supplies.

Photosynthesis also occurs in certain autotrophic bacteria, which possess pigments (bacteriochlorophylls) enabling them to trap radiant energy. Thus the green sulphur bacteria are autotrophic forms which use CO_2 as their carbon source and H_2S instead of H_2O as hydrogen donor, oxidizing the H_2S to sulphur. They do not evolve oxygen, and function as strict anaerobes. The purple sulphur bacteria are similar, except that they can carry the oxidation further to sulphate, and may sometimes make use of organic acids, so that they are not exclusively autotrophic. The purple non-sulphur bacteria are also photosynthetic forms, but preferentially use simple organic compounds, and depend upon these for their growth, so that they, too, are not strictly autotrophic. Commonly they are anaerobic, but some can function aerobically in the dark. These various types of bacteria may perhaps represent stages in the evolution of the photosynthetic mechanism of algae and green plants, but it is always possible that bacterial modes of photosynthesis may be secondary developments.

The establishment of photosynthesis was a major advance, but one further step was still needed to give the pattern of life that we find today. This was the emergence of secondary heterotrophs; organisms incapable of photosynthesis, and therefore dependent for their energy sources upon the oxidation of organic compounds formed by phototrophs. The primitive heterotrophs that we have earlier considered are visualized as a transient phase in the evolution of life, primitively dependent upon molecules formed in peculiar chemical conditions that soon passed away. Secondary heterotrophs are so called because they are visualized as having been derived from primitive phototrophs by the loss of the capacity for photosynthesis; a loss which, as we shall see later, can still occur in phototrophs at the present day. This step in evolution was thus a regressive one. But it was a case of *reculer pour mieux sauter*: it provided not only the heterotrophic bacteria and the fungi, but also the whole of the animal world, with all its immensely diversified organization that we are about to study. If we need a simple definition of animals it may be said that they 'eat'; by which is meant that they take in (or ingest) complex carbon compounds as solid food. This process, which is called phagotrophy, is commonly associated with the active pursuit or trapping of food, and it is because of this that structure and function in animals have evolved along lines profoundly different from those that characterize plants. However, heterotrophy and phagotrophy do not always go together. Some animals (often those that live in the alimentary tract of others) take in energy-rich molecules through their body surface in dissolved form; they are said to be osmotrophs.

We have postulated a complex sequence of events. Whether or not it is correct, it is certain that in this or in some other way there were established the great cycles of flow of the chemical elements that are vital to the maintenance of life, and that involve living organisms in complex webs of nutritional interrelationships (Fig. 1-7). One of these is the carbon cycle, in which complex organic compounds are built up by the phototrophs and are ultimately decomposed by heterotrophic bacterial activity, with or without prior transformation by animals. Another is the cycle of nitrogen, the element upon which all organisms depend for the building-up of their proteins. Some organisms, including the green plants, are autotrophic as regards their nitrogen supply, taking it up in an inorganic form. Usually these organisms require nitrates, but some bacteria and fungi can utilize atmospheric nitrogen. From these resources are built up the organic nitrogenous compounds that are incorporated into the bodies of the autotrophs, and that in due course are ingested by the heterotrophic animals. Finally, the supply of nitrates is maintained by heterotrophic bacteria decomposing the bodies and products of plants and animals. Phosphorus and sulphur pass through comparable cycles in which an organic phase alternates with an inorganic one. It is part of the human predicament that death is a necessary condition of life. 'Nous mourrons tous', warned the preacher, 'moi aussi peut-être.'

Finally, we must take account of another aspect of nutritional dependence which does not emerge from consideration of these simplified cycles. This is the dependence of many organisms upon the substances known as accessory growth factors, or vitamins. This dependence first became apparent from studies of human and mammalian nutrition, but its significance ranges much more extensively than that. A wide variety of vitamins are now known, classified by a conventional literal system. Some of them (those included within the A, C, and D groups) are primarily, and perhaps exclusively, vertebrate requirements. The B vitamins, on the contrary, are essential requirements for many organisms, and some are probably needed by all forms of life.

A widely accepted explanation for this is that during the chemical phase of evolution there were established metabolic patterns that were dependent upon the presence of certain enzymes and coenzymes. An example of these substances would be thiamine (vitamin B_1), which is believed to be a requirement of all living organisms. The explanation of this dependence is that thiamine gives rise to a substance called cocarboxylase (thiamine pyrophosphate). This functions as a coenzyme of carboxylase in the citric acid cycle, which we shall later see to be a sequence of reactions essential for the oxidative metabolism of living organisms. Dependence on

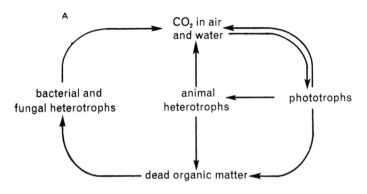

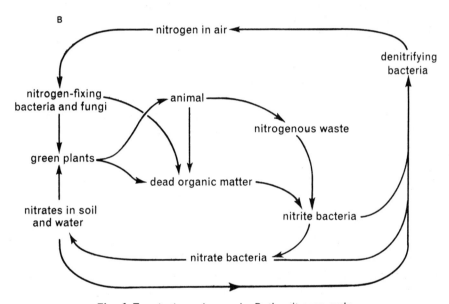

Fig. 1-7. A, the carbon cycle. B, the nitrogen cycle.

vitamin B_1, therefore, is a consequence of the dependence of living organisms on the citric acid cycle.

What is at first sight surprising in this situation is that animals, and certain other organisms also, are unable to synthesize these essential substances for themselves, but must obtain them as part of their food. This, like the origin of animals, is believed to be a consequence of regressive chemical evolution. We know that in the mould *Neurospora*, for example, there can arise mutant strains that are deficient in certain biosynthetic capacities, and it is supposed that similar mutations have been a constant accompaniment of evolution. Certain of these mutations would have resulted in a loss of the ability to synthesize particular compounds, but these defects of metabolism (for that is what they are) have persisted because the organisms concerned were able to obtain the required substances from their external environment, where they were being synthesized by other organisms. Thiamine, for example, is synthesized by plants and many bacteria, and it is because of this that animals can obtain it from their environment or their food, without having to manufacture it themselves. For them it has become a vitamin. This we can define as an essential nutrient which a particular organism cannot synthesize for itself; which it requires in only very small amounts; and which contributes nothing either to its energy supplies or to its permanent structure.

The dependence of organisms upon these metabolites thus introduces a further complication into an analysis of their nutritional relationships. A particular species may be autotrophic as regards its carbon and nitrogen sources, and yet be heterotrophic in its dependence upon an exogenous source of certain vitamins. In fact, a species can only be said to be autotrophic in the complete sense of that term when it can secure chemical energy through photosynthesis, and when it can synthesize all other essential metabolites from inorganic sources in its environment. No animal, of course, is in this position. When we consider the nutrition of the Protozoa, however, we shall find that we still have with us today organisms that to some extent bridge the nutritional gap between animals and plants; this gap may well have been repeatedly bridged in the early stages of evolution.

Anaerobic release of energy from carbohydrate is widespread today in organisms, where it takes one of two forms. One of these is the process called fermentation, well known in yeast and bacteria. This involves a complex sequence of reactions which first provides for the activation of glucose by the incorporation of phosphate radicals into its molecule. One molecule of glucose forms one molecule of fructose diphosphate, the phosphate being donated by ATP; two molecules of the latter are thus converted at this stage into two molecules of ADP. The molecule of six-carbon fructose diphosphate is next split into two three-carbon molecules of triose phosphate, and each of these is then converted into a molecule of pyruvic acid. Four molecules of ATP are generated from ADP during these stages. Thus, allowing for the initial loss of two ATP molecules, the breakdown of one molecule of glucose has so far resulted in a net gain to the organism of two molecules of ATP. In yeast the pyruvic acid is then further broken down to alcohol, so that the overall equation can be written thus:

$$C_6H_{12}O_6 \rightarrow 2CH_3CH_2OH + 2CO_2 \quad \text{free energy change} = -209.2 \text{ kJ mol}^{-1}$$

It will be noted that the phosphate radicals, which are vital for the biological significance of the reaction, do not appear at all in this representation. The 209.2 kJ mol^{-1} shown in the equation is the total free-energy change of the reaction. Of this total, however, the organism only derives the energy associated with the two molecules of ATP, amounting to 62 kJ mol^{-1}.

The anaerobic breakdown of carbohydrate in animal tissues differs from this process in two respects. It begins with glycogen, for which reason it is called glycolysis, and it ends with lactic acid. The intervening stages, however, from the phosphorylation of the glucose molecule to the production of pyruvic acid, are identical. This resemblance to fermentation is in itself a striking testimony to the essential unity of living organisms, and a powerful indication of their origin from a common ancestral pattern of metabolic organization. The overall equation for glycolysis can be written thus:

$$C_6H_{12}O_6 \rightarrow 2CH_3CH(OH)COOH + \text{free-energy change} = 150.6 \text{ kJ mol}^{-1}$$

As with fermentation, the organism gains 62 kJ mol^{-1}. It will be observed that once again the peculiar biological properties of the process are unrepresented; only a detailed analysis of the component steps can bring out the essential identity of glycolysis and fermentation.

The aerobic breakdown of carbohydrate, termed oxidation, may be represented by the following overall equation:

$$C_6H_{12}O_6 + 6O_2 \rightarrow 6CO_2 + 6H_2O + \text{free-energy change} = 2870.2 \text{ kJ mol}^{-1}$$

During this process pyruvic acid is first formed by the glycolytic pathway. The energy still present in its molecule is then released through a sequential and self-renewing series of oxidative reactions that make up the citric or tricarboxylic acid (TCA) cycle. The pyruvic acid molecule is oxidatively decarboxylated by coenzyme A with the formation of acetyl coenzyme A and the loss of CO_2 and two hydrogen atoms (Fig. 1–8). The acetyl coenzyme A enters the TCA cycle, first condensing with oxaloacetate to give citrate; this is finally oxidized to carbon dioxide, oxaloacetate being regenerated during the process and thus becoming available for a further turn of the cycle. During the passage of one molecule through a complete cycle, four pairs of hydrogen atoms are removed, the overall equation for the cycle

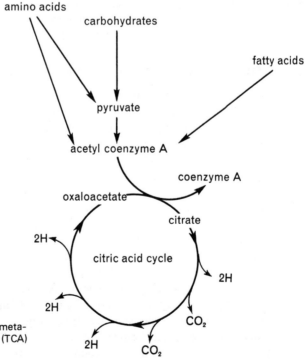

Fig. 1-8. Interrelationship of metabolic pathways with the citric acid (TCA) cycle.

being:

$$2\text{-acetyl CoA} + 2\text{-oxaloacetic acid} + 4H_2O \rightarrow 2CoA +$$
$$2\text{-oxaloacetic acid} + 16H + 4CO_2$$

Thus, for one initial molecule of glucose, 12 pairs of hydrogen atoms have been released: two pairs during glycolysis, two pairs in the conversion of pyruvic acid into acetyl coenzyme A, and eight pairs in the TCA cycle.

These hydrogen atoms are enzymically split into electrons and hydrogen ions. The electrons are passed along the system called the respiratory chain, in which the pigments known as cytochromes are prominent, and are used, through the mediation of cytochrome oxidase, to activate oxygen atoms. These combine with hydrogen ions to form water. During this electron transport sequence 36 more ATP molecules are formed, giving, in conjunction with the two formed during glycolysis, a net gain to the organism of 38 ATP molecules, amounting to about 1,178 kJ. Reference to the overall equation shows that this represents an efficiency of less than 50%. This, however, is only a measure of the efficiency with which energy is captured by the organism. The efficiency with which the energy is used in performing biological work is very much lower, and is little different from the efficiency of heat engines designed by man.

The TCA cycle serves also to complete the oxidation of fats (probably little used as an energy source by most invertebrate groups) and proteins. Glycerol is converted to triosephosphate and fatty acids to acetyl coenzyme A, while amino acids are deaminated to yield products which either enter the TCA cycle directly or via acetyl coenzyme A. Thus the cycle functions as a final common path for the katabolism of the three main categories of organic compounds, and for the efficient release of their contained energy at a temperature compatible with the maintenance of life, and in a form available, at least in part, for chemical work.

1-4 ORGANISMS AND ECOSYSTEMS

In passing from these preliminaries to the study of invertebrate structure and function it is only too easy to regard this as primarily a study of the bodies of whole animals, or of the separate (and sometimes very fragmented) parts of those bodies. But the biologist is concerned with very much more than this. We have now seen that the origin and establishment of life involved the interaction of different types of organism, and that the maintenance of life at the present day continues to be dependent upon such interactions. These are delicately balanced; so much so, for example, that the proportion of carbon dioxide in the atmosphere remains constant despite the continuous removal and replacement of this gas by the processes of photosynthesis and respiration.

Recognition of this principle of balanced interaction makes it possible to extend the analysis of adaptation in such a way that the study of animal structure and function becomes part of the wider field of ecology: a field which may be described, in Odum's phrase, as the study of the structure and function of nature. This is based upon the concept that organisms are linked together into associations which, in conjunction with the physical environment in which they exist, form what are called ecosystems. At this level of analysis each individual organism is a member not only of a species, but also of an integrated community which is regulated by processes developed by natural selection. We can here draw some analogy with what we shall see later of the evolution of the organization of cells and tissues to form bodies of increasing complexity. This organization depends upon the coordination and integration of the cells; it involves a measure of competition between them for the limited resources that are available for their maintenance; but it permits the cooperation that allows them an influence far greater than any one cell could exert on its own.

The relationships of organisms with their environment are no less complex. As we have suggested, the environment in the broadest sense of the term comprises inorganic or physical factors, and also the biotic factors that arise from the activities of the organisms themselves. All of these factors are involved in the 'struggle for existence', as was clearly perceived by Darwin, who emphasized that the term struggle is used in this context in 'a large and metaphorical sense'. The struggle is partly against the physical conditions of the environment, and partly against predators, parasites, and disease, and it involves also a testing competition for essential requirements that are limited in availability. Against this background the analytical study of invertebrate structure and function appears as more than an end in itself, absorbing though that end may be. It has also a broader aim, making an essential contribution to our understanding of the ways in which animals achieve those *éléments de prosperité*, or parameters of well-being, which are the measure of their capacity for survival in their ecosystems: the skills with which they move, feed, and digest, for example, or the vigour with which they grow and reproduce. But if we are to achieve this broader aim, it is necessary to keep in mind some simple formulation of the principles governing the structure and functioning of ecosystems.

One approach to this is to regard a community of organisms as being essentially a producer–consumer system, a simple example of this being the 'sulphuretum' of microbial ecologists. This is a stable system which depends on the metabolism of sulphur by sulphur bacteria, and which is found on black mud in environments where there is little oxygen and an abundance of reduced sulphur compounds. Carbon fixation in this miniature ecosystem depends on those autotrophic bacteria which we have earlier seen to effect the photolysis of H_2S instead of H_2O. The free sulphur which is released can be oxidized to sulphate, and this can then be reduced again to H_2S by heterotrophic sulphate-reducing bacteria. This ecosystem might well have been a dominant feature of life on the earth during its early anaerobic phase. At present, the sulphuretum makes an essential contribution to the sulphur cycle, but its further development has been prevented by the establishment of an oxygen-rich atmosphere. It is not certain, however, that this must always be an inevitable consequence of the development of life: it is conceivable that elsewhere in the universe there are oxygen-free atmospheres in which such an anaerobic system might flourish in greater elaboration.

Our concern here, however, is not with exobiology but with the life of this planet, dependent upon the obligatory interrelationships of the carbon and oxygen cycles, and the aerobic ecosystems in which they are expressed. Here the producer–consumer relationships can be stated in nutritional terms as food chains. At one end of any such chain are autotrophs. Succeeding members of the chain are the heterotrophs, which exploit the materials synthesized by the autotrophs; herbivores feed directly upon the latter, and are eaten in their turn by carnivores. A chain may include more than one group of carnivores (the first-stage ones being eaten by second-stage ones, with the top carnivores at the end), but it is unusual to find more than five stages in any one chain. At each of these stages there is typically an increase in the size of the individuals with a concomitant decrease in their number, an ecological principle formulated by Elton as the pyramid of numbers.

In practice, food chains are abstractions from a more complex system in which many different chains are interconnected to form a food web. Those species that occupy corresponding positions in their respective chains (herbivores, for example, or first-stage carnivores) are said to occupy the same trophic level. An example of such a web, illustrating the dependence of the herring (and hence of those that fish for it) upon the floating invertebrate life of the sea (plankton), is shown in Fig. 1-9.

There is, however, another way of expressing the organization of a community. This to state it in terms of its utilization of energy. Communities, like the individual organisms which compose them, are open systems, receiving a continuous input of materials and energy from an outside source. Those materials that do not yield energy (carbon, for example, or nitrogen) can circulate in the ways already mentioned, and can thus be used many times over. The fate of energy, however, is different, for it must conform to the laws of thermodynamics. These state that energy can be transformed but never created or destroyed, and that all such transformations result in energy-rich material being degraded into a less organized form, with an accompanying reduction in the amount of energy available for chemical work (free energy). Because of these limitations, the passage of energy through an ecosystem is a one-way flow and not a circulation. From this point of view, therefore, the functioning of the system must be represented as a flow diagram.

Fig. 1-10 illustrates such a diagram, prepared from studies of Silver Springs, Florida, an ecosystem in which continuous replenishment from underground main-

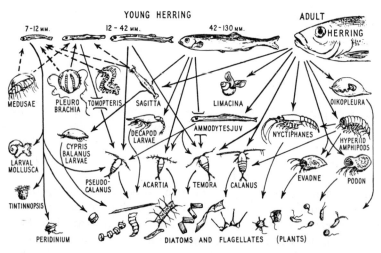

Fig. 1-9. The feeding relationships between herring of different ages and the members of the plankton. From Hardy, 1959. *The Open Sea II*: *Fish and Fisheries*: Collins, London.

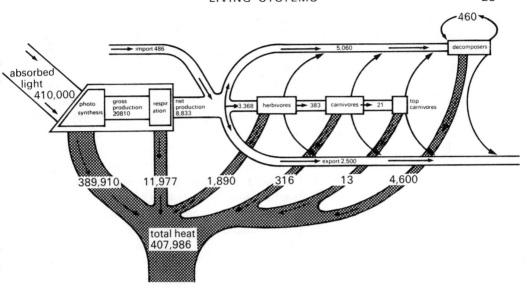

Fig. 1-10. Energy flow diagram with estimates of energy flow in kilocalories per square meter per year in the Silver Springs (Florida) community. From Phillipson, 1966. *Ecological Energetics.* Edward Arnold, London. After Teal, 1957. *Ecol. Monog.*, 283–302.

tains a marked constancy in the temperature and composition of the water. Solar radiation constitutes the main input of energy (410,000 kcal yr^{-1} m^{-2}). Some 5% of this (20,810 kcal g^{-1} m^{-2}) is incorporated into the tissues of the autotrophs, the remainder being lost as heat, either through evaporation or as a by-product of photosynthesis. This incorporation constitutes the primary production of the ecosystem, the total amount of energy so taken up being termed the gross primary production. Much of this is lost as heat in the respiration of the primary producers, the amount left after providing for this metabolic activity being the net primary production (8,833 kcal g^{-1} m^{-2}).

The stored energy of the autotrophs (supplemented by a small import of bread fed to the fish) is taken up by the heterotrophs, which are the herbivores, carnivores and top carnivores that constitute the grazing food chain. Part of this energy is assimilated by them, while the rest is lost in their faeces. This material, together with the dead bodies of the autotrophs and heterotrophs, gives rise to organic debris which is acted upon by bacterial and other decomposers, and is lost from the ecosystem by the downstream export of particulate organic matter. In some ecosystems, however, it would be deposited as the sedimentary waste called detritus. This provides a separate source of food which is exploited by the specialized detritus feeders and by those carnivores that prey upon them. Thus, argues Odum, there are two main categories of food chains; the grazing chain and the detritus chain. Phillipson has emphasized the importance of the distinction between the two, pointing out that the detritus or decomposer chain may often be of greater importance than the grazing one as regards energy flow. In Silver Springs, for example (Fig. 1–10), more energy flows through the decomposers than through the grazers (5,060 kcal g^{-1} m^{-2} as compared with 3,368 kcal g^{-1} m^{-2}). We shall see later how animals have become diversely adapted to exploit one or other of these two patterns of life, and how this has affected the course of evolution. Detritus and deposit feeding, for example, have been important in determining the early history of annelid worms and, probably, the establishment of metameric segmentation, while the grazing habit, by contrast, has greatly influenced crustacean organization.

The evidence presented in summary form in Fig. 1–10 permits the drawing up of a balance sheet of the organic matter in Silver Springs (Table 1–1). This illustrates the subjection of ecosystems to the first and second laws of thermodynamics. The total influx of energy is seen to equal the total efflux (with a small discrepances inevitable in such estimates), while most of the assimilated energy is lost as heat of metabolism, which, as we have explained, is then unavailable for further use by the organisms that have released it. In other types of system, however, some will be stored in the tissues of the heterotrophs. This storage, which results in an increase of organic material, constitutes the secondary production of the system.

Secondary production, and the importance of detritus feeders, are both illustrated by the flow diagram (Fig. 1–11) for the simple ecosystem of Root Spring, a

Table 1–1 Estimated annual balance sheet of organic matter (dry weight in g yr^{-1} m^{-2}) for Silver Springs, Florida. From Teal, 1957. *Ecol. Monog.*, **27**, 283–302.

	Credit		Debit
Primary production	6,390	Community metabolism	6,000
Bread fed to fish	120	Downstream export	766
Total income	6,510		6,966
Discrepancy 256			

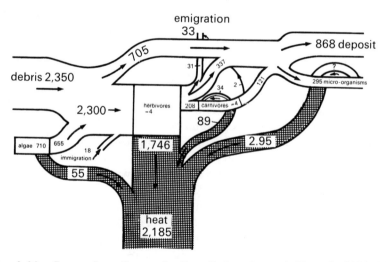

Fig. 1-11. Energy flow diagram for Root Spring. Concord, Mass. in 1953–54. Figures in kilocalories per square metre per year. Numbers inside boxes indicate changes in standing crops. Arrows indicate direction of flow. From Phillipson, 1966. *op. cit.* After Odum, 1957. *Ecol. Monog.*, **27**, 55–112.

temperate cold spring in Concord, Massachusetts. The data refer to the small spring basin, the bottom of which is covered with the mud and debris that contains the organisms. In contrast to Silver Springs, most of the energy assimilated by the heterotrophs is here imported from outside, largely as fallen leaves of apple trees and other débris, but to a small extent as immigrant caddis larvae. The balance sheet for this system is shown in Table 1–2. Secondary production is observable, some of it being exported by emigration of adult insects, but some being temporarily deposited as detritus. Eventually, however, the deposits are washed out in the spring brook, so that there is no permanent increase in usable energy within the system.

These, then, are illustrations of the larger background against which animal organization is to be viewed, and its adaptive significance assessed. Of course, not all of the structural and functional specializations that we shall be considering are directly related to nutrition; many of them are concerned with the other problems that animals face in dealing with their physical and biotic environment. The control of water content and ionic balance are examples, as also are the regulatory activities of the nervous and endocrine systems, and the ensuring of reproduction and of the successful completion of the life history. All adaptations, in these and other ways, contribute to survival, and so it is that the full significance of invertebrate structure and function only becomes apparent when it is seen as the means by which any given species maintains in its community the characteristic role which we term its ecological niche. Elton once wrote that for a biologist to see a badger should be as meaningful, and should arouse as many associative ideas, as if he were to see the vicar. Odum extends this happy thought by suggesting that we should readily recognize in any animal its address (which is its habitat, or the place where it lives) and its profession (which is its niche within that habitat).

So the ecologist at Root Spring sees his two species of planarian (*Phagocyta gracilis woodworthi* and *P. morgani*) as cannibalistic predators, and suspects that the one with multiple pharynxes may be the more aggressive of the two. Recalling that other macroscopic animals are said to be disenchanted with planarians as food, he rates these two species as top carnivores, with the corollary that those escaping destruction by cannibalism must be broken down by micro-organisms. And then, finding that only 12% of their assimilated energy is transformed into heat of metabolism, as compared with 74% for the detritus-feeding oligochaete *Limnodrilus*, he reflects on the importance of mucus in the locomotion and feeding of planarians (Sec. 4–5), and concludes that it might be worth investigating how much of the energy assimilated by *Phagocyta* may be lost in that secretion.

Table 1–2 Estimated annual balance sheet of energy (kcal yr^{-1} m^{-2}) for Root Spring, Concord. From Odum, 1957. *Ecol. Monog.*, **27**, 55–112.

	Credit		Debit
Organic debris	2,350	Transformation to heat	2,185
Gross primary production	710	Deposition	868
Import of caddis larvae	18	Export of adult insects	33
Decrease in standing crop	8		3,086
	3,086		

One last point needs to be made at this stage. Food webs and flow diagrams are static models, which tend to give an impression of stability in the structure and function of nature. Stability is indeed a necessary element in biological organization, but organic evolution depends upon a subtle balance of stability and change. Ecosystems are balanced systems which, like the whole of the biosphere (the inhabitable zones of this planet) to which we belong, are an expression of organic cooperation. But there is also interspecific competition, and it is this, in conjunction with genetic variation, which results, as we saw earlier, in animals extending their range and exploiting new possibilities in their habitats. Dynamic thrust is as much a part of the living world as are stability and regulation. Varying in extent from one species to another, it has nowhere been better shown in recent years than by man, rocketing into space and devising artificial ecosystems to take with him. The Irishman who did not begrudge his fleas their meal, but who objected to their 'continial thramplin', had perceived a fundamental characteristic of animal life.

2
Organization and Life

2-1 HOMOLOGY AND ANALOGY IN ORGANIZATION

Life depends upon more than the mere association of complex molecules; it depends also upon the patterns in which these are arranged, for the coordinated functioning of diverse metabolic pathways could not be secured without the orderly positioning of their component parts. This has long been appreciated, and the existence of structural patterns in protoplasm long suspected. The full scope and significance of the concept, however, only became apparent when electron microscopy revealed a new world of ultrastructure, strikingly similar in its fundamental organization throughout the animal and plant kingdoms. This similarity inevitably recalls the resemblances that we have already noted in the metabolic pathways by which energy is transferred. These, we suggested, might have been established once and for all during the chemical phase of evolution, and it is clear that structural resemblances might be accounted for in the same way. This explanation is not one that can ever be tested either by experiment or by direct observation. Yet it carries, both in its biochemical and its structural contexts, important implications which are bound to influence our interpretation of the history and relationships of living organisms. For this reason it merits closer consideration.

Unity of pattern is a very familiar feature of the structure of animals at the level of gross morphological analysis, and one of the aims of the comparative anatomist must be to discover such unity. The same is no less true of the comparative biochemist and the comparative physiologist, for it is a primary aim of all the sciences to expose order in natural phenomena. The morphologist, in his analysis of unity and diversity, discriminates between two different types of resemblance, known as homology and analogy. This important distinction is a pre-Darwinian concept, and was, in fact, first clearly enunciated by Richard Owen in connection with his analysis of vertebrate organization. From this point of view an organ in one species is said to be homologous with an organ in another species when the two organs are similar in their fundamental structural plan, irrespective of the functions which they may carry out. Associated with this structural similarity will be similarity of development and of relationships with adjacent structures. In contrast to this, the two organs would be termed analogous if they carried out similar functions, but were fundamentally different in their structure, development, and anatomical relationships.

The distinction, as Hertwig argued, is based in this classical formulation upon a concept that retains its validity, regardless of how it is explained. Nevertheless, the current explanation of the distinction is inevitably an evolutionary one. The possession of homologous organs by two species is attributed to the descent of these species from a common ancestral plan of structure; analogous organs are attributed to the independent evolution of similar functions in unrelated species, with different ancestries. However, the analysis has been complicated by the recognition that what animals inherit is genetic information coded in their DNA molecules, and not the structures that develop as a result of the exploitation of this information. It must therefore be accepted that species may possess organs that are homologous in their relationships, but that need not have been present in a common ancestral form. They may be possessed by the species concerned because these have inherited genetic systems with common potentialities for developing those organs. In other words, two related species, living in similar conditions, may, under the influence of natural selection, independently evolve similar adaptations because they have inherited from a common ancestor the potentiality for reacting in similar ways to similar demands. This type of resemblance is often termed latent homology, while the organs concerned are said to be homoplastic.

The question suggested in our present context by these considerations is whether the unity of pattern found in the fundamental biochemical and ultrastructural architecture of living systems is to be interpreted in terms of classical homology, latent homology, or analogy. Naturally we cannot feel sure of the answer. We can only repeat that the widespread distribution of such similarities throughout living organisms at least establishes some probability that they were evolved at an early stage of chemical evolution, and that they then became part of the common inheritance of all later forms of life. On the other hand, it must be remembered that life as we see it on this planet depends upon the unique properties of a limited range of complex molecules. Probably, therefore, there are relatively few ways in which these molecules can be brought together to form biochemical and ultrastructural systems, so that it is not impossible that similar systems might have arisen independently in more than one line of evolution, in a way which would fail to satisfy the classical concept of homology. The fact is that this concept, of such fundamental importance in comparative studies, has become increasingly difficult to define.

If this uncertainty of interpretation exists at the fundamental level of analysis that we are at present considering, it must become increasingly pronounced when we examine the complex and specialized systems that have been erected on these foundations during later evolution. It is a doubt that underlies Gray's warning concerning the uncertainties surrounding the evolutionary interpretation of comparative physiology. He suggested that in their physiological activities animals are so opportunistic that any attempt to approach the problems of phylogeny by the physiological pathway can only lead to confusion. This view may seem unduly pessimistic, but it is a valuable reminder of the need for an unceasingly critical approach to the interpretation of animal organization at all levels of analysis.

2–2 EUCELLULAR PLAN OF STRUCTURE

The best-known feature of biological organization, and one that does not rest on electron microscopy for its demonstration, is the nucleate plan of structure: the differentiation, that is, of a unit of protoplasm into cytoplasm and nucleus, the latter being demarcated by a nuclear membrane. We know that the special significance of this pattern is that the polymerized nucleotides of DNA are organized within the nucleus into the complex bodies called chromosomes. The establishment of these must have been the result of a long evolutionary process. It is probable that during the earlier phases of this process the DNA molecules existed in a less highly organized form, not yet enclosed by a nuclear membrane. This stage is still found today in the blue-green algae and the bacteria. The simplest arrangement is seen in certain sulphur bacteria, in which the DNA is present as randomly scattered granules, and in the blue-green algae, in which it is distributed as irregular strands. Division takes place by the ingrowth of a new cell boundary; the DNA is then distributed between the daughter cells without being precisely divided. In other bacteria there may be a single chromosome forming a circle. Such chromosomes, in contrast to those of higher organisms (see later), consist solely of DNA without any protein component.

Apart from these supposedly primitive exceptions, the majority of living organisms, plant as well as animal, are composed of the units formed by nuclei and their associated masses of cytoplasm. These units are the structures that we call cells, a name that derives from the observations of seventeenth-century microscopists, notably Hooke, Grew, Malpighi, and Leeuwenhoek. The term was initially applied to the thick-walled cavities that are readily visible in plant material; not until these observations were extended to animal material was increased emphasis placed upon the living contents of these units. To this extent the term has always been somewhat unsatisfactory, while there is the further difficulty of drawing a sharp distinction between this plan of structure and that of bacteria. Both depend upon the differentiation established between the cytoplasm and the coded instructions in the DNA molecules, but they clearly represent two grades of cellular organization. It is thus helpful to regard bacteria and blue-green algae as being 'procaryote'. They lack the true nuclei and membrane-bound organelles (e.g. mitochondria and plastids) which characterize all other organisms. These, which are termed eucaryote or eucellular, are distinguished by having their DNA molecules carried as chromosomes within the nucleoplasm of a fully differentiated nucleus that is separated from the cytoplasm by means of a double nuclear membrane. It is supposed that they were preceded in evolution by procaryotes.

Further characteristics of eucellular organization are the presence in the nucleus of one or more granular nucleoli, rich in RNA, and the complexing of the DNA helix with basic proteins (typically histones). These proteins provide a structural matrix which must have been crucial in favouring the evolutionary progress of the higher organisms. The single circular chromosome of *Escherichia coli* is believed to contain about 6×10^6 base pairs, while estimates of the total length of the extended DNA fibre range from 1.5 to 3 mm. In contrast to this, a single (haploid) set of coelenterate chromosomes contains about 1.8×10^8 base pairs, and a single haploid set of human chromosomes about 1×10^{10}. These figures should not be taken too precisely. When, however, they are coupled with the estimate that the total length of the fully extended DNA of the haploid set of human chromosomes may reach at least 0.5 m, it becomes evident that advances in animal organization have demanded vastly increased stores of information in the cells. Manifestly the effective translation and replication of that information requires an orderly and stable arrangement of the DNA, and it is this that is supposedly provided by the histone framework.

Nuclei with these characteristics, and with mitotic division of the chromosomes involving the development of spindle threads, are common both to animals and to seed-bearing plants. This establishes the strong probability that they were already evolved at the level of organization that was characteristic of their common ancestral stock. We shall later see more than one reason for regarding flagellate protozoans as being the closest to that stock of all Protozoa, and as antedating in their origin the other groups of living protozoans. Because of this we may reasonably expect to find in Protozoa a nuclear organization similar to that of plants and metazoans and derived, like theirs, from that same common origin. While, however, the Protozoa clearly do possess well-differentiated nuclei, the modes of division of these are so varied, and sometimes so unlike what is seen in a metazoan cell, that it was for long believed that nuclear division in Protozoa was predominantly amitotic; it was further supposed that in these organisms we see stages in the evolution of mitosis. This view, however, is certainly erroneous; mitosis is fully established in the Protozoa, although it admittedly varies greatly in its form.

In many protozoans the process corresponds closely with that of a typical metazoan cell. The centriole divides, a spindle of microtubular fibrils forms between the two products, and the divided chromosomes form a metaphase plate. Sometimes the basal bodies of flagella play the part of centrioles, which they resemble in their ultrastructure (Sec. 3–2). Not uncommonly, however, the spindle forms within the nucleus, the nuclear membrane persisting throughout mitosis. This happens, for example, in euglenids (Fig. 2–1), where, in addition, the chromosomes do not form a metaphase plate and move at different speeds during anaphase. Apparently they lack centromeres.

The ciliates present a special case, because of the differentiation of their nuclear material into micronucleus and macronucleus. The latter is dense in appearance, because it is in a polyploid condition as a result of repeated mitosis occurring within the nuclear membrane. Presumably in correlation with this, it undergoes amitotic division, accomplished by simple elongation and constriction. This involves the participation of a bundle of intranuclear microtubular fibrils, which suggests the derivation of this process from a more typical mitotic division. It is said that protein synthesis within the cytoplasm depends on macronuclear RNA, but there is, of course, nothing anomalous in this, since the macronuclei of ciliates are formed from division products of micronuclei. Indeed, the peculiar nuclear processes of the ciliate life cycle only serve to emphasize the general rule of the dependence of protozoan organization upon mitosis, for it is the micronuclei that transmit genetic material to the next generation, and these undergo mitotic division with an intranuclear spindle.

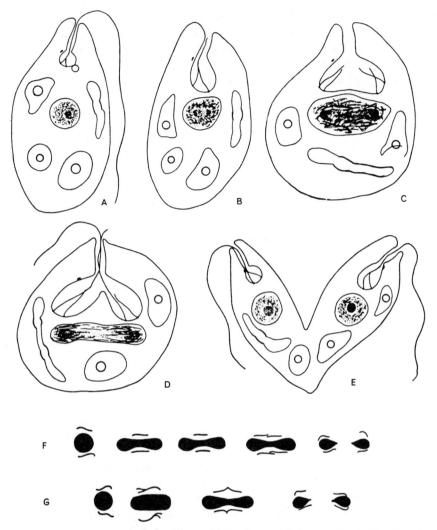

Fig. 2–1. Binary fission in *Euglena*. A-E, fission: semi-diagrammatic. Duplication of chromosomes, here shown at C (metaphase), is in some species first detectable earlier (at B, prophase) or later (at D, anaphase). F and G illustrate alternative theories of chromosome duplication and separation of chromatids. In F the chromosomes are split and their products slide one on the other towards opposite poles. In G, the chromatids are thought of as stripping apart in a widening 'V'. (A-E combined after various authors; F and G adapted from Leedale.) From MacKinnon and Hawes, 1961. *An Introduction to the Study of Protozoa*. Clarendon Press, Oxford.

2–3 CELL THEORY

The interpretation of plants and animals as being composed of cells and their products is commonly referred to as the Cell Theory. Its enunciation in the early nineteenth century by Schleiden and Schwann constituted the formulation of one of the great generalizations of biological science. Baker, in his review of the history of the theory, points out that in the form in which it was first stated by Schwann, in 1839, it was primarily a theory of the embryological development of organic structure. Schwann believed this development to take place in two stages, the first being the differentiation of a structureless substance into structural units, and the second being the differentiation of these units, or cells, into the characteristic organization of the adult. A conversation with Schleiden led Schwann to believe that the first stage in the formation of a cell was the development of a nucleus within a structureless fluid; a limiting boundary, the cell membrane, was then secreted by the nucleus, and afterwards the body of the cell appeared within the membrane.

When, later in the century, this belief was shown to be quite incorrect, the Cell Theory in Schwann's original sense became untenable. By this time, however, it had become recognized that most organisms, both plant and animal, could certainly be regarded as formed of units, consisting of nucleus and cytoplasm. This, therefore, became the fundamental tenet of the Cell Theory. It was a different theory from that originally enunciated by Schwann, yet he had observed a fundamental truth, the understanding of which had progressed with advancing knowledge. In this there is nothing unusual. It is not only organisms that evolve; so also, if in a different sense, do our descriptions of their properties, and our attempts to explain them.

To speak of the Cell Theory as based upon a fundamental truth is not to say that the issues raised by it are straightforward and its implications universally acceptable. On the contrary, it has from time to time been the centre of vigorous, not to say passionate, controversy, which has been focused on two main matters of dispute. One of these concerns the nature of individuality in living organisms. It raises the question whether eucells are individual and isolated units, or whether they should be viewed solely as parts of an organism. The other matter is closely connected with this, and raises the question of the significance of our customary recognition of two main subkingdoms of the animal kingdom, the Protozoa and the Metazoa. The bodies of protozoans each contain (with some important exceptions) a single nucleus. The metazoan body, by contrast, is composed of many eucells. Is the protozoan, then, to be regarded as unicellular, comparable with a single cell of a metazoan? Or is it to be regarded as a non-cellular organism, comparable with the whole of the metazoan body? The issue was argued with greater vigour in the past than it is ever likely to be again, yet it still deserves attention, particularly in the light of our earlier reference to the principles of scientific method.

We have seen that the progress of science depends upon the formulation of hypotheses, which are then subjected to tests designed to disprove or corroborate them. Usually these hypotheses will be limited in scope, for this makes it easier to test them. If they range over too wide a field of thought, and include too extensive a range of possibilities, it may be difficult to find ways of disproving them. It is in this respect that the predictions of the Delphic oracle differed from the assertions of the twentieth-century scientist—or so we should wish to believe. However, in our search for order in natural phenomena, and for the means of controlling them, we cannot afford to rest content with limited hypotheses. We prefer instead to systematize and extend our understanding by framing what are called theories. These are attempts to achieve a higher level of integration by grouping a series of statements into a system which is intended to provide a logical account or explanation of a range of natural phenomena. The individual statements may themselves be hypotheses, or they may be axioms (matters that are taken for granted). Or again they may be straightforward raw material of the kind called assertions or postulates, which are often just simple working ideas formulated as a basis for action. These various terms carry no sharply defined meanings. Indeed, the term 'theory' itself is sometimes used as synonymous with 'hypothesis', but the sense in which we are using it here is a common one, and it is in this sense that we can speak of the Cell Theory.

The content of this theory has been usefully analysed by Baker into seven propositions, stated here in summary form:

1. Most organisms contain or consist of a large number of microscopical bodies called 'cells'.
2. Cells have certain definable characters which show that cells are all of essentially the same nature and are units of structure.
3. Cells arise from pre-existing cells.
4. Cells sometimes become transformed into bodies no longer possessing all the characters of cells.
5. Cells are to some extent individuals, so that in most organisms there are two grades of individuality, that of the cell and that of the organism as a whole.
6. Each cell of a metazoan corresponds in certain respects to the whole body of a simple protist.
7. Metazoa and Metaphyta probably originated by the adherence of protist individuals after division.

If these propositions are viewed as a whole, it becomes clear why there has been such disagreement regarding the validity of the Cell Theory. The last of them is a statement of a hypothesis, it cannot be tested by experiment, and it certainly would not be universally accepted; there is a strongly argued view (Sec. 18–4) that Metazoa arose by the subdivision of a multinucleate protozoan. This view implies, contrary to the sixth proposition, that it is the whole body of a metazoan that corresponds with the whole body of a protist. It is because of this disagreement that one writer can homologize the protozoan body with a single cell, and another deny this homology. Proposition 5 also creates difficulties, for it raises the question of the nature and definition of individuality, while proposition 2 rests on an agreed selection of the 'definable characters' concerned, and leaves unresolved the question of what is meant by being of 'essentially the same nature'.

These difficulties, which could be pursued in much greater detail than is possible here, show that the Cell Theory in its entirety is too complex and far-ranging to provide an effective tool for an analytical approach to animal organization. The root of the problem has been clearly stated by Woodger. The classical definition of the cell as a mass of protoplasm with its associated nucleus is a highly abstract concept, and it is because of this that the concept can be so widely generalized. Not only is this concept of the cell an abstraction from the more complex reality that we perceive with the microscope, but even this more complex perceptual object is still an abstraction. This is necessarily so, because even the most perfect microscope preparation fails to reveal all the details of cell organization, and it inevitably omits the greater part of the functional relationships of the cell. Indeed, the concept of the individuality of the cell becomes progressively less meaningful as we learn more and more of the control of cells by neural and humoral processes, their dependence upon metabolites, and their regulation by internal and external feedback mechanisms.

There is much to be said for a suggestion put forward by Grimstone. Arguing from the abstract nature of the cell concept, he points out that our interpretation of cell problems, and specifically of the relationship of the cell to the protozoan body, must depend upon the context in which a particular problem is put. In some respects (and we shall see more of this below) it is undoubtedly possible and useful to compare the protozoan body with a single cell of a metazoan, and the conclusion is an empirical one (arising, that is, from observation and experiment, and not from theory). It may be regarded as a necessary consequence of the point that we have earlier made: that life is dependent upon certain patterns of molecular organization. Such a comparison is likely, therefore, to be helpful in furthering our understanding of cytological problems. In other respects, as in considerations of behaviour and life history, the individual protozoan must be compared with the whole metazoan organism, regardless of what view we may take of the way in which the Metazoa evolved from Protozoa.

Looked at in this way, there is no simple and final answer to the question whether the Protozoa are, in the classical sense, unicellular or non-cellular. The answer we give depends upon the form of the question that is being asked, and is not a complete and definitive statement. It is simply a recommendation to treat the facts in a particular way in that particular context. This flexibility of approach to problems of definition and interpretation in the biological sciences is a quality essential for the biologist to cultivate.

The need for this quality is well enough seen if we seek in a standard dictionary for a definition of 'organism'. It is, we learn, an 'organized body with connected interdependent parts sharing common life', and a 'whole with interdependent parts compared to living being'. On these definitions, the metazoan cell is clearly an organism, although it is also part of a larger organism that operates at a higher level of integration. Moreover, it is impossible to accept a lack of complete independence as an absolute and distinctive feature of a cell. The most superficial consideration of the living world shows that no organism, protozoan or metazoan, can live a life entirely independent of others, and we shall see in due course that the mechanisms by which members of communities are linked together and subjected to mutual control may be no different in fundamental principle from those that regulate the lives of cells within metazoan bodies.

2-4 MEMBRANES AND ORGANIZATION

The widespread distribution of the differentiated nucleus of the eucellular plan of structure is one aspect, and a familiar one, of uniformity in the organization of living material. Others have emerged in the more penetrating analysis made possible by electron microscopy.

One example is the widespread development of membranes. We have already referred to the membrane of the nucleus; others are found both within the cell and over its surface (Fig. 2-2). The surface membrane (plasma membrane) has a tripartite structure, detectable with electron microscopy as a central light zone

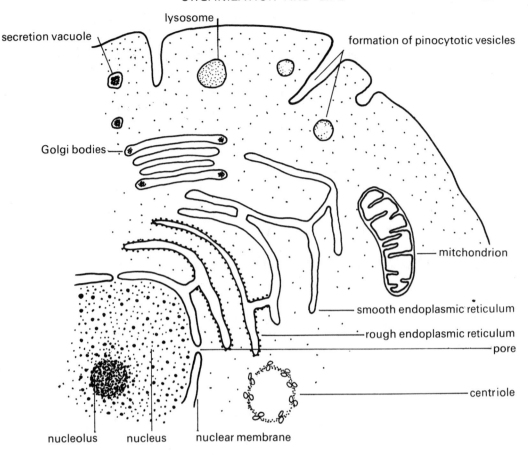

Fig. 2-2. Some features of the organization of an animal cell (not to scale).

separating two dark ones; its total thickness is about 7.5 nm, the dark zones being each about 2 nm thick and the light one 3.5 nm. This type of membrane is often called a unit membrane. Its properties include low permeability to ions and water-soluble substances, and preferential permeability to lipid-soluble ones. This suggests that it has a major lipid component, but its elasticity, mechanical strength, and low surface tension indicate the presence of protein. These considerations led Danielli and Davson to propose, prior to the use of electron microscopy, that the plasma membrane consisted of two layers of lipid (mainly phospholipid) molecules, with their polar (electrically charged) surfaces facing outwards, and bearing a covering of adsorbed protein. This interpretation (Fig. 2–3) has been widely accepted, for electron micrographs subsequently confirmed it in a striking way, the central layer corresponding with the postulated lipid zone, and the outer layers with the protein component, carbohydrate and additional protein being supposedly present over the surface. However, continuing analysis of the properties of the membrane have suggested a less homogeneous structure, stable enough for protection, yet capable of adaptive regulation of the relationships of the cell with the

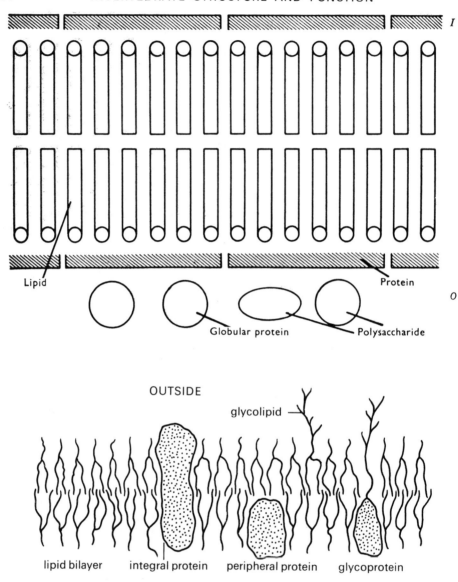

Fig. 2-3. Changing concepts of membrane structure. *A*, an earlier model, in which the membrane is regarded as essentially homogeneous. *I*, the inner surface with associated layer of enzymes, microfibrils, etc. *O*, outer surface of protein, mucoprotein and mucopolysaccharide. From Ambrose and Forrester, 1968. *Symp. Soc. exp. Biol.*, **22**, 237–248, **B**, a later model, in which the membrane is regarded as a structural, biochemical and functional mosaic. Globular integral proteins extend across the lipid bilayer, while peripheral proteins are mainly associated with the inner surface. A glycocalyx of glycolipids and glycoproteins is found on the membrane surfaces. Conformational changes in the proteins could result in the formation of channels. From Oschman *et al.*, 1974. *Symp. Soc. exp. Biol.* **28**, 305–350.

medium surrounding it. It seems certain that the evolution of primordial organisms, capable of establishing independence of the ambient medium, could not have advanced far without the development of some such membranous covering. Once the capacity for membrane formation had been established, it would also have provided for the orientation and integrated relationship of large molecules, with consequent improvement of the organization of metabolic pathways. Whether, however, the Danielli–Davson model can be generalized to apply also to intracellular membranes is uncertain, especially in view of the wide range of properties attributed to them.

The biological significance of the plasma membrane lies particularly in the opportunity that it provides for the controlled exchange of materials between the cytoplasm and its environment. This exchange is aided, however, by other factors. For example, the surface of absorptive cells is often increased by being prolonged into delicate processes called microvilli, which constitute the brush-border of light microscopy. It is probable, too, that pores are present in the membrane, for this would explain why molecules of different sizes pass through with different degrees of ease (passive pore diffusion), while the lipid component doubtless aids the movement of lipid-soluble molecules.

Many substances, however, cross the plasma membrane at speeds faster than would be possible by passive diffusion. It is supposed that they are transported after becoming attached to specific binding sites on fast-moving carrier molecules, which are estimated to be able to move to and fro across the membrane at as much as 180 times per second. Such carrier-facilitated transport may simply speed up movement down a concentration gradient (facilitated diffusion), or it may provide for movement against a gradient, from a region of low concentration to one of high concentration (active transport). The formation and movement of these carrier-solute complexes necessarily involves the expenditure of metabolic energy by the cell.

Diffusion and carrier-facilitated transport are not, however, the only means by which material can enter cells, for some can be taken in by pinocytosis. Here the plasma membrane invaginates (Fig. 2–2), engulfing droplets of the surrounding medium which are then pinched off into vacuoles. Outward (or reversed) pinocytosis is also possible, and is thought to be involved in the secretory discharge of hormones, for example, and in the discarding of unwanted substances. We see, then, that a plasma membrane has the complexity characteristic of all biological organization, involved as it is in activities that are not completely predictable in purely physical terms. Its properties, moreover, are continuously variable, and are subjected to regulation both from within the cell and from agents acting upon it from the external medium.

Unit membranes are as much characteristic of the surface of the protozoan body as they are of the surface of the metazoan cells, so that here, as in other respects to be mentioned later, direct comparison of the two types of organism is both permissible and profitable. Since, however, a protozoan depends upon this membrane for the protection of its body surface, there are necessarily various modifications and elaborations of its structure. It is found in a relatively simple form as the surface membrane (plasmalemma) in the amoeboid subclass Rhizopodea, although even here it may be modified to bear a continuous covering of filamentous molecules, as in *Amoeba*, where the filaments are thought to be mucoproteins that possibly facilitate adhesion of the animal to the substratum.

In many other protozoans the body surface is much more complex than this. It may be thickened to form a highly differentiated layer like the longitudinally striated pellicle of *Euglena*; this is formed of two separate membranes, about 8 mμ apart (1 mμ = 10 Å), the outer one being a unit membrane of the general type just described. Still more complex is the body surface of the Ciliatea. Light microscopy reveals that in *Paramecium*, for example, there is a cortex which is sculptured over its surface into a regular series of polygons, with a cilium (or sometimes a pair of cilia) arising from the middle of each (Fig. 2–4). The cortex as a whole is covered by a typical unit membrane which extends onto the cilia. The polygons are defined by a correspondingly regular series of cavities or alveoli, which form the alveolar layer that has also long been known from light microscopy. These alveoli are delimited by membranes, which are continuous from one alveolus to another. Thus in effect the body surface of *Paramecium* consists of a series of three membranes, the inner two of which are separated from each other at intervals to form the alveolar cavities.

We have seen that in the eucellular plan of structure the nucleus is separated

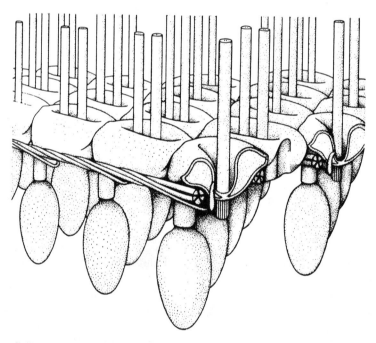

Fig. 2-4. Diagrammatic reconstruction of the cortex of *Paramecium*. Each pair of cilia emerges from the centre of a polygon; the pairs of inflated alveoli defining the polygons are shown in section at the right edge of the stereogram. Parasomal sacs are shown adjacent to the cilia in these polygons. Resting trichocysts alternate with the polygons in longitudinal rows. Kinetodesmal fibres form loose cables paralleling each kinety. (See also p. 49.) From Corliss, 1961. *The Ciliated Protozoa*. Pergamon Press, Oxford.

from the cytoplasm by a nuclear membrane. This is significantly thicker than the plasma membrane, for it commonly reaches a thickness of at least 20 mμ, and sometimes much more than this. In *Trichonympha*, for example, there are two membranes, each about 7 mμ thick, separated by a space about 23 mμ thick. The outer membrane is perforated by pores, which probably facilitate the passage of large molecules. It seems also to be common in Metazoa, and perhaps also in Protozoa, for this membrane to extend outwards at intervals and to become continuous with the system of membrane-lined cavities known as the endoplasmic reticulum (see below). It is reasonable to suppose that these extensions, like the pores, could provide pathways for the exchanges that must be continuously occurring between the nucleus and the cytoplasm.

The importance of membranes in protoplasmic organization is by no means confined to the surfaces of the cell and of its nucleus. It was suspected long before the introduction of electron microscopy that the coordinated functioning of the diverse metabolic pathways of the cell must depend on the separation of their components by some form of compartmental division of the cytoplasm. This belief has now received ample corroboration, one example being the structure of the cell organelles known as mitochondria (Fig. 2–2). The surface of these is formed of two membranes, about 10 nm apart, separated by material called the matrix. (The inner membrane has been thought to bear stalked globule-like structures, but this has been disputed.) The outer membrane is freely permeable to small molecules, but not large ones, while the inner membrane is permeable to both. It is therefore to be expected that substrates metabolized within the mitochondria may be moved by active transport through the inner membrane. Particularly characteristic of mitochondria are projections of the inner wall called cristae. In metazoans and land plants these are usually flat plates, as they are also in some flagellate protozoans. Most protozoans, however, have tubular cristae, although in some they are vesicular.

The explanation of the complex structure of mitochondria is that they provide for the functioning of the TCA cycle and respiratory (electron transport) chain. The enzymes of the former are carried in the outer membrane, while the carriers and enzymes of the transport chain are attached to the inner membrane. Because of this, dehydrogenation occurs at the mitochondrial surface; the matrix then provides for active transport of electrons and protons between the surface and the interior, ADP moving as well. ATP and water form within the mitochondrion, and are then passed out. So, to quote Palade, enzyme complexes may be thought of as 'built in, or woven into, the texture of the mitochondrial membranes in the same manner as repeated decorative patterns are woven into a sheet of damask'. The significance of the analogy lies in the element of pattern that it emphasizes, for this is what the evolution of protoplasmic membranes made possible.

We have seen reasons for believing that aerobic oxidation must have been established early in the evolution of life. The existence of a common pattern of mitochondrion in Protozoa and Metazoa is in keeping with this view, for it suggests that both groups may well have inherited this type of organelle from a common ancestry. The nature of this ancestry is a matter for speculation, but one suggestion is that mitochondria originated from aerobic bacteria which invaded anaerobic cells and, by providing them with the essentials of the TCA cycle, established a symbiotic relationship with them (Sec. 22–1). The suggestion is supported by the presence in mitochondria of DNA, ribosomes, RNA polymerase, messenger RNA and transfer RNA, all of which could well be the remains of bacterial metabolic machinery. Arguments can, however, be brought against this suggestion (which, incidentally, has also been applied on similar grounds to account for the origin of chloroplasts), and an opposing view is that mitochondria arose by the folding of the plasma membrane and the localization upon the folds of adaptively valuable sequences of enzymes. Obviously there can be no certainty here.

It should be added that conclusions regarding the functioning of mitochondria in Protozoa are to some extent inferential, for little direct evidence is available. There is, however, some indirect evidence that suggests their involvement in oxidative metabolism. For example, mitochondria are absent from *Trichonympha*, which lives within the alimentary tract of termites. This protozoan is an obligate anaerobe, presumably in adaptation to life in an environment in which the supply of oxygen is limited. In these conditions the animal is unlikely to be able to carry out the oxidative phosphorylation that is the end result of mitochondrial activity in the metazoan cell, so that the loss of mitochondria would be an understandable consequence of physiological adaptation to an entozoic mode of life.

The cytoplasm contains other sac-like structures, similar in size to mitochondria, but quite distinct from them in structure and function. These are the lysosomes (Fig. 2–2), which contain digestive enzymes that have pH optima in the acid range. The enzymes are used to digest material that has been ingested by the cell, for which reason lysosomes are particularly abundant in phagocytic cells. Their operation in this instance depends upon fusion of the lysosome membrane with the membrane of the ingestion vacuole. In addition, however, they are also responsible for self-digestion (autolysis) of tissues, a process which is best documented in the regression of the tail of the metamorphosing frog tadpole.

Another feature of the metazoan cell is the presence of groups of membranes, usually arranged as stacks of flattened sacs or cisternae, and forming a system of intercommunicating channels called the endoplasmic reticulum. Some of these membranes, which bear granules on their outer surfaces (Fig. 2–2), are the rough membranes or ergastoplasm, chemically characterized by the ribonucleoprotein composition of the granules, which are the ribosomes (Sect. 1–2). These granules consist of two unequal parts, a feature which has an important bearing on the process of protein synthesis. It is believed that the strands of messenger RNA pass through the larger part, where the code is read, and that the assembly of amino acids into protein chains takes place in the smaller part.

It is to be expected from our earlier discussion that these rough membranes would be present in Protozoa as well as in metazoan cells, and this is probably so. Granule-studded membranes similar to those of the ergastoplasm are present in Protozoa, although typically in the form of single sacs rather than as aggregates. These may be equivalent to the ergastoplasm of Metazoa, although there is no certain evidence that this is so; moreover, some Protozoa may have granules that are unattached to membranes, while in other species both granules and membranes may be missing.

In addition, metazoan cells contain unit membranes that are similar in general character to those of the ergastoplasm but which lack the granules. Such membranes, known as smooth membranes, form the Golgi bodies, which are composed of piles of flattened membranous sacs. In many cells these may be diffusely distributed through the cytoplasm, but in the sperm of most animals, and in many other cells of invertebrates, they are localized to form disc-shaped bodies called dictyosomes. The Golgi bodies may participate in the secretory activity of the cell (e.g. Fig 17–1), perhaps by concentrating and processing material that has been synthesized elsewhere under the influence of the ergastoplasm. Minute vesicles are often associated with the Golgi membranes. It is believed that secreted products are pinched off from the Golgi region in these, and so distributed to other parts of the cell, or eventually extruded from it. In this way, for example, lysosomal enzymes of phagocytic cells are conveyed in Golgi vesicles to the vacuoles that contain the material that has been ingested by the cells.

That the Golgi bodies are concerned with some fundamental activity that is common to all organized protoplasm is suggested by the fact that similar structures have now been identified by electron microscopy in many Protozoa, including flagellates, gregarines, and ciliates. They are particularly well known in flagellates, where they take the form either of dictyosomes, like those of the Metazoa, or of parabasal bodies, which are essentially similar except that they are long and cylindrical in form, and are connected by a striated fibre to a centriole. Probably these protozoan structures, too, are secretory. Good evidence comes from the Radiolaria, where the Golgi vesicles secrete scales that are later moved to the body surface.

It will be evident that membranes constitute a fundamental element in cell organization, both at the surface and within the cell body. Probably, too, they are even more intimately related than has so far been suggested in this account. In particular, it is suggested that the rough and smooth membranes are directly connected with each other and with the outer of the two nuclear membranes (Fig. 2–2), forming thereby a transport system linking all parts of the cell together. It is not certain, of course, that the various membranes are always continuous in this way, and it would be going too far to suggest that their detailed organization is always identical. Moreover, intracellular membranes are highly labile, disappearing and reforming in accordance with the demands established by the varying states of activity of the cell. In fact, the account given of them here, and the illustration of a cell shown in Fig. 2–2, are to be taken as no more than very generalized statements of cell organization. Each specialized type of cell may be expected to display its own variation upon this fundamental theme, closely adapted to the particular function for which it is responsible.

This brief review by no means exhausts the importance of the part played by unit membranes in protoplasmic organization. They are concerned, for example, in other and clearly defined functions in the Protozoa, and we shall deal with them later in connection with digestion and osmoregulation. We have seen enough, however, to suggest that the use made of these membranes in metazoan cells is very closely paralleled in protozoans. Some degree of parallelism could no doubt be acceptably explained as a consequence of independent evolution, but in these instances the parallelism seems to go beyond that limit. The complex metabolic pathways of living systems must from very early stages have been dependent for their orderly integration upon the physical properties of layers of orientated molecules. It seems very likely, therefore, that these properties were already being exploited in the early stages of chemical evolution, and that Protozoa and Metazoa have inherited their common plan of protoplasmic organization from the patterns that became stabilized at that time under the influence of natural selection.

3
Movement and Fibrils

3–1 MUSCLE FILAMENTS AND MYONEMES

The building of organic molecules into membranes is a principle of protoplasmic organization that has clearly lent itself to far-reaching exploitation. No less fertile has been another structural device, the polymerization of large protein molecules into fibrous threads. These have been widely used for effecting various types of movement in animals. Consideration of some examples will again show that living systems prove to be remarkably uniform when they are analysed at the molecular level.

The most widely studied of these fibrillar structures are those that characterize the muscle fibres of mammals, and we refer first to these since the principles involved in their organization and functioning are of wide applicability. The fibres (Fig. 3–1A) are formed of a modified cytoplasm, the sarcoplasm, surrounded by a membrane, the sarcolemma, the whole fibre consisting either of a single specialized cell or of a syncytium. The important property of a muscle fibre is its power of contraction, which, as we shall see later, can bring about movement when the fibre is associated with some form of skeletal structure. Contraction is an expression of the properties of protein molecules that are organized into threads called muscle filaments, or myofilaments (Fig. 3–1B), some of which are thick (11–14 nm diameter) and some thin (4 nm). These myofilaments are collected together into myofibrils; in which each thick filament is surrounded by an array of six thin ones, each array being shared by several thick filaments. Within the cytoplasm, and lying between the filaments, are mitochondria and a well-developed endoplasmic reticulum which is here called the sarcoplasmic reticulum.

The muscle proteins which are mainly concerned with contraction are myosin, actin, and a tropomyosin complex, the thick filaments being mainly composed of myosin and the thin ones mainly of actin. Current views of the contraction process are founded on H. E. Huxley's work on the myofibrils of the psoas muscle of the rabbit. The cross-striated appearance of the muscle results from the regular alter-

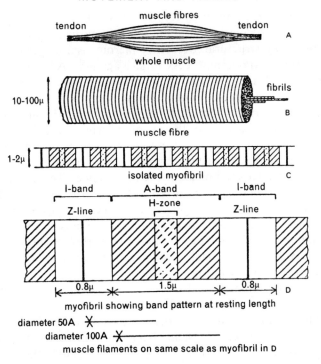

Fig. 3-1A. The structure of striated muscle at different levels of organization; dimensions shown are those for rabbit psoas muscle. From Huxley, H. E., 1960. *The Cell*, vol. 4 (Brachet and Mirsky, eds.) Academic Press, New York.

nation of strongly birefringent regions (the anisotropic bands, or A-bands) and weakly birefringent regions (the isotropic bands, or I-bands). A lighter H-zone is situated in the middle of the A-band. Of these several regions, the A-band is interpreted as being composed of overlapping thick and thin filaments, the I-band of thin filaments, and the H-zone of thick filaments. The muscle fibre itself is divided into disc-shaped units called sarcomeres, the point of division being at the Z-lines, which are formed by the interconnected ends of actin fibres.

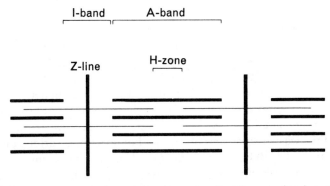

Fig. 3-1B. Diagram of the relationships of actin filaments (thin) and myosin filaments (thick) of striated muscle.

Myosin can be broken down by tryptic digestion into two components, called heavy meromyosin (molecular weight about 350,000) and light meromyosin (150,000). The latter is probably structural, but heavy meromyosin makes two fundamental contributions to the contraction process. First, it acts as an ATP-ase, which catalyses the breakdown of ATP to ADP, and thus provides the energy for the contraction. Secondly, it binds to actin (molecular weight about 60,000), the binding being effected through crossbridges formed by projections from the surfaces of the thick myofilaments. A widely accepted theory of muscle contraction, supported by electron-microscopical observations of muscle fibres in various states of tension, is the sliding filament theory, according to which the contraction of a fibre is brought about by the actin filaments sliding along the myosin ones. This sliding is believed to depend upon the rapid making and breaking of these cross-bridges, which contract each time they are formed, and so slightly displace the actin myofilaments in relation to the myosin ones. The process depends upon the presence of calcium ions, which are supposed to be released from the sarcoplasm when the muscle fibre is stimulated, and are then actively pumped back again into the reticular canals (endoplasmic reticulum) to bring about relaxation.

The tropomyosin complex is also involved in this sequence of events, which depends for its execution upon remarkably precise three-dimensional molecular architecture. This complex, which is associated with the actin filaments, contains at least two proteins, tropomyosin itself and troponin, which is a complex of globular proteins. It is supposed that the released calcium ions bind to the troponin, which then modifies the tropomyosin molecules in such a way that the myosin can now make contact with the actin molecules. Magnesium ions are also necessary, both for contraction and relaxation. However, unlike calcium ions, they cannot by themselves cause fibres to contract, whereas contraction can be evoked by injecting calcium into them.

The amount of ATP stored within muscular tissue is small, but reserves of energy are present in the substances called phosphagens, the best known of which are arginine phosphate and creatine phosphate. ADP reacts with these to form arginine or creatine as the case may be, according to the reversible equation:

$$\text{creatine phosphate} + \text{ADP} \rightleftharpoons \text{creatine} + \text{ATP}$$

These can be regenerated to high-energy phosphates by ATP. The phosphagens thus provide reserves of energy for the immediate regeneration of ATP from ADP, but the energy for continued muscular activity is provided by glycolysis, which also provides for the regeneration of the phosphagens. The main phosphagen of invertebrates is arginine phosphate, whereas creatine phosphate is characteristic of vertebrates. This distinction was thought at one time to be of phylogenetic importance. Both compounds were found in certain echinoids and in an enteropneust, and this was thought to give biochemical evidence of the origin of vertebrates from the echinoderm–protochordate line (Sec. 18–5). However, study of a wider range of species showed that this was an oversimplification. For example, various phosphagens, including creatine phosphate, have been found in annelid worms, and it is now apparent that the distribution of these substances must have been influenced by parallel biochemical evolution.

The interpretation of muscular contraction outlined above applies primarily to mammalian striated muscle. Only the future will show how universal is the mechanism thus postulated, but enough is already known to indicate some uniformity in the organization of muscular tissue in general. Throughout the animal kingdom it is characterized by the possession of myofibrils, which suggests that these must have appeared early in metazoan evolution. Thick and thin filaments, corresponding, respectively, to myosin and actin, are common, although with variation in the detailed arrangement of the myofilaments, and there is also a widespread tendency for their precise alignment, expressed as cross-striations. These, although particularly characteristic of arthropods and vertebrates, are found in certain muscle fibres of almost every phylum from the coelenterates upwards. Sometimes (in annelids and echinoderms, for example, and in some molluscs) the striations have a helical pattern, but this is a matter of fibril arrangement, and it does not necessarily imply any fundamental structural differences. There is, of course, adaptation to particular requirements, one example of this being the presence of abundant tropomyosin in the adductor muscles of bivalves, and in the anterior byssus retractor of *Mytilus*. These are muscles adapted for the sustaining of high tension over long periods, and for slow relaxation. It has been suggested that the actual contraction mechanism is of the generalized type already described, but that the tropomyosin functions as a catch mechanism, holding the muscles in the contracted state.

A comparison of rabbit psoas muscle with the muscle of insects is particularly illuminating, for there is a general similarity in ultrastructural organization in these two widely separated groups, with the insect muscle showing a regular hexagonal array of thin filaments between thick ones. One difference is that the I-band of insect flight muscles is very short. Pringle points out, however, that this is understandable if the sliding mechanism of contraction is indeed operating, for the length of this band would be expected to depend upon the amount of shortening that the muscles undergo during their normal functioning. We shall see later (Sec. 7–5) that these muscles can produce full wing movement with only a small amount of contraction. Indeed, in the extreme case of *Bombus* the natural excursion of the indirect flight muscles may amount to only 1% of their length. The shortness of the I-band is thus readily explained, and proves to be an illustration of how differences revealed by comparative studies can strengthen an hypothesis.

Further evidence for similarity is the close resemblance in amino-acid composition of the actins of the rabbit and the blowfly (*Phormia*), and also of their myosins. The same 16 amino acids are present throughout, and in the actins, for example, only eight of them show significant differences in proportion. No less striking are resemblances in physicochemical properties, additional to the ATP-ase activity which is readily demonstrable in insect muscle. Yet another point of resemblance is the presence in this muscle of granules carrying a relaxing factor which brings about a reversal of ATP-induced contraction. A similar factor, also localized in granules, is found in mammalian heart and skeletal muscle. The resemblance is illustrated in Fig. 3–2, where the activity of the relaxing factor is expressed as percentage inhibition of the ATP-ase activity of the myofibrils; the responses of the rabbit and of the locust leg muscle are very similar. The relatively weak action of this factor on the thoracic myofibrils of the locust is a consequence of the very large number of sarcosomes in this muscle. Sarcosomes, which are actually giant mitochondria, are always particularly well developed in the flight muscle of insects, where they may occupy up to

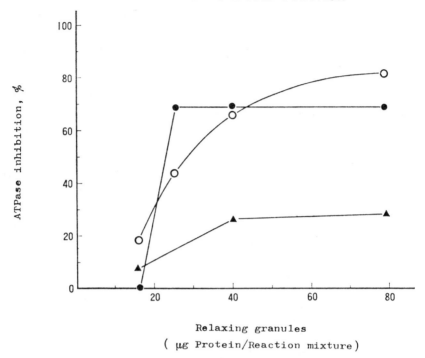

Fig. 3-2. Effects of the locust relaxing granules on myofibrillar ATPase. Assay system: 320 mM sucrose, 50 mM KCl, 2.5 mM potassium oxalate, 3 mM $MgCl_2$, 3 mM ATP, 20 mM histidine buffer (pH 7.0), varying amounts of relaxing granules as shown, and an adequate amount of myofibrils: O—O rabbit myofibri (0.36 mg protein /ml), ●—● locust leg muscle myofibril (1.2 mg protein/ml), ▲—▲ locust flight muscle myofibril (0.98 mg protein/ml). Incubated at 23°C for 10 min. From Tsukamoto et al., 1966. Comp. Biochem. (Physiol)., **17**, 569–581.

40% of the sarcoplasm volume. They contain an ATP-ase which is distinct from that of the myofibrils, and the activity of this overshadows the action of the relaxing factor upon those myofibrils.

These close resemblances are remarkable, considering that they emerge from a comparison of mammalian muscle with one of the most specialized types of invertebrate muscle. They represent very much the type of resemblance which can hardly be thought of as arising by chance. Rather, they suggest that there is a fundamental similarity in contractile mechanisms which overrides phylogenetic and adaptive diversification.

The use of muscular tissue for effecting movement is a specialized element of metazoan organization, demanding concomitant specialization in other tissues. It has not been evolved by sponges, and is necessarily absent from Protozoa. Obviously, then, the development of powers of movement in living systems preceded the evolution of muscle fibres. It is thus necessary to examine what other modes of movement have become available for animals, and, in so doing, to enquire whether the underlying biochemical mechanisms have anything in common with those of muscles. Bearing in mind what we have already learned of the widespread distribution of common metabolic pathways, we might well expect to find diverse modes of movement depending on very similar biochemical mechanisms. Again the evidence is fragmentary, yet there is some reason for believing this expectation to be well founded.

Fibrillar systems are particularly well developed within the bodies of Protozoa. One type consists of filaments, formed of molecular chains, which may be loosely aggregated into bundles when they are contractile, or cross-linked to form striated fibrils when they are concerned with support. The latter type is exemplified in the kinetodesmata of ciliate Protozoa, to be referred to later (Sec. 3–2). An example of the contractile type is seen in the ciliate *Stentor*, one of the many protozoans that are able to change their shape very quickly. In the posterior region of the body are fibrils called endoplasmic myonemes, one corresponding to each row of cilia. It is generally assumed that they are contractile organelles, for they become markedly thickened when the animal contracts. A suggestive analogy with muscular tissue is that their contraction is thought to be evoked by an increased concentration of calcium ions in the cytoplasm. It is possible that microtubular bundles in the cytoplasm are concerned with the extension of the animal that follows this contraction.

Similar myonemes occur in peritrichous ciliates (Fig. 3–3), and here there is a strong presumption that they are concerned in contraction, for in certain sessile forms, such as *Vorticella*, *Carchesium*, and *Zoothamnium*, they run together towards the base of the organism and continue into the stalk as a central structure called the spasmoneme, formed of bundles of filaments. In *Carchesium* each zooid contracts separately because it has a separate spasmoneme, but in *Zoothamnium* all the spasmonemes are continuous, so that stimulation of one individual results in contraction of the whole colony. It is of obvious significance that a spasmoneme is not found in certain other genera, such as *Epistylis* (Fig. 3–3), in which the stalk is not contractile. Our knowledge of these contractile structures is incomplete, and any interpretation must be largely inferential. Nevertheless, there is cytochemical evidence for the presence of ATP in the myonemes of vorticellids, so that there is at least some basis

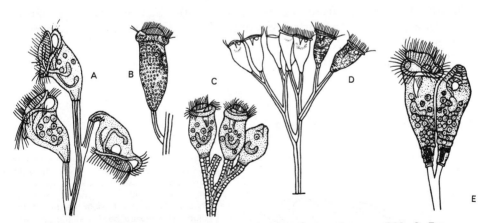

Fig. 3.3. A, *Carchesium polypinum*, ×200; B, *C. granulatum*, ×200; C, *Zoothamnium arbuscula*, ×200; D, *Z. adamis*, ×150; E, *Epistylis plicatilis*, ×200. From Kudo, 1960. *Protozoology* (4th ed.). Courtesy of Charles C. Thomas, Publisher, Springfield, Ill.

for supposing that their mode of functioning may have something in common with that of myofibrils.

Another type of fibrillar system consists of rigid structures called microtubular fibrils, with walls composed of rows of protein molecules. Examples are found in the pellicle of ciliates and of sporozoan trophozoites. They also form the axial rods of the pseudopodia (axopodia) of Heliozoia (Sec. 3–3), and they are major components of flagella and cilia.

3-2 FLAGELLA AND CILIA

More satisfactory evidence for a common plan of organization in the contractile systems of animals is to be found in the flagella and cilia which, with certain associated fibrillar systems, provide organelles of movement for many Protozoa, and which continue to function as important effector structures in most groups of animals. The exceptions are the Nematoda, from which they are entirely absent, and the Arthropoda, where they persist only in the excretory and reproductive systems of the Onychophora, in the sperms of some insects, and, in a modified form, in certain receptor organs.

Electron microscopy has shown that flagella and cilia have the same fundamental structure. The difference between them lies in their mode of beating, and in the type of movement that this produces. Both are adapted for action in a fluid medium. If they are attached to a fixed surface they set up motion in the medium relative to the surface. If, however, they arise from a movable object then this is caused to move in relation to the medium. The difference between the results of flagellar and ciliary action arises from the beat of a flagellum (Fig. 3–12) being often symmetrical, with several waves included in it at any one moment, whereas the beat of a cilium is asymmetrical and includes only one wave. In consequence, the flagellum is able to move the fluid medium continuously throughout its beat, and in such a way that the movement is at right angles to the surface of attachment. The cilium, by contrast, has an active phase, during which movement is brought about, and a recovery phase, which produces no significant movement (Fig. 3–4). The movement of the fluid is in this case parallel to the surface of attachment. Further, cilia are typically much

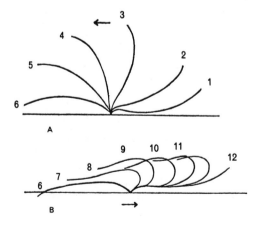

Fig. 3-4. A, forward effective stroke of frontal cilium of *Mytilus*. Note the rigid form during the whole stroke. B, backward preparatory stroke of frontal cilium of *Mytilus*. Note the flexible nature of the cilium. The flexure begins at the base and spreads to the tip. From Gray, 1928. *Ciliary Movement*. Cambridge University Press, London.

more densely massed than flagella, and their beats are coordinated in the well-known pattern of metachronal rhythm (Fig. 3–5), conventionally compared to the passage of wind over a field of wheat.

In earlier days it was natural, but incorrect, to assume that the apparently simple amoeboid movement of rhizopod Protozoa was the most primitive type of animal locomotion. We shall see reasons for supposing that the underlying mechanism of amoeboid movement may be similar in principle to that of flagellar and ciliary movement. The primitive status of flagellar movement, however, is clearly indicated by it being the characteristic mode of locomotion in the Mastigophora. This group undoubtedly occupies a basal position in protozoan phylogeny, as we shall see when we consider modes of nutrition in these animals. It is reasonable, therefore, to conclude that cilia were evolved from flagella. Pseudopodial movement must also have evolved in a stock with a flagellate ancestry, but here the structural basis can be detected only with the electron microscope, and even then with the greatest difficulty.

The close relationship of mastigophoran and sarcodine protozoans is well shown by the existence of flagellate reproductive stages in some of the latter—e.g. *Elphidium* (*Polystomella*). There are also adult forms that have both flagella and pseudopodia either simultaneously (Fig. 3–12) or in two separate phases. Examples are seen in the phytoflagellate order Chrysomonadida; many members of this group can assume an amoeboid form, and may even lose their flagella and chloroplasts so that they then become virtually indistinguishable from typical sarcodinans. An example of the reverse transformation is seen in the sarcodine *Naegleria gruberi* (Fig. 3–6). This organism normally lives in the soil, where it ingests bacteria. Transformation to a flagellate phase takes place in laboratory cultures when the medium is diluted with pure water. At first the body becomes polarized in organization, with lobose pseudopodia at one end and filiform pseudopodia at the other; flagella then appear among the latter, and the organism passes into the flagellate phase. This phase, which is freely motile, is a transient one; it has been suggested that it may be a response enabling the organism to obtain fresh food supplies or to reach an environ-

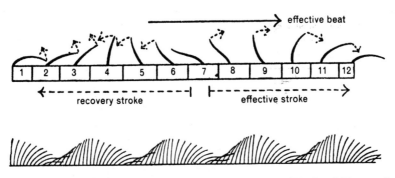

Fig. 3-5. *Above*, diagram to illustrate metachronal rhythm. Cilia *1* and *12* are at the end of the effective stroke; *2–7* indicate successive stages during the recovery stroke; *8–11* indicate stages during the effective stroke. All the cilia *1–12* are beating in sequence. *Below*, diagram illustrating the optical appearance given by a profile view of cilia beating in metachronal rhythm. From Gray, 1928. *op. cit.*

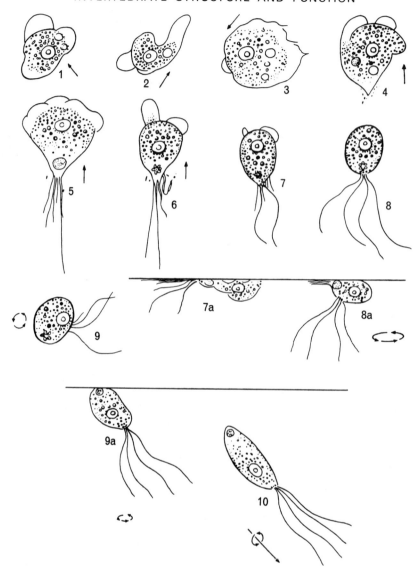

Fig. 3-6. Diagrammatic representation of the change of form of *Naegleria* when the organism is placed in distilled water. The arrows indicate the direction of motion. *1–3*, amoeboid form; *4*, 'polarized' form; *5, 6*, filiform pseudopodia present; *7–10*, acquisition of flagella. *7a–9a* are schematic of how events would appear if seen from the side when the amoeba leaves the surface of the coverslip. From Willmer, 1960. *Cytology and Evolution.* Academic Press, London.

ment which is in other respects more suitable. In short, it can get the best of both worlds.

The forces set up by flagella and cilia are small, and are totally inadequate for the production of lively movement in larger animals. They do, nevertheless, serve some locomotor function in lower metazoans such as the smaller platyhelminths. In general, however, locomotion is provided for in Metazoa by the differentiation of muscle cells. It is these, acting in conjunction with skeletal structures, that have permitted the development of the large size of so many of these animals. Flagella and cilia often persist in them, but only to carry out limited functions. These include not only movement, but also the sensory functions of non-motile cilia, which we find, for example, in the photosensitive structures of coelenterates and echinoderms, and in the rod and cone cells of vertebrates.

Flagella create the feeding currents of sponges, and they provide for the movement of sperm in most animal groups. They serve also to propel fluid along slender tubes, notably in the protonephridia of the lower Metazoa. In contrast to this, the coordinated beat of cilia is used to effect the movement of fluid over surfaces. Examples are seen in the filter-feeding mechanisms of many Metazoa, and in the lining epithelium of alimentary canals. Cilia also provide, in a stiff and relatively immobile form, the sensory processes of certain receptor cells. They are found in the scolopale sense organ of the locust, which responds to movements of the tympanic membrane, and also on the antennae of the honey bee, but in these examples the ciliary nature of the processes can only be clearly established by considering the fine structure revealed by the electron microscope.

Apart from the flagella of bacteria, which have their own characteristic features, the fine structure of flagella and cilia shows a remarkable uniformity of pattern, not only within the Protozoa but also throughout the Metazoa; it is a pattern, therefore, that may will be another example of common inheritance from a very early stage of evolution. It had already been noted in the nineteenth century that the flagellate tails of metazoan sperm might fray out into fibrils, and it is a tribute to the acuity of the observations that were made with the light microscope at that time that Ballowitz was actually able to determine in chaffinch sperm the exact number of primary fibres (11) that are now known to be typical of flagella and sperm.

A flagellum or cilium (Fig. 3–7) consists of a matrix surrounded by a membrane

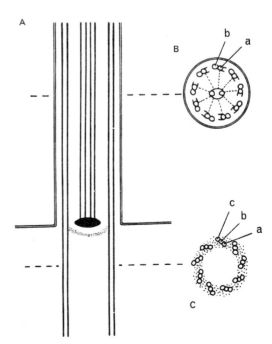

Fig. 3-7 Diagrams showing the structure of a cilium. A, structures seen in longitudinal section. B, transverse section of the ciliary shaft with the typical arrangement of internal fibrils. C, the arrangement of the fibril triplets in the basal body, showing the position of the third subfibril c, and the twist of the peripheral triplets. From Sleigh, 1962. *The Biology of Cilia and Flagella*. Pergamon Press, Oxford.

that is continuous with the plasma membrane of the cell surface. Within the matrix is an axoneme composed of 11 microtubular fibrils, each formed of rows of molecular sub-units of a globular protein (molecular weight about 55,000). This protein, called tubulin, resembles actin in certain respects. Two of the 11 fibrils are single ones (diameter about 24 nm), lying centrally, while the other nine are doublets, each composed of an A and B subfibre, and forming a circle around the central pair. One of the two subfibres of each of these peripheral doublets bears a double row of short arms which all point in the same direction. These are composed of a protein (molecular weight about 500,000) with some resemblance to myosin, but differing in certain respects; for example, it has an ATP-ase activity which is activated by Mg as well as by Ca. It is a remarkable measure of the constancy of structure in these organelles that in flagellates, a ciliate, a mollusc, an amphibian, and a mammal, these arms have been observed always to point in the same (clockwise) direction. We shall see (p. 55) that there is probably some functional significance in this. Additional features can sometimes be seen in electron micrographs of cilia and flagella. These include a sheath enclosing the central fibrils, and delicate strands which lie between the central and peripheral ones, and sometimes form spokes or radial lamellae.

Flagella and cilia take their origin in granules that are variously called basal granules, blepharoplasts, or kinetosomes (Fig. 3–7 and 3–8). Undoubtedly, there is a close relationship between these and the centrioles that produce the spindle fibres at cell division in animals (although not in plants). Not only are they both directly related to the formation of fibres, but in some instances a basal granule acts as a centriole during mitosis, dividing into two halves that move apart with only a strand connecting them. Alternatively, the centriole may be a separate granule connected with the blepharoplast by a fibril called the rhizoplast, while other bodies, such as the parabasal bodies mentioned earlier, may also be connected with it. The clearest evidence of this relationship between centrioles and basal bodies, however, is seen in their fine structure, for both contain an identical arrangement of peripheral fibres. Those of the basal body are continuous with the peripheral fibres of the flagellum or

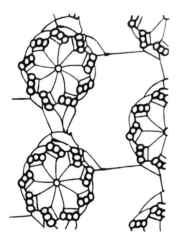

Fig. 3-8. Diagram showing the fine structure of the basal bodies of *Trichonympha* in transverse section, and the delicate system of fibres present on some of the basal bodies in this organism. Note the 9 triplet outer fibres and the 'cartwheel', which occurs in the proximal region of the basal bodies. From Grimstone, 1961. *Biol. Rev.*, 36, 97–150.

cilium associated with it, but differ from them in having a characteristic triplet pattern. Also, the triplets are twisted, and are interconnected by other fibrils. We may expect that in due course the study of basal granules and centrioles, and of the relationship between these and flagella, cilia, and spindle fibres, will shed light on the means by which living systems can produce fibrous structures by the polymerization of protein molecules. For the present we can only note the important part that this capacity evidently plays in the organization of protoplasmic activity.

We must now consider briefly how far the mode of functioning of flagella and cilia depends upon properties that they share with the contractile myofibrils of muscle cells. To what extent, in other words, is muscular contraction a development of biochemical mechanisms that were already well established before the emergence of metazoan organization? The interpretation of the action of flagella and cilia presents problems that may be approached in more than one way. It is possible, for example, to analyse the form of their movement, and to examine how this brings about movement in the organisms that depend upon it. Alternatively, we can face the more difficult problems of the nature of the cellular mechanism that is responsible for the movement of these delicate threads, and of the ways in which the necessary supplies of energy are made available. These problems are, of course, related.

The movement of flagella commonly involves the generation of waves that are transmitted along it, either in a single plane, or in a corkscrew pattern. The effect of this upon the movement of a protozoan is well exemplified by *Euglena*, an organism that presents a comparatively simple case, although one that is by no means fully understood. The waves arise at the base of the flagellum, which in this organism seems to arise from the wall of the reservoir, apparently by two roots. This appearance, however, is illusory; such an arrangement would be difficult to reconcile with what is now known of the fine structure of flagella, and it is probable that there are actually two flagella in this organism, one main one and another short and vestigial one that is fused with it. The waves pass to the tip of the main flagellum, which beats at a rate of about 12 beats per second, and which also shows a movement of rotation. This rotation causes the tip of the organism to rotate (Fig. 3–9), while at the same time pushing it to one side (Fig. 3–10). Because of this, *Euglena* rotates as it swims (at a rate of about one turn per second), and it also follows a corkscrew course. The

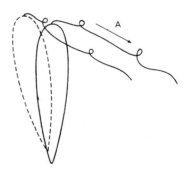

Fig. 3-9. Waves passing down the flagellum of *Euglena viridis* in the direction *A* produce a force in the opposite direction, causing the organism to take up the position indicated by the dotted figure. Adapted from Lowndes, 1941. *Proc. zool. Soc. Lond., A,* **111**, 111–134.

54 INVERTEBRATE STRUCTURE AND FUNCTION

Fig. 3-10. The varying positions taken up by *Euglena viridis* in swimming from A to B. From Lowndes, 1941. op. cit.

movement of its body is thus comparable with that of a propeller, for it sets up forces on the water that bring about forward displacement. It is not essential, therefore, for the flagellum itself to provide a forward component in such circumstances; it may well do so, however, in the particular case of *Euglena*, since in that organism the flagellum is directed backwards along the side of the body (one reason, incidentally, why students find it so difficult to observe).

While the flagellum is moving it must be dissipating energy as a result of work done against the surrounding water; this means that if it were solely dependent upon energy propagated along it from some source in the organism's body, its pulse would necessarily become diminished towards its free end. High-speed cinematography reveals that this does not happen. On the contrary, as the wave passes along the flagellum it actually develops an increase both in velocity and in amplitude (Figs. 3–11 and 3–12). From this we are bound to conclude that the flagellum must itself contribute energy to its movement. Understandably enough, we cannot yet give a complete explanation of how this contribution is contrived, but there is at least some evidence that the underlying biochemical mechanism is similar in principle to that of muscular contraction, in that ATP is the source of the required energy.

The evidence for this conclusion depends in part upon studies of the responses of intact cilia to the presence of ATP in the medium; cilia for this purpose being as relevant as flagella, in view of the similarity of their organization. Cilia from both the oyster and the frog show increased activity in the presence of ATP, the effect disappearing if the ATP is destroyed by hydrolysis. This evidence has been supplemented by the use of the glycerol extraction procedures discussed earlier in connection with the analysis of muscular contraction. Frog cilia that have been extracted with

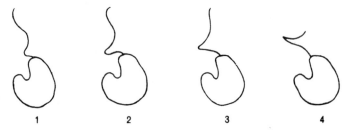

Fig. 3-11. *Peranema trichophora*, drawn from four consecutive frames of a film. The wave of movement of the flagellum is passing from the base to the tip. From Lowndes, 1941. op. cit.

glycerol for several days are inactive, but their activity can be restored if ATP is added to the medium; indeed, they will beat in such circumstances even if they have been separated from any protoplasmic connection. It might be expected, by analogy with myofibrils, that an ATP-ase would be present in flagella and cilia, and its presence has indeed been demonstrated, by enzymatic and histochemical procedures, in ciliated cells and in the tails of sperm. This ATP-ase, called dynein, is believed to form the arms that project from one side of the A subfibrils of the cilium (p. 52). These arms point towards the B subfibril of the next doublet, an arrangement which suggests that they might connect with this, much as the ATP-ase of myosin muscle filaments can connect them with the thin actin filaments. This would provide a basis for sliding to take place in the cilium, just as it is believed to do in muscle fibres, and evidence from measurements suggests that such displacement does actually take place.

Finally, ciliary activity can be arrested by the use of reagents such as sodium fluoride, sodium azide, and sodium cyanide, which selectively inhibit specific stages in the metabolic pathways of anaerobic glycolysis, the TCA cycle, and the cytochrome system respectively. We may suppose then, that flagellar and ciliary activity, like muscular contraction, depends upon a pattern of biochemical organization that we have already seen to play a fundamentally important part in the energetics of living systems.

The study of animal morphology at the macroscopic level readily convinces the observer that the design of moving parts is precisely and delicately adapted to the functions that they have to perform; we shall see examples of this in due course. It cannot be doubted, therefore, that the same must be true of the ultrastructure of flagella and cilia, provided we are justified in our assumption that all levels of animal organization have been built up under the influence of natural selection. At present, however, our information is too limited to enable us to link satisfactorily the fine structure of these organelles with our conclusions regarding their biochemical mechanisms. Analyses of isolated flagella and sperm show that protein predominates in them. This suggests that the nine peripheral fibres may well be protein macromolecules, and that movement results from contractile changes in these. Their position is ideal for this action, for they are well placed to exert a bending moment around the axis of the organelle. It has been suggested that the two central fibrils might act as compression elements in the system, providing mechanical resistance to the bending couple, but calculations indicate that they are unlikely to be strong enough to carry out this function. Possibly, therefore, the rigidity of flagella and cilia depends upon internal turgor, a situation that would present a striking analogy with the use of hydrostatic skeletons by Metazoa. Sleigh has pointed out that a cylinder in which peripheral contractile structures are antagonized by internal turgor would be transformed into a close spiral by contraction, a response that is certainly seen in the contractile stalks of peritrichous ciliates but that is not characteristic of flagella or cilia. He has suggested, therefore, that the central fibrils, and the radial strands that link them with the peripheral ones, may prevent this spiralization by restricting the effects of contraction to the sites at which it is taking place at any particular moment.

Indications of adaptive specialization are no less apparent at the level of light microscopy, as may be judged from a comparison of the three forms illustrated (Fig. 3–12). All are sessile, using their flagella to set up feeding currents in the surround-

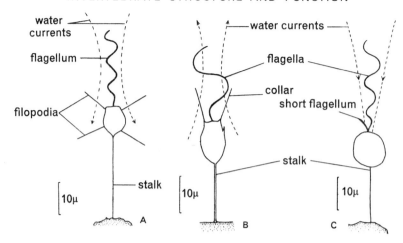

Fig. 3-12. *Actinomonas*, showing body form and the water currents caused by flagellar beating. B, *Codonosiga*. C, *Monas*. From Sleigh, 1964. *Q. Jl. microsc. Sci.*, **105**, 405–414.

ing water. *Actinomonas* is a fresh-water protozoan with a spherical body and contractile stalk. It provides an example of the situation mentioned earlier: the coexistence of a flagellum with pseudopodia, which in this case comprise a group of filopodia (see later). The wave motion of the beating flagellum passes upwards from the base, but the currents that are set up in the surrounding water pass downwards. As a result, food particles are carried towards the filopodia which catch and engulf them.

Codonosiga is one of the fresh-water choanoflagellates, organisms that resemble the choanocytes of sponges in having a flagellum surrounded by a protoplasmic collar. Here again the waves of the flagellum travel distally, but in this instance they draw water currents towards the collar from below. Food particles are trapped on the outside of the collar, which, like that of choanocytes, is composed of microvilli; these strain the water, and the trapped particles are carried downwards by a flow of protoplasm to the base of the collar, where they are ingested into food vacuoles.

Monas is another fresh-water form with two flagella. The larger one sends waves from its base to its tip, their amplitude increasing as they pass distally. As with *Actinomonas*, the water currents pass downwards, but, in contrast to what happens in the latter organism, the food particles are here brought to a focus at the base of the flagellum, the small one, with its flickering movement, perhaps helping to trap them. The broadly dispersed pattern of currents in *Actinomonas* is elegantly adapted to the trapping of food particles by the wide-ranging filopodia; in *Monas* they must be ingested at the tip of the body, near to the origin of the flagella, and the concentration of the currents provides for this.

It will be noted in these examples (and others would show the same phenomenon) that the direction of passage of the waves of the flagellum is no guide to the direction of the water currents that they produce. There is a somewhat analogous situation in free-swimming polychaete worms and fish. The eel is propelled by sinusoidal waves of the body that pass backwards, whereas *Nereis* is propelled by waves that pass forwards. We shall see later that this difference is a result of the polychaete worm having projections of its body surface in the form of parapodia, whereas no such structures disturb the smooth outline of the eel. Sleigh, in discussing this situation, has suggested that a similar explanation might conceivably account for the differing effects of the wave motion of flagella. Electron microscopy shows that these organelles sometimes have filamentous projections of their surface. Some of these (called flimmer filaments) are long and slender (about 5 nm thick). Others (called mastigonemes) are shorter, thicker (about 20 nm) and rigid. Mastigonemes bring about a reversal of flow of water, so that flagellar waves propagated from base to tip cause a flow from tip to base. The analogy with wave propagation in *Nereis* is impressive (Sec. 6–3), and shows that concepts derived from macromorphological studies may have some relevance at the micromorphological level of analysis.

One special problem, closely associated with the functioning of fibre systems, is presented by the existence in ciliate Protozoa of patterns of fibres, called kinetodesmata, which lie in the ectoplasm and which are closely associated with the basal bodies of the cilia. The cilia of these animals (Fig. 2–4, p. 38) are arranged in rows, each row, or kinety, consisting of a complex which comprises the cilia themselves, their basal granules, and the kinetodesma, which lies just to the right of each row (Fig. 3–13). The kinetodesma is visible as a fibre with the light microscope, and can be well demonstrated by silver impregnation, but electron microscopy is needed to elucidate fully its complex relationships. A single kinetodesma in *Paramecium* is composed of a number of overlapping fibrils, each of which arises from a basal body. The thickness of the whole fibre in any animal depends upon the length of the constituent fibrils; in *Stentor*, for example, they are long, extending for about half the length of the row of some 1,000 cilia, so that at any one point of its length the kinetodesma is composed of about 500 fibrils. These fibrils, which are often striated, with a periodicity of about 40 nm, may provide support or anchorage for the cilia. Other types are also demonstrable. Beneath the pellicle of *Paramecium*, for example, there are many very fine fibres, with no surface striation. These link together the basal bodies, or run towards the body surface, or pass inwards to meet in the motorium, a conspicuous body lying near the cytopharynx.

There is evidence that the motorium, together with the associated fibres, is a conducting and coordinating mechanism comparable to some extent with the

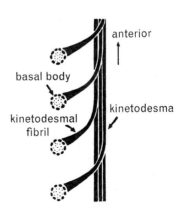

Fig. 3-13. Diagram showing the formation of the ciliate kinetodesma from overlapping fibrils arising from each basal body. From Grimstone, 1961, *op. cit.*

nervous system of Metazoa. The best evidence for this comes from the studies of Taylor on *Euplotes*. This is a ciliate in which the ventral surface of the body bears locomotor organelles called cirri, formed by the fusion of cilia. The movement of these is well coordinated, but the coordination is lost, and locomotion correspondingly disturbed, if fibres that run from the motorium to the five anal cirri are transected. It is far from certain, however, that kinetodesmata and associated fibre systems can all be accounted for in this way. It has been supposed that they might be responsible for the coordination that provides for metachronal rhythms, but it is doubtful whether such a system could possibly account for the wide range of patterns of movement that are actually observed.

Advances in knowledge since Taylor's work have opened up other lines of explanation of flagellar and ciliary coordination. It has been plausibly argued, for example, that coordination of ciliary beat in *Opalina* could be accounted for in terms of orderly depolarization of the surface membrane. Another important factor may be viscous interaction between them in the fluid medium in which they operate. There may also be some form of transmission through the cytoplasm that is independent of any visible fibres. In fact, some at least of these fibres may have nothing at all to do with conduction. They may equally well have supporting functions, or they may be concerned, like the basal bodies, with the morphogenesis of the constituent fibrils of flagella and cilia. We have accepted the existence of these latter structures without considering how they are laid down in their orderly patterns, for this again is a problem for the future to resolve.

3–3 AMOEBOID MOVEMENT

It remains now to consider amoeboid movement, which, as we have earlier seen, must have arisen later than flagellar movement. It is a form of locomotion particularly characteristic of many of the sarcodine Protozoa, but it is found also in a wide variety of metazoan cells, ranging from the oocytes of sponges to the white blood corpuscles of vertebrates. The familiar accounts of it are based largely upon studies of the lobose pseudopodia of *Amoeba* and its relatives; it is thus easy to lose sight of the range of variety within this type of movement. In *Amoeba* itself, as observed with the light microscope, amoeboid movement involves the well-known streaming movement of the protoplasm, associated with the outgrowth of pseudopodia and the progressive displacement of the organism. But as Bovee emphasizes, amoebae are not shapeless organisms, moving at random and with erratic alteration of their course. On the contrary, they are closely adapted to their particular environment, responding to it by their own characteristic patterns of reactions, as with *Subulamoeba* (Fig. 3–14). Different species, in fact, may have pseudopodia which, although lobose, are yet morphologically distinct; their differences probably reflect differences in protein composition, for they can be shown by immunological studies to be antigenically distinct from each other.

Contrasting with the typically lobose forms are the foraminiferans, which have a net-like system of branching and fusing pseudopodia, thread-like in form, and termed filopodia. These have a mechanical rigidity due to loose bundles of microtubular fibrils. A two-way flow of granules can be seen along these filopodia, the granules being largely mitochondria. It is important also not to overlook the highly

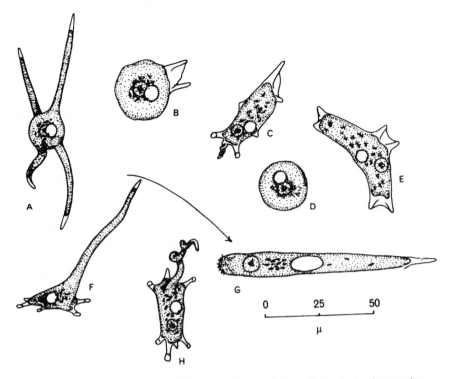

Fig. 3-14. *Subulamoeba saphirina.* A, radiate and afloat. B, beginning locomotion. C, in moderate progress, showing conical, mayorellid, determinate pseudopods. D, at rest, spherical. E, changing direction, with a new pair of pseudopods formed near the nucleus indicative of the new route. F, with a long tubular pseudopod thrown forwards, to become G, the rapidly locomotive organism. H, dorsal view of the organism, with the elongate pseudopod. From Bovee, 1964. *Primitive Motile Systems in Cell Biology* (Allen and Kamiya, eds.). Academic Press, New York.

specialized types of pseudopodia found in the Heliozoia and Radiolaria, groups that are sometimes separated from the other amoeboid protozoans as a distinct class. Thus in the Heliozoia, such as *Actinophrys sol*, there are radiating pseudopodia called axopodia. These have a central axis of microtubular fibrils, which can lengthen and shorten, probably by growth or reduction of their molecular framework.

These latter types of pseudopodia have been little studied, so that theories of the nature of amoeboid movement are largely based upon observations of the lobose pseudopodia of amoebae. This is understandable and acceptable, provided that these theories can eventually take account of the conditions in other types of pseudopodia; a particularly difficult problem for example, is presented by the occurrence of bi-directional streaming in filopodia, with an outward and an inward flow taking place along each pseudopodium.

As de Bruyn has shown, the theories have varied from time to time in obvious conformity with the prevailing interpretation of cytoplasmic structure, but this situation is by no means peculiar to this particular problem. Scientific theories must inevitably be moulded by the climate of opinion in which they are developed, and it is precisely because of this, if for no other reason, that they need periodic re-examination.

The early belief that protozoans contained organs like those of high animals led at one time to the supposition that a pseudopodium was a hernia-like protrusion developed at a point of weakness of the body surface. Later recognition of the cell-like character of the protozoan body led to views of amoeboid movement that were based upon somewhat more realistic interpretations of cytoplasmic structure. Such was Heitzmann's advocacy of a living three-dimensional network of contractile fibres which he supposed to be embedded in a non-living and non-contractile fluid. On this view amoeboid movement was ascribed to contraction of the reticulum, the substance of the fibres being transferred during contraction to the nodes of the reticulum, where it accumulated in the form of granules.

In due course, by the end of the nineteenth century, this reticular theory of cytoplasmic structure was replaced by other views, according to which the fundamental structural elements were thought to be droplets, filaments, or granules, dispersed in a structureless fluid. These views made it difficult to account for amoeboid movement in purely structural terms. Instead, it was suggested that it might depend upon local changes in surface tension; a view that acquired a certain plausibility from Bütschli's success in constructing a physical model of it. Acting on the assumption that cytoplasm had a foam-like structure, he prepared emulsions of water, olive oil, and potassium carbonate, and placed drops of these in water or in dilute glycerol. In these media the drops proceeded to move in a superficially amoeba-like way, with a streaming movement of their contents. The model was an ingenious one, no better and no worse, perhaps, than some of the others that biologists have delighted to invent in other fields of speculation, but it stands remote from the contemporary approach to the problem.

This approach derives from the recognition of the colloidal character of cytoplasm, and of the differences between the properties of the ectoplasm and those of the endoplasm. It is now suggested that the former is a colloidal gel, the plasmagel, and the endoplasm a sol, the plasmasol. Amoeboid movement, from this point of view, can be interpreted as a result of coordinated gelation and solation, the contraction of the superficial layer of gel directing the streaming movements of the plasmasol (Fig. 3–15), and perhaps forcing out a pseudopodium at some point where local conditions are evoking solation at the surface. This interpretation, which remains in principle the most favoured one, has been especially developed by Mast and Pantin. According to their views, *Amoeba* can be thought of as a tube of plasmagel containing the more fluid plasmasol. At the end that is posterior at any particular moment the gel undergoes solation; contraction forces the fluid sol forwards, and there is a transformation of sol to gel at the anterior end, so that solation is balanced by gelation. It has been suggested, however, that this interpretation may be oversimple. Allen argues that the endoplasm is not uniformly solated, so that its properties are not those of a Newtonian fluid. He thus supposes that contraction of gel at the posterior end cannot by itself account for the forward flow. Instead, he believes that contraction occurs at the anterior end, in a 'fountain zone', and that this process actively pulls the axial endoplasm forwards. This frontal-contraction model seems to account well for the complex details of movement of the giant carnivorous amoeba *Chaos carolinensis*, but Allen concedes that the tail-contraction one could serve for the giant herbivorous

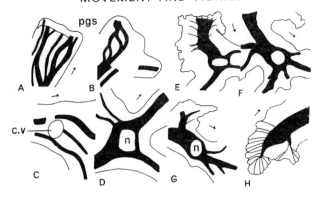

Fig. 3-15. *Above,* drawings showing position of plasmasol-stream network (black) and plasmagel network (white) in *Amoeba.* Note how the nucleus (*n*) is held back in D, E, F, and G, and the contractile vacuole (*c.v.*) in C, by the narrower channels in front. *Below,* composite drawing showing distribution of plasmagel (white) and plasmasol (black), based on photographs, cine films, and on direct observation. The smallest plasmasol streams are omitted for clarity. This distribution of plasmagel explains why the nucleus and the contractile vacuole maintain respectively a central and a posterior position. From Goldacre, 1964, *Primitive Motile Systems in Cell Biology* (Allen and Kamiya, eds.). Academic Press, New York.

species *Pelomyxa palustris*, which is a monopodial organism that progresses steadily and with little change of shape. These two species are adapted to exploit different ecological niches (it is said that *Pelomyxa* often contains 'small jagged rocks' which keep it weighted down in the mud at the bottom of lakes), but even greater diversity is to be found amongst the smaller amoebae, in comparison with which the giant ones have been described by Bovee as 'stodgy'.

Not the least of the difficulties in these interpretations is that the role of the surface membrane in amoeboid movement remains obscure; yet we cannot feel that any interpretation of amoeboid movement is satisfactory if it does not take account of the existence of the membrane, and of its elastic properites. One suggestion is that its function is essentially a physico-chemical one, changes in membrane potential perhaps initiating the outgrowth of a new pseudopodium. Another view is that it is closely associated with the plasmagel, contributing to the contractile force exerted by it, and undergoing continuous renewal from the interior. A third suggestion is that it aids the adhesion between the animal and the substratum which must be an essential factor in forward movement, although the nature of the adhesive forces is not known, nor is it understood how contact is made and broken. On the whole, it seems unlikely that the membrane provides the motive force. More probably, its value lies in its capacity to undergo deformation imposed from within the cytoplasm, the resulting tensions helping to regulate pseudopodial extension. If this be so, there would be some analogy between its function and that of the geodesic fibre systems of nemertines and nematodes (Sec. 4–6).

Allen suggests that more than one model may be needed to account completely for amoeboid movement, and that to apply the single term 'pseudopod' to the locomotor extensions of amoebae, foraminiferans, and radiolarians may be to construct a semantic trap for ourselves. This is a warning that merits close attention because of its wider implications, and we may conveniently amplify it a little at this stage.

According to Dante, everything superfluous is displeasing to God and Nature (*omne superfluum Deo et Naturae displiceat*). This has been regarded as a foreshadowing of the maximum of parsimony, a principle of logic associated particularly with the great and controversial fourteenth-century philosopher, William of Ockham, born about 1290 at Ockham, in Surrey. One of the tenets of this stormy figure was that it is vain to do with more what can be done with fewer; a proposition, known as Ockham's Razor, which is more commonly encountered in a seventeenth century formulation, *Entia non sunt multiplicanda praeter necessitatem*. Applying this principle with a ruthlessness which has been compared with that of his contemporaries in the field of politics, he developed a system of logic that led him to deny the possibility of attaining to metaphysical knowledge, and hence of demonstrating the existence of God. Not surprisingly, this brought him into conflict with the Pope. In 1324/5 he was dismissed from Oxford, where he was lecturing as a friar, and was required to proceed to Avignon to confront the papal court. Compelled eventually to flee from here to avoid arrest, he maintained his views, supported by a stream of writings, until he died, expatriate and excommunicate, in about 1349–50. Very probably he was a victim of the Black Death, which at that time was destroying a third of the population of Europe. Bertrand Russell has testified to the continuing value of the maximum of parsimony in logical analysis, and it is commonly held up as an example to biologists, with the implication that they should seek always for the simplest hypothesis and avoid unnecessary assumptions. But the recommendation needs to be handled with caution, for the logic of nature is deployed with a subtlety which makes it difficult for the biologist to know when it is safe to apply Ockham's Razor. Who can judge how relevant it is to the unravelling of the evolution of organic diversity?

Whatever interpretation of amoeboid movement may eventually win general acceptance, it must account for the force that determines the forward streaming of the plasmasol. In this connection there has been an interesting reversion to earlier views that it may depend upon the contractility of the cytoplasm. Indeed, Heitzmann might well feel that his reticular theory has now been translated into molecular terms. In these terms the structure of cytoplasm, apart from its membranes and fibrils, is commonly visualized as based upon the cross-linking of polypeptide chains, with water and metabolites present in the interstices of the meshwork. Within this system contraction could occur either by the development of additional cross-linkages, which would reduce the size of the mesh, or by the folding of the polypeptide chains, which would have a similar effect. Such contraction would seem to provide some theoretical basis for amoeboid movement, for it would result in the elimination of fluid from the contracting meshwork; this fluid might then be taken up elsewhere, to produce solation, with forward displacement of the protoplasm. The sites of these changes would doubtless be determined by a combination of internal and external influences, and in *Amoeba limax*, which moves steadily in one direction, the internal ones must presumably be particularly well established.

This type of interpretation is clearly based upon analogy with the part played by fibrillar structures in muscular contraction. As far as flagellar and ciliary movements are concerned we have seen some justification for drawing such an analogy, but it must be admitted that the justification is less clear for amoeboid movement. Current views on the organization of cytoplasm, with their emphasis on the significance of membranes and fibrils, are certainly helpful. In particular, there are thin actin-like filaments and thicker myosin-like ones in *Amoeba*, while there is evidence that a fibrillar structure may exist in slime moulds and in the testacean *Difflugia*. This animal pulls its heavy test forwards by rapid contractions of the pseudopodia after these have attached to the substratum, the action being accompanied by the temporary appearance within them of positively birefringent fibrils. These fibrils, which have been observed also with interference and electron miscroscopy, extend from the attachment points into the cell body. Flashes of positive birefringence are also visible at the tips of the pseudopodia when they are sporadically streaming. Suggestions that these filaments can contract, or take part in sliding movements similar to those of metazoan myofilaments, are very attractive, but they await further investigation.

In any case, present information suggests a simplicity of organization in the protoplasm of *Amoeba*, judged against the complexity of its behaviour. There seems to be no clearly defined ultra-structural difference between the ectoplasm and the endoplasm, nor can the distribution of visible structures be readily related to any particular phase of amoeboid movement. Mitochondria are identifiable, which confirms our estimate of the fundamental importance of these structures and of the enzyme systems that they bear, but there is no evidence of more continuous membrane systems. Instead, there is a great abundance of vacuoles, large and small; many contain food or its digestive products, while others are empty or contain particles of unidentified nature. It is not surprising, therefore, that some of the earlier investigators based their views upon the supposedly foam-like structure of cytoplasm. Indeed, it would be fair to recognize that, within the limits of the facilities available to them, this type of interpretation was not ill founded, at least as regards the structure of the cytoplasm of *Amoeba*.

It may be, of course, that highly organized systems of fibrils and membranes are not reconcilable with demands for flowing movements to develop in all parts of a unit mass of protoplasm. Whether the organization of the body of *Amoeba* is primitive or specialized in this regard it is impossible to say, but at least there is some evidence that the biochemical mechanisms underlying its activity have something in common with those involved in other forms of movement. This evidence comes from studies of extracts prepared from amoebae by controlled centrifugation and homogenization. In the presence of ATP these extracts show four significant properties: gelation, a streaming of particles, a capacity for contraction and for the transmission of tension, and an extrusion of water. Moreover, it can be shown by electron microscopy that oriented fibres develop in the gel phase. These properties could, in theory, provide at least part of the physical basis for amoeboid movement. Certainly their manifestation in the presence of ATP suggests that this movement must depend ultimately on energy exchanges similar in principle to those operating in muscular contraction and in the movement of flagella and cilia. At this stage, however, we should clearly be unjustified in pressing these comparisons any further.

4
Movement and Hydrostatics

4-1 PRINCIPLE OF THE HYDROSTATIC SKELETON

The differentiation of muscular tissue was one of the consequences of the establishment of metazoan organization, but it was not sufficient in itself to bring about effective movement of the body. The reason for this is that the contraction of muscle fibres is an active process whereas their relaxation is not. If the form of the body is to be maintained the contracted fibres must be restored to their original length, and this can only be achieved by applying an external force to them. In many higher animals this force is applied through a jointed skeleton. The muscles are attached to the movable parts of this in such a way that the contraction of one muscle brings about the relaxation of another; the two are then said to be antagonists, or to antagonize each other. This, of course, is not the only significance of a jointed skeleton; by providing a system of levers it secures the economical application of energy, while it also contributes to the general support and protection of the body.

Rigid skeletons made a comparatively late appearance in the history of animal evolution, and there are still many metazoan organisms today that do not possess them. We have suggested that life may have appeared some 2,000 million years ago. But the first remains of calcareous skeletal structures are not found until the beginning of the Cambrian period, perhaps 600 million years ago. These remains include parts of the skeletons of trilobites and echinoderms, and of the shells of molluscs and brachiopods. A great deal of evolutionary history must have preceded the emergence of these groups, but unfortunately we know tantalizingly little of that history. That there are remains of living organisms in Pre-Cambrian strata is agreed, but their nature is obscure. Stromatolites represent the primitive algae, but there is some evidence also of coelenterates and annelids, and perhaps of worm burrows. There is, however, a remarkable absence of skeletal structures, which seem to have been evolved relatively suddenly at about the onset of Cambrian times. It has been suggested that this was a new step in chemical evolution, resulting, perhaps, from animals having at last achieved a level of complexity at which the metabolism and excretion of calcium and phosphate could be adapted to the strengthening and protection of the body.

Whatever may be the truth of this, it seems clear that in animals of the remote past, as in many at the present time, movement must have depended upon something other than jointed skeletons for providing antagonistic relationships in muscular systems. The need must have been met, as it is today, by the ready availability of water. This shares with other fluids two properties that are of crucial importance in this connection: incompressibility, and the capacity for transmitting pressure changes equally in all directions. Add to this that the low viscosity of water allows it to be readily deformed, and we have the physical basis of a type of skeletal system known as the hydrostatic skeleton. The functioning of such a system depends upon the musculature being so arranged that it surrounds an enclosed volume of fluid. In these circumstances the contraction of any one part of the muscular system sets up a pressure in the fluid which is then transmitted in all directions to the rest of the body.

The consequences of this can be considered by reference to a hypothetical and highly simplified organism (Fig. 4–1A). This has the form of a continuous tube, closed at both ends, and containing a fluid which is enclosed by a body wall; the latter possesses a layer of circular muscle, but no longitudinal muscle and no elasticity. Contraction of the circular muscles at the right-hand end of the tube produces an increase of pressure in the fluid; this increase is transmitted throughout the length of the tube, the result of this depending upon whether or not the organism is free to elongate. If it is (Fig. 4–1B), the body can lengthen to the left, the diameter remaining the same at this end. If the organism is not free to elongate, the left-hand end will respond to the increase of pressure by increasing in diameter (Fig. 4–1C). There is an important difference between these two possibilities. In Fig. 4–1C the body can be restored to its original shape by contraction of the muscles of the left-hand end, for this will restore the original pressure distribution and will promote relaxation of the muscles of the right-hand end. In Fig. 4–1B the original form cannot be restored, for no change in pressure distribution can bring about shortening of the body to its original length.

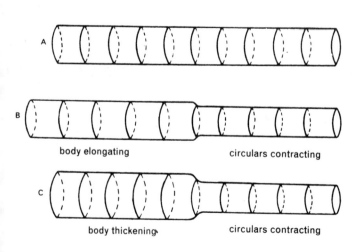

Fig. 4–1. Diagram to illustrate the action of a hypothetical animal with circular muscles only. In A all the muscles are relaxed. In B the muscles of the right-hand end have contracted but the length of this end has remained the same; the original length of the right-hand muscles cannot be restored by contraction of the left-hand muscles. In C the muscles of the right-hand end have contracted, but the muscles of the left-hand end have relaxed, and there has been no change in length; the right-hand muscles can be restored to their original length by contraction of the left-hand ones. From Chapman, 1950. *J. exp. Biol.*, **27**, 29–39.

This limitation can be overcome by the addition of a layer of longitudinal muscle to the body wall (Fig. 4–2A). Some of the consequent possible results are shown in the other illustrations. In Fig. 4–2B the circular muscles to the right have contracted and have brought about elongation of that end. In Fig. 4–2C they have contracted but the body has remained the same length; the left-hand end has consequently thickened. In Fig. 4–2D the circular muscles to the right have contracted; this end has again remained the same length, but now the left end has lengthened, without any change in diameter. From all of these positions the original form of the body can be restored. This restoration can be effected in Fig. 4–2C by contraction of the circular muscles at the left-hand end, exactly as in Fig. 4–1C. In Figs. 4–2B and 4–2D, however, it is only possible because longitudinal muscles are present. Their contraction can establish the pressure needed to bring about relaxation of the circular muscles and the restoration of their original shape. Thus the presence of the hydrostatic skeleton enables the two sets of muscles to act antagonistically to each other. Moreover, it makes locomotion possible. If, in Fig. 4–2D, there is adequate friction between the organism and the substratum, the left-hand end may remain stationary while contraction of the longitudinal muscles draws the right-hand end towards it. Thus the organism could become displaced to the left, in comparison with its original position in Fig. 4–2A.

These highly simplified examples show that antagonistic muscles, operating in conjunction with a hydrostatic skeleton, can bring about changes in shape and can also effect locomotion. The system has serious disadvantages, however, in comparison with those using jointed skeletons. Firstly, contraction at one point affects the pressure throughout the body and so influences all of the other muscles. Contraction of the latter will now demand more work than usual, for it has to be carried out against an increased pressure. This may result in a wasteful distribution of energy, and it makes it very difficult for one part of the body to respond independently of other parts. Further, contraction of any part of the body demands great change in length of

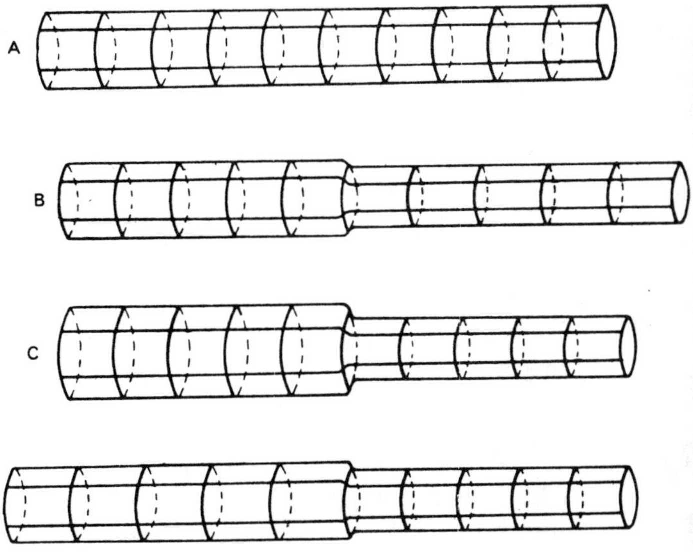

Fig. 4.2. Diagram to illustrate three possible results of the contraction of circular muscles at one end of a cylindrical animal. In A the muscles are all relaxed. In B the circular muscles of the right-hand end have contracted and this end has elongated; the left-hand end has remained unaltered. In C the length of the right-hand end has also remained the same; in D the length of the left-hand end has increased, but not its diameter. The effect of the longitudinal muscle is explained in the text. From Chapman, 1950. *op. cit.*

the muscles concerned; indeed, they may have to contract to an extent little, if at all, less than the contraction of the part of the body that they are moving. Because of this the reactions must be slow ones, for there is bound to be great waste of energy if such extensive deformations are carried out quickly.

All of these difficulties are overcome when muscle can be attached in antagonistic sets to parts of jointed skeletons. The application of force is localized, and can be quite independent of what is happening in other parts of the body. Moreover, the development of systems of levers allows very small contractions to bring about large degrees of movement; reactions can thus be rapid, and can be affected with economy of energy. Nevertheless, hydrostatic skeletons are widespread in animals, and the groups making use of them have achieved considerable success in the exploitation of their environment. For this there are two main reasons. In the simpler organisms the muscle fibres are arranged in two main layers, one circular and the other longitudinal; this facilitates reactions which, while limited in variety, can yet be of great adaptive value. In more complex organisms the relationship of the hydrostatic skeleton to the associated muscles and nerves has been elaborated in such a way as to increase both localization and complexity of response. We shall consider here only a few illustrations of these principles, but they will be sufficient to show how structural and physiological limitations have been overcome, or even turned to positive advantage.

4–2 COELENTERATE ORGANIZATION AND THE HYDROSTATIC SKELETON

The Phylum Cnidaria (Coelenterata), with some 9,000 living species, has achieved success within the limits of a simple type of organization, in which cells are grouped into tissues forming two main body layers (diploblastic structure), but the tissues are not grouped into organs. There is a single body cavity, with a mouth but no anus, and it is this cavity which contains fluid derived from the surrounding medium, that functions as a hydrostatic skeleton. This simplicity, however, is to some extent compensated for by the widespread development within the group of polymorphism (defined as the existence within a species of two or more forms of individual in numbers greater than can be accounted for solely by recurrent mutation). The primary polymorphism, which profoundly affects the functioning of the hydrostatic skeleton, is the existence of a sessile form, the polyp, and a motile one, the medusa. These may be produced in a regular alternation of reproductive generations (metagenesis), in which the polyp is the asexually reproducing phase and the medusa the sexually reproducing one, developing from a bud called the gonophore (Fig. 4–3). In many species, however, only one of these forms is present. There may also be a secondary polymorphism, associated with the colonial habit (Sec. 21–2), another feature which compensates for simplicity of organization. A single colony may thus bear two or more types of individual, differing in function (e.g. non-feeding gonozoids bearing the gonophores), and derived from either the polyp or medusa form.

Polymorphism is distributed in a characteristic pattern among the three classes which make up the phylum. The Class Hydrozoa comprises solitary or colonial forms with a simple coelenteron lacking subdivisions, and without a stomodaeum (see

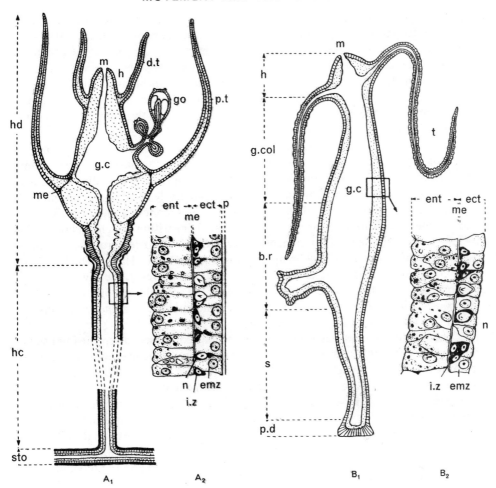

Fig. 4.3. Anatomy of two hydrozoa. A$_1$, *Tubularia larynx*; A$_2$, section through its body wall, B$_1$, B$_2$, *Hydra vulgaris*. *b.r*, budding region; *d.t*, distal tentacles; *ect*, ectoderm; *emz*, epitheliomuscular cells; *ent*, endoderm; *g.c*, gastric cavity; *g-col*, gastric column; *go*, gonophores; *h*, hypostome, peristomium; *hc*, hydrocaulus; *hd*, hydranth; *i.z*, interstitial cells; *m*, mouth; *me*, mesogloea; *n*, nematocytes; *p*, perisarc; *p.d*, pedal disc; *p.t*, proximal tentacles; *s*, stalk; *sto*, stolon; *t*, tentacle. From Tardent, 1963. *Biol. Rev.*, **38**, 293–333.

below). Colonies and metagenesis are common but not invariable, and the polyp phase tends to predominate. The medusa typically has a peripheral muscular fold, the velum. The second class, the Scyphozoa, comprises solitary forms in which the medusa, which has no velum, is the predominating phase, the polyp phase being often missing. There is no stomodaeum, but the coelenteron is often subdivided into four chambers (gastric pouches) and carries gastric tentacles which are armed with nematocysts. The third class, the Anthozoa (Actinozoa), comprises exclusively polypoid forms. Here are found the sea-anemones and the true corals. The coelenteron is subdivided by radial septa (mesenteries), and the mouth leads into a stomodaeum formed by an ingrowth of the body wall which is lined by epidermis.

Much argument has centred around the phylogenetic relations of these three classes, and, as part of the problem, around the evolutionary relationship of the polyp and medusa. These matters will be referred to later (Sec. 18–4). For the present, it should be remembered that, in discussing functional organization and modes of movement in the several classes, we are comparing plans of structure that are the products of evolutionary lines that have long been separated. Moreover, and as part of the problem of the origin of the phylum, even its apparently primitive status in the animal kingdom has been questioned.

The organization of the coelenterate polyp (Fig. 4–3) is seen at its simplest in the hydrozoan *Hydra*. (There is no medusa, which is assumed to have been lost without trace, in adaptation to freshwater life.) The body comprises a pedal disc, for attachment; a stem, and a column, distinguishable histologically because the column is the region of active digestion, so that its lining is rich in secretory and adsorptive cells; and a conical hypostome, bearing the mouth, and lying within a circlet of hollow tentacles. As we have seen, a fundamental characteristic of coelenterates is this diploblastic structure, the body wall being composed of the two cell layers which Allman, in 1853, called ectoderm and endoderm. These terms, however, are now used for the two primary germ layers of embryos throughout the animal kingdom; the two body layers of coelenterates are thus better termed the epidermis and gastrodermis. Between them lies a form of connective tissue, the mesogloea.

The epidermis and gastrodermis are largely composed of epitheliomuscular cells, which are highly characteristic of coelenterates, although they are present also in some primitive platyhelminths. These cells (Fig. 4.6, p. 73), which may each bear one or two flagella at their free surfaces, are drawn out at their bases into contractile processes or fibres ('muscle tails'). The processes of the epidermis are orientated to form a longitudinal muscle layer, while those of the gastrodermis form a circular muscle layer. Between the bases of the cells lie groups of reserve cells, the interstitial cells. Another characteristic feature are the cnidoblasts, cells which secrete the nematocysts that are used in feeding (Sec. 9–4), and which give the phylum its preferred name.

The nervous system is composed of bipolar or multipolar ganglion cells, lying just above the contractile processes in both the epidermis and the gastrodermis, and forming a nerve net. Interneuronal synapses are found, together with neuromuscular junctions and neurocnidoblast junctions. These synapses and junctions are often unpolarized (i.e. they can be shown by electron microscopy to have synaptic vesicles, presumably containing a chemical transmitter, on both sides of the junctions), or they may be polarized, with vesicles on only one side. Slender sensory cells are present in both the epidermis and the gastrodermis, but more particularly in the epidermis of the tentacles. Distally they bear a sensory flagellum-like process, while proximally each is drawn out into a conducting fibre which synapses with an adjacent ganglion cell of the nerve net.

In the various respects outlined above the structure of *Hydra* satisfies in principle the minimum requirements needed for the movements involved in the adaptive responses of a soft-bodied animal; in particular, the circular and longitudinal muscle layers, with the fluid of the coelenteron providing the hydrostatic skeleton, and a controlling nervous system, some aspects of which will be discussed later (Sec. 15–2). We may assume, from what is known of scyphozoans and anthozoans, that the mesogloea contributes visco-elastic properties that help to maintain body form, and to facilitate the antagonism of the muscle layers that are applied to its two surfaces. This is certainly suggested by the presence in it of delicate fibrils, revealed by electron miscroscopy. (In contrast to the mesogloea of the other two classes, however, it is devoid of cells.) *Hydra*, nevertheless, is too small for adequate study of the functioning of the coelenterate body. Its various properties, and their potentialities for providing adaptively specialized movements, can be much better appreciated in the anthozoan polyp, as an example of which we shall consider mainly the much studied sea-anemone *Metridium senile* (Fig. 4–4).

The structure of an anemone is more complex than that of *Hydra* in a number of features, several of which have a direct bearing on movement. Firstly, entry into the coelenteron is through a stomodaeum (actinopharynx), which, being an ingrowth of the body wall, is lined with epidermis. It bears one or two grooves (siphonoglyphs, alternatively called the sulcus and sulculus) which, in conjunction with the lateral flattening of the mouth and stomodaeum, establish a sagittal plane of bilateral symmetry. (This becomes biradial symmetry when there are two siphonoglyphs, so that the two ends of the sagittal axis are identical.) The siphonoglyphs are strongly ciliated, and maintain a current of water into the coelenteron.

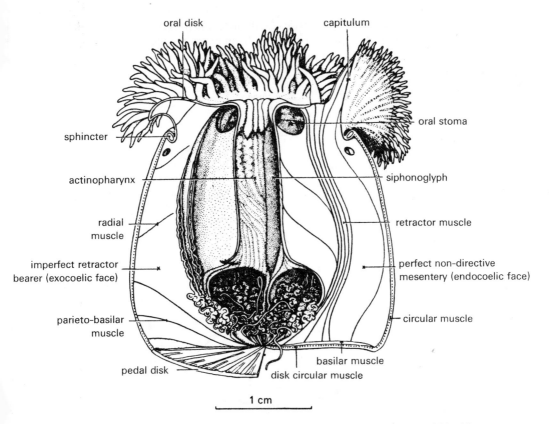

Fig. 4–4. General muscular organization and anatomical nomenclature of *Metridium*. From Batham and Pantin, 1951. *Q. Jl microsc. Sci.*, **92**, 27–54.

A second major feature is that the coelenteron is subdivided by radial septa (mesenteries), composed of mesogloea with a covering of gastrodermis. The arrangement of these septa in anemones is highly variable, even within the single genus *Metridium*, because of frequent asexual reproduction, but the general principles of the arrangement are well defined (Fig. 4–5). Some septa reach across to the stomodaeum, and are termed perfect septa; the others, which do not reach it, are termed imperfect. In either case, the septa are arranged in pairs, with one pair (directives) associated with each siphonoglyph and the others lying between the directives. The space between the two members of a pair is an enterocoele; that between adjacent pairs is an exocoele. In further emphasis of the element of bilateral symmetry, the septa are coupled, which means that a septum on one side of the plane of symmetry always corresponds to one on the other side. This coupling, in association with the septal musculature to be mentioned below, ensures symmetrical contraction of the polyp body.

Another important feature concerns the arrangement of the muscle fibres. The more superficial ones remain associated with epithelial cell bodies, as in *Hydra*. Such an arrangement is seen, for example, on the mesenteries of *Metridium*, where cell bodies of the gastrodermal epithelium form a closely knit mosaic, with their bases extending into contractile fibres (Fig. 4–6). These fibres form a lattice work connected with the cell bodies by protoplasmic strands. the interstices of which contain fluid. This perhaps has hydrostatic properties facilitating the movements of the cell bodies during contraction, and so protecting them from excessive local strain. No doubt it also aids in the diffusion of metabolites.

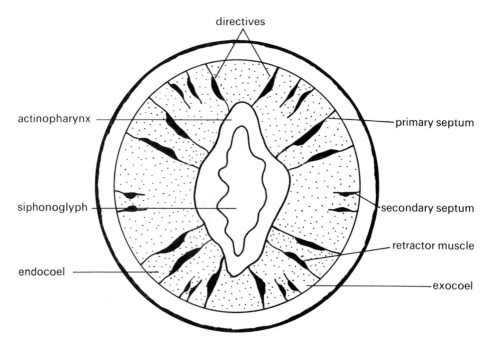

Fig. 4-5. Diagrammatic transverse section through the actinopharynx region of a sea-anemone.

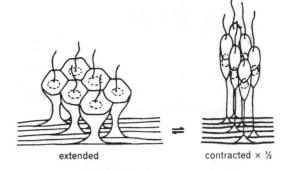

Fig. 4-6. Diagram representing movements of the musculo-epithelium (not to scale). From Robson, 1957. *Q. Jl microsc. Sci.*, **98**, 265-278.

In contrast to these superficial muscle fibres, others lie deeper as a result of their cells moving away from the epithelium. Each cell body now forms a small mass of nucleated cytoplasm lying to one side of its fibre. One consequence of this is that the layers of muscle fibres can become folded, to form almost discrete muscles that have considerable independence of action. The arrangement clearly foreshadows the appearance of the mesoderm of higher phyla.

The mesogloea of anemones, which is cellular and richly fibrous, contains mucoproteins in its matrix. These substances, with their visco-elastic properties, and their ability to hold a great deal of water, make an important contribution to the mechanical properties of this layer, and enable large amounts of it to be formed with relatively little demand for organic material. The fibres are arranged systematically to form a three-dimensional lattice (Fig. 4–7). Their source is unknown, as also is the origin of their regular arrangement, for there is no clear evidence that they are secreted by the cells of the mesogloea; indeed, the functions of these cells is unknown. However, mechanical forces are imposed on the mesogloea by the activities of the animals, and it is possible that these forces determine the arrangement of the fibres. In any event, the functions of the mesogloea are not in doubt. It holds the cell layers together, and it contributes by its elasticity to antagonize the internal hydrostatic pressure. Further, it provides a flexible bed to which muscle fibres can be attached, while its fibrillar structure enables it to accommodate muscular contraction and relaxation with varying degrees of buckling.

To the uninstructed eye *Metridium*, like other anemones, shows little movement except when it is directly stimulated, but prolonged observation, combined with time-lapse photography, reveals a slow and continuous activity, with marked changes of shape (Fig. 4–8). This activity seems to be spontaneous, in the sense that it does not arise as a consequence of external stimulation.

To understand how these movements are effected, and how the animal is able to respond adaptively to stimulation, we must consider the arrangement of the muscular system in some detail. Its essential features are shown in Fig. 4–4. Three main regions of the body can be distinguished. These are functionally differentiated, and they are to some extent able to act independently of each other. The first of these regions comprises the tentacles, capitulum, upper end of column, and oral disc, which are primarily concerned with feeding. The second region is the pedal disc, concerned with adhesion and with the effecting of a slow, creeping locomotion. The third region is the column, which is responsible for the main changes in shape of

74   INVERTEBRATE STRUCTURE AND FUNCTION

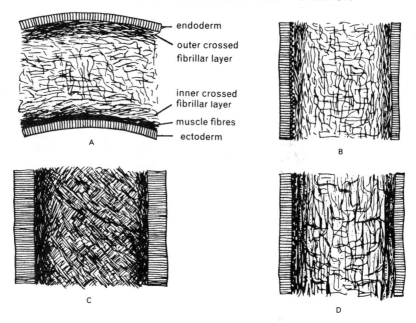

Fig. 4-7. Diagrams illustrating the general appearance of the body wall of *Calliactis* as seen in A, horizontal section, B, radial section, C, tangential section, and D, section at 45° to the long axis. Note in A and B the similar appearance of the inner and outer crossed fibrillar layers and the middle, less highly orientated region. Note in C the outer crossed fibrillar layer showing warp and weft arrangement, and in D the crossed fibrillar arrangement in which the fibres are cut either along their length or in transverse section, and are represented by dots. From Chapman, 1953. *Q. Jl. microsc. Sci.*, **94**, 155–176.

the body, and for its defensive contraction. The first two regions, however, are also involved in the movements of the column, for they close over, above and below, the central body cavity or coelenteron. This cavity contains a fluid, the coelenteric fluid. Because this can be completely enclosed the animal is able to build up in it a level of pressure that makes possible the translation of muscular contraction into movement.

The relationship of coelenteric fluid with surrounding layers of epidermal and gastrodermal muscle fibres can constitute a simple form of hydrostatic skeleton (in *Hydra*, for example), but the arrangement of muscles in *Metridium* is much more complex than this (Figs. 4-4, 4-9). The circular muscle of the column consists of a single layer of fibres, lying at the base of the gastrodermal epithelium and in contact with the mesogloea to which the fibres are attached. They run as an almost continuous layer around the body, passing underneath the radial mesenteries. In the oral disc there is an additional sheet lying at the bases of the epidermis.

In more primitive anemones the longitudinal musculature is supplied by the epidermis, as it is in *Hydra*. So it is also in the tentacles of *Metridium*, but in the

MOVEMENT AND HYDROSTATICS 75

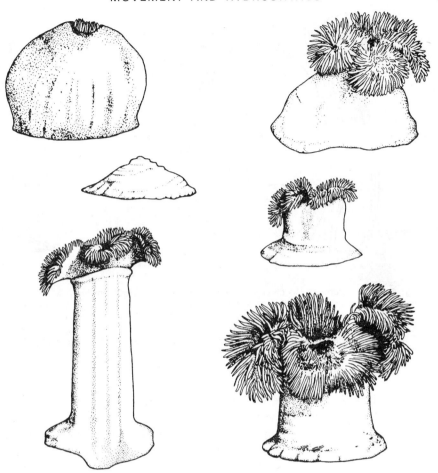

Fig. 4-8. Changes of shape of *Metridium senile*. All drawings are of the same individual on different occasions, and all are to the same scale. From Clark, 1964. *Dynamics in Metazoan Evolution.* Clarendon Press, Oxford.

column of this animal it is derived from the gastrodermis, and is borne on the mesenteries as bands of fibres forming parietal, parieto-basilar, radial, and retractor muscles.

The parietal muscles, located mainly on the endocoelic faces of the mesentery pairs, and best developed in the youngest cycles, have a broad origin on the marginal sphincter of the oral disc, and run down to fan-like radial insertions on the pedal disc, to which they transmit vertical stresses. The parieto-basilar muscles, mainly on the exocoelic faces, run from the lower parts of the column to the centre of the pedal disc, and carry stresses that are normal to the surface of the lower part of the body wall. The radial muscles, situated in the perfect mesenteries, enlarge the stomodaeum when they contract, and thus regulate the loss of fluid from the coelenteron. The retractor muscles, borne on the perfect and the larger imperfect

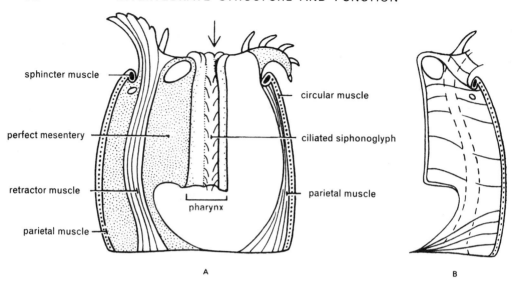

Fig. 4-9. Diagrams of the chief musculature of the column of *Metridium*. A, section through whole specimen, showing endocoelic faces of a non-directive perfect mesentery on left, of a young imperfect mesentery on right. B, exocoelic face of a non-directive perfect mesentery, showing direction of radial muscles. Broken lines indicate position of retractor muscle on opposite face. From Batham and Pantin, 1950. *J. exp. Biol.*, **27**, 264–288.

mesenteries, face away from each other on the directive pairs, but towards each other on the other pairs. These muscles, which are strong and conspicuous as a result of overgrowth and foldings, transmit stresses from the oral disc to the pedal disc. Their tonic contraction, in conjunction with the action of the parietals, maintains the shape of the column, but they are also responsible for the fast retraction response which leads to shortening and closing-over of the anemone as a result of sudden stimulation.

The muscle fibres are exceedingly slender, being about 0.5 μm thick, but they can reach a length of over 1 mm. In this they surpass the maximum recorded length of mammalian smooth muscle fibres, which may extend to 500 μm in the uterus of the pregnant female. Even so, the fibres are short in relation to the total length of the muscles. These are therefore composed of slender fibres which are placed end to end and side by side, and which form what are called muscle fields. The fibres, like all smooth muscle fibres, have a great capacity for deformation, much greater than that of striated muscle. Thus smooth muscle generally may contract by as much as 400–500%, whereas a corresponding figure for striated muscle fibres would be 20% or less. This property contributes to the substantial changes of shape that are so marked a feature of coelenterate responses, but these depend also upon the support given to the muscles by the mesogloea, with which they are closely associated (Fig. 4–10). This association becomes apparent when the column contracts. As the body then shortens, much of the mesogloea thickens, its capacity for this marked deformation being presumably dependent upon the lattice structure to which we have referred. The layer of the mesogloea that immediately adjoins the muscle

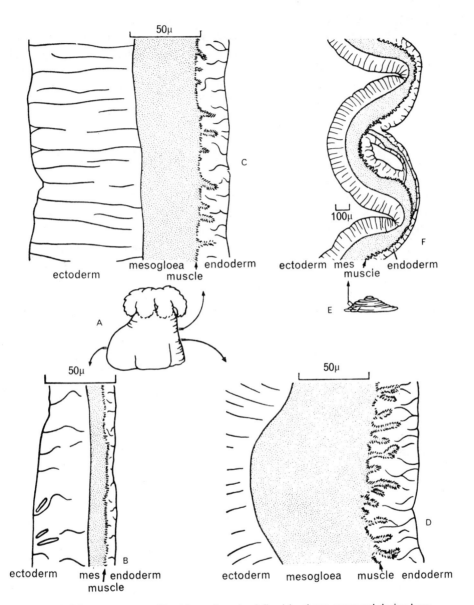

Fig. 4-10. Buckling in *Metridium*. A, animal fixed in shape assumed during locomotion. Arrows indicate regions whence sections B, C, and D were taken. B, C, D, vertical sections of body wall showing circular muscle layer unbuckled in complete expansion (B); and in increasing states of contraction, partially buckled (C) and highly buckled (D). Note great thickening of mesogloea. E, maximally contracted animal, showing in the low-power vertical section (F) the appearance of buckling of the second order in the whole body wall. From Batham and Pantin, 1951. *op. cit.*

fibres is intimately bound to these, and virtually forms part of the muscle layer. Because of this, the shortening of the muscle causes a buckling of this region of the mesogloea, the muscle layer being thereby thrown into the folds that are so familiar in sections of the contracted coelenterate body (Figs. 4–10c and D). Further contraction eventually throws the body wall itself into folds; these constitute a second stage of buckling (Fig. 4–10F), which marks the maximum degree of contraction that can be attained. Thus the close association of muscle fibres and mesoglea imposes limits on the degree of deformation that the body can undergo.

We have already noted that one of the consequences of possessing this type of hydrostatic skeleton is a great waste of energy if a contraction is performed quickly. This is doubtless one reason why coelenterates are characteristically slow-moving in their sessile phase, and it explains why the movements of *Metridium* are sluggish and imperceptible. A corollary of this is that maximum efficiency of muscular action, by which is meant in this context the least waste of energy, requires that the pressure in the coelenteron shall be kept as low as is consonant with prevention of the collapse of the body. This condition seems to be achieved in *Metridium*, for in the unstimulated animal the pressure is of the order of only 2–3 mm of water. This value is naturally influenced by the condition of the musculature of the body wall; it increases to a value of 6–7 mm when the animal contracts to some 30% of its body length.

An inescapable point of weakness in the coelenterate hydrostatic skeleton is that the coelenteron is open to the outside at the mouth, so that there must necessarily be some exchange of fluid with the external medium. To some extent this is countered in anemones by the pharynx projecting down into the coelenteron (Figs. 4–4 and 4–9) and functioning as a valve, which helps to retain the fluid in the body cavity. However, in extreme cases, often at certain intervals after feeding, the animal may collapse and empty itself of coelenteric fluid, and thereafter proceed to refill itself. Even at other times it must have some means of replenishing the fluid. The agents for this are the ciliated siphonoglyphs (Fig. 4–9), which provide a channel down which a gentle stream of water can be continuously passed into the coelenteron. The pressure that it can build up is necessarily a small one, but, as we have seen, this is all that the animal needs, and the siphonoglyphs are evidently well adapted to meet this requirement.

Pantin has pointed out that the existence of a hydrostatic skeleton is responsible for introducing an element of bilaterality into actinian organization, for when the pharynx collapses under lateral pressure it will establish an axis of bilateral symmetry; this may well have influenced the situation of the siphonoglyph. It is common to think of bilateral symmetry as a consequence of a free-moving mode of life, and of radial symmetry as associated with sessile habits; the combination of both types of symmetry in the coelenterates is thus an interesting feature of their organization, and one to which some phylogenetic importance has been attached. Pantin's suggestion reminds us, however, that a particular plan of symmetry may arise in more than one way, and that its significance needs to be carefully assessed in relation to all aspects of the mode of life of a group.

We have mentioned that there is considerable functional differentiation between the column, the foot, and the tentacles and oral disc. In some anemones waves of muscular contraction pass across the foot and provide for a slow creeping that is probably similar in principle to that seen in flatworms and gastropods, although it has not been closely analyzed. The tentacles and disc, however, are concerned with feeding, and here we find a particularly good illustration of the relationship between form and function in the actinian body. This region differs from the column in retaining what is presumably a more primitive arrangement of the muscular system (Fig. 4–11). The circular muscles are provided by the endoderm, as in the column, but the longitudinal and radial muscles are provided by the ectoderm. One advantage of this concerns the bending of the tentacles that is involved in the ingestion of food; this is carried out by local contraction of the ectodermal longitudinal muscles, and is thus functionally isolated from the contractions of the column. This ectodermal relationship is also of mechanical advantage, for it enables the longitudinal muscles of the tentacles to exert a greater bending moment than they could if they lay internally to the mesogloea. Moreover, in the latter position their efficiency would be diminished by the resistance of the mesogloea, which would antagonize their action.

It is noticeable that in the tentacles the longitudinal muscle are much better developed than are the circular ones (Fig. 4–11). This feature, too, is of functional significance. The longitudinal muscles are responsible for transporting food to the mouth, and they must be able to carry out massive contractions, which may involve them shortening to an extent similar in magnitude to the actual movement of the tentacle. By contrast, the circular muscles are concerned only with maintaining the shape of the tentacles, and with ensuring that when they are fully extended their internal pressure is not uselessly dissipated in local bulging and deformation. As we have seen, the pressure changes that are actually transmitted through the coelenteric fluid are very small; a comparatively slight tonic action of the circular muscles will thus be sufficient to control the distribution of these changes through the length of the tentacle.

Paradoxically, the significance of these circular muscles can readily be judged from the condition in many of the hydrozoan polyps, where they are absent. In *Obelia*, for example, the endoderm is transformed into a flexible supporting rod that is somewhat reminiscent of a notochord; the movements of the tentacles are effected by longitudinal muscles acting directly against this rod, which is effectively a flexible endoskeleton. Such coelenterates are exploiting a mechanical system that is different in principle from that operating in the tentacles of *Metridium*, and that is curiously similar to the relation between muscles and an endoskeleton. It is because of this that circular muscles are not developed in the tentacles of such forms.

Finally, it is worth noting that the muscles of the tentacles and oral disc, despite their generally primitive arrangement, provide an example of muscle fibres moving inwards and carrying reduced cell bodies with them, while separate epithelial cells provide the outer covering layer (Sec. 4–5).

So far we have considered the movements of coelenterates in relation to the lives of sessile polyps. Actually, their hydrostatic mechanisms do not preclude some freedom of movement, although the species exhibiting this are few and quite exceptional. An example is the sand-burrowing anemone, *Peachia*, which is adapted for its mode of life by having its base modified to form an inflatable lobe, the physa. Inflation of this is brought about by peristaltic waves which pass down the body and drive the coelenteric fluid into the physa. The consequent eversion of the latter disperses the sand and creates a cavity into which the anemone can gradually pass. The principle is not dissimilar in principle from the digging action of the proboscis

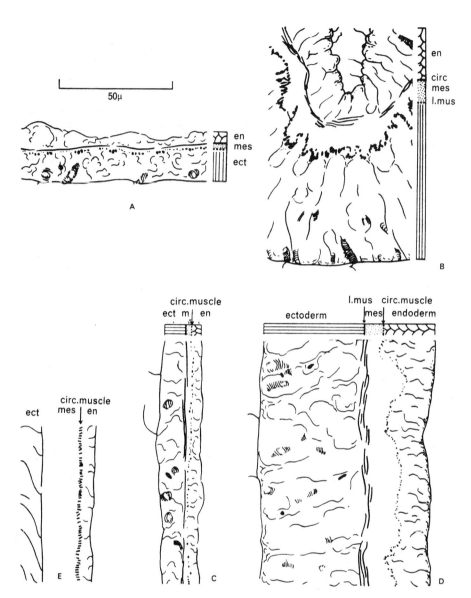

Fig. 4-11. A–D, sections of *Metridium* tentacles. Mesogloea is left unstippled. A and B, transverse to tentacular axis, A extended, B contracted. Note simple buckling of ectodermal (longitudinal) muscle field. C and D, longitudinal sections, C extended, D contracted. Note relatively weak circular muscle field, and simple buckling on contraction. A and C are from tentacle from one of the outer cycles. B and D are from the same tentacle from the innermost cycle. E, longitudinal section of extended body wall on same scale. Compare this more uniform and compact muscle layer with tentacle musculature. *circ*, circular muscle; *en*, endoderm; *ect*, ectoderm; *mes*, mesogloea; *l.mus*, longitudinal muscle. From Batham and Pantin, 1951. *op. cit.*

of *Arenicola* which we shall encounter later. It is, however, much less efficient, for it takes *Peachia* some 15 min to achieve a degree of anchorage that can be attained by *Arenicola* in about 20 sec.

Certain other anemones have evolved a much freer degree of movement than this. One example of these is a swimming anemone, *Stomphia*, which, as we shall see (Sec. 15–2), swims away by means of writhing contractions after certain starfish have touched it. Another is a small anemone, *Gonactinea prolifera*, a mere 6 mm high, which detaches itself and shortens its column when certain nudibranchs come in contact with it (Fig. 4–12). It then rows itself away with its tentacles, although it can only manage some 10–30 strokes before it tires. Yet another anemone of eccen-

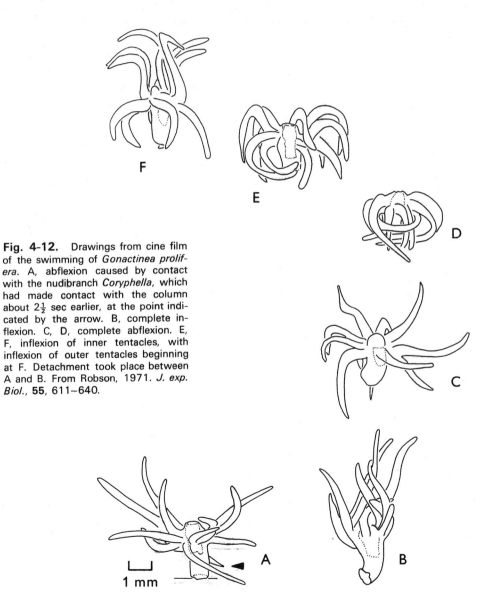

Fig. 4-12. Drawings from cine film of the swimming of *Gonactinea prolifera*. A, abflexion caused by contact with the nudibranch *Coryphella*, which had made contact with the column about $2\frac{1}{2}$ sec earlier, at the point indicated by the arrow. B, complete inflexion. C, D, complete abflexion. E, F, inflexion of inner tentacles, with inflexion of outer tentacles beginning at F. Detachment took place between A and B. From Robson, 1971. *J. exp. Biol.*, **55**, 611–640.

tric habits is *Boloceroides*, which swims away in response to mechanical stimulation. The means of locomotion are again the tentacles, which are exceptionally well developed, constituting some 90% of the body weight.

The adaptive significance of these reactions is far from clear (the nudibranchs which stimulate *Gonactinea*, for example, do not necessarily prey upon it), while the responses are so limited that they serve to emphasize how ill-suited for locomotion is the design of the polyp. Hence the fundamental importance in coelenterates of the polymorphic differentiation of sessile polyp and motile medusa. We may now consider how movement is provided for in the latter.

It has been customary in elementary teaching to derive one of these two forms from the other by changing the proportion of cell layers and mesogloea. Functional considerations, however, show that more than changes in proportion must be involved, for the movement of medusae is a form of jet propulsion, dependent upon rapid rhythmic contractions that drive water from under the subumbrella, or ventral surface. This activity, which depends upon antagonism between circular musculature and a richly fibrous mesogloea, contrasts markedly with the generally slow and largely tonic contractions of the polyp. In hydrozoan medusae (Fig. 4–13) it depends primarily upon striated circular muscle in the marginal velum (which, however, is rudimentary in *Obelia*; an example of the way in which favoured instructional 'types' depart from the typical). Similar muscle lies also in the subumbrella, together with weakly developed radial muscles. All are developed from epithelio-muscular cells of the epidermis. In the Scyphozoa there is no true velum, but merely an internal flange, called the pseudovelum, which is devoid of muscle. Swimming is here dependent upon the coronal muscle, a broad band of circular muscle lying in the subumbrella, its fibres formed by the almost complete conversion of the cell protoplasm into contractile substance. Longitudinal muscle is present in all types of medusae, mainly in the tentacles and in the manubrium.

Two other factors further emphasize the differences in functional organization of polyp and medusa. One of these is the condition of the coelenteron which, in hydrozoan medusae, is small in relation to the size of the body. The gastrodermis extends out from the central gastric cavity to form four radial canals, which fuse at their tips to form a circular canal, the whole being connected by a gastrodermal lamella. Such a system, while well adapted for nutrition, cannot act as a hydrostatic skeleton. The same is true in principle of scyphozoan medusae, although here there is much variation in the form of the coelenteron. Characteristic of this group are ingrowths of the body wall, which form septa subdividing the coelenteron into four gastric pouches. In *Aurelia*, however, the septa degenerate during development, and the adult possesses a complex system of radial canals, with a circular canal as well, through which a digestive circulation is maintained (Sec. 9–4).

The other factor influencing the mode of movement of medusae is the extensive development of the mesogloea. This is of value in more ways than one. Its transparency makes for concealment in the surface waters, while its ionic composition may contribute to buoyancy (Sec 13–1). As already pointed out, its mass is achieved with only minimal demands for an organic contribution, while it provides support for the substantial areas of epidermal and gastrodermal epithelium which are necessary in large animals that have no differentiated organs. Finally, its elasticity makes

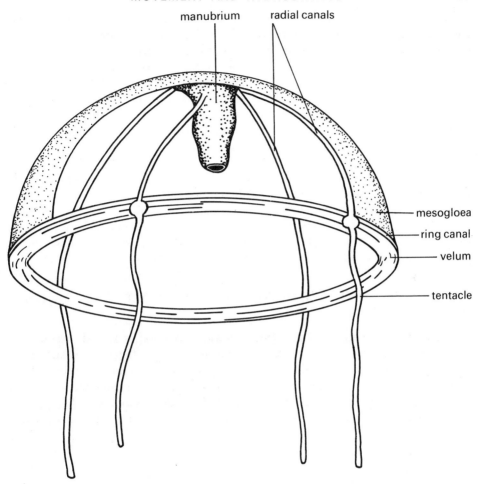

Fig. 4.13. Diagram of a hydrozoan medusa. After Hyman, 1940. *The Invertebrates.* McGraw-Hill, New York.

an essential contribution to the locomotor mechanism, by providing for the restoration of the medusa to its flattened shape after contraction of the bell. This is the more important because exumbrellar muscles are little developed, and are, indeed, absent from most scyphozoans, so that there is no muscular antagonist for the subumbrellar muscles.

The significance of these features is well seen in the large jelly-fish of the class Scyphozoa, one species of which (*Cyanea arctica*) reaches the impressive diameter of 2 m. Thick fibres run vertically through the mesogloea of the scyphozoan medusa *Pelagia noctiluca*, and are continuous with an exumbrellar and subumbrellar meshwork of finer fibres. The resting shape of the medusa is like that of a plano-convex lens. Distortion of this shape by muscular contractions sets up tensions within the mesogloea, and it is these that antagonize the subumbrellar musculature. The value of this is seen when the tension created by the fibres is weakened by the making of eight vertical cuts through the umbrella. The swimming movements are still carried out to some extent, but the animal cannot relax fully to its normal shape.

The characteristic disc shape of the scyphozoan medusa has the effect of retarding fall through the water by increasing drag. Low pressure pulses are thus probably adequate to maintain position and movement. Different conditions obtain, however, when the medusa is bowl-shaped, for this shape may be correlated with more powerful swimming movements, as exemplified in the hydrozoan medusa *Polyorchis montereyensis* (Fig. 4–14). Contraction of the subumbrellar muscles here forces out a jet of water through the narrowed velar opening, while the upper part of the bell alters little in diameter. The consequent distortion is facilitated in this species by the mesogloea being subdivided into eight vertical segments which are connected by flexible joints. During the propulsive phase the radially arranged fibres are stretched, so that part of the energy of the muscular contraction is stored in the mesogloea. The elasticity of the mesogloea promotes the recovery phase, during which water is drawn again into the subumbrellar cavity.

4–3 CTENOPHORE ORGANIZATION AND THE EVASION OF HYDROSTATICS

The term 'coelenterate' originally carried a wider connotation than its present one, for it was introduced by Leuckart in 1847 to include also the sponges and ctenophores, with the object of distinguishing these three types of animal from the echinoderms. Hatschek, in 1888, first clearly separated Leuckart's coelenterates into three phyla, comprising respectively sponges, cnidarians, and ctenophores. This practice is now generally followed, except that the designation of cnidarians as coelenterates

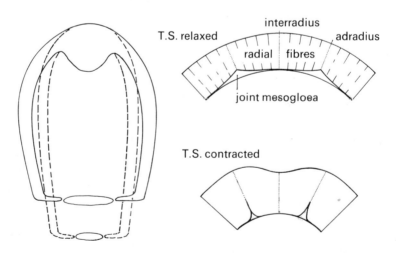

Fig. 4-14. Outlines of the bell of the medusa *Polyorchis montereyensis*, and sections of a quadrant in the relaxed and contracted state, showing the fibrous mesogloea and, at each adradius, the deformable joint mesogloea. After Gladfelter, 1972; from Chapman, 1975. *J. exp. Zool.* **194**, 249–270.

sensu stricto has been sanctioned by usage, while the sponges are often regarded as constituting a separate subkingdom, the Parazoa. Ctenophores clearly resemble coelenterates in certain features, but there are also important differences, which, taken in conjunction with a fundamentally different approach to locomotion, justify the separation of the two, and show how rich are the potentialities of diploblastic structure. Ctenophores and coelenterates may have had a common origin, and the production of nematocysts by one ctenophore, *Euchlora rubra*, possibly justifies this view, although it is said that there is nothing in the ultrastructure of the two groups to indicate a particularly close relationship.

Ctenophores (Fig. 4–15) are biradially symmetrical and free-swimming animals

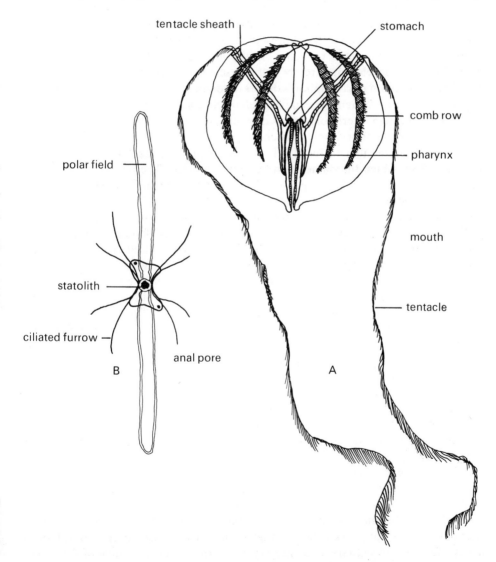

Fig. 4-15. A, ctenophore (*Pleurobrachia*), from life. B, aboral sense organ and polar fields of the same. From Hyman, 1940 *op. cit.*

which are usually transparent and gelatinous in appearance, the more primitive members having something of the general form of medusae. There is an outer epidermis and inner gastrodermis, an epidermal nerve net, and a mesogloea containing cells and fibres. An ectodermal stomodaeum (pharynx) leads into a stomach; this extends into a complex system of gastrovascular canals, which opens to the outside at two aboral anal pores. The gonads develop in the walls of the canals.

These could all be accepted as coelenterate-like features in principle, despite differences in detail. Against them must be set the absence of polymorphism and colonial life, the occurrence of determinate cleavage and mosaic development (Sec. 18–4), and the formation of a distinctive cydippid larva, resembling the adult. (There is no larval form comparable to the planula of coelenterates, except in *Gastrodes*, a minute parasite of *Salpa*.) Further, the place of nematocysts in feeding is taken by adhesive (lasso) cells (colloblasts), present in large numbers in the surface epithelium of two solid and retractile tentacles, while the muscle cells differ from those of coelenterates in developing directly from mesenchyme cells in the mesogloea.

The locomotion of ctenophores differs fundamentally from that of coelenterates in depending upon a highly specialized ciliary mechanism instead of upon a hydrostatic skeleton. Eight meridional rows of plates extend over the surface of the body, beginning near the aboral pole and ending near the oral pole. Each plate is composed of a large number of fused cilia and consequently looks like a comb (ctena), a feature from which the name of the group is derived. The beat of these combs, which propel the animal, is regulated by a highly specialized apical sense organ (Fig. 4–15), more advanced in structure than anything found in coelenterates. This organ, which is protected by a covering membrane, consists of four groups of 'balancer' cilia supporting a statolith composed of several hundred minute calcareous grains. Each balancer consists of 100–200 cilia, partly fused together, their cells being connected by ciliary tracts to the rows of combs. The whole complex forms a functionally continuous system, the ciliated cells composing it being varied specializations of the same cell family. This is shown by their uniformly upward power stroke, and by a curious ultrastructural feature of their cilia, all of which have strands (compartmenting lamellae) connecting two opposite fibrils of the typical $9+2$ system to the surface membrane.

The locomotor combs can be inhibited by the nerve net, but they are activated by the balancer cilia, which beat with a rhythm controlled by the direction of the pressure exerted on them by the statolith. Each beat of the balancer is propagated along the ciliated grooves to the combs, orientation of the animal being determined by appropriate changes in the beat of individual rows. It is believed that the propagation depends upon cell-to-cell flow of electric current, the beat of the cilia in each cell being triggered by depolarization of the cell membrane.

The difference from coelenterates in respect of locomotion is clearly profound, but there is, nevertheless, some involvement of the muscle cells and the nerve net in ctenophores. This is well shown in *Beroë*, a thimble-shaped form, with no tentacles or tentacle sheaths, and with its interior largely taken up by an immense stomodaeum. This animal can bend by means of local muscular contractions. Indeed, newly metamorphosed larvae, at stages prior to the full development of the combs, are continually bending and constricting, and can even progress by peristaltic movements. *Beroë* also gives a highly specific feeding response when its lips come in contact with another ctenophore, *Pleurobrachia*. The response, which cannot be evoked by the juices of any other animal, is described as a 'great gulp', effected by a propagated wave of contraction of radial muscle fibres which are embedded in the mesogloea around the mouth. Another element of the response is closure of the lips when they are stimulated by touch on the inside. There are also radial muscle fibres in the mesogloea which, on contraction, increase the volume of the stomach and bring about an inspiration of sea-water. That the nervous system is concerned in these movements can be shown by cutting appropriate conducting tracts, but direct transmission from one muscle cell to another seems also to be involved.

An involvement of muscle cells in movement is also seen in some of the more specialized ctenophores, an example being *Cestum (Cestus) veneris* (Venus's girdle), a ribbon-like member of the order Cestida which reaches a length of $1\frac{1}{2}$ m. It moves by means of combs, together with wave-like undulations of the whole body.

Even more specialized are the remarkable flattened forms placed in the order Platyctenea, and represented by *Ctenoplana* and *Coeloplana*. *Ctenoplana* (Fig. 4–16) is a planktonic animal with a body composed of a thicker central region and two thin lateral lobes. The flattened oral surface consists of the everted pharynx, but apart from this the general features of ctenophores, including tentacles and combs, are readily apparent. There is, however, some modification of locomotion; the combs are still used, but are aided by flapping of the lateral lobes. Moreover, *Ctenoplana* may rest on the bottom and move a little by creeping. This habit has been developed

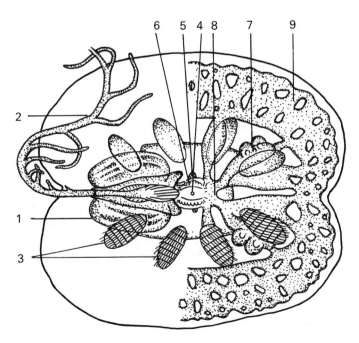

Fig. 4-16. *Ctenoplana*. 1, tentacle sheath; 2, tentacle; 3, comb rows; 4, statocyst; 5, anal pores; 6, pharynx; 7, gonads; 8, gastrovascular canals; 9, peripheral network of the same. From Hyman, 1949. *The Invertebrates: Protozoa through Ctenophora.* McGraw-Hill, New York and London.

further in *Coeloplana*, which lives as an ectocommensal on certain alcyonarians. Tentacles and statocyst are retained, but the combs are lost, and the animal creeps by the cilia of the stomodaeum which, as in *Ctenoplana*, forms the ventral surface of the body.

One might suppose, by analogy with the coelenterates, that tensile forces in the mesogloea would play some part in those ctenophores in which muscular contractions contribute to movement. Nothing, however, is known of this, and in any case the general patterns of locomotion in the two phyla emphasize their divergent evolutionary history.

4–4 SOME LIMITATIONS OF SPONGES

Sponges, like coelenterates and ctenophores, are diploblastic animals, but their organization is very different. So much so, that (as already seen) they are commonly regarded as constituting a separate subkingdom, the Parazoa, with the implication that they may have originated from protozoan ancestors quite independently of the main metazoan line. Indeed, even the animal nature of the sponges was in doubt for a long time, although the water currents that they produce had been recognized in the eighteenth century. The doubt lasted well into the nineteenth century; when Fyfe reported in 1819 that iodine was present in their bodies, he held this to confirm the view that they must be plants.

What is of particular interest about sponges is that their plan of structure constitutes one of the major blind alleys of evolution, and has evidently not lent itself to further elaboration. This does not mean, however, that sponges are to be lightly dismissed as a minor group. Already present in the Lower Cambrian, they must certainly have been evolving in the Pre-Cambrian, while their abundance in the past is testified by strata of chert, largely composed of sponge spicules, and reaching a thickness of 105 m. Even today they still comprise some 5,000 surviving species. Yet with all this they present a striking combination of long-standing adaptations with other characteristics that have limited their evolutionary potentialities, and it is of interest to enquire what these limitations are.

In part, and especially significant in the present context, they arise from the lack either of differentiated muscular tissue or of a hydrostatic skeleton, associated with which is the absence of mouth and digestive cavity. A more fundamental limitation, however, is that the body is composed of a population of cells which have little tendency to form tissues, and in which the function of any particular cell may be influenced as much by its position within the body as by its earlier differentiation. There is no reserve comparable with the interstitial cells of coelenterates. Instead, it is homeostasis within the total cell population that ensures the functions upon which the lives of sponges depend: skeletal secretion, flagellar action, phagocytosis, and reproduction.

These and other features of sponge organization are best understood by reference to the so-called olynthus stage, through which calcareous sponges commonly pass in their development (Sec. 18–4), and which in a few cases (asconoid sponges) constitutes also the adult form. This is a radially symmetrical vase-shaped body, of which the outer layer (Fig. 4–17), the dermal epithelium or pinacoderm, consists of flattened, non-ciliated cells called pinacocytes. These are highly contractile, as is the

Fig. 4-17. Diagram of the simplest type of sponge, the asconid type. *1,* osculum; *2,* layer of choanocytres; *3,* spongocoel; *4,* epidermis; *5,* pore through porocyte; *6,* porocyte; *7,* mesenchyme; *8,* amoebocyte; *9,* spicule. From Hyman, 1940. *The Invertebrata*, vol. 1. McGraw-Hill, New York. Used by permission.

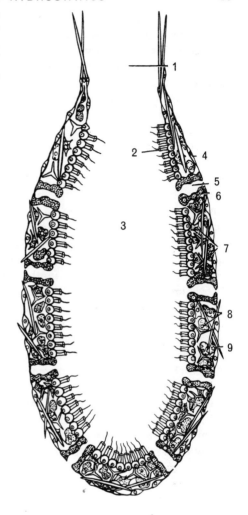

sponge body as a whole, but the movements of these cells are amoeboid rather than muscular. They are not, however, the only contractile elements in sponges. Some species have stellate connective tissue cells (collencytes) which extend across the spaces through which the water passes, and which serve to reduce the size of these spaces when the need for this arises. Spindle-shaped cells called myocytes are also often present. These resemble the smooth muscle cells of higher animals in their general form, and they have been thought in consequence to represent specialized contractile tissue; it seems possible, however, that they may often be supporting or secretory in function.

The inner layer of the body is formed of peculiar cells called choanocytes, each possessing a single flagellum which is surrounded at its base by a protoplasmic collar composed of very thin pseudopodia or microvilli held together by delicate microfibrils. These cells have long been regarded as restricted to sponges, and to imply a derivation of the group from the choanoflagellate protozoans, which have a similar collar, and which are sometimes colonial, although without any differentiation of cell types. More recently, however, cells of similar nature have been found elsewhere; in echinoderm larvae, for example, and in the planula larva of a coral. It has also been suggested that the terminal organs of protonephridia might have evolved from a choanoflagellate type of cell (Sec. 12–1). Whether this indicates a possible derivation of all metazoans, as well as of sponges, from choanoflagellates is too uncertain a proposition to be usefully discussed here, although it has been put forward as a possibility.

In any case, the two main layers of the sponge body are highly characteristic of the group, and very different from the epidermis and gastrodermis of coelenterates. They lack basement membranes, and their functions are not rigidly determined. Thus the pinacocytes of calcerous sponges give rise to porocytes, which are cells with an intracellular lumen that form the surface pores in that group. Choanocytes, primarily concerned with feeding and the maintenance of water current, can transform into sperm (at least in the Calcarea) or can act as nurse cells for sperm received from another individual.

No less characteristic of sponges is the gelatinous mesogloea, which lies between the two main layers, and which contains many free-moving amoeboid cells. These are another indication of the low level of differentiation in the sponge body, for they play a part in a diversity of functions that in higher animals would be assigned to specific tissues. Some of them, with lobose pseudopodia and large nuclei, are called archaeocytes, for they seem to be a reserve from which other types of cell can be differentiated. Others, called scleroblasts, are responsible for secreting the skeleton, which consists either of spicules, fibres of spongin, or of both these. Others again give rise to the reproductive cells, or may be concerned with transport of sperm (Sec. 18–4), or with the transport, digestion and storage of food material.

But it is particularly the absence of nerve cells that testifies more than any other single character to the lowly organization of sponges. Of course, there must be some means of communication within the body, but probably it is provided by chemical diffusion, foreshadowing the advanced chemical communication systems of higher animals. Electron microscopy shows direct contacts between individual cells; perhaps these facilitate the transmission of chemical signals direct from one cell to another.

The impression of loose organization and wide diversification of cellular functioning in the sponge body is sufficiently clear, yet order can also be discerned, and not least in the skeletal materials which provide for support and protection. A very characteristic feature of the group are the spicules (Fig. 4–18), which may be formed either of calcite or of opaline silica. They are often divisible into larger ones (megascleres), which contribute to the framework of the skeleton, and smaller ones (microscleres), which, together with some of the megascleres, lie loose in the tissues. Their form, which is very precisely determined, is designated according to the number of their axes (suffix *-axon*) and of their rays (suffix *-actine*). The spicules provide the main basis for the classification of sponges into three classes.

Of these, the Class Calcarea comprises the calcareous sponges, mainly littoral or sublittoral forms, with exclusively calcareous spicules that are not found in the other classes. These spicules may be monaxon, triactine, or tetractine. The Class

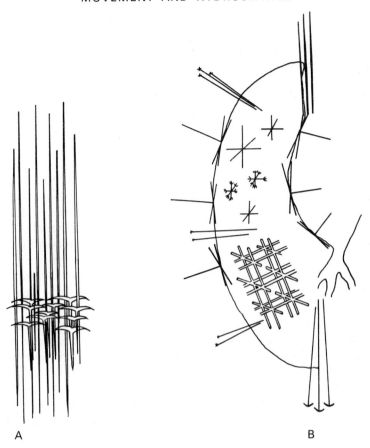

Fig. 4-18. A, monaxon and triactine spicules at the osculum of *Sycon*, from life. B, diagram of the spicule arrangement of a hexactinellid sponge, in vertical section. From Hyman, 1940. *op. cit.*

Hexactinellida comprises the glass sponges, which are deep-sea forms with fundamentally triaxon (hexactine) siliceous spicules that may be separate, or united to form skeletal networks of great elegance. There is no surface epithelium, and the choanocytes are confined to simple layers of finger-shaped flagellated chambers (Sec. 21–2). The Class Demospongia comprises forms in which the skeleton is variable in composition, being formed of siliceous spicules (which are not triaxon), or of a tough but flexible secretion called spongin (related chemically to mammalian hair and horn), or of both. Sometimes neither spicules nor spongin are present, the only skeletal material being then a soft colloid-like substance.

What these various types of skeletal structures can do is to provide supporting frameworks (Fig. 4–18), often complex and sometimes very beautiful, yet their function is limited to the strengthening of the delicate body. The skeletons are unjointed, lacking the locomotor potentialities of the skeletons of higher animals, and unable to establish functional relationships in this respect with the contractile cells, such as they are. These cells can provide effectively enough for protective contraction of the sponge body, and closure of the canal system (Sec. 9–3) through which the water passes. What they cannot do is to effect either locomotion of the whole body or controlled movements of individual parts of it. We have seen already that such movements demand well-differentiated muscles that can antagonize each other, and that can exert force either on a jointed skeleton or on an enclosed fluid. The necessary conditions are totally lacking in sponges. Quite apart from the absence of an integrating nervous system and of a jointed skeleton, contractile tissue is diffuse and unspecialized. No less important, the water in the spongocoel is contained in a network of channels open at many points to the outside, so that pressure changes cannot be transmitted in the manner required of a hydrostatic skeleton.

These limitations impose on sponges an immobility that must be a major factor in the evolutionary sterility of their organization. Paradoxically, it has been said that littoral sponges are 'continually on the move', but the movement is partly an illusion resulting from differential growth, and partly a consequence of the amoeboid activity of the pinacocytes. They are certainly not motile in the way that we find in most metazoans, and their behaviour is consequently of the most extreme simplicity. There are no well-differentiated defence responses. The animals depend for protection upon the spicules in the body wall, or the tough spongin; upon the force of the exhalent jet of water; and upon their capacity to contract the inhalent openings and thereby keep out intruders. Locomotion is confined for practical purposes to the flagellated larvae. The adult sponge is a sessile animal, unable to move in search of food, and dependent for its nutrition, as we shall see, upon the feebly organized hydraulic systems of its canals and its flagellated chambers. These it is unable to supplement by such devices as the extension of the body or the movement of tentacles which are so familiar in other sessile invertebrates, notably the coelenterates.

One may ask what advantages were gained by cells aggregating at such a low level of integration, for many of the Protozoa exemplify the high degree of differentiation that can be achieved within the limits of a uninucleate body. The answer is doubtless to be found in the larger size of the resulting structure, which must increase the capacity for resisting the physical stresses of the environment. Moreover, cell differentiation, combined with the organization of the different cell types into tissues, increases the efficiency with which functions can be discharged, so that already in sponges we find a protective external dermal epithelium, and an internal epithelium which is concerned, with other cells, in digestion. Complex reproductive processes are also favoured by multicellular organization, not only because the cells concerned can be set apart from the others (a complete segregation of this sort is not, in fact, found in sponges), but also because the maturation and safeguarding of the gametes is very much more easily achieved.

4–5 TRIPLOBLASTIC STRUCTURE: PLATYHELMINTHES

The coelenterates provide an illustration of the hydrostatic skeleton functioning within the limits of a comparatively simple organization of body structure. In other invertebrate groups, at more advanced levels of morphological and histological differentiation, the fundamental principles of the hydrostatic skeleton continue to operate, although they are expressed in different and more elaborate patterns of structure. This is already apparent in the Platyhelminthes. The relationship of this group to the coelenterates is a matter for interesting speculation (Sec. 18–4), but in any case it is clear that the organization of the platyhelminth body is more complex than that of the coelenterate one, and that this makes possible the exploitation of a greater range of action and of habitat. Differentiation in coelenterates, as in sponges, is a differentiation of tissues. In the flatworms these have become further elaborated and integrated to form compact organs, between which there is a division of labour far more extensive than is possible between one tissue and another.

Closely associated with the development of organs, and probably an essential factor in the promotion of this, is the appearance during development of a mesoderm, an extensive mass of cells which separates the ectoderm from the endoderm, and which is derived from one or other of those two germ layers. It largely arises from the endoderm, and is then known as endomesoderm; ectomesoderm, however, derived from the ectoderm, also has an important part to play in many animals. But whatever its origin, the mesoderm as a whole is regarded as constituting a third germ layer. Those animals possessing it are therefore referred to as triploblastic, in contrast to the diploblastic sponges and coelenterates. No doubt the evolution of the mesoderm is foreshadowed in both of those groups in the wandering cells that move through the mesogloea, particularly in sponges, and in the increasing differentiation of muscle layers that we have noted in anthozoans. Never in those groups, however, does this trend in development reach as far as in the platyhelminths, where for the first time we find a true mesodermal parenchyma. This has been regarded as a syncytium with interstitial fluid, but electron microscopy shows that it is fully cellular in composition. Functionally, it constitutes an advance that greatly aids the morphological and functional differentiation of the organs that we have noted as a new feature of the group. It relieves the derivatives of the ectoderm and endoderm of some of their primitive functions, gives them space in which to extend, and provides a measure of transport and communication between them by diffusion and by the passage of fluid.

The platyhelminths constitute five classes, one of which, the Class Turbellaria, comprises predominantly free-living forms which retain a ciliated epidermis. The others (Trematoda, Monogerea, Cestodaria and Cestoda) contain exclusively parasitic forms which have lost this ciliation; they will be dealt with later in other contexts. The ciliated epidermis of the free-living forms makes an important contribution to their locomotion, the very small ones being able to swim or glide entirely by ciliary action, much like ciliate protozoans. With increase in size, this method becomes inadequate, and is largely replaced by muscular action, which can generate much greater forces. The muscles operate in conjunction with a hydrostatic skeleton, but there is an important difference from coelenterates in that the mesenchymatous parenchyma replaces the coelenteric fluid as the deformable mass. Associated with this is a different arrangement of the musculature. As Pantin remarked, these worms present us with a new invention: the development of muscle fibres freely, and in depth, at any point in the parenchyma, instead of their restriction to the ectodermal and endodermal epithelia. We have seen, however, that this invention is already foreshadowed in the coelenterates.

Some Acoela possess musculo-epithelial cells in the ectoderm, a feature that is suggestive of close relationship with coelenterates, although it might have been independently evolved. The very active movements of the Turbellaria, however, depend in general upon an effector system that is more complex than that of coelenterates, and that is arranged according to a fundamentally different plan (Fig. 4–19). Immediately beneath the epidermis lies a well-developed basement membrane containing inextensible fibres. These form a flexible lattice structure which can extend and retract and yet limit the extensibility of the body. A layer of longitudinal muscle and another of circular muscle are present beneath the basement membrane; both layers are separate from the epidermis, and both are several or many fibres thick. Between these layers lie diagonal muscle fibres, while dorso-ventral fibres run vertically between the dorsal and ventral surfaces of the body.

The functioning of the muscles depends upon the circular and longitudinal layers, with their associated diagonal fibres, being situated immediately beneath the basement membrane. The significance of this position is that all of these effector structures thus lie externally to the deformable parenchyma. In coelenterates, by contrast, one muscle layer is external to the mesogloea and the other typically internal. The difference is correlated with the use of the parenchyma as the deformable hydrostatic skeleton. Only because the two antagonistic systems of circular and longitudinal musculature lie externally to that skeleton can they bring direct pressure to bear upon it. Exactly what contribution the diagonal muscle fibres make is not clear, although it has been suggested that they may contribute to the total strength of the body musculature and assist also in the maintenance of internal turgor. The dorso-ventral muscles are presumably of particular importance in these animals in ensuring the maintenance of the flattened body form.

The total effect of this arrangement of musculature and hydrostatic skeleton is that flatworms can generate delicate muscular waves passing backwards along their ventral surface, and, in addition, can reversibly deform their bodies by twisting and

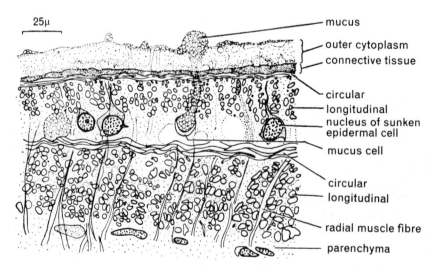

Fig. 4-19. Drawing of part of a transverse section of *Prorhynchus putealis* Haswell, to show position of surface layers. From Robson, 1957. *op. cit.*

flattening to an extent that cannot be achieved by coelenterates. This makes possible several modes of progression, additional to ciliary movement.

Large aquatic planarians glide by means of these muscular waves. They provide the prime propulsive force, but they are accompanied by ciliary action (Fig. 4–20) in association with an abundant epidermal mucus secretion (Sec. 1–4), in which the cilia beat, and which is left behind as a slime trail (revealed by treatment of the substrate with powdered charcoal). The waves, which may involve the whole surface, or may be confined to the edges of the body, are produced by localized contraction of the longitudinal muscles. They result in localized swelling of the body, which adheres temporarily to the substratum by the epidermal mucus. This provides the friction that we have earlier mentioned as essential for progression. Displacement of the body here results from its extension anterior to the point of contact, an extension that is brought about by contraction of the circular muscles, with coordinated relaxation of the longitudinal ones. Coordination, indeed, is an essential factor; transmission of pressure changes through the parenchyma is a fundamental requirement, but the activity of these animals must depend also upon control of the musculature exerted through the nervous system.

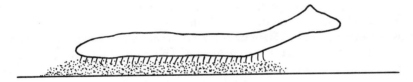

Fig. 4-20. Diagram of a planarian moving forward with its cilia beating in its mucus secretion. After Pearl, 1903, *Qt. Jl microsc. Sci.*, **46**, 509–714.

Locomotion is similar in the common British land planarian, *Rhynchodemus terrestris*, except that here the cilia are restricted to the ventral surface which forms a creeping 'sole', while the muscular waves are few in number. The presence of mucus in these land planarians thus makes possible a mode of locomotion which is, in principle, of an aquatic type. Another terrestrial species, *R. bilineatus*, which moves faster than *R. terrestris* (5 cm per minute for a 12–13 mm animal at 18°C), throws the ventral sole into large protuberances (myopodia) which leave separate footmarks instead of the continuous mucus trail left by the other animals. The implications of this are of interest, both functionally and ecologically. Functionally, the myopodia are produced by waves of relaxation which permit local extensions of the sole, under tonic pressure of the muscles of the body wall. Ecologically, this method is much more economical of mucus than is that of *R. terrestris*, and, in correlation with this, *R. bilineatus* is found in less damp situations. The continuous production of mucus inevitably makes great demands upon water reserves, and is ill-suited for habitats other than very damp ones.

4–6 TRIPLOBLASTIC STRUCTURE: NEMERTINEA

Nemertine worms are so obviously related to platyhelminths in many features of their organization that they have sometimes been regarded as a class of that phylum. However, it is more usual to regard their advances as being sufficiently substantial to justify placing them in a Phylum Nemertinea (Rhynchocoela), the characteristics of which clearly foreshadow other improvements which appear in annelids and arthropods.

Nemertines are typically marine littoral animals (occasionally fresh-water or terrestrial), ranging in length from a few millimetres to as much as 30 m (credited to *Lineus longissimus*, and suggesting some intriguing problems of physiological organization). Turbellarian-like features include the flattened body, ciliated epidermis, the lack of a coelom, and the possession of a parenchyma and an excretory system of protonephridia. In general, however, they have carried further the differentiation of tissues and organs. Thus, the epidermis is more complex than that of platyhelminths, having ciliated columnar cells, together with interstitial, glandular, and sensory cells. Underneath it is a dermis, which may be gelatinous or fibrous, resting on a basement membrane with inextensible fibres similar to those of platyhelminths. Two or more layers of muscle are present, including at least an outer circular and an inner longitudinal. The parenchyma is a complex connective tissue, continuous with the dermis, and varying greatly in development. In general, it comprises a ground substance, free cells of diverse functions, and a continuous system of filaments and fibrils. Other features of nemertines include the presence of an anus, a circulatory system (sometimes with haemoglobin) and a proboscis of a design unique to the phylum, in which use is made of a fluid hydrostatic skeleton.

This proboscis (Fig. 4-21), in its resting position, is invaginated within the body in a fluid-filled cavity called the rhynchocoel. The walls of this cavity are provided with longitudinal and circular muscle, contractions of which bring about pressure changes in the fluid, with consequent eversion of the proboscis to a length that may greatly exceed that of the worm itself. This organ is used both for feeding and for burrowing. The latter function depends on the proboscis being inserted into the sand of the substratum, and then dilated at its tip. Contraction of its retractor muscles will now draw the rest of the body after it, the dilated tip serving as the anchorage during this movement. The speed, precision, and power of the movements of this organ are a good testimony to the efficiency of a completely fluid hydrostatic skeleton, in association with fully differentiated muscle layers. This is the device that has been so fully exploited in annelid worms and many other invertebrates, but, as we shall see when dealing with these animals, the exploitation was only possible after further advances in organization.

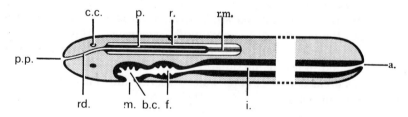

Fig. 4-21. Diagrammatic vertical longitudinal section of *Lineus*. *a.*, anus; *b.c.*, buccal cavity: *c.c.*, cerebral commissure; *f.*, fore gut; *i.*, intestine; *m.*, mouth; *p.*, proboscis; *p.p.*, proboscis pore; *r.*, rhynchocoel; *rb.*, rhynchodaeum; *r.m.*, retractor muscle. From Jennings and Gibson, 1969. *Biol. Bull. mar. biol. Lab., Woods Hole*, **136**, 405–433.

Nemertines are typically creeping or burrowing animals, using cilia and muscles in essentially the same ways as already described for turbellarians. As in the latter, a purely ciliary method is typical of young stages or small adults. However, the terrestrial nemertine, *Geonemertes dendyi*, which can glide over a glass plate at 7–10 mm per minute, does so without using muscular waves. It secretes abundant mucus, which is carried back by cilia, and the animal swims forwards in the mucus tube, in an essentially aquatic type of movement. As with terrestrial planarians, the movement is ill-adapted for dry conditions. In dry air the animal soon ceases to move, but it is able to survive by toughening the mucus into a cocoon, within which it can twist and turn, and from which, when need be, it can escape.

Pantin, pointing out how unsuitable terrestrial life is for such an animal in a world of fast-moving predators, showed that it has another protective adaptation. This involves everting the proboscis to two or three times the length of the body. The distal end of the proboscis adheres to the substratum, the animal shoots forwards over it by sudden longitudinal contraction and circular relaxation, and the proboscis is then withdrawn. The actual forward escape flight is completed in $\frac{3}{4}$–1 sec.

Other movements of nemertines depend upon the muscles of the body wall, often involving substantial waves of peristalsis which are made possible by the advanced muscular organization. In principle this is not different from planarian movement, but another type of locomotion is found which is not shown by those animals. This is swimming by dorso-ventral undulations, the body (e.g. in *Cerebratulus marginatus*) being dorso-ventrally compressed, with consequent increase of locomotor efficiency. As may be seen from Fig. 4–22, peristalsis and undulatory movement may both occur at the same time in this species, the type of movement depending on whether or not the region concerned is free in the water or is in contact with the substratum.

One feature of planarian and nemertine organization that has an important bearing on the functioning of their hydrostatic skeleton is the lack of a cuticle, which in higher worms plays an important part in strengthening the body surface and restricting deformation. How, then, is deformation kept under control? The answer, at least for nemertines, is found in the existence in the body wall of the inextensible fibres to which we have already referred. They are arranged as a lattice of alternately left- and right-handed spirals around the body, forming a multilayered system in the basement membrane, and also a myoseptum in the muscle layers (Fig. 4–23). These fibres, because of their spiral arrangement, form a geodesic system, which is to say that each one marks the shortest distance between two points of the body surface. Moreover, they are attached to each other where they cross, so that they cannot slip, but can move over each other with consequent increase or decrease of the lattice angle, much like the effect of handling 'lazy tongs' or a garden trellis.

The influence of these fibres on movement is illustrated in general terms in Fig. 4–24. This shows (A) a diagram of part of the body of a cylindrical animal (i.e. with circular cross-section) with a single turn of a single geodesic fibre, and (B) this same part after it has been cut open along the top and spread out flat. The fibre, by definition, now forms a straight line of length D. It can be seen that increase in the angle θ, between the fibre and the longitudinal body axis, changes the shape of the body towards a thin disc with a circumference formed by D; in the limit, the volume approaches zero. Conversely, as the angle θ decreases, the shape changes towards

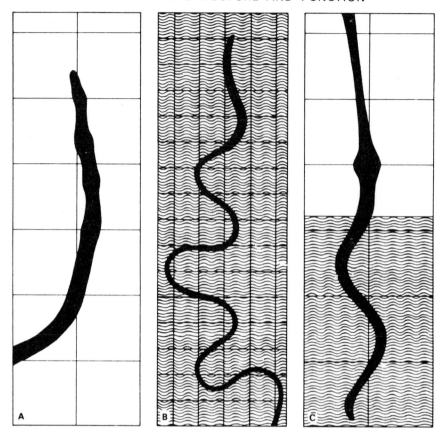

Fig. 4–22. *Cerebratulus marginatus.* A, creeping over solid surface; note peristaltic waves. B, swimming freely on water; note undulatory movement. C, a specimen suspended vertically with posterior end only in water; note peristalsis in anterior region in air and undulatory movement of posterior end in water (drawn from photographs by H. W. Lissmann). From Gray, 1939. *Proc. R. Soc. Lond., B,* **128**, 28–62.

a thin thread of length D and again in the limit the volume approaches zero. Between these two limits, as shown in Fig. 4–24c, the cylindrical body will, in theory, have maximum volume, and it can be calculated that this is reached when the angle θ is 54°44′ (i.e. $\tan^2 \theta = 2$).

In fact, however, an animal cannot live like this, for its volume must remain constant in order to maintain a proper relationship between the circular and longitudinal muscles of the body wall and the fluid of the hydrostatic skeleton. This constant volume is indicated by the horizontal line in Fig. 4–24. Comparison of it with the theoretical volume indicated by the curve shows that when $\theta = 54°55′$ the volume of the body must be less than the theoretically maximum volume. This condition the animal achieves by departing from a cylindrical shape and adopting an elliptical cross-section. It thus reduces its volume, making use of the principle that the volume of a body at any given length is greatest when its cross-section is circular. As the body shortens or lengthens, towards the two limits, the cross-section approaches to the

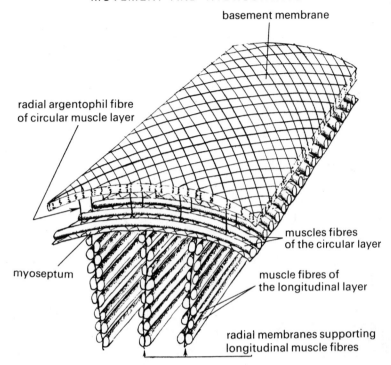

Fig. 4-23. Diagram of the structure of the body wall of *Amphiporus lactifloreus*. The weak diagonal muscle, which lies between the circular and longitudinal muscle, is not shown. From Cowey, 1952. *Qt. Jl. microsc Sci.*, **93**, 1–15.

circular. When it becomes circular, at the intersects F and G, the volume is at the theoretical maximum for that particular length. Lengthening beyond the intersect F, and shortening beyond the intersect G, are both impossible because, as may be seen from the diagram, they would require the animal to decrease its volume. And this, by definition, it cannot do.

The geodesic lattice thus plays, to some extent, the part of a cuticle, not only by strengthening the body surface, but also by permitting changes of shape and yet restricting them in accordance with the above principles. But this is not all, for here, as so often, animals set us all an example in overriding theoretical considerations. Clark and Cowey, in an analysis of the situation in a number of nemertine species, show that some of them depart from theoretical predictions to an extent which is correlated with the mode of life and the requirements of the habitat. For example, the terrestrial nemertine *Geonemertes* always approximates to a circular cross-section, and achieves this by limiting its extensibility. This peculiarity is correlated with the need of this animal to conserve water by reducing evaporation; an end which is better secured with a circular cross-section, for this gives a low surface area/volume ratio. *Lineus*, by contrast, is a marine creeping form, and therefore has no need to conserve water. Thus it shows considerable extensibility and also some flattening, a feature which facilitates movement over the substratum by cilia (see below). *Cere-*

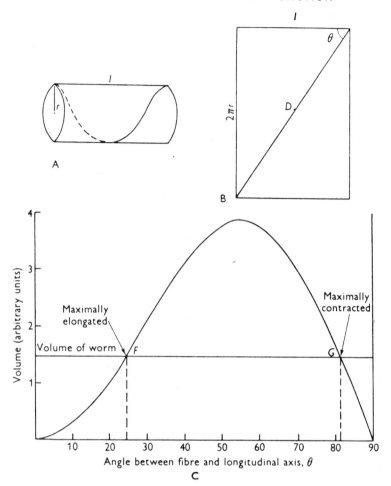

Fig. 4-24. A, unit length of a cylindrical 'worm' bounded by a single turn of the geodesic fibre system. (Fibres running in the opposite sense have been omitted for clarity.) B, the same unit length of 'worm' slit along the top and flattened out. C, the curve represents the theoretical relationship between the volume contained by the fibre system and the inclination of the fibres to the longitudinal axis. The horizontal line represents the actual and constant volume of the nemertean *Amphiporus lactifloreus*. It intersects the curve at F and G, which are the limiting position of elongation and contraction, respectively, for the species. From Clark and Cowey, 1958. *J. exp. Biol.*, **35**, 731–748.

bratulus is markedly flattened, but has only a low extensibility, because it swims and needs a flattened body to facilitate this. Here the flattening takes precedence of extensibility, and does so because there are inextensible reticulin fibres in the muscle layers, while the powerful longitudinal musculature, together with dorso-ventral muscles, further tends to impede changes of shape.

Similar considerations must apply in principle to turbellarians, where there is also no cuticle. These worms, however, are always flattened, so that, despite possessing the lattice of fibres that we have already noted, they make little use of the changes

MOVEMENT AND HYDROSTATICS 101

of shape which are exploited so effectively in the nemertines. There are several reasons for this. One is that a flat body enables the animal to deal more easily with the problem of gaseous diffusion; less compelling in nemertines, no doubt, because they have a blood system. But another reason is certainly the burden imposed by ciliary locomotion. This demands as large a surface area of ciliated epithelium as possible, a condition which can be achieved by keeping the body as thin as possible, with a high surface/volume ratio.

4-7 TRIPLOBLASTIC STRUCTURE: NEMATODA

The principles of the hydrostatic skeleton are fundamentally so simple that they can operate within more than one type of structural organization. This is sufficiently evident from a comparison of coelenterates with platyhelminths, but it is illustrated no less well by the Nematoda. These animals form a very isolated assemblage, sharing a combination of features so characteristic of the group that it is difficult to determine either its relationships or its evolutionary background. The body wall (Fig. 4-25) is covered by a cuticle, secreted by a non-cellular epidermis, and consisting of protein, not of chitin. Underneath the epidermis is a single layer of peculiar muscle cells, which have an outer contractile portion and an inner core of unmodified and non-

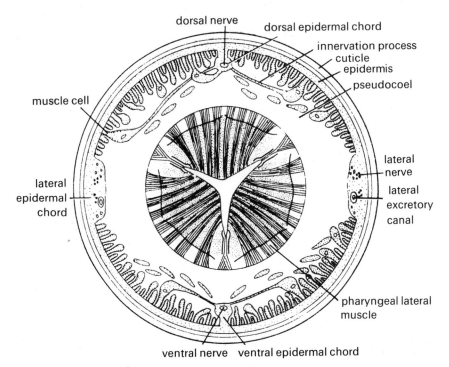

Fig. 4-25. Diagrammatic transverse section through *Ascaris* in the region of the pharynx. After Hirschmann, from Meglitsch, 1972. *Invertebrate Zoology* (2nd ed.) University Press, Oxford.

contractile protoplasm. These cells are arranged as a layer of longitudinal muscle, circular muscle being entirely absent from the body wall. Between this muscle layer and the alimentary canal lie highly vacuolated cells, the vacuoles running together to form a pseudocoel, filled with fluid, which constitutes a hydrostatic skeleton. Here, however, there is no antagonism between longitudinal and circular muscles. Instead, the longitudinal muscles are antagonized by the force exerted upon the cuticle by the pressure of the body fluid.

This is a unique type of hydrostatic skeletal system, depending for its operation upon the highly specialized structure of the cuticle. In *Ascaris* (and this is probably true of most nematodes) the cuticle contains sets of diagonally crossed fibres, which are sufficiently thick to be seen in fixed material. There are three layers of these fibres, creating a geodesic lattice like that of the nemertines that we have just discussed. Just as in nemertines, these fibres can move with reference to each other in such a way that the worm can become either shorter and thicker, or longer and thinner. Because of this the cuticle, in association with the longitudinal musculature, forms a system capable both of distortion, and of exerting tension against the internal pressure. The fibres themselves are probably inextensible; it is supposed, therefore, that the cuticle must also contain an elastic component.

The internal hydrostatic pressure of *Ascaris* is considerable (Fig. 4–26), although it shows a wide range of variation, from 16 to 125 mm of mercury, with a mean value of 70 mm of mercury (equivalent to 95 cm of water). This pressure, acting in conjunction with the tension exercised by the body wall, and with some degree of muscular tone, accounts for the constant shape, cross-section, and length which is so characteristic of the nematode body. It provides also for the simple undulating movement of these animals; a type of movement that, by fitting them so well for life

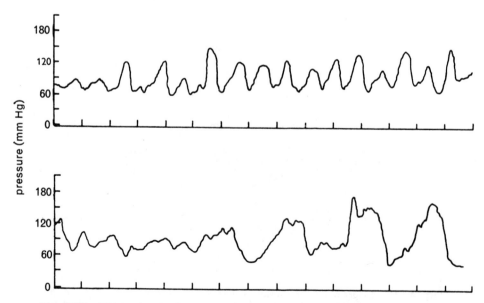

Fig. 4-26. Tracing from a photographic record of pressure changes in *Ascaris*, made with a glass helix pressure gauge. The two halves of the record are consecutive. Time marks are at 30-second intervals. From Harris and Crofton, 1957. *op. cit.*

in viscous media, is one factor that pre-adapts them for the parasitic habit (Sec. 23–4). It depends on the fact that when muscular contraction at one point produces local shortening, this is compensated for by lengthening elsewhere, because the displaced body fluid exerts pressure against the elastic force maintained by the body wall. Usually the lengthening will be on the side of the body opposite to the point where contraction is occurring, with the consequent production of sinusoidal waves. A variant is found in the chrysanthemum eelworm, *Aphelenchoides*, much smaller than *Ascaris* and with virtually no pseudocoelomic fluid. Here the deformable material is provided by the muscles themselves, together with the alimentary tract and its contents. Contractions of the muscles of the body wall occurs in only short lengths at a time, the result being a remarkable economy of action. Gut contents can be moved by simultaneous contractions of the dorsal and ventral halves of the musculature, while sinusoidal movement of the worm is effected when the dorsal and ventral waves are in opposite phase (see also below).

Undulatory propulsion is common in the animal kingdom, and is produced in more than one way, but its principles are essentially the same whether we are concerned with a snake or a fish, with a nematode or a sperm. They are, however, particularly well illustrated by nematodes because their shape conforms so closely with that of a smooth and elongated cylinder.

A nematode moving in a dense medium uses its head as a wedge. The propulsive force applied to this is derived from the resistance of the medium to displacement by the animal's body in a direction normal to the surface of the body at any given point. Under ideal conditions, this resistance is sufficiently great to prevent the track of the animal being wider than the body, and it is then that maximum speed is attained. Such conditions obtain when the small nematode (*Haemonchus contortus*) glides over the surface of agar jelly (Fig. 4–27). The sinusoidal locomotor waves are stationary relative to the ground, and the animal leaves a sinusoidal track behind it. If, however, the resistance of the medium is not sufficiently great, the body is displaced laterally,

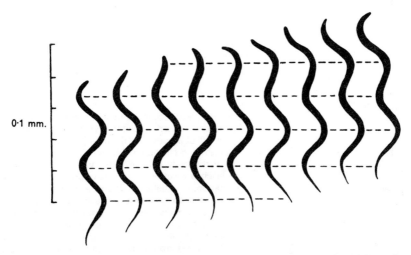

Fig. 4-27. *Haemonchus contortus* creeping with zero slip over the surface of an agar gel. Tracings from successive photographs taken at approximately 1/3 sec intervals. Note that the waves are stationary relative to the ground. From Gray and Lissmann, 1964. *J. exp. Biol.*, **41**, 135–154.

energy is no longer applied entirely to forward movement, and the sinusoidal waves of the body move backwards relative to the substratum, the speed of progression equalling the speed at which they pass backward relative to the head. This backward movement of the waves is called their slip, which is given by the expression

$$\% \text{ slip} = [(V_w - V_x)/V_w] \times 100$$

where V_x is the speed of progression and V_w the speed of propagation of the waves.

Conditions in a homogeneous aqueous medium differ from those found in a dense suspension, and this affects the propulsive waves. Their amplitude, frequency, and wavelength are greatly increased, while the speed of progression is reduced relative to the speed of propagation of the waves. With nematodes, as with fish, the amplitude of the waves increases as they pass backwards, an arrangement which makes it possible for the resultant of the transverse waves to be zero, while lateral displacement is minimal at the anterior end. This reduces side-to-side movement to a minimum, and thereby conserves the energy which would otherwise be needed to overcome the forces resisting the lateral displacement. To make this balancing of transverse forces possible, however, the body must be long enough to provide for more than one complete wave (Fig. 4.28). This condition is satisfied in the vinegar worm, *Turbatrix*

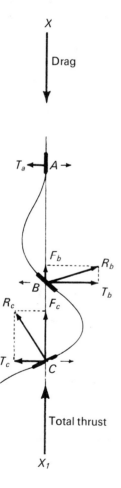

Fig. 4-28. Diagrammatic representation of the forces exerted by water against the undulatory movement of a body. If a distal region (C), is moving to the right and eliciting a reaction R_c, the motion of the rest of the body relative to the water must be such as to elicit a reaction equal but opposite to R_c. This condition is satisfied if there is an anterior region (A) moving to the right and eliciting a reaction I_a, and an intermediate region (B) moving to left and eliciting a reaction R_b. The resultant of the forward components (F_b and F_c) of R_b and R_c is equal to the total backward drag of the whole body, while the resultant of the transverse forces (T_a, T_b, T_c) is zero. The small arrows show the direction of lateral movement of the three regions named. After Gray, 1958. *J. exp. Biol.*, **35**, 96–108.

aceti, when it swims actively to the surface of vinegar (Fig. 4–29). *Haemonchus contortus*, with a shorter body (547 μm, as compared with 840 μm for *Turbatrix*), displays less than one complete wave while swimming freely, and exhibits marked periodic yawing.

Nematode locomotion can be elegantly adapted to the demands of particular habitats, as has been well shown by Wallace in a comparison of the beet eelworm, *Heterodera schachtii*, with the chrysanthemum eelworm, *Aphelenchoides ritzema-bosi*, to which some reference has already been made. *Heterodera* can move at speeds of up to 300 mm/hr, optimum conditions being attained when the channels between the soil particles are about the same diameter as the eelworm. Studies in which the movements were observed in single layers of particles showed that progress was inhibited in wider channels (e.g. 100 μm), because there was now less restriction on lateral movement of the body. This led to a reduction in forward movement, according to the principle outlined earlier. Pores with a diameter less than 20 μm also provided a barrier. This is because the nematode locomotor mechanism, dependent, as explained above, on a high internal pressure opposed to the tension of the body wall, does not permit the animal to force its way through narrow channels by constriction. The consequences are further illustrated by the data of Fig. 4–30. The distance travelled is seen to be greatest when the particles are 150–250 μm in size, and also when the pores are only partly filled with water. Saturation with water encourages the disadvantageous lateral movement of the body.

The chrysanthemum eelworm is more active than *Heterodera*, in that its propulsive waves have a higher frequency and speed. Moreover, it can swim readily in thick films of water, whereas *Heterodera* is most efficient in thin ones. The significance of this difference is that it enables *Aphelenchoides* to move up the stem of the plant immediately after rain instead of having to wait until the plant has partially dried. Eventually, however, it has to penetrate the leaf of the plant through the stomata.

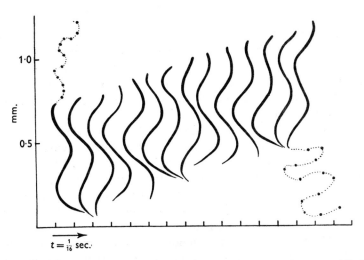

Fig. 4-29. Successive positions at intervals of 1/16 sec of a specimen of *Turbatrix aceti* in which the amplitude of transverse movements of the tail was about four times that of the head. From Gray and Lissmann, 1964. *J. exp. Biol.*, **41**, 135–154.

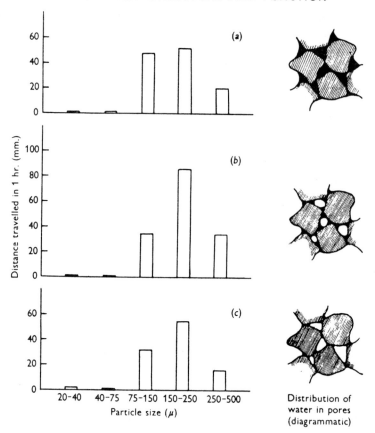

Fig. 4-30. Migration of *Heterodera schachtii* in single-particle layers of different faction of soil at three different moisture levels. The diagrams to the right of the histograms indicate the distribution of water (black) in pores at three moisture levels. From Wallace, 1958. *Ann. appl. Biol.*, **46**, 74–85.

This can only take place when movement has been slowed down, and this is provided for during the drying of the plant. As the film of water becomes thinner, the movements become slower, until conditions for penetration are attained. Such is the adaptive subtlety which can be revealed by the functional analysis of a particular component of animal organization.

We have been concerned here with locomotion, but Harris and Crofton show that it is possible to extend the analysis to other aspects of nematode functioning. Commenting on the constant form of these animals, they remark that 'the elementary student may be forgiven at times for thinking that there is only one nematode but that the model comes in different sizes and with a great variety of life histories.' This constant form, they suggest, may well be determined by limitations inherent in the plan of structure that we have outlined.

For example, transfer of fluid from point to point within the body provides for an essentially mechanical coordination of movement that eliminates the need for local reflex pathways. Hence the simplicity of the nematode nervous system, which has anterior and posterior groups of nerve cells linked by nerve cords that contain few if any cells. The high internal pressure would cause collapse of the lumen of the alimentary tract if it were not antagonized in some way. This antagonism is ensured by the existence of a muscular pumping pharynx, which, incidentally, is another pre-adaptation for parasitic life. The absence of cilia is a feature of nematodes, and it is probably significant that ciliary action would be quite inadequate for maintaining sufficient pressure within the intestine of these worms. Maintenance of this pressure is aided by the 'self-sealing' of the anus, defaecation being provided for by the action of dilator muscles of the hind gut. Then, to give one further example, it is characteristic of nematodes that the excretory canals are embedded in lateral thickenings called the lateral lines. This may well be an adaptation for ensuring that these canals are not closed by the internal pressure.

This interpretation of the form of nematodes as being determined by mechanical factors is somewhat theoretical, yet very plausible. No less plausible is the interpretation of the form of platyhelminths as being determined partly by the factors that we have discussed earlier, and partly by considerations of transport and diffusion (Sec. 11–1). In animals, as in the structures that we design ourselves, external appearance reflects functional requirements, and is often all the more satisfying the more closely that it does so.

5
Movement, Hydrostatics, and the Coelom

5-1 SIGNIFICANCE OF THE COELOM

An advance in animal organization, no less important than the development of a mesoderm and internal musculature, is the appearance within the mesoderm of the cavity known as the coelom. This can be defined as a cavity that arises within the endomesoderm, and that is therefore covered on its outer surface by the somatic mesoderm and on its inner surface by the splanchnic mesoderm. It contains a fluid, the coelomic fluid, and is lined by an epithelium, the coelomic epithelium or peritoneum.

It is usually referred to as the secondary body cavity, because it is preceded in development by the primary body cavity or blastocoel, which is the cavity of the blastula. This cavity is usually obliterated by the development of the archenteron, the primitive digestive cavity. The archenteron persists in coelenterates as the only cavity, but since its lumen is strictly part of the outside world it is not regarded as a body cavity *in sensu stricto*. It also persists as the only cavity in the platyhelminths, for in them, too, the blastocoel is obliterated and the coelom has not yet appeared. This group is triploblastic but acoelomate. Sometimes the blastocoel persist to form in the adult a body cavity that is superficially like a coelom in that it contains a fluid, but differs from a typical coelomic cavity in lacking a lining epithelium. We have already encountered this type of cavity, called a pseudocoel, in the Nematoda. It is also found in the Nematomorpha, Rotifera, Gastrotricha, Kinorhyncha, Acanthocephala and Entoprocta. The body cavity of the Priapulida may also be a pseudocoel.

The question of the evolutionary origin of the coelom has been much debated, but there is a lack of definite information, and conclusions can only be reached by inference. Four main theories have been proposed: that it originates from outgrowths of the alimentary canal (enterocoel theory), from the enlarged cavities of nephridia (nephrocoel theory), from splits in the mesoderm (schizocoel theory), or from the enlarged cavities of gonads (gonocoel theory).

As Hyman points out in her analysis of these several theories, none is wholly free from difficulties. The nephrocoel theory is only of historical interest. It was first proposed by Lankester in 1874 at a time when the relationships of the coelom with nephridia and coelomoducts (Sec. 12–1) had still to be clarified; it has since found few supporters. The schizocoel and enterocoel theories derive from the two methods of development of the coelom, which may arise either by splitting of the mesoderm, in which case the cavity is called a schizocoel, or by the evagination of pouches from the wall of the archenteron, in which case it is called an enterocoel. A schizocoel is found in the Annelida, Arthropoda, Mollusca, and Pogonophora—groups that are believed to be closely related and that can be placed, with the Platyhelminthes and Nemertinea, in a group called the Protostomia (Sec. 18–5). An enterocoel is found in the Echinodermata, Hemichordata, Cephlochordata, and perhaps in the Urochordata—groups that are believed to have close relationships with the Vertebrata and that can be placed, with them, in the Deuterostomia.

This may seem to suggest that the mode of development of the coelom is of phylogenetic importance, but in fact it is by no means certain that this is so. Even within the echinoderms and hemichordates, for example, there is marked variation in this development. Moreover, the vertebrates have a schizocoel, whereas the cephalochordates, to which they are undoubtedly very closely related, have an enterocoel. We see here an illustration of a fact that cannot be over-emphasized: the pathways of early development are greatly subject to evolutionary modification, and therefore form an insecure basis for phylogenetic speculation unless they are carefully considered in relation to other aspects of organization.

The enterocoel theory was associated by Sedgwick in 1884 with a specific aspect of coelenterate organization, for he suggested that the coelom might have been derived from the coelenteric pouches enclosed by the mesenteries in the Scyphozoa and Anthozoa. This version of the theory, however, demands the derivation of triploblastic animals from forms that are already so highly specialized that they would seem unlikely to have been the starting points of a major evolutionary advance. Nor is it easy to explain the selective advantage of such a step. These coelenteric pouches serve to increase digestive efficiency through the enlarged surface area that they provide, and through the specialization of the cells along their free edges (Sec. 8–5), while their mesenteries help to regulate the water content of the hydrostatic skeleton. These advantages would be lost if the pouches separated off as separate coelomic cavities.

The schizocoel theory is at first sight more plausible than either of the preceding theories, yet it has received little support. This is probably because of its essentially indefinite character, and particularly because of its inability to account for the clear-cut morphological difference between the coelom and the spaces of the blood vascular system.

The most favoured theory of the four has certainly been the gonocoel theory, illustrated in diagrammatic form in Fig. 5–1. Among the points in its favour is the fact that some acoelomate nemertines pass through a stage in which they possess a series of gonads with large and sometimes empty cavities; it is possible to visualize the further enlargement of these into the spacious coelomic cavities of annelids. Moreover, this particular theory would account for gonads commonly developing from the coelomic wall. In this respect, however, it fails to explain why the germ cells

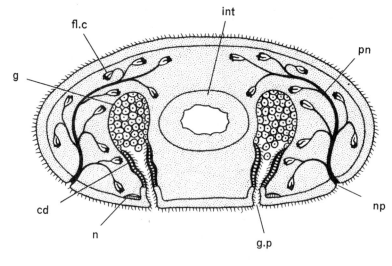

A

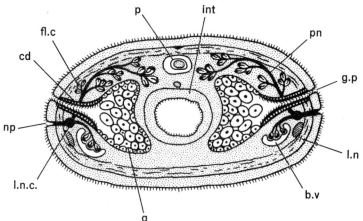

B

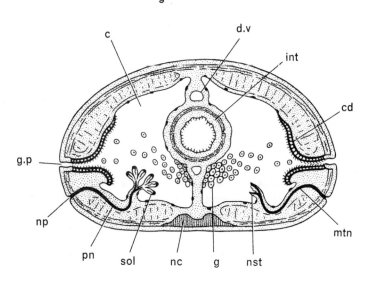

C

often make their first appearance in regions other than the coelomic wall, and only reach this after considerable passage through the body. Nor is this the only difficulty presented by the gonocoel theory. To derive the coelom from serially repeated gonads is to imply that its origin was closely connected with the establishment of metamerism. Clearly this entirely fails to account for the existence of many unsegmented coelomate groups, for these cannot be dismissed with the assumption that they have lost their segmentation after derivation from a metamerically segmented ancestor. We shall return to this point when we deal with the origin of metamerism, although theories relating to this do little, if anything, to clarify the present question. It is wiser, in fact, not to seek a formal answer to it.

The difficulties outlined above—and they could be pursued in much greater detail—suggest that a discussion of the origin of the coelom couched in terms of classical morphology is bound to be sterile. Rather, we should examine the uses to which the coelom can be put, in order to understand the conditions that may have favoured its appearance and early evolution. We shall see that it must have proved so advantageous that it might have arisen independently on more than one occasion.

One obvious advantage derived from a coelomic cavity is that it establishes a cavity around the alimentary tract, which is now surrounded by the peritoneum and, in many animals, suspended in the cavity by a double fold of peritoneum called a mesentery. This, together with the development of a visceral musculature from the associated mesoderm, permits free movement of the tract, an advantage conferred also upon the heart and other mobile organs. As regards the alimentary canal, this freedom improves efficiency in the handling of the food material. In microphagous animals, which tend to rely largely upon cilia for the handling of the food particles, this is less important, but it must have been a crucial factor in promoting the development of macrophagy, with the consequent need to break up food and to mix it with the enzymes of the alimentary lumen.

Another advantage is concerned with transport. We have noted that the presence of fluid in the interstices of the parenchyma of the acoelomate plathyelminths must help in the transmission of metabolites. For this purpose the coelomic fluid is more efficient, as is apparent from the way in which excretory organs, whether nephridia or coelomoducts, open into the cavity and abstract dissolved nitrogenous waste from it. At the same time, and through the same pathway, the coelomic fluid has an important part to play in osmoregulation. We shall be considering both of these aspects later, and we shall see then how the development of a blood vascular system has become involved in their execution.

In addition, and arising from what we have said above regarding the origin of the germ cells, the coelom is an important factor in reproduction, for the germ cells mature in it, either in the cavity as a whole or in restricted parts of it, and are even-

Fig. 5-1. Diagrams of transverse sections illustrating the gonocoel theory of the origin of the coelom. A, the platyhelminth stage; B, the nemertine stage; C, the annelid stage. *b.v.*, lateral blood vessel; *c*, coelom; *cd*, coelomoduct; *d.v.*, dorsal longitudinal blood vessel; *fl.c*, flame-cell; *g*, gonad; *g.p.*, genital pore; *int*, intestine; *l.n*, lateral nerve; *l.n.c*, longitudinal nephridial canal; *mtn*, metanephridium; *n* and *nc*, ventral longitudinal nerve; *np*, nephridiopore; *nst*, nephridiostome; *p*, proboscis; *pn*, protonephridium; *sol*, solenocyte. Adapted from Goodrich, 1945. Q. Jl microsc. Sci., **86**, 113–392.

tually discharged through the coelomoducts. These ducts were very probably genital in their initial function, and only later became concerned with excretion and osmoregulation, displacing in consequence the nephridia, which, in their closed protonephridial form, are the most primitive excretory organs (Sec. 12–1).

Lastly, but perhaps most important of all, the coelom with its contained fluid provides the structural basis for a hydrostatic skeleton more highly organized than that of coelenterates and platyhelminths. In this respect it probably played a major part in ensuring the successful survival of the lower coelomates during the period when calcium and phosphorus metabolism had not been elaborated to a point at which rigid and jointed skeletons could be constructed. The movements of annelid worms, echinoderms, and many of the smaller groups, are based upon hydrostatic principles in essentially the same way as are the movements of coelenterates, but they show a greatly increased flexibility and speed of response. This is partly, of course, because of the more highly differentiated structure of the body as a whole, and in particular because of the much greater elaboration of the nervous system. A major contribution to this improvement, however, is certainly attributable to the presence of the coelom, and it may well be that this aspect of the function of the coelom was the most important single factor determining its appearance. This possibility will be more readily appreciated if we examine some examples of its functioning as a hydrostatic skeleton.

5–2 SIPUNCULA, ENTOPROCTA, POLYZOA (ECTOPROCTA)

Coelomic hydrostatic skeletons are used to provide for locomotion, or, in sessile microphagous animals, for the extension of tentacles used in feeding. In echinoderms, as we shall see later, these two possibilities have been combined in the exploitation of benthic life.

The Phylum Sipuncula, a coelomate group probably related to annelids, illustrates the use of a coelomic hydrostatic skeleton in burrowing, at a relatively simple level of organisation with an undivided coelomic cavity. *Sipunculus nudus* is a marine animal inhabiting galleries in clean and well-oxygenated sand, which it can readily re-enter if it is removed from it and laid on the surface. Its burrowing is carried out in a series of digging cycles, each involving the eversion and withdrawal of a proboscis or introvert which carries the mouth at its tip, surrounded by tentacles.

Each cycle, lasting about 30 sec and affecting a forward movement averaging some 1.66 cm, begins with longitudinal contraction and thickening of the trunk (Fig. 5–2,*a*,*b*). This brings about a rise in coelomic pressure which causes a rapid eversion of the proboscis, the swollen end of the trunk providing anchorage to the substratum (*b*,*c*). The proboscis now increases in volume by about 50%, and its anterior end dilates to provide a new anchorage. The trunk then narrows, its former anchorage is lost, and its anterior end is drawn forwards by slow retraction of the proboscis (*d*,*e*). The anterior end of the trunk now thickens and again forms an anchorage (*f*) which permits the rest of the trunk to be drawn forwards by longitudinal contraction. The cycle then ends with the complete retraction of the proboscis (*g*).

Direct pressure measurements, obtained by inserting a cannula into the coelom and connecting it to a pressure transducer, confirm that pressure is uniform throughout an undivided coelomic cavity, and show that when the animal is on the surface

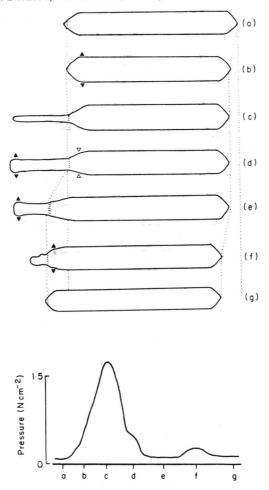

Fig. 5-2. Schematic representation of a digging cycle of *Sipunculus nudus* in relation to the internal pressure (below). Closed arrowheads indicate anchorage; open arrowheads indicate loss of anchorage. For further explanation, see text. From Trueman and Foster-Smith, 1976, *J. Zool.* **179**. 373–386.

it everts the proboscis at a coelomic pressure of about 1.8N cm^{-2}. Below the surface of the sand the pressure peak is lower, presumably because receptors are monitoring the condition of the substrate. Such pressure records, in conjunction with measurements of the proboscis and of the trunk musculature, and of the rate of forward movement, make it possible to calculate the energy expenditure of the animal. Trueman and Foster-Smith have thus calculated that the work done by an individual of about 1,000 mg dry weight, in effecting proboscis and trunk movements during a single digging cycle, is between 1.9 and 4.2×10^{-5} J/mg. Comparable figures for a bivalve mollusc, *Donax denticulatus*, will be given later (p. 144).

This analysis, simple enough in principle, illustrates one important aspect of the functioning of hydrostatic skeletons in burrowing: the use of two alternating anchors. The posterior one (at the anterior end of the trunk in *Sipunculus*) can conveniently be termed the penetration anchor, in the terminology suggested by Trueman, and the anterior one (at the anterior end of the proboscis in *Sipunculus*) the terminal anchor. We shall see other examples of this later, in a diversity of animals.

The importance of a fluid-filled body cavity in the quite different circumstances of sessile life is well illustrated by two phyla, the Entoprocta and the Polyzoa (Ectoprocta). (Their names are confusing; they used to be included together in a single group called the Phylum Polyzoa (Bryozoa), but their resemblances are only superficial, and there cannot be any close relationship between them.)

We have already mentioned the Entoprocta as an example of a group in which the blastocoel persists as the permanent body cavity or pseudocoel. These animals (Fig. 5–3) are microphagous feeders (Sec. 9–2), creating a feeding current by means of circlet of tentacles which encloses both the mouth and anus, and which can be withdrawn for protection within a vestibule formed by an extension of the body wall. The opening of the vestibule to the exterior can be closed by a sphincter muscle; when this muscle is relaxed the tentacles spread outwards, presumably because of the hydrostatic pressure of the fluid of the pseudocoel.

The Polyzoa (Ectoprocta), like the Entoprocta, are microphagous, with feeding tentacles, but these, because they are born on a ridge which encircles only the mouth and not the anus, constitute a true lophophore in the strict definition of that term (Sec. 11–3), unlike the circlet of Entoprocta. Moreover, the Ectoprocta are coelomate, so that extrusion is here brought about by the hydrostatic pressure of the coelomic fluid. Quite apart from this difference, the mechanics of extrusion in many ectoproctans are much more complex than in the Entoprocta. This is because the body wall in the trunk region secretes an exoskeleton called the zooecium (the two together are termed the cystid) which often develops a degree of rigidity that does not seem likely at first sight to favour hydraulically mediated movements. In correlation with this the tentacles are not simply folded inwards, but are withdrawn into a space formed by the introversion of the flexible anterior part of the body. This introvert, which is termed the polypide, is everted when the tentacles are to be exposed, and takes part of the body contents with it. Introversion, by contrast, is a simpler proposition, since it merely depends upon the contraction of specialized retractor muscles which run from the body wall to their insertions on the polypide.

The organization of the fresh-water forms (such as *Plumatella* and *Cristatella*), which constitute the Class Phylactolaemata, does not present any special difficulty. The cystid is flexible, and is well provided with circular and longitudinal muscle, so that contraction of the musculature is sufficient, in cooperation with the coelomic fluid, to bring about extrusion or withdrawal of the polypide and its tentacles. In the remaining Ectoprocta (the typically marine Class Gymnolaemata) the situation is more complicated. In these the cystid develops a varying degree of rigidity, and it is in connection with this that elegant hydrostatic devices have been evolved (Fig. 5–4).

The difficulty here lies in varying the pressure in the coelomic body cavity when this is surrounded by a body wall that is not flexible. *Membranipora* exemplifies one solution of this problem. In this animal, as in many other members of the Order Cheilostomata, the individuals (zooids) are associated in an encrusting colony. The zooecium is largely a rigid box, with its base attached to the substratum, and with four sides attached to neighbouring zooecia. The remaining wall, which is exposed

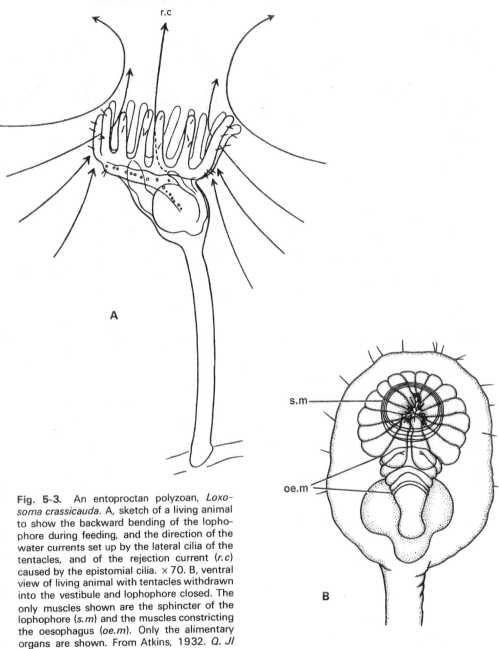

Fig. 5-3. An entoproctan polyzoan, *Loxosoma crassicauda*. A, sketch of a living animal to show the backward bending of the lophophore during feeding, and the direction of the water currents set up by the lateral cilia of the tentacles, and of the rejection current (*r.c*) caused by the epistomial cilia. × 70. B, ventral view of living animal with tentacles withdrawn into the vestibule and lophophore closed. The only muscles shown are the sphincter of the lophophore (*s.m*) and the muscles constricting the oesophagus (*oe.m*). Only the alimentary organs are shown. From Atkins, 1932. *Q. Jl microsc. Sci.*, **75**, 393–423.

to the outside of the colony, is flexible, and it is upon this, the frontal membrane, that the solution of the hydrostatic problem depends. Muscles, termed the parietal muscles, are inserted upon the membrane, their origins being on the rigid lateral walls. Contraction of these muscles pulls the membrane inwards, causing an increase of pressure that is sufficient to evert the polypide.

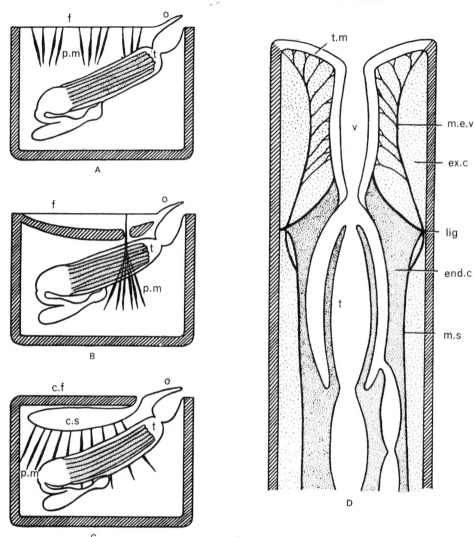

Fig. 5-4. Diagrams to illustrate the mechanism of protrusion of the tentacles in various types of polyzoan. A, *Membranipora*. B, *Micropora*. C, the Ascophora. D, a cyclostomous polyzoan. *c.f*, calcified frontal wall; *c.s*, compensation sac; *end.c*, endosaccal coelom; *ex.c*, exosaccal coelom; *f*, frontal wall of zooecium; *lig*, ligament of membranous sac; *m.e.v*, extensor muscles of the vestibule; *m.s*, membranous sac; *o*, operculum; *p.m*, parietal muscles; *t*, tentacles; *t.m*, terminal membrane; *v*, vestibule. After Harmer. From Chapman, 1958. *Biol. Rev.*, **33**, 338–371.

In other genera, such as members of the group Ascophora, the exposed wall of the zooecium is calcified, so that it loses its flexibility. This modification, which increases the protective value of the zooecium, may seem to destroy the possibility of increasing internal pressure, yet the difficulty is overcome by the development of a separate internal cavity, the compensation sac. The parietal muscles are here attached to the floor of this sac in such a way that when they contract they enlarge the lumen of the sac, and thus produce the required change of pressure in the coelomic fluid. Actually, the difference between this arrangement and the preceding one is probably less than it may seem, for it is likely that the floor of the compensation sac represents the original frontal membrane, which we can regard as having become arched over by an external secondary wall.

One other example of these elegant devices may be illustrated by *Micropora*. Here the frontal membrane remains flexible, but an internal calcified wall called the cryptocyst becomes deposited under the membrane; as with the previous example, this probably gives some protective advantage. The parietal muscles extend through the cryptocyst to reach the frontal membrane; their contraction depresses the membrane and thereby increases the coelomic pressure.

5-3 ECHINODERMS

The Echinodermata, known from the earliest Cambrian, constitute a well-characterized phylum in which the exploitation of hydrostatic principles for locomotion and feeding has evolved in association with the functioning of a mesodermal endoskeleton. To understand how this has been brought about, it is necessary to consider briefly some aspects of the history and classification of the phylum.

An important feature of echinoderm embryology and larval organization (Sec. 18–5) is the early appearance of a paired coelom, enterocoelic in origin, which becomes divided on each side, partially or completely, into three regions: anterior (axocoel), middle (hydrocoel) and posterior (somatocoel). These regions, however, are asymmetrical from the beginning, the axocoel and hydrocoel being better developed on the left side than on the right. The appearance of this asymmetry so early in development suggests that it must have originated correspondingly early in the history of the phylum, and it is commonly supposed that this was correlated with the assumption of sessile life by settlement on the left side. This sessile habit, by inference from living forms, would have been associated with microphagous feeding by means of ciliated tentacles which were distended by coelomic pressure.

In all of these respects there are close resemblances between echinoderms and hemichordates, for the latter, too, develop a tripartite coelom, associated in the adult with a corresponding division of the body into prosoma, mesosoma and metasoma (Sec. 19–4). The resemblance is heightened by the sessile habit of the pterobranch hemichordates, and their possession of hollow ciliated tentacles, used for feeding, and containing extensions of the mesosomatic coelom. This region of the coelom corresponds with the hydrocoel of echinoderms, which, as we shall see, is also involved in the functioning of the tentacles.

These resemblances are among the many good reasons justifying the view that echinoderms and hemichordates are phylogenetically related, and that they belong with the protochordates and vertebrates to the assemblage called the Deuterostomia. It may be, therefore, that the pterobranchs (Fig. 10–17, p. 278) give us the best idea today of the supposed sessile stage of the early echinoderms. But however this may be, the assumption that the echinoderms passed through a sessile phase common to the whole phylum remains entirely hypothetical, for already at their earliest appearance in the Lower Cambrian they show a distinction between sessile and free-

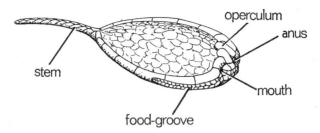

Fig. 5-5. Restoration of the Middle Cambrian 'carpoid' *Trochocystites*, in which there are food grooves in the plates of the margin. From Nichols, 1972. *Palaeontology*, **15** (4), 519–538.

living forms. Indeed, it has long been customary to divide the phylum into two assemblages based upon this distinction. One of these, termed the Subphylum Pelmatozoa (*pelma*, sole of foot), includes, in addition to many fossil groups, living forms that are attached aborally during their whole life (sea-lilies) or during the early part of it (feather-stars), with their oral surfaces directed upwards, and using ciliated tentacles for microphagous feeding. The other assemblage, termed the Subphylum Eleutherozoa (*eleutheros*, free), today comprises the asteroids (starfish or sea-stars), ophiuroids (brittle-stars), echinoids (sea-urchins) and holothurians (sea-cucumbers), all of them free-living.

However, it has become apparent that the possibility of convergence has been too readily disregarded in this classification, and that there is compelling evidence that pelmatozoan and eleutherozoan modes of life must have evolved independently along several lines. This point of view has been expressed by Fell in an alternative classification based in part upon an analysis of patterns of growth gradients. In this classification, which will be used here, the following four subphyla are recognized (Fig. 5–6).

Subphylum Homalozoa. Fossil forms, including the 'carpoids' (Fig. 5–5), known from the Middle Cambrian to the Middle or Upper Devonian. They were not attached, they lacked radial symmetry, and they were fundamentally asymmetrical. They perhaps represent an early adaptive radiation, preceding the development of the pentamerous radial symmetry found in the other three subphyla.

Subphylum Crinozoa. Radially symmetrical forms (Fig. 5–13), known from the Lower Cambrian to the present time, including the present-day crinoids (known from the Lower Ordovician) and a range of exclusively Palaeozoic fossils. The growth pattern is partly meridional, producing a cup-shaped or globoid theca, and partly

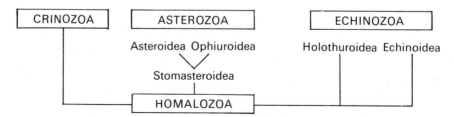

Fig. 5-6. Suggested classification of the main echinoderm groups.

radial, producing food-gathering arms. These animals, as already mentioned, are attached aborally for at least part, and usually for the whole, of their lives.

Subphylum Asterozoa. Radially symmetrical free-living forms (from the Lower Ordovician to the present time), in which a radial growth pattern produces a star-shaped body with rays. There is a single class, the Stelleroidea, which includes, according to one view, three subclasses. These are the Somasteroidea (perhaps derived from the crinoids, and represented by a few Palaeozoic genera and a single surviving one, *Platasterias*), the Asteroidea, and the Ophiuroidea. The latter two, which presumably diverged from somasteroid ancestors, are readily distinguishable as adults, not least by the snake-like locomotor movement of the ophiuroid arms (*ophis*, snake). However, the ophiuroids pass through an asteroid-like stage after metamorphosis, and the two groups have much in common in their fundamental organization. (An alternative view contrasts a Class Asteroidea with a Class Ophiuroidea, the somasteroids forming a subclass of the former.)

Subphylum Echinozoa. Radially symmetrical free-living forms (from the Lower Cambrian to the present time), in which a meridional growth pattern gives rise to a globoid body without separate rays or feeding arms. The Class Echinoidea and Class Holothuroidea are included here, together with various imperfectly understood fossil groups (e.g. Edriasteroidea), some of which were probably fixed, with an upwardly directed mouth, to the sea-bed or to other animals.

The calcareous endoskeleton of echinoderms, already present in the earliest fossils, differs markedly from that of other animals, even of the vertebrates, despite the relationship of the two groups. Its chemical composition is unique, consisting as it does of 71–95% calcium carbonate and 13–15% magnesium carbonate. Moreover it is spongy in appearance as a result of the mineral matter (which probably has little, if any, protein matrix) containing a network of interconnected spaces occupied by connective tissue. This mode of association of mineral matter (stereom) with organic matter (stroma), which differs from that discussed later (p. 134) in connection with the other skeletons, has advantages. The open structure combines lightness with strength, and conserves calcium, while the penetration of the skeletal elements by living tissue provides for nutrition, growth, and repair, aids their binding and articulation, and lessens the need for complex interdigitating surfaces.

This skeletal material may form minute spicules strengthening a flexible body wall, as in holothurians; large ossicles moved by muscles, as in asteroids and ophiuroids; and plates united as a protective test, as in echinoids. Its versatility is shown in the uses to which it is put, additional to its close association with the coelomic hydrostatic skeleton. For example, the internal row of vertebrals in the ophiuroid arms, formed by the fusion of pairs of ambulacrals (see later), have ball-and-socket joints, and, in conjunction with muscle action, provide an interesting convergence with the axial skeleton of vertebrates. Other examples of versatility are the jaws (Aristotle's lantern) of echinoids (a structure already present in the Ordovician), the articulated spines which contribute to feeding, locomotion, and burrowing in those animals (also present in the Ordovician), and the pedicellariae of asteroids and echinoids.

Pedicellariae, composed of two or three opposed ossicles which act like the blades of scissors or forceps, are used for protection, for cleaning the body surface, and for seizing prey. In asteroids they are either sessile or have a fleshy stalk, the ossicles of both types articulating with a basal ossicle. In echinoids the pedicellariae have stalks which are strengthened with skeletal elements, and the ossicles articulate with each other. These differences support the view that these two groups of echinoderms are not closely related, their pedicellariae having evolved independently.

The development of the echinoderm skeleton begins in the larva with the secretion of small triradiate spicules in an extracellular matrix immediately adjacent to the single mesenchyme cells. These spicules, each of which behaves optically as a crystal unit, present an instructive contrast to those of sponges, for they grow and branch, while others are added from other cells throughout ontogeny. The skeletal elements so formed are thus composed of an aggregation of minute units, but nevertheless each element itself behaves as a single crystal of calcite, and retains this optical property throughout its growth. This crystallographic property is one determinant of skeletal form, for it gives the initial triradiate shape to the spicules, the branches tending thereafter to grow straight. The direction of their growth, however, is modulated by the pseudopodia of the mesenchyme cells. The triradiate spicules of the larva are initially orientated at random, but this is rapidly corrected so that each element becomes appropriately orientated in accordance with the requirements of the larval form. The subsequent development of the skeleton, with all its variety, is a result of interaction between the pseudopodia and the ectoderm cells. The skeletal material is laid down within a pseudopodial complex, from which fine cable-like structures extend to the ectoderm. The pseudopodia, which are highly contractile, explore the ectoderm and become attached to it at certain points which are determined by differences in its adhesiveness. The direction of growth of the spicule arms, which may involve considerable bending, is then guided by the pseudopodial cables.

The interrelationship of skeleton and coelom in the movement and feeding of echinoderms depends upon the subdivision of the coelom into several well-defined components. These components, already foreshadowed in the development of the larva (Sec. 19–4), comprise the main perivisceral coelom (derived from the left and right somatocoels), the water-vascular system (derived from the left hydrocoel), the hyponeural sinus system (probably derived mainly from the left somatocoel, and alternatively called the perihaemal system, since it surrounds the haemal or blood system), the axial sinus (left axocoel), and the madreporic vesicle or dorsal sac (right axocoel). All of these components are lined by an endothelium, usually ciliated, and they are filled with coelomic fluid containing various types of coelomocytes which are involved in the transport of food and metabolites. It is from this fluid that are derived the hydrostatic properties of the coelom, which are of prime importance in the functioning of the perivisceral coelom and the water-vascular system.

The relative importance of the calcareous and hydrostatic skeletons in echinoderm movement varies from group to group. A primarily hydrostatic system operates in the holothurians, which are in some respects very worm-like in their mechanism of locomotion. The metasomatic coelom here forms a spacious and fluid-filled body cavity which, in conjunction with the musculature of the body wall, permits peristalsis. Some of them have thus been able to develop burrowing habits. An example is the holothurian *Leptosynapta*, which excavates sand with its tentacles and moves along its burrow by forwardly-directed peristaltic waves. In echinoids and asteroids, by contrast, the body wall is so firm that peristaltic movements are impossible, Locomotion in these animals depends upon the transformation of the ancestral ciliary feeding mechanism, with its tentacles and contained coelomic spaces, into the tube-feet and water-vascular system, and the interaction of this system, especially in asteroids, with movable elements of the calcareous skeleton. Tube-feet are also important in holothurians (although some specialized burrowers have lost them), but the spicular skeleton here does no more than strengthen the body wall.

The principles involved in the use of tube-feet for locomotion can be illustrated by a consideration of starfish, which have been particularly well studied from this point of view. Along each arm of these animals extends an open ambulacral groove, arched by a series of ambulacral ossicles. From each groove arises a series of tube-feet (podia). These are arranged in two rows, and usually have suckers at their tips. Each tube-foot (Fig. 5–7) has a coelomic cavity which communicates by a narrow neck with an ampulla lying in the main perivisceral coelom. This ampulla, which may be divided into a median lobe and two lateral lobes, forms with the tube-foot a functional unit. The coelomic cavity of this unit is connected by a lateral water vessel with the main radial water vessel that extends along the length of each arm, and that connects with a circumoral ring vessel in the central disc of the body. From this the stone canal runs upwards to open to the outside at the madreporite. This structure, derived from the hydropore of the larva, is perforated by many small apertures, through which water is drawn into the stone canal, and thence into the vessels of the water-vascular system. Its significance will be explained later.

The functioning of the system depends upon differences between the muscula-

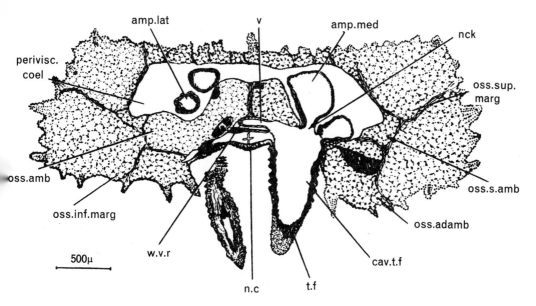

Fig. 5-7. Transverse section through the arm of *Astropecten irregularis*. *amp.lat*, lateral lobe of the ampulla; *amp.med*, medial lobe of the ampulla; *cav.t.f*, tube-foot cavity; *nck*, neck of the ampulla; *n.c*, radial nerve cord; *oss.adamb*, adambulacral ossicle; *oss.amb*, ambulacral ossicle; *oss.inf.marg*, infra-marginal ossicle; *oss.s.amb*, supra-ambulacral ossicle; *oss.sup.marg*, supra-marginal ossicle; *perivisc.coel*, perivisceral coelom; *t.f*, tube foot; *v*, valve. *w.v.r*, radial water vessel. From Smith, 1946. *Phil. Trans. R. Soc. B*, **232**, 279–310.

ture of the ampulla and the tube-foot (Fig. 5–8). In the ampulla the muscles consist mainly of rings of smooth muscle fibres which are set vertically and which lie parallel to the long axis of the arm. Protraction of the tube-foot is brought about by the contraction of these muscles, the effect of this being to drive fluid out of the ampulla into the foot. The increase in pressure is wholly translated into elongation of the foot, any wasteful lateral bulging of this organ being prevented by a collagenous sheath of connective tissue in its wall. This sheath allows extension but resists lateral pressure, so that it may be compared, from a functional point of view, with the circular muscle fibres in the tentacles of anemones. The musculature of the tube-foot (Fig. 5–9), in contrast to that of the ampulla, consists of longitudinal muscles, which are bounded on the inside by the ciliated epithelium of the coelom and on the outside by the collagenous connective tissue, by the ectoderm, and by the cuticle. The hydrostatic pressure of protrusion brings about relaxation of these muscles (Fig. 5–10). Withdrawal of the foot is accomplished by their contraction, the connective tissue fibres becoming pushed together into layers during this process. Bending of the foot is achieved by localized contraction of the longitudinal muscles, while postural muscles provide for its orientation. This makes possible the highly organized stepping movements by which these animals pull themselves along. The whole system is a striking example of the complexity of behaviour that can be mediated by a hydrostatic skeleton, given adequate specialization of the muscles and nervous system.

The importance of the calcareous skeleton in asteroid feeding and movement is well exemplified by the burrowing form, *Astropecten*. This animal has no suckers on its tube-feet, and is perhaps primitive in this respect. It uses these tube-feet to pass large numbers of molluscs into its mouth, and subsequently rejects their shells. An asteroid from the Upper Ordovician, distended with food in just this way, shows how old the method is. *Astropecten* is an expert burrower in sand. Digging is effected by the tube-feet pushing sand away from beneath the body, which sinks down and is finally covered by the excavated material falling onto its aboral surface. Heddle shows how the skeletal elements permit an arm to be used in either a feeding or a digging position, the former with the tube-feet brought together, and the latter with

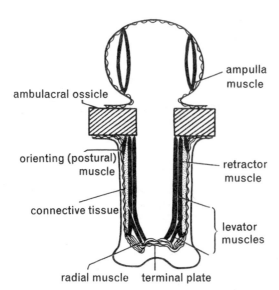

Fig. 5-8. Diagram of a longitudinal section through a foot and ampulla of *Asterias rubens*, showing the arrangement of the chief muscle systems. From Smith, 1947. *Q. Jl microsc. Sci.*, **88**, 1–14.

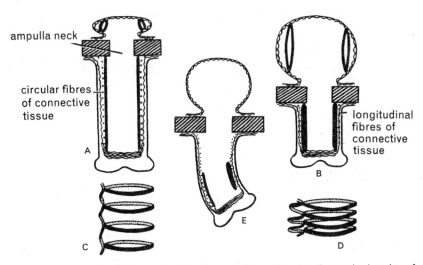

Fig. 5-9. A, B, E, diagrams showing the conditions of contraction and relaxation of the ampulla muscles and the retractor fibres of the foot of *Asterias rubens* during A, protraction, B, retraction, and E, localized bending of the podium. Muscles in contraction are represented by the thicker, and relaxed muscles by the thinner, of the black bands. C and D show how the longitudinal and circular fibres of the connective tissue sheath within the column of the foot are arranged when the foot is C, protracted, and D, retracted. From Smith, 1947. *op. cit.*

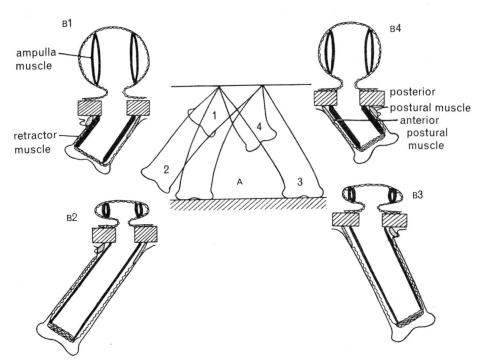

Fig. 5-10. A*1*–*4*. The successive phases of the ambulatory step. B*1*–*4* show the conditions of contraction and relaxation of the protractor, retractor, and postural muscles of the foot during the successive phases of static posture of the 'ideal' step; the protractor and retractor muscles are shown in black, the orienting (postural) fibres are stippled. The anterior postural fibres orientate the foot in the forward direction, the posterior fibres in the backward direction of the step. From Smith, 1947. *op. cit.*

them more widely separated (Fig. 5–11). The adambulacrals act as pivots for these changes in position, while the superambulacrals provide for internal muscle attachments, and also brace the marginals against the ambulacrals to resist the pressure of the sand. *Asterias* has tube-feet with suckers, and so, unlike *Astropecten*, it can climb up rocks. It can also use a different mode of feeding, placing itself over a bivalve mollusc, raising its body into a dome shape, and forcing the two valves apart with the suctorial podia. The edges of the stomach are then inserted between the valves, and the prey digested externally. The strains involved when the body is raised are met primarily by the dorsal musculature and by the series of adambulacral ossicles, which are themselves joined by muscle. An asteroid preserved in this attitude has been found in the Ordovician; this, and the association of asteroids with a bed of bivalves in the Middle Devonian, again suggest the antiquity of these echinoderm habits. An elegant refinement of them has been noted in *Acanthaster planci*, the Crown of Thorns starfish which feeds upon corals. This animal extrudes over its prey, by coelomic pressure, folds of the cardiac stomach, which are then pulled and pushed by the tube-feet until they conform with the contours of the coral. The reaction is readily elicited by inserting extracts of the coral *Acropora formosa* into its mouth.

In dealing with coelenterates we have noted how these animals avoid undue loss of fluid from their hydrostatic skeletal system, and this is an important consideration also in echinoderms. Protrusion is an entirely passive process, as far as the foot is concerned; thus in *Asterias rubens* the volume of the ampulla (which in this genus is undivided) is about the same as the maximum extension attained by the foot. In this situation there must be no serious loss of fluid during protrusion and withdrawal, for the effect would be as disastrous as having a leaking hydraulic brake system. Loss is prevented by a valve at the junction of the tube-foot with the lateral vessel. This valve is essentially an extension of the lateral water vessel, arranged to permit fluid to pass from the vessel into the tube but not in the reverse direction. Its importance may be judged from the fact that the total volume of fluid in the stone canal, circumoral vessel, and radial vessels amounts to only 1–2% of the estimated volume of the contents of the tube-feet and ampullae; it would thus be quite impossible for the latter organs to be replenished at all rapidly from the fluid in the vessels. Some loss, however, is inevitable, and what those vessels do is to replace it with a trickle of fluid which enters through the madreporite and stone canal; an arrangement that is functionally comparable, although at a higher level of complexity, with the action of the siphonoglyphs of anemones.

Clearly, it is important that the path of entry of the fluid should not be blocked by foreign material. Thus, the surface of the madreporite is kept clean by strong ciliary currents, while removal of any obstruction is further ensured by the outward-

Fig. 5-11. A, diagram of a transverse section of an arm of *Astropecten* showing features of the musculo-skeletal system. B, transverse section of an arm of *Astropecten* in the walking posture. C, the same in the digging posture. Key: i.a.m., infra-ambulacral muscle; i.l.m., inner lateral transverse muscle; m.m., marginal muscle; o.l.m., outer lateral transverse muscle; p.m., pericoelomic transverse muscle; s.a.m., supra-ambulacral muscle; s.m., superambulacral muscle; a.d.o., adambulacral ossicle; a.g., infra-marginal plate; p., paxilla; s.o., superambulacral ossicle; s.p., supramarginal plate; t.f., tube-foot. From Heddle, 1967. *Symp. zool. Soc. Lond.*, **20**, 125–141.

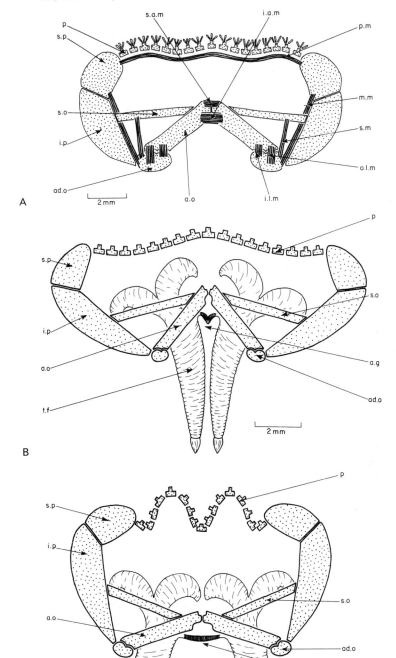

126 INVERTEBRATE STRUCTURE AND FUNCTION

beating ciliation of the lining of the pores. Inward passage of water is effected partly by suction and partly by the inward beat of the cilia of the stone canal. Many of the pores open into this canal, at least in asteroids, but others open into the axial sinus, in which the cilia beat aborally. The madreporite, however, is more than a device for replenishment. A further important function has been shown for echinoids by Fechter's analysis of its action in *Echinus esculentus* (Fig. 5–12). He attached a manometer tube to the madreporic plate, and showed that raising or lowering the external

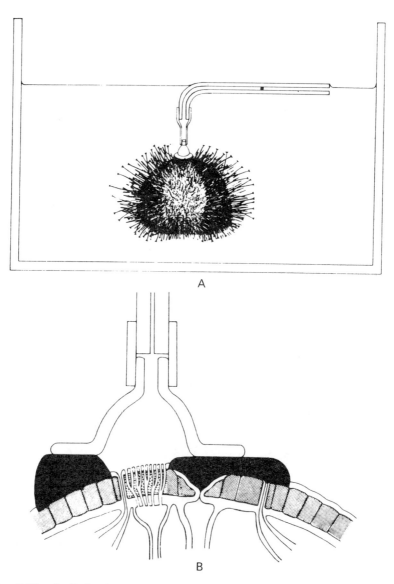

Fig. 5-12. A, Fechter's manometric experiment to demonstrate pressure relationships in the water-vascular system of an echinoid. B, the attachment of the manometer to the madreporite. From Fechter 1965. *Zeitschr. vergl. Physiol.*, **51**.

pressure relative to the internal, prevented the normal operation of the tube-feet. This is because in *Echinus*, as in *Asterias*, movement of the tube-feet depends upon the external and internal water pressure being equalized. Fechter's experiment indicates that this equalization takes place in the madreporite and stone canal.

In other groups the same principle operates but in a somewhat different way. Crinoids have a number of short stone canals opening from the ring canal into the coelom. This, however, is connected with the outside by hundreds of ciliated funnels which conduct water through the oral surface of the body wall into the coelomic fluid. Holothurians have secondarily lost the external opening of the stone canal, although a hydropore is present in the larva. The stone canal here ends within the coelom as a perforated bulb, the madreporic body, the pores of which are lined by cilia that beat inwards. It is thus coelomic fluid that maintains the pressure in the water-vascular system.

In a number of genera the structures called Polian vesicles probably make a contribution to this regulatory mechanism. They are muscular sacs, resembling ampullae in histological structure, which are found as outgrowths of the ring canal in many asteroids (although not in *Asterias*), and also in ophiuroids and holothurians. They are thought to provide reservoirs for the temporary storage of water, which can then be made available for the tube-feet at short notice. They may also help to equilibrate the pressure within the system when the tube-feet are moving.

We have so far been dealing with the reciprocating system of ampullae and tube-feet which is found in asteroids, echinoids, and holothurians. However, ampullae are absent from crinoids and from ophiuroids, and it is an instructive

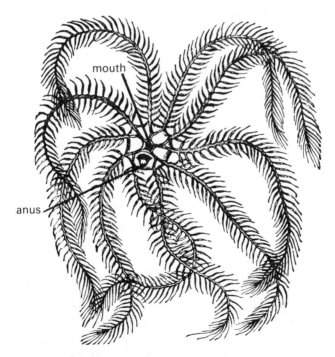

Fig. 5-13. *Antedon bifida* in oral view. From Sedgwick, after Claus, from Borradaile *et al.*, 1958. *The Invertebrata*, Cambridge University Press.

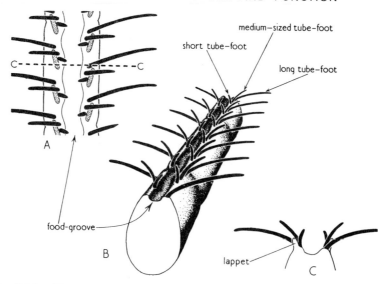

Fig. 5-14. Diagrams to show the arrangement of the tube-feet on a pinnule of *Antedon*. A, plan view of part of a pinnule to show the relationship between the tube-feet (silhouetted) and the lappets (stippled). B, perspective view, looking distally, of part of a pinnule, showing the angle at which the tube-feet are held during feeding. The lappets have been omitted for clarity. C, transverse section across the line CC in A, showing the angle at which the tube-feet are held in relation to the lappets during feeding. From Nichols, 1960. *Qr. Jl microsc. Sci.*, **101**, 105–117.

demonstration of the flexibility of natural selection to see how their place is taken by devices that are similar in principle, although quite different in organization. To illustrate this, we shall briefly consider the crinoid *Antedon bifida* (Fig. 5–13).

Crinoids, the most primitive of living echinoderms, continue today the ancient microphagy of the early forms. Thus in *Antedon* the tube-feet (tentacles) are mainly concerned with ciliary feeding, although they also have some sensory function. They are borne on 10 arms, and also on the numerous pinnules which project on each side of the arms. A food-groove extends along both arms and pinnules, and the tube-feet are arranged on each side of this in groups of three, each group being associated with a lobe (lappet) which closes over the groove when the tube-feet retract; this they can do to about one-half of their fully extended length.

The tube-feet of the pinnules (Fig. 5–14) are of three different lengths: the outermost of each group of three, projecting almost at right angles, is the longest; the intermediate-sized one projects outwards at 45°; the smallest one projects upwards. They differ also in the adaptive organization of their musculature. The longest ones, which are especially concerned with catching food, have muscles only on their oral surface, which enables them to bend inwards for transferring the food. The medium-sized ones have muscle on both their oral and aboral surfaces; this enables them to bend outwards to pick up food strands, and also to direct them inwards. The shortest ones have their muscle arranged as a complete cylinder; they are thus able to bend in all directions and are particularly effective in pushing the mucus strands into the food-grooves. On the main arms, the members of each group of three tube-feet are orientated in the same way as those on the pinnules, but they are all of the same size, and all have a complete circular musculature.

MOVEMENT, HYDROSTATICS, AND THE COELOM

These tube-feet are probably stimulated to action when they are touched by food organisms in the plankton. If the animal is violently disturbed, however, all the tube-feet are rapidly retracted and return only slowly to the feeding position. These varied movements, of food trapping and transfer, and of protective retraction, are made possible by the hydrostatic mechanism of the water-vascular system, working in cooperation with the elegantly adapted musculature of the tube-feet, but without the aid of ampullae. They depend upon the structure of the radial canals and their branches in the pinnules (Fig. 5-15). In the arms, but not in the pinnules, each canal is expanded centrally to form a central tube. On either side of this, and in corresponding positions along the canals of the pinnules, there are muscle fibres which stretch across the lumen. In addition, the canals, both in the arms and in the pinnules, are regularly constricted into chambers, at points lying between the successive groups of tube-feet. Thus six tube-feet, one group on each side, arise from the chambers between successive constrictions.

Nichols's interpretation of the functioning of this system is that protraction of the tube-feet depends first upon closure of the constriction, the canal muscle fibres at the same time closing off the lateral parts of the canal from the median region (that is, from the central tube in the arm, and the less clearly defined central chamber in the pinnules). This isolates the corresponding group of tube-feet; these can now be protruded by continued contraction of the muscle fibres, which will force water into the

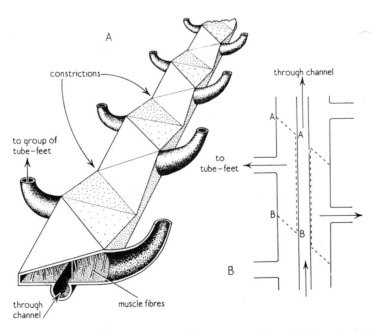

Fig. 5-15. Schematic diagrams to show the structure and suggested mode of action of the water-vascular canal in *Antedon*. A, perspective view of the canal of an arm, showing the relationship between the constrictions, which cross the canal diagonally, and the side-branches to the tube-feet. B, a plan view of the canal of an arm or pinnule, showing the suggested mode of protraction of the tube-feet. The dotted lines show the supposed delimitations of one cavity on each side. For further explanation, see text. From Nichols, 1960. *op. cit.*

tube-feet. The clear central tube or channel ensures that water continues to reach the more distal tube-feet, which would otherwise become inoperable when the more proximal ones were active. The only limitation imposed by this arrangement is that all six tube-feet associated with one chamber must be protracted together when that chamber is isolated by muscular constriction, but this is no disadvantage, since the tube-feet are functionally associated in the feeding process. Each tube-foot can, however, be withdrawn independently of the others, presumably because their intrinsic musculature is separately innervated.

As already mentioned, ophiuroids also lack ampullae, but they, too, have a device for constricting the radial canals, formed in this instance by muscle sphincters. In addition, their tube-feet have muscular and swollen bases, which, although external to the body, are to some extent functionally equivalent to ampullae in that they provide pressure for protraction. This suggests a possible step in the evolution of true internal ampullae, but evidence from fossil asteroids and echinoids is not clear on this point, and we are again frustrated by the early diversification of these animals.

However, the extinct somasteroid, *Chinianaster* (Fig. 5–16), from the Lower Ordovician of the vineyards of St Chinian, in France, is relevant to the argument. This genus, probably the most primitive of the known somaseroids, had large ambulacral elements on the oral surface, corresponding to the brachials of crinoids with series of smaller ossicles, the virgalia, lying on either side of them. There could not have been any internal ampullae, for there were no pores through which their necks could have passed. There was however, a depression (cupule) for the lodgement of each tube-foot, and thus could have accommodated a swollen base like that found today in ophiuroid tube-feet. In theory, the base could then have moved in to form a true internal ampulla. This would give the condition found in the only surviving somasteroid, *Platasterias latiradiata*, a species known since 1870 but not recognized as a surviving somasteroid until some 90 years later. In this living fossil, in which small internal ampullae are present, the tube-feet are still used mainly for collecting food, supplementing ciliary tracts that provide for suspension feeding, and for this the small size of the tube-feet is adequate. If, however, this suggested explanation of the origin of internal ampullae is well founded, the inward movement must have occured very early, for openings in the ambulacral grooves show that internal ampullae were present in other extinct somasteroids, and were probably present in all Palaeozoic asteroids. It is significant that the radial (ambulacral) channels are still shallow in somasteroids. Invagination of the ambulacral groove, as seen in modern asteroids, would have permitted enlargement of the ampullae. This in its turn has allowed enlargement and increased activity of the tube-feet, for the volume of the protracting fluid increases, and so also does the muscular force brought to bear upon it. The tube-feet can therefore be extended to greater lengths, and can function with greater strength.

As regards echinoids, internal ampullae were already present in the Ordovician. This emphasizes yet again the remarkably early differentiation of echinoderm characters, and leaves entirely open the question whether or not internal ampullae evolved independently in asteroids and echinoids, and perhaps in holothurians as well. One can only say that the manifest advantages of the reciprocating system in the exploitation of benthic life could readily account for the independent establishment of

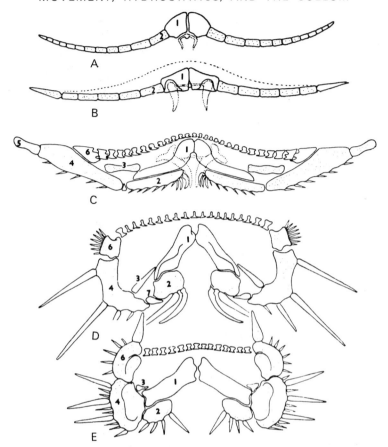

Fig. 5-16. Transverse sections of arms of astroradiate echinoderms, showing phases in evolution prior to development of suctorial tube-feet. A, biserial crinoid; B, *Chinianaster* (× 8); C, *Platasterias* (× 4), ambulacral ossicles becoming erected; D, *Luidia* (× 4); E, *Astropecten* (× 4); here and in *Luidia* the buttress skeleton is becoming perfected. 1, ambulacral (brachial in A); 2, virgalium-1, adambulacral; 3, virgalium-2, superambulacral; 4, virgalium-3, inframarginal; 5, virgalium-4, marginal radiole; 6, superomarginal paxilla or plate; 7, actinal intermediate (interalary) plate. From Fell, 1963. *Phil. Trans. R. Soc. Lond. B*, **246**, 381–435.

this feature along more than one line. Differences in the organization of the suckers of the tube-feet of asteroids and echinoids support this view. Where the suckers of asteroids have discs of connective tissue, those of echinoids have a system of ossicles which contribute in several ways to the efficiency of the tube-feet; e.g., by maintaining the shape of the suckers during adhesion and by providing muscle attachments.

The adaptive potentiality of these tube-feet is particularly well shown in the spatangoid sea-urchin *Echinocardium* (Fig. 5–17). This animal, which burrows in sand to a depth of 18 cm, has highly extensile dorsal tube-feet that build and maintain (in conjunction with the articulated spines which contribute to feeding, locomotion, and burrowing) a respiratory funnel which leads down from the surface of the sand. Water is drawn down this funnel by cilia over the body surface and on the fascioles, which are bands of the spatulate spines called clavulae. Suckers have been second-

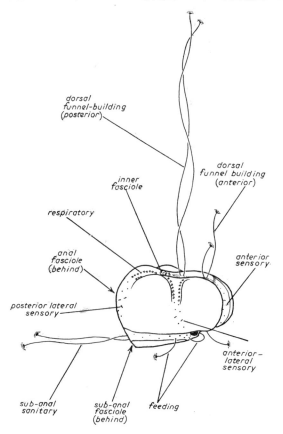

Fig. 5-17. Diagram showing the position of the various tube-feet and clavulae (the spines of the fascioles) on the test of *Echinocardium cordatum*. A few of each group of tube-feet are drawn extended; the others are represented by their pores. From Nichols, 1959. *Qt. Jl microsc. Sci.*, **100**, 73–87.

arily lost from the tube-feet. Instead, discs bear glandular and muscular papillae, each supported by a skeletal rod, while just behind the disc is a large curved spicule which is used for the scraping of material from the walls of the burrow. Posteriorly, the animal bears other tube-feet, similar to the building ones, but lacking the scraper. These maintain a drain for the removal of waste, and to permit the outflow of the respiratory current. Feeding is carried out mainly by a third category of tube-feet situated around the mouth. These, which also bear glandular papillae, take up material and pass it into the mouth.

The subtlety of adaptive organization of these structures can be illustrated by one feature. Muscle fibres run between the mucous glands of the funnel-building and drain-building tube-feet, but are not present in the feeding tube-feet. Nichols suggests that the difference is correlated with the need for the building tube-feet to discharge mucus on to the walls of the burrow, whereas the secretion must remain on the feeding tube-feet in order to ensure efficient transport of the food. The action of the feeding tube-feet is supplemented by that of the dorsal funnel-building ones, which pick up material from the surface of the substratum and transfer it to the mouth region when they are retracted. Finally, there are dorsal respiratory tube-feet, with thin and folded walls, while various parts of the body bear sensory tube-feet with a rich supply of nervous tissue. Such are the almost inexhaustible biological potentialities of hydrostatics.

5-4 MOLLUSCS

The ground plan of the molluscs (Fig. 5-18) is a body divided into two functionally distinct regions. One of these, the upper part, comprises the visceral hump and mantle; it functions largely through mucus secretion and ciliary action. The other, the lower part, comprises the head and foot, and is largely muscular, although it too is provided with cilia and mucus cells. This plan of structure is not easily relatable to that of any other phylum; one idea is that it might have evolved from a platyhelminth-like stock in which enlargement of the alimentary system led to the development of the characteristic molluscan visceral hump.

Whatever the truth of this, it is certainly very difficult to interpret the history of the coelom (Fig. 12-9). This is typically small in molluscs, and it is usually described as consisting of a pericardial coelom around the heart, a gonadal coelom, and paired coelomic ducts which, together with the pericardial wall, serve as excretory organs. The complication here is that there are good grounds for considering that the molluscs are closely associated with the Annelida and the Arthropoda as members of the Protostomia. Has, then, the small coelom of molluscs been directly derived from a stage of evolution preceding the appearance of the spacious coelom of annelids, or has it been secondarily reduced from a more extensively developed condition? If the former, its appearance would have antedated the establishment of its hydrostatic

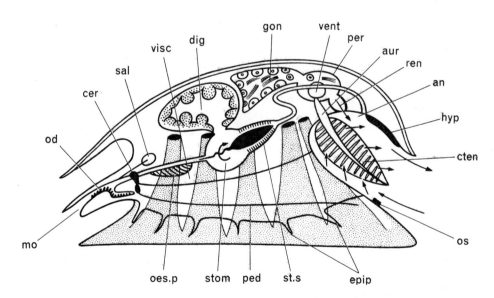

Fig. 5-18. Schematic view of an early mollusc. *an*, anus; *aur*, auricle; *cer*, cerebral ganglion; *cten*, ctenidium; *dig*, digestive gland; *epip*, epipodium; *gon*, gonad; *hyp*, hypobranchial gland; *mo*, mouth; *od*, odontophore; *oes. p*, oesophageal pouches; *os*, osphradium; *ped*, pedal cord; *per*, pericardium; *ren*, renal organ; *sal*, salivary gland; *stom*, stomach; *st.s*, style sac; *vent*, ventricle; *visc*, visceral loop. Certain structures, such as the organs of the head and nerve cords, are turned so as to appear slightly in dorsal view, and the digestive diverticula have been displaced from their respective left and right positions. The odontophore supports the radula. Adapted from Morton, 1958. *Molluscs*. Hutchinson, London.

functions, which in any case are often of minor importance in molluscs. Indeed, Salvini-Flaven suggests that molluscs inherited their locomotor mechanisms from platyhelminth-like ancestors, and so had no need for a coelomic hydrostatic skeleton. Perhaps, therefore, their coelom at first functioned simply to protect the heart and facilitate its movements. If the molluscan coelom has been secondarily reduced it is surprising that there is no trace of this in the development of present-day forms, for the undoubted reduction of the coelom of arthropods has left clear embryological evidence of the process, at least in *Peripatus*. However, there is no comparable evidence in the embryology of the Urochordata, although reduction of the coelom must certainly have taken place in this group.

A puzzling element of the problem is that the archaic mollusc *Neopilina*, discovered in deep-sea dredgings from the Pacific Ocean, contains two extensive dorsal coeloms (Fig. 5–19). Surprisingly, this animal has a metameric-like repetition of gills, nephridia, and retractor muscles, but the lack of any sign of metamerism in other molluscs suggests that this repetition is a secondary specialization and not a primitive characteristic. If this be the correct explanation, then the large coeloms of *Neopilina* may also be a secondary feature. The fact is that the origin and evolutionary relationships of the molluscan coelom are another example of a biological problem that eludes our understanding because of the lack of adequate evidence. The possibility that it may have evolved independently of the annelidan coelom certainly cannot be excluded.

Movement in mollucs is primarily dependent upon the foot, but the shell may also play an important part, especially in the bivalves. The molluscan shell is essentially an exoskeleton, lying outside the tissues and secreted by the mantle; contrary to what was at one time believed, it is stronger than echinoderm skeletal material. The chitons (Polyplacophora) are exceptional in having an eight-valved shell with an upper layer (tegumentum) formed of protein (conchiolin) and calcium carbonate, and an inner layer (articulamentum) which is completely calcareous. In other groups there is usually an outer layer (periostracum), formed by conchiolin, and an inner layer (ostracum), containing crystals of calcium carbonate in the form of calcite or aragonite, held in a matrix of conchiolin. Both types of crystal may be present in the same shell. Typically, the outer region of the ostracum forms the prismatic layer, in which the crystals lie vertically to the shell surface. Beneath this layer is the pearly nacrcous layer, a stronger material in which the crystals are arranged as sheets parallel to the surface. (Commercial pearls are formed of this material; which is laid down around foreign particles lying between the mantle and the inner surface of the shell). There may also be a third type of structure called crossed lamellar, in which additional sheets of crystals are arranged at an angle to the main one, and in which the protein component is sparse.

The functional significance of these different types of material is somewhat obscure, although the crossed lamellar type, which is the hardest, is perhaps of value in some species in increasing resistance to erosion. However, the close association of protein with mineral crystals is in no way peculiar to molluscs; it is a common feature of skeletal structures, in ectoproctans, brachiopods and arthropods, for example, as well as in the bone of vertebrates (but not, as we have seen, in echinoderms). In general, the crystals provide stiffness in tension and compression, and hence give support for bending moments, while the protein provides a matrix that binds the crystals toge-

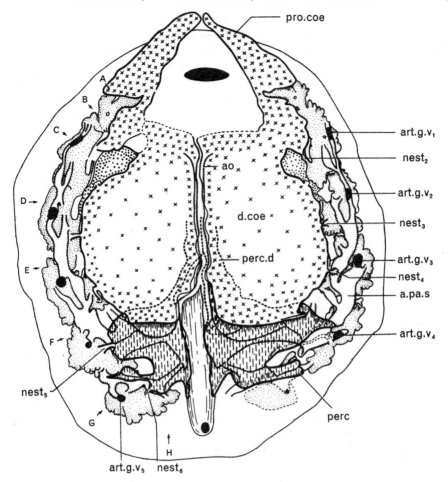

Fig. 5-19. Graphic reconstruction of *Neopilina*, showing the urogenital system, vessels, and coelomic cavities. A–H, positions of pedal retractors (not drawn); *a.pa.s*, arterial pallial sinus; *ao*, aorta; $art.g.v_{1-5}$, entrance of arterial vessels from 1st to 5th gill; *d.coe*, dorsal coelom (with crosses); $nest_{2-6}$, communications between coelom and kidneys; *perc*, pericardium; *perc.d*, pericardial diverticula along the aorta; *pro.coe*, preoral diverticula of dorsal coelom. From Lemche and Wingstrad, 1959. *Galathea Rep.*, vol. 3. Danish Science Press, Copenhagen.

ther, contributes to resistance to tension, confers pliancy, and tends to prevent the spreading of cracks.

This analysis of structural properties can be taken much further, as Currey shows, but he also points out that the total adaptation of a species may lead to more than purely mechanical factors being involved in skeletal design. For example, the secretion of a skeleton sufficiently massive to resist all predatory attack might be metabolically so expensive for some sedentary species that it could be better policy from the point of view of survival of the species, to secrete a lighter shell, with only partial protection, and to apply the surplus energy to achieving a larger output of offspring. We shall encounter a similar balancing of considerations when we deal, in a later chapter, with the significance of life-history patterns.

The use of the molluscan foot in locomotion depends upon it being always a very muscular organ, and upon the musculature operating to some extent in conjunction with the hydrostatic properties of the blood system. This is possible because the blood of molluscs, although conveyed from the heart in a system of arteries, passes eventually into a haemocoel, a complex of rather indefinite vascular cavities lying in a spongy meshwork. The relative degree of development of haemocoel and of well-defined vessels varies a good deal, but in principle possession of a haemocoelic body cavity is characteristic of the whole phylum. To this extent it resembles the Phylum Arthropoda, but it does not follow that the haemocoel has arisen in the same way in the two groups. It is clear from the embryology of *Peripatus* that in the arthropods the haemocoel has evolved by the encroachment of the blood system upon an originally well-developed and metamerically segmented coelom. No comparable information is available for the molluscs, and the earlier history of their haemocoel is as unknown as is that of their coelom.

Small gastropods, and more particularly aquatic ones, can glide by means of cilia on the ventral surface of the foot, but the locomotor power of cilia is inadequate for larger animals. These move by muscular waves passing over the ventral surface of the foot, with their concavities downwards. This involves interaction of muscular contractions with the pressure of the haemocoelic fluid, and the secretion of mucus over the pedal surface. The waves may pass forwards (direct waves) as in most pulmonates, or backwards (retrograde waves) as in chitons and the limpet *Patella*. Further, they may extend across the whole width of the foot (monotaxic), or two sets of waves may pass independently on either side of the middle line (ditaxic).

Patella moves (Fig. 5–20) by retrograde ditaxic waves, the ditaxy, which facilitates turning, being perhaps a useful factor in its precise homing. The foot musculature is complex, but the elements mainly concerned are the dorso-ventral muscles, passing from the shell to the sole of the foot on the same side, and the transverse muscles, passing from the shell to the sole and lateral margin on the opposite side. They represent respectively some 70% and 25% of the total pedal musculature. The blood, which provides hydrostatic support for the foot, is derived from two pedal sinuses that discharge into many small spaces, about 10 μm diameter, lying about 0.5 mm above the sole.

The leading edge of a wave is established by contraction of the dorsal-ventral muscles at that point; as a result, the surface of the foot is concave during the passage of the wave, and the pedal epithelium in the wave is stretched. In consequence, the overlying blood spaces are squeezed. Since the posterior part of the foot is anchored by mucus (usually an important factor in gastropod movement), and the lateral muscles prevent lateral expansions of the blood spaces, these must necessarily extend forwards, and so the retrograde movement of the waves is translated into forward progression.

At the trailing edge of the wave the dorso-ventral muscles relax, being thus antagonistic to the waves at the leading edge. We have seen that the function of a skeleton is to provide for such antagonism and thus to restore the relaxed muscle to normal. One factor here in this antagonism is presumably the haemocoelic pressure; it has not actually been measured, because the blood spaces are so small, but its

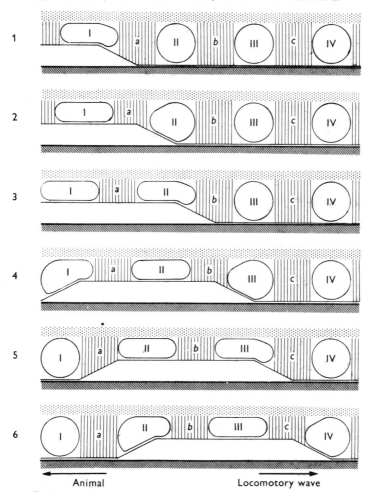

Fig. 5-20. Sequence of six diagrams of a model to show the progression of a retrograde locomotor wave in *Patella*. Haemocoel spaces (I–IV) are distorted by contraction of the dorsoventral muscles (a–c). This contraction also draws up the sole, creating negative pressure beneath it; haemocoelic pressure, in conjunction with this negative pressure, antagonizes the dorso-ventral muscles when they relax. From Jones and Trueman, 1970. *J. exp. Biol.* **52**, 201–216.

effect may reasonably be assumed. This will tend to restore the sole of the foot to the substratum. Another and less obvious factor, and one that has been directly recorded, is that lifting of the foot during the passage of the wave (and it is only raised by about 0.2 mm) results in the pressure beneath it falling by about the equivalent of 6 cm of water. This negative pressure will supplement the action of the blood pressure in promoting the lowering of the foot.

Patella may be contrasted with a slug, *Agriolimax reticulatus*, where the pedal waves are direct and monotaxic. The principles involved in the operation of these may be assumed to apply to many other gastropods. At the upper boundary of the foot is a compact layer of longitudinal muscle, about 0.1 mm thick. This is connected with the pedal epithelium by a meshwork of oblique fibres lying in the haemocoel. Some fibres run forwards and upwards from the epithelium; these, the anterior oblique fibres, exert an upward and forwardly directed force on the sole of the foot. The remaining fibres run backwards and upwards from the epithelium; these, the posterior oblique fibres, exert an upward and posteriorly directed force. The contraction of these two sets of muscle fibres, at different stages, brings about the formation of the waves and the forward progression of the body.

The pedal waves, as in *Patella*, are concave downwards, but differ in being regions of shortening of the epithelium (Fig. 5–21). Each is produced by contraction of the anterior oblique muscles, which pull the pedal epithelium upwards and forwards. Thus the sole moves forwards relatively to the body, the regions of the sole between the waves stretching so that the slug remains more or less of the same length. In general, terrestrial molluscs seem to require high blood pressure: cephalic pressure in *Helix*, for example, has been found by direct measurement to be of the order of 17 cm of water. Presumably, then, it is haemocoelic pressure which reapplies the pedal epithelium of the wave to the substratum. Contraction of the posterior oblique fibres then pulls the longitudinal muscle forwards, and the rest

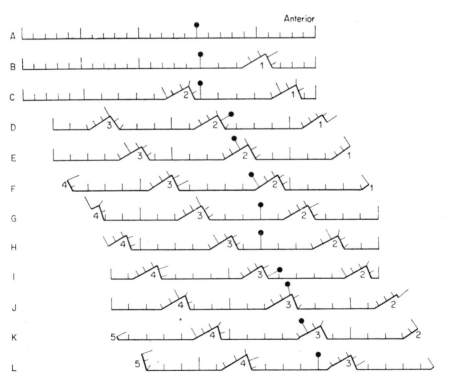

Fig. 5-21. A series of hypothetical diagrams to illustrate the beginning of locomotion and subsequent movement in a slug, using direct locomotory waves. For further explanation, see text. From Jones 1973. *J. Zool.*, **171**, 489–498.

of the body with it. The pedal epithelium is thus, as it were, overtaken, and so the slug moves steadily on.

There is a certain amount of inference in this interpretation, but it is based also on the study of sections of the foot of slugs that were killed while in the act of crawling. This was achieved by allowing animals to crawl on filter paper suspended by a single cotton thread, which was then burned through with a Bunsen burner, the animal and its paper substratum being received into a vacuum flask of liquid nitrogen. Death, it is surmised, was instantaneous, since the optic tentacles were only about one-third retracted.

The forces involved in this type of movement have been demonstrated in snails by encouraging them to crawl over a pivoted bridge (Fig. 5–22). When the head pushes forward it is resisted by sliding friction, which is overcome by the hind end applying a backwardly directed force (static reaction). These two components can be separated if the bridge is sufficiently wide, and the opposing forces can then be recorded through a transducer as first the head and then the tail pass on to the bridge. The sliding friction of the head displaces the bridge forward, and then the static reaction of the tail displaces it backward (Fig. 5–22). Mucus is important for facilitating these reactions. In some species (*Limax*, for example) it may be of two kinds: a more fluid type needed for the locomotory waves, and a more viscous one for anchoring the stationary parts of the foot.

The potentialities of the gastropod foot are so great that its uses thus far described are by no means the only ones. For example, it plays an important part in burrowing, examples of which are found in both prosobranchs and opisthobranchs.

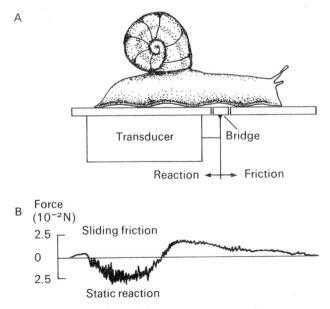

Fig. 5-22. A, arrangement for recording forces developed in locomotion between foot of *Helix pomatia* and substrate. B, recording obtained using a wide bridge (about 1 cm). From Trueman, 1975.

The important feature here is the formation of a mucus-covered plough. A head-shield develops from the anterior end of the foot in prosobranchs, or from the head itself in opisthobranchs, while in both groups the sides of the foot grow up as parapodia, the general result being to create a measure of streamlining.

Pelagic life is another field of action for the foot of gastropods. In the heteropod prosobranchs the foot is drawn out into a highly mobile fin, which in *Carinaria*, for example, is held upwards, the animal swimming upside-down. In the opisthobranchs the tendency to streamlining, already noted in connection with burrowing, has found an application to pelagic life. One can perhaps see the process beginning in *Aplysia* (Fig. 5–23). During creeping the anterior and posterior ends of its foot are lifted away from the ground in alternation, with the remainder of the body arching between them. But the animal can even lift itself wholly away from the substratum by flapping its parapodia, and *Pleurobranchus* actually swims in this way (Fig. 5–24). The extreme development of this trend occurs in the pteropods (sea-butterflies). These move by the steady beat of the lateral epipodia (Fig. 5–25), which are presumably derived from the parapodia of less modified forms. The extent of the locomotor adaptation is well seen in *Clione* (Fig. 5–26); here the paired 'wings' have only a narrow base of

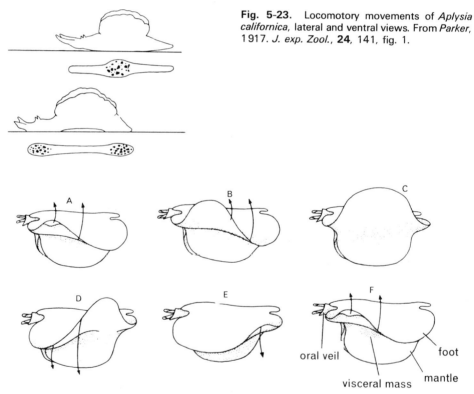

Fig. 5-23. Locomotory movements of *Aplysia californica*, lateral and ventral views. From *Parker*, 1917. *J. exp. Zool.*, **24**, 141, fig. 1.

Fig. 5-24. The swimming mechanism of *Pleurobranchus*; the animals are shown as viewed from the morphological right side, but the gill is not illustrated. The swimming lobe of the left side is here shown, for simplicity, to be stationary in a relaxed position. The visceral mass is visible through the body wall. From Thompson and Slinn, 1959. *J. mar. biol. Ass. U.K.*, **38**, 507–524.

Fig. 5-25. *Limacina retroversa*. Outline drawings showing successive positions of the wings in swimming upwards. *(1–4)*, and in descending *(5)*. *6* shows the wings held motionless and vertical, in a different view from *5*. From Morton, 1954. *J. mar. biol. Ass. U.K.*, **33**, 297–312.

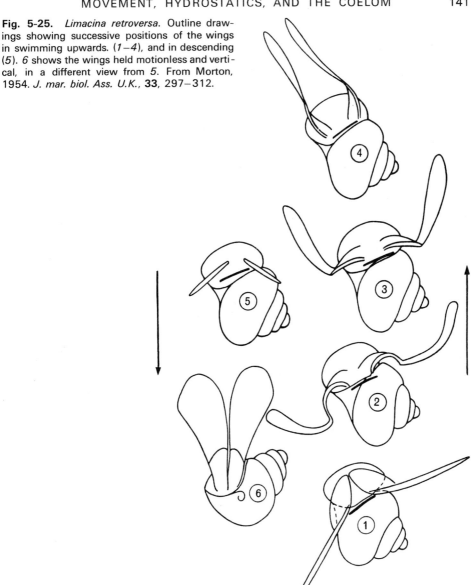

attachment, and are twisted at that point during both upward and downward beats so as to increase the power of their stroke in both directions.

A different pattern of use of the molluscan foot is found in the Bivalvia (lamellibranchs), where the interaction of shell, muscles, and hydrostatic skeleton has evolved along lines very different from those of gastropods, and has been the basis of their remarkably successful exploitation of burrowing in soft substrata. In the more primitive *Nucula* the foot has a flat ventral surface not unlike that of a gastropod, although it already differs in its mode of operation. It is thrust into the mud, and

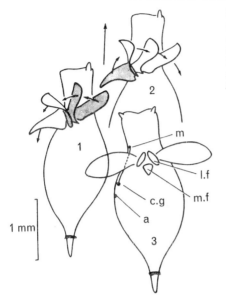

Fig. 5-26. Outline sketches of *Clione* showing the successive positions of the wing in downward (*1*) and upward (*2*) strokes (left-side view), and (*3*) ventral view. *a*, anus; *c.g*, common genital aperture; *l.f*, lateral lobe of the vestigial foot; *m*, male aperture; *m.f*, median lobe of the foot. From Morton, 1958. *J. mar biol. Ass. U.K.*, **37**, 287–297.

then expanded into a holdfast; the body is then drawn after it by contraction of the pedal retractor muscles. In more advanced forms (*Tellina* and *Donax* are among those that have been well studied), the foot has become a narrow wedge or tongue. In principle, this is thrust forward while the shell acts as a penetration anchor; swelling of the foot then enables this to act as a terminal anchor, and the body is drawn forwards. This type of progression has been developed into a highly effective boring mechanism, as in the piddock (*Pholas*), which bores into rock, and the shipworm (*Teredo*), which bores into wood. In both animals the edges of the shell valves form the boring tools, the motive power arising from the interaction of the muscles and the adhesive power of the foot.

The use of electronic transducers of force, pressure, and displacement, in conjunction with multichannel recorders and film, has made it possible to analyze in some detail the cyclical burrowing activities of bivalves when they are below the surface and out of sight. The main structural elements involved, additional to the bivalved shell, are the extrinsic muscles of the foot (mainly anterior and posterior retractors) and the intrinsic muscles (mainly transverse). These are antagonized by the blood of the haemocoel. This may achieve a pressure of 30–50 cm of water, the force being withstood by the transverse muscle and by the sand of the burrow.

The burrowing of *Tellina* (which exemplifies that of a generalized bivalve Fig. 5–27) is initiated from the surface by probing and scraping movements of the foot. These movements are brought about by the actions of the retractors and transverse muscles, with the blood mediating their antagonism, but blood pressure in the foot is low at this stage, which, as we have seen earlier, is always a difficult one for burrowers. Once the foot is below the surface, and can gain some purchase, it begins to dilate (stage i). The siphons now close, to prevent egress of water (stage ii), and this is followed (stage iii) by a rapid adduction of the valves of the shell (0.1 sec). This raises the blood pressure (sometimes as high as 30 cm

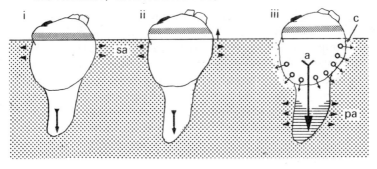
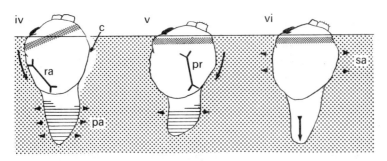

Fig. 5-27. Diagrams of a generalized bivalve burrowing into sand (shaded area) at different stages of the digging cycle. The dotted band across the valve indicates movements of the animal with reference to the surface of the sand. The horizontal shading on the foot indicates the region of the pedal anchor (*pa*). Stages i–vi, are referred to in the text. ←—Movement of the shell; ←O water ejection from the mantle cavity loosening the sand around the shell (*c*); ←—◄ probing and extension of the foot; →< hydrostatic pressure produced by adduction (*a*) of the valves causing pedal dilation; >—< contraction of anterior (*ra*) and posterior (*pr*) retractor muscles. Arrowheads indicate pedal (*pa*) and shell (*sa*) anchors. From Trueman, 1968. *Symp. zool. Soc. Lond.* **22**, 167–186.

of water in the mantle cavity) and brings about maximum foot dilation, together with the ejection of water from the ventral mantle margins. This ejection, facilitated by the prior closure of the siphons, loosens the sand, and thus reduces resistance to the subsequent movement of the shell. We shall later see this principle exploited by *Arenicola* (Sec. 6–3).

With the dilated foot now acting as a terminal anchor, the retractors contract, the siphons reopen, and the shell is drawn forward, the adduction of its valves having reduced its profile, thus aiding its movement through the substratum. The pedal dilation is then reduced (stage iv). The adductor muscles of the shell now relax, and the gape of the valves increases (stage v). The shell can now act as a penetration anchor during the early stages of the next cycle, with the foot probing forwards, and the elastic hinge ligament helping to maintain pressure of the valves on the burrow. There may, however, be a static period (stage vi) before the actual beginning of a fresh cycle, with only slight movements of foot and shell taking place.

This beautifully coordinated sequence of actions takes longer to describe than it does for the animal to execute it; the clam *Donax denticulatus* burrows into the substratum within 4 sec in a rapid series of digging cycles, at a rate of 0.4 cm s^{-1}. The energy requirements have been calculated by Ansell and Trueman, using data derived from animals that have been restrained in their progress by threads attached to force transducers. This makes it possible to measure the maximum force that an animal can exert, and hence the maximum drag forces that it will be able to overcome during its burrowing. Calculated in this way, the energy requirement by *D. denticulatus* for each of its digging cycles is found to be 4.4×10^{-6} J/mg dry weight. This may be compared with figures given earlier for *Sipunculus nudus* (p. 113). The comparison is not straightforward, since allowance must be made for the greater length of *Sipunculus* (15 cm as compared with 2 cm for *Donax*) and its slower rate of burial (0.05 cm sec^{-1} as compared with 0.4 cm sec^{-1}), but it does seem that the burrowing mechanism of *Donax* may be the more economical of the two.

Certainly there are two ingenious features of the bivalve mechanism which must have helped these animals to become among the most successful exploiters of sandy and muddy substrates. One is that the profile of the shell can be enlarged for anchorage and then reduced for movement. This suggests that the evolution of the bivalved shell was primarily an adaptation for burrowing. The other ingenious and economical feature is that bivalves integrate the use of two fluid systems, that of the mantle and that of the haemocoel. But it is not only in the burrowing process that the hydrostatic properties of sea water are exploited by these animals. A good example of another pattern of use is given by the protrusible siphons. Consider *Mya arenaria* (Fig. 5–28) which is characteristically found in the tidal mud of estuaries, living as a suspension feeder at a depth of 6–8 in, with the openings of its siphons lying level with the surface of the mud. Into these it draws currents of water which bear suspended debris and micro-organisms, the subsequent fate of which we consider elsewhere. The siphons are extended by water pressure, contractions of the adductor muscles of the shell forcing water into them from the mantle cavity. During extension the siphon apertures and the foot opening remain closed, forming a watertight system in which pressure can be built up. This process has to proceed in stages, elongation taking place about 1 cm at a time. Between each successive elongation more water is taken into the mantle cavity so that an increased volume of water can be provided to match the increasing volume of the siphons. In this way a degree of extension can be secured that would be unattainable if the animal had to rely upon pressure increases in a body cavity containing only a restricted volume of fluid.

An interesting contrast to *Mya*, and one that well illustrates the elegant way in which hydrostatic mechanisms can be adapted to the requirements of different modes of life, is provided by *Scrobicularia plana* (Fig. 5–29), another bivalve which lives in mud between the tide marks, where there is some lowering of salinity by a flow of fresh water. It differs from *Mya* in being a deposit feeder, with siphons that are separate from each other; an arrangement that permits the inhalent siphon to extend over the surface of the mud flat and thus to draw in the rich organic matter that accumulates there. These siphons may elongate to a length of 10 cm or more, and they do so in a way different from that found in *Mya*, for the movement is a steady and smooth one, unrelated to any pumping action of the valves of the shell. This indicates that extension cannot depend upon the forcing of water into the siphons, as occurs in *Mya*. Indeed, it can proceed even after one valve has been completely removed.

Fig. 5-28. *Mya arenaria*, with siphons and foot fully extended. Half natural size. From Yonge, 1949. *The Sea Shore*. Collins, London.

Fig. 5-29. *Scrobicularia plana*, showing mode of feeding. The animal normally lives much deeper, up to 6 or 8 inches below the surface. (From Hesse, Allee, and Schmidt, 1951. *Ecological Animal Geography* (2nd ed.). Wiley, New York.) From Yonge, 1949, *op. cit.*

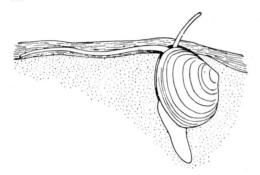

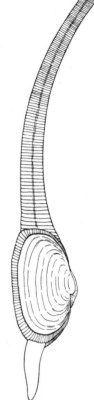

The structures actually involved in extension are the intrinsic muscles of the siphons, with the blood acting in this instance as the fluid component of the system. Elongation of the siphons is brought about by the contraction of strands of radial muscle, this making the walls thinner and displacing some of the blood they contain. The fluid acts as an antagonist of the muscles, while collagen fibres resist deformation, their lattice-like arrangement allowing the siphon wall to become thinner and longer, but resisting outward bulging. Because of these fibres the total cross-sectional area of the siphons remains unaltered during extension, while the walls become thinner; an arrangement that has the great advantage of ensuring that the passage of incurrent water is not hindered. The withdrawal of the siphons is a consequence of muscular action alone, the structures here concerned being longitudinal muscles that take their origin from the shell.

5–5 MOLLUSCS: CEPHALOPODA

Cephalopods are certainly the most remarkable of all the molluscs for the high level of adaptive efficiency that they have achieved, not only in locomotion but in other respects as well. For example, their elaborate three-dimensional musculature, working in conjunction with a highly deformable body, makes possible a variety of postures (Fig. 5–30). These are deployed in conjunction with the action of chromatophores (pigment cells) which, unlike those of crustaceans and vertebrates, are controlled by muscles and are regulated exclusively by the nervous system, without the partici-

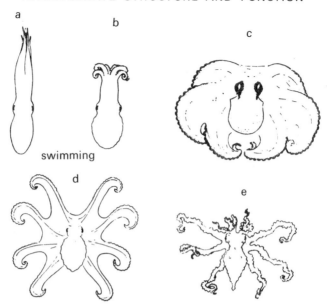

Fig. 5-30. Some of the postures adopted by *Octopus vulgaris* (not to scale). From Packard and Sanders, 1969. *Endeavour*, **28**, 92–99.

pation of hormones. As a result of all this, cephalopods have an unequalled faculty for changing their appearance, with a repertoire of patterns emerging as reflex responses that are fully specified and programmed in the central nervous system. Thus, write Packard and Sanders, they can become from moment to moment 'seaweed, a sponge, a rock—a devil!'

As regards their locomotion, they, too, have exploited the potentialities of the foot, but they differ from the rest of the phylum in having developed a design involving also the mantle, which is related primarily to efficient swimming by jet propulsion. Actually, this method of movement is also found elsewhere in the molluscs. It is seen in the scallops, where, against the whole evolutionary trend of the bivalves, we find pelagic animals that derive their motive force by expelling water from between the clapping valves of their shell. The device, however, is not one that can be used for long periods, and the cephalopods have achieved an altogether more efficient variant of it. Their interest from this point of view is all the greater because of what has been called the 'operational parallelism' between fish and cephalopods. Aristotle knew that cephalopods are molluscs, but, writes Packard, 'ours is a functional and democratic age; no longer is genealogy of primary importance. And Cephalopods functionally are 'fish'. They live in the same medium as fish, they are as large, as mobile, and as speedy, they have ingenious methods of regulating buoyancy that recall teleostean ones (p. 148), and they often occupy the same trophic levels as fish.

This convergence in mode of life between two groups so diverse in their origin owes much to the progressive improvement of their respective modes of locomotion. The establishment of jet propulsion was clearly the determining event in cephalopod history. The principle, startlingly misunderstood in a sixteenth century illustration

Fig. 5-31. The Paper Nautilus, *Argonauta argo*, from Belon (1551). The animal is incorrectly drawn using its arms as oars and its membrane as a sail. From Singer, 1931. *A Short History of Biology.* Clarendon Press, Oxford.

of the pelagic octopod *Argonauta* in supposedly full sail (Fig. 5–31), depends upon the modification of the posterior part of the foot into an exhalent funnel (Figs. 5–32, 11–12), through which water is discharged from the mantle cavity with great force. The remainder of the foot forms a crown of tentacles around the mouth, the animals being so orientated in the water that these lie anteriorly, the visceral hump posteriorly, and the mantle cavity ventral and lateral to the viscera. The funnel, being flexible, can be

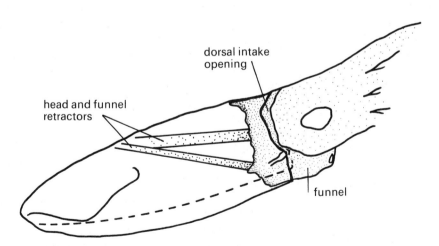

Fig. 5-32. Lateral view of *Loligo* at start of power stroke showing anatomy of the squid and the change of mantle shape in swimming. The broken lines shows the position of the ventral mantle surface at the end of the power stroke. The stippled areas are attached to the head; in the collar area, the mantle wraps around and over these parts. From Ward, 1972. *J. Zool.*, **167**, 487–499.

adjusted to point in various directions, promoting corresponding flexibility of movement.

Cephalopod jet propulsion is seen in a primitive form in *Nautilus*. This animal, the only surviving genus of the Subclass Nautiloidea (known from the Cambrian), is tetrabranchiate (with two pairs of gills), it has a bilobed funnel and it retains the chambered external shell, of which it occupies only the last chamber. The mantle is bound to the shell, and because of this the water has to be expelled by contraction of the muscles of the funnel, together with retraction of the body, the mantle wall being unable to contribute. Even so, the movements of *Nautilus* are rapid, although they do not compete in speed with the most efficient of the other surviving cephalopods. The animal is aided by the buoyancy of the unoccupied chambers of the shell, which are filled with gas, and serve to make the animal very nearly the same density as sea water.

In all other recent cephalopods, which are dibranchiate (with one pair of gills), the design is improved by the funnel becoming a complete tube, while the shell becomes internal (covered, that is, by the mantle) and is then either reduced or completely lost (as it is in *Octopus*). This reduction seems to have taken place along several lines, doubtless because of the functional advantage that results from the increased flexibility of the mantle. This is well provided with circular and radial muscle fibres, and a variable amount of longitudinal ones. Contraction of the circular muscle drives water out of the funnel with great force, which is increased by closure of the mantle edges around the head during this phase. Other adaptive improvements, leading to a maximum speed of 200 cm sec^{-1} in the squid *Loligo* for a single jet cycle, include the development of stabilizing fins, streamlining of the body, and the use of stud and socket fasteners to hold the mantle to the head so as to prevent the egress of water at this point during the locomotor jet cycle. Loss of the supporting and protective function of the shell is counterbalanced by the development of an internal skeleton of cartilage, closely resembling in its histology that of vertebrates. The convergence is enhanced by the use of this cartilage, as in the vertebrates, for the protection of the highly developed brain.

What is left of the shell in squids and cuttlefish also contributes to locomotor efficiency. It prevents energy being wasted by distortion of the body during the jet propulsion cycle, while in *Sepia* it contributes to buoyancy, the shell (cuttlebone) containing gases which account (as in *Nautilus*) for the density of cuttlefish being about the same as that of sea water. This helps the animals to hover while stalking their prey, which they catch with their long tentacles. Control of density in cuttlefish depends on the balance between the hydrostatic pressure, which drives water into the cuttlebone, and the osmotic pressure of the body fluids, which draws it out, gases diffusing into the chambers of the shell to fill the spaces that are left. Here is an example of convergence with fish, this time with the swimbladder of teleosts.

The freeing of the mantle of dibranchiates has evidently made a major contribution to their locomotor success, but its functioning is still not fully understood, for it has been studied less than might have been expected. This is surprising, considering, as Ward and Wainwright remark, that in the squid 'it effects the fastest locomotion known in aquatic invertebrates, generates the highest hydrostatic pressures measured in any animal, contains functional giant axons and tastes good'. Only a few points from the fascinating picture of adaptation that is now emerging from squid studies can be mentioned here. The mantle contains circular and radial muscle, arranged in alternating rings, but not, in these animals, any longitudinal muscle. In addition, there are abundant collagen fibres, forming, in part, a geodesic system. During the intake of water, the mantle diameter increases, with thinning of its wall, largely through contraction of the radial muscles; the funnel is closed at this stage, and valves at the intake apertures are opened. When the water is expelled, the circular muscles contract, the thickness of the mantle increases, and its circumference decreases; as a result, it looks smaller from the outside. During locomotion, water can only leave through the funnel; during respiratory movements, which are similar, but with smaller amplitude, water can also leave around the mantle edges. Throughout all of these movements, the length of the mantle remains constant, and there is no sign of peristalsis, or of any other asynchrony of the muscles.

These actions are easy to describe, but they raise questions. For example, constancy in length of the mantle has an obvious advantage in that thrust is not wasted in passing water into a lengthened cavity, but how is this constancy achieved? Probably by the resistance set up by the collagen fibres, with the geodesic component facilitating smooth changes in shape. Another question arises from the implied mutual antagonism of the circular and radial muscles. This is comparable to the antagonism of the circular and longitudinal muscles of coelenterates and worms, but in the mantle there is no fluid-filled cavity to act as a hydrostatic skeleton. How, then, is the antagonism achieved? Presumably, suggests Trueman, the circular muscle bands are interacting with the adjacent radial muscle; in other words, the muscles themselves represent the hydrostatic skeleton, as they do also in the arms. This argument, however, would only provide for local application of force without any transfer over the mantle wall; yet the rapid jet cycles demand simultaneous contraction of the whole wall, which is, in fact, one of the remarkable features of the system. How is this achieved? Here the giant nerve fibres (Sec. 16–4) are of prime importance, providing for rapid conduction of nerve impulses from the stellate ganglion to the mantle muscles, with the diameter of the fibres being elegantly adjusted to the distances over which they have to conduct. The thinner fibres, which conduct more slowly, pass to the anterior end, while the thicker ones, conducting more quickly, pass to the posterior end.

The attainment of high speed during jet propulsion can be shown, from Newton's laws of motion, to depend in part on the mass (m) and velocity (v) of the water ejected. This imparts momentum (mv) to the body, but the actual velocity of the animal will also depend on its own mass and upon its drag. The latter, which is the force resisting the animal's motion, will be reduced by the streamlining of the body, which is an obvious feature of the fast-swimming cephalopods. Jet velocity is inversely proportional to the cross-sectional area of the funnel, and is further influenced by the thickness of the circular muscle of the mantle, while the mass of water ejected is determined by the volume of the mantle cavity. It follows that jet propulsion favours large size, and Packard shows that in *Loligo* velocity is proportional to body weight. It is not difficult, therefore, to see why the giant deep-sea squid (*Architeucthis*) reaches a weight of several tons. And it may be that this accounts for the rapid growth rate of cephalopods, for this would be strongly favoured by selection.

However, the exact pattern of these relationships varies with mode of life, as may be seen from Table 5–1, which shows that there are significant differences in the mantle properties of three genera, but the locomotor efficiency of these animals

Table 5-1 Comparison of factors concerned with jet propulsion in certain cephalopods. From Trueman and Packard, 1968, *J. exp. Biol.*, **49**, 495–507.

species	Wet weight (g)	Cross-sectional area of jet (cm²)	Mantle capacity (ml)	Pressure pulse duration (msec)	Velocity of jet (cm/sec)	Mantle water all expelled	Theoretical maximal velocity of animal during single jet cycle (cm/sec)		
							2/3 water expelled	1/2 water expelled	1/3 water expelled
Octopus vulgaris	370	0.7	53	600	126	18	—	—	—
Eledone moschata	600	1.6	300	200	940	470	177	94	40
Loligo vulgaris	350	1.5	200	180	730	420	155	81	33
Sepia officinalis	250	1.5	62	150	275	68	27	15	0.6

is not in doubt. Some species of squid can swim at 15 knots (790 cm sec^{-1}), which is the greatest known speed of any aquatic invertebrate, while the efficiency of the muscle is apparent from the fact that the complete muscular cycle (inhalent and exhalent phases) occurs in as little as 0.45 sec in *Loligo*.

The functional organization of *Octopus*, however, is very different from that of squids and cuttlefish, in accordance with its benthic life. The Decapoda concentrate particularly on speed, with stabilizing fins, a streamlined body, a well-defined visual sense, and tentacles adapted for seizing food and for copulating. Octopoda, by contrast, have a closer association with the substratum. They have lost the internal shell, generally lack fins, and use their tentacles for locomotion and for the recognition of food, revealing in this a well-balanced provision of visual and tactile receptors. Recordings of jet thrust can be made by tethering the animal with a thread, and attaching this, through a force transducer, to a recorder. Simultaneous records of pressure in the mantle cavity can be made by inserting a cannula into the cavity. Using this type of procedure, Trueman and Packard found that squids developed a thrust approximately equivalent to their body weight, whereas in *Octopus* it was equivalent to not more than one-half of their weight.

Power in *Octopus* has gone to the arms. An animal of about 1 g weight can produce a jet thrust of less than 1×10^{-2} N. Holding on with five arms, however, it can resist a pull of 1 N (Fig. 5–33). The limitations of octopod jet propulsion are reflected in the more globular body, as compared with that of squids, and in the more limited inhalent apertures. The power of the tentacles, however, facilitates crawling, and a

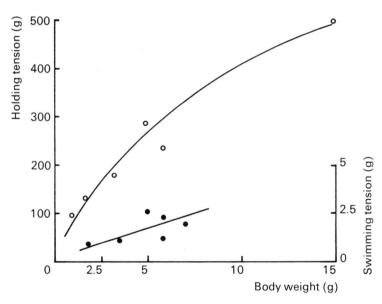

Fig. 5-33. Graphs showing relationship between the increase in the body weight of young *Octopus*, the maximum tensions developed by production of jets (swimming tension, ●) and the maximum holding pull (holding tension, ○) with five arms attached to the bottom of the tank. The holding tension is represented by points at which a thread stitched through the cephalic cartilage broke free. From Trueman and Packard, 1968. *op. cit.*

report of these animals keeping up with a briskly walking man provides agreeable testimony to this.

Octopods are thus more versatile than the decapods, being able, as Morton reminds us, to 'perform an elaborate courtship, fondle and repeatedly cleanse and ventilate their eggs, capture a moving crab with a fine precision, and even transport stones to build houses'. Given the complex brain, the elaborate nervous system, and the highly organized neural pathways with which these animals are endowed, it is natural that they should have provided such excellent material for the study of advanced forms of behaviour (Sec. 16–6).

6
Movement and Metamerism

6–1 SIGNIFICANCE OF METAMERISM

Important though the evolution of the coelom has proved, there is another innovation, closely associated with it, that has had no less profound an influence upon the history of many of the invertebrates, and of the whole of the vertebrates. This is the development of metamerism, the plan of structure in which the body is differentiated along its longitudinal axis into a series of units or segments, each of which contains elements of some of the chief systems of organs. The characteristics of a typical segment were usefully defined by Goodrich. In invertebrates each segment is demarcated externally by an anterior and a posterior groove. This, however, applies primarily to those animals in which the primitive pattern of segmentation remains unmodified and unobscured. In arthropods, particularly in the head and thorax, grooves may have a functional significance unrelated to segmentation. Ideally, each such segment will contain a pair of mesodermal somites with coelomic cavities, and a pair of coelomoducts leading from these to the outside. In addition, there will be a pair of nephridia (in those species that have retained this particular system), a pair of nerve ganglia borne on the paired ventral nerve cord, and often a pair of appendages.

In principle a similar situation exists in the vertebrates, but with variations imposed by certain fundamental differences in the organization of this group as compared with invertebrates. For example, the muscles of the body wall of fish are associated with an internal axial skeleton, so that a hydrostatic skeleton plays no part in movement. Here the criteria of a segment are the presence of paired mesodermal somites and coelomic cavities, units of striated muscle (myomeres) derived from the somites, paired coelomic cavities, paired coelomoducts in the form of kidney tubules, and paired dorsal and ventral nerve roots. In these animals the paired limbs are not metamerically repeated, a difference that sheds some light on the circumstances that have influenced the origin of metameric organization (see below).

Metamerism, as we have here defined it, is a characteristic of the Annelida, the

Arthropoda, and the Vertebrata (together with the protochordate amphioxus). This means that it must certainly have evolved independently at least twice, for the vertebrates probably share a common origin with the echinoderms within the Deuterostomia, and certainly show no close relationship with the metameric invertebrates. Moreover, the protochordates include the unsegmented Urochordata, which in itself indicates that metamerism must have appeared within the vertebrate line quite independently of the metamerism of annelids and arthropods. In addition to these major expressions of metamerism, however, a division of the body, with repetition of parts, is also found in the tapeworms. One can thus argue—as does Hyman, for example—that these acoelomate animals should also be regarded as metamerically segmented. Nevertheless, many zoologists have been unwilling to accept this interpretation, preferring to consider the annelids, arthropods, and vertebrates as the only truly metameric animals. For this attitude there is good justification, if we consider the functional significance of metamerism.

We shall see that this significance is undoubtedly to be found in the modes of locomotion of metameric animals, and especially of the more primitive ones, which illustrate something of the circumstances in which metamerism first became established. In tapeworms the repetition of the body structure is a parasitic and not a locomotor specialization; it serves primarily for the construction of a reproductive machine, each proglottid containing a complete set of hermaphroditic reproductive organs. Locomotion is merely a matter of movement of individual proglottids after they have broken off from the main body. One feature which is often emphasized is that the new segments of a worm or arthropod appear at the hind end of the body, while in vertebrates the centre of growth is also posterior, situated at the junction of body and tail. In tapeworms, on the other hand, the growth zone is at the anterior end, immediately behind the scolex. This has led some supporters of the metameric interpretation of the structure of these animals to argue that the scolex is, in fact, posterior, so that the worm is attached to its host back to front. Such an argument is perhaps more revealing of the mental processes of its devisor than of the evolution of the cestodes; it illustrates the danger of placing undue emphasis upon form to the exclusion of functional considerations. Bearing in mind, then, the close relationship of annelids and arthropods, we may take it that metameric segmentation must have appeared twice, once in the annelidan–arthropodan stock and once in the chordate stock.

The problem of the origin of metamerism is no less obscure than is the similar problem of the origin of the coelom. It may be relevant that the lower groups of animals show a marked tendency for repeating certain organ systems. The platyhelminths are clearly unsegmented animals (Fig. 6–1), but some have multiple gonads alternating with pouches of the alimentary canal, and this is seen also in nemertines (Fig. 6–2). In the echinoderm–chordate line a well-known example is the repetition of gill slits, gonads, and alimentary pouches in the hemichordates. The advantage of such repetition probably lies in the resultant increase in surface area of the organs concerned. It is likely to be of particular significance in the acoelomate invertebrates, where the absence both of a coelom and often of a vascular system must limit the efficiency of transport of metabolites. This type of repetition, sometimes termed pseudometamerism, is, therefore, an understandable phase of the evolution of animal organization, but whether it was a determining factor in the

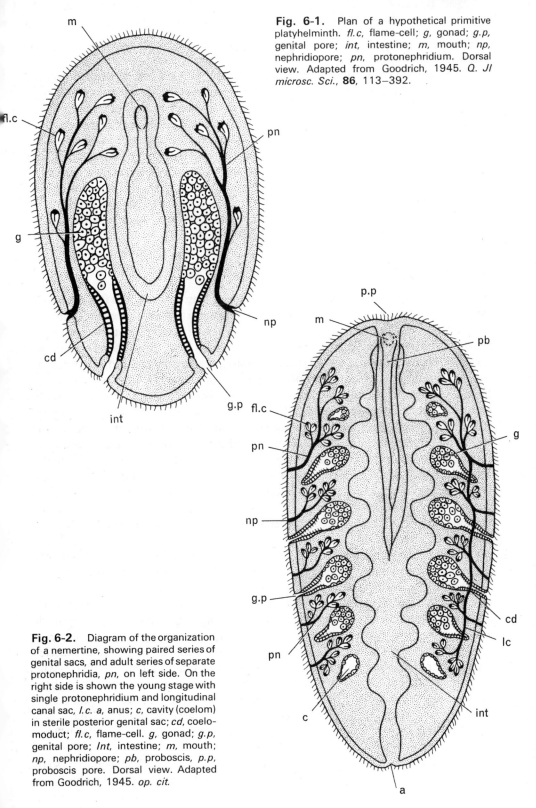

Fig. 6-1. Plan of a hypothetical primitive platyhelminth. *fl.c*, flame-cell; *g*, gonad; *g.p*, genital pore; *int*, intestine; *m*, mouth; *np*, nephridiopore; *pn*, protonephridium. Dorsal view. Adapted from Goodrich, 1945. *Q. Jl microsc. Sci.*, **86**, 113–392.

Fig. 6-2. Diagram of the organization of a nemertine, showing paired series of genital sacs, and adult series of separate protonephridia, *pn*, on left side. On the right side is shown the young stage with single protonephridium and longitudinal canal sac, *l.c. a*, anus; *c*, cavity (coelom) in sterile posterior genital sac; *cd*, coelomoduct; *fl.c*, flame-cell. *g*, gonad; *g.p*, genital pore; *Int*, intestine; *m*, mouth; *np*, nephridiopore; *pb*, proboscis, *p.p*, proboscis pore. Dorsal view. Adapted from Goodrich, 1945. *op. cit.*

evolution of metamerism is more doubtful. Certainly a much more important one must have been locomotion.

We shall be considering below how far the movement of annelids and arthropods depends upon the segmentation of the musculature, coelom, and nervous system. So important is the subdivision of the musculature in annelids and vertebrates, and of the coelom in annelids, that it becomes highly probable that metamerism in these animals was determined primarily by these structures, the repetition of other systems becoming arranged in conformity with them. Some of these systems may already have been subdivided for reasons just indicated, while some may have become subdivided later as a matter of structural convenience. The interrelationship of form and function is here somewhat involved, but it should become easier to grasp if we examine some aspects of the organization of locomotion in the metameric invertebrates, and the contribution made to it by the metamerism of the body.

6-2 LOCOMOTION OF OLIGOCHAETES

Annelid worms are essentially creeping and burrowing animals, and it is reasonable to accept their present-day modes of life as being those that influenced the origin and early evolution of their metamerism. The lower vertebrates, by contrast, are characteristically free-swimming and often pelagic animals, and this, in conjunction with the early development of an axial skeleton, must have determined the form that metamerism took in them and in their protochordate ancestors. Vertebrate limbs seem to have arisen, probably on more than one line of advance, as lateral folds. These met certain hydrodynamic requirements, involving the stability of a body moving in a fluid medium, and in doing so they became restricted to two pairs of appendages. It was therefore swimming that laid the foundations for the eventual emergence of tetrapod vertebrates, and of human beings with two hands and two feet. In complete contrast to this, it was probably the creeping and burrowing of unknown worm-like ancestors, exploiting the rich food resources of bottom deposits, that determined the subsequent evolution of arthropods, with their serially repeated and remarkably versatile limbs, and their feeding methods which have drawn heavily upon this versatility.

It will be convenient to look to the oligochaetes for our first example of annelidan locomotion. They are by no means the most primitive of living annelids in their habits, but the earthworm possesses a plan of structure that conforms very closely in some respects with the idealized type of hydrostatic skeleton we considered earlier. The body wall is composed of continuous layers of longitudinal muscles (on the inside) and circular muscles (on the outside), these being placed so that they can exert pressure upon the coelomic fluid which they surround. The longitudinal muscle is often said to be segmented in conformity with the external annulation, but this is not strictly correct. The longitudinal muscle fibres are long enough to extend through two or three segments, so that in this respect the segments are linked in small groups. Each segment, however, has its own complement of segmental nerves, so that precise and localized control of muscular contraction can certainly be exercised by the central nervous system.

There is, however, another factor that makes an essential contribution to the localization of muscular activity: the subdivision of the coelomic cavity by the transverse septa, which are developed uniformly throughout the greater part of the body. It is in this respect that the hydrostatic skeleton of the earthworm differs in a very important way from that of the unsegmented animals (and of the idealized model) that we have previously considered. We saw that in these the pressure changes set up at one point are distributed throughout the body, with a consequent dissipation of energy. This must impose limits upon the speed and efficiency of the locomotor responses of unsegmented animals. In the earthworm, by contrast, the septa damp down the pressure changes and tend to limit them to particular regions of the body. This they are the better able to do because they are well provided with intrinsic muscle fibres which enable them to resist stress. Nevertheless, they cannot provide complete barriers. They are penetrated by the ventral cord, and a foramen at this point (Fig. 6–3) provides for continuity of coelomic fluid from one segment to the next. In actively moving worms, however, these foramina are closed by sphincter muscles; because of this, and because of the intrinsic musculature of the septa, increase of pressure in the fluid of one segment is substantially isolated from adjacent segments. This makes possible the application of thrust at individual points called *points d'appui*.

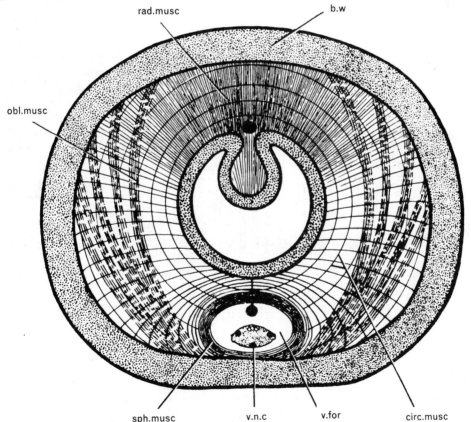

Fig. 6-3. Diagram to show the arrangement of the septal muscles in the earthworm. *b.w*, body wall; *circ. musc*, circular muscle; *obl. musc*, oblique muscles; *rad. musc*, radial muscles; *sph. musc*, sphincter of ventral foramen; *v. for*, ventral foramen; *v.n.c*, ventral nerve cord. From Newell, 1950. *J. exp. Biol.*, **27**, 110-121.

158 INVERTEBRATE STRUCTURE AND FUNCTION

Metameric segmentation, as we find it in the earthworm, therefore provides an escape from an important limitation inherent in unsegmented hydrostatic skeletal systems. The influence of this becomes apparent when we examine how the animal moves. According to the analysis of Gray and Lissmann, locomotion begins with a contraction of the circular muscles in a limited region of the anterior end of the body (Fig. 6–4). This contraction passes backwards down the body as a peristaltic wave, each wave being followed by a wave of contraction of the longitudinal muscle; this continues in regular alternation. At those regions where the longitudinal muscle is contracting, the body surface bulges outwards and the chaetae are protruded; here the worm can exert a thrust against the substratum. Where the circular muscle is contracting the body becomes thinner, and the chaetae are withdrawn. Here the segments extend forwards, aided by the thrust from the swollen regions. Thus the extension of the thinner parts of the body can be translated into forward movement. It is a method of locomotion particularly well suited for a burrowing animal, since the swollen parts of the body press against the whole circumference of the burrow and can thus act according to circumstances as penetration or terminal anchors, just

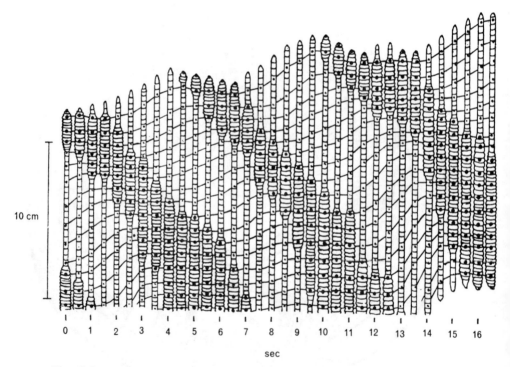

Fig. 6-4. Diagram (prepared from a cinematograph record) showing mode of progression of an earthworm. Regions of the body showing longitudinal contraction are drawn twice as wide as those undergoing circular contraction and are also marked by larger dots. As long as a segment is longitudinally contracted it remains at rest relative to the ground; it moves forwards during all other phases of the cycle. The track of individual points on the worm's body and their movements relative to each other are shown by the lines running obliquely forwards from left to right of the diagram. From Gray and Lissmann, 1938. *J. exp. Biol.*, **15**, 506–517.

as in the unsegmented sipunculids. In those earthworms, such as *Lumbricus* and *Allolobophora*, where the four pairs of chaetae are situated in the ventral half of the segments, an effective thrust can also be exerted against the substratum when the animal is moving over the surface. On a polished surface such movement is difficult because of backward slip, but the worm can still pull itself forward by raising its body and using its anterior end as a sucker; thus it can climb out of a glass beaker.

Peristaltic waves of the body wall are not confined to segmented animals. They occur in unsegmented ones as well, but in these the waves are irregular, and the movement that they produce is comparatively sluggish and is rarely prolonged. The locomotion of earthworms is based upon a much more advanced physiological organization. It depends upon a refinement of integration, which is exerted through the metamerically segmented nervous system and acts upon correspondingly segmented effectors. We shall see later something of the importance in this connection of the organization of patterns of activity within the nervous system. The locomotion depends equally upon the segmental partitioning of the coelomic fluid, which results in force being exerted sequentially at restricted points of the body surface, and which allows one part of the body to act in this respect independently of other parts.

An indication of the efficiency of the mechanism is given by measurements of the degrees of force that can be built up within the coelom. Pressure recordings show that resting pressure is very low, and may even be negative, ranging from -2 to 2×10^{-2} N cm^{-2}. This is due to several factors. The body wall has an inherent stiffness and elasticity which, in conjunction with the septa, tend to maintain the form of the body, while stability is further enhanced by feedback from the stretch receptors of the muscles. Pressure changes within a single segment during locomotion (Fig. 6-5) involve, in principle, the exertion of maximum force during contraction of the circular muscles, while a smaller peak occurs when these muscles are antagonized by contraction of the longitudinal muscles. The first peak provides the force for the forward thrust of the head; the second peak, which coincides with protrusion of the chaetae, provides the *point d'appui*. These changes vary in detail in different regions of the body, and according to whether the worm is burrowing or moving over the surface. During burrowing, contraction of the longitudinal muscles can bring about very high pressures at the hind end, amounting to as much as 3.75 N cm^{-2}. This provides for enlarging the burrow and consolidating its wall. High pressures are similarly attained during the digging cycles of *Arenicola* (see later), probably for the same reason.

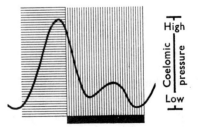

Fig. 6-5. Diagram summarizing coelomic pressure fluctuations (vertical scale) in a single segment of *Lumbricus* during the passage of a locomotor wave. Horizontal hatching: thinning and elongation of the recording segment by contraction of the circular muscles. Vertical hatching: thickening and shortening by contraction of the longitudinal muscles. Horizontal bar: duration of chaetal protrusion. The large and small pressure pulses are associated respectively with elongation and shortening. From Seymour, 1969. *J. exp. Biol.* **51**, 47–58.

6-3 LOCOMOTION OF POLYCHAETES

We have dealt first with oligochaete worms because they illustrate in a comparatively simple way the interaction between metameric segmentation and the functioning of a hydrostatic skeleton. Polychaete worms are generally held to be more primitive than the earthworms, and are clearly so if they are judged by their modes of life, for creeping and burrowing in a marine environment must have preceded the fresh-water and terrestrial habits that characterize the oligochaetes. Yet polychaetes are certainly highly specialized animals within the limits of their general plan of organization, and their mode of locomotion introduces certain features that are lacking in the oligochaetes. These we shall illustrate mainly by a consideration of *Nereis*, although it must not be forgotten that this animal provides only one example of the widely diversified modes of life of polychaete worms.

Two structural elements are especially important in the locomotion of *Nereis*: the longitudinal muscles of the body wall, and the parapodia (Fig. 6–6). As in the earthworm, the segmentation of the muscles and of their nerve supply facilitates the passage of waves of contraction down the body. An important difference, however, is that in *Nereis* the longitudinal muscle does not form a continuous layer. Instead, it is broken up into two pairs of blocks, one pair dorsal and the other ventral. Because of this the muscles of the two sides of a segment can be in opposite phases, one contracted and the other relaxed, so that the passage of waves of contraction along the body can throw it into lateral undulations. These are a characteristic feature of the locomotion of *Nereis*, and contrast markedly with the peristaltic waves of the earthworm. Another point of difference from the later animal is that in *Nereis* the waves of contraction pass forwards; we shall see later the explanation of this.

So far in our consideration of hydrostatic skeletons we have found that the musculature of body walls is commonly arranged so that circular muscle can be opposed to longitudinal muscle. The nematodes provided an exception to this, and the principle is also somewhat departed from in *Nereis*. In this animal the circular muscle layer is relatively weak, more particularly because it is interruped laterally where muscles derived from the circular layer run into the parapodia. This means that the body wall itself is weaker than that of the earthworm. Some support is probably given by fibres that are present in it, and a contribution is also made by the oblique muscles that run upwards and outwards from the region of the ventral nerve cord. Despite this supplementary strengthening, however, the body is clearly ill adapted for the peristaltic movements that are so well provided for in earthworms. It is the lateral undulations that are important in the locomotion of *Nereis*, together with the parapodia that translate these into forward displacement of the body. We see here a form of movement different from that of the earthworm, and one that anticipates the locomotory mechanisms of arthropods.

A parapodium is a hollow extension of the body, typically divided into dorsal and ventral components called the notopodium and neuropodium (cf. Fig. 6–6). Each lobe carries a bundle of bristles, or chaetae, strengthened by a supporting aciculum. The parapodium can be moved forwards and backwards, its point of attachment to the body wall acting as a hinge, while it is sufficiently flexible to undergo a considerable measure of protrusion and withdrawal. Protrusion takes place in part as a result of hydrostatic pressure exerted through the coelomic fluid, this being

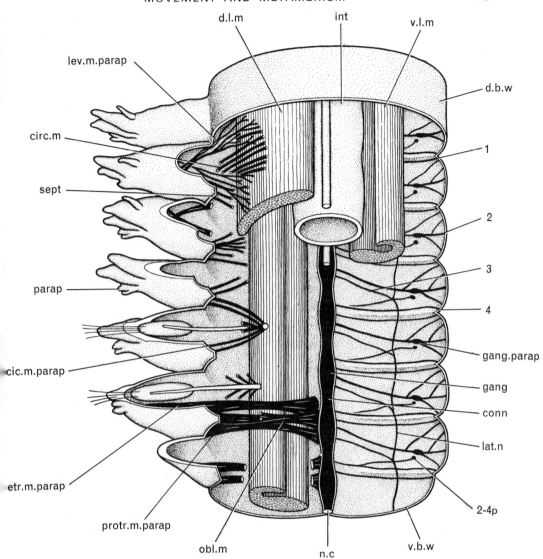

Fig. 6-6. Stereogram of seven body segments of *Nereis virens* dissected at various levels to show the nerve cord and segmental nerves and the muscles of the body wall and parapodia. The anterior end is at the top of the figure. *1–4*, the segmental nerves in antero-posterior succession; *2–4p*, peripheral connexion between nerves 2 and 4; *acic.m.parap*, acicular muscle of the parapodium; *circ.m*, circular muscles of the body wall. *conn*, nerve cord connective; *d.b.w*, dorsal body wall; *d.l.m*, dorsal longitudinal muscle; *gang*, nerve cord ganglion; *gang.parap*, ganglion of the parapodial nerve (II); *int*, intestine; *lat.n*, lateral nerve; *lev.m.parap*, levator muscle of the parapodium; *n.c*, nerve cord; *obl.m*, oblique muscle; *parap*, parapodium; *protr.m.parap*, protractor muscle of the parapodium; *retr.m.parap*, retractor muscle of the parapodium; *sept*, intersegmental septum; *v.b.w*, ventral body wall; *v.l.m*, ventral longitudinal muscle. From Smith, 1957. *Phil. Trans. R. Soc. B.*, **240**, 135–196.

162 INVERTEBRATE STRUCTURE AND FUNCTION

possible because the coelomic cavity extends into the parapodium. The parapodial musculature, however, also contributes to this, and is responsible for other movements as well. The arrangement of the muscles is remarkably complex (Figs. 6-6, 6-7), but, according to Mettam's analysis of *Nereis diversicolor*, those mainly concerned with movement are as follows.

Oblique muscles, which originate mid-ventrally close to the nerve cord. These run in part to the lateral body wall, where they resist transverse stress, and in part to the parapodia, where they are inserted at the base of the neuropodium, some on its anterior face and some on the posterior.

Dorsal and ventral parapodial muscles, which originate externally to the longitudinal muscle of the body wall (they are perhaps derived from the circular muscle) and are inserted at various points on the notopodium and neuropodium.

Intrinsic parapodial muscles, which have both their origin and their insertion within the parapodium.

Muscles of the aciculum and chaetae sacs, constituting protractor and retractor muscles, some of the acicular protractors passing deep into the parapodium.

Nereis diversicolor lives in permanent or semi-permanent burrows, moving by creeping, or by a partial extension of its proboscis which is effected by coelomic pressure. It is aided in its burrowing by its ability to retract the parapodium and chaetae. However, the properties of the musculature and parapodia of nereids in

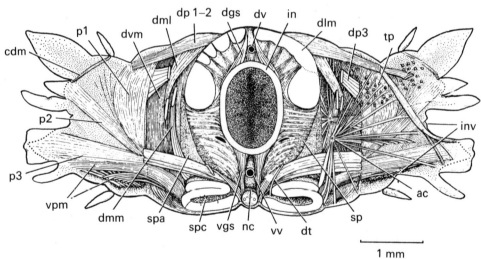

Fig. 6-7. Anterior view of posterior half of segment from mid-body region of *Nereis diversicolor*. Acicula and acicular muscles removed from right side; acicula removed from left. *ac*, acicular muscles of neuropodium; *cdm*, dorsal cirrus muscle; *dgs*, dorsal gut suspensor; *dlm*, dorsal longitudinal muscle; *dml*, lateral diagonal muscle; *dmm*, median diagonal muscle; dp 1, dp 2, dp 3, dorsal parapodial muscles; *dt*, transspetal diagonal muscle; *dv*, dorsal vessel; *dvm*, dorsoventral muscle of body wall; *in*, intestine; *inv*, invagination line of neuropodium; *nc*, nerve cord and supraneural muscle; p 1, p 2, p 3, parapodial intrinsic muscles; *sp*, septum; *spa*, anterior fold of septum; *spc*, septal connective tissue; *tp*, transparapodial muscles; *vgs*, ventral gut suspensor; *vpm*, ventral parapodial muscles; *vv*, ventral vessel. From Mettam, 1967. *J. Zool.* **153**, 245-275.

general interact to provide for several different types of movement. One of these is a slow creeping, in which only the parapodia are active. The two members of a segmental pair alternate in phase, while the movement of any one parapodium begins slightly after that of the one behind it. Waves of parapodial movement thus travel forwards over the length of the body. Initially, a parapodium moves inwards, upwards, and forwards, with its tip lifted from the ground, and with the aciculum withdrawn. This phase of limb movement, called protraction, is the preparatory stroke; it is effected mainly through the action of the intrinsic muscles of the parapodial wall and of the dorsal parapodial muscles. Following this, the aciculum is protruded and the parapodium, having made contact with the ground, is able to form the *point d'appui*, through which power is applied. This power stroke (called retraction) involves rotation and protraction of the parapodium, which is carried out mainly by the acicular muscles. During this stroke, the body is pulled forwards, while the parapodium, remaining stationary relative to the ground, becomes directed backwards. (The extension of the neuropodium is of particular importance during this phase, the extreme lability of this lobe being an adaptation for creeping.) The distance through which the parapodium moves back relative to the body during the complete cycle is termed its span, in Gray's nomenclature, while the distance through which the body moves forward during the cycle is the stride.

Slow creeping readily passes into rapid creeping. This involves a similar rhythmic pattern of waves, but with the difference that the longitudinal muscles of the body wall are now of primary importance. They contract serially in parallel with the movements of the parapodia. The oblique muscles of the latter are now probably of much less importance than in slow creeping, for rapid creeping is mainly effected by the longitudinal muscles pulling against points of friction established between the parapodia and the substratum. The contractions of the longitudinal muscle throw the body into lateral sinusoidal waves that pass anteriorly along its length (Fig. 6–8). At any given moment a parapodium on the crest of a wave, where the longitudinal muscles are relaxed, will be stationary with respect to the substratum. Its partner on the opposite side of the segment is in the trough of the wave, having moved there as the longitudinal muscles contract (Fig. 6–8). During this contraction the muscles exert a thrust on the substratum, but this is transmitted through the stationary parapodium on the opposite side. The parapodia on the crests are not entirely inactive, for they exert their own power stroke. The force of this stroke is thus added to that generated by the contraction of the longitudinal muscles of the opposite side; yet it is the latter muscles that make the major contribution, and that permit the rapid creeping type of locomotion.

It is because of this relationship of muscles and parapodia that the sinusoidal locomotory waves of *Nereis* pass forwards over the body, in the opposite direction to the peristaltic locomotory waves of the earthworm. Were they to travel backwards the worm would also move backwards. This consideration applies equally in the third type of movement displayed by *Nereis*, which occurs in the free-swimming phase. The pattern of movement is essentially the same as that of rapid creeping; sinusoidal waves of the body interact with parapodial movement, but there is a marked increase in the length of the waves, and also in their amplitude and frequency (Fig. 6–9). Here again the worm is propelled forwards by body waves that also pass forwards; a situation that presents a striking contrast with fish, which swim forwards through the

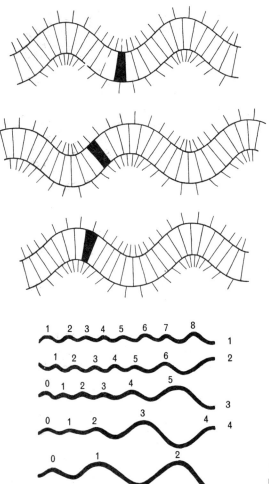

Fig. 6-8. Diagrams showing (A, B, C) forward rapid creeping movement (to the left) of *Nereis* relative to the ground. Note that the two sides of a segment (black) move forward alternately, one side pivoting on the other. Each parapodium on the crest of a wave is lifted from the ground as it completes its effective stroke, and is carried forwards and inwards as the underlying longitudinal muscles contract. It is then carried forwards and outwards as these muscles relax, the early part of this phase constituting the preparatory stroke. The motive power is derived almost entirely from the longitudinal muscles pulling against *points d'appui* established on the opposite sides of the body by the bases of the parapodia at the crests of the waves. Adapted from Gray, 1939. *J. exp. Biol.*, 16, 9–17.

Fig. 6-9. Diagram showing the transition from rapid creeping pattern of movement of *Nereis* to that of swimming. From Gray, 1939. *op. cit.*

agency of waves that pass backwards. The difference is again a result of the presence of the parapodia. These move backwards during their active stroke, and so, by creating a backward flow of water, give a forward thrust to the body. The efficiency of their action is probably very low; indeed, the progress of the animal is slow as compared with the frequency and speed of the waves of propulsion. Nevertheless, without this parapodial action the worm would swim backwards. The deflection of the parapodia is effected largely by the parapodial oblique muscles, the forward movement by the anterior components of these and the backward movement by the posterior ones, while the aciculum keeps the parapodium in a semi-rigid condition. On this analysis, therefore, the muscle system involved in swimming must be different from that involved in slow creeping.

Nereids are examples of the polychaetes referred to as 'errant' forms, so called because their mode of life is typically free-moving, and contrasts with the relatively inactive, burrowing life of the 'sedentary' forms. Errant forms may, of course, be active burrowers, as we have seen in the case of *Nereis diversicolor*. Indeed, this mode of life is so advantageous, both in the protection that it provides and in the rich supplies of organic deposits that are thus made available as food, that it may well have been a major factor in stimulating the evolution of metamerism with its associated improvements in locomotor mechanisms. We have seen that these improvements powerfully contribute to the efficiency of the earthworm's burrowing, and it is likely that they also did so during the evolution of the polychaete locomotor mechanisms.

As already mentioned, the burrowing of *N. diversicolor* is aided by protrusion of the proboscis, which comprises essentially the buccal cavity and pharynx. The thrust exerted by this is a result of pressure exerted by the muscular body wall and transmitted through the coelomic fluid; a simple application of the operation of the hydrostatic skeleton. The same principle is even better seen in *Nephtys*, an errant form that is a particularly active burrower. As described by Clark, the animal makes an opening in the sand by vigorously everting its proboscis, meanwhile anchoring its body at its widest region, which lies between segments 15 and 45. After this the proboscis is withdrawn into the body and the worm crawls forwards into the space thereby exposed, using for its locomotion lateral undulations of the same general type as those described above. These undulations are elegantly adapted to the variation in width of the body at different points of its length. The anterior segments are particularly narrow, and here the undulations are large enough to bring the sides of the body into contact with the wall of the burrow. Farther back their amplitude is reduced, in correlation with the greater width of the body. The form of the parapodium in *Nephtys*, with the characteristic divergence of the notopodium and neuropodium, is probably another aid to burrowing, for it enables the worm to grip the dorso- and ventro-lateral walls of the burrow and so stabilize the body during the violent eversion of the proboscis.

It is thought that in some circumstances burrowing polychaetes make use of thixotropy, a property of colloidal systems that allows an increased rate of shear to produce a reduction in resistance in the system. This is seen in certain types of sand, particularly when clay is present. It has been observed, for example, that when *Nephtys* begins to burrow in such a medium it may hold its head against the sand and set this into vibration by lateral waves of movement of the body. Because of its thixotropic property the sand becomes semi-fluid, permitting the entry of the head; thereafter, the worm continues to burrow by the method just outlined.

The refinement of adaptive organization that underlies polychaete movements is well brought out by further comparison of *Nephtys* with *Nereis*. *Nephtys* cannot crawl by its parapodia alone, but, in contrast to *Nereis diversicolor*, it is a frequent and efficient swimmer. As in *Nereis*, this movement involves backward deflection of the parapodia, but in *Nephtys* this deflection takes place just when the parapodium reaches the crest of the undulatory wave of the body. In *Nereis* the deflection begins on the leading edge of the wave, and is thus applied with less precision, so that the power stroke is less efficient. Probably the swimming of *Nephtys* is also aided by the muscular lobes of the parapodium, for similar lobes develop on the parapodia of those nereids (which do not include *Nereis diversicolor*) that develop a heteronereid phase (Sec. 18–2).

In the sedentary polychaetes we commonly find a marked reduction of the parapodia and of their associated musculature. This is correlated with an improved development of the longitudinal and circular muscles, which come to form much more complete layers than in *Nereis*, and which permit a peristaltic type of movement similar in some respects to that of the earthworm. As always, there is (Sec. 10–1) much variation in the degree of specialization. *Sabella*, for example, is capable of peristalsis but can also move forwards or backwards in its tube through the activity of its parapodia. More commonly, movement by peristaltic waves seems to be incompatible with the exertion of positive thrust through the parapodia.

Another adaptive modification in the locomotor mechanism of polychaetes, closely associated (but not exclusively so) with a burrowing habit, is the reduction of the transverse septa. This is a particular example of a phenomenon that is widespread in those groups of animals that are metameric: the reduction or complete loss of the metameric repetition of certain of their organ systems. Presumably those particular facets of metamerism lost their functional advantage in the particular modes of life adopted by these animals. By considering this phenomenon we should therefore learn more about the advantages that were initially gained by a metameric organization. From this point of view a study of the habits of lugworms has proved to be particularly illuminating, for these animals, living in their familiar U-shaped burrows (Sec. 11–3), have achieved a remarkable complex of adaptive movements providing for consolidating the burrow, digging the sand and ingesting it for its contained food, irrigating the burrow with sea water for respiration, and dealing with faeces.

The burrowing of *Arenicola*, like that of earthworms, involves forwardly directed peristaltic waves of the body wall, permitting its use as a penetration and terminal anchor, but these act in *Arenicola* in conjunction with eversion and retraction of the proboscis, which comprises the papillate buccal mass and pharynx. Eversion is brought about by coelomic pressure (Fig. 6–10), applied gently and not explosively as it is in *Nephtys*; retraction is effected by the sheath of retractor muscles which suspends the proboscis from the body wall. The body wall of *Arenicola* is less firm than that of the earthworm, so that its form is maintained by a higher coelomic pressure, which fluctuates around 5×10^{-2} N cm^{-2}. Entry into the sand is effected by repeated scraping actions of the proboscis, which continue until some four chaetigerous segments have been inserted into the sand. Intracoelomic pressure remains low up to this stage, but from now on the anterior end can act as terminal anchor, and pressure is rapidly built up (Fig. 6–11) to as much as 1.5 N cm^{-2} by cyclical contractions of the circular and longitudinal muscles. The head can now be forced forwards, while the hinder segments are drawn after it. During this process, anchorage of the body to the surface of the sand is aided by the chaetae and by the erection of flanges of the body wall (Fig. 6–10), especially at the anterior end. These flanges contain rings of coelomic cavity that are isolated from the main coelom. They can thus be erected by hydraulic pressure, but, because of their isolation, they can remain erected when the pressure in the main coelom falls, and the body become thinner. It has been thought paradoxical that the calculated maximum pressure in the proboscis of a worm on the surface of the sand is of the order of 16×10^{-2} N cm^{-2}, whereas a human being may exert a pressure of 1.5 N cm^{-2} without sinking into it. How does the worm achieve its entry? One suggestion is that burrowing depends upon the proboscis movements liquefying the mixture sand and water because of its thixotropic property already mentioned. Probably, however, it is unnecessary to appeal to this; the effect of the repeated scraping action of the anterior end could be sufficient to resolve the apparent paradox.

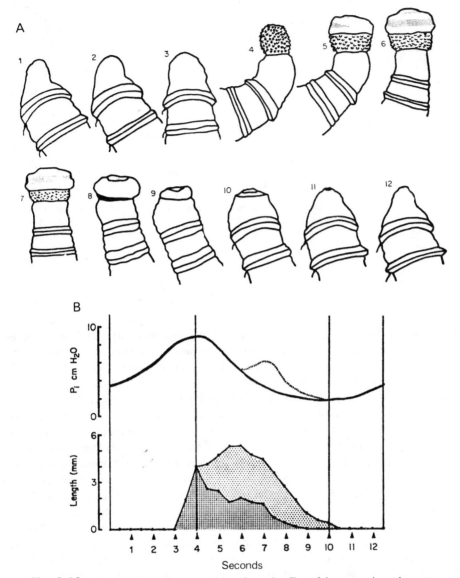

Fig. 6-10. **A**, successive diagrams, drawn from cine film, of the ventro-lateral aspect of the anterior end of a specimen of *Arenicola marina* at 1/6 sec intervals, showing proboscis eversion and retraction and the flanging of the anterior two chaetigerous annuli.
B, graphs showing (*above*) the pressure recorded from the trunk coelom, the dotted line representing the extra pressure pulse on withdrawal of the buccal mass, and (*below*) the length of the buccal mass (fine stipple) and pharynx (coarse stipple) extruded during the same period. From Seymour, 1971. *J. Zool.*, **164**, 93-132.

Such softening action does, however, seem to play a part once the worm is fully beneath the sand. Burrowing involves cyclical contractions of the muscles of the body wall, reflected in corresponding cycles of pressure change within the coelom. The cycles lengthen once the worm is covered (Fig. 6–11), the explanation of this being

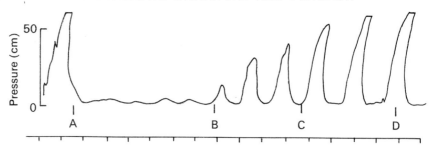

Fig. 6-11. Pressure changes in the body cavity of *Arenicola* during the beginning of burrowing into sand. The initial peak is due to insertion of the cannula. Subsequent major peaks are only seen after several segments have entered the burrow. A, head down to sand, followed by repeated proboscis eversion and penetration. B, C, D, two, three, and four chaetigerous annuli, respectively beneath the surface. The flat top of the last two peaks is due to limitation of the travel of the pen. Time mark: 5 sec. From Trueman, 1966. *J. exp. Biol.*, **44**, 93–118.

that a more complex pattern of activity now develops to provide for digging. The sequence may be considered to begin with a shortening of the body, during which the everted proboscis scrapes sand away from the end of the burrow. The proboscis is then suddenly withdrawn, so that water is drawn into the cavity that it has made, and the sand is consequently liquefied. This permits elongation of the trunk, and penetration of the head into the water space so created, a penetration anchor being formed by pressure of chaetae and flanges on the wall of the burrow. Thus an efficient digging tool is formed by interaction of proboscis, parapodia, and body wall.

The successful exploitation of the hydrostatic skeleton by *Arenicola* depends, in contrast to what we have seen of the superficially similar mode of life of earthworms, upon the reduction of the transverse septa. These have largely been lost from the trunk region, the only ones remaining being the three anterior ones, usually termed diaphragms. Lugworms are thus no longer able to confine pressure changes to those regions of the trunk in which localized muscular contraction is taking place. Against this, however, they have gained the advantage of being able to generate a large thrust through the combined action of a substantial length of trunk musculature. The gain is specifically of advantage in burrowing, provided always that the internal pressure remains below the danger limit. Outside the burrow it becomes apparent that this loss of septa sets limits on the versatility of movement of the animal. Lugworms have little capacity for swimming, or even for crawling on the substratum; in this latter respect the advantage lies with the earthworm, and its capacity for localized application of thrust.

Some of the hydrostatic adaptations of lugworms show a degree of interspecific variation (Fig. 6–12). In *Arenicola ecaudata*, for example, the anterior or head coelom is substantially isolated from the rest of the coelom, apart from the existence of a dorsal valve and ventral foramen, which are probably closed during the extrusion of the pharynx. The pressure for this extrusion is exerted through contraction of the first diaphragm (differentiated into a retractor sheath and gular membrane), so that the process is independent of pressure changes in the main coelom. In *A. marina*, on the other hand, the first diaphragm is less well differentiated, and here the general

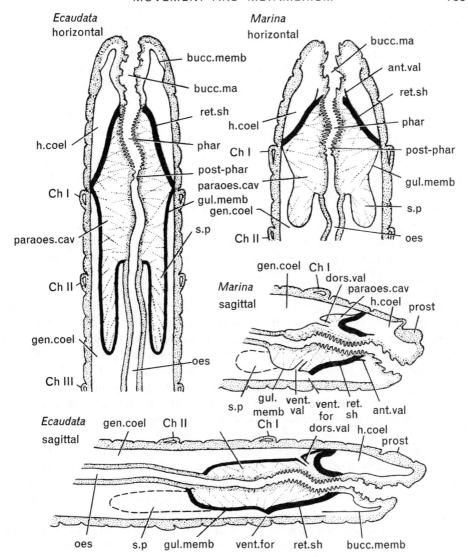

Fig. 6-12. Diagrams of the structure of the proboscis apparatus of *Arenicola ecaudata*, left and *A. marina*, right. *ant. val*, anterior valve; *bucc.ma*, buccal mass; *bucc. memb*, buccal membrane; *Ch*, chaetigerous annulus; *dors. val*, dorsal valve; *gen. coel*, general coelom; *gul. memb*, gular membrane; *h.coel*, head coelom; *oes*, oesophagus; *paraoes. cav*, paraoesophageal cavity; *phar*, pharynx; *post-phar*, post-pharynx; *prost*, prostomium; *ret.sh*, retractor sheath; *s.p*, septal sheath; *vent. val*, ventral valve. From Wells, 1954. Q. Jl microsc. Sci., **95**, 251–270.

coelomic pressure is of greater importance in extrusion, for the second and third diaphragms are incomplete and exert little restraining effect upon movements of the fluid. The structure of the tail, however, shows that the animal makes use of the advantages of metameric septa. They are well developed in this region, perhaps because they help to control the extrusion of faeces. The supposition is that they isolate the tail from the effect of changes of pressure in the continuous trunk coelom, and aid localized increases of pressure when discharge of the faeces is to take place. There could be no better illustration of the versatility of action of hydrostatic skeletons than the adaptive variations of organization that have thus become established within the limits of the body of a single polychaete worm.

6-4 LOCOMOTION OF LEECHES

That the Hirudinea are closely related to oligochaetes, and are probably derived from them, is not in doubt. It is shown in the exploitation of aquatic and damp terrestrial habitats; in the hermaphrodite reproductive system, the clitellum, and the cocoon; in the persistence of chaetae (otherwise lost in the class) in the primitive Russian leech *Acanthobdella*; and in the loss of chaetae and the development of leech-like jaws and anterior and posterior suckers in a group of parasitic oligochaetes, the Branchiobdellida. But in certain respects leeches have advanced beyond the level of oligochaete organization, and not least in features related to locomotion.

The muscle cells are peculiar, with their central sarcoplasm and peripheral contractile myoplasm, although it is not clear whether this is relevant to locomotor efficiency; possibly the isolation of the contractile region of the cell is of value in this respect. What is clearly significant, however, is the high level of development of the musculature as a whole. Outer circular and inner longitudinal muscle layers are well developed, while between them lies a system of oblique muscle fibres. These form a double layer, arranged in left and right geodesic spirals in essentially the same way as the non-extensible fibres of the body wall of nemertines and nematodes. Further, there are well-developed vertical muscle fibres, running through the other muscle layers and connecting the dorsal and ventral body walls.

Some other features are reminiscent of arthropod organization. Metamerism is reduced internally and is difficult to detect externally because of the secondary annulation, while, in contrast to other annelids, the number of segments is fixed at 33. The coelom (Fig. 6–13), well developed and still segmented in *Acanthobdella*, is reduced in other leeches as a result of invasion by connective tissue (botryoidal tissue) to a system of sinuses, of which the lateral ones are muscular vessels in the Gnathobdellida (the jawed leeches). This interconnected system, with its coelomic fluid (which contains haemoglobin in gnathobdellids), forms the blood vascular system. The original (true) blood vessels, still represented in the Rhynchobdellida (the jawless leeches with a proboscis), are completely absent in gnathobdellids.

If we pursue the logic of our analysis, it is possible to interpret most of these features as an integrated specialization for the modes of live of leeches, with their characteristically agile locomotion. These animals can move in two ways. One is by creeping, which involves alternating use of the anterior and posterior suckers, and which facilitates predation, terrestrial life, and the sucking of blood. The other is swimming, which is effected by dorso-ventral undulations. Efficient application of power in this method demands a flattened body, and this is ensured by the vertical muscle. Creeping demands efficiency in lengthening and shortening the body. This is provided for in part by the circular and longitudinal muscle fibres, but Mann argues that the spirally arranged oblique muscle fibres also make an important contribution,

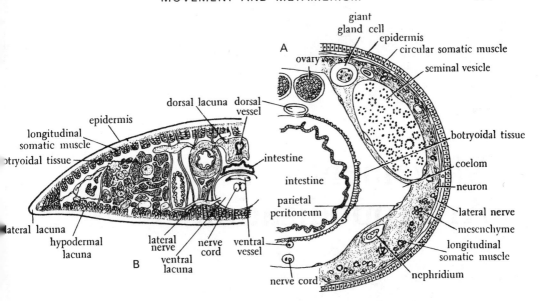

Fig. 6-13. **A** transverse section through *Acanthobdella*, showing the open coelom. From Meglitsch, 1972. *Invertebrate Zoology* (2nd ed). Oxford University Press. **B**, transverse section through a glossiphonid leech, *Placobdella*, in which the coelom is reduced to narrow passages in the botryoidal tissue. From Meglitsch, 1972. *op. cit.*, after Livanoff, from Meglitsch, 1972. *op. cit.*

for when the body is elongated they will reinforce the subsequent contraction of the longitudinal muscle, while when it is shortened they will reinforce the subsequent contraction of the circular muscles. Furthermore, they may serve to increase the hydrostatic pressure and thereby make the body more rigid. Mann suggests that this may help leeches to stand erect upon their posterior sucker and carry out the characteristic swaying movements with which they explore their environment.

Why the coelom has been reduced in leeches is not clear, although the reduction is perhaps foreshadowed in oligochaetes by the tendency to break up the coelom with the septal outgrowths that form the seminal vesicles. But one underlying factor (which has certainly determined the loss of chaetae, and may well have influenced the restriction in the number of segments as well) is that the body moves as a single unit, and is not dependent upon subdivision of the hydrostatic skeleton. The necessary deformability is provided by the botryoidal tissue and the remains of the coelom, while the segmented nervous system provides the neural basis for the agility of these animals. This last aspect we shall examine later.

7
Movement and Arthropodization

7-1 SKELETON OF ARTHROPODS

By the standards that we have discussed earlier, the Arthropoda are outstandingly successful. They have adapted themselves to water, land, and air, sometimes under most extreme conditions, in a way that bears comparison with the achievements of vertebrates, while the Class Insecta alone, with some 700,000 species, contains more species than all the rest of the animal kingdom (Fig. 1–2). Their success is due to a process that is sometimes called 'arthropodization'. This is essentially the exploitation within the group of the far-reaching potentialities that are latent in the plan of structure exemplified in annelid worms. This does not mean that the arthropods must have evolved from primitive annelids. Their origins are, in fact, unknown, for they must lie far back in Pre-Cambrian times, but we are justified in assuming that the group arose from some worm-like stock. A comparison of them with modern annelids is thus a useful basis for assessing the significance of the features to which they owe their success, some of which, as we shall see, must have been achieved independently along more than one line of evolution.

One of the most important of those features, and the key to all their subsequent history, is the development of a firm exoskeleton which can resist deformation, and which is suitable for the construction of systems of levers. The movements of worms are restricted in scope and wasteful of energy, being dependent upon contractions of a muscular body wall and the distribution of the resultant pressure changes. Often they involve the development of lateral undulations, which represent an expenditure of considerable effort in return for a comparatively slow rate of forward progression. With a firm exoskeleton, and with the jointed appendages that this allows, the muscular body wall can be broken down into separate muscles, arranged so as to ensure that their contractions bring about responses that are localized both in time and space. In principle, however, these results are also attained with the endoskeleton of echinoderms and vertebrates. What, then, is the biological significance of an exoskeleton, and what contribution has it made to the success of the arthropods?

One line of thought, which was developed by D'Arcy Thompson, bears on the use of tubular construction in the endoskeleton of vertebrates as well as in the arthropodan exoskeleton. It derives from the fact that a hollow tube, whether it be a wheat-straw, a feather quill, or a girder of the Forth Bridge, has a greater resistance to bending than has a solid rod of the same cross-sectional area. A thin sheet of paper can be rolled up into a relatively stiff cylinder. But there are many other considerations, some of which differ in their consequences for the two groups. These considerations have been analyzed in some depth by Currey. One factor is that the rigidity and, to a less extent, the strength of a hollow cylinder increase as its external diameter increases. If we consider an arthropod as a series of such cylinders, and if we accept that there must be a limit to the weight of skeletal material that an animal can manufacture and carry, we can see that there is some advantage in having the skeleton external to the soft tissues, where it can attain maximum diameter. If, however, there must be economy of material, increase in size may be expected to lead to a relatively thinner skeleton, and eventually this will become liable to buckle. This will tend to limit the size of the body.

Another factor limiting size, and one that is critical for anthropods, is that the exoskeleton cannot grow, so that, unlike an endoskeleton, it must be periodically shed by moulting. At this stage the animal must support its body with minimal skeletal aid. Aquatic forms, and particularly marine ones, will be aided in this by the buoyancy of the medium, but terrestrial ones will not. We may therefore expect that marine arthropods will reach larger sizes than terrestrial ones, and this is exactly what we find. Yet another limiting factor is that stresses resulting from impact, which are prone to injure the exoskeleton, can be shown to increase with the size of the body. The vertebrate endoskeleton is protected from such injury by the plasticity of the overlying soft tissues, which is one reason why vertebrates have evolved into large and active animals. An exoskeleton is better suited to small and active ones.

Other factors, too, must be considered in evaluating the contribution of the exoskeleton to arthropod organization. We shall see later that its composition lends itself well to waterproofing, to the achievement of strength and flexibility by tanning and calcification, and to the development of a degree of elasticity which is an important element in the success of insect flight. It seems fair to conclude, therefore, that the exoskeleton has contributed greatly, through its structural and chemical adaptability, to the success of the arthropods, while tending to limit their size, especially when they are living on land. We shall see, too, that their respiratory system, which is related in its form to the presence of an exoskeleton, has been another limiting factor in influencing the size of insects.

Arthropods, like annelids, are metamerically segmented, and so their appendages, like parapodia, form a metamerically repeated series. Within this series there has evolved a regional specialization and division of labour, leading to an increased complexity of reaction and greater efficiency of performance. A contributing factor is the extensive development of cross-striations in the muscle fibres. Except in the Onychophora, the muscles are almost entirely striated, a feature that is believed to secure increased speed of contraction and hence of response.

Two other features are also very characteristic of the arthropod group. One is the almost complete loss of cilia, presumably a result of the increased development of cuticle over the external surfaces and to some extent over the internal ones also. The other is the loss of the coelom as the main body cavity, and its restriction to vestigial spaces in the gonads and excretory organs. Consideration of the development of *Peripatus* (Fig. 12–8) suggests that this loss is closely connected with the breakdown of the blood vessels to form a continuous blood-filled cavity called the haemocoel, the heart coming to lie in this and taking in blood through openings called ostia. The haemocoel provides for a hydrostatic skeleton that maintains turgor in animals in which the cuticle remains flexible, as it does, for example, in the Branchiopoda, and in centipedes. It is also of particular importance during the moulting of animals with rigid exoskeletons, for it contributes to the distension of the body, which ensures that the newly formed skeletal covering is adapted to the shape of the growing animal. It has been suggested that in this respect the haemocoel has some advantage over a segmented coelom in that its fluid can be driven through it by the force of the heart beat, and without obstruction by metameric septa.

However this may be, it is clear that arthropods have lost much of the internal metamerism that we assume was present in their annelid-like ancestors. We have earlier suggested that the hydrostatic function of the coelomic fluid was an important factor influencing the initial establishment of metamerism. The evolution of arthropod locomotor mechanisms seems to have brought about a decreased dependence on hydrostatic properties. We may think of the haemocoel as being adaptively correlated with other factors of arthropodan organization. It has certainly evolved quite independently of the similar cavity of molluscs.

It is common to find that the body surface of invertebrates is protected by an epidermal secretion (the cuticle). This is true of arthropods, but the secretory process in these animals is of exceptional complexity, while their epidermis shows an impressive range of potentialities. Typically there is a firm basement membrane to which muscles can be attached, and which is strengthened by being linked through intracellular fibrils to the cuticle. The muscle attachments may consist of thickenings of the membrane, or of outgrowths of it that take the form of fibrillar bars. Further, the ectoderm may give rise to subepithelial connective tissue fibrils, while striated muscle may sometimes arise from it.

Fundamentally the arthropodan cuticle is a complex of protein and chitin (see below), but lipids also play an important part in its composition, as do the polyphenols and phenolases (polyphenol oxidases) that are involved in the tanning process to be described later. These various substances are distributed in two zones, an inner chitinous procuticle (often differentiated into an outer exocuticle and an inner endocuticle) and an outer non-chitinous epicuticle. In general the significance of the differentiation of an epicuticle in arthropods is that is both protects the underlying endocuticle and also provides an important physiological barrier between the animals and their external environment. The epicuticle has been of the utmost value in supporting their exploitation of life on dry land. It is already recognizable in the very primitive cuticle of the Onychophora, where electron microscopy reveals an epicuticle of four distinct layers. Its chemical composition in this particular group is uncertain, but it is probable that it contains both protein and lipid components, and it is likely that the molecular arrangement of these components contributes to the markedly hydrofuge (or 'unwettable') properties of the body surface of *Peripatus*—which do not, however, make it waterproof, so that the Onychophora cannot resist desiccation.

In other groups the epicuticle provides more complete protection. Its composition is best understood in insects (Fig. 7–1), where, as a layer which is only about 1 μm thick, it provides impermeability, but enough is known of the crustacean cuticle to show that in that group its organization is very similar (Fig. 7–2). The proteins of the insectan epicuticle form a cuticulin layer, composed of polymerized lipoprotein. According to Wigglesworth's analysis of events preceding the moult in the bug *Rhodnius*, the cuticulin is the first layer of the epicuticle to be secreted. It is penetrated by pore canals containing processes of the epidermal cells, from which a secretion is passed on to the epicuticular surface. Out of this secretion there separates, just before the moulting of the old cuticle, a crystalline wax layer which waterproofs the cuticle. Finally, just after moulting, a protective cement layer is secreted by dermal glands. This layer, which may sometimes be resinous, forms a protective outer covering. Thus the innermost layer of the epicuticle is the first to be formed, and the outermost layer the last, a mode of development that clearly depends upon the epidermal cells having channels of communication with the outside. This applies also in principle to the crustacean epicuticle, for here the cuticle is penetrated by pore canals communicating with the epidermis, and by ducts arising from tegumental glands that lie below the epidermis. The precise contribution that these make to the development of the cuticle is not, however, clearly understood. In the Onychophora there are no such communications. Here the layers of the epicuticle are presumably

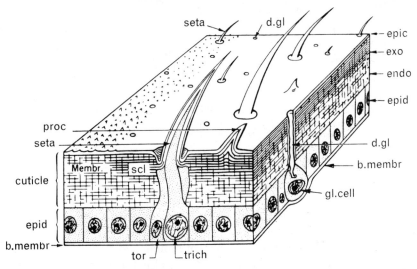

Fig. 7-1. Diagrammatic reconstruction of a block of the integument of an insect to show the general organization, drawn at boundary between a sclerite and membrane to show the fundamental similarity of soft and hard cuticles. *b.membr*, basement membrane; *d.gl*, duct of unicellular gland; *endo*, endocuticle or soft inner portion of procuticle; *epic*, epicuticle (subdivisions not indicated); *epid*, epidermal cell layer; *exo*, exocuticle or hard and usually darkened outer portion of procuticle; *gl.cell*, gland cell; *membr*, membrane; *proc*, a type of immovable non-cellular process; *scl*, sclerite or hardened and darkened area of cuticle; *seta*, the commonest type of movable (tactile) projection; *tor*, tormogen or socket-forming cell; *trich*, trichogen or seta-forming cell. From Richards, 1951. *The Integument of Arthropods*. University of Minnesota Press, Minneapolis. Copyright 1951 by the University of Minnesota, University of Minnesota Press.

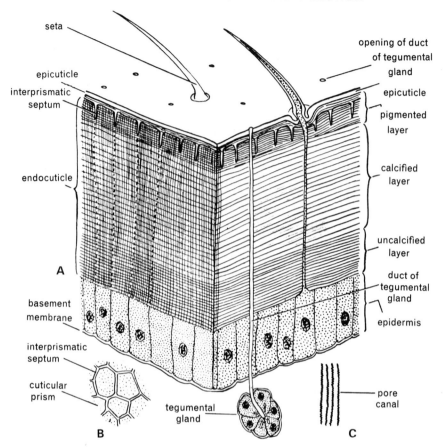

Fig. 7-2. A, diagram to illustrate the structure of the decapod cuticle as seen in vertical section. B, horizontal section through pigmented layer of endocuticle. C, pore canals as they appear in vertical sections of the epicuticle under higher magnification than (A) and (B). From Waterman (ed.), 1960. *The Physiology of Crustacea*, vol. 1. Academic Press, New York.

laid down in the order that they occupy, an illustration of its more primitive organization.

The endocuticle consists of a complex of protein and chitin. Chitin (Fig. 7–3) is a polysaccharide which was first discovered in 1811 in fungi, and which, at that time, was named fungine. It can be chemically defined as a polymer of high molecular weight, consisting of acetyl-glucosamine residues which are linked together by 1,4-β-glycosidic bonds to form long and unbranched molecules. Like cellulose, which it resembles in some of its properties, chitin is a 'structural' polysaccharide, well suited by the aggregation of its linear chains to contribute to the protection and support of the body. Its value can be measured by the prodigious quantities in which it is secreted. According to Tracey, copepods alone are estimated to produce several times 10^9 tons per year, whereas the world production of natural cellulose amounts to 10^{11} tons per year.

The term chitin has in the past been used very loosely, to refer to various types of

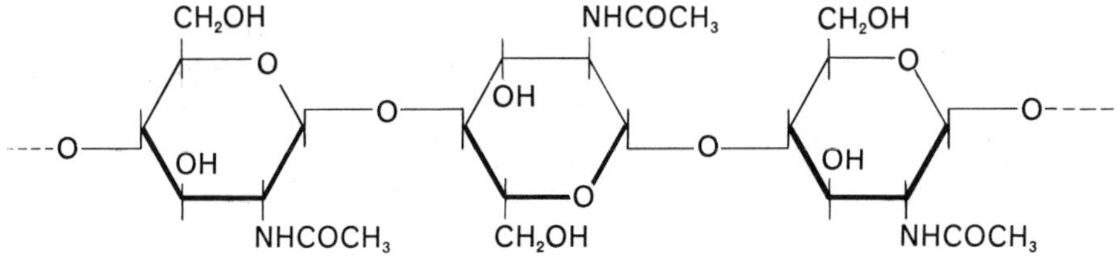

Fig. 7-3. Chitin.

hard cuticle which, regardless of their chemical constitution, have been described as 'chitinized'. This is a misleading terminology, for the hardening of the arthropod cuticle does not primarily involve chitin at all, nor is this substance confined to the arthropods. Its presence, in a precise chemical sense, can be established by specific procedures such as colour tests and X-ray diffraction studies. From such studies it is known to be widely distributed in the animal kingdom, although it does not always have the same crystallographic properties. One form, referred to as β-chitin, occurs in association with collagen, as, for example, in the chaetae of polychaetes and in the reduced shell (pen) of *Loligo*. The other form, α-chitin, is characteristic of arthropods, and is found also in fungi.

The distribution of chitin, determined by methods involving the use of purified chitinase, shows that its biosynthesis must have been a primitive property of animal cells. It is found in the cyst walls, shells, or spores of ciliates, rhizopods, and cnidosporans; in hydrozoans and other coelenterates; and in most of the protostomian groups, with the exception of platyhelminths and nemertines. In striking contrast, it is completely absent from the deuterostomian line, comprising the echinoderms, hemichordates, and chordates (Sec. 18–5). A possible exception is its presence in chaetognathans, which, according to some authorities, are perhaps related to chordates. Presumably there was gene repression at the start of the deuterostomian line of evolution. In the protostomian line, by contrast, the production of chitin has been developed to a point at which it has become a major key to arthropodan success. Chitin has not been identified in the cuticle of annelids, despite its presence in their chaetae and jaws. The contrast between annelids and arthropods is in this respect clear-cut, for in the latter group it is universally present as a continuous layer. This advance in chemical complexity of the protective covering of the body is thus an important difference between the two groups.

The cuticle may remain a soft and flexible material, with a protein component, termed arthropodin, that is predominantly soluble. This is seen, for example, in the Onychophora and centipedes, in the cuticle of larvae, in many branchiopod crustaceans, and in the intersegmental membranes that form the mobile junctions between the hard parts of toughened exoskeletons. Associated with arthropodin is another protein, resilin, which provides the elasticity that makes an essential contribution to movement, as, for example, in the hinges at the base of the wings of insects. Cuticle of this type is thus very flexible. Normally, however, much of it, particularly in adults, undergoes hardening. This is a progressive process, upon which, and upon the varying extent to which it is developed, the mechanical properties of the exoskeleton necessarily depend.

Hardening may be brought about in two ways. One of these is by the deposition of calcium carbonate, mainly as calcite but to some extent also in an amorphous form. The process is well known in the decapod crustaceans, where calcification starts in the epicuticle and then extends inwards through the endocuticle. Resorption of the calcium salts takes place in these animals prior to the moult, so that this method of hardening involves a regular cycle in calcium metabolism. Calcification is also characteristic of diplopods, in which its development in the outer endocuticle confers upon their exoskeleton a rigidity and inflexibility that is functionally correlated with their burrowing habits. Its adaptive significance is emphasized by its absence in *Polyxenus*, a diplopod that is not a burrower, but that exploits instead the roofs of crevices.

The other method of hardening, called sclerotization, involves a change in the molecular organization of the protein component of the cuticle, accompanied by a characteristic darkening in colour. This process, when maximally developed, produces an exoskeletal material that differs from unmodified chitin in being highly elastic and essentially non-plastic, so that it returns to its original shape when deforming forces are removed. Sclerotization, like the calcification of the decapod cuticle, begins in the epicuticle and is a progressive process. Local differences in the degree of hardening often reflect different mechanical requirements at specific regions of the body.

In *Peripatus*, where the exoskeleton remains very flexible, sclerotization is only weakly developed; over the body as a whole it is restricted to the epicuticle, where it may help to prevent the non-elastic cuticle from being pulled out of shape. It does, however, extend inwards to toughen the procuticle of the hard jaws and claws, and we shall see that its restriction elsewhere is adaptively correlated with the mode of life of the Onychophora. In insects this inward extension is more fully developed, resulting in an external dark sclerotized region, the exocuticle, overlying an inner light endocuticle, the latter being unsclerotized and thus more or less transparent (Fig. 7–1). A similar extension takes place in many crustaceans, but full uniformity of nomenclature of insectan and crustacean cuticles has not yet been generally agreed, nor is it certain that it could be justified by the actual composition of the two types of exoskeleton. Considerable sclerotization is also found in chilopods. The exoskeleton of these myriapods thus has an elasticity that contrasts with the rigidity of the calcified exoskeleton of the diplopods. We shall later consider the adaptive significance of this contrast.

To understand the nature of sclerotization it is necessary to appreciate its structural consequences. The unhardened endocuticle already has certain valuable properties, for the highly polymerized molecular chains of the chitin and protein complex are very strong, and provide a covering which is both firm and flexible. These latter properties would be adequate for an animal that still relied upon hydrostatic forces for its movements, and we shall shortly see these forces operating in *Peripatus* and to some extent also in centipedes. The strength of the endocuticle results from the molecular chains being grouped in lamellae that have been compared to the veneers of plywood. Within each lamella the chains are parallel with each other, but those of one lamella lie at an angle with those of adjacent ones. The flexibility of this complex of lamellae is a result of there being little bonding between adjacent molecules, and still less between adjacent lamellae, so that the cuticle can be readily bent. Stiffness can only be secured by the insertion of appropriate bonds, and it is this bonding that is effected by sclerotization. The result is the production of a highly resistant, and insoluble protein called sclerotin.

The stabilization of proteins by the introduction of cross-linkages between their molecules has been achieved by animals in two distinct ways. One of these, characteristic of the vertebrates, consists in the establishment of disulphide bonds between the cysteine residues of the molecules. This method, which is comparable with the vulcanization of rubber, yields the product called keratin. This was perhaps first evolved as a waterproofing device, although its mechanical properties have been well utilized by the vertebrates in such skeletal products as beaks, feathers, and claws.

The other method, which is particularly characteristic of arthropods, involves the use of orthoquinones to form cross-linkages between the free imino or amino groups of the protein molecules (Fig. 7-4). This process, which yields sclerotin, may be compared with the commercial tanning of collagen by benzoquinone. Under natural conditions the quinone may be formed by dehydrogenation of an *ortho*-dihydroxyphenol, the reaction being catalysed by a phenolase (or polyphenol oxidase) (Fig. 7-5). One example of this is the formation of the egg case (ootheca) of the cockroach. The two colleterial glands secrete the several components that are needed. Included in the secretion of the left gland are three substances: a protein, a phenolase, and a glucoside of protocatechuic acid. The secretion of the right gland is a clear fluid containing a glucosidase.

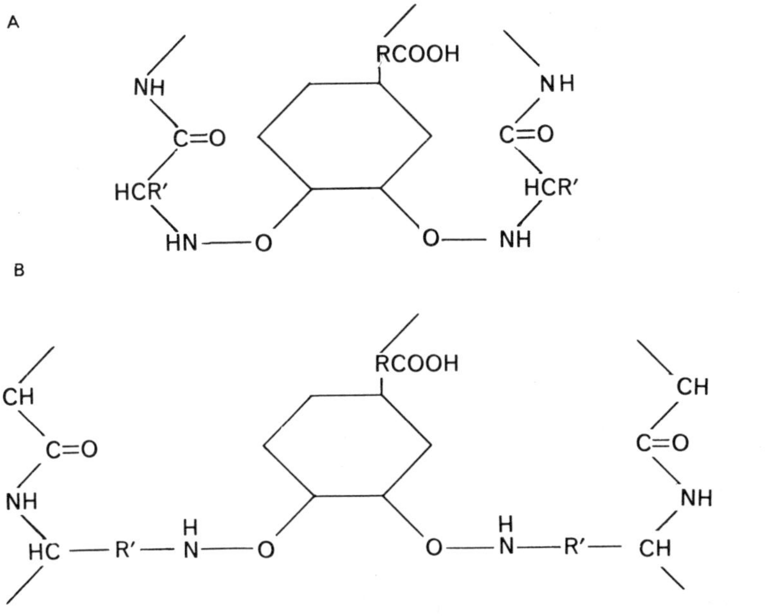

Fig. 7-4. Quinone-tanning reactions. A, quinone tanning by reaction with imino group of the main protein chain. B, quinone tanning by reaction with amino group of basic side chains.

The two secretions are mixed when they are discharged, and the glucosidase hydrolyzes the glucoside, with the release of protocatechuic acid, which is an *ortho*-dihydroxyphenol. The protocatechuic acid is oxidized to an orthoquinone by the phenolase, and the orthoquinone then reacts with the protein molecules, binding them together to form the tough sclerotin of the fully formed ootheca. The ootheca of *Locusta migratoria* is formed by a similar mechanism, with 3,4-dihydroxyphenylacetic acid (Fig. 7–5) taking the place of protocatechuic acid.

It has been thought that reactions similar to these occur in all insects during the sclerotization of the cuticle. However, the difficulties of analysis and interpretation are here greater than in the case of the cockroach because the secretions are produced in such small quantities, and it is not surprising that many features of the process are still obscure. It is certain that the pattern of events has become much diversified in its details, but it must be sufficient here to describe one other example as an illustration of this. At the end of the last larval instar of the blowfly, *Calliphora erythrocephala*, the larval cuticle becomes hardened to form the characteristic puparium. Prior to this stage the tyrosine metabolism mainly takes the form of deamination (Fig. 7–6A). Puparium formation is believed to be initiated by a metabolic change evoked by the secretion of the moulting hormone, ecdysone (Sec. 17–4). Tyrosine is now metabolized along a pathway (Fig. 7–6B) that leads to the formation of N-acetyldopamine quinone. Phenolase is active, because it can catalyze the *ortho*-hydroxylation of monophenols as well as the dehydrogenation of *o*-dihydroxyphenols; dopa decarboxylase and acetylcoenzyme A are the other enzymes involved. The quinone is believed to act as the tanning agent in the formation of the puparium; a conclusion that is confirmed by the demonstration that if isotopically labelled N-acetyldopamine is injected into the larva towards the end of the last larval instar, it is incorporated into the puparium.

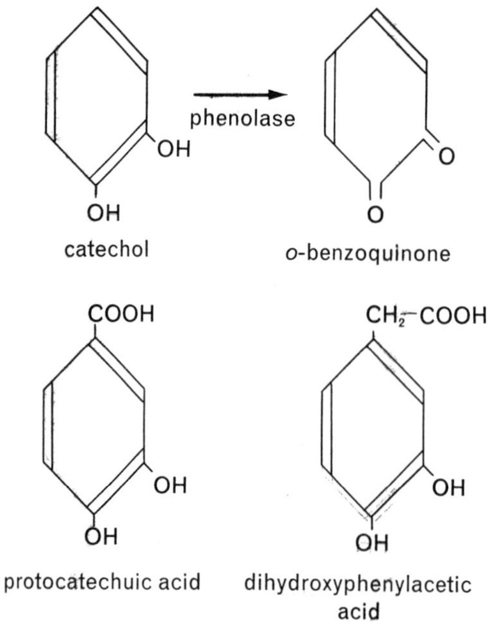

Fig. 7-5. A quinone and some polyphenols.

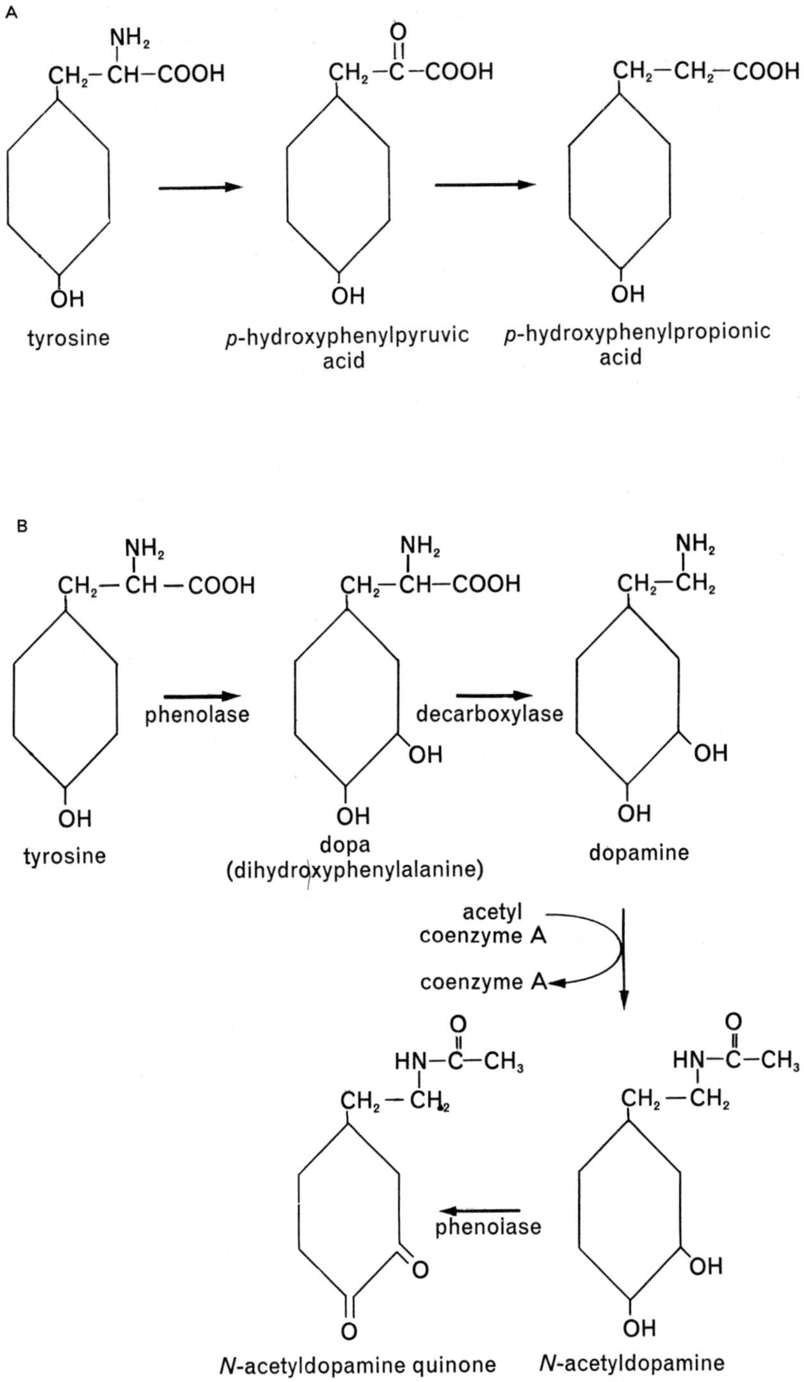

Fig. 7-6. Pathways for the metabolism of tyrosine.

Quite apart from the variation found within the insects themselves, it is probable that different groups of animals differ in their tanning mechanisms. There is certainly scope for such differences, for there is good evidence that sclerotization, in the general sense of tanning through the agency of aromatic cross-linkages, is by no means restricted to arthropods. It is not found in echinoderms or chordates, but it has been reported in the central capsule of the radiolarian *Thalassicola*, in the cysts of the eelworm *Heterodera*, in the egg capsules of the liver-fluke *Fasciola*, and in the chaetae of polychaete and oligochaete worms. It is probably widespread also in molluscs, for the byssus of *Mytilus*, the hinges of lamellibranch shells, and the radulae of gastropods, are all thought to be hardened by sclerotization.

It must be emphasized that the tests used to justify these conclusions are not specific in the sense of demonstrating that the products are identical with arthropod sclerotin; nor is the evidence sufficient to justify any firm conclusion regarding the distribution of the process in the various groups. We probably see here one of many examples of the way in which a highly advantageous biochemical feature is based upon materials of common occurrence, which in this instance are proteins containing phenolic amino acids, and the enzymes capable of oxidizing them. It is to be expected that sclerotization could arise in many different evolutionary lines, and it is all the more interesting that in the annelids the process has been restricted to isolated structures like the chaetae. The lack of any extension of it into the general cuticle is correlated with a locomotor mechanism that depends upon hydrostatic pressure and a contractile body wall.

7–2 TAGMATA AND THE HEAD

The metamerism of arthropod limbs has introduced the possibility of division of labour among them, and this has led to a great diversity of specialization. This adaptive diversification is associated with a grouping of the appendages so that those carrying out similar functions lie on adjacent segments, for this ensures efficiency of operation. The resulting regional differentiation is reflected not only in the limbs themselves, but also in the form of the segments, which is necessarily influenced by functional demands. Because of this the arthropod body becomes demarcated into clearly defined regions called tagmata. In general there is a well-defined head, where the appendages are predominantly concerned with sensory and alimentary functions, while over the rest of the body the appendages, where present, are primarily locomotory and often also respiratory. This broad subdivision of functions is by no means invariable or absolute, for there is much overlap of function between the several tagmata. Each major group, however, has its characteristic pattern of 'tagmosis', and retains this with remarkable constancy during its adaptive radiation. The composition of the insect head, for example, has remained unchanged since Devonian times. Because of this we might expect the various patterns to provide a helpful guide to the mutual affinities of the arthropod classes. To some extent they do this, but the evidence, not only from tagmosis but also from other aspects of their organization, is far from being clear-cut and conclusive.

Heads are characteristic of free-living animals that are bilaterally symmetrical. It is in this region that new stimuli are usually first encountered, so that paired receptor systems tend to become concentrated here, together with the associated regulatory centres, both neural and neurosecretory.

> Something I owe to the soil that grew—
> More to the life that fed—
> But most to Allah who gave me two
> Separate sides to my head.

Here, too, the food will be ingested, so that the mouth is located in this region, often in association with structures used in seizing and manipulating the food. The resultant specializations are expressed in the process called cephalization, or head formation. This leads in metameric animals, both vertebrate and invertebrate, to fusion and modification of originally separate and uniform segments, as a result of which the segmental composition of the head becomes increasingly difficult to determine. We may thus expect (Sec. 6–1) a segment to comprise a pair of mesodermal somites (often with coelomic cavities), a pair of coelomoducts, and a pair of nerve ganglia, while externally it should bear a pair of appendages and be delimited anteriorly and posteriorly by a transverse groove. Nevertheless, one or other of these features may be unrecognizable in the specialized body form of the adult; grooves, moreover, may be deceptive, for they sometimes have a functional explanation that is unrelated to any history of segmentation. It is thus commonly necessary to seek evidence for segmental composition in early stages of development.

Signs of cephalization are already apparent in the Platyhelminthes, where nerve ganglia (often called cerebral ganglia), eyes, and tentacles are localized at the anterior end. As might be expected, the process is still at a very early stage in these animals, and in the Acoela the cerebral ganglia are represented by no more than thickenings of the nerve plexus. In annelids the process has advanced further. In an errant polychaete such as *Nereis* a morphologically recognizable head region is formed by two regions called the prostomium, the former lying anterior to the mouth while the peristomium surrounds it.

The prostomium, which primitively contains the cerebral ganglion ('brain'), lies in front of the first true segment. In polychaetes it develops directly from the anterior (upper) region of the trochophore larva. The interpretation of the peristomium is less clear, although it has commonly been assumed that it is composed of one or more cephalized segments. In *Nereis*, for example, it bears four pairs of cirri, which are probably the remains of two pairs of parapodia belonging to two cephalized segments. It is possible, however, that more than two may be present in some polychaetes. In certain ariciid polychaetes no less than three pairs of mesodermal somites have been identified in the mouth region during early development, although no ganglia are differentiated in relation to them nor is there a corresponding annulation of the ectoderm. We can conclude that cephalization is well established in polychaetes, but that there is variation within the group, so that a final interpretation of the structure of their head is not yet possible.

The head of arthropods, which is very much more complex than that of annelids, has evolved along several lines, each characteristic of one of the major groupings within the phylum (but regarding this term, see p. 186). Analysis of the structure of the head in this phylum should therefore provide important clues to an understanding of the relationships of those groupings. On the whole, this expectation is justified, but, as in other aspects of arthropod organization, it is essential to take careful account of the possibility of parallel and convergent having occurred.

It is characteristic of arthropods that the segmentation of the anterior end of the embryonic head is restricted to the ventral and ventro-lateral regions, where developmental stages of somites, limbs, and ganglia are recognizable to varying degrees. The dorsal region is thus formed only of unsegmented tissue, which is sometimes termed the acron. Another characteristic is that the most anterior of the true segments have come to lie in front of the mouth. This is often expressed in the statement that the mouth appears to have moved backwards in the course of evolution. Tiegs and Manton have shown, however, that this situation is actually a consequence of morphogenetic movements that carry the anterior head segments forwards, while at the same time there is a backgrowth of the labrum and prostomium relative to the somite region. This brings the antennae into a position in which they can function as pre-oral sense organs, which is a highly advantageous adaptation in active animals. The pre-oral region formed in this way may be called the procephalon.

Behind the procephalon there is commonly a post-oral region, also belonging to the head. This region may be called the gnathocephalon, for here the appendages are primarily concerned with feeding. Usually it comprises three segments, so that the head is then composed altogether of six segments. There are, however, exceptions to this plan. Moreover, even when the same plan is found in different animals, this cannot be taken as necessarily establishing common ancestry. The difficulty is well illustrated by Manton's analysis of the functional morphology of the mandibles, which are commonly developed in arthropods out of a pair of appendages associated with the mouth.

Two types of mandible can be distinguished from the developmental point of view. In one of these (type A of Manton's analysis) the biting jaw is formed from the base of the appendage, while the distal part is reduced in the adult to a biramous or uniramous palp (cf. Fig. 7–9). The biting region develops out of the proximal part of the coxa (coxopodite), together with the proximal endite, a component that may be enlarged on other crustacean appendages to form a gnathobase (Fig. 10–24). Mandibles of this type are found in Crustacea and, in a somewhat different form, in Chelicerata. In the second type of mandible (type B), the jaw is formed from the whole of the embryonic limb, and not merely from its base (cf. Fig. 7–11). This type is found in the Onychophora, and also in the myriapods (where it is usually jointed), and in Insecta (where it is unjointed).

Analysis of the functioning of the arthropod mandible discloses a further complication, for it exposes a second dichotomy that cuts across the dichotomy of development. In the Onychophora the two mandibles carry out a backward slicing movement that is essentially like the movement of the walking limbs. The mandibles of other groups are more complex in their movements than this, and present two modes of action, referred to by Manton as types I and II. Type I is a rotatory and counter-rotary movement; it is carried out around a dorso-ventral axis, and is similar in principle to the movement of a coxa on a locomotor appendage of the trunk, from which movement it is doubtless derived. This type is found in the Branchiopoda and less specialized Malacostraca, and in *Petrobius*, for example, among the insects.

Type II is a stronger movement. It is carried out transversely (at right angles, that is, to the roll of type I), and is more effective for crushing, gripping, and cutting. The achievement of the type II movement must have presented mechanical difficulties, for it is necessary to secure abduction of the mandible by muscles, and this is not easily contrived when the appendage already lies at the side of the head. The problem appears to have been solved in two ways. The myriapods and Chelicerata have derived this transverse movement directly from a transversely gripping movement of the trunk limbs, whereas in Crustacea and Insecta it has been secondarily developed by independent modification of the rolling movement of type I.

We have seen that the structure of the crustacean mandible differs so fundamentally from that of insects that the jaws of these two groups (respectively belonging to type A and type B) must have been independently evolved. It follows that the establishment in these same two groups of the type II movement must also have been achieved independently. The situation is illustrated in Fig. 7–7. This shows that four types of mandible must have been evolved independently in the arthropods, and that convergent similarities have developed between crustaceans and insects (as regards direction of bite) and between certain insects and myriapods (as regards the evolution

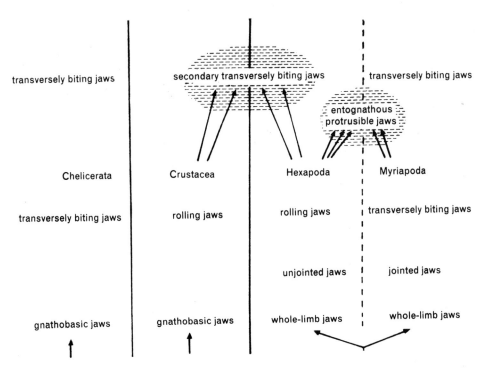

Fig. 7-7. Diagrams showing the conclusions reached by Manton concerning the distribution of the principle types of mandibles or jaws. The heavy vertical lines indicate an entire absence of common ancestry between the jaws referred to on either side; an interrupted vertical line indicates separate evolutions of the jaw mechanisms of Hexapoda (Insecta) and Myriapoda which probably had a common origin; and the shaded areas indicate mandibular mechanisms showing convergent similarities derived from unlike origins. From Manton, 1964. *Phil. Trans. R. Soc. B*, **247**, 1-183.

of protrusible jaws). This is a good illustration of parallel and convergent evolution in operation, and we shall later find that similar considerations apply in the analysis of other aspects of arthropodan evolution. So we are led to the conclusion that this vast group must be polyphyletic. Recent support for monophyletic arthropodan phylogeny, argues Manton, has been founded on 'abundant phantasies and errors of fact'. Her interpretation, which will be adopted here as a basis for analysis, is that there are three main lines, each perhaps equivalent to a phylum, and comprising respectively the Crustacea, the Uniramia (Onychophora, Myriapoda, and Insecta), and the Chelicerata (Merostomata and Arachnida). The Pycnogonida are probably chelicerates, while the positions of the Tardigrada, Pentastomida, and Trilobita remain uncertain. Arguments in favour of this point of view will emerge in the subsequent discussion, and it will prove helpful to keep it in mind throughout.

We can now consider the arrangement of the segments of the head in the main groups, as shown in Table 7–1. A primitive form of head is to be expected in the Trilobita (Fig. 7–8), marine forms that were dominant in the early Palaezoic but finally became extinct in the Permian. The expectation is justified, although the data give little help in determining the affinities of the group. One pair of antenna-like pre-oral appendages is present, probably associated with the second segment, and therefore homologous with the antennules of crustaceans. The remaining appendages of the body are biramous, the outer ramus bearing gill filaments, while the inner is a walking leg. They differ very little from each other, so that they give no clear basis for establishing the posterior limit of the head. Certain external features, including a facial suture and genal spines, do, however, make it possible to delimit a presumed head from a trunk, and on this basis four pairs of post-antennary appendages are regarded as cephalic ones. They pass into the series of trunk appendages without interruption or sudden change of form, so that from the functional aspect the head of these animals is very imperfectly defined.

A case for associating the trilobites with the crustaceans has been based largely upon supposed resemblances in their appendages. The evidence, however, is by no means convincing, as we shall see later, and in any case the head has evolved very much further in crustaceans. In these animals it includes a procephalon of three pre-oral segments, the first of which, the pre-antennary segment, has no appendages; its coelomic sacs, however, have been detected during development. The remaining two pre-oral segments bear sensory appendages: the antennules (first antennae) and the antennae (second antennae). This procephalon may be thought of

Table 7–1 Appendages of the first seven segments of arthropods

Somite	Onychophora	Trilobita	Crustacea (Malacostraca)	Myriapoda (*Scolopendra*)	Insecta	Arachnida
1	Preantennae	?	Embryonic	Embryonic	Embryonic	Embryonic
2	Mandibles	Antennae	Antennules	Antennae	Antennae	Chelicerae
3	Oral papillae	Biramous limbs	Antennae	Embryonic	Embryonic	Pedipalpi
4	Legs	Biramous limbs	Mandibles	Mandibles	Mandibles	Legs
5	Legs	Biramous limbs	Maxillae (1st)	Maxillae (1st)	Maxillae (1st)	Legs
6	Legs	Biramous limbs	Maxillae (2nd)	Maxillae (2nd)	Maxillae (2nd)	Legs
7	Legs	Biramous limbs	Maxillipeds	Maxillipeds	Legs	Legs

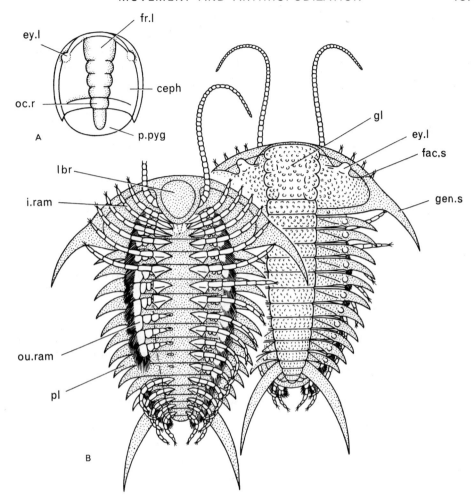

Fig. 7-8. A, diagram of protaspis larva after Whittington, 1957. B, reconstruction of the trilobite *Ceraurus pleurexanthemus* Green, Ordovician, 27 mm, after Størmer, 1951. *ceph*, cephalon; *ey.l*, eye lobe; *fac.s*, facial suture; *fr.l*, frontal lobe; *gen.s*, genal spine; *gl*, glabella; *i.ram*, inner ramus of leg; *lbr*, hypostome (labrum); *oc.r*, occipital ring; *ou.ram*, outer ramus of leg; *pl*, pleural fold; *p.pyg*, protopygidium. Adapted from Tiegs and Manton, 1958. *Biol. Rev.*, **33**, 255-337. Used by courtesy of the Cambridge Philosophical Society.

as the primitive head of crustaceans, with which other segments become associated to form a gnathocephalon of variable constitution. This variability suggests a certain parallel with vertebrates; in both groups the progress of cephalization has tended to shift the posterior limit of the head backwards.

A well-defined feature of crustaceans is the development of mandibles on the fourth segment. This is the first segment of the gnathocephalon, which is typically completed by two more segments, bearing the maxillules (first maxillae) and the maxillae (second maxillae). The form of these appendages varies in relation to the

188 INVERTEBRATE STRUCTURE AND FUNCTION

feeding habits of the group concerned, but they are commonly somewhat lobate. In some crustaceans (e.g. *Mysis*) there is a distinct mandibular groove immediately behind the mandible; this has been thought to suggest that the primitive crustacean head may at one stage have ended at this level, but in fact the groove is a mechanical necessity correlated with the mandibular musculature. In other forms (e.g. amphipods, isopods) the first thoracic segment fuses with the head, a situation that is functionally associated with the use of the appendages of this segment (maxillipeds) in feeding (Fig. 7–9). In the Eucarida there are as many as three pairs of maxillipeds, all associated in some way with the feeding activities of the head appendages. Since the carapace of these animals arises from the maxillary (sixth) segment it is generally accepted that this segment marks the true morphological limit of the head region. However, there is some uncertainty because the division of labour among the appendages of crustaceans has not established a sharp functional differentiation between the head and the anterior segments of the remainder of the body. This will become even more apparent later when we consider the filter-feeding mechanisms of these animals.

Cephalization within the Uniramia assemblage has clearly advanced least in the Onychophora (Fig. 7–10), for in these animals the pre-oral shift of segments is minimal. Only the lateral parts of the first (antennal, or, strictly, pre-antennal) segment become pre-oral, while the jaws, which are on the second segment, remain lateral to the mouth. With only one pair of feeding appendages, the head may be termed monognathan. All of this is in accord with the generally primitive level of organization of the group, which we shall see illustrated further when we examine its locomotor mechanisms. The oral (slime) papillae are a specialization, but other primitive features are the persistence of appendages on the first segment, and the

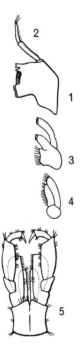

Fig. 7-9. *Left*, The mouth appendages of *Gammarus*. *1*, left mandible; *2*, palp; *3*, 1st maxilla of left side; *4*, 2nd maxilla of left side; *5*, maxilliped of each side together forming an underlip. From Shipley and MacBride, 1904. *op. cit.*

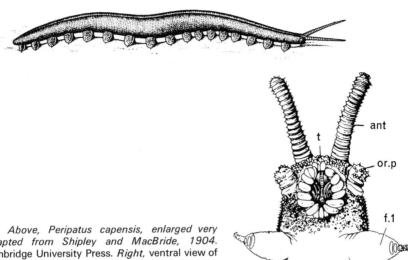

Fig. 7-10. Above, Peripatus capensis, enlarged very slightly. Adapted from Shipley and MacBride, 1904. Zoology. Cambridge University Press. Right, ventral view of the head of P. capensis. ant, pre-antenna; or.p, oral papilla; f.l, first leg; t, tongue. From Sedgwick, 1922. Cambridge Natural History, vol. 5. Macmillan, London.

form of the mandibles, which bear some resemblance to the claws of the trunk limbs in biting at the tips instead of at the sides. The leg-like movement of these appendages, to which we have earlier referred, may also be primitive, but is adaptively correlated with the onychophoran habit of squeezing crevices. The same may be said, incidentally, of the smooth muscle of the body wall. By suffering great changes of length, it permits correspondingly great changes of shape, while, because of its slow responses, it serves well to moderate changes in hydrostatic pressure. But the jaws have striated muscle, precisely suited for their functioning, and showing how readily the nature of the muscular tissue can be adapted to local needs.

Myriapoda and insects have a plan of head structure that in certain respects resembles that of crustaceans, notably in having a procephalon of three segments, of which the pre-antennary segment bears no appendages and the second segment bears antennae. The third segment, however, in contrast to that of crustaceans, bears no appendages. The head is completed by a gnathocephalon of two (dignathan) or three (trignathan) segments in Myriapoda, and there are three also in insects (Fig. 7–11), bearing respectively the mandibles, the first maxillae and the second maxillae (fused in insects to form the labium). However, despite the presence of mandibles on the fourth segment in all these groups, the fundamental difference between the mandibles of Crustacea and Uniramia means that the development of a three-segmented procephalon in both lines must be ascribed to convergence. It is thus no longer possible to sustain the view that crustaceans, myriapods, and insects should be classified together as the 'Mandibulata'. In any case, the remainder of the crustacean gnathocephalon is not closely comparable with the myriapod and insectan type, from which it differs in the form of its appendages and in the degree of its variability. The head appendages of isopods and amphipods admittedly show a superficial resemblance to those of insects, particularly in the form of the maxillipeds, which recalls the insect labium Figs. 7–9, 7–11. These appendages, however, arise in the crustaceans from the seventh segment, and it is certain that this particular resemblance must be

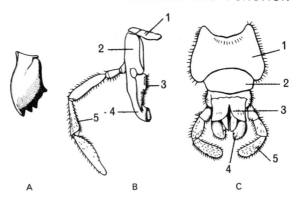

Fig. 7-11. *Below*, Mouth appendages of *Blatta*. A, mandible. B, first maxilla; *1*, cardo; *2*, stipes; *3*, lacinia; *4*, galea; *5*, palp. C, right and left second maxillae fused to form the labium; *1*, submentum; *2*, mentum; *3*, ligula, corresponding to the lacinia; *4*, paraglossa, corresponding to the galea; *5*, palp. From Shipley and MacBride, 1904, *op. cit.*

another result of convergence. It has been attributed to the influence of a bottom-dwelling mode of life upon the feeding mechanism.

As regards the three groups of Uniramia, each has a characteristic type of whole-limb mandible (Fig. 7–7). According to Manton, no one type could have given rise to either of the others, if functional continuity were to be maintained during the transition. Even within the Symphyla, a small group of myriapods often regarded as particularly close to insects, the mandible is myriapodan in type, and does nothing to bridge the gap between myriapods and insects. The logical conclusion must, therefore, be that the uniramian line had initially the common feature of a whole-limb mandible, differing in this respect from the crustacean line. It must then have diverged very early into the onychophoran, myriapodan, and insectan lines, in which jaw structure was one of the many differentiating features.

Finally, the organization of the head region of the Chelicerata is so anomalous as to suggest that these animals form a natural group with no obvious affinity with any of the others. There is no clearly defined head in the sense in which we have so far described it. Instead, the head region is incorporated into an anterior tagma, called the prosoma, which comprises eight segments. No appendages are found on the first segment, but there is some evidence for coelomic sacs in the embryo. Prehensile chelicerae are developed on the second segment, taking the place of the antennae that are found here in all other arthropods except the Onychophora.

A distinctive peculiarity of the Chelicerata, although one that is seen also at a more primitive level of organization in the trilobites, is the absence of mandibles. In the chelicerates these are replaced functionally by gnathobases, borne on the anterior appendages. A gnathobase may be defined as a jaw process that has the function of assisting, by apposition to its fellow on the opposite side, in seizing and moving particles that may be introduced into the mouth. Most chelicerates are fluid feeders, and in some of them the crushing action of the gnathobases is used to extract juices from the prey. An exception to this is *Limulus* (Fig. 7–12), the only living member of the Subclass Xiphosura; it ingests solid food, and, through the agency of the gnathobases, can make use of either soft or hard material. Soft food is held by this animal

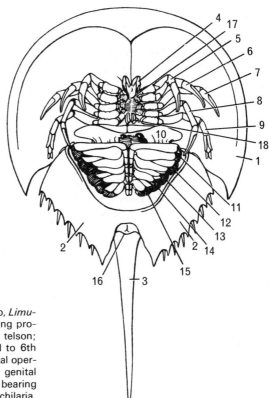

Fig. 7-12. Ventral view of the King-crab, *Limulus polyphemus*, ×½. *1*, carapace covering prosoma; *2*, mesosoma and metasoma; *3*, telson; *4*, chelicera; *5*, pedipalp; *6, 7, 8, 9*, 3rd to 6th appendages, ambulatory limbs; *10*, genital operculum turned forward to show the genital aperture; *11, 12, 13, 14, 15*, appendages bearing gill-books; *16*, anus; *17*, mouth; *18*, chilaria. From Shipley and MacBride, 1904, *op. cit.*

in a ventral depression of the body; there it is chewed by the gnathobases (borne on all of the walking limbs), and the shredded material is passed forwards to the mouth. Hard material is gripped between the chilaria, pressed forwards by the genital operculum, and cracked by the gnathobases.

Manton, in her account of *Limulus*, emphasizes the unusual features of this method of feeding. Contrary to the general tendency of arthropods to develop mandibles adjacent to the mouth, it is the more posterior gnathobases of *Limulus* that are the strongest and most jaw-like. Further, as we have noted, the feeding movements are directed transversely, at right angles to the antero-posterior swing that provides for locomotion, so that feeding and locomotor movements are mutually exclusive. The feeding methods of *Limulus*, like so much else in its organization, may be regarded as an adaptation to burrowing into the upper layers of the substratum. Their efficiency may be judged from the report that when a single animal, only 3 in. wide, was placed on a limited area of substratum with 100 ½-in. clams, it removed 99 of them in 72 hr.

The third segment of chelicerates, which lies behind the mouth, bears appendages called pedipalps, In *Limulus* these are sensory or prehensile in function, while in spiders they are also involved in reproductive activity. They are followed by four pairs of limbs which are primarily ambulatory, although, as we have just seen, they are important also in feeding. The eighth segment of the prosoma is visible only in the embryo, but it is represented in the adult by the chilaria.

7-3 LOCOMOTION AND THE CRUSTACEAN LIMB

Tagmosis in the post-cephalic region of the arthropod body is closely related to modes of locomotion. Its interpretation raises complex issues, not only of phylogenetic history but also of the factors that influence the functioning of locomotor appendages as individual units and in relationship to each other. These factors are best considered in the predominantly terrestrial groups, which have been analyzed in depth by Manton. It will be convenient first, however, to refer to the Crustacea, which have mostly retained the aquatic mode of life that must have characterized the first arthropods. Within this habitat they have attained none the less a remarkable degree of diversity, and their limbs show the versatility with which metameric structures can respond to the demands of feeding and locomotion.

Many smaller crustaceans are filter feeders. We shall see later that this type of activity depends upon the simultaneous use of limbs for swimming and for filtering water. These limbs, which move like paddles, are typically of the kind called phyllopodia (Fig. 10–22), with an axis that bears foliaceous outgrowths, arranged as a row of endites on the medial side, and one or more exites on the outer side. Often the cuticle of these limbs is thin, and their shape is largely maintained by blood turgor, The movement of these delicate structures is integrated in a metachronal rhythm, and it is partly because of this that the filter-feeding crustaceans are small, for the locomotor efficiency of this type of limb is at its highest in small organisms. The larger size that is found in many malacostracans seems likely to have emerged from patterns of adaptation associated with predation and the utilization of bulky food material (macrophagy, Sec. 9–2). Weapons have been acquired by the development of powerful chelipeds, and the protective value of the cuticle has been increased by heavy calcification. Also associated with this mode of life is the evolution of walking by means of legs, in the form of slender, biramous, non-foliaceous appendages called stenopodia.

Phyllopodia (Fig. 7–13) bear a certain resemblance to parapodia in their lobate form, delicate structure, and dependence upon turgor. Because of this, and also because of their presence in the primitive Branchiopoda, it has been argued that they represent the primitive type of crustacean limb. An alternative view, however, is that stenopodia may be the more primitive of the two types, one reason for this idea being that they occur in the nauplius larva (Fig. 19–7). By itself this consideration could not carry much weight, for, as we shall see later, larval organization is a very uncertain basis for phylogenetic speculation. Another and more substantial reason is that the stenopodial type of limb is present in trilobites, where (as mentioned earlier) it consists of an inner seven-jointed walking ramus, and an outer branchial one with a fringe of respiratory filaments (Fig. 7–8). Clearly it differs in detail from the typical crustacean stenopodium; furthermore, there is no good evidence that it bore gnathobases, which is another point of difference. Nevertheless, there are certain Mid-Cambrian fossils that combine a generally crustacean appearance with a trilobite type of limb, so that the possibility of some relationship between trilobites and crustaceans cannot be wholly dismissed. But even if the relationship be admitted (and many palaeontologists have favoured it), it still does not justify the conclusion that the trilobite limb is a primitive forerunner of the crustacean one.

The fact is that the question whether the phyllopodium or the stenopodium is

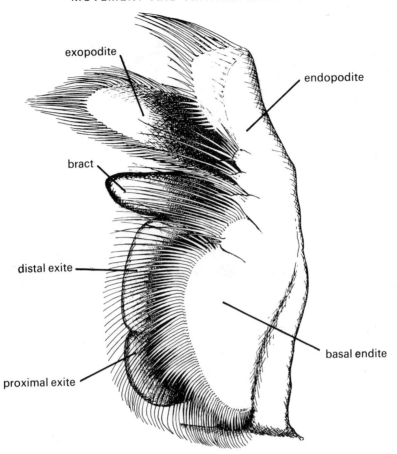

Fig. 7-13. Median view of a trunk limb of *Chirocephalus* to show its normal arrangement on the body. From Cannon, 1928. *Trans. R. Soc. Edinb.* **55**, 807–822.

the more primitive is the type of zoological dispute which presents a problem in an artificial and oversimplified way. Common sense suggests that the ancestral limb is more likely to have been of a generalized and simple type, giving a basis for the subsequent adaptive radiation of the microphagous and macrophagous ways of life. Happily, common sense is supported by evidence from two sources.

One of these is *Lepidocaris rhyniensis* (fig. 7–14), an extinct crustacean from the mid-Devonian, which had both types of appendage. The anterior ones were foliaceous with gnathobases, but further back on the thorax there was a change to biramous appendages without these (figs. 7–14, 7–16). Gnathobases develop in phyllopodia out of the proximal endite. In many living crustaceans they make an important contribution to filter feeding, and Cannon, in his analysis of the well-preserved material of *Lepidocaris*, showed that they may have done so in that animal as well.

The other source of evidence is the cephalocarid *Hutchinsoniella macracantha* (Fig. 7–15). This is a primitive living crustacean, discovered in 1955 in soft subtidal

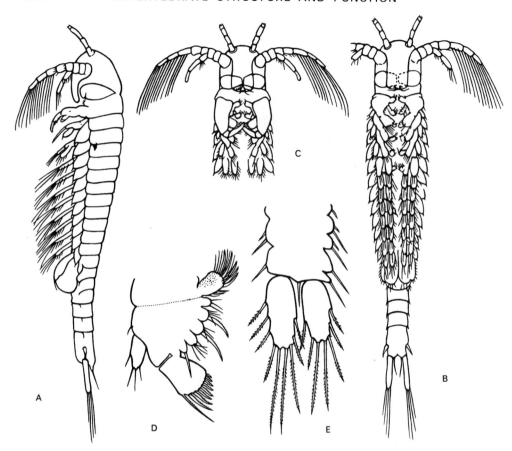

Fig. 7-14. Reconstructions of the crustacean *Lepidocaris rhyniensis* Scourfield, 3 mm, Devonian. A, female from the side; B. female from below; C, head of male, ventral view; D, one of the first pair of trunk limbs; E, one of the posterior (7th? to 11th) pairs. After Scourfield, from Tiegs and Manton, 1958. *op. cit.* Used by courtesy of the Cambridge Philosophical Society.

mud of the North American Atlantic seaboard, and of exceptional importance because *Lepidocaris* is its closest known relative. The trunk limbs of *Hutchinsoniella*, borne on only the first nine segments of the trunk, provide some additional evidence for a relationship between trilobites and crustaceans. This is the only known crustacean which resembles trilobites in having limbs with seven-jointed endopodites, although it differs in having a flattened accessory lobe (pseudoepipodite) and a coxal gnathobase. Furthermore, these limbs, which are similar in form, which beat metachronally, and which are used for feeding and locomotion, meet very well the requirement for a generalized structure from which more specialized types could be derived. Sanders shows that their evolution can be visualized along two possible paths (Fig. 7–16). One could lead to a primitive type of biramous malacostracan limb, by the loss of the gnathobase and of a joint of the endopodite. The other line could lead to the limbs of *Lepidocaris* and to the foliaceous ones of branchiopods such as *Branchionecta*.

Fig. 7-15. *Hutchinsoniella macracantha.* From Sanders, 1957. *Syst. Zool.,* **6**, 112–128.

One other aspect of crustacean tagmosis that merits attention is the establishment in the Malacostraca of the plan of organization called the caridoid facies. This is the body form in which the head is followed by a thorax of eight segments, with limbs that are used for walking, and that also bear natatory exopodites and respiratory epipodites. The body is completed by a third tagma, the abdomen. This is composed of six abdominal segments, five of which bear swimming pleopods, used as paddles, while the sixth bears uropods that form, with the post-segmental telson, the tail fan.

The origin of the caridoid facies is unknown. We shall see later, however, that certain functional considerations operative in terrestrial arthropods may lead to a reduction in the number of walking legs as these become longer. Because of this, reduction has taken place in chelicerates, insects, and in certain myriapods, and the malacostracans probably provide another example. This interpretation implies that the caridoid facies first evolved as an adaptation to the bottom-dwelling life and ambulatory movement that we have already envisaged, but evidently it did not restrict the group to that habitat. Mysids and euphausiaceans, in which the caridoid facies is well defined, are free-swimming and filter-feeding forms. Probably, then, there was a divergence of habit at an early stage of malacostracan evolution, with some lines perfecting the bottom-dwelling mode of life while others exploited the possibilities of pelagic life and filter feeding. In the later stages of this divergence the abdomen of bottom dwellers, with its natatory appendages, became less important, as is shown by the great reduction that it has undergone in crabs.

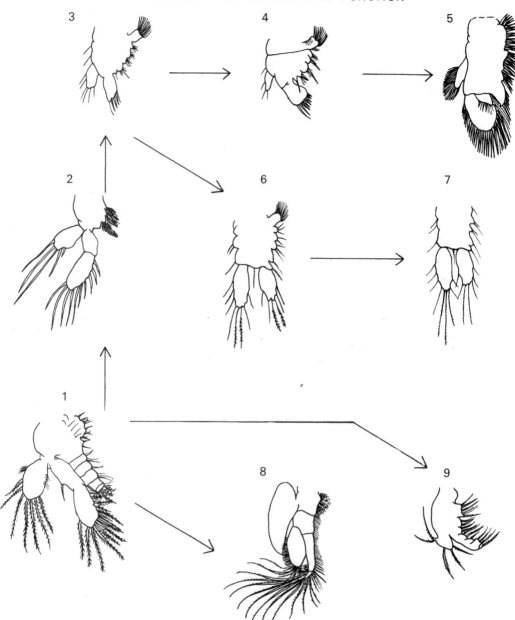

Fig. 7-16. Scheme showing the possible evolution of the various crustacean limbs from the cephalocarid limb. 1, Fourth thoracic limb of *Hutchinsoniella*; 2, eighth thoracic limb of the same; 3, second or third thoracic limb of *Lepidocaris*; 4, first thoracic limb of the same; 5, foliaceous thoracic limb of *Branchionecta paludosa*; 6, fourth to sixth (?) trunk limb of *Lepidocaris*; 7, biramous copepodan seventh (?) to eleventh trunk limb of the same; 8, malacostracan thoracic limb as exemplified by *Nebalia bipes*; 9, second maxilla of a penaeid larva. All limbs oriented with endopodite on the right. From Sanders, 1957. *op. cit.*

Crustaceans, although well able to develop walking legs, have shown remarkably little facility for the exploitation of the land. At least four independent attempts have been made from the littoral zone, originating in the isopods, amphipods, anomurans, and brachyurans. The mechanical practicability of the adventure is sufficiently illustrated by the amphibious isopod *Ligia*, which is the common shore slater that lives just above the tidal zone. This animal uses essentially the same walking mechanism on land as in the sea, so that the mode of functioning of the appendages of the aquatic members of the group is presumably pre-adapted to terrestrial life. There is here a parallel with the evolutionary history of the vertebrates, for the structure of the rhipidistian fin showed the essential pattern of the pentadactyl limb before the emergence of the Amphibia.

We shall see later that the terrestrial limitations of crustaceans are essentially physiological ones, involving, for example, failure to achieve adequate modification of nitrogen excretion or of water balance. Any explanation of this can only be hypothetical, but it may be significant that these various terrestrial lines all originated in specialized shore-dwelling groups. Perhaps the very success of crustaceans as aquatic animals has militated against the establishment in the group of any alternative mode of life; probably those that did attempt life on land were already too specialized to develop an adequate range of new adaptations. Another possibility is that limitations in their gene complexes resulted in a lack of suitable mutations, but this is difficult to reconcile with the widespread occurrence of parallel and convergent evolution in the arthropods.

7-4 LOCOMOTION IN TERRESTRIAL ARTHROPODS

Our understanding of the functioning of the locomotor mechanisms of terrestrial arthropods reveals *Peripatus* to be an animal of key importance, showing potentialities that were to be realized more fully as the 'arthropodization' of the phylum advanced further. Sclerotization is only feebly developed over much of the cuticle of this animal. As we have shown, this is not due to any inability to develop the necessary biochemical mechanism, for a tough and well-sclerotized procuticle is present in the jaws and claws. In fact, the retention of a soft and flexible cuticle over the general body surface is part of an adaptive complex that includes also the retention of smooth muscle, with its capacity for extensive deformation. Because of this the cuticle is highly permeable, and the animal is consequently unable to resist desiccation. In one sense this permeability, like that of Amphibia, is a serious limitation in a terrestrial animal. It undoubtedly restricts the range of habitats that are open to the Onychophora, yet within their chosen range they nonetheless show a high level of adaptive organization. A much-favoured habitat of *Peripatus* is the deeper parts of decaying logs. Here it profits from its soft cuticle and the deformability of its smooth muscle, for these features enable it to penetrate through narrow passages into concealed cavities where it is protected both from loss of water and also from the pursuit of predatory arthropods. Surprisingly, it can pass through holes as small as one-ninth of the area of its resting cross-section. We have seen in other invertebrates how deformability is associated with some form of fibrous mesoglea. This is true also of *Peripatus*, which has as one of its characteristic specializations a well-developed connective tissue layer underlying the ectoderm, and providing for the insertions of the muscles. A similar adaptation is also found in myriapods.

The Onychopora are sometimes thought of as a primitive group that has managed to survive 'in spite of' its primitive features, as though they were an example of 'Mir war auf dieser Welt das Glück nicht hold'. But to think in this way is to raise a difficult question: why has selection pressure not led to the replacement of primitive features by more advanced and less limiting ones? The fact is that this point of view, as Manton has emphasized, is a mistaken one. It is self-evident that *Peripatus* represents an early stage of arthropodization, yet its life is not a struggle against structural and physiological limitations. We may rather say that these have been accepted, and that the animal is successfully exploiting them through the agency of a highly specialized behaviour pattern. After all 'if a man does not keep pace with his companions, perhaps it is because he hears a different drummer'.

It is reasonable to derive the Onychophora and hence the uniramian line, from worm-like animals in which parapodia-like limbs became directed more ventrally instead of laterally. The mid-Cambrian fossil *Aysheaia* (Fig. 7–17) is significant in this regard, for it closely resembles onychophorans in the form of its body and limbs. It was marine in habit, so that it may very well represent a stock, possibly ancestral to the Onychophora, in which the limbs had already become pre-adapted for terrestrial locomotion.

However, the possible relationship between onychophoran limbs and the parapodia of worms needs critical consideration, as Manton has emphasized. The limb is well provided with muscles that are responsible for its forward, backward, upward and downward movement (e.g. the remotor muscles in Fig. 7–18), and also for changes in its shape and for movements of the claws. But these muscles have to work against hydrostatic pressure, and this probably accounts for the form of the haemocoel, which is subdivided in a way permitting a flow of blood (arrows in Fig. 7–18) between the perivisceral haemocoel and the paired lateral haemocoels, and between these and the haemocoels of the legs. This flow can be regulated by various

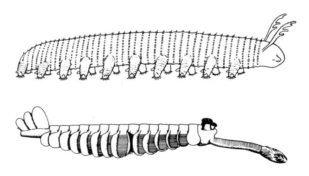

Fig. 7-17. *Above, Aysheaia pedunculata* Walcott, 50 mm, from the Middle Cambrian. A conjectural restoration, after Hutchinson. From Tiegs and Manton, 1958. *op. cit.* Used by courtesy of the Cambridge Philosophical Society. *Below, Opabinia regalis* Walcott, also from the Middle Cambrian. A reconstruction, showing the animal in lateral view, with the lateral lobe and gill of segment 7 cut off proximally to show the lateral lobe and gill of segment 8. From Whittington, 1975, *Phil. Trans. Roy. Soc. B,* **271**, 1–43.

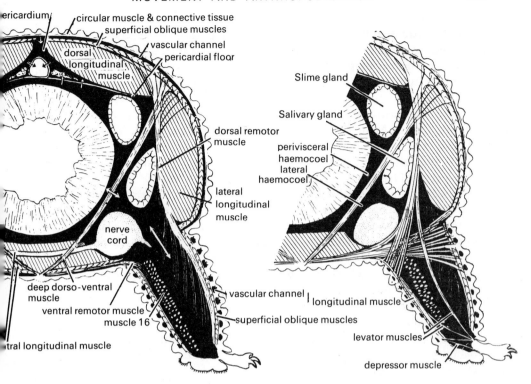

Fig. 7-18. Diagram, based upon *Peripatopsis*, to show features associated with the mode of action of the lobopodial limb. Cut longitudinal muscles are cross-hatched, the nerve cord is stippled, and haemocoelic spaces are black. The superficial ectoderm, subcutaneous connective tissue and circular muscles are white. The segmental organs and leg glands, which lie in the haemocoelic space dorso-lateral to muscle 16, are omitted, except for the exit ducts. The figure shows a transverse thick section passing through a leg and viewed from in front. From Manton, 1967. *J. nat. Hist.*, **1**, 1–22.

devices, including the arrangement of the deep dorso-ventral muscle. This particular muscle contains ostia, and it can also isolate the limb haemocoels so that they can be hydraulically independent of each other.

The polychaete parapodium, although resembling the onychophoran limb in being hollow and possessing intrinsic musculature, differs fundamentally from it in having the aciculum and its muscles as the primary effectors in a power stroke. Unlike the onychophoran limb, it has little capacity for change in shape or in length, although its action may depend in part upon changes in coelomic pressure.

These considerations have led Manton to support and develop Snodgrass's concept of two types of worm: lobopods and chaetopods. The latter comprise polychaete-like animals, with a coelomic body cavity, and with parapodia (chaetopodia) depending upon the acicular mechanism. The lobopods are a hypothetical type of worm, conceived to have had a haemocoelic body cavity, and parapodia (lobopodia) that made use of changes in hydrostatic pressure in the haemocoel. The Onychophora could have evolved from such a group, whereas they seem unlikely to have done so from chaetopod worms, for lobopodia have not developed in polychaetes, even amongst those (such as *Harmothöe*) which are able to walk upon their neuropodia. We are therefore led, on this interpretation, to the conclusion that the Onychophora, and presumably the Uniramia as a whole, must have arisen from a hypothetical group of segmented, worm-like ancestors, perhaps ancestral also to annelids, which have left no fossils or living record. The relationship of Crustacea and Chelicerata can only be a matter for speculation. However, on the general argument adopted here, we must conclude that a hard and jointed exoskeleton evolved in these two groups independently of the Uniramia. This conclusion is in accord with what we have learned earlier of the selective advantages of this type of skeleton, for these could well have promoted such parallel evolution.

What might these hypothetical ancestors of arthropods have looked like? A possible clue is provided by a strange Cambrian fossil, *Opabinia regalis*. This enigmatic animal (an earlier version of the reconstruction shown in Fig. 7–17 is said to have been greeted by the Palaeontological Association with loud laughter), which was supposedly benthic in habit, was covered by a very thin cuticle and had a body composed of two main regions. One of these was a head with two pairs of stalked eyes and a sessile one, and with a frontal process that perhaps trapped food and conveyed it to the mouth. The other region comprised a trunk of 15 segments, ending in a tail fan. Each of the segments bore a pair of lateral lobes to which were attached (expect in the first segment) lamellate gills. *Opabinia* was formerly interpreted as a branchiopod crustacean, of the order Anostraca. It is now thought to have been neither arthropod nor annelid, but perhaps a representative of a group ancestral to annelids, or to arthropods, or to both.

Returning now to firmer ground, another difference between polychaetes and onychophorans is that in worms the body musculature plays a large part in movement, although aided by the action of the parapodia. In onychophorans, by contrast, movement is carried out by the protraction and retraction of the limbs, effected by their own extrinsic muscles. The main function of the longitudinal musculature (Fig. 7–18) is to maintain, in conjunction with the haemocoelic fluid, a rigid body wall upon which the extrinsic muscles of the limbs are inserted. This change in function of the muscles must have been of crucial importance in the transformation of worm-like forms into arthropods; progressive sclerotization and hardening of the exoskeleton in higher arthropods would then have improved the efficiency of the limb muscles, while releasing them from dependence upon the hydrostatic pressure of the haemocoelic fluid. This use of extrinsic limb muscles (Fig. 7–18) instead of the musculature of the body wall, helps to reduce body undulations, which seem to be so wasteful of energy (although Gray points out that they are of some value in increasing the stride). A further advantage gained by onychophorans is that the length of the limbs can be regulated by their intrinsic muscles. This regulation is essential for efficient walking, since the distance between the base and the tip of the limb must be shortest at mid-stroke. In other arthropods, the jointing of the limbs provides for this regulation.

Evidently the locomotion of an arthropod, even at the relatively primitive level of onychophoran organization, involves a remarkable refinement of adaptation and control. Nor is this all, for subtle variations in the pattern of limb movement permit changes of speed and corresponding changes in the application of driving force. One way in which animals can increase their speed is by increasing the frequency of limb movement. This device is found in myriapods and insects, not to mention vertebrates, but the Onychophora make little use of it. They mainly rely upon two other procedures, which are also used in the other groups as well: an increase in the span of the limbs, and an increase in the ratio of protraction time to retraction time (T_p/T_r). These two factors determine the number of legs that are actually propelling the animal at a given moment, and hence the driving force that they apply, for retraction, it will be recalled, is the power stroke. On this basis, Manton distinguishes three main gaits in onychophorans: bottom gear, middle gear, and top gear (Fig. 7–19).

In bottom gear the span is small and so also is the T_p/T_r ratio; as a result, more than half of the legs are simultaneously in contact with the ground at the same time, so that a powerful driving force is exerted. This gait is therefore used for starting and for gaining speed. In middle gear the protraction and retraction times are equal, but half of the legs are still propelling at the same time. This gait is for walking when

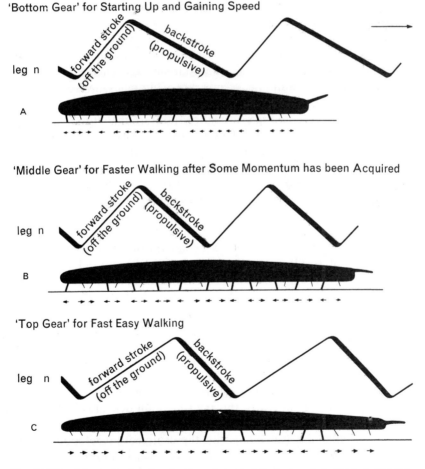

Fig. 7-19. Diagrams showing the three 'gears' of *Peripatus*. Adapted from Manton, 1950. *J. Linn. Soc. (Zool.)*, **41**, 529–570. Reproduced with permission.

some momentum has been secured. In top gear the T_p/T_r ratio is still higher; it may approach 2/1, whereas in bottom gear it may be 1/2. Less than half of the legs are now in contact with the ground at once, so that this gait is for fast walking, when speed has been established and a powerful driving force is less needed.

In the more advanced terrestrial arthropods, with their more rigid skeleton, new problems are presented. To some extent these animals tend to specialize upon one particular gait, suited for a particular mode of life. Yet this is not the only solution; flexibility can still be secured provided that the number and interrelationship of the limbs is suitably modified. These aspects of locomotor adaptation are well illustrated in the myriapods and Insecta.

Sedgwick, in a discussion of myriapods, suggested in 1909 that the Onychophora, Myriapoda, and Insecta might be the survivors of a once great and continuous group of land arthropods, a large number of which had become extinct. We have now seen that, in principle, this view is still acceptable. Features suggestive of a relationship between onychophorans and myriapods, apart from jaw structure, are the terrestrial habit, the restriction of tagmosis to the head, the development of tracheae throughout the body, and the presence in certain species of eversible sacs that absorb moisture, and that resemble comparable structures in the Symphyla. Significant also in the presence, in the embryos of certain myriapods, of large coelomic vesicles recalling those of the onychophoran embryo. In both groups the midgut lacks the digestive gland found in crustaceans and chelicerates, while other features in common are the absence of any trace of biramous limbs or of a suppressed nauplius stage. The absence of Malpighian tubules from *Peripatus* results in the intestine of this animal secreting uric acid crystals throughout its length, in physiological adaptation to terrestrial life. This intestinal activity does not occur in myriapods to the same extent, for these animals do possess Malpighian tubules; nevertheless there is some particulate excretion from the intestine in certain species.

Evidence for a relationship between the myriapods and insects is no less convincing. Especially striking, in addition to the fundamentally terrestrial mode of life, is the uniform structure of the head, with a similar distribution of appendages and with the same mode of development of the mandible. Other features shared by the two groups are the absence of a digestive gland, the presence of tracheae throughout the body and of a well-developed fat body in the haemocoel, and considerable similarity in the structure and development of the heart and aorta.

It is now possible to see, from Manton's studies, that this interpretation of the relationship between the three groups conforms well with the organization of their locomotor mechanisms. What seems to have happened, in general principle, is that the myriapods, in adaptation to contrasting modes of life, have specialized in one or other of the several gaits seen in the more generalized Onychophora, and have to some extent reduced the number of their legs, while in insects a high degree of flexibility of movement is correlated with a reduction in the number of the legs to three pairs. A full illustration of this interpretation demands a closer attention to complex detail than is practicable here, but the argument can be illustrated by contrasting the millipedes (Diplopa) with the centipedes (Chilopoda). These two groups are differentiated from each other by very marked differences in mode of life. The differences are reflected in many details of structural organization, yet not so profoundly as to preclude the possibility of a common ancestry. This is the justification for continuing to give centipedes and millipedes the common designation of myriapods.

Diplopods are burrowers, which push head foremost into loose soil and crevices, or underneath stones; they have also developed the protective reaction of coiling their body. We have seen that the structure of their cuticle is influenced by the burrowing habit; no less characteristic is the effect of this habit upon the mode of locomotion. This depends upon a specialization of the low-gear type of gait. The back stroke is of relatively long duration, which makes possible a strong and slow movement of the legs. Further, the legs are very numerous; thus a large number can be included in each metachronal wave, so that at any one moment many legs are exerting the driving stroke, and a correspondingly large force is transmitted to the anterior end of the body.

Burrowing requires the elimination of unnecessary projections. Thus the integument of diplopods is smooth as well as hard, the contours of the head are rounded, and the antennae can be folded flat. The walking limbs also make an important contribution to this aspect of adaptation. They are short, and can therefore exert a large force, because the muscles have a considerable mechanical advantage. Moreover, they arise midventrally, so that they project little, if at all, from the lateral body surface, and do not obstruct movement within the burrow. Associated with this is the peculiar form of the individual segments. These are unusually wide and deep, aiding the covering over of the legs from above. In addition, they are fused together in pairs to form what are termed diplosegments. Manton has shown that this fusion is an adaptation that can be explained in terms of the functional demands of diplopod movement and mode of life. The individual segments are very short, presumably because if they were of more normal length the body, with its large number of limbs, would be too long for effective transmission of force for head-on burrowing. A large number of short segments, however, would also create problems. The body would now be unnecessarily flexible, and this flexibility would have to be regulated by muscles. Further, it would be difficult to provide an adequate area for the insertion of the segmental muscles. These difficulties are overcome by the formation of diplosegments, without losing the advantage of the propulsive force derived from the large number of appendages. Moreover, the provision of muscle insertions is facilitated, and the body is more rigid. This rigidity is further secured by the development of ball and socket joints between the individual rings; these discourage longitudinal telescoping and contraction of the body.

Millipedes eat as they push their way through decaying vegetable matter. Centipedes, by contrast, are active carnivores, searching widely for their food, which they pursue in the open and in crevices. Manton remarks that in consequence of this difference of habit, chilopods are nearly always hungry when they are caught, while diplopods are seldom in need of a meal. Chilopods, as we have seen, possess an elastic cuticle in contrast to the rigid one of diplopods, and, in correlation with this, they have not exploited the mechanical principles involved in the association of muscles with systems of rigid fulcra and levers. Large trunk sclerites are present, so that chilopods are clearly more advanced in arthropodization than are the Onychophora, but these sclerites do not articulate. Thus the hydrostatic pressure of the haemocoel assumes great importance in these animals, and the muscles tend to act indirectly on the sclerites through the mediation of this fluid system. The result of all this is that the chilopod body is highly deformable, and is capable of much dorsoventral flattening, and of shortening and extension. Because of this, the animals can readily insert themselves into crevices, but by a method quite different from the burrowing action of the rigid diplopod body; thus, the head is flattened, the antennae are inserted anteriorly instead of dorsally, while the poison claws and second maxillae can move horizontally, so that they can manipulate prey in shallow spaces.

The legs of chilopods, in association with the free-running mode of life, may be long, and, in contrast to the diplopod legs, they exploit the top-gear type of gait. Longer legs make greater speed possible, for the stride can now be longer and the muscles, although losing mechanical advantage in comparison with those of millipedes, have a considerable velocity advantage, so that speed is correspondingly increased, always provided that the duration of pace is not increased. Exploitation of top gear will also contribute to greater speed, for in this type of gait the back stroke is relatively shorter than in bottom gear, so that the duration of pace can actually be decreased.

Scolopendra is a chilopod in which the legs are relatively short, in correlation with life in confined spaces. Despite this limitation, however, it still manages to secure a long stride by adopting gaits in which there is a large phase difference between successive legs. This animal exploits the top-gear type of gait, with a high T_p/T_r ratio, while the movement of the legs is so precisely coordinated that groups of three propulsive legs use virtually the same footprint. As may be seen from Fig. 7–20, increasing speed, with increasing stride, leads to these footprints being more widely separated, so that at top speed the animal may be momentarily supported at only three points.

Figure 7–20 shows that with increasing stride the body tends to undulate, as a consequence of the legs in any one pair being in opposite phase. These undulations are mechanically disadvatageous, for they tend to reduce forward speed, and it is thus essential that they should be reduced as much as possible. This reduction is secured by the presence of some long terga, which, in conjunction with specialized muscle insertions, give points of relative stability at which the undulations can be damped down. In *Scolopendra* tergites 7 and 8 are particularly important in this respect.

Lithobius (Fig. 7–20) is a chilopod that is adapted for hunting in the open, and for hiding and eating in wide crevices. In correlation with this it has longer legs than *Scolopendra*, but it has to pay a price for this by accepting a more limited range of gaits. This is because long legs tend to get in the way of each other. In order to avoid this difficulty, and the stumbling that would result, locomotor movements must be executed with very great precision, which means a corresponding loss in flexibility of gait. In *Lithobius* the gaits are adjusted so that crossing of the legs occurs only during protraction, when they are not exerting any driving force. During retraction (propulsion) they diverge, and because of this, and also because of their length, there is a great tendency for the body to undulate. The devices to prevent this are therefore even better developed than in *Scolopendra*, although they are still based upon the presence of long terga; these provide specialized muscle insertions for most of the dorsoventral and oblique muscles of the body.

Lithobius, thanks to these and other adaptation, can attain a speed of 280 mm/sec, but is exceeded in accomplishment by *Scutigera* (Fig. 7–20), which, at 420 mm/sec for an animal 22 mm in length, is the fleetest of centipedes. The success of

Scutigera, within the limits of so short a body, is secured in part by having very long legs which differ in length. They are so arranged that each is longer than the one in front, an adaptation already foreshadowed in *Lithbius* (Fig. 7–20). Each, therefore, can be put down just outside the tip of the preceding one (Fig. 7–21), and in this way a maximum stride is ensured without the mechanical interference of one leg with another. This device, however, can only operate with success if there are relatively few legs. This is because there is a limit to which the legs can vary in length, since at any one speed the stride must be the same throughout the body. It would be impossible to achieve this with a large number of legs, all different in length, and so in

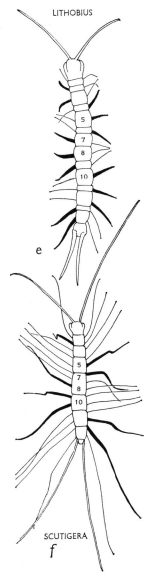

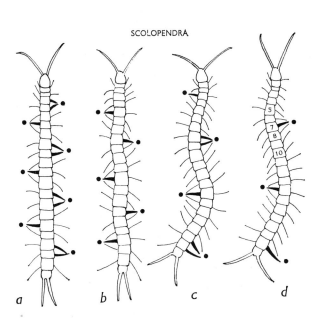

Fig. 7.20. Tracings from photographs showing the running of Chilopoda. *a–d*, *Scolopendra cingulata* running progressively faster. Legs with their tips in contact with the ground and performing the propulsive backstroke are shown in heavy lines; legs off the ground performing the recovery forward swing are shown by thin lines. The points of support of the body against the ground are indicated by black spots. The distances between the spots show the stride lengths. Note that body undulations are absent in *a*, but present in progressive measure in *b*, *c*, and *d*. A shorter body could not accommodate the fastest gait shown in *d*. *e* and *f* are similar tracings of *Lithobius forficatus* and *Scutigera coleoptera* respectively. Compared with *Scolopendra*, the legs are progressively longer, but undulations smaller owing to heteronomy of segments and their associated morphology. The short tergites 2, 4, 6, 9, 11, and 13 are minute, and are covered by the long tergites in *Scutigera*, where 7 and 8 are fused at the zone of maximum stability. *Scutigera* is the fleetest of all centipedes, and lack of control of body undulations would make speedy running with long legs an impossibility. From Manton, 1965. *J. Linn. Soc. Zool.* **45**, 251-483.

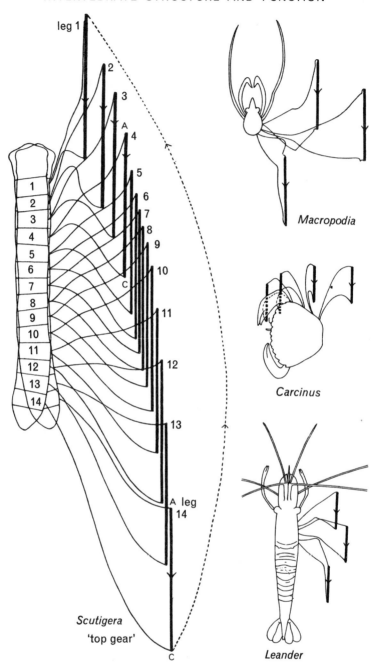

Scutigera (as, indeed, in *Lithobius*) they have been reduced in number. With such long legs the tendency to undulate is still further enhanced, and so we find an even greater reliance upon the development of some long terga. The shorter tergites are greatly reduced, and the centre of stability, which is still at tergites 7 and 8, is further strengthened by the fusion of these two sclerites (Fig. 7–20).

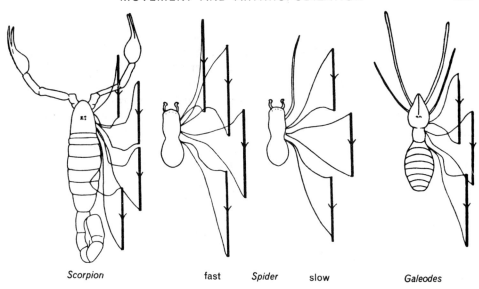
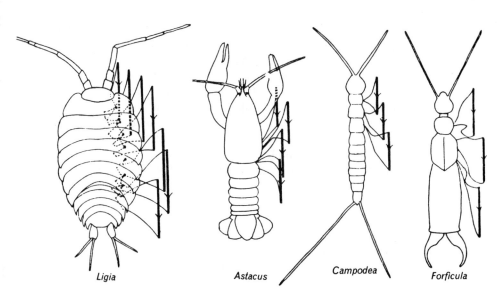

Fig. 7-21. The field of movement of the legs of a series of Arthropoda. The heavy vertical lines represent the movements of the tip of the leg relative to the body during the propulsive backstroke. In *Scutigera* the limb-tips are swung forwards approximately on the common dotted line, and the forward stroke for the other animals is not shown. The other animals are the spider-crab *Macropodia rostratus*, the prawn (*Palaemon*) *Leander serratus*, the shore-crab *Carcinus maenas*, the scorpion *Bluthus australis*, a British lycosid spider progressing at 36 and 250 mm/sec in the two diagrams, *Galeodes arabs*, *Ligia oceanica*, *Astacus fluviatilis*, a British species of *Campodea*, and the earwig *Forficula auricularia*. From Manton, 1952. *J. Linn. Soc. (Zool.)*, **42**, 93–117.

The principles of chilopod locomotion have been stated here in very general terms, but within the group there has been much adaptive divergence in matters of detail. One example illustrates the application of the principles within a somewhat different pattern of life than the one that we have been discussing. *Geophilus* is a burrowing chilopod, but it burrows like an earthworm and certainly not like a diplopod. We have noted the importance of the hydrostatic functions of the haemocoel in chilopods, and they are well exploited by *Geophilus*, which makes use during burrowing of the deformability of its hydrostatic skeleton. The legs are necessarily short. During burrowing they serve for anchorage, while when the animal is walking it uses a large angle of swing to compensate for their shortness. The burrowing habit is here associated with a comparatively long body. The body wall is elastic, as usual in chilopods, and the dorsal and ventral sclerities are able to slide over each other, with intercalary sclerites increasing the capacity for telescoping. Further, much lateral expansion is made possible by the elasticity of the pleural region, which possesses only isolated sclerites. The contrast with diplopods is fundamental, and illustrates well the consequences that flow from the adoption of two entirely different approaches to a burrowing habit.

The principles that we have outlined can reasonably be applied to the history of insect locomotion as well. The characteristic tagma of insects is the three-segmented thorax with its three pairs of elongated legs (Fig. 7–21) that are often longer than those of myriapods. This reduction in the number of appendages carries to an extreme the principle outlined earlier. Mutual interference of the legs is avoided, while their very small number leaves the animals free to use a flexibility of gait without stumbling. The localization of these three segments immediately behind the head illustrates another consideration. The action of the legs, when they are so few in number, will be more effective if the leg-bearing segments are adjacent to each other, for this abolishes any tendency to the lateral undulations that we have seen to be wasteful of energy and confusing to movement. It is thus not surprising to find these appendages borne close together on one tagma. Further, the localization of this tagma directly behind the head aids in the support and manipulation of that region.

The modes of operation of the legs of insects vary a good deal, both between species and also within a species in relation to the speed of movement, but they conform to common principles. For one thing, maintenance of stability must depend on at least three feet being on the ground at once, and this is what is normally found. However, some insects, when moving at speed, do very briefly find themselves in a position of instability with only two legs supporting them.

Hughes has shown that two rules underlie the variation in pattern of insectan limb movement: (i) a foreleg or middle-leg is not lifted until the leg behind it is touching the ground; (ii) each leg alternates in phase with the opposite leg of the same segment. (Jumps, such as those executed by grasshoppers, are naturally exceptions to these rules, for they may be executed by two legs of a pair moving together.) An illustration of the precise control that is consequently required is seen in Fig. 7–22, which shows the cockroach (*Blatta*) during normal walking. The positions of the legs conform to the rules just stated, but it will be observed that four or five legs are supporting the body at any one time, and at one stage all six legs are doing so. Given this general pattern of movement, an insect can inscrease its speed by the procedures stated earlier: increases in span, in frequency, and in the protraction/retraction ratio.

MOVEMENT AND ARTHROPODIZATION 209

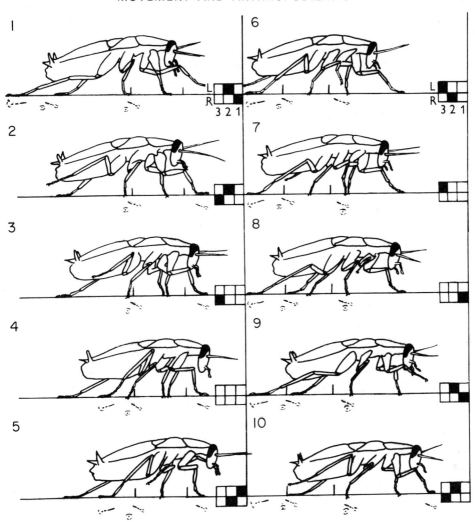

Fig. 7-22. A series of drawings of *Blatta* taken from successive frames of a film (30 frames/sec) to show the sequence of leg movements in side view during normal walking. The legs which are off the ground during each phase are indicated by the black squares on the right of each drawing, and the markings left by the insect when walking over smoked paper are shown beneath. From Hughes, 1952. *J. exp. Biol.*, **29**, 267–284.

In practice, the last possibility is the most favoured, for variation in gait is aided, as we have indicated, by the presence of so few legs. Speed in running is not, however, of prime importance to insects (chilopods, for example, achieve higher speeds relative to the size of the segments). This is because the evolution of wings on the insectan thorax made possible the development of flight, and thus of an entirely new method of locomotion. Diverse in its expression, which ranges from the fast forward flight of larger insects to the hovering of small ones, this has established for insects a dominance of air space which has been little disturbed by the subsequent evolution of birds and men. Before we consider this, however, we may note that the principles operating in the evolution of ambulatory movement in the Onychophora, myriapods, and Insecta can also be applied to an understanding of the ambulatory locomotion of crustaceans and chelicerates. Thus it becomes easier to understand the origin of the caridoid facies of malacostracans, provided that we accept, as suggested earlier, that this arose in association with a bottom-dwelling habit. The development behind the head of these animals of an eight-segmented thorax, with some of the segments bearing long walking limbs, can be interpreted as illustrating the same principles as are thought to have governed the origin of the insect thorax, although the principles have not been carried to quite such an extreme conclusion here.

As for Chelicerata, this group is so different in organization from other living arthropods that, as we have seen in considering the head region, it must have followed an independent path. It has been suggested that the class may have had some remote relationship with the Trilobita, but there is a substantial gap between the two groups which we are unable to bridge with any assurance. The Chelicerata began their history as an aquatic group, the Merostomata, represented today by *Limulus*; the scorpions are terrestrial derivatives of these, and are found as early as the Upper Silurian, when the vertebrates had not yet passed beyond the agnathan phase of their evolution. The characteristic tagmata of the scorpion (Fig. 7–21) are the prosoma, bearing chelicerae, pedipalps, and walking legs, and the opisthosoma, the latter divided into a mesosoma and a metasoma.

As in crustaceans, the concentration of locomotion in a few pairs of walking legs leaves no ambulatory function for the more posterior appendages, but the evolution of the latter has taken a manifestly different course in the two groups. Characteristic of the Chelicerata is the distinction between prosomatic locomotor appendages and mesosomatic respiratory ones, a characteristic that is already becoming apparent in the aquatic fossil forms. Some of these were active swimmers, while others were crawling forms, but from a very early stage there was a tendency for the adaptive evolution of relatively few locomotor limbs. The eventual establishment of four pairs of long walking legs in the terrestrial forms is in accord with the general principles suggested earlier. The mesosoma of scorpions includes four pairs of respiratory appendages, forming the lung books which are peculiar to the Arachnida. The six-segmented metasoma is reduced, serving primarily in scorpions for manipulation of the terminal sting.

7–5 FLIGHT IN INSECTS

In the course of evolution there have been a number of attempts to exploit the possibilities of flight, outstanding success having been achieved by insects, birds, and man. The human solution has depended upon the use of wings as aerofoils, power being applied through a moving airscrew (propeller), or by the use of jet propulsion. The latter device has been extensively applied by other animals to secure movement in water, but has not been used for flight. Instead, movement through the air has been achieved by applying power to mobile wings in order to establish a flow of air over them; they thereby serve both as airscrews and aerofoils. In birds the mechanics of the system are remarkably uniform, but in insects the situation is otherwise. This is partly because flight in insects, unlike that of the vertebrates, does not depend upon the transformation of limbs. Outgrowths of the thorax that foreshadow wings are first seen in fossil insects from the Upper Carboniferous. Probably they were first used in gliding, but later these expansions, together with their supporting musculature, became transformed into a flight system which was added to the already existing locomotor mechanism, instead of being a transformation of it.

Another factor is the exoskeleton. We have considered earlier (Sec. 7–1) some of the advantages of this, and here we find another, for Weis-Fogh points out that its toughness and flexibility make it superior to the brittle, calcified endoskeleton of birds in the absorption and release of elastic energy. In conjunction with the associated specialized muscles, it makes an invaluable contribution to the flight mechanism by allowing the kinetic energy of the upstroke of the wings to be stored for use in the propulsive downstroke. An important factor in this economical use of energy is a protein called resilin, which is extensively used in the joints of insects, and particularly in the hinges of the wings. This substance, because of the mobility of its amino acid chains, has almost perfect elasticity, being able to return about 97% of the energy which is stored in it by compression or extension.

The elasticity and flexibility of the exoskeleton, used with particular ingenuity in the 'click' mechanism to be described later, goes a long way to avoid the waste of power that would otherwise result from inertial torque, and which could be particularly serious when there is a very rapid stroke frequency, such as is found in hovering flight. Weis-Fogh shows that in this type of flight the power demands are particularly heavy because, in contrast to the more primitive type of forwardly-directed flapping flight, there is no accumulation of kinetic energy, and the wings can only produce lift when they are actively moved. But diversity of adaptation, so characteristic of insects, is found even in the exploitation of elastic structures in flight. Thus in the dragon fly, *Aeschna*, the elasticity is chiefly located in the wing muscle, with the thoracic exoskeleton contributing less than 25%. In the privet hawk moth, *Sphinx ligustri*, by contrast, it is mainly in the scelerotized cuticle of the thoracic box and in the ligaments, which are composed largely of resilin.

Diversity of adaptation is seen also in the variety of wing movements found in insects. The complexity of these movements becomes evident when the path traversed by the wing of an insect is analysed. A wing has to translate the energy derived from the muscles into forces that will provide thrust for forward movement in opposition to the drag exerted by the air, and lift to counteract the effect of gravity. In addition, the insect must be stable in the three planes of space, despite lacking secondary stabilizing agents comparable with the tail of birds. Reflex changes in the path of the wing-tips may contribute to this stability, but functional analysis of wing movements suggests that their subtle pattern contributes in itself to stability in yaw, pitch, and roll. Failure to appreciate this complexity of wing functions has led to misunderstanding of their mode of operation. In an extreme case, as Chadwick points out, it has even been concluded that it is mathematically impossible for the bee to fly!

Of course, the insect wing does not move exactly like an airscrew. (The wheel and axle is one of the very few human inventions that have not been anticipated in earlier stages of evolution.) It actually moves in a sinuous sweep, so that if the insect is stationary, the tip of the wing describes a figure of eight (Fig. 7–23). This may be a broad one, as in the fly *Phormia*, for example, or a narrow one, as in the bee. The forces acting upon the wing of *Phormia*, which beats at about 120 times per second, have been analysed with the help of high-speed photography. During the downbeat,

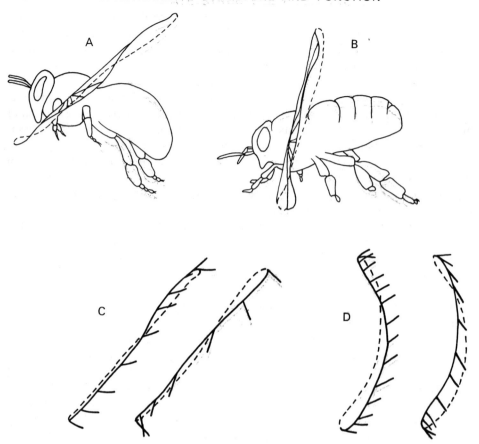

Fig. 7-23. *Apis mellifera*. Body attitude and path of the wing-tip during flight, A at 3 m/sec airflow, and B, when distributing secretion from abdominal scent gland. C, path of the wing-tip and angle of attack of the widest part of the wing during A; D, the same during B. From Pringle, 1968. *Adv. Ins. Physiol.*, **5**, 163–227.

the true ventral surface of the wing is the aerodynamic surface, the resultant force acting upon it being nearly vertical. During this phase most of the lift is obtained. At the end of the downstroke the wing is turned actively at its base so that its true dorsal surface is now the aerodynamic lower surface. The resultant force is now directed forward, and it is this phase therefore, that provides the main thrust. At the end of the upstroke the wing again rotates rapidly. This rotation, like the preceding one, leads to changes in the balance of forces that seem to serve mainly to decelerate the wing in advance of the changes in stroke.

Insect wing movements naturally vary greatly in detail, as may be judged from the bee, *Apis mellifera*, in which, as already noted, the loop described by the wing is narrow in contrast to that of the fly (Fig. 7–23). The angle of attack during the downstroke (that is, the angle of incidence of the flow of air) is about 15°, but during the upstroke it is zero. Thus the upstroke, in contrast to that of the fly, performs little useful aerodynamic work. But bees use their wings for purposes other than flight. For example, they are used to waft away the secretion from their scent gland, an important contribution to the organization of their social life (Sec. 21–4). During flight, the plane of the wing stroke lies at about 120° to the longitudinal axis of the body. During scent movement, however, when the bee raises its abdomen and clings with its legs, the stroke plane is nearly vertical and the angle of attack is large in both upstroke and downstroke. This pattern provides for efficient circulation of the scent, at the cost of high muscle loading during both phases of wing beat, reflected in the oscillating movements of the body during the fanning action.

One consequence of the insectan mode of flight is that in so far as the wing functions as an aerofoil, it can do so with efficiency only if the air stream flowing towards it is free of turbulence. This condition can operate for the anterior pair of wings, but the movement of these creates a turbulence that could impede the functioning of the posterior pair when, as in most insects, both pairs move in flight. Various adaptations have been evolved to overcome this problem. In dragonflies both pairs of wings are functionally equivalent in their contribution to flight, but they beat out of phase with each other. The effect of this is that the posterior pair meets the backward stream of air before the latter has been disturbed by the anterior pair. Often, however, the two pairs are moved as nearly as possible as a single pair. Bees achieve this by the hooking together of the anterior and posterior wings while Lepidoptera obtain the same result without hooks because the posterior wings are pressed down by the anterior ones. In some insects only one of the two pairs of wings is used for propulsion. This is so in some Coleoptera in which the anterior stiffened pair (the elytra) do not beat. Even so, they probably do make some contribution to movement by providing lift, and by controlling direction; perhaps, too, they can control power by regulating the flow of air over the hind wings. In the Diptera, by contrast, propulsion is effected by the anterior wings, the posterior ones being reduced to halteres. These structures, which vibrate in flight at the same frequency as the wings, have groups of campaniform sensilla (Sec. 14–3) at their bases; information provided by these sensilla enables the insect to maintain its equilibrium during flight.

Following Boettiger, we can analyse the machinery of insect flight into four components. The first of these comprises the wings, which we have already briefly described. The second component is the complex of parts that articulates the wings to the thorax; included here are the base of the wing and the structures relating it to the thorax, together with the direct flight muscles. The latter are so called because they run from the pleural and sternal regions of the thorax to be inserted directly on the sclerites at the base of the wing; they thus control the setting of these parts. The third component comprises the well-developed and powerful indirect (or driving) flight muscles (Fig. 7–24). One group of these runs dorsally and logitudinally between the mesothorax and the metathorax. The other group runs dorso-ventrally between the tergum and the sternum. These muscles are termed indirect because they do not run direct to insertions on the wings. They are, however, coupled functionally to the wings by parts of the thorax; the latter constitute the fourth component of the flight machinery.

These four components are not always used. In the more primitive insects (for example, the dragonflies) power is applied to the wings through the direct muscles, each of the four wings of a dragonfly possessing elevator and depressor muscles.

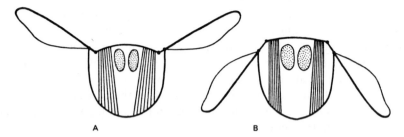

Fig. 7–24. Diagrams illustrating the early view of the mode of action of the indirect flight muscles of insects. From Imms, 1959. *Outlines of Entomology* (5th ed.). Methuen, London.

These are called lamellar muscles, because the protofibrils of the muscle cells are grouped together to form sheets or lamellae, with sarcoplasm and mitochondria lying between them. Lamellar muscles are found also in cockroaches and mantises. Little is known of the physiology of this muscular tissue, but it may be a primitive type of flight muscle. Its important characteristic is that each contraction of the muscle is the result of stimulation by a single nerve impulse; in this respect the muscle behaves like the general body musculature of insects. This type of flight mechanism has been called the synchronous type, because the stroke of the wings is synchronous with the stimulation of their musculature by the nervous system.

In many insects, including the Ephemeroptera, Locustidae, and Lepidoptera, the protofibrils are grouped to form delicate fibrils with a diameter of about 1.5 μm. Muscles of this type, called microfibrillar muscles, are characteristic of insects with comparatively soft bodies. Those with harder cuticles, such as Hymenoptera, Coleoptera, and Diptera, have fibrillar muscles in which the fibrils are larger, with a diameter of 3.0 μm. These are the insects with the most highly developed powers of flight; powers that are believed to depend in part upon the complexity of the wing articulation, and in part upon the physiological properties of the fibrillar muscular tissue that forms the indirect flight muscles. In such insects important parts are played by all four of the components that we have mentioned. Their mode of operation, however, is very complex, and it will be impossible to do more here than describe it in brief outline.

One important characteristic of this second type of flight mechanism is that it is asynchronous, by which is meant that the wing stroke is not synchronous with the nervous stimulation of the muscles. The rate of stimulation can be determined by direct recording of the action potentials in the muscles, and it can thus be shown that it bears no relation to the rate of beat of the wings. Instead, it appears that each nerve impulse sets up in the muscle an active state, during the maintenance of which a variable number of contractions can occur. It is because of this that insects possessing as asynchronous flight mechanism can maintain such a remarkably high frequency of wing beat. Whereas the wings of a butterfly may beat at a frequency of 9 per second, and those of a dragonfly at 28, the frequency in the housefly may be 180–200 per second, and in the mosquito 1024.

The mode of operation of an asynchronous flight mechanism has been analysed in the fly, *Sarcophaga bullata*, by Boettiger. His analysis depended on the fortunate circumstance that when the fly is exposed to fumes of carbon tetrachloride its wings become set in either an up or a down position corresponding to the positions during the normal flight movements. Study of these positions, and pushing of the wings from one position to another, thus made possible the elucidation of the normal functioning of the relevant parts of the skeleton. Previously it had been supposed that the downstroke was produced by a lateral expansion of the notum brought about by contraction of the longitudinal indirect muscles; then, it was thought, contraction of the vertical indirect muscles would draw the notum inwards and bring about the upstroke (Fig. 7–24). Boettiger's analysis showed that this cannot be so, and that in fact contraction of both longitudinal and vertical muscles will tend to produce lateral expansion of the notum. The true course of events is complex, and reference must be made to his account for a full description. In principle, however, tension in the indirect muscles produces an outward thrust at the point f (Fig. 7–25), the effect of this being to force apart the mesopleural process (a) and the hinge (h) which connects the parascutum to the lateral border of the tergum. This thrust strains the tergum and the mesopleural process and results in the building-up of a store of potential energy. The

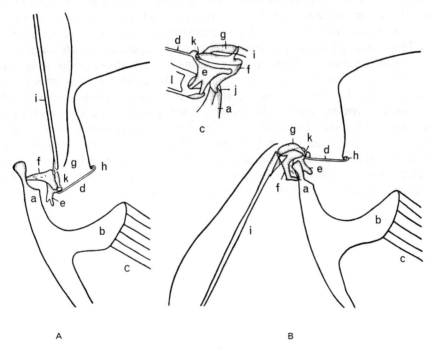

Fig. 7-25. Cross-sectional view of thorax of *Sarcophaga bullata*, showing details of the articulation of the right wing. A, wing in up position, anterior view. B, wing in the down position, anterior view. C, posterior view of the axillary sclerites of right wing showing their relation to the mesopleural process, the lever arm, and the anterior parascutum. *a*. mesopleural process; *b*, pleural apophysis; *c*, anterior pleurosternal muscle; *d*, anterior parascutum; *e*, first axillary sclerite; *f*, second axillary sclerite; *g*, base of radial vein; *h*, hinge; *i*, radial vein; *j*, hook articulation; *k*, point of articulation of anterior notal process, first axillary sclerite, base of radial vein and second axillary sclerite; *1*, end of the lever arm. From Boettiger and Furshpan, 1952. *Biol. Bull. mar. biol. Lab., Woods Hole*, **102**, 200–211.

relationship of adjacent parts of the skeleton are such that the point k is removed. A critical stage is reached in this movement when k moves past the line running from the hinge (h) to the point of articulation of the second axillary sclerite (f) to the mesopleural process. At this stage the point k is forced towards its extreme position, either upwards or downwards, because there is a recoil of the skeletal elements that are under strain. The stored potential energy is thus released, and is manifested in wing movement. The whole process is referred to as the click (or snap) mechanism, because the wing 'clicks' from one position to another when the parts are subjected to appropriate pressure in flies that are under the influence of carbon tetrachloride.

The result of this sequence of events is that changes in length of the indirect flight muscles are magnified some 400–600 times in wing movement. Thus these muscles can operate almost isometrically, with minimum change of length. No less important is the peculiar physiological property of fibrillar muscles, which, as we have noted earlier, permits them to respond to nervous excitation by the maintenance of prolonged tension. It would further appear that they are able to contract as a direct response to the stimulus of being stretched; contractions can thus occur without the tissue being dependent upon a corresponding inflow of nerve impulses. The operation of the 'click' mechanism depends upon muscular tension being applied to the wing articulation. Movement of the wing into one position (up or down, as the case may be) produces a quick release of the particular muscles concerned in that movement. This causes an immediate stretching of the antagonistic muscles, which then contract and bring about, through the resulting tension, the opposite wing movement. Such a 'click' mechanism is not necessarily associated with asynchronous flight muscle, however. It is found in the metathorax of the locust, *Schistocerca*, where the flight muscle is synchronous, and of beetles, where it is asynchronous; on the other hand, it has not been identified in bees, where the muscle is also asynchronous. There is certainly much to unravel before the evolution of the fully developed flight mechanism can be understood. Fibrillar muscle seems to have evolved along a number of lines, which leads Pringle to suggest that its behaviour must be latent in insect muscles in general, and perhaps in all muscular tissue.

8
Nutrition of Protozoa

8–1 FEEDING

Flagellate Protozoa of the Superclass Mastigophora show great diversity in their modes of nutrition. Some are plant-like, some animal-like, and there is a marked tendency for the appearance of holozoic or saprophytic nutrition in groups that are primarily phototrophic. This is an accord with our discussion of the course of biochemical evolution, and of the origins of green plants and animals, for the Mastigophora stand very close indeed to the common stock from which both plants and animals have diverged.

At one time it was suggested that a flagellate could be regarded as an animal if it lacked a cellulose cell wall, and if it divided by longitudinal and not by transverse fission, but these criteria cannot be sustained. No doubt flagellates are a polyphyletic assemblage, but, subject to this limitation, modes of nutrition are a convenient guide to relationships. Many flagellates are phototrophs (phytoflagellates), and it is not sensible to try to separate these from unicellular algae. It is best to group them all within a Class Phytomastigophorea, to which must be added certain colourless forms that are closely related to them. Other flagellates, which lack chloroplasts, and which have no algal relationships, are then contrasted with them in a Class Zoomastigophorea (zooflagellates). Some zoomastigophoreans may lose their flagella and become amoeboid. It is because of this that the amoeboid Protozoa (Superclass Sarcodina) are held to be closely related to the flagellates, and are placed with them in a Subphylum Sarcomastigophora.

Several species of *Euglena* provide good examples of the tendency for the appearance of heterotrophy in groups that are primarily phototrophic. This genus is typically phototrophic, but *E. gracilis* can survive and grow in the dark, in conditions in which it cannot possibly be deriving energy from photosynthesis. Its survival is due to osmotrophic heterotrophy: in the absence of solar radiation it can obtain energy by oxidizing acetic acid. Indeed, many flagellates flourish in the presence of this substance. They seem to have little capacity for utilizing carbohydrate, and the acetate provides them with what is probably their sole source of carbon.

The osmotrophy of *E. gracilis* provides the organism with a source of energy, but it also meets other needs, for euglenids, despite being phototrophic, have certain specific vitamin demands. This was first demonstrated for *E. pisciformis*, which cannot grow without an exogenous supply of thiamine. This dependence seems to be a common situation in euglenids, and thiamine is not the only substance so required. *E. gracilis* and *E. viridis*, together with many other phytoflagellates, need an exogenous supply of vitamin B_{12}, although closely related cobalt compounds can sometimes be substituted for this.

We have already suggested that vitamin requirements may be a result of regressive biochemical evolution. Another example of such regression is the tendency for certain phytoflagellates to lose their chlorophyll. *E. deses* does so after prolonged culture in the dark, while *E. gracilis* may give rise spontaneously to colourless forms; these are identical with a flagellate that was known at one time as *Astasia longa*. The appearance of colourless forms can also be induced by treatment with streptomycin, *E. gracilis* being a species that responds in this way. Not all euglenid species, however, can abandon photosynthesis; *E. pisciformis*, for example, will die if it is maintained in the dark. It is thus an obligatory phototroph, whereas *E. gracilis* is a facultative one.

The species that can survive without chlorophyll depend upon their capacity for osmotrophy, but we might expect that during evolution the establishment of colourless forms through mutation would be followed by the development of phagotrophy. This must surely account for the presence in the Order Euglenida of truly animal-like forms, colourless and phagotrophic, of which *Peranema* is an example. This organism (Fig. 8–1), like other euglenids, possesses a reservoir and a contractile vacuole, but in addition there is a rod organ, consisting of two parallel rods formed of microtubular fibrils. This organ, which can be protruded or withdrawn during

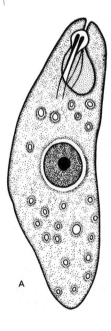
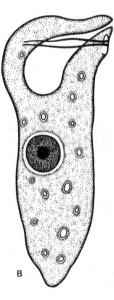

Fig. 8-1. *Peranema*. A, viewed from ventral surface; B, twisted, and viewed from the side. From Chen, 1950. *Q. Jl. microsc. Sci.*, **91**, 279–308.

feeding, is associated with a mouth or cytostome. *Peranema* takes a variety of food material, including bacteria and detritus, and readily attacks other euglenids provided that they are stationary. When it touches the anterior end of a *Euglena* the rod organ protrudes and becomes attached to the surface of the prey (Fig. 8–2). The anterior end of the *Peranema* then dilates, its body moves around the prey, and the rod organ begins to push the latter through the cytostome. Through continual detachment and re-attachment of the organ, combined with forward movement of the *Peranema*, the prey is gradually swallowed, the process being completed within an average time of 8 min.

We shall see that suctorial feeding has evolved in the Metazoa as a specialized process, often dependent upon the possession of piercing structures, and it is thus of particular interest to find the same development occurring here. Sometimes the rod organ rasps at the surface of the attacked *Euglena* until this has been cut open. The cytoplasm and chromatophores then flow into the body of the *Peranema*, and accumulate there in food vacuoles. As with other euglenoids, paramylon is a reserve product of this organism, and so also is oil. Observations of the fate of starch, oil, and casein taken into the food vacuoles show that it can digest all three types of material, and that it is able to convert them into those reserve substances.

The ease with which colourless flagellates arise today suggests that there may in the past have been repeated evolution of heterotrophic flagellates from phototrophic ones. Certainly the coexistence of phagotrophy and phototrophy within the same order is not peculiar to the Euglenida, for we find this situation also in another group of flagellates, the Order Dinoflagellida. Members of this group often have an external covering of cellulose plates, and are further characterized by the possession of two flagella arranged in grooves in a clearly defined way; one lies transversely to the main axis of the body, in a groove called the cingulum, while the other is trailed along that axis in another groove, the sulculus. The group includes both fresh-water and marine organisms, but it is particularly prevalent in the open sea, where its members are among the most plentiful of micro-organisms.

Chloroplasts are present in dinoflagellids, which are a typically phototrophic group, but their photosynthetic activity is sometimes combined with the capacity for ingesting small organisms. One example of this is the brackish-water form *Gyrodinium*, which is able to immobilize large ciliates and to engulf them through the

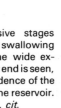
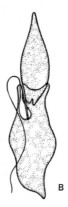

Fig. 8-2. Successive stages (A–D) of *Peranema* swallowing a *Euglena viridis*. The wide extension of the anterior end is seen, and also the independence of the food entrance from the reservoir. From Chen, 1950. *op. cit.*

posterior end of the sulculus (Fig. 8–3). Another example is *Ceratium hirundinella*, some of which are colourless while others possess chromatophores. Such colourless individuals can feed on solid prey, which they seize either in a fine network of cytoplasm extended over the body surface, or in a single pseudopodium. Whether individuals with chromatophores also feed in this way is uncertain, but it is at least likely that they do so. Dinoflagellids certainly occur in the sea at depths too great to permit photosynthesis, and they may play a significant part as heterotrophs in the cycle of marine nutrients.

The complete abandonment of phototrophy, analogous to the situation in *Peranema*, is seen in *Noctiluca*. This is an aberrant dinoflagellid, often present near the surface of the sea in enormous numbers, and especially conspicuous because of its bioluminescence. *Noctiluca*, which lacks chromatophores, is exclusively phagotrophic, and, although itself only about 1.5 mm in diameter, is able to feed on animals as large as copepod larvae. These are drawn in through a permanent mouth (cytostome), after having become attached to a movable adhesive structure called the tentacle, which is found also in some other dinoflagellids. The longitudinal and transverse flagella of the latter may be represented in *Noctiluca* by structures called, respectively, the cilium and the tooth. Their function is not clear, but they probably play some part in manipulating the food material. Clearly, then, the development of phagotrophy has involved considerable specialization of the feeding mechanism in dinoflagellids, as it has in *Peranema*.

The remaining groups of Protozoa completely lack the power of photosynthesis, so that they can be regarded without equivocation as true animals, Protozoa *in sensu stricto*. We have no direct evidence regarding their origin, but bearing in mind the many lines of evolution that are discernible in the algae and phytoflagellates, there is certainly no reason to suppose that they are monophyletic. They include many free-living forms, but also show a marked tendency towards entozoic habits, leading in the case of the trypanosomes both to harmless commensals and to highly important pathogenic forms, parasites of man and his domestic animals (Sec. 23–4). Thus *Bodo*, which is both free-living and entozoic, contains food vacuoles and is phagotrophic, whereas *Trypanosoma* is osmotrophic. In many other zooflagellates an entozoic mode of life is combined with an exceedingly complex body form which involves the development of an elaborate system of flagella, ranging in number from a score or so to many thousands. These animals may be phagotrophic (Fig. 8–4) or osmotrophic. Within the hind-gut of the cockroach, for example, are found *Lophomonas striata* and *L. blattarum*. The former is osmotrophic, but the latter ingests bacteria through the soft posterior region of the body wall. We shall be returning later to the habits and significance of some of these remarkable organisms.

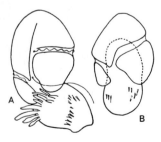

Fig. 8-3. *Gyrodinium pavillardi* capturing (A) and ingesting (B) a specimen of *Strombidium*. The ciliate is held in the region of the frontal membranellae, and ingested in the ventral part of the sulcus, the lips of which are distended. From MacKinnon and Hawes, 1961. *An Introduction to the Study of Protozoa*. Clarendon Press, Oxford.

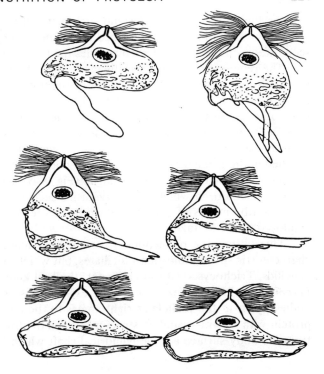

Fig. 8-4. *Trichonympha* sp. from the gut of a termite showing mode of ingesting fragments of wood. From Baer, 1951. *Ecology of Animal Parasites*. University of Illinois Press, Urbana Ill.

The Superclass Sarcodina comprises organisms that are characterized, during at least part of their life cycle, by the use of pseudopodia for movement and feeding. They are typically phagotrophic, the food being surrounded with cytoplasm so that it becomes enclosed in a food vacuole. We have earlier noted the close relationship of these animals with the flagellates. The facts are fully in accord with the view that the rhizopods have been derived from flagellate ancestors, with the complete abandonment of phototrophy. In the light of our earlier arguments, and the diversity of their pseudopodia, they are likely to represent more than one line of evolution.

The origin of the ciliates (Subphylum Ciliophora) is more difficult to determine, for the ciliation of these animals, their remarkable nuclear specialization, and the phenomenon of conjugation, set them very much apart from the preceding groups. However, we have seen that there is no fundamental distinction to be made between flagella and cilia. Moreover, since some of the most specialized flagellates have developed a complete covering of flagella, the ciliation of the Ciliophora does not preclude them from being derived from some unknown flagellate ancestry. The existence of *Opalina*, a ciliated organism with a unique arrangement of nuclei, and with some flagellate-like features (for example, division between the ciliary rows instead of across them), supports this view. This animal seems to represent an independent line of ciliate-like evolution, and can reasonably be classified in a separate superclass of the Sarcomastigophora.

The more elaborate organization of ciliates in comparison with the Sarcodina is seen in the way they secure their food. The Suctoria, generally considered to be an aberrant group of ciliates, have an exceptional method of feeding in that they capture their food by tentacles and suck material down these into the body. The means by which this is accomplished is not understood, although electron microscope studies have shown that sometimes the tentacles possess internal canals.

The remaining ciliates typically ingest food through a mouth, but they use this in two distinct ways that foreshadow the distinction between microphagy and macrophagy which we shall encounter in the Metazoa. Some seize and ingest large animals, often other ciliates, with the mouth (cytostome). This may be set on a proboscis, as it is in *Didinium*, which can eat ciliates several times larger than itself. The cytostome leads into a cytopharynx, which may be strengthened by rods in a manner closely paralleling the arrangement that we have noted in *Peranema*. An example is *Chilodonella*, which ingests filamentous algae (Fig. 8–5) as well as bacteria. Others use a different method, in which small organisms such as bacteria are directed towards the mouth by means of currents set up by their cilia, a principle that is also widely employed in the Metazoa.

This latter type of feeding is assisted by trichocysts, which are particularly characteristic of the holotrichous ciliates, but which are also found in certain dinoflagellids. Trichocysts vary in character, the best known being of the type found in *Paramecium*. Prior to extrusion, they exist in this animal as oval or rod-shaped bodies, lying in the pellicle at right angles to the body surface, and composed of protein with a paracrystalline structure. When the trichocyst discharges it extrudes from the body surface a long cylindrical shaft which bears a short spine at its tip.

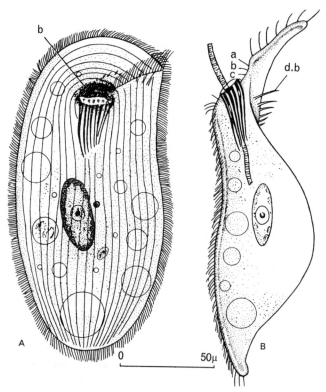

Fig. 8-5. *Chilodonella cucullulus*. × 500. A, from the ventral side; B, from the left, the organism feeding on a filamentous alga. *a, b, c*, pre-oral and circumoral cilia or ciliary lines; *d.b*, dorsal brush. From Mackinnon and Hawes, 1961. *op. cit.*

In so doing it must undergo some considerable molecular reorganization which is said to be completed in a few thousandths of a second. The effect of this reorganization is detectable under the electron microscope by the appearance of cross-striations and fibrillar patterning in the body of the trichocyst; these extend also into the fully formed shaft. One factor in the discharge of the trichocyst is thought to be a considerable extension of the paracrystalline lattice. Some trichocysts, such as those of *Prorodon* (Order Gymnostomatida) are different. They have a curious resemblance to the nematocysts of the coelenterates, and are, for this reason, called cnidotrichocysts. The body consists of a capsule containing a delicate tube, coiled up inside it, that is turned inside out on stimulation. Both the function and the evolutionary origin of trichocysts remain obscure, but it seems likely that they serve to attach the animals to accumulations of bacteria. These structures, like the locomotor mechanisms of Protozoa, illustrate very well the specialized development to which fibrous macromolecules can be subjected.

There is much variation in the details of the ciliary feeding mechanisms of ciliates, but the primitive arrangement is probably to be seen in certain gymnostomatids where rows of body cilia converge upon an apical cytostome. Food particles are driven into the cytostome and then passed on into the endoplasm through the cytopharynx, which is an unciliated passage. An elaboration of this arrangement leads to the cytostome being situated at the base of a depression, called the buccal cavity; the somatic cilia extend into this, and may fuse together to form an undulating membrane and smaller membranelles. In *Paramecium* (Fig. 8–6) the buccal cavity lies at the inner end of a vestibulum, but in the Spirotrichia, such as *Vorticella*, it is situated at the body surface, where it forms the peristome; its membranellae may thus play a part in locomotion as well as in feeding.

Ciliary feeding can readily be observed in *Paramecium*. This animal does not normally ingest any food while it is swimming actively, feeding being restricted to those periods when it is swimming slowly or when it is at rest. At such times the cilia set up currents down the vestibulum and small particles are thus driven towards the buccal cavity. These particles may vary a great deal in character, but they seem to be subjected to some form of selection, for many of them may be passed out again before they enter the buccal cavity. The mechanism of this selection is not understood, but its existence in a protozoan foreshadows the great importance of selective mechanisms in many metazoans that feed in this way upon small particles.

Ciliates have lent themselves well to nutritional studies, and particularly to the development of methods of axenic culture, in which the animals are reared in cultures that are free of all other organisms, including bacteria. In this way the extent of their dependence on preformed organic materials has been determined. For example, we know the complete requirements of *Tetrahymena pyriformis*, which is a common small ciliate of fresh-water and other habitats. This animal requires 10 amino acids, 6 vitamins (including thiamine and riboflavin), and 2 nucleic acid derivatives (guanine and either uracil or cytidine), together with various inorganic ions including magnesium, potassium, iron, copper, and phosphate. Thus far have the Protozoa moved from their autotrophic ancestors; further, indeed, than the higher vertebrates in at least one respect. The rat and pigeon, for example, can synthesize purines and pyrimidines, and can incorporate them into nucleic acids, whereas *Tetrahymena* presumably lacks certain of the enzymes needed to achieve this synthesis. The

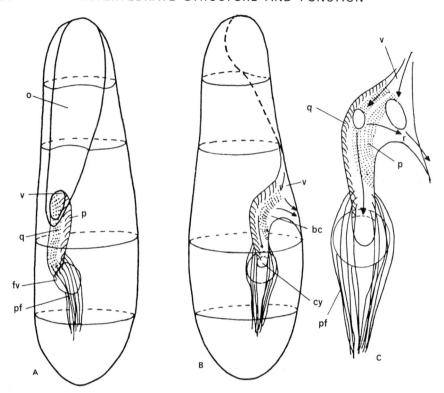

Fig. 8-6. *Paramecium:* the feeding apparatus (diagrammatic). A, ventral view, the animal turned slightly to its right; B, from the right side, with the feeding and rejection paths marked by arrows; C, part of B enlarged, *bc*, buccal cavity; *cy*, cytostome; *fv*, food vacuole; *o*, oral groove; *p, q,* ciliated bands; *r*, rejection path; *v*, vestibulum. (After Mast.) From Mackinnon and Hawes, 1961. *op. cit.*

needs of *Paramecium* have been less completely established, but they are known to include 10 amino acids, and also acetate, phosphate, a sterol, and several vitamins, including thiamine and riboflavin.

Finally, and in complete contrast to the Ciliata, the Subphyla Apicomplexa, Myxospora, and Microspora are exclusively parasites. They have in common an osmotrophic nutrition that is an adaptation to their parasitic life, but they show marked differences among themselves in their structure and life-cycles. The flagellated gametes of certain sporozoans (the coccidian *Eimeria*, for example) suggest that these may have had a flagellate origin, while an amoeboid stage in myxosporan and microsporan life histories suggests an amoeboid origin for these groups. The evidence is tenuous, however, and we are hardly justified in doing more than drawing the obvious conclusion that these osmotrophs have had a history that is unknown but which was certainly polyphyletic.

In considering the feeding of Protozoa, as well as their locomotion, we have inevitably referred to the phylogenetic relationships of these organisms. One possible representation of these, which gives general expression to the views stated here, is shown in Fig. 8–7. The basal position of the flagellates is not in doubt. The amoeboid

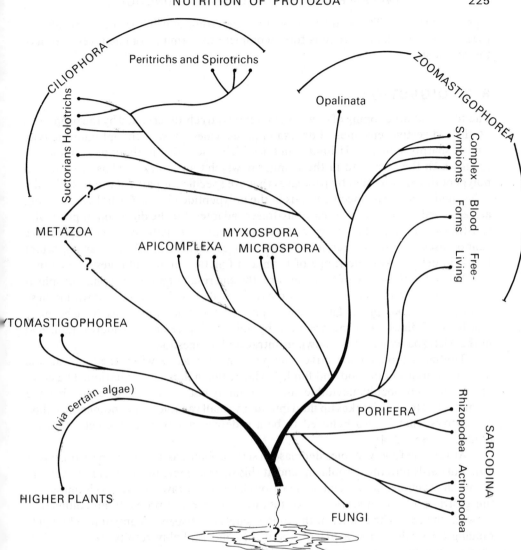

Fig. 8-7. A proposed phylogenetic free for the phylum Protozoa. Other interpretations are possible. Modified from Corliss, 1959. *Syst. Zool.*, **8**, 169–190.

Protozoa probably branched off from them at an early stage, while the ciliates evolved later, possibly as a derivative of the complex zoomastigophoreans. The relationships of the Apicomplexa, Myxospora, and Microspora are, as already emphasized, uncertain, but they must surely represent several evolutionary lines. The higher plants, the sponges, and the Metazoa are shown as products of three offshoots of the early flagellates. That sponges arose independently of true metazoans is generally accepted. Less clear is the path of origin of the metazoans themselves, and the figure expresses this uncertainty. One thought is that they may not have a direct relationship with the flagellates, but possibly arose by cellularization of a ciliate type of organism. These are matters for interesting and informative speculation (Sec. 18–4), but the evidence is too incomplete to permit a confident resolution of the problem.

8–2 DIGESTION

The form and functioning of digestive systems is largely determined by the properties of their digestive enzymes. For example, enzymes show their optimum activity within sharply defined pH limits, and are highly selective in the type of molecule upon which they act, and in the manner in which they attack it. Thus proteins and polypeptides are digested by proteases that are specific for particular types of molecules, and that break specific bonds in the polypeptide chains. Carbohydrates, too, are attacked by a variety of carbohydrases, adapted for the digestion of particular substrates, although the degree of their specificity is a matter of some dispute. We cannot discuss these biochemical aspects of digestion here, but they are an important factor in determining the range of form and function in animal digestive systems.

Whatever the mode of food capture, the end result of protozoan phagotrophy is the ingestion of the food material into cavities called food vacuoles; it is within these that digestion takes place. This intracytoplasmic digestion, which must have been the first form of digestion to be practised by animals, has remained widely established in the Metazoa, where it is known as intracellular digestion.

This occurs in a comparatively simple form in *Amoeba*, where the food vacuole initially contains both food and fluid, the latter taken in during ingestion. The entry of this fluid is probably inevitable, and is certainly not surprising, for even when the animal is not feeding it takes up fluid into many small vacuoles, a phenomenon, called pinocytosis, that is also believed to be a widespread activity of the cell of higher organisms (Sec. 2–4).

At first the food vacuoles decrease in size, probably as a result of the diffusion of fluid outwards into the cytoplasm, and at this stage an increased acidity is detectable within the vacuole. Superficial comparisons have been drawn between this phase and the gastric phase of digestion in vertebrates, but there is probably no connection between the two. The acidity of the food vacuoles in *Amoeba* is comparatively slight, ranging no further than a pH of about 5.6, and probably results from chemical changes associated with the death of the prey, which ceases its movements during this phase. Later the acidity decreases, the pH increasing to a value of about 7.3; this is probably because of the inflow of fluid from the cytoplasm, a movement that may conceivably be aided by a rise of osmotic pressure in the vacuole. During this later, more alkaline, phase, digestion of the prey continues, for it now becomes broken down and diminished in mass. Presumably, then, digestive enzymes are now being secreted into the food vacuoles, and under their influence digestion continues until there remains only a small residue of indigestible material which is discarded from the body. This sequence of events demands a functional organization in the food vacuole which is not fully detectable by light microscopy. The electron microscope, however, reveals features that clearly demonstrate secretory and absorptive phenomena (Fig. 8–8). The enzymes are initially enclosed in small vacuoles surrounding the food vacuole, and are discharged into the latter by fusion of the small vacuoles with it. These small vacuoles are examples of the organelles called lyso-

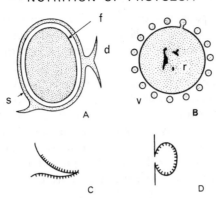

Fig. 8-8. The structural features of food vacuoles. A, a recently formed vacuole in a stage thought to coincide with the secretion of enzymes into the vacuole, and rapid digestion; *f* is the food organism, *d* are tongue-shaped diverticula which may facilitate secretion by increasing the area of the enclosing membrane, *s* is a layer of material which may be either products of digestion or the secreted enzymes. B, an older type of vacuole, probably in the absorptive phase; small vacuoles may be budding off to carry away digested material; *r* represents indigestible fragments. C, tongue-like diverticula of the membrane possibly represent absorption phenomena. D, possibly the budding off of vesicles during absorption. Adapted from Mercer, 1959. *Proc. R. Soc. B*, **150**, 216–237.

somes, which we have seen to be membrane-bound bodies of common occurrence in protozoans and also in metazoan cells. Their enzymes are agents in intracellular digestion in Metazoa (to be referred to later), and also in the phagocytic activity of certain types of blood cell. They also provide for the autolysis of dead cells, a process that facilitates the replacement and repair mechanisms of the body.

Another protozoan in which digestion has been closely observed is *Paramecium*. We have seen how the food of this animal is driven into the buccal cavity. Once it is there the fluid pressure exerted by the ciliary beat forces the food particles towards the distal end, which becomes distended to form a small sac. This sac is then constricted off as a food vacuole, probably in part through the agency of fibres that extend inwards at this point (Fig. 8–6). It is perhaps also a consequence of the action of these fibres that the newly formed vacuole passes quickly backwards towards the hind end of the body. Thereafter it moves more slowly through the body, probably through the influence of streaming movements in the cytoplasm; eventually it reaches the permanent anus (cytoproct), through which the indigestible residues are discharged.

As in *Amoeba*, there are changes in the pH of the fluid contents of the food vacuoles, but they are here much more marked; an acidity as high as pH 1.4 has been recorded by means of indicator dyes that have been ingested with the food. We cannot be sure whether or not this acidity arises in the same way as has been suggested for that found in the vacuole of *Amoeba*; the possibility that there is a true secretion of acid cannot be excluded, but even so this is not to be compared with the peptic digestive phase of the vertebrate stomach. In *Paramecium*, as in *Amoeba*, the death of the prey takes place during the acid phase, but later the acidity decreases, and digestion takes place at a pH of about 7.8. Further parallels with the digestive mechanism of *Amoeba* are seen in the diminution in size of the food vacuole during the acid phase, and its subsequent enlargement when the acidity decreases; this latter change is presumably associated with the passage into the vacuole of digestive secretion from the surrounding cytoplasm.

Information regarding the nature of the digestive enzymes of the Protozoa can be obtained partly by observing what types of food materials are actually broken down within the food vacuoles, as we have already seen in *Peranema*. It is possible also to identify the enzymes by chemical investigation of extracts prepared from cultures of the animals. By these means it has been shown that the enzyme equipment of these organisms compares well with that present in the digestive systems of Metazoa. This does not mean, however, that all Protozoa can necessarily digest all types of food material.

One element in the adaptive specializations of digestive mechanisms is certainly an emphasis upon the production of those enzymes that are most needed for the characteristic diet of a particular species. How far this principle operates at the protozoan level of evolution is not clear, for our information is still too limited, but it would appear that *Paramecium* can readily digest protein and fat, while only small amounts of ingested starch are broken down. It is said, too, that *Amoeba* is unable to digest starch, but its capacity for digesting protein and fat is well attested, the presence of a protease having been demonstrated by chemical methods. In the Metazoa the digestion of protein occurs in stages, the process being completed by peptidases that act specifically upon the short amino acid chains of dipeptides and tripeptides; this seems to be true also of Protozoa, for a dipeptidase has been demonstrated chemically in *Pelomyxa*. Indeed, the presence of peptidases is clearly indicated by the capacity of ciliates for maintaining growth in media that contain dipeptides and tripeptides; these are presumably broken down by the appropriate enzymes within the body.

As regards the relation between the pH in the food vacuole and the optimum pH of the enzymes, extracts of *Paramecium* show optimum proteolytic activity at pH 5.7 to 5.8, concordant with the usual optimum values for lysosomal enzymes. The significance of this result is admittedly limited, for the value may well represent the combined activities of a number of enzymes, including cytoplasmic enzymes that are not necessarily involved in digestion at all. Nevertheless, conditions within the vacuole are evidently quite adequate for the action of digestive enzymes, since the majority of these, as far as we can judge from studies on Metazoa, are likely to have pH optima close to neutrality.

The digestive activity of the Protozoa is confined within a small space, and has no multicellular differentiation to support it; yet it is clearly specialized to an extent that is as adequate for the requirements of these animals as is the alimentary tract of metazoan forms. In this, as in so many other respects, the Protozoa have a complexity of structure and function that is fully comparable with that of the whole body of a metazoan (p. 34).

9
Nutrition of Some Lower Metazoa

9–1 FOOD RESOURCES

The food resources available for exploitation by animals may be broadly classified as liquid or particulate, and, if particulate, as microparticulate or macroparticulate (massive). Of these categories, the first might seem especially valuable, for the waters in which so many species live are rich in dissolved nutrients. Inorganic and organic substances are continuously draining into them from the land, and are also in continuous production through the breakdown and decay of plant and animal life. As a result of this, the shallow seas contain dissolved organic matter to a concentration of about 4.5 mg/l, while the content of fresh-water lakes may be even greater. Aquatic animals would thus be amply provided with nutrients, if only they were able to make direct use of them. This, however, they cannot do to any great extent. At one time it was suggested that they might absorb dissolved nutrients through their gills and other permeable surfaces, but it is doubtful whether they usually do so in significant amounts. However, the Phylum Pogonophora provides an important exception to this.

Pogonophorans are benthic animals which live in the sediment of the sea floor in individual tubes made of chitin and protein. First described in 1914, they were thought at one time to possess a tripartite coelom; because of this, and because they have anterior tentacles, it was suggested that they were related to the deuterostomes. This view had to be discarded when, in 1964, it was shown that nobody, up to that time, had actually described a complete individual (Fig. 9–1). The missing part was the hind end (opisthosoma), which readily breaks off from the delicate trunk. This region is segmented, with muscular septa and with segmentally arranged chitinous chaetae, blood vessels, and nerve ganglia. Perhaps, as has been plausibly suggested, it serves as a contractile anchor in burrowing, or for moving up and down in its tube. But however this may be, the description of the opisthosoma, in

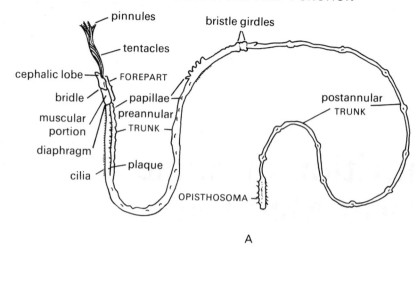
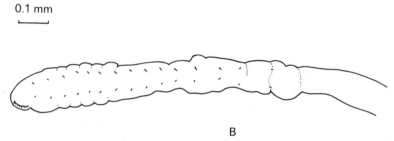

Fig. 9-1. A, diagram of a pogonophore, much shortened, showing the regions of the body. B opisthosoma of *Siboglinum ekmani*. From Southward, 1975. *Symp. zool. Soc. Lond.*, **36**, 235–251.

conjunction with certain ultrastructural features, has led to the view that the phylum may be related to annelids; the affinities, however, remain uncertain.

In the present context, the group is of unique interest because pogonophorans, although free-living animals, have completely lost their mouth, alimentary tract and anus. How, then, do they feed? There is no evidence for external digestion, but radiochemical studies have shown that they can take up amino acids and proteins present in the ambient sea-water. Sometimes this uptake is slower when they are inside their tube than when they are outside it, but this may not be true for all species. Ultrastructural studies reveal epidermal microvilli penetrating the overlying collagenous cuticle, and it is likely that absorption is effected by these, in conjunction with pinocytosis and an active uptake mechanism. Much more remains to be learned of the nutrition of these remarkable animals, but already enough is known to demonstrate once again how every available environmental resource is exploited by some form or other of animal life. Pogonophorans, of course, are exceptional, but many groups of animals of most diverse habits have found the sea to be highly favourable for life. Of the 50 or so classes of animals recognized in current taxonomy, 19 are purely marine, and only the Onychophora, myriapods, and Amphibia are wholly absent from the sea. The explanation of this, however, lies in something other than the direct utilization of dissolved nutrients; we shall see later what that explanation is.

Fluid feeding, in fact, is a highly specialized mode of metazoan life. It is seen in the osmotrophy of tape-worms, which take up material in solution from the intestinal contents of their hosts, making use of the properties of a complex and active body surface. Electron microscopy has shown that the supposedly inert and protective cuticle which was formerly thought to cover the surface is actually a living cytoplasmic layer, underlying a polysaccharide-containing layer, the glycocalyx. The two form part of a complex body surface which, in addition to making possible the passage of nutrients, presumably contributes to biochemical and physiological responses to fluctuating conditions in the host. There are also many animals that feed by suction (recalling the suctorial ciliates), a habit that is often aided by piercing structures. These are seen in a simple form in the jaws of the gnathobdellid leeches, which draw their food directly from the tissues of their host or prey, a habit that recalls the feeding mechanism of *Peranema*.

The exploitation of liquid food is particularly characteristic of nematodes, arachnids, and certain insects. Many free-living nematodes feed upon decaying products of plant and animal material; they are said to be saprophagous. It is a short step from this to making use of the fluid contents of living organisms, for the transition is facilitated by a characteristic feature of nematode organization, the buccal capsule. This structure, lying immediately behind the mouth, is commonly lined by a sclerotized cuticle, which can be elaborated into varying patterns of teeth and jaws, or may form a protrusible stylet. No less elaborate and variable is the next region of the alimentary tract, the pharynx, which can form an efficient muscular sucking organ. We have already noted that this may be a direct consequence of the peculiar way in which nematodes exploit hydrostatic principles. But whatever the reason, these animals are well equipped for the development of suctorial feeding, either as carnivores or herbivores. This capacity, together with the saprophagous habit that is widespread among them, has no doubt been an important factor in the evolution of the parasitic habit within the group. It contributes also to their success as members of the soil fauna, in which, numerically speaking, they are second in importance only to the Protozoa. Many of these soil nematodes are saprophagous, and many are predatory, the latter applying their lips to their prey, piercing it with their stylet, and sucking out its fluid content. This is a feeding method well suited to soil dwellers, for it depends upon the animals making use of a resistant substratum while they are thrusting with the stylet.

Fluid feeding among the arthropods depends upon the adaptive versatility of the arthropod limb, which has lent itself to the evolution of complex assemblies of crushing, piercing, and sucking appendages. It is particularly prevalent in the Arachnida, which typically employ a suctorial pharynx to remove the juices from their prey. The anterior appendages cooperate in this by crushing the prey, for instance, as in scorpions and spiders. In ticks the chelicerae form cutting and piercing organs, the fluids being drawn up through a channel formed by these organs and the median hypostome. In spiders a further adaptive device may be brought into use.

Some of these animals secrete enzymes into their prey, so that external digestion occurs before the fluid food is taken into the alimentary canal.

Fluid feeding in insects is particularly well developed in the Hemiptera and in certain flies. Some of the latter feed upon decaying organic matter, others upon the nectar and other fluids of plants, while the development of blood-sucking habits has led to dipterans becoming important vectors of disease-producing organisms. In accordance with this diversity of habit, the mouth parts of sucking flies are highly variable in form. Those of Hemiptera are less so, for in this group the members are all adapted for feeding by piercing and sucking. In principle, their maxillae and mandibles are modified to form slender stylets which are protected in a grooved labium. A sucking action draws in the food material, probably aided by capillarity and, in the case of plant feeders, by the pressure of the plant sap. As with flies, this feeding mechanism has led in some instances to the development of close relationships with other organisms. For example, plant-feeding aphids may have to ingest large amounts of fluid in order to obtain the protein that they need; the surplus is extruded as honeydew, which, being rich in carbohydrates, is attractive to ants. This leads in certain species to the establishment of remarkable symbiotic relationships between the two types of insect.

Animals, such as nematodes and hemipterans, that have evolved methods for feeding upon the fluids of plants, have circumvented a major barrier that impedes the exploitation of higher plants by the animal kingdom. This barrier, which is created by the resistant qualities of cellulose and lignin, is one that has not been easily overcome by digestive specialization. We shall refer to this matter later in the context of symbiotic relationships, for it is the use made of micro-organisms by higher animals that has provided another solution to the problem. For many animals, however, the immense food resources represented by the higher plants have remained unavailable as a direct source of nutrition.

These higher plants appeared relatively late in the evolution of life, but prior to their appearance animals must have been exploiting, as they still do today, the vast assemblage of phototrophs that live in the surface layers of the seas and inland waters. Sunlight penetrates water to an average depth of 100 m, so that within this limit it is possible for phototrophs to carry out photosynthesis and to absorb the dissolved nutrients that they require. Some of these plants are the seaweeds of the seashore and shallow waters, but these are comparatively unimportant. The most important component of this aquatic pasturage, both marine and fresh-water, is composed of micro-organisms that swarm in vast numbers in the upper levels of open waters. It is these that provide the energy-rich compounds upon which the vast aquatic fauna depends for its food supplies; these are provided in the sea to an extent that is fully comparable with the productivity of terrestrial grassland and forest.

That micro-organisms are concerned here, rather than multicellular plants with flotation devices, can be understood in the light of the physical factors that influence life in a fluid medium. There is an advantage in presenting to the medium a surface area that is large in proportion to the body volume, for this facilitates the absorption of nutrients through the body surface. Such a ratio of surface area to volume is found in small organisms, and particularly in unicellular ones; the development of multi-cellular structure would tend to reduce this benefit. There is also another advantage that phototrophs gain from their small size. They must remain in the upper layers of the sea, since it is only there that sunlight is able to penetrate. Consequently they will be aided by any property that retards their rate of sinking through water; small size is exactly such a property, for the relatively large surface area means that frictional resistance between the organism and its surroundings medium counteracts to a considerable extent the effect of gravity.

The phototrophic organisms that compose this floating and drifting pasturage are termed the phytoplankton (*planktos*, wandering). Certain bacteria are included in this assemblage, but first and foremost are the diatoms. These are unicellular algae that are characterized by the possession of a siliceous covering, a feature that in itself has an important limiting effect upon the action of the digestive enzymes of animals that ingest them. The rest of the phytoplankton are flagellates, and of these the Dinoflagellida are usually the most abundant. In addition there are the coccolithophores, which are flagellates with calcareous plates, and there are also enormous numbers of minute flagellates that constitute what is called the nanoplankton (*nanos*, dwarf). It is a remarkable fact, and a striking tribute to the efficiency with which animals have exploited these sources of food, that the existence of this nanoplankton was for a long time unknown, because the organisms were too small to be retained within the usual type of net used for collecting plankton samples. The first demonstration of its presence in the sea came from a study of the filtering activity of the urochordates known as Larvacea, which have developed a means of separating these minute organisms out of the surrounding water.

The exploitation of this aquatic pasturage depends largely upon its ingestion by herbivorous animals, which are ingested in their turn by carnivorous forms. Many of these animals are themselves small drifting organisms, living in continuous association with the phytoplankton, and constituting the zooplankton. Others are large, and may feed directly upon the plankton, or they may devour relatively large prey that have themselves fed upon it. In this way are built up the food chains which link organisms of diverse sizes into ecosystems (Sect. 4–1). Metazoans predominate in these communities, and have always attracted most attention, but protozoans also play a part. One example of this is the freshwater *Enteromyxa paludosa*, an amoeba-like organism which lives amongst the slowly gliding filaments of the alga *Oscillatoria* (Fig. 9–2). It engulfs these filaments, sometimes three at a time, and rapidly digests them (Fig. 9–3). Another example is a flagellate, *Pseudospora*, which enters an algal colony such as that of *Eudorina*, and feeds upon it, moving through it in an amoeboid way with passive flagella. These and other remarkable and still little-known protozoans are thought to be important in the control of planktonic algae in lakes. By feeding upon the larger algae, they supplement the action of the metazoan herbivores such as rotifers and crustaceans which tend to feed upon the smaller members of the phytoplankton. In addition to this ingestion of the plankton, there is also the catabolic activity of heterotrophic micro-organisms which bring about the breakdown of organic material without subjecting it to phagotrophy. This results in the production of inorganic nutrients which can then be taken up by the phototrophs in the cycling of nutrients that we discussed earlier. Bacteria play an important part in this, but major contributions are also made by diatoms, flagellates, yeasts, and fungi. All of these heterotrophs are added to the enormous numbers of minute organisms that form such an important part of the food resources available for animals. Those that feed directly upon this micro-

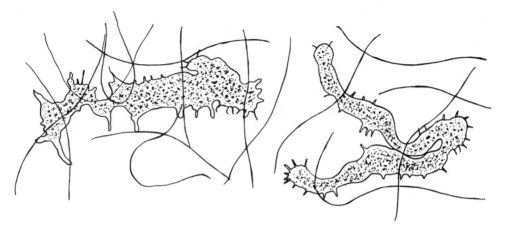

Fig. 9-2. Two individuals of *Enteromyxa paludosa* amongst filaments of the alga *Oscillatoria*. From Canter, 1966. *J. Linn. Soc. (Zool.)* **46**, 143–154.

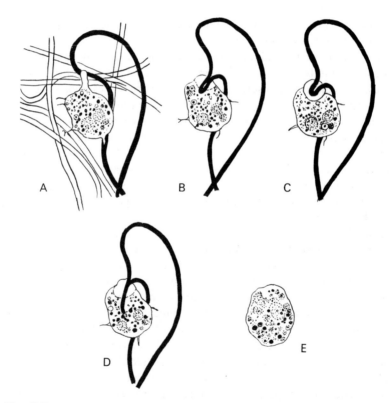

Fig. 9-3. Stages in the capture and ingestion of an algal filament by *Enteromyxa*. A thread is captured (A) by a hyaline process which is then slowly retracted, the filament being carried into a swollen area of clear ctyoplasm (B-D). Fifteen minutes elapsed between stages D and E; in this time the contents of the thread had been liberated into the body of the protozoan. From Canter, 1966, *op. cit.*

scopic life are said to be microphagous; alternatively, because of the small size of the particles that they ingest, they are said to be particulate feeders. Microphagy is so important in the economy of aquatic life that we must now consider it in some detail.

9–2 MICROPHAGY AND MACROPHAGY

Most generalizations that can be made about animal feeding mechanisms will have their exceptions, but we can begin by recognizing two types of particulate feeding. Both types often depend upon the use of cilia, the animals concerned being then termed ciliary feeders. One of the types, called suspension feeding, makes use of the minute organisms and other particulate material that are suspended in water; this usually requires the filtering of the water and the extraction of the food particles from it, so that organisms feeding in this way are often described as filter feeders. They are highly diversified in form, for the habit is distributed among many groups of the animal kingdom, but they share certain common requirements. Usually filtration demands the setting-up of a current in the surrounding water. This is commonly created by cilia, with mucus aiding the trapping and sometimes also the filtering, although the crustaceans provide an important exception to this. A great deal of the suspended material may be inedible or positively harmful, so that it is desirable to have some form of rejection mechanism. Further, even if the material is edible, the animal concerned may be limited as regards the size of the particles that it can deal with, so that some sorting mechanism is also desirable, combined, perhaps, with the rejection mechanism. Then, of course, food particles must usually be directed towards a mouth, although there are exceptions even to this apparently self-evident generalization, as, for example, in sponges and pogonophorans. Finally, when the particles have entered the alimentary tract, they must be suitably manipulated to ensure digestion and absorption, and the elimination of waste. Clearly the possibilities here for complexity of adaptation are immense, and under the pressure of natural selection, which tends constantly towards the improvement of the ways in which animals exploit their environment, they have been very fully realized.

Filter feeders include sessile and free-swimming forms, in both of which the capacity for movement, which is so characteristic of animals, is turned to advantage; indeed, the primary reason why movement is so much more typical of animals than of plants is that the former depend upon it so greatly for the securing of their food. The free-swimming filter feeder has the advantage of being able to move directly among its food, but the sessile animal uses cilia or appendages, which are themselves often locomotor structures in their origin. The sessile animal is aided by the existence of natural currents in the water, which help to replenish its food supplies and to remove its waste products. The extent to which these currents are exploited varies, but an extreme example is presented by the hexactinellid sponges, which make use of the steady current of their deep-sea habitat and spread the net-like structure of their bodies across it.

The second type of particulate feeding makes use of the deposits of organic material (detritus) that accumulate on the substratum; this is known as deposit feeding. Animals exploiting this method of feeding usually depend, like suspension feeders, upon ciliary action; indeed, there may not necessarily be a sharp distinction between the two modes of feeding, for one animal may make use of both sources of supply. However, the accumulation of detritus is not confined to the surface of the substratum; material also accumulates within the sand or mud, and so deposit feeding in the sense defined above grades into the swallowing of the substratum itself. This is seen in the lug-worm, *Arenicola*, for example, and in the echiuroids; these animals occupy very much the same ecological niche in the marine environment as do the earthworms on land.

The swallowing of sand, mud, or soil, while relatable to deposit feeding and thus to particulate feeding, may also be regarded as an example of the third method of feeding enumerated earlier; the exploitation of massive food material. Other and very different examples of this are seen in those animals that feed upon encrusting organisms such as algae, Polyzoa, and sponges. Such are the limpets (*Patella*), which use their finely toothed radula to browse upon algae, a method of feeding that is probably primitive in molluscs, and that has had important consequences in the group, as we shall see later. The rasping of the radula results in a stream of fine particles entering the alimentary tract, so that in one sense this, too, is particulate feeding. Here, as so often, is seen the difficulty of framing biological categories in such a way as to avoid any overlap between them. Another example of browsing habits is found in the regular echinoids (sea-urchins), the organ involved in this instance being Aristotle's lantern. In its fully developed form this comprises five teeth and a number of supporting parts amounting to some 40 pieces altogether. With the teeth growing continuously from sacs, and with complex associated muscles, this organ has a wide range of use; thus echinoids can act as scavengers, feeding on many types of dead animal material, and even upon live animals when these are sufficiently slow-moving to be captured by them.

Feeding upon large masses (often referred to as macrophagy) has the advantage of opening up for a group a wider choice both of food and of habitat, and, on the whole, it favours the attainment of larger size. On the other hand, it commonly involves active predation, or the pursuit of living prey, which demands the development of specialized behaviour patterns and thus makes considerable demands upon muscular and nervous organization. It might be expected, then, that microphagy would be found among the smaller and less highly organized species, and that macrophagy would tend to replace it in the larger and more complex ones. Ciliary feeding certainly occurs today in some decidedly primitive groups (the sponges and pterobranchs, for example), and it probably occurred in the earliest echinoderms. Moreover, it is found also in various other groups of small animals such as the Ectoprocta, the Entoprocta, and the Brachiopoda—groups that are of uncertain relationships, but which are primitive in many features of their organization.

Yet the feeding of many phyla show that events have not always followed such a simple course. For example, in the polychaetes it is the more primitive errant forms that are macrophagous and predatory, while the sedentary species are highly specialized for microphagy. Within the Phylum Mollusca we find ciliary feeding being exploited by the highly specialized lamellibranchs. Here its evolution can be seen as a consequence of the existence of the molluscan gill, or ctenidium, and of the occurrence of rasping feeding in primitive molluscs; both of these factors, as we shall explain later, can be regarded as pre-adaptations favouring the elaboration of filter and suspension feeding. Indeed, the cephalopods are the only molluscs in which predation has become fully developed, and it is significant that these are the members of the phylum that have broken away most completely from the old-established molluscan habits and organization.

The Crustacea form another group in which microphagy is associated with a high level of structural organization. Here it is to be attributed to the remarkable versatility of the arthropod limb. It is particularly characteristic of the smaller crustaceans, and it may well have been practised by primitive aquatic arthropods. The larger members of the class, and, in general, the remainder of the living arthropods, are macrophagous, although some, as already mentioned, practise suctorial feeding. These feeding habits are associated in part with the diversification of modes of locomotion, and in part with the assumption of terrestrial life, which, of course, disposes of the possibility of filter feeding.

Finally, the Deuterostomia began their history, as far as we can judge, with microphagy practised by sessile and tentaculate forms. Within the chordate line the ciliary feeding of the protochordates is succeeded by the macrophagy of vertebrates, a transformation associated with a unique invention called pharyngotremy, which is the development of perforations in the wall of the pharynx.

Already we have encountered an important difficulty in generalizing about feeding or any other aspect of the functional organization of invertebrates. Each major group, endowed with its own characteristic plan of structure, exploits the resources of its environment in its own particular way. The history of one group provides no basis for predicting the history of another. But this does not mean that no general principles can emerge from invertebrate studies. In the present instance we can see that both microphagy and macrophagy are widely distributed in the animal kingdom. This is to be expected, for both contribute to the establishment of organized biotic communities. Moreover, microphagy is not only appropriate for the nutrition of the simpler and smaller animals; it is also well suited to the specialized organization of more advanced groups. The degree of persistence of microphagy, the extent of its replacement by macrophagy, the secondary development of microphagy in primitively macrophagous groups—all of these are ultimately determined by the possibilities latent in the fundamental plan of organization of any particular phylum. We must therefore consider the more important groups individually if we are to appreciate the success with which animals exploit their food resources. A few examples will serve to show something of the elegance and complexity of the structural and functional adaptations of feeding mechanisms, as well as the principles that have influenced their evolution.

9-3 FILTER FEEDING AND DIGESTION IN SPONGES

Sponges, sessile and weaponless, live by drawing in an unceasing flow of water through minute pores. This flow is nutritive, it brings oxygen and it removes the waste products of metabolism, in so far as they do not diffuse direct to the outside. In the calcareous sponges (Class Calcarea) the pores are intracellular canals in specialized cells called porocytes, which are derived from the dermal epithelium. Porocytes may not be present in other sponges, but pores certainly are, and they are always small; consequently they constitute a simple but effective sorting device that permits the passage of only the smallest particles. The course that the water follows may be very complex, but in principle it passes through the pores into the central spongocoel, and leaves this by a large opening, the osculum (Fig. 4–17, p. 89). Each individual initially possesses one such opening. As the body grows, additional oscula may form, and we then regard the sponge as a colony (Sec. 21–2), each osculum representing one individual. The definition is admittedly somewhat arbitrary; individuality is much less well defined in these organisms than it is, for example, in coelenterate colonies.

This weak individualization, like the feeble capacity for movement that we have considered earlier, is indicative of the exceptionally low level of organization at which sponges exist. Another illustration of this, the limited coordination between individual cells, is well seen in the feeding mechanism. The current of water depends upon the completely uncoordinated beating of the flagella of the choanocytes (Sec. 4–4). As we shall see later, these cells, together with wandering amoebocytes, abstract the food particles from the water and immediately ingest them, so that no structural specialization for the manipulation of the food is required. What has influenced the body form, however, is the dependence on uncoordinated flagella for the creation of the water flow. This influence is expressed in three main levels of organisation.

The simplest arrangement is the asconoid type (Fig. 9–4), consisting of a single ascon chamber (Fig. 4–17, p. 89) lined by choanocytes. The uncoordinated and circular beat of the flagella contributes nothing to directional movement of the water, but a flow from the porocytes towards the osculum is encouraged by the small size of the former and the larger size of the latter, since the water will tend to flow towards the larger opening. The flagella, however, can only establish a very small head of pressure, which is a very inefficient way of keeping a central mass of water in continuous movement. (Records for siliceous sponges show that when a tube of 1 mm bore is inserted into the osculum the level of water in the tube only rises to a height of 2–4 mm.) This hydraulic difficulty has conditioned the increasingly complex canal system of the other two levels of organisation: the syconoid and leuconoid.

The syconoid type (Fig. 9–4) results from a folding of the body wall so that the choanocytes are restricted to radially arranged canals which are essentially ascontype chambers, the pores of which are now termed prosopyles. The radial canals discharge through internal ostia into a large excurrent chamber that opens at the osculum. This is the simpler type of syconoid structure, seen in *Sycon*. It becomes further elaborated by the development of a cortex, supported by spicules and overlying subdermal cavities. A more continuous body surface is thus formed, penetrated by dermal ostia which lead into a system of incurrent canals.

Fig. 9.4. Diagrams of various types of sponge structure. A, asconoid type. B, syconoid type, early stage without cortex. C, final syconoid stage, with cortex. D, leuconoid type with eurypylous chambers. E, leuconoid type with aphodal chambers. F, leuconoid type with diplodal chambers. Choanocyte layer in heavy black, mesenchyme stippled. *1*, mesenchyme; *2*, choanocyte layer; *3*, incurrent pore; *4*, prosopyles; *5*, radial canal; *6*, incurrent canal; *7*, osculum; *8*, spongocoel; *9*, internal ostium; *10*, dermal ostium; *11*, excurrent channel; *12*, flagellated chamber; *13*, aphodus; *14*, apopyle; *15*, prosodus. Adapted from Hyman, 1940. *The Invertebrates: Protozoa through Ctenophora.* McGraw-Hill, New York. Used by permission.

NUTRITION OF SOME LOWER METAZOA

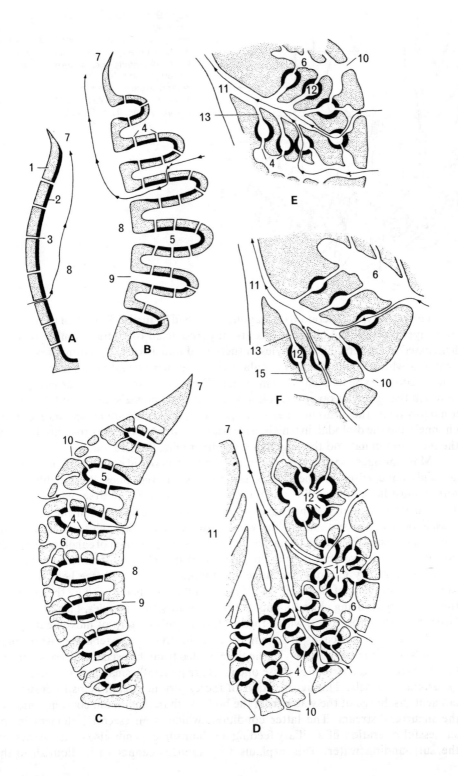

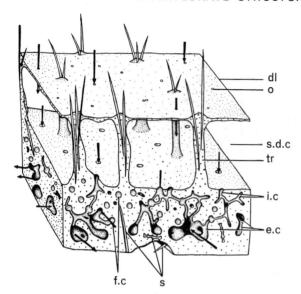

Fig. 9-5. Scheme of the outer layer of the body of *Spongilla*, reconstructed from living and fixed sections. *dl*, pseudo-epithelium; *e.c*, excurrent canal; *f.c*, flagellated chamber; *i.c*, incurrent canal; *o*, ostium; *s*, spicule; *s.d.c*, subdermal cavity; *tr*, trabeculum. Arrows indicate the water flow. From van Weel, 1949. *Physiologia comp. Oecol.*, **1**, 110–126.

The third type (leuconoid or rhagon) is reached by further folding of the choanocyte layer, so that the choanocytes are now restricted to a large number of small flagellated chambers. The apopyles of these lead into a complex system of excurrent channels which finally open at oscula without discharging into a main spongocoel. The leuconoid type of canal system (Fig. 9–5) exists in three grades: the eurypylous, in which the apopyles open direct into the excurrent channels; the aphodal, in which a narrow duct, the aphodus, lies between the flagellated chamber and its excurrent channel; and the diplodal, in which there is also a narrow duct, the prosodus, between the incurrent canal and the flagellated chamber (Fig. 9–4).

Most sponges (perhaps 98%) conform to the leuconoid plan, which must almost certainly have arisen independently along many different lines. Presumably, therefore, it must have outstanding advantages, and it is not difficult to see what these are. The leuconoid canal systems have an effect somewhat like that of the branching vessels of a blood system, for the water, after entering through the dermal ostia, moves increasingly slowly as it passes into an increasing number of canals of diminishing diameter. Movement within the flagellated chambers is thus slow enough to permit respiratory gaseous exchange and the filtering of food (see below). At the same time, movement of water out of the chambers is encouraged by the collars of the choanocytes being close to, and directed towards, the apopyles. These tend to be larger than the prosopyles, which further favours this pattern of movement.

When the water leaves the chambers it begins to move faster as it flows through channels of increasing size, and it is finally expelled from the oscula with considerable force, since the diameters of these openings are generally smaller than those of the excurrent channels. The acceleration of the excurrent stream helps to ensure the efficient discharge of the water from the body, with the minimum contamination of the incurrent stream. The latter condition, which is an essential element in the successful operation of a ciliary feeding mechanism, depends also on movement of the surrounding water. This explains why sponges cannot easily flourish in the absence of currents. It is further ensured in sponges by the form of the body, which often contributes to the efficient separation of the incurrent and excurrent streams.

Bidder, in his analysis of this aspect of sponge organization, called the angle between the intake and outflow currents the angle of supply. Between the two currents there is established a re-entrant vortex, the diameter of which he called the diameter of supply. This must be large enough to provide a good chance that surrounding currents or drift will carry away the outgoing water. In non-stalked sponges (Fig. 9–6A) the angle of supply is 90°. The presence of a stalk increases the angle, and thereby reduces the risk of contamination of the intake current. This means that the osculum can be opened out, since the water need not be ejected with as great a force; thus evolves the type of sponge body represented by Neptune's Cup (Fig. 9–6B). If the cup is set on one side, the angle of supply becomes 180°. Oscular velocity is no longer an important consideration, and so the body can open out into a flattened

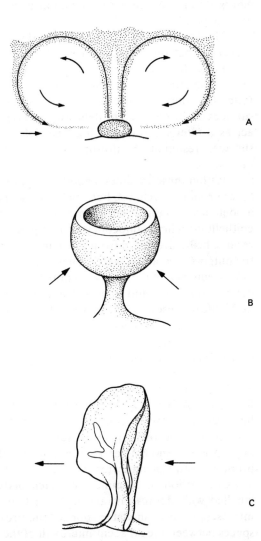

Fig. 9-6. Relationship of the form of sponges to their feeding currents. A, the bath sponge (*Euspongia*), a sessile form in which the angle between the inflow and outflow currents is 90°. B, Neptune's Cup, in which a stalk increases the angle of supply, so that there is less risk of mixing the two currents; the osculum therefore opens out. C, *Phakellia*, in which the body is a flattened fan, with one inflow face and one outflow face. Adapted from Bidder, 1923. Q. Jl microsc. Sci., **67**, 293–323.

form (Fig. 9–6C); this has the advantage of permitting a maximum flow of water through the body, and hence affords an improved opportunity of securing food.

The achievements of sponges are thus not to be underestimated. They are animals with the minimum of cell differentiation, and with so little coordination that they can almost be regarded as colonies of cells. Nevertheless, by the evolution of their form along lines that can be interpreted in simple hydraulic terms, an effective feeding mechanism is undoubtedly attained. However, the elaboration of a complex hydraulic system is not a necessary condition of this achievement. As we have earlier noted, some of the deep-sea hexactinellid sponges rely upon a body is shaped as a flat or curved net, with water entering on one side and leaving on the other. In this simplicity of structure they are exploiting, with a minimal expenditure of energy, the steady currents of the abyss. Setting their lattice-like structure across the direction of flow, and filtering particles that are brought to them, they are, in Bidder's words, 'a moment of active metabolism between the unknown future and the exhausted past'.

The low level of coordination in the Porifera is further shown in the digestive processes that follow the filtering of their food material. It would appear, from the few species that have been studied, that the choanocytes are responsible not only for maintaining the food current, but also for the initial ingestion of the particles. In the Calcarea, where these cells are particularly large, the food adheres to the outer surfaces of the collars. The delicate processes of which these are composed (Sec. 4–4) act as a filter for trapping food particles, which are then passed down the collar surface, presumably by protoplasmic streaming, to be taken up into food vacuoles at its base.

Within these vacuoles a good deal of digestion may take place, very much as in heterotrophic flagellates, but some at least of the material is transferred to wandering amoebocytes in the mesenchyme that separates the flagellated cells from the dermal epithelium. These amoebocytes complete the digestive process, the indigestible residue being discarded from them and eventually removed from the sponge body in the outgoing stream of water. These same cells also store reserve material, and thus, in consequence of their movement through the organism, provide the equivalent of a combined storage and transport system. Amoebocytes therefore play a central part in the life of these animals; as we shall see, they contribute also in an important way to the reproductive processes.

In the other groups of sponges, the choanocytes (Fig. 9–7) are smaller, and they seem to be concerned more with ingestion than with digestion, for food material is transferred more immediately from them to the wandering cells. This happens, for example, in the fresh-water sponge, *Spongilla*, where the amoebocytes both digest the food and also pass it on to other cells which may complete the breakdown. It is said, however, that they never pass material into the choanocytes, which implies that these cells must have their own self-sufficient digestive mechanism. Nevertheless, and appropriately enough for such a cell republic, even ingestion is not exclusively a property of the choanocytes. This can be well shown by feeding sponges with suspensions of carbon or carmine, or (much better) with dead bacteria that have been labelled with fluorescent antibody, so that their fate can be traced by ultraviolet microscopy. In fresh-water sponges the prosopyles are 5 μm in diameter, while the spaces between the adjacent microvilli of the collars are of the order of 0.1 μm. Thus,

Fig. 9-7. Scheme of water flow in the flagellated chamber of *Spongilla*. *p*, prosopyle, lying between choanocytes. From van Weel, 1949. *op. cit.*

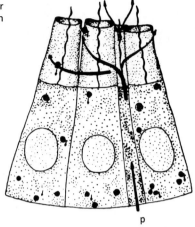

the flagellated chambers are well adapted for filtering individual bacteria, which fall within this range. Larger particles, however, such as clusters of bacteria, can still be captured. They can be phagocytosed by the dermal cells if they are too large to enter the ostia, or by amoebocytes and other cells if they are trapped in the incurrent spaces.

Whether or not sponges can make use of soluble organic material is uncertain, although the possibility cannot be excluded. A complicating factor in studies of the nutrition of these animals is the presence within the mesenchyme of many species of a rich flora of bacteria and unicellular algae. The physiological significance of the bacteria is not clear, but the algae are thought to be symbionts (Sec. 23–2).

9-4 FEEDING AND DIGESTION IN COELENTERATES

Coelenterates, with a level of cellular differentiation that is so much more advanced than that of sponges, show a correspondingly greater complexity in their feeding and digestive mechanisms. They are all specialized for a carnivorous diet, capturing their food by a highly discriminatory sensory and motor system. Tentacles play a prominent part in this, used in conjunction with various combinations of mucus secretion, ciliary movement and the action of cnidoblast cells.

Cnidoblasts, so characteristic of coelenterates, secrete within their cell bodies the structures called nematocysts (Fig. 9–8), which consist of a pear-shaped capsule containing an inverted tube attached to one pole of the capsule wall. On appropriate stimulation the tube is extruded by inversion, through a mechanism not fully understood, to form a thread that fulfils a function determined by its form. The range of form (no less than 17 different types have been described) is illustrated by the four types found in *Hydra* (Fig. 9–8). One of these is the desmoneme or volvent, which has a coiled thread, with minute bristles, that winds round any surface projections of the prey. Another type is the penetrant or stenotele (it seems profligate to carry precision to the point of describing this as a stenotelic rhopaloidic heteronemic stomocridic nematocyst, although such is indeed its designation in one system of classification). This has a basal enlargement called the butt, on which are situated three rows of

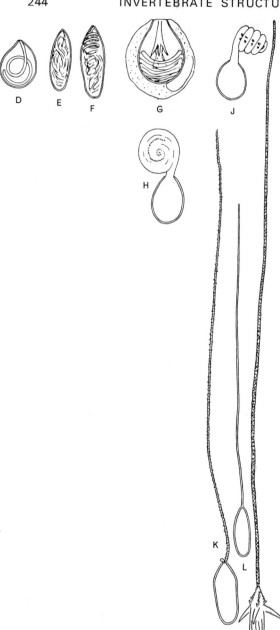

Fig. 9-8. The nematocysts of *Hydra littoralis*, from life. D–G, undischarged: D, desmoneme; E, atrichous hydrorhiza; F, holotrichous isorhiza; G, stenotele inside its cnidoblast. H–M, the same, discharged; H, desmoneme seen from end view showing spiral of thorns; J, desmoneme seen from the side; K, holotrichous isorhiza; L, atrichous isorhiza; M, stenotele. From Hyman, 1940. *op. cit.* Used by permission.

spines, the lowest members of each row being much enlarged to form a stylet. This nematocyst is a piercing structure, which is believed to inject poison into the prey. Also present in *Hydra* are atrichous isorhizas, with a smooth thread, and holotrichous isorhizas, with barbs throughout its length. These barbs are folded within the capsule;

as the tube everts, they point forwards, and then flick backwards like blades of a penknife. They give the impression of being formidable weapons, but the force of discharge is apparently not very great, for their movement can be arrested by a drop of paraffin. Both types of isorhiza are probably concerned with the injection of poison.

By contrast, the anthozoan Subclass Octocovallia has only a single type of nematocyst, which is said to be absent from the Order Actiniaria of the Subclass Zoantharia, or at least rare in it. Further, the Zoantharia have structures called spirocysts, which have been thought to differ from true nematocysts in certain respects, including the nature of the wall, staining properties, and permeability to water. These structures are not found in either the Hydrozoa or Scyphozoa. Some phylogenetic significance has been attached to such matters, but the difficulty of observing these small structures is so great that there are understandable disagreements in this field, and phylogenetic implications should certainly be handled with caution.

Considering the exceedingly small size of nematocysts, the capsules of which commonly range in length from only 5 to 50 μm, their structural specializations are sufficiently remarkable, but this is not the whole story. The toxins that they inject are sometimes very distressing. The siphonophore *Physalia* catches and paralyzes fish with its 9 m long tentacles; its nematocysts are potentially dangerous to man, and can raise inflamed welts on the skin. Much more deadly are certain scyphozoan medusae (e.g. *Chironex fleckeri*). These have been responsible for some 50 recorded deaths in Australian tropical waters, although most of their stings are usually much less dangerous than this implies. Nevertheless, severe pain and frenzied agony, followed within a few minutes by sudden and complete collapse, may be the melancholy consequences of extensive entanglement with the tentacles of these virulent animals.

Another aspect of these specializations is the remarkable flexibility and subtle adaptations of the cnidoblast action system. Nematocyst discharge is influenced by various factors, including temperature and the physiological condition of the animal, and individual cnidoblasts differ in these respects. For example, the discharge of atrichous isorhizas in *Hydra* is evoked by mechanical stimulation and inhibited by an extract of copepods, while the discharge of stenoteles is augmented by this extract. The influence of physiological conditions is well shown also by the reaction of a satiated *Hydra* to living *Artemia*. It discharges its nematocysts when the prey touches the tentacles, but, in contrast to the behaviour of a starved animal, quickly drops the prey because the cnidoblasts are squeezed out of the epidermal cells.

Fundamental to the specialization of cnidoblast responses is the fact that each can only discharge one nematocyst; they would therefore be largely wasted if their discharge was indiscriminate. It is thus important that their response should be so controlled that it is only evoked in the presence of suitable food material. This control is, in fact, ensured, but it is not exerted by the nervous system, as can readily be demonstrated in larger coelenterates such as the sea-anemone, *Anemonia sulcata*. Electrical stimulation of the tentacles of this animal through micro-electrodes results in the discharge of nematocysts only in the immediate region of stimulation; there is no evidence at all of any conduction to other parts of the body. This is because cnidoblasts are able to respond independently of any other tissue element; they are independent effectors, containing within themselves both the sensory and the motor properties needed to evoke discharge, although the nervous system may sometimes determine the threshold at which they are excited. It is commonly supposed that the projecting cnidocil of cnidoblasts of *Hydra* is the sensory element, but the presence of this structure is not essential. None is associated with the spirocysts of anemones, nor, probably, are they possessed by all of the true cnidoblasts of these animals.

No less important than the functional independence of the cnidoblast is the nature of the stimulus that triggers its response. Mere physical contact of prey and tentacle would not be sufficiently precise, for such contact stimulus could equally well be given by inanimate objects. A chemical stimulus would be another possibility, but this by itself would again be unsatisfactory, for chemical substances, diffusing from the prey might evoke discharge before the tentacles could make the essential contact. These difficulties have been overcome by the evolution of a specialized sensory mechanism requiring mechanical stimulation. The response, however, depends upon the source of the stimulus. If it originates from animal material a discharge of nematocysts is readily evoked, but if the source is inorganic a response will only result from a considerable degree of stimulation. For example, a smooth capillary ball produces no discharge (Fig. 9–9), but some discharge, although still very limited, is seen if the tentacle of an anemone is scratched with a broken capillary rod.

Another requirement is that there must be chemical stimulation by material present in the prey. The effect of this is shown in the copious discharge that takes place when a tentacle is touched by a capillary that has been smeared with molluscan extract. The nature of this stimulating material is not wholly clear, but it appears to be a lipid that is strongly adsorbed on to protein, from which it can be removed by treatment with ethanol or acetone. Its effect is to sensitize the cnidoblasts so that the threshold of their response to mechanical stimulation is markedly lowered. Material with this property seems to be widely distributed in animal products, for a human hair can also evoke discharge of the nematocysts (Fig. 9–9). Human saliva will similarly produce this effect. A tentacle immersed in sea water to which a small amount of saliva has been added will discharge a few nematocysts without any mechanical stimulation at all, and will produce a copious discharge if it is then touched with a capillary ball.

The result of this adaptation is that nematocysts will normally be discharged only in the immediate and close presence of food; diffusion of material from the latter, or direct contact with it, will produce the chemical sensitization, and the mechanical stimulus of contact will complete the process of discharge. The cnidoblast, then, is not merely an independent effector, but an independent effector with a dual sensory mechanism. As Pantin points out, there is no obvious analogy to this situation in any of the tissues of higher Metazoa. It is a good demonstration of the way in which precise adaptation enables animals with an inherently simple organization to secure from their environment the essential requirements for their nutrition.

It is not enough, however, for a coelenterate to capture and paralyse its prey. The food must next be swallowed. This process is well exemplified in *Hydra*, where it involves a characteristic feeding reaction (Fig. 9–10), in which the tentacles, after first extending and writhing, contract and bend towards the mouth, which enlarges to receive the food; ingestion follows when the food touches it. In this response, as in the capture of prey, the animal is able to discriminate. Inert objects are ingested, provided that they have touched the mouth, but the initial feeding reaction is evoked only by living animal prey, although it can also be evoked by inert animal material if this is first moistened with the juice of dead animals.

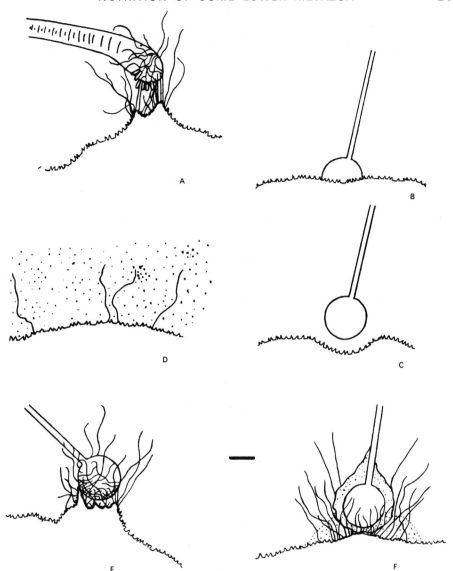

Fig. 9-9. Responses of tentacles of *Anemonia*. Bar = 100 μm. A, response of cnidae to touch by human hair; B, C, lack of response to clean glass bead; D, response to immersion in 1% dry weight of human saliva in sea water; E, sensitization of cnidoblasts to glass bead by 5 min immersion in 0.1% dry weight of saliva in sea water; F, response to glass bead smeared with alcoholic extract of *Pecten* mantle. From Pantin, 1942. *J. exp. Biol.*, **19**, 294–310.

Chemical and biological studies have shown that the feeding response of *Hydra littoralis* (which can be expressed quantitatively as the time during which the mouth is kept open) is specifically activated by the reduced form of the tripeptide glutathione,

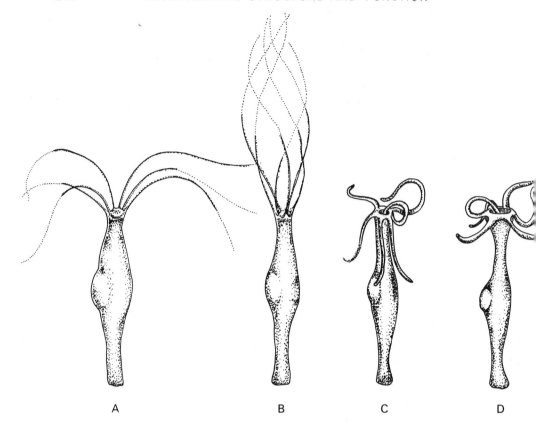

Fig. 9-10. Stages of the normal feeding reflex of *Hydra*. The tentacles first extend and writhe (A, B), and then contract and bend towards the enlarged mouth (C, D). From Lenhoff, 1961. *J. gen. Physiol*, **45**, 331–344.

and that the animal shows remarkable discrimination in its recognition of this substance:

$$H_3N-\underset{CO_2}{CH}-CH_2-CH_2-CO-NH-\underset{CH_2SH}{CH}-CO-NH-CH_2-COOH$$

Glutamyl Cysteinyl Glycine

Glutathione

Even asparthione, which differs from glutathione in being only one methylene group smaller, is unable to substitute for it. This capacity is not peculiar to this particular genus. The siphonophore *Physalia physalis*, widely separated taxonomically from *Hydra*, also shows a glutathione-mediated feeding reaction, while proline functions in this role in *Cordylophora*.

The testing of *Hydra* with related compounds shows that the thiol grouping of glutathione is not required, and that it is the tripeptide chain that the animal recognizes. The glutamyl region is particularly important, for the action of glutathione can be competitively inhibited by glutamine and by glutamic acid, but not by any other amino acids. No doubt the molecule is recognized by a specific receptor site, which might, for example, possess charged groups that become bound to complementary groups on the activator. Apparently glutathione is not chemically altered during the binding, and must be constantly present at the site if the feeding reaction is to be maintained. This has led to the suggestion that the activator causes a reversible allosteric modification of the tertiary structure of the receptor molecule. A similar view has been advocated as one explanation of the reaction of hormonal molecules with their target cells. Such similar viewpoints are not surprising, for both phenomena can be seen as aspects of the very wide spectrum of activities encompassed by chemical communication (Sec. 17–8).

In the present instance, the recognition of the activator molecule certainly seems to be more than an example of simple chemical olfaction, for the result of the activation is expressed through neuromuscular pathways as a coordinated manipulation of the prey. Glutathione is particularly well suited for this function. It is a common molecule, but is only present in living animals, from which it is readily released when they are pierced by the barbed penetrants. Thus the feeding reaction is only brought into use when living prey has been captured by the nematocysts. Cnidoblasts, their nematocysts, and a highly specific chemical sensitivity thus work together to ensure that the feeding response of *Hydra* is focused upon material that will be of the maximum benefit to the animal.

9–5 INTRACELLULAR AND EXTRACELLULAR DIGESTION: COELENTERATES

The mode of digestion in sponges, in which the food is broken down inside the cell within food vacuoles, is termed intracellular digestion. We must suppose that it has been directly taken over from protozoan ancestors. Indeed, there is at least one striking point of resemblance between the two groups, for the food vacuoles of the amoebocytes are at first acid and later alkaline in reaction, exactly as in the protozoan examples considered earlier.

Obviously, intracellular digestion greatly restricts the size and variety of the food that an animal can utilize. Because of this there has been a widespread tendency for the increase in size of animals to be associated with the establishment of extracellular digestion, in which the digestive processes take place largely within the lumen of an alimentary canal under the action of digestive enzymes that are extruded from the cells of its lining. Such extrusion is already foreshadowed in the Protozoa; the heliozoan *Vampyrella*, for example, is able to extract the cytoplasm of algae by boring through their cell walls with its pseudopodia, apparently by extracellular action.

In the coelenterates digestion is a combination of the intracellular and extracellular methods. At first sight it may seem that the coexistence of the two is an intermediate stage of evolution, associated with the generally simple organization of those animals. Consideration of other groups, however, shows that the relationship between intracellular and extracellular digestion is by no means always so straightforward as this. If it were, we should expect to see the extracellular method soon replacing the more primitive and restricted intracellular one. This certainly does happen in some groups, but not in all; on the contrary, the intracellular method persists in such highly specialized forms that we must assume that it has been retained because it confers some positive advantage.

The significant fact here is that the presence of well-developed intracellular digestion is commonly associated in higher animals with filter feeding, although this association is not invariable. It is well suited to provide for the digestion of small particles, but only if certain contingent requirements are met. The trapping of food particles during their passage through the alimentary tract demands a relatively large area of ingestive epithelium, which in its turn creates the need for extensive morphological and functional specialization of the alimentary tract. Nor is this all. We shall see examples of collecting mechanisms playing an important part in the sorting of food particles, but often the alimentary tract has also to make a contribution to this. It has further to provide for mixing the food with an extracellular secretion that may be produced by the digestive epithelium, and to delay the passage of the mixture through the alimentary tract. This delay is needed both to allow time for any preliminary extracellular digestion that may be provided for, and also to permit the ingestion of an adequate proportion of the filtered particles.

Extracellular digestion, by bringing a concentration of secretion to bear upon the food, undoubtedly facilitates the breakdown of the latter, but this method of digestion also makes its own demands. It requires a control of pH within the lumen of the alimentary tract, and, in consequence, some degree of regional differentiation. There must also be provision for the movement and mixing of the food and secretion, and for ensuring that the enzymes are economically used by being discharged only when they are actually required. There is, in fact, a balance of advantage and disadvantage to be weighed in comparing one method with the other, and different groups have, so to say, assessed the situation in different ways.

The persistence of intracellular digestion in the coelenterates is what we might expect from their comparatively lowly position, but extracellular digestion is also well established in them, permitting them to deal with relatively large prey. *Hydra* is one example of this, capturing small crustaceans and annelids by the tentacles and transferring them into the mouth, which can be enormously distended to receive the prey in the feeding response that we have just examined. In this instance the food material, although small, is large in relation to the size of the hydroid, which may therefore be described as macrophagous. So effective is the extracellular phase of digestion in *Hydra* that a *Daphnia* is broken down into small particles within 4 hr of its ingestion, as a result of the secretion of enzymes into the coelenteron. Undigestible remnants, such as the exoskeletons of the prey, are ejected from the mouth.

Two types of cell (secretory and absorptive) are concerned in these digestive processes. The enzyme-secreting cells, which are club-shaped, are most abundant in the hypostome, plentiful also in the column, but rare in the tentacles, peduncle, and base. Their zymogen granules are released by apocrine secretion, the tip of the cell being pinched off and the contents replaced from the more basal region. The absorptive cells are columnar epitheliomuscular cells, with one or more pairs of flagella projecting into the coelenteron, together with many branching or anastomosing microvilli. Partially digested food is taken up by pinocytosis, so that cells of the hypostome and stomach regions, in well-fed animals, contain many pinocytotic vesicles and digestive vacuoles, as well as reserve materials in the form of lipid droplets and glycogen. The corresponding cells of the peduncle and tentacles are much less active in digestion. They do, however, contain large intracellular cavities, and probably function in storage. The disposal of intracellular waste, always a requirement in this mode of digestion, is dealt with by the fragmentation of the absorptive cells, so that portions containing indigestible residues can be extruded into the coelenteron.

Macrophagy is not restricted to polyps. It is found also in some sycphozoan medusae, the tentacles catching the food and conveying it to the mouth. *Haliclystus*, for example, which is essentially a sessile medusa, has been observed feeding upon the amphipod *Caprella*; the prey is trapped with the bunches of capitate tentacles borne on the margin in the umbrella, and transferred directly to the short quadrangular manubrium.

In contrast to this, the scyphozoan jelly-fish *Aurelia* is microphagous, making use of mucus and ciliary movement (Fig. 9–11). Placed in a suspension of plankton, a medusa 10 cm in diameter can clear the larger planktonic organisms from 700 ml of water in less than an hour. Food collecting, which is aided by the normal swimming movements, takes place mainly at the margin of the umbrella, on the oral arms, and on the subumbrellar surface. The food is trapped in mucus and moved centrifugally, by ciliary action, over both the exumbrellar and subumbrellar surfaces. Much of the exumbrellar material is lost, but the rest of the food collects in the marginal groove at the edge of the umbrella. From here it is licked off by the oral arms.

Ciliary currents convey the mucus and food from these into the gastric pouches,

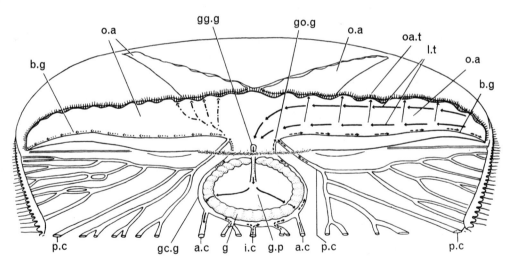

Fig. 9-11. *Aurelia:* oblique view of part of the disc from the sub-umbrellar surface, showing some of the currents in two of the four oral arms and in one of the four gastric pouches. The right side shows the path of food (solid arrows) in the lateral tract, and of excretory matter (broken arrows) in the basal groove; the vertical arrows represent the rejection reaction in the lateral tract. The left side illustrates the main paths of the gametes at spawning (path of sperm in male shown by dotted arrows, path of eggs in female shown by dotted and dashed arrows). *a.c*, adradial canal; *b.g*, basal groove tract; *g*, gonad; *gc.g*, gastro-circular groove; *gg.g*, gastro-genital groove; *go.g*, gastro-oral groove; *g.p*, gastric pouch; *i.c*, interradial canal; *l.t*, lateral tract; *o.a*, oral arm; *o.a.t*, oral arm tentacle; *p.c*, perradial canal. From Southward, 1955. *J. mar. biol. Ass. U.K.*, **34**, 201–216.

where any living prey is killed by nematocysts borne on filaments. It then enters the coelenteric canal system, passing outwards along the unbranched adradial canals to the marginal canal, returning along the branched perradial and interradial canals, with ciliary mechanisms operating at various stages of this progress to provide for selection of material. Waste material is finally removed by outwardly directed currents on the oral arms. Digestion in *Aurelia* begins in the stomach, and, according to the usual coelenterate pattern, is completed after ingestion by gastrodermal cells, which can occur at any part of the canal system. The circulation in this system, which takes about 20 min for the complete round, is clearly well adapted for the shape and size of the body, and provides for the distribution of nutrients which, in higher forms, is effected by a vascular system.

A comparable diversity of feeding methods is found in anthozoan polyps. Many feed on relatively large prey, caught on the tentacles, and transferred by these to the mouth which opens to receive it. Other anemones, however, make use of smaller organisms and rely more on mucus and ciliary tracts. An example is *Metridium*. Here, food particles and minute organisms are caught by the branched tentacles in a mucous secretion, and then moved across the oral disc by ciliary action.

A combination of intracellular and extracellular digestion is found in the Anthozoa, as in other coelenterates, but the structural adaptations for digestion are more complex than in hydrozoan polyps because of the subdivision of the coelenteron by the septa. The free edges of these are thickened to form mesenterial filaments, the organization of which differs as between the Alcyonacea and the Zoantharia. In the former the two asulcal mesenteries (i.e. the two lying on the side opposite to the single sulcus) bear ectodermal filaments which are heavily ciliated; these create an upwardly directed current which balances the inward current of the sulcus. The other septa bear endodermal filaments which are richly provided with gland cells.

In the Zoantharia, the mesenterial filaments, borne on all but the smallest mesenteries, have a trilobed structure in their upper regions. The median lobe (cnidoglandular lobe) is provided with gland cells and cnidoblasts, while the lateral lobes are strongly ciliated. Intermediate tracts (zooxanthella tracts) may lie between the median and lateral ones, packed with zooxanthellae (Sec. 23–2), and having also an excretory function. The lower regions of the mesenteries may bear only the cnidoglandular tract, and in the more primitive anemones this may be the condition throughout. In some species the lower ends of the mesenterial filaments are prolonged into filaments called acontia. These are similar in their histology to the cnidoglandular tracts, and presumably contribute further to the efficiency of digestion. They can also be extruded through the mouth, particularly when the animal contracts, and so have a defensive function as well.

Preliminary digestion in Zoantharia takes place, as in *Hydra*, in the general lumen of the coelenteron, but here the process is aided, at least in macrophagous forms, by the grasping of the food by the mesenterial filaments. This is possible because the filaments are armed with many nematocysts. In this way the digestive secretion is directly applied to the food, an arrangement that doubtless increases the efficiency of action of the enzymes. Ingestion of the partially digested food, and the completion of digestion, takes place over the whole extent of the gastrodermis, including that of the septa, but excluding the epithelium of the cnidoglandular tracts.

9-6 FEEDING AND DIGESTION IN PLATYHELMINTHES AND NEMERTINEA

The free-living platyhelminths provide another example of intracellular digestion playing an important part in a comparatively lowly group of animals, which yet depend on specialized feeding mechanisms. Of particular interest here is the remarkable way in which particulate feeding and intracellular digestion are organized in certain species so as to permit the handling of large prey.

In the acoelan *Convoluta*, the nutrition of which will be referred to later in the context of symbiosis (Sec. 23-2), the endoderm forms a solid syncitium which is protruded through the ventral mouth. Digestion is necessarily intracellular, and the food is engulfed into food vacuoles much as though the syncitium were a giant pseudopodium.

Digestion is largely intracellular in the triclad *Polycelis*, but not entirely so. This animal feeds on bulky invertebrate prey, such as oligochaetes and crustaceans, which it traps in a mucous secretion, holding fast to the substratum to resist the prey's escape. Its long (plicate) protrusible pharynx (Fig. 9-12) is inserted into the body of the prey so that the whole of its soft contents can be withdrawn, the process being aided by the production by the pharynx of a peptidase. The withdrawal is effected by the passage of waves of contraction over the proboscis; these, in conjunction with the enzyme action, break up the food into small particles, while it is passing to the alimentary canal, and thus prepare it for intracellular digestion.

The alimentary canal is lined by gland cells and phagocytic cells. The gland cells secrete a peptidase, similar to that of the pharynx, which continues extracellular

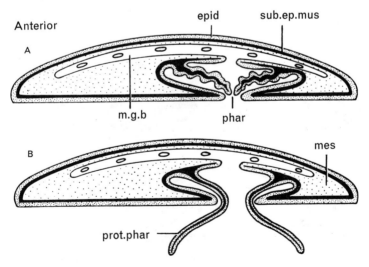

Fig. 9-12. Diagrammatic longitudinal sections of *Polycelis* to show the cylindrical plicate pharynx. A, the normal condition with pharynx retracted; B, pharynx protruded for feeding; *a.g.b*, anterior gut branch; *epid*, epidermis; *m.g.b*, median gut branch; *mes*, mesenchyme; *phar*, pharynx; *p.g.b*, origin of posterior gut branches; *prot.phar*, protruded pharynx; *sub.ep.mus*, sub-epidermal muscles. From Jennings, 1957. *Biol. Bull. mar. biol. Lab., Woods Hole*, **112**, 63-80.

digestion in an acid medium, at an optimum pH of about 5. Thereafter, the food particles are taken up into food vacuoles within the absorptive cells for intracellular digestion. The first phase of this is proteolytic, again in an acid medium, after which the pH of the vacuolar contents rises to pH 7.0–7.5. In this alkaline phase, proteolytic digestion is completed by further peptidases, and carbohydrates and fats are broken down. Indigestible material is voided from the cells into the gut lumen; in due course this is washed out by the taking in of water through the mouth and the flushing out of the gut contents by contraction of the general musculature. Strictly the animal is microphagous, judged by the size of the material that it ingests. It is thus practicable for the animal to rely upon this, and yet be able to exploit large prey. In precisely the same way the land planarian *Orthodemus terrestris* is able to feed upon slugs and worms, although its digestion, too, is entirely intracellular.

The diversity shown by closely related forms is apparent when we compare these animals with the polyclad *Cycloporus papillosus*. This is said to feed exclusively upon colonies of *Botryllus* and *Botrylloides* in a very selective manner, sucking the individual zooids into its alimentary tract through its protrusible pharynx. The plicate pharynx of a polyclad (Fig. 9–13) differs from that of triclads in having the form of a curtain hanging from the wall of the pharyngeal cavity. Perhaps it is because of this that the zoids are still undamaged when they arrive within the body. In this instance, and in complete contrast to *Polycelis*, digestion is extracellular; the food is homogenized and broken down within the gut, with no sign of intracellular digestion.

Feeding in nemertines shows advances upon the corresponding processes in platyhelminths, both in food capture and in digestion, although, in principle, digestion follows similar lines. Decaying material can be swallowed directly, but living animals are caught by eversion of the proboscis, which wraps around them. Some

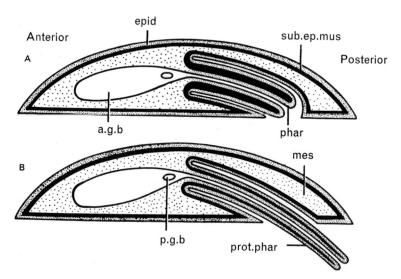

Fig. 9-13. Diagrammatic longitudinal section of *Leptoplana* to show the ruffled plicate pharynx. A, the normal condition with pharynx retracted. B, the pharynx protruded to envelop food. Abbreviations an in Fig. 9–12. From Jennings, 1957. *op. cit.*

nemertines (Subclass Enopla) have a proboscis armed with piercing stylets and a poisonous secretion that kills the prey. In *Lineus*, which is one of the unarmoured heteronemertines (Subclass Anopla), the prey is killed in the foregut by acid secretions, the chemical basis for this being an abundance of carbonic anhydrase in the gland cells of this region. This enzyme brings about the release of hydrogen ions, as it does in the production of acid in the vertebrate stomach; the nature of the acid released by nemertines, however, is not known.

Extracellular proteolytic digestion takes place in an acid medium (pH about 5) in the intestine, and this is followed by phagocytosis and intracellular digestion in an alkaline medium. Digestion is thus completed much as in platyhelminth triclads, but with the intracellular acid phase being lost.

10
Filter Feeding

10-1 FILTER FEEDING IN POLYCHAETES

With the increasing development of cell differentiation and coordination that marks the history of the Metazoa, there arose methods of filter feeding much more elaborate than those of sponges. This type of feeding is found only in aquatic animals, the absence of filter feeding from terrestrial forms being largely related to the much lower density of air as compared with that of water, and to the consequent lack of inert organic particles floating in it. However, air does contain an abundant supply of flying insects, the trapping of which by birds and web-spinning spiders can be regarded as the nearest approach to filter feeding on land. Filter feeding in aquatic animals is not easy to understand, yet it is worth while trying to do so, for the mechanisms involved are unexcelled for the precision and beauty of their adaptive organization. The principles involved show a considerable degree of uniformity over a wide range of species; this is to be expected, having regard to the uniform character of the food material and to the limited range of structures that animals can deploy in capturing it.

Good examples occur among the polychaetes. Primitively, these worms were probably free-moving and macrophagous, following the mode of life familiar in such animals as *Nereis* and *Nephtys*. These have an eversible pharynx that is used both for burrowing (p. 165) and for feeding. Prey is seized by the hooked jaws that arise from the lining of the pharynx; these are situated at its tip when the pharynx is fully everted. Microphagous feeding is characteristic of the sedentary species, and involves mechanisms so specialized that it is difficult to see immediately any close relationship between the two modes of life. Yet with feeding, as with other functions, we cannot suppose that such specializations arose already fully organized—they must have evolved stage by stage, under the influence of natural selection. Closer analysis often suggests in such instances that the more primitive forms possessed structures suitable for adaptive modification in a particular direction, even though they may have initially served some quite different function. We have earlier referred to this as pre-adaptation; it is a phenomenon that must have been important in facilitating evolutionary change.

Examples of it occur in the errant polychaetes. Like so many creeping and burrowing animals, they produce over their body surface a mucous secretion which protects the surface and forms temporary linings to burrows. *Nereis diversicolor* sometimes forms this secretion into a net within its burrow; water can then be pumped through this net, so that it can be used for a simple form of filter feeding. Particles collect in the secretion as though in a bag, and from time to time the material is swallowed. We can visualize that the further elaboration of some such mechanism might have been aided by the presence on the head of tentacles and palps, which are used by errant worms for sensory purposes and to assist in the manipulation of food. These structures have given rise in sedentary worms to complex and beautiful systems of tentacular outgrowths, often called gills, or branchial crowns, because they were at one time regarded as primarily respiratory in function. No doubt they do play a part in respiration, but they also provide mechanisms for the collection and sorting of food particles; they are aided in this by the production of mucus, which is distributed over tracts of ciliated epithelium.

With this potentiality as a starting point, ciliary feeding mechanisms could have evolved in the sedentary polychaetes along many independent lines. Terebellid worms (Fig. 10–1), for example, which live in permanent tubes in mud, are deposit feeders; they obtain detritus by extending long ciliated tentacles from their head over the surface of the substratum, the food particles being trapped in mucus and swept along ciliated grooves into the mouth (Fig. 10–1). The tentacles, which are highly mobile, are beautifully adapted for this function. Their shape, as seen in cross-section, varies at different points at any particular moment according to the use to which they are being put. At one point a tentacle will be flattened to form a zone of attachment, distal to which the remainder of the tentacle explores the substratum. Proximally to the attachment point the surface of the tentacle will be folded to form a ciliated groove along which the food is propelled. This propulsion is effected in part by ciliary action, but in part also by muscular contraction.

In complete contrast to this, the lugworm obtains detritus by swallowing the mud in which it is contained. This requires active cyclical movements (p. 305), by which the animal draws into its burrow the detritus deposited on the surface; these animals, therefore, although taxonomically sedentary polychaetes, are more active than most worms of this type. They rely upon the sucking action of the anterior end of the alimentary tract to enable them to swallow, and this, coupled with their burrowing activity, accounts for the absence of tentacles.

Sabella, which presents a contrast to both of the above worms, provides an example of the elegance of adaptation that has been achieved in the feeding mechanisms of the polychaetes. It is a comparatively large animal, found in the littoral zone, where it builds tubes that project in large numbers from the surface of the mud. It is a suspension feeder, extracting its food from water currents, which are created by coordinated cilia. These are set upon the branchial crown (Fig. 10–2), which consists of a large number of cephalic tentacles or filaments, differing from those of *Terebella* in being stiff and pinnate, and in being arranged in two groups of about 30. Each group is united towards its base into a lateral lobe, the two lobes being joined together dorsally but extending back independently on the ventral surface.

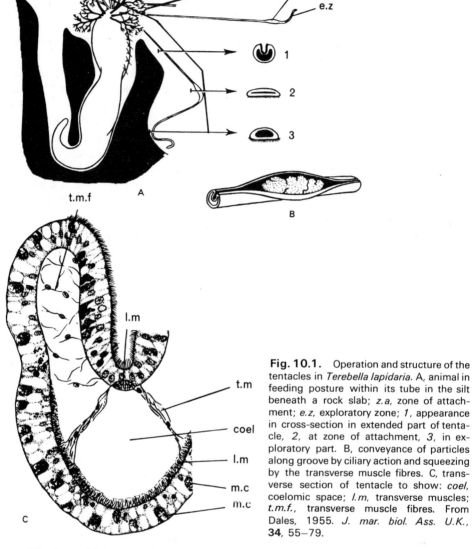

Fig. 10.1. Operation and structure of the tentacles in *Terebella lapidaria*. A, animal in feeding posture within its tube in the silt beneath a rock slab; *z.a*, zone of attachment; *e.z*, exploratory zone; *1*, appearance in cross-section in extended part of tentacle, *2*, at zone of attachment, *3*, in exploratory part. B, conveyance of particles along groove by ciliary action and squeezing by the transverse muscle fibres. C, transverse section of tentacle to show: *coel*, coelomic space; *l.m*, transverse muscles; *t.m.f.*, transverse muscle fibres. From Dales, 1955. *J. mar. biol. Ass. U.K.*, **34**, 55–79.

The branchial crown forms a wide funnel, with the mouth of the animal lying at the base. The problem, therefore, is to secure the food and to direct it towards that point, and it is for this purpose that the cilia are used. Their operation in feeding depends upon two rows of outgrowths, the pinnules, along each filament. These are set in such a way (Fig. 10–3) that a pinnule in one row makes an angle of rather more than 90° with the corresponding pinnule in the other row. The pinnules at the distal end of a filament are separated from the ones on the next adjacent filament, but towards the lower part of the branchial funnel they are brought closer together and finally interlock. As a result, they form a filtering system upon which food particles can be trapped.

FILTER FEEDING 259

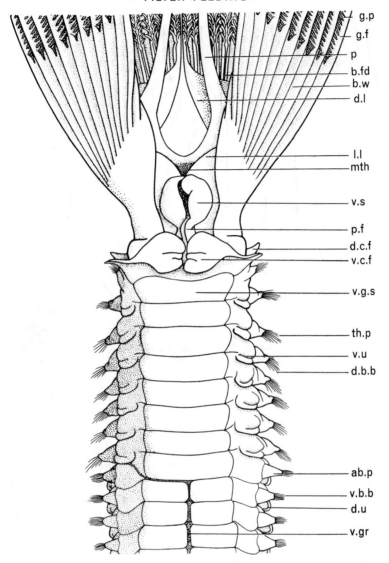

Fig. 10-2. Ventral view of the base of the branchial crown and the first 12 body segments of *Sabella pavonina* to show the external features × 12. *ab.p*, abdominal parapodium; *b.fd*, basal fold; *b.w*, basal web; *d.b.b*, dorsal bristle bundle; *d.c.f*, dorsal collar fold; *d.l*, dorsal lip; *d.u*, dorsal uncini; *g.f*, gill filament; *g.p*, gill pinnule; *l.l*, lateral lip; *mth*, mouth; *p*, palp; *p.f*, parallel folds; *th.p*, thoracic parapodium; *v.b.b*, ventral bristle bundle; *v.c.f*, ventral collar fold; *v.gr*, ventral groove; *v.g.s*, ventral gland shield; *v.s*, ventral sac; *v.u*, ventral uncini. From Nicol, 1930. *Trans. R. Soc. Edinb.*, **56**, 537–598.

The outer surfaces of the pinnules (Fig. 10–3) bear cilia, the abfrontal cilia, which beat strongly towards the tip of each pinnule. These cilia draw water into the funnel, the process being completed by latero-frontal cilia, which beat inwards, at right angles to the beat of the abfrontal ones. Food particles enter the funnel with

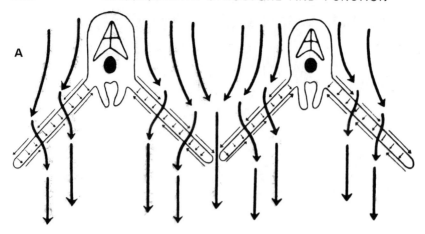

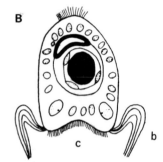

Fig. 10-3. A, diagrammatic section of two gill filaments of *Sabella*, to show the direction of flow of the water entering the branchial funnel, and the direction of beat of the cilia which cause the current. The small arrows indicate the direction of beat of the cilia; the large arrows indicate the direction of flow of the water. B, transverse section through a pinnule to show the ciliation. × 500. *a*, abfrontal cilia; *b*, latero-frontal cilia; *c*, frontal cilia. From Nicol, 1930. *op. cit.*

the steam of water, and are thrown, partly by eddies and partly by the beat of the latero-frontal cilia, onto a groove that runs along the inner edge of each pinnule. In this groove are cilia, the frontal cilia, that beat towards the base of the pinnule in the opposite direction to the beat of the abfrontal ones. The frontal cilia thus drive the food particles to the bases of the pinnules. From here they enter a longitudinal groove than runs down the whole length of each filament, and they are driven along this by the cilia that line it.

Towards the base of each filament the two rows of pinnules pass into two continuous folds, the gill folds or basal folds (Fig. 10–4), which are ciliated on both their outer and inner faces. The cilia mostly beat upwards towards the free edge of the folds, but on their inner surfaces the folds bear three ciliary tracts in which the beat is downwards. This downward beat is directed towards the mouth, which is bordered by a dorsal and two lateral lips (Fig. 10–2); these are formed in part from the bases of a pair of long palps, and in part from the bases of the branchial crown. The lips are ciliated, and, in particular, they bear three ciliary tracts that correspond with the three tracts of the gill folds. These various structures constitute part of the sorting mechanism; we have remarked that this is an essential feature of a highly organized ciliary feeding mechanism, ensuring that only suitable particles are directed into the digestive system.

Rejection in *Sabella* depends upon the capacity of the sorting mechanism to

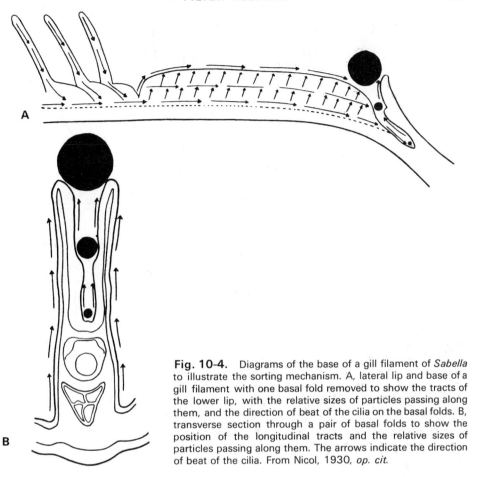

Fig. 10-4. Diagrams of the base of a gill filament of *Sabella* to illustrate the sorting mechanism. A, lateral lip and base of a gill filament with one basal fold removed to show the tracts of the lower lip, with the relative sizes of particles passing along them, and the direction of beat of the cilia on the basal folds. B, transverse section through a pair of basal folds to show the position of the longitudinal tracts and the relative sizes of particles passing along them. The arrows indicate the direction of beat of the cilia. From Nicol, 1930, *op. cit.*

differentiate between particles of various sizes. The pinnules play some part in this, because large particles falling from above cannot enter the longitudinal grooves of the filaments since these are protected by the overarching of the expanded bases of the pinnules. Lower down a more subtle differentiation is found, permitting the sorting of large, medium, and small particles (Fig. 10–4). The gill folds are arranged in pairs; the members of each pair enclose a basal groove, but their inner surfaces are so close together that large particles cannot enter between them. Such particles, therefore, do not come under the influence of the ciliary tracts that beat downwards towards the mouth. Instead, they are moved away from the mouth towards the edges of the lips. The smallest particles, however, can pass between the inner surfaces of the gill folds into the basal groove, and they are then carried towards the mouth. The medium-sized ones follow yet another course, because, although they can enter between the inner surfaces of the gill folds, they are prevented from proceeding as far as the basal groove owing to the existence of a longitudinal ridge. They thus come under the control of a different ciliary tract, which transports them into two expansions of the lateral lips called the ventral sacs (Fig. 10–5). They do not, therefore,

262 INVERTEBRATE STRUCTURE AND FUNCTION

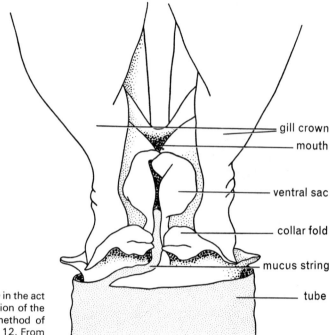

Fig. 10-5. Ventral view of *Sabella* in the act of tube-building, to show the formation of the mucus and sand string and the method of applying it to the edge of the tube. × 12. From Nicol, 1930. *op. cit.*

enter the mouth, but their selection is none the less important. They are used, after being mixed with mucus, for the building of the mud tube. This is formed through the activity of the collar folds that lie just posterior to the mouth.

The large particles are removed from the animal through the action of rejection currents, the cilia of the lips and palps being important in this. Other débris is removed by the same route, together with the faeces that are passed forward from the hind end of the animal in a ciliated groove. Mucus is extensively used in these rejection movements, where it helps the cilia to pass along the discarded material, but it would not be practicable for it to be used in the transport of food, for this depends upon the sorting of individual particles by size. In correlation with this the collection and sorting processes depend largely upon the cilia alone, and upon the currents and vortices that are established by their beat.

There are other polychaetes with feeding mechanisms similar in general principle to that of *Sabella*, but with great variation in detail. One example is *Pomatoceros*, common on rocky shores in its calcareous tubes. The branchial crown, which is very similar to that of *Sabella*, consists of two halves, each of which comprises some 14–20 pinnate tentacles borne on an outgrowth of the prostomium. The two halves are joined by the dorsal and ventral lips, between which lies the mouth. The collection and transport of food is similar in principle to that in *Sabella*, but simpler in detail, perhaps because the animal is smaller and possesses a calcareous tube. In particular, there is no specialized sorting mechanism, so that all particles transported by the pinnules are likely to reach the mouth. Some degree of sorting, however, presumably results automatically from the small size of the animal, for this ensures that only small particles will be transported by the pinnules in the first instance. If, as may sometimes happen, the filaments become clogged with an excess of material, the tip of a filament will bend over and remove the obstruction; further protection is afforded by rejection currents on the palps, lips, and basal folds.

One other example, which shows the variety of these feeding mechanisms even within the limits of one class, is *Chaetopterus*, a highly specialized worm of bizarre form that lives in sand or mud within a U-shaped tube of parchment-like consistency (Fig. 10-6). In this animal there is no branchial crown. Instead, water is drawn through the tube by the beating of three pairs of fans that are presumably derived from the parapodia of related forms. Farther forward another pair of outgrowths forms two wings that are pressed against the wall of the tube. Mucus secreted by these is drawn backwards by cilia in a ventral groove, and is formed into a conical bag, the apex of which lies within a small cup. Food particles are strained out by this mucous bag, the substance of which is continuously secreted by the wings and rolled up into a pellet in the cup. At intervals the secretory process stops and the cilia in the ventral groove move in reverse; as a result the pellet from the cup, with its contained food particles, is transported to the mouth and swallowed.

In the more advanced invertebrates extracellular digestion tends to replace the intracellular method, for reasons that we have earlier indicated, but the mode of feeding may be decisive in determining how far this tendency proceeds. In annelid worms the situation has been comparatively little studied, but there is evidence that

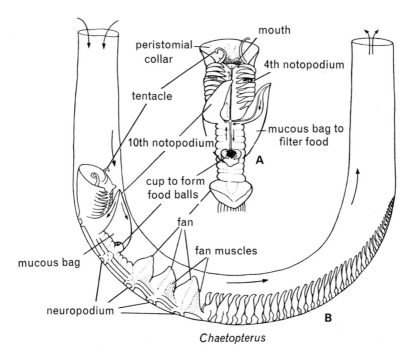

Fig. 10-6. *Chaetopterus*. In B, the direction of water is indicated by arrows. From Borradaile *et al.*, 1958. *The Invertebrata* (3rd ed., ed. Kerkut). Cambridge University Press, London.

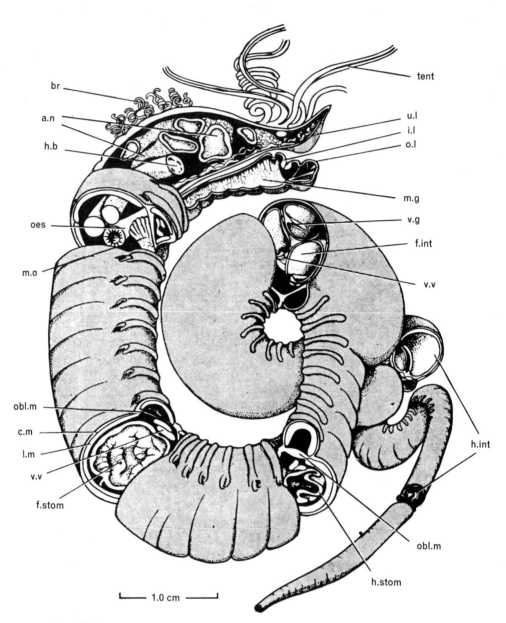

Fig. 10-7. *Amphitrite johnstoni*, illustrating the main regions of the gut in relation to the rest of the body. *a.n*, anterior nephridia; *br*, branchiae; *c.m*, circular muscles; *f.int*, fore-intestine; *f.stom*, fore-stomach; *h.b*, heart body; *h.int*, hind-intestine; *h.stom*, hind-stomach; *i.l*, inner lips; *l.m*, longitudinal muscle; *m.g*, mucous glands; *obl.m*, oblique muscles; *oes*, oesophagus; *o.l*, outer lips; *tent*, tentacles; *u.l*, upper lip; *v.g*, ventral gutter; *v.v*, ventral vessels. From Dales, 1955. *J. mar. biol. Ass. U.K.*, **34**, 55–79.

the extracellular method predominates. Some phagocytosis may, however, take place, as, for example, in *Arenicola marina*, where digestion is completed in wandering amoebocytes that take up from the alimentary epithelium particles that its cells have ingested. It is surprising that digestion appears to be largely extracellular in terebellids, for their filter-feeding habits would seem to favour the persistence of intracellular digestion. In fact, their gut shows considerable regional specialization for extracellular digestion (Fig. 10–7), being differentiated into an oesophagus, a fore-stomach, a muscular hind-stomach which serves as a mixing region, and an intestine. Enzymes are secreted in the fore-stomach and fore-intestine, absorption taking place in the intestine. Arthropods also, incidentally, rely almost completely upon extracellular digestion, even in microphagous forms; intracellular digestion is restricted to the final stages in the digestion of proteins, more especially in arachnids. This contrasts strikingly with the situation that we shall find in the Bivalvia, where microphagy is associated with the retention of a highly specialized form of intracellular digestion.

10–2 FILTER FEEDING AND DIGESTION IN MOLLUSCS

No animals provide better illustrations of filter feeding than do the bivalve (lamellibranch) molluscs, for all the members of this class, with the exception of the secondarily modified septibranchs, obtain their food in this way. As with the polychaetes, we can see that a substantial element of pre-adaptation has been involved, although the course of events has been quite different in the two groups. To judge from the type of feeding found in the chitons and in the most primitive living gastropods, the earliest molluscs must have been microphagous, using a rasping radula to break up encrusting algae, and then transferring particles of these organisms, together with deposits of detritus, into the mouth. We shall see later that the molluscan alimentary tract is highly specialized to deal with the intake of particulate material. What the bivalves have done is to exploit certain potentialities inherent in this molluscan plan of organization. Making use of the protective value of the shell, and of the muscular power of the foot, they have become typically semi-sessile animals, inhabitants of sandy or muddy substrata. Here, with an alimentary system already adapted for microphagy, they have utilized to the full the particulate food resources of their habitats.

Probably ciliated labial palps, lying on either side of the mouth, were the feeding organs of the earliest bivalves, as they are to-day in the members of the order Protobranchia, in which the palps extend into tentacle-like proboscides which probe the substratum. The ciliated gills (ctenidia), initially respiratory in function, would probably have been incorporated later into the feeding mechanism, but it is uncertain how far they are involved in this in the protobranchs, for in these animals the palps are responsible for some quantitative sorting of the particles as well as for their collection. In most other bivalves, however, the ctenidia have become highly specialized for filtering suspended and deposited particles from a current of water, maintained by ctenidial cilia, which enters through an inhalent siphon and leaves through an exhalent one.

The elongated ctenidia of bivalves (Fig. 10–8) are organized around an axis which bears two demibranchs, each of these being composed of a parallel row of

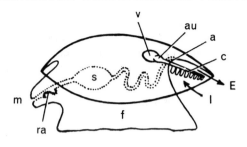
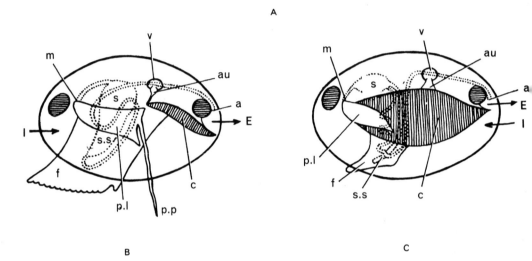

Fig. 10-8. Diagrams illustrating stages in the evolution of the Lamellibranchia, showing significant changes in the orientation of the body, form of the shell, foot, ctenidia, and heart, alimentary canal (stippled), and labial palps. A, structure of hypothetical primitive Mollusca (modified after Pelseneer); B, protobranch stage; C, final eulamellibranch condition without siphons. *a*, anus; *au*, auricle; *c*. ctenidium; *f*, foot; *m*, mouth; *p.l*, palp lamellae; *p.p*, palp proboscides; *ra*, radula; *s*, stomach; *s.s*, style sac region; *v*, ventricle. From Yonge, 1939. *Phil. Trans. R. Soc. B*, **230**, 79-147.

filaments. It has been customary to distinguish three grades of arrangement of these filaments (protobranch, filibranch, and eulamellibranch), and to use them as a basis for classification. A different taxonomic treatment is now favoured, but the three terms retain their descriptive value. In the protobranch type the filaments are unfolded, but in the filibranch and eulamellibranch types they are folded so as to form ascending and descending limbs (Fig. 10–9). In the filibranch type (e.g. *Mytilus*) adjacent filaments are joined by ciliary junctions; in the eulamellibranch types (e.g. *Anodonta*) they are joined to each other by vascular interfilamental junctions (Fig. 10–10). Each demibranch thus forms a folded lamella, and the ascending and descending plates of this are joined to each other by interlamellar junctions.

In bivalves, as in polychaetes, use is made of cilia that are arranged in frontal

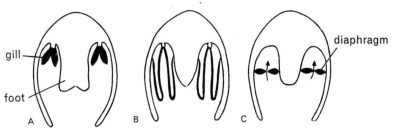

Fig. 10-9. Ventrical section of Bivalvia to show different arrangements of the ctenidia. A, protobranch; B, filibranch and eulamellibranch; C, septibranch, with gills modified into muscular septa. The arrows in C show the direction of water flow through the septum when the latter moves downwards. From Borradaile et al., 1958. op. cit.

and lateral series, but their action is complex, and shows much interspecific variation. In principle, lateral cilia (Fig. 10–10) draw water into the mantle chamber, and from there into the interlamellar and suprabranchial cavities. As the water passes between the filaments, the food particles are filtered out by latero-frontal cirri. These were formerly called cilia, but ultrastructural studies have shown that they are actually ciliary complexes, each consisting of 20–25 pairs of cilia, arranged as a double row, all beating together at right angles to the long axis of the gill filament. The distance between adjacent cirri is about 2.0 to 3.5 μm, but the effective space is smaller than this, since the cilia bend at regular intervals along the cirrus so that they form a meshwork between the cirri and also between adjacent filaments. This remarkable arrangement, which seems to be unique to the bivalves, accounts for the extreme efficiency of their filtration, which permits the retention of a high proportion of incoming particles in the size range 1–3 μm, and virtually a 100% retention of those of 4 μm.

The particles thus trapped by the latero-frontal cirri are thrown onto the frontal cilia, which then sweep them, entangled in mucus, over the surface of the gill lamellae. From here they may pass either into a ventral marginal groove, or into a dorsal groove along the axis of the gill, at the base of the demibranchs. Along one or other of these grooves, depending upon the particular species concerned, the food material is carried to the two pairs of labial palps; these are triangular structures, one pair lying on either side of the mouth.

It would take too long to describe the possible courses of these particles in detail, so complex are the specialized ciliary mechanisms involved, but as far as sorting is concerned this occurs partly on the gills and partly on the palps. It is common to find a differentiation of ciliary tracts on the gills. Some tracts are provided with fine cilia, and are adapted for conveying the fine particles required for food; other tracts have coarse cilia and deal with the rejection of larger particles. Coupled with this simple differentiation, however, are complex and highly diversified patterns of ciliation, associated in their turn with variations in the form of the gills.

For example, the gill lamellae are sometimes folded (plicate), as they are in *Pecten* and *Ostrea*, with the ciliary beat on the crests of the folds differing from that in the grooves. The smaller particles required for food are carried chiefly upwards by fine cilia to the dorsal grooves, while coarse particles, such as sand grains, are

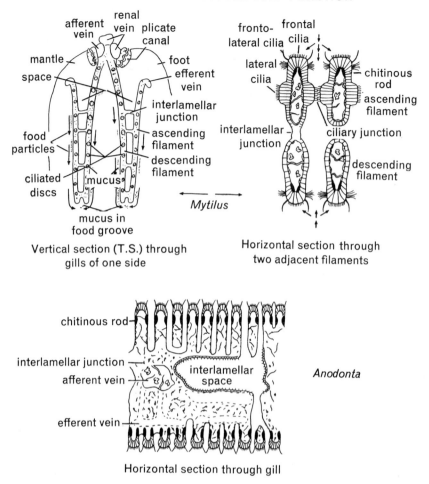

Fig. 10-10. The ctenidia of the Bivalvia. The arrows indicate the direction of the food current and the path of the food particles it contains. *Mytilus* (filibranch); *Anodonta* (eulamellibranch). From Borradaile *et al.*, 1958. *op. cit.*

carried downwards. The latter tend to drop off from the gill edge (Fig. 10–11), or are shaken off by muscular movements of the demibranchs.

In other forms the ventro-marginal grooves may contribute to the sorting in a way reminiscent of that which we have seen in *Sabella*. For example, they may be divided into a deep channel lined by fine cilia and a superficial one lined by coarse cilia. Fine particles may then be carried into the bottom of the groove, whereas closure of this may cause large ones to be conducted only to its edge, so that they are eventually rejected. As another possibility, long cilia on the edges of the marginal groove may permit the entry of small particles, but prevent the entry of larger ones; this principle operates in *Mya*.

The labial palps (Fig. 10–12) are ciliated structures; their sorting function depends particularly on the structure of their internal faces, which are crossed by a

FILTER FEEDING

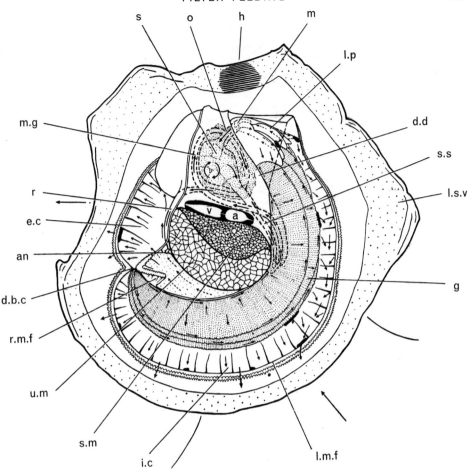

Fig. 10-11. *Ostrea edulis*, right shell valve and mantle removed. *a*, auricle; *an*, anus; *d.b.c*, division between inhalent and exhalent chambers; *d.d*, digestive diverticula; *e.c*, exhalent chamber; *g*, gills; *h*, hinge; *i.c*, inhalent chamber; *l.m.f*, left mantle fold; *l.p*, labial palps; *l.s.v*, left shell valve; *m*, mouth. *m.g*, mid-gut; *o*, oesophagus; *r*, rectum; *r.m.f*, right mantle fold; *s*, stomach; *s.m*, adductor muscle, portion with striated fibres; *s.s*, style sac; *u.m*, adductor muscle, portion with smooth fibres; *v*, ventricle. Large arrows external to shell indicate direction of ingoing and outgoing currents. Broken arrows denote currents on under surfaces. From Yonge, 1926. *J. mar. biol. Ass. U.K.*, **14**, 295-386.

series of diagonal folds. These folds overlap each other in the direction of the mouth, all but the uppermost part of one fold being covered by the next adjacent one. The sorting mechanism is here said to depend solely on the weight and not on the size of the particles. Of the particles that are carried over the top of the surface, the heavier ones settle down into the grooves between the slope of one fold and the crest of the next; in this position they come under the influence of a powerful ciliary current that sweeps them to the upper margin of the palp. Lighter ones avoid this current because they do not sink in the same way; as a result, they are swept from one slope to the

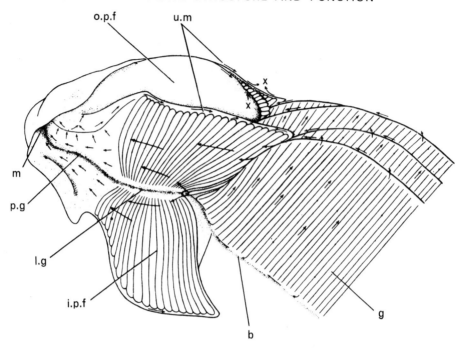

Fig. 10-12. Junction of palps and gills of *Ostrea*, right palps opened out so as to expose inner, ridged surfaces. ×8. *b*, base of demibranch; *g*, gill; *i.p.f*, inner palp face; *l.g*, lateral oral groove; *m*, mouth; *o.p.f*, outer palp face; *p.g*, proximal oral groove; *u.m*, upper margin of palps; *x*, point where material is rejected from palps. From Yonge, 1926. *op. cit.*

next and pass towards the mouth. A similar sorting takes place among particles that pass down between the folds, heavier ones being carried by another current towards the upper margin and lighter ones towards the mouth. Thus these structures have their cilia arranged so as to produce a diversity of ingeniously integrated currents.

We shall follow later the fate of those particles that finally enter the mouth and pass into the alimentary canal. The rejected particles pass onto the ciliated epithelia of the mantle or visceral mass, where the strong ciliary currents produced by these epithelia give rise to vortices. These entangle the particles in mucus to form masses that eventually accumulate below the internal opening of the inhalent siphon. Periodically the animal closes its exhalent siphon and, by a sudden movement of contraction, forces the rejected material (pseudofaeces) out of the inhalent siphon in a current of water.

We have suggested that the rasping method of feeding of primitive molluscs was a pre-adaptation to the evolution of the specialized filter feeding of bivalves. This rasping method is essentially a form of microphagy. Thus the requirements for the handling of small particles must have influenced the organization of the alimentary canal of molluscs from the beginning of their history, and must further have aided the establishment of bivalve feeding methods. Moreover, the ingestion of small particles probably determined the retention of the intracellular method of digestion, which is so widespread in molluscs. Indeed, the group as a whole provides a good illustration of the supplanting of the intracellular method by the more advanced extracellular one, and the retention of the former in microphagous forms.

Particularly characteristic of the molluscan alimentary tract is the stomach, an organ that demonstrates very strikingly the problems of combining microphagy with intracellular digestion. We have earlier mentioned that intracellular digestion requires a large area of phagocytic epithelium. To take full advantage of this, the food must be delayed in its passage through the alimentary tract and distributed over the epithelium. The satisfying of these two requirements has determined the evolution of the molluscan stomach. In bivalves there projects into the lumen of this organ a long and flexible rod, composed of layers of mucoprotein. This structure, the crystalline style, is secreted by a style sac (Fig. 10-13); this is an extension of the

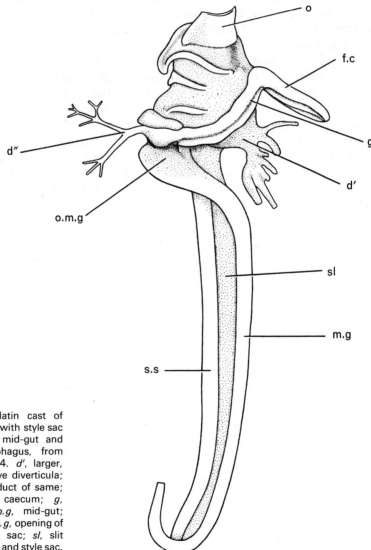

Fig. 10-13. Gelatin cast of stomach of *Ostrea* with style sac and first part of mid-gut and portion of oesophagus, from ventral aspect. ×4. *d'*, larger, left duct of digestive diverticula; *d''*, smaller, right duct of same; *f.c*, food-sorting caecum; *g*, ventral groove; *m.g*, mid-gut; *o*, oesophagus; *o.m.g*, opening of mid-gut; *s.s*, style sac; *sl*, slit connecting mid-gut and style sac. From Yonge, 1926. *op. cit.*

stomach that may open widely into that organ, or be more or less completely cut off from it. Cilia in the style sac cause the style to rotate, and at the same time drive it forwards into the stomach. Here its free end is worn away by friction against the gastric shield (Fig. 10–14), which is a thickening of the cuticular lining of the stomach wall. The wearing away is aided by the alkaline stomach contents, which cause the style substance to dissolve. This substance contains a digestive amylase adsorbed to its mucoprotein base, while in some species a cellulase is perhaps also present. Thus the dissolution of the style results in these enzymes being added to the contents of the gastric lumen.

The effect of all this is that the food-bearing strands of mucus that enter the stomach are caught up and wound into a spiral mass by the rotation of the style, and are simultaneously mixed with its digestive secretion, so that the extracellular digestion of carbohydrate is initiated. Furthermore, particles of food and mucus are continually being broken off from the main mass, partly because of its rotation, and partly because the pH of the stomach contents (about 5–6) lowers the viscosity of the mucus. These detached particles are now sorted by the stomach wall (Fig. 10–14), much of which is lined with ciliated ridges and grooves that have an action similar in principle to that of the labial palps. The larger and heavier particles enter the deeper grooves, and are transported by cilia into the intestine; there the more alkaline pH of the contents has the effect of forming them into faecal pellets that are eventually extruded through the anus. Finer particles, however, take a different route, which promotes their further digestion; they are borne over the cilia on the ridges of the stomach wall towards the openings of the digestive diverticula (or glands), which constitute another characteristic feature of the molluscan alimentary tract.

There are two of these diverticula, each consisting of a highly branched system of blind tubules opening into the stomach by a ciliated duct. The epithelium of the tubules is also ciliated, and in bivalves is usually composed of a single type of highly vacuolated cell. This cell is phagocytic, capable of ingesting fine particles into food vacuoles, and it is within these vacuoles that the digestive process is completed. In fact, digestion is very largely intracellular in bivalves, except in so far as the secretion of the style sac initiates the digestion of carbohydrates. We see here very clearly how the retention of the intracellular method demands a sorting mechanism of the type provided by the stomach. Only fine particles can be taken up into the phago-

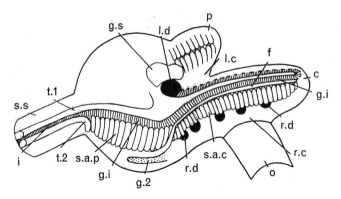

Fig. 10-14. Diagram of the anatomy of the stomach of *Glycymeris*. *c*, caecum; *f*, fold; *g.i*, intestinal groove; *g.s*, gastric shield; *g2*, groove; *i*, intestine; *l.c*, left wall of caecum; *l.d*, opening of duct from left lobe of digestive gland; *o*, oesophagus; *r.c*, right wall of caecum; *r.d*, opening of duct from right lobe of digestive gland; *s.a.c*, sorting area of caecum; *s.a.p*, posterior sorting area of stomach; *s.s*, style sac; *t1*, *t2*, typhlosoles. From Graham, 1949. *Trans R. Soc. Edinb.*, **61**, 737-761.

cytic epithelial cells, and the entry of large ones into the digestive diverticula would clog their delicate ducts. A supplementary means of ingestion is, however, available; larger particles, including whole diatoms or blood corpuscles that have been artificially fed to the animals, can be ingested by wandering amoebocytes in the stomach, a procedure that recalls the part played by the amoebocytes of sponges.

We have already seen in our discussion of coelenterates that intracellular digestion results in the phagocytic cells becoming loaded with waste material. This is very evident in the digestive diverticula. It leads to the epithelial cells undergoing fragmentation; spherical masses with vacuoles, ingested material, and waste are thus given off from the epithelium, and are conveyed out of the diverticula into the stomach and thence into the intestine. Such masses must contain enzymes that have been secreted by the phagocytic cells, and this probably accounts for the fact that weak enzyme activity can be detected in the stomach even though it does not itself possess a digestive epithelium. Whether such residual enzymes are of any importance in normal digestion may be doubted. On the other hand, it is quite conceivable that their discharge from phagocytic epithelia, an inevitable consequence of the course of intracellular digestion, might have been a factor promoting the evolution of the extracellular method.

The bivalves are specialized for the exploitation of the food sources provided by suspended organic matter and detritus, and their feeding and digestive mechanisms are, with one exception, very uniform. The exception is provided by the Septibranchia, an order of carnivorous lamellibranchs that draw in dead or dying animals by means of contractions of muscular septa derived from the gills (Fig. 10–9). In correlation with this, the crystalline style is reduced in size.

The gastropods, by contrast, are altogether more varied in their feeding habits. In consequence, there is a corresponding diversity in the structure of their alimentary canal, with a marked tendency for intracellular digestion to be replaced by the extracellular method. Some gastropods are herbivores, and of these there are genera in which the organization of the digestive system shows a general resemblance to that of the bivalves. *Crepidula*, for example, is a ciliary feeder, with amylase as its only extracellular digestive enzyme, and with digestive diverticula that are solely absorptive in function and that contribute no external digestive secretion. What is particularly interesting, however, is that this animal, like bivalves, possesses a crystalline style, a structure that is also found in a number of the other herbivorous gastropods, especially in the Taenioglossa but also in the Rhipidoglossa and the thecosomatous pteropods.

The distribution of the style in gastropods raises some interesting considerations, for it is not present in all of the herbivorous forms. Those that do possess it are microphagous herbivores, and this is significant, for it is possible to see two factors that would promote the establishment of a style in such forms. Only some of them are ciliary feeders; indeed, the majority depend upon their radula for the securing of their food, but this method of feeding, like ciliary feeding, results in a continuous stream of small particles passing into the alimentary canal. It is in precisely such conditions that the crystalline style confers an advantage; not only does it wind the food into a spiral mass and mix it with the amylase, but it also provides a means for the continuous release of small quantities of that enzyme.

A further factor is the absence of an extracellular protease. If this were present it would, of course, digest the style. No doubt this is one reason why this structure is lacking in carnivorous forms, for many of these, including carnivorous prosobranches such as *Murex*, have both an amylase and a protease present in their digestive secretions. Another consideration is that gastropods possess oesophageal glands, which provide an alternative source of extracellular enzymes. Carnivorous species commonly obtain their food at irregular intervals, so that glands like these, which may well be under nervous or hormonal control, probably meet their needs better than would the continuous release of enzymes from a style. It has been suggested that the two modes of secretion, by a crystalline style and by oesophageal glands, are mutually inconsistent; their distribution in prosobranch gastropods certainly goes some way to confirm this, for in that group they do not coexist in the same species, but seem rather to be developed as alternative mechanisms.

We have mentioned that the style is not universally present in herbivorous gastropods. *Patella*, *Haliotis*, *Aplysia*, and *Helix* are among those forms that lack it, but there seems to be no single reason for this. A tendency for the replacement of intracellular digestion by extracellular is doubtless one factor that determines it. This situation exists in *Helix*. In this herbivore no solid particles at all enter the cells of the alimentary tract, although the digestion of protein is probably completed within them after they have absorbed soluble peptides. *Patella*, on the other hand, has its digestive system organized at a more primitive level, in that most of its digestion is intracellular, with amylase as the only extracellular enzyme. This is not produced, however, through the mediation of a crystalline style, but is secreted from the oesophageal glands, known in these animals as sugar glands. It has been suggested that in this instance the lack of the style is conditioned by the mode of life of this herbivore in the intertidal zone. Its feeding is restricted to those periods when the tide is in and it can move without danger of desiccation; it could well be, therefore, that a style mechanism would be ill adapted for sporadic feeding of this sort, much as it is in carnivores.

The molluscs generally display a remarkable diversity and elaboration of digestive specialization. Among them the cephalopods, as might be predicted of such a highly organized group of predators, have an exclusively extracellular digestive mechanism. This (Fig. 10–15) bears some superficial likeness to the alimentary system of vertebrates, for the first stage of digestion occurs in a so-called 'stomach', and the second stage in the 'caecum', while the digestive gland has become subdivided into two regions which have been called the 'liver' and the 'pancreas'. The terms, however, are misleading. The pH of the stomach contents is 6.2, with no indication of an acid peptic phase such as is found in most vertebrates, while the 'liver', at least in *Octopus*, is a typically molluscan organ, carrying out the functions of secretion, absorption, storage, and excretion. The functions of the 'pancreas' have still to be clarified; however, in *Sepia* it certainly produces no enzymes, and it is capable of absorbing amino acids.

An interesting adaptive feature is said to be involved in the functioning of the 'liver' in *Octopus*, in that each cell performs each of these three functions in a cycle. The organ cannot, therefore, release enzymes into the 'stomach' while absorption is in progress. The squid, *Loligo*, however, differs in that its 'liver' is not concerned with absorption, so that this organ can secrete enzymes without interruption. This difference may be correlated with the bottom-dwelling habit of *Octopus* and the active

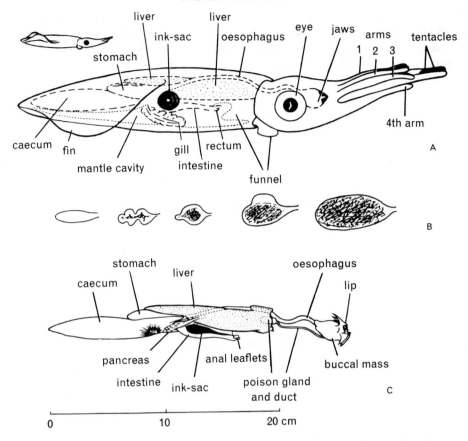

Fig. 10-15. A, young living *Loligo vulgaris* as seen swimming in tank. Optical section of mantle-cavity in dotted outline. Pancreas and details of gills not visible in life. Oesophagus, stomach, and caecum only visible when containing food. B, stomach empty, half-full and actively churning, and fully distended. C, digestive system from animal's right side. A, B, and C to same scale. From Bidder, 1950. *Q. Jl microsc. Sci.*, **91**, 1–43.

swimming and predation of the squid. Digestion in a sedentary *Octopus* can, perhaps, be successfully completed over a considerable period of time, whereas in *Loligo* the ceaseless activity demands a more rapid handling of the food.

10–3 FILTER FEEDING AND DIGESTION IN DEUTEROSMIA

Consideration of annelids and molluscs has already shown us how systems of microphagy, not fundamentally dissimilar in their ciliary mechanics, may yet differ widely in the circumstances that condition their origin and in their subsequent evolutionary history. In annelids they arise as a later development in a group that seems initially to have been more predatory in its habits. In molluscs they appear as a logical development from an initially microphagous habit, although one that was of a peculiar character and that was not based upon ciliary mechanisms. One other group that merits attention from this point of view is the Echinodermata, particularly because the history of its ciliary feeding has a bearing upon the wider issue of the evolutionary history of the chordates.

The crinoids are all ciliary feeders, as might be expected of this largely sessile group, the process having a hydrostatic basis which we have already examined in the secondarily free-swimming *Antedon*. This animal attaches to the substratum by its cirri when feeding. Its tube-feet then secrete a mucus net for the trapping of plankton, moving vigorously while they do so. The mucus is largely derived from papillae on the tube-feet, each papilla bearing a group of four to six sensory processes, a single muscle fibre, and several mucus and nerve cells (Fig. 10–16). It is supposed that the secretion is shot out by the contraction of the muscle fibre. The long tube-feet of the pinnules (p. 127) throw the food towards the food groove,

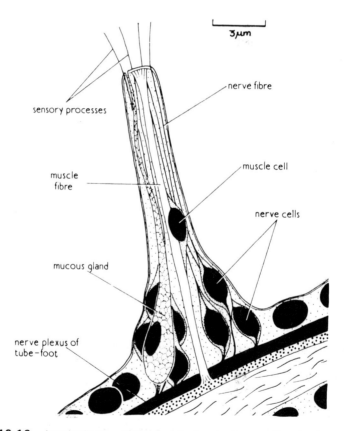

Fig. 10-16. Longitudinal section of a single tube-foot papilla showing, on the left, one of the four or so mucous glands, and, on the right, three of the eight or so nerve-cells, most of whose processes protrude from the tip of the papilla. The cytoplasmic envelope, which can sometimes be seen surrounding the length of the central muscle fibre, has been omitted for clarity, except in the region of the nucleus, where it is normally clearly visible. From Nichols, 1960. *Q. Jl microsc. Sci.*, **101**, 105-117.

further transfer being effected by the other tube-feet in the nicely adapted movements that we have earlier analysed.

Presumed food grooves are outlined in fossils of early sessile echinoderms, and it is therefore reasonable to suppose that ciliary feeding is primitive in this group, and that crinoid-like tentacles were early established. Subsequently, as we have seen, the sessile life was abandoned for movement, this advance being accompanied by progressive modifications of the water-vascular system along more than one line of evolution. These modifications, which led to the elaboration of the ciliated tentacles into the reciprocating complex of ampullae and tube-feet, permitted the replacement of microphagy by macrophagy. Even so, some modern forms, such as the heart urchins (p. 131) and certain holothurians, ingest mud and sand for their organic content, and are to this extent microphagous still.

We have already commented on the association of the Echinodermata with the vertebrates, and their relations with the assemblage termed the Deuterostomia. One justification for grouping several phyla in this association is that the history of their feeding mechanisms, despite their diversity of specialization, can be interpreted within a logical and consistent framework.

Within the Phylum Hemichordata the use of ciliated tentacles as a feeding mechanism persists in the pterobranchs *Cephalodiscus* and *Rhabdopleura*. In *Cephalodiscus* (Fig. 10–17) it is associated with the pharyngotremy mentioned earlier, for this animal draws into its mouth a stream of water which leaves by a pair of gill slits in the wall of the pharynx. Judging the value of this device is difficult, because pterobranchs have almost certainly become secondarily simplified in correlation with a reduction in size. Perforation of the pharynx may facilitate the expulsion of the excess water that is driven in by the ciliary feeding feeding mechanism, but possibly it was initially developed as a respiratory adaptation. This is suggested by the pharyngotremy found in the second group of hemichordates, the Enteropneusta, where the pharynx is perforated by a series of gill pores that seem to be primarily respiratory in function. Enteropneusts, which are mostly burrowing forms, feed on detritus and sand, engulfing vast quantities of the substratum and discarding it at the surface of their burrows as sand castings. They probably evolved from pterobranch-like ancestors, and their feeding mechanism shows some trace of this history. The ciliated proboscis can trap food particles in mucus and pass them towards the mouth—a process clearly reminiscent of the food-trapping action of ciliated tentacles.

The relationship of the Phylum Chordata to these several groups is demonstrated by two chordate subphyla, the Urochordata (Fig. 10–18) and the Cephalochordata. In these two groups, collectively termed the Protochordata, pharyngotremy has become the basis of a new type of ciliary feeding mechanism; the pharynx is used both for feeding and for respiration, and there is no longer any sign of ciliated tentacles. Despite the novelty of this device, however, the ciliary mechanics of the pharynx prove similar in principle to those of the branchial crown of the polychaetes and of the gills of lamellibranchs, for the feeding of amphioxus and the ascidians depends upon groups of cilia beating at right angles to each other.

Lateral cilia on the delicate fenestration of the pharynx draw water through the pharyngeal wall into an atrial cavity that has been secondarily developed as protection for this vulnerable structure. The filtering of this water depends upon the presence in the floor of the pharynx of a longitudinal groove, the endostyle. A secretion

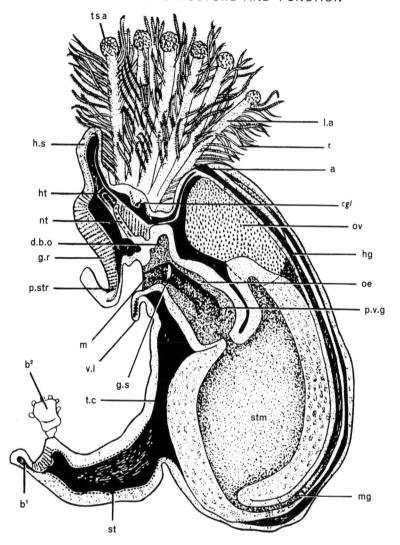

Fig. 10-17. Anatomy of *Cephalodiscus*, seen in median L.S. *a*, anus; b^1, b^2, buds; *c.gl*, cerebral ganglion; *d.b.o*, dorsal median blind projection of oesophagus; *g.r*, glandular region; *g.s*, gill slit; *hg*, hind-gut; *h.s*, head shield; *ht*, heart; *l.a*, lophophore arm; *m*, mouth; *mg*, mid-gut; *nt*, notochord; *oe*, oesophagus; *ov*, ovary; *p. str.* pigment streak; *p.v.g*, posterior vacuolized groove; *st*, stolon; *stm*, stomach; *t*, tentacle; *t.c*, trunk coelom; *tsa*, terminal swelling of arm; *v.l*, ventral lip. From Schepotieff, 1907. *Zool. Jb. Abt. Anat.*, **24**, 553-608.

produced by this organ is moved up the pharyngeal wall by frontal cilia, which are the functional equivalent of the frontal cilia of worms and molluscs; thus the food particles are caught up in a moving filter formed by this secretion. The material is moved up to the mid-dorsal line, and from there is passed backwards as a food cord which is digested in a manner to be briefly considered below. This feeding mechanism

FILTER FEEDING

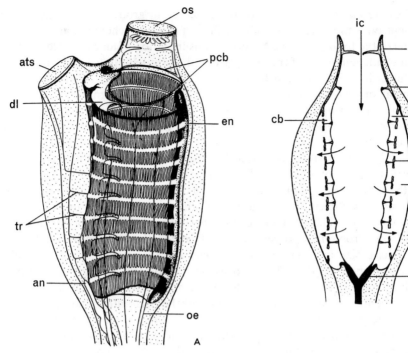

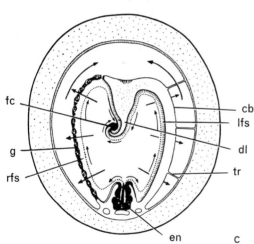

Fig. 10-18. Food collection in *Clavelina lepadiformis*. A, lateral view of anterior region of body. B, diagrammatic frontal section. C, diagrammatic transverse section through the pharynx, passing on the left through a row of stigmata, and on the right through a ciliated bar. *ac*, atrial cavity; *an*, anus; *ats*, exhalant siphon; *cb*, ciliated bar; *dl*, dorsal languet; *en*, endostyle; *fc*, food cord; *g*, gill; *ic*, inhalent current; *lfs*, left mucous filtering sheet; *mf*, funnel of mucus; *oe*, oesophagus; *os*, inhalant siphon; *pcb*, peripharyngeal ciliated bands; *rfs*, right mucous filtering sheet; *te*, tentacle; *tr*, supporting trabecula. In *B*, the arrows indicate the direction of the water currents. In *C* the small arrows indicate the direction of movement of the filtering sheets, and the large arrows the direction of the water currents. Adapted from E. and B. Werner, 1954. *Helgoländer wiss. Meeresunters.*, **5**, 57-92.

is well suited for sessile and bottom-dwelling animals, but is much modified in the pelagic urochordates, such as the salps and doliolids. In the latter, for example, the flow of water is maintained by pulsations of the body wall, and the food is trapped by strands of mucus in the lumen of the pharynx.

The endostyle is very characteristic of the protochordates, but a curious parallel to it is found in the gastropod *Crepidula*. This animal, which uses its gills for ciliary feeding, has developed an organ that seems to fulfil a function very similar to that of the endostyle, and which, in consequence, has been given the same name. This organ of *Crepidula* is a ciliated and mucus-secreting groove, from which a secretion is swept onto the food-collecting surface of the gills, where it contributes to the trapping of the food particles. Naturally, this imitation of an endostyle has been evolved quite independently of the endostyle of the protochordates. Indeed, this is where its interest lies, for it reveals how similar requirements may condition the evolution of similar organs in wholly diverse groups. Like the patterns of ciliation that we have been considering, it is an example of how the common factors that underlie the diverse organization of animal groups may, from time to time, result in the building of closely similar types of adaptation.

The ciliary and pharyngeal feeding mechanisms of the early Deuterostomia certainly had potentialities of exceptional importance, for they contributed powerfully to the evolution of vertebrates. We cannot follow this aspect in any detail here. It must be sufficient to say that the substitution of muscular for ciliary action led on to the type of microphagous feeding that we see today in the ammocoete larva of the lamprey. Probably it proved a more efficient method of securing food, and permitted increased size and activity. The crucial stage, however, was the development of macrophagous feeding. This process was absolutely dependent upon the initial establishment of pharyngotremy, for it was made possible by the development of jaws out of parts of the supporting skeleton of the perforated pharynx.

The digestive processes of the Deuterostomia are less well known than could be wished. They have been most closely studied in amphioxus, which provides another example of the retention of intracellular digestion in association with ciliary filter feeding, showing also some remarkable parallelisms with the digestive processes of lamellibranchs and of certain other invertebrate filter feeders. The food cord formed in the pharynx is passed back by ciliary action into the mid-gut, at the hind end of which its passage is arrested in a specialized region called the ilio-colon ring (Fig. 10–19). Here cilia set it into rotation around its longitudinal axis. This rotation ensures the necessary delay in the backward movement of the food cord; at the same

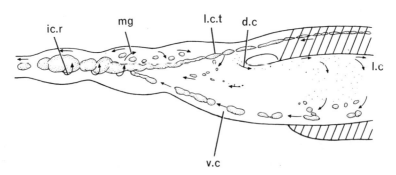

Fig. 10-19. Diagram to show the currents and movements of material in the alimentary tract of amphioxus. *d.c*, dorsal current carrying small particles into the caecum; *ic.r*, 110-colon ring, containing a rotating mass of food and secretion; *l.c*, small particles deposited and ingested on the lateral wall of the caecum; *l.c.t*, lateral ciliated tract; *mg*, mid gut, in which material is thrown forwards; *v.c*, ventral current carrying secretion from the caecum into the mid-gut. From Barrington, 1965. *The Biology of Hemichordata and Protochordata*. Oliver and Boyd, Edinburgh.

time it mixes the cord with digestive enzymes that are passed into the ilio-colon ring, again by ciliary action, from the mid-gut caecum where they are secreted. The rotation also breaks up the mass of food and secretion into fragments; the smallest of these are swept forwards by ciliary currents into the caecum, while the larger ones drop back into the mid-gut and are returned to the rotating mass. Within the caecum the small particles come to rest on the epithelial surface and are ingested by the cells, so that while the initial stages of digestion are extracellular, the process is finally intracellular.

The rotatory action of the ilio-colon ring presents a remarkable analogy with the style mechanism of molluscs. The style probably originated in a mechanism similar to that seen today in the protobranchiate *Malletia*, where food is mixed by ciliary action with a mass of secretion produced by a style sac. In more advanced bivalves this device has evolved into the firm rotating style already considered, but this has not happened in the protochordates. The two groups show, however, the independent evolution of a fundamentally similar device for delaying the passage of the food, for mixing it with secretion, and for distributing it to the ingestive regions of the alimentary epithelium. The digestive gland of bivalves also resembles in principle the mid-gut caecum of amphioxus. Both provide for an increase in surface area for secretory and ingestive activity, and in both groups the distribution of food particles is so controlled that only the finer particles enter these blind alimentary diverticula. The two groups thus illustrate yet again how similar needs, acting within the inevitable limits of animal organization, must evoke similar, yet independent, solutions to functional problems. Nor are bivalves and cephalochordates the only example of this. A rotation of material in the alimentary canal, essentially similar in principle to that described above, occurs also in the Polyzoa (Ectoprocta), the Entoprocta, the Brachiopoda, and the Phoronidea.

10–4 FILTER FEEDING IN CRUSTACEA

Although cilia lend themselves particularly well to the organization of filter-feeding mechanisms, they are not the only means available for this, as is shown in the microphagy practised by Crustacea. This group, sharing with other arthropods a general lack of cilia, has exploited instead the potentialities of the arthropod limb. The mechanisms involved are quite different from those that we have so far considered, and they are the more interesting because of this. They show how similar ends may sometimes be attained by quite different means, determined by differences in the fundamental plan of structure of the groups concerned.

We have already seen something of the ways in which the crustacean limb has become diversified in relation to habitat and to method of locomotion. It has also been profoundly influenced by its involvement in crustacean feeding mechanisms. We can only speculate as to the starting point for its diversification, but among living forms the Branchiopoda seem to shed the most light on this problem. Branchiopods have a variable number of segments and an elongated tubular heart with segmental ostia, and they often use their antennae for swimming. The first two of these features are certainly primitive, and the last one may well be so also, for the Devonian *Lepidocaris* probably used its antennae in the same way. Indeed, it may be that initially the antennae were the chief locomotor organs, with the abdominal appendages serving primarily for respiration, and that these were later drawn into use for swimming and feeding, as they are in branchiopods today. At first they would doubtless have been simpler in form than the limbs of present-day species; we have seen that they might have been of a simple foliaceous type, already with filtering gnathobases, such as were possessed by *Lepidocaris*. Metachronal rhythm would have prevented the limbs interfering with each other, and their orderly beat would have provided for an economical expenditure of energy, in that locomotor movements could have contributed also to the requirements of filter feeding, and would have aided respiratory ventilation.

In this way the early crustaceans could have begun to exploit the rich food resources of open waters. However, this was not necessarily the primitive mode of life of the group. Filter feeding has certainly evolved along many independent lines in the course of crustacean history (in malacostracans, for example, quite independently of branchiopods), while many crustaceans have also exploited the nutritive resources of the substratum. We shall see that the phylogenetic interrelationships of these diverse methods of feeding (bottom-dwelling and pelagic, raptatory and suspension feeding) are complex, and that the course of history has not always followed the same pattern in the various lines. One conclusion, however, is self-evident: the ability of crustaceans to achieve in their feeding mechanisms the perfection of mechanical adaptation that we shall now be examining must have been a major factor in establishing the success of the group as aquatic animals.

As a first example of these filter-feeding mechanisms we may consider the anostracan Branchiopoda. These animals swim by means of the metachronal beat of thoracic appendages which are of the foliaceous type that we have called phyllopodia (Fig. 10–20). The inner edge of each of these bears a series of endites, the most basal of which is much larger than the others and is possibly formed by the fusion of two. These endites in their turn bear a fringe of large setae, which, together with the endites themselves, are directed somewhat backwards. The outer edge of the limb bears several lobes; the most distal of these, which is also provided with setae, is the exopodite, while the more basal ones are the epipodite and the proepipodites.

The beating of these limbs produces currents in the surrounding water, and it is these that are used in feeding. Thus the animal economically employs the limbs simultaneously for feeding as well as for locomotion, while their delicate structure enables them also to serve for respiratory exchange. An important factor in the feeding mechanism is the existence of an inter-limb space between any two limbs of one side (Fig. 10–21). This space is delimited on the outside by the exites of the more anterior of the two limbs, and on the inside by its endites. Normally the animal swims upon its back, so that the inter-limb space will be closed above by its large distal endite (or endopodite), and below by the ventral body wall. All the inter-limb spaces communicate with the continuous mid-ventral space that extends between the two rows of limbs.

The metachronal rhythm of locomotion is so organized that the phase of each limb is slightly in advance of the one immediately in front of it. At the end of the backstroke the limbs are inclined back at such an angle that their inter-limb spaces are reduced to their minimum volume; as they move forwards in succession (Fig. 10–22, limbs 6–10) their inter-limb spaces are correspondigly enlarged. This enlargement creates a suction which results in the endites and exites of one

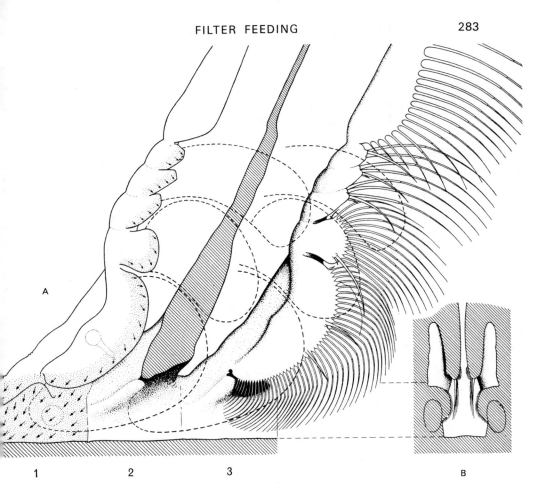

Fig. 10-20. A, median view of three consecutive trunk limbs of the branchiopod *Branchinella australiensis*. Only the median parts of the limbs are shown. *Limb 1:* The setae are omitted. The setules on the edges of the endites and on the wall of the food grooves are indicated by stippling, their direction being shown by arrows. *Limb 2:* The endites have been cut away to show the shape of the inter-limb space behind and its exit channel. *Limb 3:* The endites complete with setae are figured. The extent of the setae on limbs 1 and 2 is indicated by dotted lines. B, posterior view of lower edge of basal endites of a pair of trunk limbs of *Branchinecta gaini*, showing the exit grooves opening into the food groove between the filter setae and the main axis of the limb. From Cannon, 1933. *Phil. Trans. R. Soc. B,* **222**, 267-352.

limb pressing backwards on the limb behind it, and so closing off the upper and side entrances to the inter-limb space. The suction draws water into the enlarged inter-limb space from the mid-ventral space between the pairs of limbs. Because of the closing action of the endites and exites, however, this water can only enter the inter-limb space by passing through the fringe of setae along the edges of the endites. When the limbs beat backwards (Fig. 10–22, limbs 1–5) the situation is different. It is in the backward phase that the animal is driven forwards. During this phase the limbs are comparatively rigid, and they become separated from each other during their extension; this opens up the inter-limb spaces and allows water to

284   INVERTEBRATE STRUCTURE AND FUNCTION

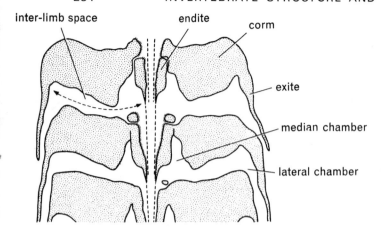

Fig. 10-21. Frontal section through three consecutive pairs of trunk limbs of *Branchinecta gaini* to show the shape of the inter-limb spaces and the valvular arrangements of the exite series. Adapted from Cannon, 1933, *op. cit.*

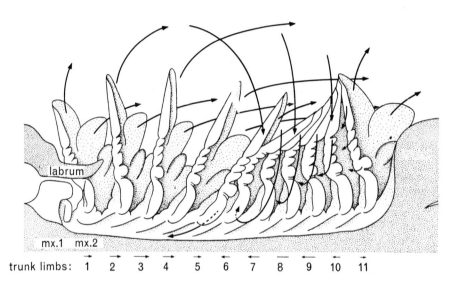

Fig. 10-22. Outline sketch of left half of an Anostracan (based on *Branchinella australiensis*) to show swimming and feeding currents. The arrows below the trunk limbs indicate their relative movements; limb 5 is completing its backward stroke, while limb 6 is beginning to move forward. All setae have been omitted. Adapted from Cannon, 1933. *op. cit.*

pass out backwards from them. It is, in fact, this stream of water that causes swimming.

Thus the swimming movements of the limbs result in water passing from the mid-ventral space between the limbs into the inter-limb spaces. To replace it more water must enter the mid-ventral space from the water surrounding the animal, and it is this inflow that brings the supply of food material. As the water passes into the inter-limb space, this material is trapped on the setae of the endites, which thus act as the filtering elements of the mechanism.

Another important factor in this complex process is the presence of very fine setules (Fig. 10–20) that form a feltwork on the median faces of the endites close to the main filtering setae. As the limbs move forwards, and each one presses against the one behind it in the way already explained, the setules of any one limb will project between the setae of the limb in front of it; as a result, they comb off the accumulated food material, which is then drawn by the water currents into a food groove. This groove is a channel that runs forwards along the mid-ventral line of the body wall, immediately between the bases of the limbs. In ciliary feeders the passage of food material along a corresponding path would be effected by cilia. Here it is probably brought about by anteriorly directed spurts of water which leave the inter-limb spaces during certain phases of the limb movements. In this way the food reaches the labrum, where it becomes entangled in a mucous secretion discharged from labral glands. This secretion facilitates its passage into the mouth, into which it is pushed by the maxillules, a pair of small appendages bearing long setae.

All branchiopods, and many filter-feeding Malacostraca, exploit essentially the same mechanical principle in their feeding mechanisms: the use of limbs with filtering setae for simultaneous feeding and respiration, and forward transmission of food material to the mouth. The principle has been deployed, however, with much diversification of detail. To take another example, the mechanism found in *Daphnia* (Fig. 10–23) shows the effect of a reduction in the number of limbs, and of some degree of division of labour among them, while the development of the bivalved carapace means that this can act as the outer wall of the water chamber. The limbs of *Daphnia* also differ from those of *Chirocephalus* in the reduction of the distal parts and in the development of a prominent gnathobase from an enlarged basal endite.

In all there are five pairs of trunk limbs in *Daphnia* (Fig. 10–24), and of these it is the third and fourth pairs that act as a filtering mechanism similar in principle to that of the Anostraca. This they are able to do because of the presence of a fringe of setae, representing their gnathobases. The fifth limb lacks these setae, and plays no direct part in filtering, but it contributes by its movements to the maintenance of the feeding currents; its outer part hinges forward and backwards on its inner part, so that it closes the inter-limb space at one stage and opens it at another. The first limb probably aids the forward passage of food along the ventral food groove by increasing the suction in this region, while its setae, together with those of the second limb, probably prevent over-large particles from entering the groove. The gnathobase of the second limb possesses a fringe of setae, one of which is particularly elongated; these seem to play some part in assisting the removal of food particles from the filtering setae of the more posterior limbs, and in influencing their movement towards the mouth. There is some disagreement regarding the exact operation of the anterior limbs, but their action results eventually in the food being entangled with a labral secretion as in anostracans; the material is then passed into the mouth by the action of the maxillules and mandibles.

So common is *Daphnia* in the fresh-water plankton that the Daphniidae are easily thought of as being entirely planktonic, but this is not so. Some are littoral and even benthic in habit, but the degree of diversity permitted by cladoceran organization is much better illustrated by a related family, the Chydoridae. Here the modes of life include scraping and filtering, crawling and scrambling, burrowing and mud-dwelling, scavenging and parasitism. But the niches that these animals occupy are even more diverse than this list may suggest. As Fryer points out in his

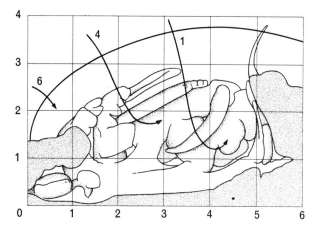
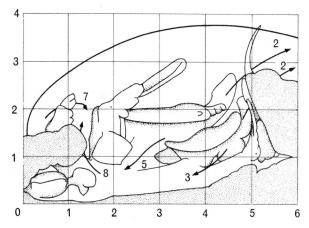

Fig. 10-23. Outline sketches of left half of *Daphnia magna*, to show movements of limbs and feeding currents. *Above*, the third and fourth trunk limbs are approximately at the end of their forestroke and *below*, at the end of their backstroke. A coordinate frame has been traced over the figures so that the movements of the limbs can be seen. Adapted from Cannon, 1933, *op. cit.*

analysis of the group, an expression such as 'weed-frequenting' is too generalized to be very informative, for two species may frequent the same frond and may yet be exploiting different ways of life. This is a measure of the precision of adaptation which is forced upon potentially competing species by the pressure of natural selection.

The possibilities of chydorid specialization can be illustrated by *Graptolebris testudinaria* (Fig. 10–25); a cladoceran which, within the limits of a pin's head (it averages 0.7 mm in length), has achieved a remarkable convergence with browsing gastropods, demonstrating thereby the power of natural selection in securing the most intimate relationship between organization and mode of life. The particular expertise of this animal is adhesion to surfaces of the Canadian pond-weed (*Elodea canadensis*) by its carapace chamber, which it achieves by applying the edges of this and lowering the pressure within it to establish suction. This demands sealing the anterior, posterior, and ventral openings of the chamber, and these it plugs with parts of its appendages and with numerous setae. The seal, which depends on the presence of a film of water over the leaf, is readily released when the animal elects to move away, but it 'swims only with reluctance', being ill-designed for that purpose.

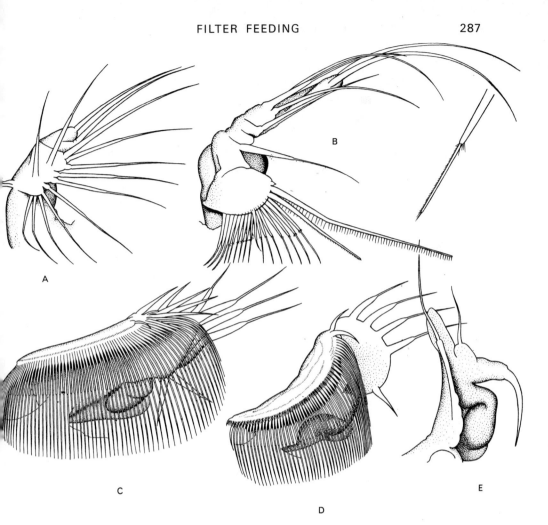

Fig. 10-24. Median view of left trunk limbs of *Daphnia magna*. A–E first to fifth trunk limbs respectively. The gnathobase at the lower end of limb B bears a number of setae, one of which is shown separately at the right. Adapted from Cannon, 1933. *op. cit.*

Movement over the leaf is effected mainly by the diphasic manipulation of the first pair of trunk limbs (Fig. 10–26). These slide forwards and then, pressed against the substratum, exert a drag which pulls forwards the carapace; this glides on its edges much as though they were the runners of a sledge. An equal drag by both limbs pulls the animal straight forward; an unequal drag moves it to left or right, the antennae cooperating. Food (which is probably mainly bacterial) is collected by scraping setae on the first two pairs of trunk limbs, the actions of which are very closely coordinated with each other and hence with the movement of the body. As the first pair move forwards, the second pair move inwards (Fig. 10–26), carrying food with them. As the first pair drags the body along, their scrapers swing inwards and collect food, while the second pair carry out their non-working recovery stroke. In the

288   INVERTEBRATE STRUCTURE AND FUNCTION

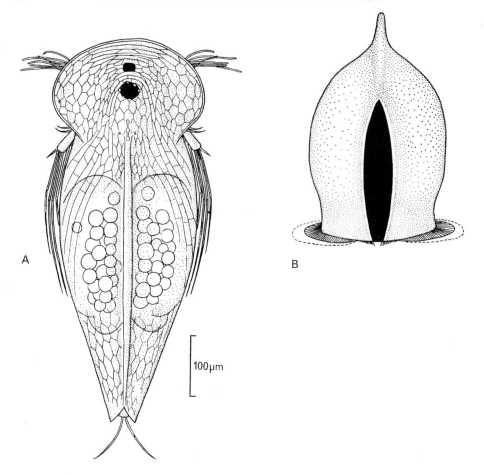

Fig. 10-25. *Graptoleberis testudinaria*, a gliding animal, seen A, from above. Note the array of tactile sense organs both around the head and posteriorly.
B, *Graptoleberis testudinaria* seen from behind, the observer being slightly above the level of the substratum. The outline of the animal is taken from a photograph focused somewhere near the middle which gave an accurate silhouette, and the posterior aperture, shown in black, was added from another photograph. The general appearance of the median ventral setae is correct, but the setae themselves are shown only diagrammatically. From Fryer, 1968. *Phil. Trans. R. Soc. Lond.*, **254**, 221–385.

course of these movements the food is directed into the median food-groove and passed forwards to the mouth. Correlated with this adaptive pattern is a reduction in the number of limbs used in feeding, a loss of filtration in collecting and handling the food, and a reduction in respiratory pumping, the last feature being probably aided by the sedentary mode of life.

Filter feeding of a different type from that so far described, in that it is associated with a considerable degree of differentiation and divergent specialization among the limbs, is seen in copepods such as *Diaptomus* and *Calanus*, in which the feeding current is created out of vortices produced by the swimming movements (Fig. 10–27). The swimming of *Calanus* depends mainly on the backward thrust produced by the

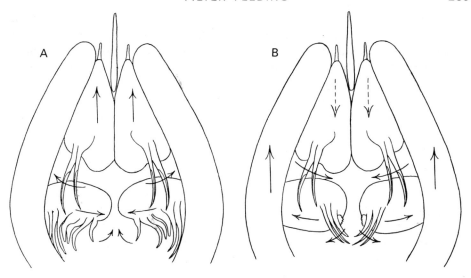

Fig. 10-26. *Graptoleberis testudinaria.* Diagram showing the coordinated sequence of movements of the trunk limbs and of the entire animal in gliding over and feeding from a flat surface. In A the carapace remains stationary while the first trunk limbs slide forward (and have here reached their anterior limit) and their scrapers swing outward. At the same time, the scrapers of the second trunk limbs (the major food collectors) move medially and deeper into the food chamber, thus being swept over the substratum while the animal is stationary. Having completed these movements, the sequence is, in essence, reversed (B). The first trunk limbs grip the substratum and pull. Because their 'soles' remain stationary, the carapace, and therefore the animal as a whole, slides forward. Simultaneously, the scrapers of the first trunk limbs swing towards the mid line (perhaps collecting material for the next working stroke of the second trunk limbs), and the second trunk limbs, here seen near the end of their working stroke, reverse their previous movement, thus making their non-working movement as their scrapers move forward over the substratum. From Fryer 1968. *op. cit.*

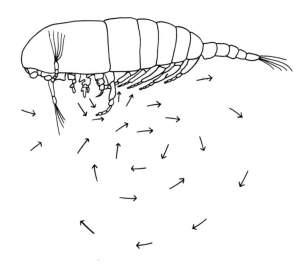

Fig. 10-27. Swimming vortex of *Calanus*, from the side. After Gauld, from Marshall, 1973. *Adv. mar. Biol.* **11**, 57–120.

antennae, the exopodites and endopodites of which beat in alternation and thus provide a smooth movement. As regards the other limbs, the first thoracic segment, which is here fused with the head (cf. p. 188), bears a pair of maxillipeds, while the remaining five free thoracic segments each bear a pair of swimming limbs. These latter, while contributing to the vortices, are not primarily involved in the feeding mechanism, which is mainly the concern of the more anterior appendages. The mandibles, maxillules and maxillipeds, which, like the antennae, bear long plumose setae, are kept in rapid vibration, and it is these movements, less important for swimming than those of the antennae, which are the basis of the filter-feeding.

The filter chamber of *Calanus* (Fig. 10-28) is bounded dorsally by the body wall, ventrally by the tips of the more anterior swimming limbs, medially and anteriorly by the maxillary setae and the labrum, and laterally by the maxillulary setae. Water is first drawn into this chamber by the suction created by the outward swing of the setae of the maxillipeds, and is then drawn out of it, through the maxillary setae, by the forward beat of the maxillules. The effect of this is that food is collected on the fringe of the maxillary setae, which thus act as the filter; the maxillae form a stationary filter, which does not show the rhythmical movement that is displayed by the other limbs. Finally, the food is removed from the maxillary setae and passed towards the mouth by setae on the maxillular endites and on the maxillipeds.

Fryer has suggested that this use of swimming vortices to produce feeding currents may be a secondary development of filter feeding in a group that was primitively raptatory. However that may be, there is certainly much diversity of feeding habits in copepods. Some are herbivores, some carnivores, some omnivores, with mandibles and other appendages correspondingly diversified in size and structure. The maxillipeds of carnivores, for example, may be very large and possess grasping spines. There are reports of copepods seizing fish larvae, and even filter feeders may take individual organisms, or follow them when they are offered to them in the laboratory. Moreover, not all copepods are pelagic. Some are suctorial parasites, ranging in size from the 32 cm *Pennella balaenopterae*, parasitic on whales, to *Sphaeronellopsis monothrix*, parasitic on marine ostracods, with a male only 0.11 mm in length. Some are benthic, these including species which live in sand, where they are said to scrape off the food material attached to the sand grains. The irrespressible resources of animal adaptation are well exemplified in this context by *Macrostella gracilis*, which, although adapted for benthic life, succeeds in escaping to the plankton by attaching itself to the floating algal filaments on which it feeds.

Maxillary filtering plates are not peculiar to copepods. They are also used in the Malacostraca, as, for example, in *Hemimysis*; in this animal they act both as a suction pump and as a filter, drawing a food-bearing stream forwards from a ventral food groove by their vibration. The food stream in this animal is produced by the swimming movements of the thoracic exopodites, much as in anostracans, but there is an important difference in their mode of action, for the limbs of Malacostraca bend forwards, whereas those of the Branchiopoda bend backwards. This suggests that the method of producing the food stream in the two groups evolved independently, for any stage intermediate between them would have been hydraulically inefficient.

Many other complications are revealed when we attempt to trace out the evolutionary pattern underlying these various systems. For example, *Nebalia bipes*, a malacostracan of the group Phyllocarida, is a mud-living form that is common under

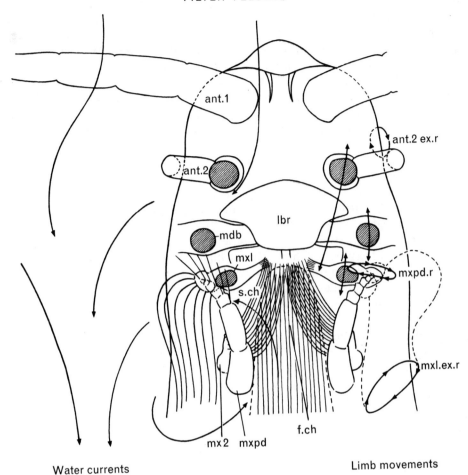

Fig. 10-28. Diagram of anterior region of *Calanus finmarchicus*. The endopodite of the antenna, the mandibular palps, and the distal parts of the maxillules have been removed. The position of the swimming trunk limbs is indicated by the shaded area inside the dotted line. On the *right* side of the figure the limb movements are indicated, on the *left* the water currents. *ant.1*, antennule; *ant.2*, antenna; *ant.2 ex.r*, rotation path of tip of exopod of antenna; *f.ch*, filter chamber; *lbr*, labrum; *mdb*, mandible; *mxl*, maxillule; *mxl.ex.r*, rotation path of tips of setae of maxillulary exite; *mx.2*, maxilla; *mxpd*, maxilliped; *mxpd.r*, rotation path of tip of maxilliped; *s.ch*, suction chamber. From Marshall and Orr, 1955. *The Biology of a Marine Copepod*. Oliver and Boyd, Edinburgh.

stone in the littoral zone. It favours areas that are rich in organic débris, and feeds on particles that are filtered from a food stream produced by the oscillatory movements of the foliaceous trunk limbs. This stream, however, differs from that described in our other examples in that it enters the food-collecting zone anteriorly and leaves at the posterior end of the carapace. Moreover, the filter is here formed by the setae on the endopodites of the limbs.

One explanation of the origin of this unusual mechanism is that the Phyllocarida may be derived from primitive mysid-like forms that took to a burrowing life, as a result of which they abandoned their swimming movements and, in consequence, their primitive filter-feeding mechanism. The use of oar-like movements of the antennules to pull the animal through the mud might then have tended to suck particles into the anterior opening of the carapace. An increasingly foliaceous development of the trunk limbs might have encouraged this backward current, and have led eventually to the evolution of a new type of filtering mechanism. Mechanical principles similar to those employed in more primitive forms would in this way have found a new structural expression.

The complexity of the evolutionary interrelationships of crustacean feeding mechanisms and modes of life will be sufficiently evident, even from this brief survey. To some extent they are an expression of the parallelism and convergence found in so many aspects of arthropod organization. Thus, the caridoid facies of the Malacostraca, although probably developed in bottom-dwelling forms, has nevertheless lent itself to the exploitation of pelagic life, and filter feeding has evolved as a natural accompaniment. Despite this, however, the higher Malacostraca have come to rely on crawling or burrowing in the littoral zone and on the sea bottom. Increasing emphasis upon macrophagy is natural in these conditions, and is aided by the versatility of the crustacean limb, and notably by the development of chelae on the walking limbs. This in turn favours continuing increase in size, and so we arrive at the large and varied raptatorial decapod fauna of today.

11
Respiration

11–1 SOME PHYSICAL FACTORS

The processes involved in securing and utilizing oxygen are grouped under the general category of respiration, but this term comprises two distinct aspects of those processes. Firstly, there is the exchange of oxygen and carbon dioxide between the organism and the external environment: this is external respiration. Secondly, there is the complex of reactions that takes place within the cells, and that results in the release of energy by the oxidation of the energy-rich molecules derived from the food: this is internal respiration. Linking these two are the transport mechanisms that convey the oxygen and carbon dioxide between the external respiratory surfaces and the metabolizing tissues.

We are here concerned with external respiration and transport. The organization of these is influenced by certain physical considerations that were clearly stated by Krogh, whose analysis we may conveniently follow. The availability of oxygen for living organisms depends upon its concentration in the surrounding medium, and in this respect conditions are more favourable in air than in water. Air consists of 20.95% oxygen and 0.03% carbon dioxide, the remainder being composed of inert gases, mainly nitrogen (78%) and argon (0.94%). In respiratory uptake, however, the pressure rather than the volume of these gases is important; consequently it is more meaningful to express their concentrations in terms of pressure. Thus at sea level, at a normal barometric pressure of 760 mm of mercury, the oxygen pressure will be of the order of $20.95/100 \times 760$ mm, i.e. 155 mm, whereas at 16,000 ft it will have fallen to 88 mm, and life will be correspondingly more difficult to maintain.

The situation in water is very different from this. The component gases of the air are soluble, but only to a limited extent, and less so in salt solutions than in pure water. Solubility is expressed as the absorption coefficient. This, in Krogh's definition, is the quantity of gas, measured dry at 0°C and at atmospheric pressure, which can be taken up by one volume of water from an atmosphere of that gas at normal pressure (760 mm); it can conveniently be expressed as a percentage of the water volume. Absorption coefficients vary with temperature. At 15°C, to give one example, the values for oxygen, nitrogen, and carbon dioxide in pure water are respectively 3.5, 1.7, and 100%. Normally, of course, organisms are concerned with the mixed gases of the atmosphere, each of which is independently absorbed according to its partial pressure. Thus 100 volumes of water at sea level and at 15°C will take up $155/760 \times 3.5 = 0.72$ vol of oxygen, and $605/760 \times 1.7 = 1.35$ vol of nitrogen. It follows from all of these considerations that oxygen is very much less available in water than in air, and that it is less available in sea water than in fresh water (Table 11–1).

Other factors, however, are relevant in judging the availability of oxygen for the organisms living in any particular habitat. In the sea, despite the relatively low solubility of the gas, there is certainly enough in most regions to meet the requirements of animal life—not only at the surface, where it is taken up directly from the atmosphere, but in the deep waters of the abyssal zone, where it is continuously replenished by the massive currents of the oceans. The same is true for fresh water, provided that there is ample circulation of the water. Where circulation is reduced, as in swamps and ponds, or in lakes at a level below the summer thermocline (where contact with the surface is lost), there may be oxygen deprivation, particularly where abundant organic material is undergoing bacterial decay. Conversely, the occurrence of much photosynthesis may raise the oxygen content of water above what could be secured by diffusion from the atmosphere; the water is then said to be supersaturated.

The rate of uptake of oxygen by an organism is limited by the gradient of concentration between the organism and the layer of the medium that immediately surrounds it. Regardless of whether an animal is living in water or in the air, this layer will be aqueous, for even in terrestrial forms the respiratory surfaces are kept moist by a thin film of fluid; it is this that maintains their permeability. Efficient respiratory exchange depends also upon the continuous replacement of the oxygen that is taken up from the medium, a replacement which is effected partly by diffusion of oxygen through the medium, and partly by movement of the medium itself. This movement may be partially provided for by convection currents, but the more important factor will usually be ventilation, which comprises the movements set up in the medium by the organism itself. It is essential for air-breathing animals to keep their respiratory surfaces moist, and to prevent an undue loss of water through them. Provided that these conditions are satisfied, the balance of respiratory advantage lies with them rather than with those animals that take their oxygen from water. One reason for this

Table 11–1 Volume of oxygen (ml/l) in fresh and salt water saturated with atmospheric air at the stated temperature. From Krogh, A., 1941. *The Comparative Physiology of Respiratory Mechanisms.* Pennsylvania University Press.

Temp. °C	fresh water	10% Cl	20% Cl
0	10.29	9.13	7.97
10	8.02	7.19	6.35
15	7.22	6.50	5.79
20	6.57	5.95	5.31
30	5.57	5.01	4.46

is that oxygen diffuses much more readily in air than in water. This has been expressed by Krogh in terms of diffusion constants, which are based upon the difference in partial pressure of the gas per unit length of the phase through which it is diffusing. The diffusion constants for oxygen at 20°C are 11.0 in air and only 0.000034 in water. A further advantage for air-breathing animals is the lower density of air as compared with water; the specific gravity of air at 20°C, when it is dry and at atmospheric pressure, is 0.0012. This means that convection currents are more readily established in air than in water, and also that ventilation movements in the former demand a smaller output of energy.

We shall see many ways in which the form and organization of the animal body have been influenced by the factors outlined above. In addition, they have been influenced by the problem of ensuring the transport of oxygen from the surrounding medium into the area of metabolic activity within the body. Small size brings the advantage of a high ratio of surface area to volume. Because of this the smallest animals need rely on no more than passive diffusion of the oxygen into and around their body. This is most obvious in Protozoa, but it holds also in sponges and coelenterates, where thin layers of epithelium are directly bathed by the external medium.

There are limits, however, to the value of unaided diffusion. These can be appreciated from the calculation that the difference in oxygen tension between the surface and the centre of an organism, if constant metabolic rate is to be maintained, will be proportional to the square of the radius of the organism. Krogh has drawn from this the conclusion that organisms relying solely upon diffusion can only maintain a high rate of metabolism if they are of 1 mm diameter or less. Larger animals that continue to rely upon diffusion alone must necessarily, therefore, have a low metabolic rate. It is true that animals have some room for manoeuvre in this regard, for their relations with the environment can be improved by appropriate modifications of their shape. The point here is the well-known one, fundamental to the design both of the whole body of an animal and of its constituent parts, that a sphere provides the smallest possible ratio of surface area to volume. Departures from a strictly spherical shape will, therefore, benefit the organism by increasing this ratio and so facilitating respiratory exchange by diffusion. This factor must have influenced the evolution of the shape of platyhelminths and nemertines, for their flattened form both increases the ratio and reduces the distances over which diffusion has to take place.

Two other developments have been important in attaining large size. Both are related to the need for providing internally some system of convection to replace diffusion, and to provide an equivalent to the external convection secured by the ventilation movements. The first has been the establishment of a circulatory system providing for the movement of a fluid transport medium through the body. To some extent the movement of tissue fluids resulting from locomotion and feeding goes a little way to meet this need; but the crucial advance in this connection was the evolution of a blood system, which provides for a continuous and comparatively rapid circulation of fluid to all parts of the body.

We have already seen, however, that a saline medium, which is what the body fluids necessarily are, can only absorb comparatively small amounts of oxygen in simple solution. Consequently many animals have a circulatory system which is fortified for respiratory functioning by the presence in it of oxygen carriers, or respiratory pigments. These have properties of remarkable adaptive advantage, being able to take up oxygen where it is abundant, to release it where it is scarce, and to transport it in concentrations greater than could be secured by its simple solution. By developing these substances many animals have obtained an invaluable basis for metabolic efficiency. Curiously enough, however, one of the major groups, the insects, together with certain of the other members of the Arthropoda, have not adopted this device. Instead, they have secured a high level of metabolic efficiency through the aid of physical diffusion, sometimes, although not always, supplemented by ventilation movements. For this they have paid the price of small size.

11–2 RESPIRATORY PIGMENTS

We can only account for the distribution of oxygen carriers in the animal kingdom if we assume that these compounds have been independently evolved in many different lines, for more than one type of compound has been employed, and the distribution of these follows no simple phylogenetic plan. The most widespread of these substances is the group of compounds called haemoglobins; these have certainly evolved quite independently in unrelated groups, doubtless as a consequence of the capacity of living organisms for synthesizing the compounds known as porphyrins, which may be regarded theoretically as derivatives of a parent compound called porphin (Fig. 11–1). This substance, known only from laboratory synthesis, consists of four pyrrole rings linked by four methane bridges into a cyclic system which is evidently of great stability, for porphyrins of biological origin occur in mineral deposits such as coal and oil. They are widely distributed in living organisms, in bacteria and plants as well as in animals. In fact, the capacity for synthesizing them is universal in aerobic organisms, with the exception of certain bacteria, and they are probably another example of a molecular pattern that appeared very early in the history of evolution.

A particularly important property of porphyrins is their ability to associate with

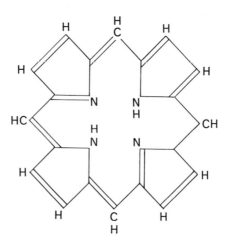

Fig. 11-1. Porphin.

metals to form coordinated compounds known as metallo-porphyrins. These compounds serve as the prosthetic groups of proteins that have a wide range of catalytic functions in the cell. Thus iron porphyrins are the prosthetic groups of the peroxidases that are found largely in plant tissues, and also of the catalases of bacteria and animal tissues, while chlorophyll is a magnesium-porphyrin complex (Sec. 1–3). Furthermore, the cytochromes, which we have seen to be essential components of intracellular oxidation mechanisms, are composed of proteins having iron porphyrins as their prosthetic groups. Probably the capacity for synthesising this type of molecule led to the evolution of adaptively valuable compounds, with the consequent development of certain oxygen carriers.

One of the porphyrins that can be derived from porphin is protoporphyrin. The addition to this of one ferrous iron atom produces ferrous protoporphyrin (haem, Fig. 11–2), in which the iron atom is joined by four of its coordination bonds to the four nitrogen atoms of the protoporphyrin. One of the remaining six coordination bonds can then be joined to a molecule of protein called globin. The result is the formation of the compounds known as haemoglobins, which are the best known of all of the oxygen carriers. The haemoglobin molecule can carry an oxygen molecule attached to the remaining coordination bond, with the iron remaining in the ferrous state (Fig. 11–2); this is of peculiar value in oxygen transport. The oxygen is readily taken up at the respiratory surface, and is equally readily given up within the tissues at regions of low oxygen concentration; this latter process, called dissociation, involves the replacement of a molecule of oxygen by a molecule of water. The two forms of the compound, oxygenated and deoxygenated, are known respectively as oxyhaemoglobin and deoxyhaemoglobin. They must be distinguished from the oxidized form, methaemoglobin; in this the iron is in the ferric form, and the sixth bond carries a hydroxyl group, with the result that the molecule is no longer available as an oxygen carrier.

Haemoglobins occur in all vertebrate animals, with some very rare exceptions (the leptocephalus larva of the eel is one), so that they can be regarded as a biochemical characteristic of the group. In invertebrates the situation is quite otherwise. It has been known since the early investigations of Ray Lankester in the nineteenth century that haemoglobin is widely distributed in these animals, but its appearance is curiously sporadic, and of no obvious phylogenetic significance. As suggested above, this is most readily explained as a consequence of the widespread capacity for synthesizing protoporphyrin. Conceivably haemoglobin could have arisen because of changes in the cytochromes resulting from mutation. Cytochromes, however, depend for their functioning upon oxidation, with an associated change in the valence state of the iron from the ferrous to the ferric form. Thus, as Munro Fox and Vevers remark, it is probably easier to imagine haemoglobins arising from the chance association of protoporphyrin with various globins.

Among the invertebrates haemoglobin is particularly characteristic of the Annelida and the entomostracan Crustacea, where it is typically dissolved in the blood fluid, although it may also be present in the tissues. It occurs only rarely in other arthropods, the blood of the midge larva, *Chironomus*, providing one example. In molluscs its occurrence is sporadic; when present it is usually in the tissues, but it is dissolved in the blood of one pulmonate, *Planorbis*, and it occurs in blood corpuscles in a few bivalves. It has been identified also in animals as diverse as *Para-*

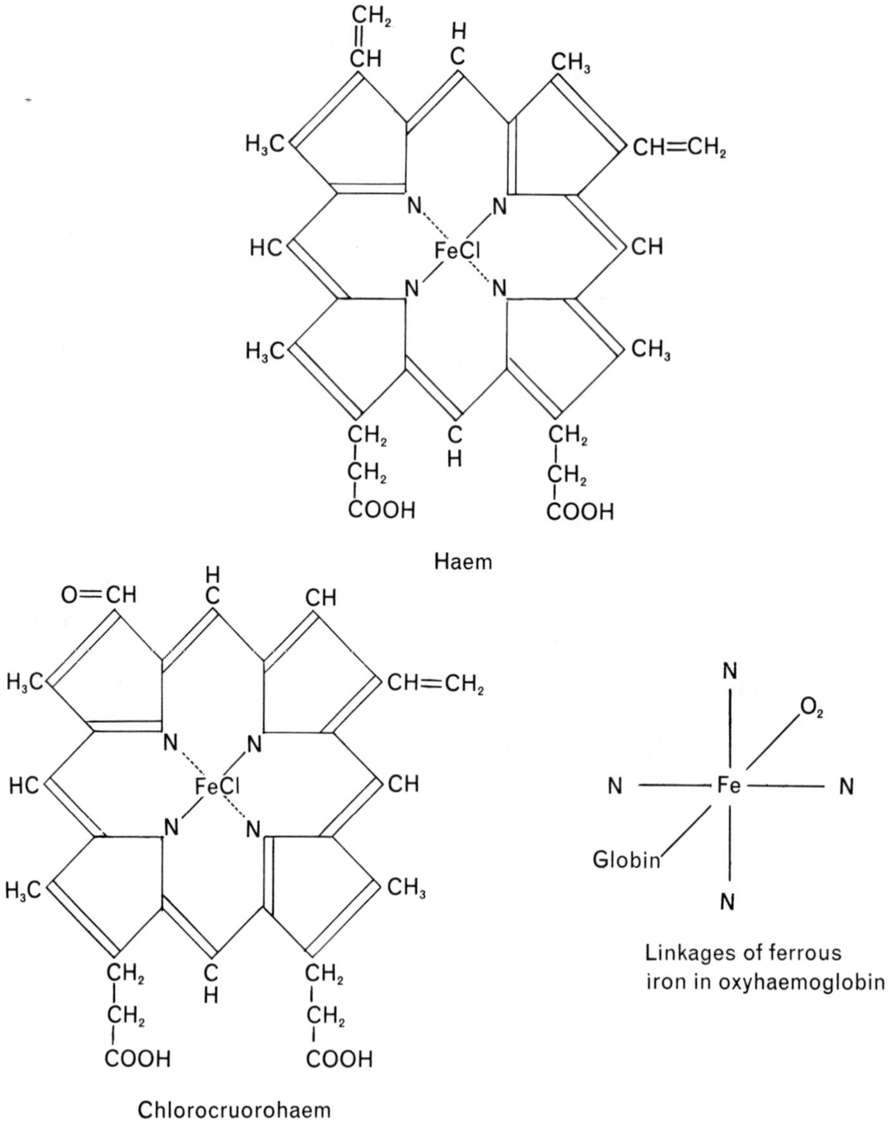

Fig. 11-2. Haem and related compounds.

mecium, trematodes, nemertines, *Ascaris*, and *Phoronis*. Finally, a closely similar substance (leghaemoglobin) is found in the root nodules of Leguminosae, where it results from the interaction between the plant itself and its symbiotic micro-organism, *Rhizobium*; neither of these can produce it unaided.

Oxyhaemoglobin and deoxyhaemoglobin are respectively red and purple in colour, and are characterized by well-defined absorption bands in the visible spectrum. Oxyhaemoglobin has alpha and beta bands in the yellow and green regions (the alpha nearer the red end) and a larger gamma band in the violet region. Deoxyhaemoglobin has the gamma band, although at a slightly different position, together with only one other band, lying in the green region. The positions of these bands vary in haemoglobins from different sources, which means, of course, that there must be many different haemoglobins. Indeed, spectroscopic characteristics are not the only differences between them. They are distinguishable also by physical properties, such as their isoelectric points and their amino acid composition, and these are reflected in differences in the kinetics and equilibria of their interactions with oxygen. These characteristics arise solely from differences in the globin portion of their molecules, the haem portion being always identical. This can be demonstrated by taking advantage of the capacity of pyridine for displacing the globin and combining with the haem to form a pyridine–haem complex. The spectroscopic properties of this complex are always precisely the same, irrespective of the properties of the haemoglobin from which it is derived.

All haemoglobins consist of unit molecules, each with one haem and its globin, and having a molecular weight of 17,000–18,000. In general, the invertebrate haemoglobins differ from the vertebrate ones in having less histidine and lysine, and more arginine and cystine, but they also differ among themselves in the number of units that they contain. These range from two in the larva of *Chironomus* (m.w. 31,400), for example, to several hundred in *Arenicola* (m.w. 3 million). This variability has doubtless helped to provide for the adaptive molecular evolution of haemoglobins that are particularly suited for use in different types of environment. There are, however, other differences to be found in respiratory pigments, showing that substances that are similar in principle in their oxygen-transporting properties may yet be very different in their molecular structure.

One example of these is chlorocruorin, a green respiratory pigment found only in four families of the Polychaeta: the Ampharetidae, Chlorhaemidae, Sabellidae, and Serpulidae. It is closely related chemically to haemoglobin (Fig. 11–2). The protein is different, as is to be expected, but the iron porphyrin of its prosthetic group is a haem which differs only in having one of the two vinyl groups ($-CH=CH_2$) substituted by a formyl (aldehyde, $-CHO$) group. Its molecular weight, however, is extremely high, amounting to about 3 million, and it contains some 190 iron atoms in its complex molecule. Chlorocruorin is described as dichroic, for it has two colours which are dependent on its concentration, green in dilute solutions and red in concentrated ones. Worms with this pigment are not necessarily green, however, for the colour may be masked by other pigments, as it is, for example, in the tentacles of serpulids. More confusing is the fact that closely related species may differ in the presence or absence of the pigment, as in the genus *Spirorbis*. One species, *S. borealis*, possesses chlorocruorin; another, *S. corrugatus*, has haemoglobin; a third, *S. militaris*, has no respiratory pigment at all.

The close similarity of chlorocruorin to haemoglobin makes it likely enough that chlorocruorin could evolve from haemoglobin by genetic mutation, and that it might well have arisen independently in different species. Thus the presence of this pigment is not necessarily evidence of the close phylogenetic relationship of the groups concerned. The need for caution in making such interpretations is shown by the presence of chlorocruorohaem, which is the prosthetic group of chlorocruorin, in two starfish, *Luidia* and *Astropecten*; these animals have not, however, developed the respiratory pigment itself.

The devious pathways of biochemical evolution are further illustrated by the haemerythrins. These are reddish-violet iron-containing respiratory pigments that are known only in the sipunculids (including *Sipunculus* and *Phascolosoma*), the polychaete *Magelona*, the priapulids *Halicryptus* and *Priapulus*, and the brachiopod *Lingula*. Haemerythrin is thus a rare pigment, and, apart from the fact that it has been identified in all sipunculids examined, its distribution is sporadic and without any phylogenetic significance. The sipunculids are possibly related to annelids, but the priapulids are non-coelomate forms, while the brachiopods are perhaps related through the structure of their lophophore to *Phoronis* and the Ectoprocta. There are other distinguishing features of this pigment. For one thing, it is always present in cells, which are usually in the coelomic fluid; in *Magelona*, however, they are true blood corpuscles, this animal being the only polychaete to carry a respiratory pigment in this particular way. This intracellular location of haemerythrin is doubtless correlated with its molecular weight (66,000 in *Sipunculus*), which is low in comparison with that of most invertebrate respiratory pigments, and more comparable with that of vertebrate haemoglobin (68,000). Retention of respiratory pigments of low molecular weight within cells is thought to be an adaptation preventing their loss through the excretory system. Another feature of haemerythrin is that, despite its oxygen-carrying capacity and its red colour, it is not closely related to haemoglobin. Its spectrum shows no strong absorption bands, and in molecular structure it is a protein, with a high percentage of iron in its molecule, but with no associated porphyrin, the iron being attached directly to the protein. Thus in haemoglobin and haemerythrin we find a similar result achieved by different chemical means.

Haemerythrin is misleadingly named, and so also is haemocyanin, another respiratory pigment in which the prosthetic group is not a haem, and in which the metal is probably attached directly to the protein. Here, however, the metal is copper instead of iron. In this respect haemocyanin differs from all of the previously mentioned respiratory pigments, and shows some relationship to the copper-containing phenol oxidases, a relationship analogous to that of the haemoglobins to the iron-containing oxidases and peroxidases. The oxygen combines with two atoms of copper in haemocyanin, as compared with the one iron atom of the haemoglobin, and in so doing produces a blue compound, a deoxygenated form of the pigment being colourless. The molecular weights are considerable—units with two copper atoms range from 50,000 to 74,000, and are associated into complex molecules with molecular weights as high as 6,650,000 in *Helix pomatia*. Not surprisingly there is variation here, as with the haemoglobins, in the properties of the pigment.

Haemocyanin, like chlorocruorin, occurs only in solution, in this instance in the haemolymph of certain arthropods and molluscs, where its distribution shows a certain phylogenetic pattern. It is the only respiratory pigment of malacostracan Crustacea, occurring in the decapods and the stomatopods. It is found in no other crustaceans, nor does it occur at all in insects, but it appears in some chelicerates: *Limulus*, *Euscorpius*, and spiders. As regards molluscs, it is found in chitons, in cephalopods, and in many gastropods, including particularly the prosobranchs and the pulmonates, but it has not yet been identified in the bivalves. Molluscan haemocyanins have molecular weights of several million, while those of crustaceans are of the order of several hundred thousand; the difference doubtless reflects the independent origin of these substances in the two groups.

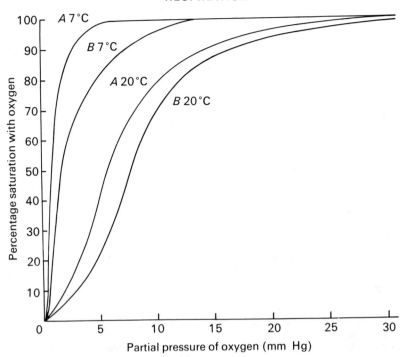

Fig. 11-3. Oxygen dissociation curves for the haemoglobins of two earthworms, *Allolobophora terrestris* (A) and *Lumbricus terrestris* (B) at the temperatures shown. These are high oxygen affinity pigments. From Haughton et al., 1958. *J. exp. Biol.* **35**, 360–368.

The wide distribution of respiratory pigments, and their occurrence in such different groups, is sufficient indication that they must confer some respiratory advantages. Yet it needs to be emphasized that, as in the case of *Spirorbis*, one species may exist without any of these pigments, while related species possess them. Indeed, it is not always easy to decide just what their function is in certain species. As we have already suggested, the molecules of these pigments seem well suited for undergoing adaptive modification. Yet we must judge their function not only from the properties of the particular pigment, but also with an understanding of the conditions of life of the species concerned.

In general, respiratory pigments function either as oxygen carriers, or as oxygen stores, providing reserves to be used at times of shortage. The former function may reasonably be attributed to those pigments that are present within the blood stream, but where they are present in other tissues (as, for example, in the muscles of *Arenicola*) they are more probably serving for oxygen storage. However, storage is not sharply separable from transport, for the tissue pigments, by combining with the oxygen arriving in the blood stream, must accelerate the passage of oxygen into the tissues. This is likely to be particularly important in organs that carry out rhythmical bursts of activity; it is significant, therefore, that there is evidence of haemoglobin being particularly abundant in the muscles of gizzards and radulae. Moreover, its distribution may be adapted to the mode of feeding, as it appears to be in *Aplysia*,

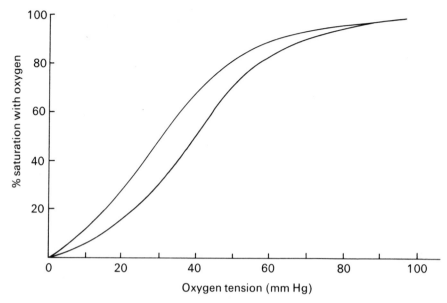

Fig. 11-4. The Bohr effect, illustrated by a pigment of low oxygen affinity, typical of man and other mammals. The curve is shifted to the right by an increase of CO_2 tension in the blood stream; such an increase, which may be expected to occur in regions of high metabolic activity, will thus facilitate the release of oxygen to the tissues.

which feeds continuously, and the related *Navanax inermis*, which feeds sporadically. The former has muscle haemoglobin (myoglobin) in its buccal musculature, whereas the latter lacks it. It is often supposed that storage is also the main function of the respiratory pigment of the body cavities, such as the coelomic haemoglobins of some annelids and the coelomic haemerythrins of sipunculids. This may be so, but we have noted that the use of such fluids in hydrostatic skeletons results in a good deal of movement in them, and certainly brings them into close functional relationship with contractile tissues; some transport function cannot, therefore, be excluded in such instances.

One key factor in the mode of functioning of these pigments is the relationship between the amount of oxygen that they can take up and the pressure of oxygen (oxygen tension) to which they are exposed. This relationship is expressed in an oxygen dissociation curve (or oxygen equilibrium curve), which is obtained by exposing the blood to oxygen at a series of different tensions. Examples of such curves are seen in Fig. 11–3, which shows the dissociation curves for two species of earthworms. These have haemoglobins of high oxygen affinity, which are well saturated at a wide range of environmental oxygen tensions. Their low unloading tensions imply that oxygen tensions are correspondingly low in the tissues. This, it is suggested, may be due to the lack of respiratory organs, combined with a body surface that has a high resistance to diffusion.

It will be apparent that a useful index of the mode of functioning of these pigments is given by determining the partial pressure of oxygen at which they are half-saturated with oxygen. This value, expressed as p_{50}, varies with temperature, lower temperatures shifting the curves to the left, and vice versa (Fig. 11–3). It is also

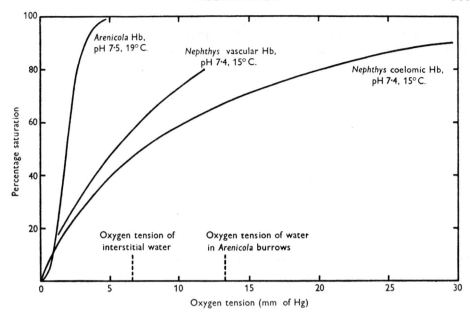

Fig. 11-5. Comparison of the oxygen dissociation curves of the vascular and coelomic haemoglobins of *Nephtys hombergii* at pH 7.4 and 15°C and of the haemoglobin of *Arenicola marina* at pH 7.5 and 19°C. The levels of interstitial oxygen tension and of oxygen tension in the residual water of exposed *Arenicola* burrows are indicated by vertical lines above the abscissa. From Jones, 1955. *J. exp. Biol.*, **32**, 110-125.

affected by acidity. If the pH decreases, so does the oxygen affinity, and the curve is moved to the right (Fig. 11–4). An increase in carbon dioxide tension will have this effect (the Bohr effect), and since this tension will be high where tissues are active, the result is to facilitate the release of oxygen to the tissues that most need it. A reversed Bohr effect is sometimes found, the dissociation curve shifting to the left with a rise in pH.

It has often been suggested that a low oxygen affinity indicates an oxygen transport function for the pigment, but that a high affinity may indicate a storage function, allowing oxygen to become available in conditions in which transport has become impossible. This, however, is an oversimplification, as may be illustrated by reference to *Arenicola*. The oxygen affinity of the haemoglobin of this animal is high (Fig. 11–5), with a p_{50} ranging from 2.0 to 8.3 mm, while its oxygen-carrying capacity is exceptionally high for an invertebrate. Even so, however, this capacity is low in relation to the known metabolic rate of the animal. Even if the pigment were functioning solely as a store, the oxygen would only be sufficient for about 21 min of activity. Admittedly even as small a store as this could be of value in the intervals between the rhythmic irrigation movements which, as we shall see later (Sec. 11–3), the worms carry out in their burrows, and which provide a regular replenishment of the oxygen supply. However, it is probable that the pigment does, in fact, provide a mechanism for transporting oxygen. The oxygen tension of the water in the sand around the burrow at low tide will contain an oxygen concentration equivalent to about 6.7 mm of mercury. Reference to Fig. 11–5 shows that the pigment can continue to take up oxygen at even lower tensions than this, so that it is well able to transport it to the tissues and release it to the haemoglobin of the muscle. We shall see that the respiration of *Arenicola* is further aided by the animal's ability to draw oxygen into its burrow when the tide is out; presumably, therefore, it is not dependent for long periods upon stored oxygen. We may conclude that the transport capacity of the haemoglobin of *Arenicola* is well adapted to the needs of an animal which experiences a wide range of oxygen tensions in its burrow. At the same time, some storage function is not excluded.

Another burrowing polychaete, *Nephtys hombergii*, presents an interesting contrast to *Arenicola*. It has haemoglobin in the coelom as well as in the blood, the p_{50} values being, respectively, 5.5 and 7.5 mm. These are within the range of values for *Arenicola*, but the shape of the curve is very different (Fig. 11–5), showing that oxygen is released over a much wider range of tensions. This, therefore, is a low-affinity pigment. The difference can be ascribed to *Nephtys* moving vigorously through the substratum in temporary burrows that collapse when the tide is out. Unlike *Arenicola*, therefore, it has no substantial store of environmental oxygen, while it can get little benefit from the sparse interstitial water, for the oxygen tension of this is so low as to be virtually unusable. A high-affinity pigment would be of little value for transport when the tide is in. When it is out, the animal must reduce activity, and its metabolism then is probably largely anaerobic. Where so many invertebrates differ from mammals is in their ability to maintain metabolism at very low internal oxygen tensions. This is probably the situation in *Arenicola*, and it is probably true also of *Tubifex* and earthworms, which can continue their metabolism even when their haemoglobin is completely immobilized by the presence of carbon monoxide.

11–3 GILLS AND LOPHOPHORES

The thin flat body of platyhelminths facilitates respiratory exchange, and also ensures that metabolites can readily be transmitted throughout the body of diffusion; in correlation with this there is no blood system in these animals. A simple type of circulatory system is, however, present in nemertines, and is probably associated with the elongation of the body. In its simplest and presumably primitive form, as seen in *Cephalothrix*, it consists of no more than a pair of longitudinal vessels, united in front and behind; in other genera this plan is further elaborated by the development of transverse vessels and a mid-dorsal one. The larger vessels are contractile, with circular muscle fibres in their walls, and there is a true circulation of the blood. This movement can be seen in smaller animals, the flow being forwards in the dorsal vessel, as it commonly is in invertebrates.

The blood of nemertines is usually colourless, with cells suspended in it, but sometimes these cells contain pigment, and in a few instances this has been shown by spectroscopic examination to be haemoglobin. Presumably, then, the vascular system of nemertines may sometimes be of significance for respiration. But the vessels lie below much or all of the well-developed muscle layers of the body wall; here they are so far from the body surface that they must often be unable to play much part in respiratory exchanges with the external medium. Probably the system is chiefly important as a transporter of metabolites; we would expect, however, that once established a circulatory system would become associated, sooner or later, with the uptake and movement of oxygen and carbon dioxide; in fact, we find a system functioning in this way in the Phylum Annelida.

In many annelids, particularly in oligochaetes and leeches, respiratory exchanges take place through the body surface. In earthworms the epidermis is sufficiently thin for diffusion to take place through it, and the vessels run at its base; in leeches, with a more specialized body wall, capillaries penetrate among the epidermal cells. Even within this simple type of arrangement there is room for adaptation to specialized habitats, as may be seen in *Tubifex*. This oligochaete is well adapted for withstanding oxygen shortage. It is, in fact, the first species to become re-established in rivers below points where animal life has been destroyed by the discharge of effluents, and for this reason it is a useful biological indicator of high pollution. It lives head downwards in burrows in the mud, and obtains its oxygen by uptake through the posterior body wall. Ventilation is effected by a rhythmic waving of its tail at a frequency that increases as the amount of available oxygen declines. The arrangement of the posterior segmental vessels is influenced by this; they are long and convoluted, and are closely applied to the body wall, so that they are well suited for the uptake of oxygen. A similar arrangement at the anterior end, however, would result in the loss of oxygen to the deoxygenated mud, and it is presumably in adaptation to this difficulty that the more anterior segmental vessels are shorter, and are not so closely applied to the body surface.

The respiratory exchange of errant polychaetes is aided by the parapodia; these are hollow and mobile extensions of the body surface, and are also well vascularized. True gills or branchiae are also common in polychaetes. An outgrowth of the parapodium of *Nephtys* is probably respiratory in function, but gills are more particularly characteristic of sedentary worms such as *Amphitrite* (Fig. 10–7), where the parapodia are of a different form. In *Arenicola* gills are developed as branched and highly vascularized outgrowths on a number of the trunk segments, and the adaptations of this animal for life within a burrow (Fig. 11–6) include a mechanism for ensuring efficient ventilation. This is carried out by outbursts of irrigation movements, which recur with great regularity at intervals of about 40 min. The movements are initiated by a tailward locomotion of the animal, which takes it towards the posterior end of the burrow. This is followed by the chief phase of irrigation, which consists of anteriorly directed peristaltic waves of the body surface, combined with some headward creeping. During this phase water is drawn forwards and over the gills. Finally, there is a brief phase of tailward irrigation.

One means for bringing about such irrigation movements would, in theory, be for them to arise as reflex responses to shortage of oxygen, but this is not the means adopted in *Arenicola*. Instead, they are the product of an innate spontaneous rhythm with a period of about 40 min; this rhythm is uninfluenced by oxygen shortage, and is probably evoked by a pacemaker situated in the central nervous system. Similar outbursts of activity occur if the worm is pinned out in a dish of well-aerated sea water (Fig. 11–7); they are even shown by strips of its body wall, provided that these include a piece of the ventral nerve cord. Innate rhythms of this type are examples of the functioning of what are called 'animal clocks'. They are of great importance in animal behaviour patterns, and the irrigation movements are not the only example of their action in *Arenicola*. There is also a periodicity of feeding movements, smaller in amp-

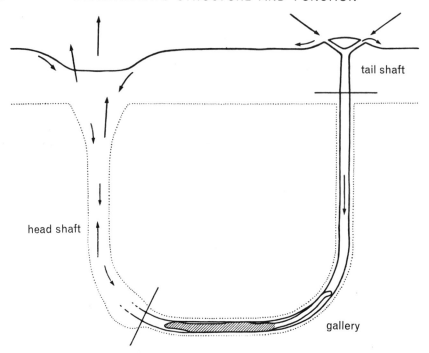

Fig. 11-6. Generalized diagram of a lugworm burrow, with the worm lying quietly in the gallery. The cross lines are drawn at the boundaries between head shaft, gallery, and tail shaft. The dotted line is the boundary between yellow and black sand. The long, thin arrows show the movement of water, and the short, thick ones that of sand. From Wells, 1950. *Symp. Soc. exp. Biol.,* **4**, 127–142.

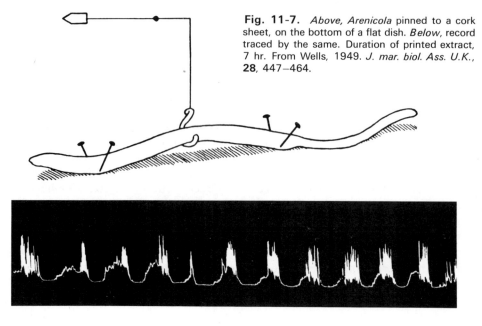

Fig. 11-7. *Above, Arenicola* pinned to a cork sheet, on the bottom of a flat dish. *Below,* record traced by the same. Duration of printed extract, 7 hr. From Wells, 1949. *J. mar. biol. Ass. U.K.,* **28**, 447–464.

litude and with a period of about 7 min; it probably arises from a pacemaker located in a nerve plexus on the oesophagus (Sec. 16–2).

The potentialities of this mode of behaviour, and its suitability for life in a burrow, are illustrated by the fact that the irrigation movements can be used when the tide is in and the burrow filled with water, as well as when the tide is out and the burrow deprived of fresh supplies of water. In the latter circumstance the same rhythmicity results in the worm extending its tail to the surface of such water as may be present over the sand. The forward-directed irrigation movements then draw air bubbles over the dorsal body surface to the gills. This behaviour allows a worm to spend as much as 120 h in an artificial burrow in the laboratory, with the same 15 ml of water, and still be in good condition at the end of that period. It may be expected from our discussion of blood pigments that the haemoglobin of *Arenicola*, which is present in its muscle as well as in its blood stream, will further increase the efficiency of its respiration.

Many sedentary polychaetes have a crown of tentacles at the anterior end. These, as we have seen, can serve as filter-feeding mechanisms. Before this function was understood the tentacles were generally termed gills, or branchial crowns, and the term is not necessarily inappropriate, for respiratory exchange may be one of their functions. It is possible that here, as with the gills of molluscs, respiration was actually the primary function of these tentacular crowns, and that their incorporation into the feeding mechanism was a later step. The polychaetes, however, provide no special evidence regarding this.

The development of branchial crowns at the oral end of the body is not restricted to polychaetes. A similar principle is seen in a number of the smaller invertebrata phyla, including the Phoronidea, the Polyzoa (Ectoprocta) (Fig. 5–4), the Brachiopoda, and the Entoprocta (Fig. 5–3). Of these, the first three share common features of organization in that the ring of tentacles surrounds the mouth but not the anus, and the tentacles are hollow and contain an extension of the coelom. To this specific type of branchial crown the term lophophore has been applied. The Entoprocta, which have no close relationship with the Ectoprocta, are of a more archaic organization, and lack a coelom (Sec. 5–2). In them the tentacles enclose the anus as well as the mouth, and they are not retractile. Nevertheless, the term lophophore is sometimes loosely applied to their tentacular crown also, and, indeed, to all such rings of tentacles, irrespective of their morphological organization, and including those of the polychaetes.

In all of the groups mentioned above the tentacles serve as ciliary feeding organs; they can reasonably be assumed also to assist in respiration, by providing a large surface area for gaseous exchange. In the Phoronidea and the Brachiopoda there is the additional advantage of a blood system. In the former this includes a ring vessel running in the coelom at the base of the lophophore, and giving off a single vessel into each tentacle. The circulation of blood passes up and down the tentacular vessels, and must considerably increase the efficiency of the respiratory exchange, particularly since haemoglobin is present.

11–4 GILLS AND LUNGS IN MOLLUSCS

The permeable body surface of molluscs permits cutaneous respiration, which is, indeed, obligatory in the Scaphopoda and certain other forms that lack differentiated respiratory organs. Most aquatic molluscs, however, possess gills or ctenidia, the history of which, like that of the organs just considered, shows how close a relationship exists between respiration and ciliary feeding. Respiratory surfaces must be kept clean, and a flow of water maintained over them; needs that are often met in animals by the secretion of mucus, and by the development of cilia to maintain movement over the surface. Clearly this is precisely the situation that also best lends itself to the collection and transport of food particles, but unfortunately we often lack the evidence to determine whether or not events have followed this course. In the molluscs, however, we have a group of animals in which the close interlocking of respiratory and nutritional requirements is particularly well shown, and in which there is evidence that enables us to trace something of the history of the organs concerned.

As already noted, we can regard the body of a mollusc as composed of two components: a ventral portion (the head and foot), in which the activity is predominantly muscular, and a dorsal portion (the visceral hump), in which ciliary action and mucus secretion predominate (Fig. 11–8). The surface of the visceral hump extends into an overhanging fold, the mantle, which secretes the shell, and which encloses between itself and the visceral hump a space called the mantle cavity. This cavity is the centre of external respiration, for protected within it are the gills or ctenidia. It has been generally assumed that, in the primitive state, there were two of these, but some doubt now attaches to this since the discovery of *Neopilina*, with its five pairs of ctenidia (Fig. 11–9). However, even if the earliest molluscs did have serially repeated ctenidiá (and we cannot be sure that this was so), these organs must have been reduced to a single pair at a very early stage of molluscan history. Their condition at that stage was the basis for the later history of the respiratory organs in the group, and we can develop our analysis from that point.

The structure of the early molluscan ctenidium, which can be deduced from its condition in primitive archaeogastropod prosobranchs, may be thought of as consisting of a longitudinal axis from which triangular filaments were given off alternately on its two sides (cf. Fig. 11–8). Essentially it was a hollow outgrowth of the body, supplied with branchial muscles, nerves, and blood vessels; the blood travelled up and down the axis in a dorsal (afferent) and ventral (efferent) vessel, and flowed from one to the other of these through the filaments. If we assume that only two ctenidia were present, oxygenated blood would have flowed from them into a single pair of auricles, and so into the ventricle for general distribution (Fig. 11–8A). The condition in *Neopilina* could then have been derived from this by secondary increase in number (Fig. 11–8B).

The functional relationships of this system, so simple in principle, are subtly adapted in present-day molluscs to ensure the maximum efficiency of respiratory exchange. The protection of the filaments within the mantle cavity makes ventilation the more essential, and this is secured by lateral cilia on the face of each filament. These cilia, which draw the respiratory stream of water into the mantle cavity, lie near the efferent limb of the circulation. Here they create a current of water that is directed upwards and inwards, in the opposite direction to the flow of the blood. This arrangement therefore involves counterflow; its effect is to increase the efficiency of gaseous exchange between the blood and the water. The gill filaments must be supported against the flow of water, and chitinous supporting rods provide for this. Further, the surface of the filaments needs to be kept clear of foreign material which

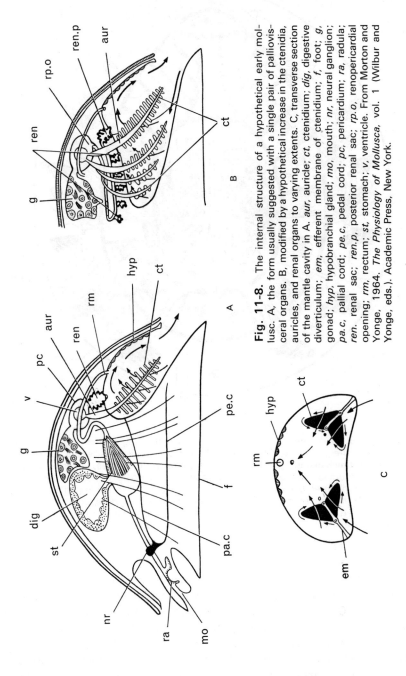

Fig. 11-8. The internal structure of a hypothetical early mollusc. A, the form usually suggested with a single pair of pallioviseral organs. B, modified by a hypothetical increase in the ctenidia, auricles, and renal organs to varying extents. C, transverse section of the mantle cavity in A. *aur,* auricle; *ct,* ctenidium; *dig,* digestive diverticulum; *em,* efferent membrane of ctenidium; *f,* foot; *g,* gonad; *hyp,* hypobranchial gland; *mo,* mouth; *nr,* neural ganglion; *pa.c,* pallial cord; *pe.c,* pedal cord; *pc,* pericardium; *ra,* radula; *ren,* renal sac; *ren.p,* posterior renal sac; *rp.o,* renopericardial opening; *rm,* rectum; *st,* stomach; *v,* ventricle. From Morton and Yonge, 1964. *The Physiology of Mollusca,* vol. 1 (Wilbur and Yonge, eds.). Academic Press, New York.

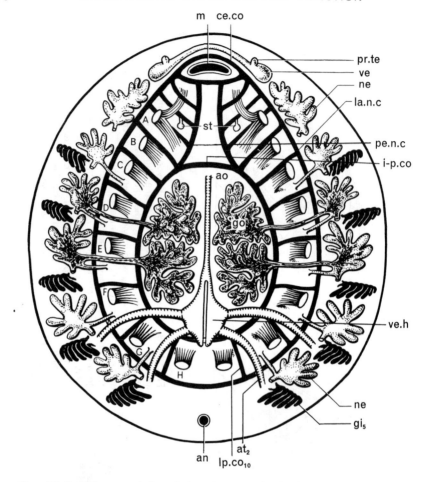

Fig. 11-9. Diagram of the relations between the 'segmented' organ systems in *Neopilina*. The gill nerves, the gill vessels, and many smaller muscles are also repeated, but are not included in the drawing. A–H, foot retractor muscles; *an*, anus; *ao*, aorta; at_2, 2nd atrium (auricle) of heart; *ce.co*, cerebral commissure; gi_5, 5th gill; *i-p.co*, interpedal commissure; *la.n.c*, lateral nerve cord; $lp.co_{10}$, 10th latero-pedal connective; *m*, mouth; *ne*, excretory organs; *pe.n.c*, pedal nerve cord; *pr.te*, preoral tentacle; *st*, statocyst; *ve*, velum; *ve.h*, ventricle of heart. From Lemche and Wingstrand, 1959. *Galathea Rep.*, vol. 3, 9–71. Danish Science Press, Copenhagen.

will tend to settle on them from the water, and in adaptation to this requirement there are frontal and abfrontal cilia situated on the afferent and efferent edges of the filaments (cf. Fig. 10-10, p. 268). These sweep material towards the central axis, where it is removed by a ciliary current along the afferent surface.

As Yonge points out, these arrangements will not in themselves ensure the protection of the ctenidia from sediment, particularly as these organs increase in size, for the ciliary currents may prove too weak to remove heavy deposits of material. It is probably because of this that two other paired structures, the osphradia and the hypobranchial glands, are present in the primitive mantle cavity; these structures,

Fig. 11-10. The process of torsion in a prosobranch veliger larva, showing the position of the asymmetrically developed shell muscle. Asterisk indicates position of the mantle cavity. *rl*, left retractor muscles; *rr*, right retractor muscles; *vel*, velum. From Morton and Yonge, 1964. *op. cit.*

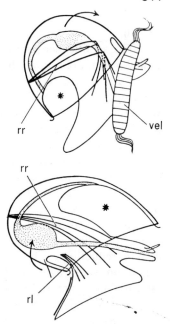

with the ctenidia, form a functionally associated system which can be called the pallial complex. The osphriada, which are universally present in the mantle cavities of aquatic gastropods, irrespective of their habitat or the nature of their food, are receptors, consisting of raised areas of epithelia that are rich in mucus, and in ciliated and sensory cells. They test the quality of the incoming water, perhaps by chemical sensitivity, for they are extremely large in carnivores such as *Buccinum*. Perhaps they also estimate through mechanoreception the amount of sediment entering the mantle cavity. The hypobranchial glands (Fig. 11–8), present in the gastropods and in some bivalves, are folds of mucus-secreting epithelium lying on the roof of the mantle cavity. They vary in size and complexity, apparently in relation to the amount of sediment that is likely to enter the mantle cavity in the particular habitat favoured by the species; their function seems to be to aid the removal of the sediment by consolidating it into larger masses. Their absence from cephalopods is thought to be correlated with the exceptionally powerful currents in those animals, which remove the need for such consolidation.

Not surprisingly, the highly organized respiratory systems of the pallial cavity reflect a great deal of adaptive evolution in relation to changes in the habits and organization of the members of this highly diversified phylum. This is already apparent in the primitive chitons. In these animals the number of gills is secondarily increased in correlation with the forward extension of the mantle cavity, and the osphradia lie posteriorly, where the outgoing current of water leaves. Within the gastropods the respiratory process has been profoundly affected by two characteristic features of the group. The first of these is the asymmetrical coiling of the visceral mass, which is an adaptation to secure a more compact arrangement of the internal organs: The second is torsion, a process that takes place early in development (Sec. 19–2), quite independently of coiling, and which brings the mantle cavity of the veliger larva from its primitively posterior position round to the anterior end (Fig. 11–10).

The significance of torsion for the respiratory process is that the visceral and pallial organs are rotated through 180°, so that the mantle cavity is now in more sensitive touch with its surroundings. Additional advantages are that the forward movement of the animal will now reinforce the incurrent stream instead of opposing it, while the gills receive water direct from in front of the animal, and the osphradium is in the anterior position that is suitable for a major receptor organ. On the other hand, a serious problem arises in disposing of the excurrent water stream with its contained faeces. It would be unsatisfactory for these to be discharged anteriorly into the path of the animal, and so, as a primitive solution to this problem, openings are formed in the shell to direct this stream away from the incurrent one. *Haliotis* is an example of this.

More advanced solutions to the problem are influenced by the fact that asymmetrical coiling produces on one side of the mantle cavity a compression that favours reduction of the corresponding part of the pallial complex. Thus the right ctenidium and its associated auricle are lost, the original two rows of gill filaments of the ctenidium are reduced to one, and the axis of the ctenidium now becomes attached along its whole length to the wall of the mantle cavity. With the water current entering on the left side of the head and leaving on the right side, close to where the rectal opening has moved, an efficient respiratory circulation is ensured for most of the prosobranchs, without the need to have openings in the shell. Only the Order Archaeogastropoda retain the primitve symmetry of paired gills and auricles, together with the shell perforations. The Subclass Opisthobranchia shows varying degrees of detorsion, accompanied by reduction of the shell and mantle, until in the Order Nudibranchia a secondary symmetry is achieved with loss of the shell, mantle cavity and ctenidium. Respiration now takes place through the body surface, or through secondarily developed gills.

Many prosobranchs, particularly in the tropics, have become terrestrial, and have converted their mantle cavity into a lung for aerial respiration, its vascularized epithelium forming the respiratory surface. A similar adaptation is found in the Subclass Pulmonata, which probably evolved from prosobranchs. The pulmonate lung usually retains its anterior position, but it has only a single external opening, the pneumostome, which can be opened and closed. Because of this change of function, the hypobranchial gland is lost, as would be expected from what we have said above regarding its supposed role in the removal of small particles, while the osphradium is usually outside the mantle cavity. The respiratory surface is now the lining of that cavity, and, as in other types of lung, its area is greatly increased by ridges that are richly supplied with blood vessels. These ridges are said to increase the respiratory surface of air-breathing snails by as much as two or three times.

We are accustomed to regard lungs as being organs that require ventilation if they are to function satisfactorily, and this is obviously true of vertebrates. Probably because of this, it has been thought that the pulmonate lung also requires ventilation, but Krogh showed that this was not necessarily so. In slugs (*Arion*) of about 10 g weight, with a pneumostome of 4–6 mm diameter, and with a respiratory surface of 6–7 cm^2, a pressure difference of only 2 mm will ensure diffusion of atmospheric oxygen to the wall of the mantle cavity; even larger animals could therefore be satisfactorily supplied in this way. This conclusion applies, of course, to animals with a relatively low metabolic rate; ventilation of the lungs certainly becomes necessary with increasing oxygen consumption, and with increase in size of the body.

In any case, ventilation movements do occur in *Helix*, and have been demonstrated by recording pressure changes in the lung (Fig. 11–11). When the pneumostome is open, as it may be when the animal is crawling slowly, the pressure remains unchanged and respiration depends upon diffusion. When the penumostome is closed the pressure begins to rise, reaching a maximum in 4–5 sec, and then falling again. These pressure changes, which are brought about by muscular movements of the floor of the lung, result in air movement and hence redistribution of oxygen within the lung cavity; at maximal pressure, they probably facilitate the passage of oxygen into the blood stream.

The pulmonate lung is by no means purely a terrestrial respiratory organ, nor is it restricted to aerial respiration. We shall deal elsewhere with some of the factors that operate as animals of different phyla move to and fro between water and air in the course of their evolutionary history. It will be sufficient here to note some examples of the varied respiratory potentialities of this organ. The siphonariid limpets are marine pulmonates that are completely aquatic in habit; the mantle cavity is filled with water, and has developed secondary pallial gills, formed by folding of the wall of the cavity. Many other examples, showing different degrees of adaptation to aquatic or amphibious habits, are provided by the fresh-water pulmonates. Thus *Lymnaea truncatula* has its lung filled with air; this animal, living an essentially aerial life in marshy habitats, plays a role of no small economic importance in

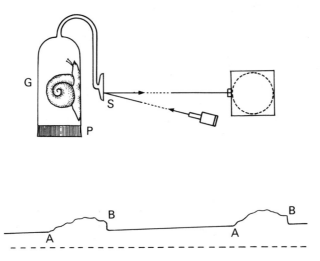

Fig. 11-11. *Above*, measurement of respiratory movements in *Helix*. The snail is closed in a vessel (G) connected to a Marey tambour (S). A small mirror on the tambour reflects a beam of light on a photokymograph. Pressure changes inside the vessel will be recorded through movements of the mirror. *Below*, modifications of the lung in *H. pomatia* during normal respiration. At A the pneumostome is closed; at B it is opened. From Ghiretti, 1966. In *Physiology of the Mollusca*, vol. 2. (Wilbur and Yonge, eds.). Academic Press, New York.

providing an intermediate host for the sheep liver fluke, *Fasciola hepatica*. At the other extreme is *L. abyssicola*, which lives in deep water without coming to the surface, while *L. stagnalis* is intermediate in habit, living in water, but returning to the surface to refill its lung with air.

Planorbis corneus resembles *Lymnaea stagnalis* in this respect, but is able to handle the situation differently because it possesses dissolved haemoglobin in its blood; a good example of the ecological implications of the presence or absence of the pigment. *Planorbis* has a secondary gill, and probably respires through its body surface as well, but it relies largely upon its air-filled lung, functioning in conjunction with the vascular haemoglobin. This pigment, because of its high oxygen affinity, makes the animal better able than *Lymnaea* to take up oxygen from its pulmonary cavity at lower concentrations during the later stage of the dive, for *Lymnaea* can only take it up into simple solution. As a result, *Planorbis* can exploit its oxygen store more effectively (Table 11-2), can dive deeper and for longer periods, and can feed by browsing upon the bottom deposits. The dives of *Lymnaea* are more restricted, and, in correlation with this, it feeds on submerged vegetation near the surface. Nevertheless, the ecological significance of vascular haemoglobins in molluscs is not always so clear, as is apparent in their absence from many species that live in conditions of oxygen shortage. Individuals of *Glycymeris nummaria*, for example, even though they may be living in the same environment, range widely in their content of haemoglobin, some of them lacking it altogether.

Probably, then, the haemoglobin of *Planorbis* functions mainly to facilitate the uptake of oxygen from the pulmonary store, and there is little evidence that the pigment itself serves any storage function. A puzzling feature is the reversed Bohr effect which is found at high carbon dioxide tensions, as it is in some haemocyanin-containing snails. The physiological significance of this is not clear, but it may perhaps facilitate the transport of oxygen from the lung to the tissues when the animal is withdrawn into its shell and the carbon dioxide tension rises.

It is to be expected that oxygen lack will be one factor in stimulating the return of these animals to the surface of the water, but another factor is the hydrostatic property of the lung. If the cavity of a fresh-water pulmonate is artificially filled with oxygen instead of air, the animal will return to the surface before it has exhausted the supply; it is driven by the reduction in the volume of the gas enclosed in the lung.

The most complex of all molluscan ctenidia are those found in the Bivalvia, where they are involved in the elaborately specialized ciliary feeding mechanisms of this group. In our earlier discussion we saw that the respiratory functioning of the ctenidium demands the presence on it of mucus glands and ciliated tracts that serve to keep its delicate surface clean from detritus, so that molluscs may be said to have

Table 11-2 Percentage of oxygen in pulmonary air (with standard errors) at beginning and end of dive. Number of samples in brackets. From Jones, 1961. *Comp. Biochem. Physiol.*, **4**, 1-29.

Venue	Initial		Final	
	Planorbis	*Lymnaea*	*Planorbis*	*Lymnaea*
Aquarium (18)	15.2 ± 0.51	19.0 ± 0.12	7.9 ± 0.84	10.2 ± 0.53
Ditch (6)	16.2 ± 0.99	18.2 ± 0.48	4.2 ± 1.47	8.8 ± 0.81

been pre-adapted for ciliary feeding. We have observed that certain gastropods have taken advantage of this to become ciliary feeders, but it is clearly the bivalves that have most fully exploited the situation.

Yonge has outlined a course of events that could have led to the elaboration of the gills of the most advanced bivalves. The first step, according to his analysis, was the overgrowth of the body by the mantle lobes, for this led to the anterior and ventral extension of the mantle cavity, and to the forward movement of the ctenidia. This movement brought them into functional association with the mouth, through the intermediation of the labial palps. Initially, while the ctenidia were primarily respiratory, they were horizontally disposed, but later the filaments bent downwards to form a V-shaped pattern. The outer filaments became fused with the mantle, and the inner ones to their partners on the other ctenidium or to the visceral mass; thus the mantle cavity became subdivided into inhalent and exhalent chambers. The filaments then became elaborated into an increasingly complex sieve by the development of interlamellar junctions, and of ciliary and tissue junctions (Sec. 10–2).

The increasing ciliation which would have accompanied these structural advances probably increased the intake of sediment with the respiratory stream, and the development of latero-frontal cilia would serve to counteract this by preventing the passage of particles between the filaments. In correlation with this the abfrontal cilia, still present in *Nucula*, would have been lost. It is obvious that all of these advances, serving primarily to improve respiratory exchange by increasing the flow of water and by elaborating the surface of the gills, would also have improved the collection of suspended food material. Indeed, the two functions were presumably elaborated side by side in this group, together also with the specialization of the sorting and transport mechanisms. At quite the opposite extreme is the situation in the anomalous and highly specialized septibranchs, where a muscular pumping organ is formed by a septum, which is perforated by ciliated pores, and which probably evolved from the fusion of modified ctenidia.

The close interrelationship between respiration and ciliary feeding, upon which we have already commented, is shown particularly clearly in studies of bivalves. Oxygen utilization can be determined by comparing the oxygen content of samples of water taken from the incurrent and the excurrent siphons. Because of technical difficulties (e.g. there is a gradient of oxygen within the exhalent current) the results are only approximate, but they certainly show a very low rate of utilization (Table 11–3). This is surprising, considering the very high rates of ventilation that are made possible in these animals by the specialized structure of the ctenidia. Galtsoff found that ventilation rates in the oyster (*Crassostrea virginica*) might be as high as $3.9 \, l \, hr^{-1}$ at $25°C$; other procedures, based on the rate of clearance of suspensions in the water in which the animals are placed, have yielded results of the same order for *Mytilus*. The energy expended by these animals in their filtering activities must presumably be justified by the food obtained, rather than by the oxygen.

In contrast to this, there is a much higher level of oxygen utilization in gastropods and cephalopods (Table 11–3). The active life of Cephalopoda, in particular, presents a remarkable contrast with that of the sedentary forms that we have just considered (Fig. 11–12). In this class the ctenidia are contained within a mantle cavity that has become elongated in a dorso-ventral direction, and that is usually itself involved in the respiratory movements. A primitive form of the respiratory

Table 11–3 Percentage of available oxygen utilized by various molluscs. Data from various authors. From Ghiretti, 1966. *Physiology of Mollusca* (K. M. Wilbur and C. M. Yonge, eds.), vol. 2, 175–208. Academic Press, New York.

	%
BIVALVIA	
Mya arenaria	3–10
Cardium tuberculatum	6–10
Solen siliqua	7–12
GASTROPODA	
Haliotis tuberculatus	48–70
Doris tuberculata	64–69
CEPHALOPODA	
Octopus vulgaris	50–80

mechanism is found in the Tetrabranchia, represented by *Nautilus*, where the presence of two pairs of ctenidia, instead of the single pair that is so generally characteristic of molluscs, has been regarded as a secondary specialization. We have seen that this may also be true of the five pairs in *Neopilina*.

In respiration the important functional characteristic of cephalopods is the production of the respiratory current by muscular action. *Nautilus* creates this ventilation by pulsations of the funnel, which is formed, as we have seen, from two halves of the foot that are not yet fused at this primitive stage of evolution. In the Coleoidea fusion is complete, and the current correspondingly more vigorous; it is now brought about by contractions of the mantle wall as well as of the muscles of the foot and head, this being made possible by the reduction of the shell and its overgrowth by the body. Cilia are thus no longer a necessary part of the respiratory mechanism, and they are absent from the gill surface, which is greatly increased in area by primary and secondary folding of the filaments.

We have seen that the cephalopods are beautifully designed for rapid locomotion by jet propulsion. Execution of this is directly and economically linked with the respiratory mechanism. As with all such active animals, there is a demand for the most efficient possible oxygen supply; the elaborate folding of the gill surface is one contribution to this, while the presence of a respiratory pigment is another. Significantly, of all the animals that possess haemocyanin, cephalopods have the blood with the highest oxygen-carrying capacity. Yet this is not as great as capacity of the blood in many animals with haemoglobin. Representative values for the latter are 21 volumes of oxygen per 100 volumes of blood in man, 5.5–7.8% for the dogfish *Mustelus*, and 8.4% to 9.7% for *Arenicola*. Corresponding values for the haemocyanin-containing blood of molluscs are 3.1% to 4.5% for *Octopus vulgaris*, 3.8% to 4.5% for *Loligo pealei*, and 1.15% to 2.2% for *Helix pomatia*. The environment of cephalopods is rich in oxygen, and in adaptation to this the blood of *Loligo* has a low oxygen affinity, and is saturated only at relatively high oxygen tensions. There is a typical and large Bohr effect, which facilitates the release of oxygen to the active tissues. While, therefore, these animals are very sensitive to oxygen lack, under normal conditions they can achieve the very high level of activity which is so characteristic of them.

The efficiency of their respiratory mechanism is further increased by a capillary

Fig. 11-12. A, *Sepia officinalis*, mantle cavity of topographically under (morphologically posterior) surface, showing disposition of organs and course of respiratory current. B–D, diagrams illustrating postulated mode of evolution of mantle cavity. B, primitive mollusc with posterior mantle cavity. C, postulated intermediate condition, broken arrows indicate alternative directions of respiratory current at this stage (i.e. as in B if produced by cilia; as in D if produced by muscular movement of pedal flaps). D, cephalopod mantle cavity. *a*, anus; *am*, afferent membrane of ctenidium; *em*, efferent membrane of ctenidium; *f*, foot; *ga*, genital aperture; *hb*, hypobranchial gland; *k*, kidney aperture; *lct*, left ctenidium; *lk*, left kidney aperture; *o*, osphradium. *rct*, right ctenidium; *rk*, right kidney aperture; *vms*, visceral mass. From Yonge, 1947. *Phil. Trans. R. Soc. B*, **232**, 443–518.

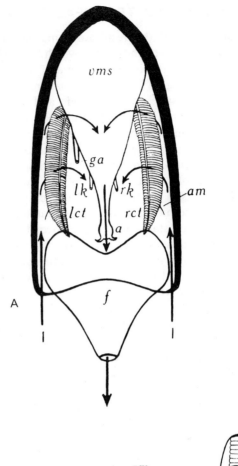

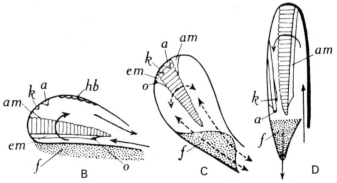

circulation in the gills, contrasting markedly with the haemocoelic type of system found in other molluscan groups. Associated with this are branchial hearts at the bases of the gills. Thus the whole system is organized for maximum efficiency of respiration within the limits of the molluscan plan of structure. The complex of respiratory, locomotor, and circulatory mechanisms in cephalopods provides an instructive example of the way in which the parts of highly specialized animals can become closely integrated, so that each contributes to the successful functioning of the others.

11-5 GILLS AND TRACHEAE IN ARTHROPODS

The organization of the respiratory mechanism of the Arthropoda reflects the evolutionary history of the group, which, as we have already shown, began in the sea and continued later in fresh water and on the land. The most primitive method of respiratory exchange is found in the smaller aquatic crustaceans, such as early larvae, adult copepods, and most ostracods and cirripedes. These animals, like many worms, respire through their general body surface. It is reasonable to assume that the thin-walled and foliaceous appendages of the branchiopods, like the parapodia of polychaetes, facilitate gaseous exchange, and that they have, therefore, a respiratory function additional to their use in feeding. The maintenance of the filter-feeding current would, of course, serve also for ventilation. In fact, it may well be that in these animals, as in the bivalves, the feeding mechanism was a specialization of a more primitive respiratory one, but we cannot be sure of this.

The larger crustaceans develop specialized outgrowths that are regarded as gills, although there is often a lack of physiological evidence for their respiratory function. Structures of this type are particularly well developed in the Malacostraca, where they take the form of foliaceous outgrowths of the coxae of the thoracic limbs. These outgrowths, which are termed podobranchs, may be supplemented, particularly in the decapods, by similar outgrowths arising from the arthrodial membrane at the base of the limb (arthrobranchs), or from the body wall (pleurobranchs). In their simplest form such gills may be only hollow, flattened outgrowths, but they become much more complex, with a central axis and various types of lateral branch. They are well vascularized, with an afferent and efferent circulation, but they do not develop either capillaries or branchial hearts such as are found in the cephalopods. Ventilation is maintained by the rhythmic beating of one or more appendages; primitively a number of these are concerned, as in *Anaspides*, but in the decapods this function is restricted to the scaphognathite of the second maxilla.

Another example of the use of limbs in respiration is probably to be seen in the Trilobita (Fig. 11-13), although here we are restricted to inferences drawn from their fossil remains. We have seen that these animals show a uniformity of structure in their biramous limbs, apart from the differentiation of the antennules. Of the two rami, the inner one supposedly functioned as a walking leg, since it bore a terminal claw. The outer ramus, with its fringe of broad filaments, presumably had some respiratory function.

A third example of the respiratory use of the limb occurs in *Limulus* (Fig. 7-12, p. 191). Five pairs of swimming paddles exist on the opisthosoma of this animal, each of these limbs consisting of a slender internal ramus and a broad external one. The latter bears a peculiar type of gill formed of as many as 200 delicate branchial leaves, for which reason it is called a gill book. Ventilation of the gill book is readily provided by the locomotor actions of the limbs. For much of its time the animal is shovelling its way through sand and mud, and the moulding of the ventral surface of the body into a trough probably provides some protection for the delicate gill books. The overhanging genital operculum, regarded as the fused limbs of the first opisthosomatic segment, provides a further safeguard.

We have seen that few crustaceans have achieved any success in the invasion of land (Sec. 7-3). Some malacostracans, however, have surmounted the difficulties,

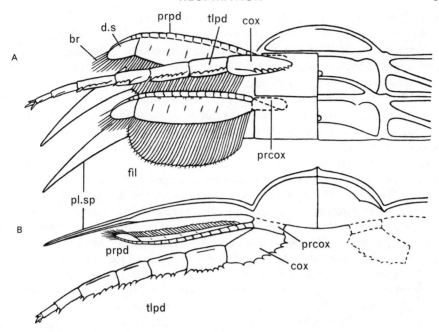

Fig. 11-13. Reconstructions of the appendages of the trilobite *Olenoides* (*Neolenus*) *serratus* Rominger, 100 mm, Middle Cambrian. After Størmer. *br*, bristle; *cox*, coxa; *d.s*, distal segment of pre-epipodite; *fil*, filaments; *pl.sp*, pleural spine; *prcox*, precoxa; *prpd*, pre-epipodite; *tlpd*, telopodite. From Tiegs and Manton, 1958. *op. cit.* Used by courtesy of the Cambridge Philosophical Society.

examples being found among the anomuran decapods, and also among the true crabs. Here the solution of the respiratory problem has been the development of vascularized folds of the wall of the branchial chamber; a development that is not greatly different in principle from the possession of pleurobranchs. These adaptations may be called lungs. Another solution is found in the Porcellionidae and Armadillidiidae, terrestrial isopods in which the respiratory endopodite of each pleopod is protected by an operculum formed by the exopodite, which also contains branched tubules (pseudotracheae, p. 383, opening to the outside by a narrow aperture. This is a simple illustration of the possibility of distributing gases through tissues by tubular ingrowths of the body surface. The principle has been widely exploited in the arthropods, where it has presumably been encouraged by the development of the hard cuticle; it provides one of the clearest indications of the widespread convergent evolution that has marked the history of the group.

It is convincingly demonstrated, for example, in the chelicerates, which, since they include the aquatic *Limulus*, must have developed their terrestrial adaptations independently of other arthropods (Sec. 7–2). The line of evolution seems likely to have passed through the scorpions, which are known from the Upper Silurian. These animals bear some resemblance to the eurypterids, an extinct group of aquatic chelicerates that were more generalized in structure than the highly specialized *Limulus*. Unfortunately the respiratory organs of the eurypterids are not satisfactorily known, but we may assume that they were of the same type as the gill books of *Limulus*, and that the aerial respiratory organs of scorpions were evolved from them. These are invaginated organs that are called lung books, because they consist of closely apposed leaflets, which are set, like the pages of a book, within a pit that opens to the outside through a narrow aperture. No special provision is made for ventilation, so that gaseous exchange must depend upon diffusion.

Limulus can survive for several days out of water; it has been suggested, therefore, that primitive chelicerates had a similar capacity, and that lung books evolved as a consequence of this. It may be, for example, that the limbs were closely applied to the ventral body surface, that the gill books became enclosed by fusion of the limbs with the body wall, and that the gill lamellae in their turn fused with the wall of the chamber. Four pairs of lung books exist in scorpions, which have remained conservative in their respiratory equipment, but other arachnids are more specialized. Thus some spiders have two pairs of lung books, some have only one pair, and in one family there are none at all. This reduction is doubtless correlated with the tendency for lung books to be replaced in these animals by tracheae, which resemble in principle those of insects and other groups to be mentioned below, but which have clearly been developed quite independently. In the spiders they may perhaps have evolved as diverticula of the missing lung books. Tracheae are also found in other arachnids: the Solifugae, the Phalangidae, and some of the Acarina. In these groups there is no evidence for a primitive lung-book phase, and it is at least possible that they may have developed tracheal respiration independently of the spiders.

We have already seen that *Peripatus*, the myriapods, and the insects may constitute a natural group of terrestrial arthropods, the product of an invasion of the land that was achieved quite independently of the arachnid line. All of these forms have developed tracheal respiration, but their tracheae are not uniform in structure, and we cannot assume that they necessarily had a common evolutionary origin. They could, in theory, be independent expressions of a common genetic potentiality in these several groups. We have no means of judging this, and the dangers of speculation are shown by the fact that the tracheae of the Solifugae show detailed resemblances to those of insects, despite their undoubted independent origin. In *Peripatus* (Fig. 11–14) the tracheae are delicate tubules, passing inwards to the organs, and arising in tufts from pits of the body surface. Each pit opens by a spiracle, the spiracles being scattered irregularly instead of showing the segmental arrangement seen in insects. A fact of great physiological and ecological importance for the Onychophora is that they are unable to close the spiracles. Because of this they cannot resist desiccation, a feature that is correlated with their occupation of sheltered and damp habitats. In the myriapods the tracheae are more complex in structure, and are commonly supported, as they are in insects, by a spiral thickening. In centipedes they usually branch and anastomose; in millipedes they may branch but anastomoses do not develop.

The operation of tracheae has been best studied in insects, where they achieve an efficiency in operation that makes a major contribution to the diversity of specialization and high level of activity that these animals attain. Tracheae have, however, the one major disadvantage that transport of gases by diffusion is suited only to small organisms. Thus they are one of the features of insect organization (the exoskeleton is another) that severely limit the size of these animals; a limitation for which the rest of the animal kingdom should be thankful.

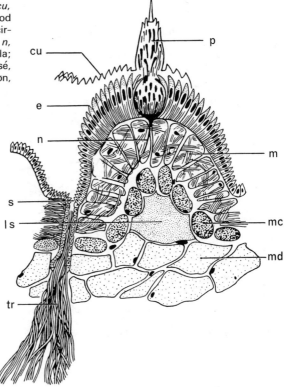

Fig. 11-14. T.S. *Peripatus trinitatis*. *cu*, cuticle, detached; *e*. epidermis; *ls*, blood lacuna; *m*, subepithelial muscles; *mc*, circular muscles; *md*, diagonal muscles; *n*, nerve of sensory papilla; *p*, sensory papilla; *s*, stigma; *tr*, bundle of tracheae. From Grassé, 1949. *Traité de Zoologie*, vol. 6. Masson, Paris.

Tracheae are ectodermal structures, formed by invagination, or ingrowth, from the surface. As a result they are lined with cuticle, called the intima, which is thickened to form delicate ridges arranged either as a continuous spiral or as separate rings. These thickenings, the taenidia, serve to maintain an open lumen throughout the tracheal system, and thereby ensure the passage of gases. Communication with the outside is by means of openings called spiracles, referred to above; they may be regarded as the sites of the original invagination. Typically these are found in the thorax and abdomen (Fig. 11–15), situated on the pleura, but there is much variation in detail, and distinctions can be made in the insects between holopneustic, hemipneustic, and apneustic systems. The first of these, with eight pairs of spiracles on the abdomen and two on the thorax, is the most primitive, and is particularly characteristic of adult stages. Hemipneustic systems, in which one or more of the pairs of spiracles are closed, are particularly characteristic of those larvae in which respiratory exchange has become localized at one end of the body; this is usually an adaptation for life in a fluid or semi-fluid medium. In apneustic systems all of the spiracles are closed, so that respiratory exchange must now occur either through the body surface, or through outgrowths of it that are called gills. This mode of functioning is particularly characteristic of endoparasites and of fully aquatic insects.

The essential feature of the fully developed tracheal system of insects is that it transports oxygen to the tissues by tracheal tubes that branch to supply all parts of the body, and form extremely fine terminations called tracheoles. These terminations

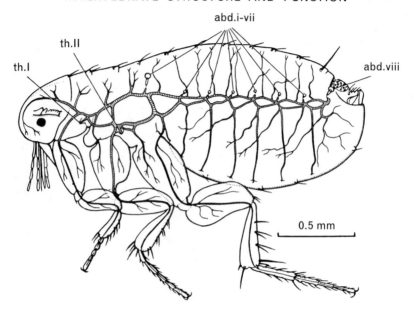

Fig. 11-15. Tracheal system of *Xenopsylla cheopsis. th.I, th.II*, thoracic spiracles; *abd. i viii*, abdominal spiracles. From Wigglesworth, 1947. *The Principles of Insect Physiology* (3rd ed.). Methuen, London.

possess thin walls that are permeable to water, and that are provided, like the tracheae, with taenidia so delicate that they can only be seen by electron microscopy. Because of the minute size of the tracheoles they enter into so close a relationship with the tissue that they surround the cells and end blindly within them. Moreover, the degree of their development and branching can be adaptively adjusted to meet fluctuations in the oxygen demands of particular tissues.

It is difficult sometimes to make a clear-cut distinction between pulmonary and tracheal respiration. As Krogh points out, the tracheae of spiders may function essentially as lungs, for they do not convey oxygen to the tissue, but merely aerate the blood in the adjacent ventral sinus. Even in insects there is a blood stream which must necessarily play some part in gas transport; the tracheoles may sometimes be found suspended in the blood, and presumably provide a source of oxygen that it can then transport. Moreover, tracheae are sometimes expanded to form thin-walled air sacs. These may sometimes serve as hydrostatic organs, but they are probably of particular importance in increasing the capacity of the respiratory system as a whole, so that the respiratory movements result in the exchange of a correspondingly greater volume of air. There is an analogy here with the function of the air sacs of birds, and a further analogy is implied in a suggestion that the air sacs of insects may aid flight by reducing specific gravity. A similar function has been suggested in birds, for in these animals the air sacs penetrate extensively into the bony skeleton.

As with other types of respiratory system, the mode of functioning of tracheae is closely governed by the physical properties of air and water. We have seen that air is a more favourable medium for respiratory exchanges than in water, and that this is particularly apparent in connection with the diffusion of oxygen, which is very much more rapid in air than in water. Krogh's calculations showed that in a large *Cossus* (goat moth) larva, 60 mm long and weighing 3.4 g, the tracheae had an average length of 6 mm, with a cross-sectional area of 6.7 mm^2. In these circumstances diffusion is ample for supplying the necessary oxygen; it requires a pressure difference of only 11 mm, so that the animal need not expend any energy at all upon ventilation movements. The same is probably true of very many tracheate arthropods, including the Onychophora, the arachnids, the myriapods, and a large number of the smaller insects, together with larvae and pupae. The efficiency of tracheal respiration, given the small size of insects (see also p. 296) is well brought out by some calculations of Alexander. These show that a larva with the same oxygen consumption as a goat moth larva (0.3 cm^3/g hr), and with the same proportion of its body occupied by tracheae, could in theory reach a radius of 0.9 cm, whereas an earthworm-like animal, relying upon a circulatory system for distributing its oxygen, could only reach 0.3 cm.

Two factors that modify this situation are size and activity. Many insects are smaller than the larva mentioned, and are in an even more favourable position, but conditions deteriorate with increased size, as Krogh showed. If, for example, the linear dimensions increase by a factor of 10, giving tracheae 6 cm long and 6.7 cm^2 in cross-section, the rate of diffusion could be increased by a factor of 10, but the animal would be 1,000 times heavier, and its metabolism at least 100 times greater. Diffusion would now be totally inadequate for meeting its needs, and this is why tracheal respiration limits the size of arthropods. Conditions are at their most difficult in insects. The expenditure of energy during flight is formidably high, amounting to 100 cm^3/cn^3 tissue hr, and transport of oxygen by diffusion is thought to become inadequate when the body weight reaches 0.1 g. Respiratory movements are now needed to ensure an adequate supply of oxygen, and so these are seen in bees, for example, which weigh 100 mg. It is, of course, the flight muscles which are mainly consuming the oxygen, and they have a specialized tracheal supply, with a main trunk running throughout the muscle, and giving off lateral branches which continue to divide. Even so, Alexander shows that the greatest diameter that a flight muscle could achieve, while still maintaining this rate of oxygen consumption, is 0.5 cm. This is, indeed, the limit normally found. But there is a tropical water bug, *Lethocerus*, which reaches a length of 11 cm, being one of the largest of known insects. This has a further adaptation in its wing, in that air is pumped through the larger of the branch tracheae as well as through the main trunk, whereas in other insects it is only pumped through the latter.

A special difficulty arises in species in which the legs are unduly long relative to the size of the body; reliance upon diffusion over such distances may result in the oxygen concentration in the tracheae of these limbs falling to very low levels. It is doubtless in adaptation to this that harvestmen (Order Opiliones), with exceptionally long legs, have spiracles on their tibiae.

With tracheae, as with lungs, there is bound to be some loss of water vapour through the respiratory membranes; this constitutes a serious drawback in the tracheal system, particularly as terrestrial animals must achieve maximum economy in the use of water. Since the general body surface is impermeable, the spiracles become the main site of water loss. Primitively this loss was doubtless accepted, as it still is today in *Peripatus*, but the extraordinarily successful exploitation of aerial life by insects has depended upon overcoming this particular limitation by developing devices for closing the spiracles and thereby controlling diffusion through them (Fig. 11–16). The spiracles thus become complex organs, varying greatly in the details of their organization, but commonly provided with muscles, with which cuticular elements may be associated. These form closing mechanisms by which the opening can be completely occluded. Further, the tracheae arise from a vestibule that is provided with filtering hairs; these serve to reduce any diffusion of water vapour through the spiracle, in addition to preventing foreign particles from entering the system.

For successful operation, these closing mechanisms must be regulated to meet the respiratory needs at any particular moment. Here again there is much variation. In air-breathing vertebrates use is made of the carbon dioxide concentration in regulating the rate of pulmonary ventilation. The respiratory centre in the medulla of the brain is stimulated by the small increases in the carbon dioxide tension in the arterial blood that result from increased metabolism—an example of the use of a by-product of metabolism as a chemical signal. A similar principle seems to be used by at least some insects. The cockroach, for example, opens its spiracles with increasing activity; this is thought to be a result of increased production of carbon dioxide, for experiments have shown that the spiracles are provided with receptors that are specifically sensitive to this gas.

In the locust (Fig. 11–17) the respiratory movements are under the control of a

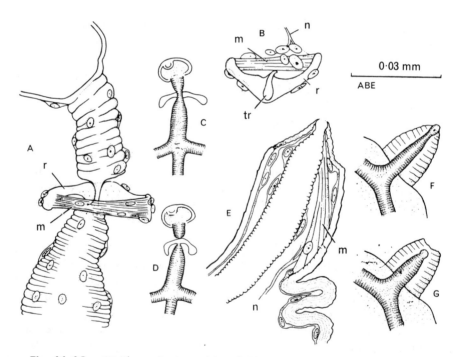

Fig. 11–16. Closing mechanisms of the spiracles of *Xenopsylla*. A. abd. viii in surface view; B. abd. vi in transverse section; C. abd. vi as seen in living insect, open; D. the same, closed. E. th. II in horizontal section; F. th. II as seen in living insect, open; G. the same, closed. *m*, muscle; *n*, nerve; *r*, chitinous rod; *tr*, trachea. From Wigglesworth, 1935. *Proc. R. Soc. B*, **118**, 397–419.

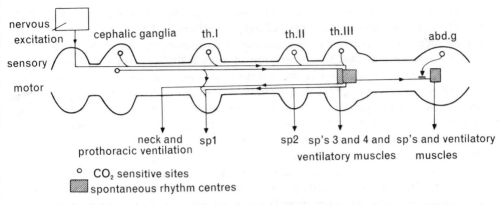

Fig. 11-17. A summary of the control of ventilation in the locust. Sensory fibres coupled to carbon dioxide receptors in each ganglion run in the nerve cord to the metathoracic ventilation centre; other fibres run from the head to the metathoracic ganglion and then to the neck and prothoracic ventilation muscles. From Miller, 1960. *J. exp. Biol*, **37**, 224–236.

ventilation centre in the metathoracic ganglion. This is regulated by impulses transmitted from carbon dioxide receptors situated in the nerve ganglia themselves. The centre can also be controlled by external stimuli acting through the cephalic ganglion; for example, respiration may be momentarily interrupted if the insect is handled.

In the flea *Xenopsylla cheopsis*, on the other hand, no specific carbon dioxide receptors are present, and in this animal the opening of the spiracles is apparently due to increased acidity of the tissues. Spiracular closing mechanisms sensitive to carbon dioxide are found also in the scorpion *Buthus* and the tick *Ornithodorus*; an example of physiological convergence in insects and arachnids. We may recall in this connection that the effect of carbon dioxide on mammalian respiration was once thought to result from the increased acidity caused by its presence, rather than from the direct action of this metabolite on the respiratory centre. The situation in insects shows that this mechanism is a feasible one, and it may also play a part in mammals although in this group direct action is probably also important.

One further remarkable specialization in tracheal systems remains to be described. This is a mechanism, discovered by Wigglesworth, that makes it possible for the insect to increase the availability of oxygen for the tissues, and thereby supplement the effect of the closing mechanisms in reducing gaseous diffusion. We have seen that the tracheae end in tracheoles that surround and penetrate the cells. These tracheoles are necessarily permeable to water, for without such permeability the oxygen could not diffuse through their walls. As a result, water passes into them from the tissues. The amount that does so is determined by the relationship between the osmotic pressure of the tissue fluids and the force of capillarity within the tracheole tubules. When the tissues are respiring with an inadequate oxygen supply, as happens at times of great activity, metabolites accumulate in the tissue fluids and raise their osmotic pressure. This results in water being withdrawn from the tracheoles into the tissue fluids, the space left in the tracheoles being filled by air moving along them.

Since oxygen diffuses much more rapidly through air than through water, the effect of this will be to improve the supply of oxygen to the cells. This result is analogous to the opening-up of the capillaries in the active muscles of the vertebrate body.

It remains to add that the possession of such a highly specialized respiratory system has not prevented insects from returning to an aquatic life, any more than the possession of lungs has prevented mammals from doing the same. Various devices are adopted for this purpose. Some larvae obtain air from aquatic plants but a more common practice is to return to the surface to breathe through the spiracles. In this case, hydrophobe structures or secretions must be present around the spiracular openings to prevent flooding of the tracheae. This procedure may be supplemented by carrying down under the water a temporary reservoir of air on the body surface. This reservoir, which serves a hydrostatic function as well as being an oxygen store, may be retained beneath the wings, as in *Dytiscus*, or among hydrofuge hairs, as in *Notonecta*.

Another device is the use of spiracular gills, mainly confined to the pupae of certain flies and beetles, in which they must have been independently evolved on many occasions. These structures, which are modifications of the spiracle, or body wall, or both, usually bear a device called a plastron. This is an air store held in place

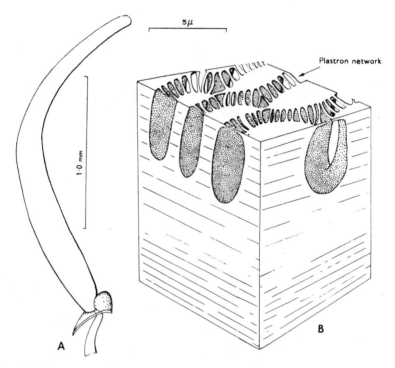

Fig. 11-18. Spiracular gill of *Eutanyderus wilsoni*. A, lateral view of right spiracular gill, which is formed by an outgrowth of the body wall adjoining the first pair of spiracles. B, diagram of the structure of the wall of the spiracular gill showing the plastron. Vertical pillars distributed over the surface of the gill are connected by horizontal struts. The interstices between the struts provide the air–water interface. From Hinton, 1968. *Adv. Ins. Physiol.*, **5**, 65–162.

by a complex system of hydrofuge structures, which may consist of hydrofuge hairs, or be formed of much more elaborate systems of tubercles, struts, and arches (Fig. 11–18). The air–water interface of this structure is very large, facilitating gaseous exchange without any increase in the permeability of the cuticle. The plastron can be regarded as a 'physical gill', depending for its operation on the air store being maintained at constant volume. Give this condition, and provided that the ambient water is well aerated, oxygen diffuses continuosly from the water into the plastron, replacing what has entered the tracheae; the animal can thus remain submerged for a long time. If, however, the oxygen tension in the water is low, the animal suffers because oxygen will diffuse out from it, through the plastron, into the water. If the animal is exposed to air, the plastron no longer functions, except in so far as it can provide channels for gaseous diffusion, but it does restrict water loss through the spiracles. These properties have important ecological implications. Plastron respiration is usually restricted to insects living in well-aerated water, subject to frequent changes in water level, with consequent periodical exposure to dry conditions.

Finally, many aquatic insects have become fully aquatic by developing tracheal gills, which make it possible to obtain oxygen by diffusion from the surrounding water. The essential feature of these is the presence of many fine tracheae immediately below the cuticle, ventilation being sometimes provided by the movements of the animal, and sometimes by movements of the gills themselves. Krogh pointed out in this connection a curious analogy between the respiratory system of the nymph of the dragonfly *Aeschna* and that of the cephalopods. The former has tracheal gills that lie in the rectum, and that are ventilated by muscular movements that also provide for the locomotion of the animal. In principle, this is the same combination of respiratory movement and locomotion that has been developed with such success in the operation of the cephalopod pedal siphon.

12
Excretion

12–1 EXCRETORY ORGANS

While zoological knowledge was advancing structures were sometimes described and named before their function was understood, and before their relationship could be determined. This is why so much confusion has in the past surrounded the study of the so-called excretory organs of animals. We say 'so-called', because the very term excretion is used in more than one sense. Strictly speaking, it refers to the removal of the waste products of metabolism, which comprise the carbon dioxide and water that are released by the oxidation of energy-rich compounds, and the nitrogenous waste that results from the metabolism of proteins and nucleic acids. But the removal of carbon dioxide is part of the respiratory process, and because of this, and because there is more than one route for the passage of water, excretion has commonly been thought of as the removal of nitrogenous waste. We shall use the term in that sense here, but with the proviso that the process is closely bound up with the regulation of the flux of water and certain electrolytes between the organism and external environment. We have, in fact, to find sites of operation of functions that were not appreciated when invertebrate organs were first described and named. This is true also of many other aspects of animal organization, but it is particularly important in studies of excretory function not to be misled by traditional nomenclature, especially since subsequent studies have revealed unexpected structural complexity.

We owe to Goodrich the foundations of our modern understanding of the history of excretory systems. It was he who first clearly formulated, in 1895, the fundamental principle that the animal body is primitively connected with the external environment by two distinct sets of tubular structures, the nephridia and the coelomoducts (Fig. 12–1) with cilia or flagella providing their motive power. He defined a nephridium as an organ that is developed centripetally, and quite independently of the coelom, being probably derived from the ectoderm. Its lumen is

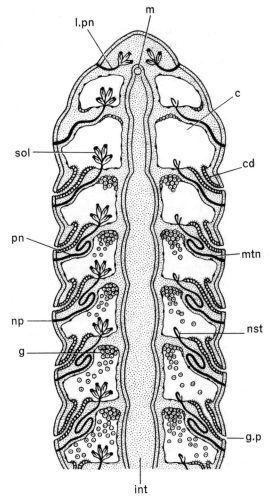

Fig. 12-1. Diagram of primitive annelid in longitudinal section, showing relations of coelomoducts and nephridia to segmental coelomic cavities. Protonephridia, *pn*, on left; metanephridia, *mtn*, on right. *c*, coelom; *cd*, coelomoduct; *g*, gonad; *g.p*, genital pore; *int*, intestine; *l.pn*, largal protonephridium; *m*, ventral mouth; *np*, nephridiopore; *nst*, nephridiostome; *sol*, solenocyte. Adapted from Goodrich, 1945. Q. Jl microsc. Sci., **86**, 113-392.

formed by the hollowing-out of the nephridial cells, and is consequently intracellular. Primitively this lumen is closed internally, in which case the organ is called a protonephridium. Frequently, however, it secondarily acquires an opening into the coelom, this opening being the nephridial funnel or nephrostome. The organ is then called a metanephridium. In complete contrast to a nephridium, a coelomoduct is developed centrifugally as a mesoblastic structure, formed as an outgrowth of the gonad or of the wall of the coelomic cavity. Its lumen, which is an extension of that cavity in coelomate animals, is not intracellular, but is bounded by a layer of epithelial cells; it opens into the coelom by a ciliated funnel, the coelomostome.

The distinction between a nephridium and a coelomoduct is not purely developmental and morphological, but involves also functional criteria. Both types of organ are present in the lower coelomate invertebrates, and in these the primary function of the coelomoducts is to convey to the outside the germ cells that develop in the coelomic epithelium. In the acoelomate platyhelminths and nemertines these ducts are represented by the gonoducts, which, on the gonocoel theory of the origin of the coelom, would therefore be the direct forerunners of coelomoducts. The nephridia, on the other hand, seem primitively to have had no connection with reproduction, although such a connection may be secondarily established in a relatively small number of species. The primitive function of these organs is usually said to be excretory; this is correct is we use the term in a broad sense to include regulatory functions, without necessarily implying that the removal of nitrogenous waste is an important aspect of their function (Sec. 12-2).

These are matters that will become clearer as we pursue our analysis. For the moment it is important to appreciate that nephridia are essentially characteristic of the earlier stages of invertebrate evolution, and, to some extent, of the larvae of higher forms, although their presence in amphioxus reminds us that they have also persisted side by side with considerable advances in other aspects of organization. What created confusion for those who first described and named these organs is that in the course of evolution the coelomoducts have extended their field of activity to take over functions originally carried out by the nephridia, so that the distinction between the two sets of organs is obscured. The distinction remains, however, of fundamental importance.

There are no excretory organs in coelenterates. In these animals excretion and the composition of the body fluids are presumably regulated by the epithelia of the body surface and coelenteron, and it must have been at the rhabdocoeloid stage of evolution (on the common assumption that coelenterates are at a more primitive level, Sec. 18-4) that increased size and solidity of the body, with the associated problems of internal transport, promoted the appearance of excretory or regulatory organs in the form of nephridia. They might well have evolved from epithelia already concerned with regulation, which would account for their development today from the ectoderm. The disappearance of nephridia later in evolution is associated with the increasing metabolic complexity of the more advanced animals. Increasing exploitation of a wider range of habitats and modes of life has demanded an increasing specialization and diversification of function from the excretory and regulatory systems, and it may well be that coelomoducts have proved better able to provide for this than have nephridia. Indeed, we shall see that in the arthropods even the coelomoducts proved insufficient for these needs, and that in this group there evolved another type of excretory organ that belongs to neither of these two primary categories.

Ultrastructural studies have shown that there are at least three main categories of protonephridia. In one of these the inner blind end of the protonephridium is formed of a terminal organ which is actually a compound structure (Fig. 12.2) composed of two cells which interdigitate by means of finger-shaped processes. One of these cells is a flame cell, bearing a bunch of motile flagella forming the flame. The other is a tubule cell, the tubule of which is formed by the wrapping of cytoplasm around an extracellular space, the two edges of the cytoplasm being bound together by a desmosome which has been aptly compared with a zip fastener. It follows that in this respect the postulate that the nephridial tube is always intracellular cannot be sustained, and in some species that applies to the tubule as a whole. Protonephridia of this type are especially characteristic of the platyhelminths, and are found also in nemertines and in the pseudocoelomate entoproctans. Interestingly

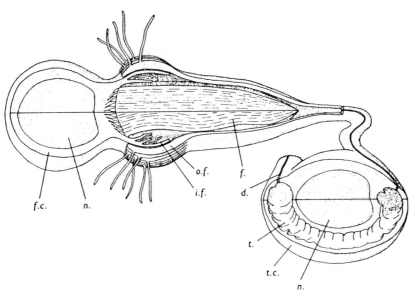

Fig. 12-2. Reconstruction of the terminal organ of the protonephridium in the miracidium larva of *Fasciola hepatica*. The flame cell (on the left) bearing a tuft of flagella, the flame, interdigitates with the first cell of the tubule. The inner fingers of cytoplasm project from the flame cell, the outer fingers from the tubule cell. d., desmosome; f., flame; f.c., flame cell; i.f., inner finger of cytoplasm; n., nucleus; o.f., outer finger of cytoplasm; t., tubules; t.c., tubule cell. From Wilson and Webster, 1974. *Biol. Rev.* **49**, 127-160.

enough, Goodrich had already concluded in 1895, from light microscopical studies, that the excretory and genital ducts of entoprocts were of a platyhelminth pattern. That acoelomates should resemble pseudocoelomates in this respect is perhaps surprising; whether the resemblance implies a common origin is, however, doubtful, as will appear later.

The second category of protonephridium, characteristic of rotifers, has a terminal structure called a flame bulb (Fig. 12–3), formed of a single cell. A flame of flagella arises from the apical cytoplasm of the flame bulb, the latter being connected with the nephridial tubule by a complex of cytoplasmic channels, columns, pillars, and microfilaments which presumably serve to anchor the flame.

The third category has a terminal structure formed of elongated tubular cells called solenocytes, with a nucleus in the apical cap and with a single flagellum. These cells are found in *Priapulus*, in the gastrotrich *Chaetonotus*, in the archiannelid *Dinophilus*, in certain polychaetes (e.g. *Glycera*), and in the protochordate *Branchiostoma* (amphioxus). In amphioxus the nucleus lies in the head of the terminal structure, from which arises a tubule composed of rods. There is no material between these rods, so that fluid can pass readily into the tubule without having to be filtered. (Strictly, this could be held to be a metanephridial rather than a protonephridial characteristic, on the grounds that the inner ending is not a blind one.) This particular solenocyte, which is closely related to the atrial epithelium and to the blood system, has been called a cyrtopodocyte (Fig. 12–4), because its base is

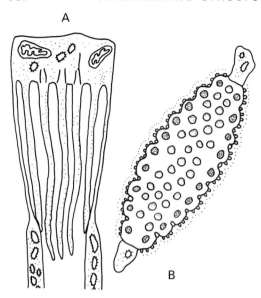

Fig. 12–3. A, longitudinal section through the flame bulb of the rotifer *Asplanchna brightwelli*; B, transverse section through the flame bulb. a.c., apical cytoplasm; c.ch., cytoplasmic channel; c.col., cytoplasmic column; f., flagellum; p., pillar. From Wilson and Webster, 1974. *op. cit.*

drawn out into branched and palisade-like feet resembling those of the podocytes of the Bowman's capsule of the vertebrate kidney. Perhaps this structural similarity between a protochordate and the vertebrates indicates no more than similarity of permeability requirements. But it has been suggested that the kidney might have evolved from such a condition by branching of the blood vessels to form the glomerulus, with the protonephridial cells forming the capsule and nephron tubules.

It will be apparent that, given these different types of protonephridia (and not all can be easily fitted into the three categories listed here), the elucidation of their phylogeny presents difficulties. It has been suggested that protonephridial terminal structures may have evolved from a choanoflagellate type of cell. We have earlier seen, however, that this has been identified in more than one group of animals other than sponges, so that its existence in excretory organs is not necessarily evidence of their common ancestry. Willmer is properly cautious in suggesting that genetic information for such cell structure might have been retained in suppressed form in many groups, and have become activated independently when conditions favoured this. There may, therefore, be an element of latent homology (Sec. 2–1) in protonephridia, but we can no longer feel sure of what was implicit in the original concept of a protonephridium: that all of these organs had a direct common origin. This, of course, is in line with what we have noted elsewhere (in relation to arthropod evolution, for example) regarding the remarkable possibilities of convergence.

Platyhelminths may be thought of as having a pair of protonephridia, their canals being much branched and bearing flame cells at the ends of the branches. These cells, therefore, are scattered throughout the parenchyma. Similar protonephridia are found in the nemertines, sometimes as a single pair situated far forwards. Sometimes they may be extended into longitudinal collecting canals, into which open many smaller efferent canals that lead from the scattered flame cells. Such is the arrangement in the terrestrial form *Geonemertes* (Fig. 12–5), where the system consists in effect of many hundreds of separate protonephridia, each dis-

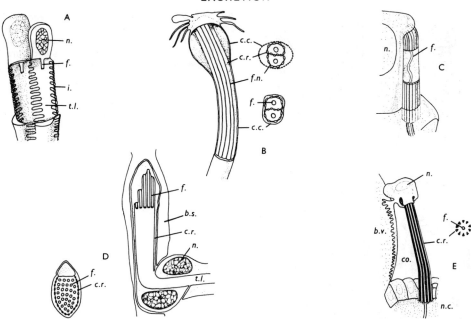

Fig. 12-4. Solenocytes and other protonephridia. A Schematic diagram of the priapulid protonephridium (after Kümmel, 1964). B Reconstruction of the terminal organ of the gastrotrich *Chaetonotus* in longitudinal and transverse section (after Brandenburg, 1962). C Structure of a solenocyte from the annelid *Glycera unicornis* (after Brandenburg and Kümmel, 1961). D Longitudinal and transverse sections through the cyrtocyte of *Dinophilus gyrociliatus* (after Brandenburg, 1970). E diagram of the cyrtopodocyte of *Branchi ostoma* (based on Bradenburg and Kümmel, 1961). *b.s.*, blood sinus; *b.v.*, blood vessel; *c.c.*, cytoplasmic covering; *co.*, coelom; *c.r.*, cytoplasmic rod; *f.*, flagellum; *f.n.*, fibrous network; *i.*, interdigitations; *n.*, nucleus; *n.c.*, nephridial canal; *t.l.*, tubule lumen. From Wilson and Webster, 1974, *op. cit.*

charging through its own efferent canal. From such an arrangement the nephridial system of annelids might have been derived. Here each typical segment primitively possesses a pair of the organs, opening independently of each other at segmental nephridiopores. The nephridia themselves are intersegmental in position, their inner end penetrating the anterior septum of the segment in which the main body of the organ lies.

The nephridia of the annelids show many variation upon this simple ground plan, and for this reason it is important to understand the history of the nephridial system of these animals. The particular complication in the polychaetes is that their nephridia frequently bear ciliated funnels; they are then termed metanephridia. It was at one time supposed, because of the presence of the funnels, that organs of this type were homologous with coelomoducts, especially since they sometimes serve for the passage of germ cells. The situation was finally clarified by Goodrich, who, in showing that these organs were nephridia, showed also that nature had set a trap for investigators by developing more than one variant of them.

The trochophore larva of certain polychaetes (e.g. *Nereis*, *Pomatoceros*, Fig. 19-2) possesses a pair of simple and typical protonephridia, the so-called head kidneys. These structures, each with a flame cell bearing a single flagellum, dis-

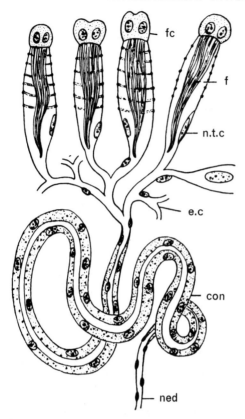

Fig. 12-5. *Geonemertes agricola.* Diagram of single protonephridium (from Coe, 1930). con, convoluted canal; *e.c.*, end canal; f, flame of cilia; *fc*, binucleate flame-cell; *ned*, efferent duct; *n.t.c*, nucleus of terminal chamber. From Goodrich, 1945. *op cit.*

appear later, and the nephridia of the adult worms are metanephridia, with open ciliated funnels. Sometimes, however, even adult polychaetes (e.g. *Glycera*, *Nephtys*, *Phyllodoce*) possess only protonephridia, which in these instances have solenocytes. Coelomoducts are also typically present in polychaetes, not necessarily in every segment, but at least in those in which germ cells develop. They give rise, however, to two modifications. One of these is found in the nereid worms, where the coelomoducts are reduced to inconspicuous areas of ciliated epithelium, discovered by Goodrich and named by him the dorsal ciliated organ. These are possibly phagocytic, but they have certainly lost their primitive function of conveying the germ cells, which in these worms escape by rupture of the body wall.

The other modification is more widespread. It is remarkable that only one family of the polychaetes, the Capitellidae, preserves the primitive arrangement in which coelomoducts and nephridia (in this instance metanephridia) are separate from each other. In the other families the two structures become associated to form a compound organ called a nephromixium, in which the nephridial tube bears a ciliated mesodermal funnel. The nephridial component of this may be a protonephridium, as, for example, in the Phyllodocidae; here the developing coelomoduct has been observed to grow backwards alongside the protonephridial canal, an open communication between the two arising at sexual maturity. The organ thus formed is a protonephromixium (Fig. 12–6). Alternatively, and more commonly, the

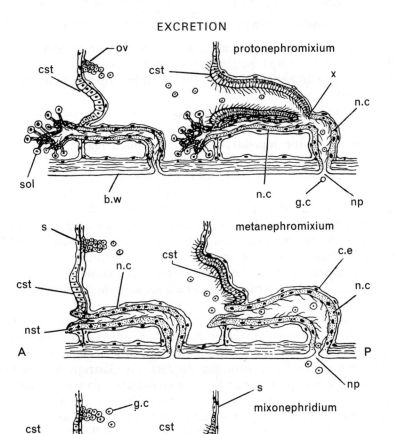

Fig. 12-6. Diagrams showing structure and formation of nephromixia by combination of coelomoduct with nephridium. A, anterior, young stage before combination. P, posterior, combination completed. *b.w*, body wall; *c.e*, coelomic epithelium; *cst*, coelomostome; *g.c*, germ cell; *n.c*, nephridial canal; *np*, nephridiopore; *nst*, nephridiostome; *ov*, ovary; *rnc*, rudiment of nephridium; *s*, intersegmental septum; *sol*, solenocyte; *x*, point where coelomoduct opens into nephridial canal. From Goodrich, 1945. *op. cit.*

nephridial component of the nephromixium is a metanephridium, the resulting compound organ being called either a metanephromixium or a mixonephridium, according to the way in which the junction is formed (Fig. 12–6). An example of the latter is seen in *Arenicola*, where there are usually six pairs of these organs, present in segments 5 to 10. These organs, conspicuous with their rich vascularization, frilled funnel lip, and attached gonad, serve both for the passage of germ cells and, presumably, for the regulatory functions that we shall be considering later. Probably the elaborate structure of the funnel aids in the selection of ripe ova and in their removal from the coelom.

Such, in outline, is the disposition of coelomoducts and nephridia in the polychaetes, although it is accompanied by much variation in detail from family to family. Elsewhere in the annelids the situation is much simpler. In the oligochaetes the nephridia are all of the metanephridial type, familiar in the nephridium of the earthworm. The evolutionary relationship between the complex funnel of this type of organ and the flame cell and simpler funnel of many polychaetes can readily be appreciated in such a form as *Enchytraeus*, where the nephridiostome is small and possesses a flame of cilia recalling the arrangement of the endings of protonephridial tubes. Coelomoducts are also present, but the tendency for the restriction of these to the genital segments, seen in polychaetes, is here carried further, one duct now being associated with each gonad. The arrangement in the Hirudinea is very similar in principle, except for the insertion of a capsule between the funnel and tubule of the metanephridium.

Other groups of animals also possess nephridia. In the Archiannelida, an artificial assemblage of worms of rather doubtful affinity, but presumably related to annelidan ancestors, we find either protonephridia or metanephridia, and probably also nephromixia. The Entoprocta, the Rotifera, the Gastrotricha, and the Kinorhyncha have a pair of protonephridia, with gonoducts representing coelomoducts. Protonephridia have also been found in the Priapulida and in the Acanthocephala. In the latter the nephridial and genital ducts join to form a median urogenital canal, giving rise, therefore, to a peculiar type of nephromixium. The actinotrocha larva of *Phoronis* has a pair of protonephridia, but the corresponding organs of the adult have wide ciliated funnels, and are probably nephromixia, since they carry the genital products to the outside. The Sipuncula, the Echiura, and the Myzostomaria are other groups in which the so-called nephridia are probably nephromixia, although this is not certain; nephromixia are probably also present in the Brachiopoda.

The generalization with which we started, that there are two types of ducts, is thus of wide applicability, and among the lower invertebrates there are only a few groups that cannot be readily fitted into it. The Chaetognatha and the Polyzoa (Ectoprocta) have no nephridia at all, but coelomoducts are represented by ciliated ducts leading to the outside. Another exceptional group is the Nematoda. The gonoducts may here be held to represent the coelomoducts of other groups, as they do in the platyhelminths, while the excretory system is supposed to be represented by a pair of longitudinal canals lying in the lateral thickenings of the body wall. These, however, show no resemblance at all to nephridia; they have no internal opening, nor do they possess cilia, which are not present at all in this group. Moreover, the whole system is believed to be formed by the enlargement and excavation of only two cells.

Another important group in which there is no sign of nephridia, either in the larval or adult stages, is the Echinodermata, and this applies also to the Hemichordata and the Urochordata, which, as we see in other contexts, are probably related to the echinoderms. Whether this means that these groups have lost these organs during their evolution is very difficult to say, but it is certainly of significance that protonephridia are present in amphioxus. It is obvious from our earlier discussion that we must be cautious in assuming a common ancestry for all protonephridia, but it would certainly be surprising if these organs had appeared in amphioxus without being represented at all in the protochordate and echinoderm line. This, then, seems to be an instance in which nephridia may have been lost without trace.

The history and relationships of the excretory systems of arthropods are no less complex than other aspects of the organization of this group. The characteristic excretory organ of many of these animals is a tubular mesodermal structure, which, before the distinction between nephridium and coelomoduct had been clearly formulated, was often referred to as a 'nephridium'. It follows from our earlier definition, however, that this was a misnomer, and that the organ must, in fact, be a coelomoduct.

The primitive arrangement is well seen in *Peripatus* (Fig. 12–7), which possesses a pair of these structures, called coxal glands, in almost every one of its segments. Their development is particularly illuminating. Hollow coelomic follicles, or somites, appear in each segment, and in the trunk region, each becoming subdivided into a dorsal, a lateral and a ventral portion (Fig. 12–8). The cavity of the ventral portion persists to form the end sac of the coxal gland; this becomes connected with the outside by a mesodermal duct which grows from it towards the ectoderm. Thus arises the prototype of the coxal gland of other arthropods. In its fully developed form, the end sac opens by a ciliated canal, which we can regard as the coelomostome, into a coiled excretory canal; the terminal portion of this is enlarged to form a vesicle or bladder. There is no sign at all of nephridia in *Peripatus*. The derivation of the coxal glands from coelomoducts is perfectly clear, and is emphasized by the mode of formation of the gonads and their ducts. The dorsal portions of the coelom become reduced anteriorly, but posteriorly they fuse on either side to form the paired gonads, which open to the outside at the last segment. At this point the coelom is not divided into dorsal and ventral portions, so that each of the gonads becomes continuous with a duct that meets its partner at a median ventral pore. As a result, no coxal glands are formed in this particular segment.

Coxal glands of similar structure are also found in the Arachnida and the Crustacea, but are greatly reduced in number (Fig. 12–7). In the Crustacea they are found only in the third and sixth segments, where they open respectively at the base of the second antenna (antennal gland) and second maxilla (maxillary or shell gland). In the Branchiopoda, Ostracoda, Copepoda, Brachiura, Cirripedia, and lower Malacostraca, the antennary gland is present in the larva and the maxillary gland in the adult. In the Amphipoda, Euphausiacea, and Decapoda the antennary gland persists into the adult, while the maxillary gland either disappears or fails to develop; but in the Mysidacea, generally conceded to be primitive animals, both of the glands may be functional in the adult. Given this facility of the coxal glands for disappearing, the restricted glands of modern crustaceans can readily enough be derived from a hypothetical continuous series such as that of *Peripatus*. Support for this interpretation is found in various crustacean species where groups of cells at the bases of other appendages can take up injected foreign material, such as indigocarmine. In doing this they resemble cells of the glands themselves, and they are probably to be regarded as vestiges of former coxal glands. The openings of the

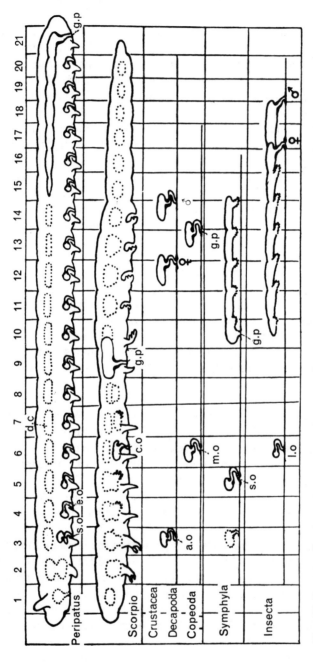

Fig. 12-7. Diagram showing excretory and genital coelomoducts in various Arthropoda. *Peripatus* and *Scorpio* drawn complete, with embryonic coelomesoblastic somites and their dorsal portions shown in dotted line. Crustacea, Symphyla, and Insecta simplified; transient vestigial coelomoducts shown in dotted line. Opening of antennary organ, *a.o*; of coxal organ, *c.o*; of excretory organ, *e.o*; of lingual gland, *l.o*; of maxillary organ, *m.o*; of salivary organ, *s.o*. Dorsal coelom, *d.c*; genital pore, *g.p* (both sexes indicated in Decapoda and Insecta). From Goodrich, 1945. *op. cit.*

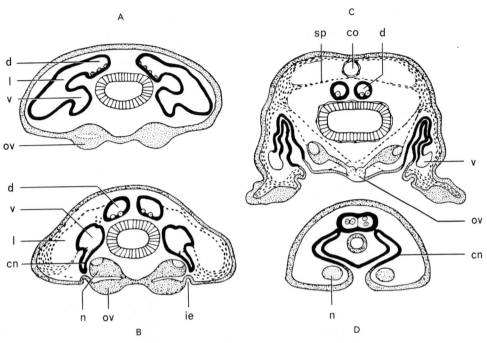

Fig. 12-8. Scheme illustrating the development of the coelom in *Peripatus*. A, each coelomic sac divides into dorsal (*d*), lateral (*l*), and ventral (*v*) cavities, the dorsal enclosing the germ cells. *ov*, ventral organ. B, the dorsal cavity forms the gonad cavity, the lateral one the lateral body cavity by disintegration of its walls, and the ventral one the sac of the excretory organ with its canal, *cn*. *ie*, epidermal invagination; *n*, ventral nerve cord. C, section at the level of the gonads. The body cavity is traversed by cords of mesenchyme. *co*, heart; *sp*, pericardial septum. D, section at level of genital opening. *cn*, excretory canal forming oviduct. From Grassé, 1949. *Traité de Zoologie*, vol. 6. Masson, Paris.

genital ducts, which vary in position from the eleventh to the nineteenth segment, would have formed part of the same series of structures. As in *Peripatus*, there is no certainty that nephridia were ever present in crustaceans. There is, however, some evidence that the terminal portion of the duct of the gland forms from ectoderm; this has led to the suggestion that it may be the vestige of a nephridial canal.

Most arachnids possess a pair of coxal glands opening on the sixth segment, at the base of the fifth pair of appendages (third pair of walking legs). Doubtless there were originally more of these, for in *Limulus* the gland also receives contributions from the three next anterior segments, while during the development of the scorpion vestiges of glands are detectable in the fourth, fifth, and seventh segments; they appear as diverticula of the coelomic cavities, and can reasonably be regarded as traces of coelomoducts. Outgrowths of the coelomic cavities in segment 9 form the mesodermal parts of the gonoducts.

In those arthropods that have become fully adapted to terrestrial life there is an important development in the organization of the excretory system: the appearance of Malpighian tubules. These are outgrowths of the alimentary canal, and are thus an entirely new introduction, owing nothing either to nephridia or to coelomoducts. We shall see that their appearance is correlated with the novel physiological situation presented by life on dry land. They are, therefore, new adaptive devices, and it is understandable, although none the less remarkable, that they have arisen by convergence along more than one evolutionary line. They are found in Myriapoda, Insecta, and Arachnida, and we are bound to conclude, in view of what has been said earlier regarding the history of these groups, that the Malpighian tubules of arachnids must have been evolved independently of those of the other two groups. This interpretation is strengthened by their development in arachnids from the end of the embryonic midgut, whereas in the other groups they are ectodermal structures, arising from the proctodaeum. The difference is a small one, but of the sort that we expect to find when comparing convergent features of organization.

We have suggested earlier the possibility that the myriapods and Insecta evolved from a remote common ancestor. Perhaps, therefore, the Malpighian tubules of these two groups had a common ancestry, but we cannot assume that this must have been so. If convergence in this respect has occurred in the arachnids, it may well have occurred in other lines also. As we have earlier emphasized, not the least interesting feature of arthropod relationships is that however we interpret them we are bound to postulate some degree of convergence in their evolutionary history. It is hardly necessary to add that no trace of nephridia remains in any of these groups. As for coelomoducts, these persist as the genital ducts, and it is possible that the 'head' or 'salivary' glands of certain myriapods may represent the coxal glands.

There remain the molluscs, a group that preserves a good deal of uniformity in the excretory system, despite the variety of form shown by other organs such as the ctenidium and the foot. As with the crustaceans, nephridia are absent from the adults; in this instance, however, we have some evidence that they have been secondarily lost, for protonephridia have been identified in the larvae of certain Pulmonata (e.g. *Ancylus*, *Planorbis*, *Lymnaea*, *Arion*, *Helix*) and Bivalvia (e.g. *Dreissensia*).

The coelomoducts of the adult (Fig. 12-9) consist typically of a pair of tubular structures, leading from the coelomic cavity to the outside, and primitively constituting the genital ducts. It has been generally assumed that the molluscs originally possessed a pair of coelomic cavities, which met dorsally to enclose the heart, and the walls of which proliferated the germ cells, but we have seen that it is not easy to decide how these cavities might have arisen in the first place. We can only suppose that some simple arrangement of paired cavities gave rise by further differentiation to an anterior region, the gonad; a central region, the pericardial coelom; and a posterior region, the gonoduct. The last of these also came to take over excretory function, presumably in correlation with the disappearance of the nephridia. Thereafter there was a progressive tendency for the renal and genital ducts to separate, just as happened also in the vertebrates; presumably there is much functional advantage in this separation. The coelomic cavities, whatever their precise origin, are clearly tripartite structures as we see them today. There is no reason, however, to suppose that this subdivision represents any form of metameric segmentation, for these cavities arise by the hollowing-out of a pair of coelomoblast masses. In so far as segmentation is to be found at all in the coelom of molluscs, it is present in the six pairs of excretory organs of *Neopilina* (Fig. 11-9, p. 310), but, as we have already seen, it is very doubtful whether this repetition is truly metameric, or even primitive.

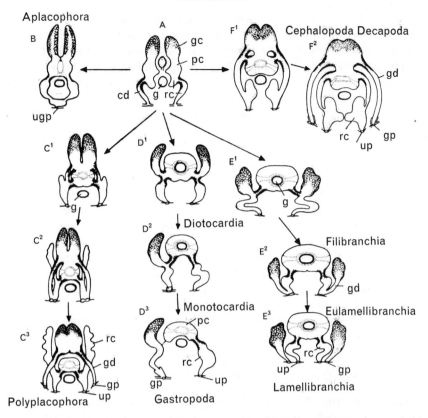

Fig. 12-9. Diagram showing specialization and subdivision of coelom and coelomoducts in Mollusca. A, primitive plan. B, Aplacophora. *cd*, coelomoduct; *g*, gut; *gc*, gonadial coelom; *gd*, genital duct; *gp*, genital pore; *h*, heart; *pc*, pericardial chamber; *rc*, renal organ; *ugp*, urinogenital pore; *up*, excretory pore. Along line C coelomoduct becomes split from coelomostome outwards and genital coelom closed off. Along line D (Gastropoda represented as untwisted) both heart and pericardial chamber may surround gut, and left genital chamber comes to open into renal organ and left coelomostome may be closed, left coelomoduct remaining as genital duct. Along line E gonadial chamber comes to open into coelomoduct which may become split into separate excretory and genital ducts. In line F coelomoduct also becomes split into two, and generally they are asymmetrically developed. From Goodrich, 1945. *op. cit.*

In the Aplacophora the arrangement of the coelomic cavities still corresponds closely to the primitive plan, but in other molluscs modifications have been developed along several evolutionary lines (Fig. 12–9). One result of these is the appearance of some degree of asymmetry; another is the separation of genital and excretory regions. This separation is achieved in the Polyplacophora by a splitting of the coelomoduct in the region of the coelomostome, while the gonadial cavities become closed off from the pericardial coelom. The excretory coeloms remain connected with the latter at the coelomostomes, but are increased in relative size by forward prolongation and by the development of branched outgrowths.

The gonad of the post-torsional left side is always lost, but both kidneys may remain well developed (e.g. *Haliotis*). In the higher forms only the post-torsional left kidney persists, the gonad opening into a coelomoduct that has lost both its renal function and its connection with the pericardial coelom. The position of the external renal opening varies in relation to the other factors influencing the organization of the pallial complex. In the pulmonates a secondary ectodermal ureter runs along the rectum and opens outside the mantle cavity (Fig. 13–5). The bivalves lack the complications of asymmetry that are such a feature of gastropod organization, but in them, too, the genital and renal ducts become separated. In the more primitive protobranchs the gonads discharge their products into the renal organs just beyond the renopericardial funnel; in filibranchs the connection is much nearer the posterior end of the kidney, while in eulamellibranchs the two organs have developed separate openings. During the course of this transformation there is also a modification in the renal organs themselves. Initially, in the protobranchs, the whole of the coelomoduct is excretory, but later it becomes bent into a U shape; the lower limb is now glandular, while the upper, and more distal, limb forms a bladder. Finally, the separation of the genital and excretory components of the coelomic complex has been achieved also in the cephalopods, where the genital duct comes to run separately from the renopericardial canal and kidney. Some degree of asymmetry may be present in this group, as, for example, in *Sepia*, where a renal papilla is present on each side, but a genital papilla on the left side only.

12–2 EXCRETION OF NITROGEN

Amino acids, absorbed from the alimentary tract or arising from the metabolism of proteins, are subjected to two main types of metabolic process. One of these, known as transamination, is a reversible process in which amino nitrogen is transferred from one compound to another:

$$\underset{NH_2CHCOOH}{\overset{R}{|}} + \underset{O = CCOOH}{\overset{R'}{|}} \rightleftharpoons \underset{O = CCOOH}{\overset{R}{|}} + \underset{NH_2CHCOOH}{\overset{R'}{|}}$$

This reaction, which is probably operative throughout living tissues, both plant and animal, involves an interaction between the amino acid and α-ketoglutaric acid, and is catalyzed by specific enzymes.

The second main type of process is known as oxidative deamination, and results in the breakdown of the amino acid to a keto acid and free ammonia:

$$\underset{NH_2CHCOOH}{\overset{R}{|}} + \tfrac{1}{2}O_2 \rightleftharpoons \underset{O = CCOOH}{\overset{R}{|}} + NH_3$$

This, too, is a reversible reaction, catalyzed by specific enzymes, and is known to occur in plants and micro-organisms as well as in animals. Both processes, therefore, are presumably part of the fundamental biochemical equipment of living tissues.

Animal tissues can assimilate some of the ammonia produced by deamination, but usually their catabolic activity produces far more than can be dealt with in this way, and it is here that the problem of nitrogenous excretion arises. Ammonia is the major waste product of protein metabolism in all animals, but it need not leave the body in that form. To determine whether or not it does so, it is necessary, ideally, to identify the organs responsible specifically for the removal of nitrogenous waste, and to establish the chemical nature of their product. There are obvious difficulties in the way of doing so, however; not the least is that the animals concerned are often small. Fortunately, much valuable information can be gained with aquatic animals by analysing the medium in which they are living, and determining the nature of the nitrogenous products that accumulate in it.

No group limits its nitrogen excretion to one product, but aquatic invertebrates commonly excrete much of their nitrogen as ammonia. This constitutes 52.7% of the total nitrogen excreted by actinians, other representative figures being 80% (*Aphrodite*), 60% (*Astacus*), 67% (*Sepia*), and 39.3% (*Asterias*); a relatively low value in echinoderms is associated with the excretion of large amounts of amino acids. All such animals, in which ammonia is the predominant form of nitrogenous waste, are termed ammonotelic. Ammonia is a substance that diffuses readily through body surfaces into surrounding water, so that its removal is unlikely to demand the development of specialized excretory organs. This is confirmed by data obtained from crustaceans, which show that ammonia may be plentiful in the blood but very scarce in the excretory fluid. In *Maia squinado*, for example, the ammonia content of the blood may be as high as 2.4 mg/100 ml, but the amount of nitrogen lost in the urine accounts for less than 10% of the total nitrogen excreted by the animal. The explanation seems to be that ammonia is lost largely through the permeable body surfaces, and particularly through the gills. In consequence of this, a considerable amount of ammonia can still be lost even after the excretory pores have been blocked. Evidently, then, some function other than nitrogenous excretion is needed to account for the invertebrate excretory organs that we have described, but before seeking for this it will be well to examine why certain animals have departed from this primitive pattern of ammonia excretion.

Compounds other than ammonia may be excreted by ammonotelic animals, including urea, uric acid, guanine, and amino nitrogen (see Table 12–3, p. 350), but these are secondary in importance to ammonia. Even *Asterias* excretes only 23.8% amino nitrogen, while for *Astacus* the figure is 10.1%. There are, however, other invertebrates in which uric acid is the predominant nitrogenous waste product; in *Rhodnius*, for example, as much as 92% is excreted in this way. These forms, which are termed uricotelic, comprise most of the insects and many gastropods. The two groups have thus developed, quite independently of each other, a specialized mode of nitrogen excretion, and one that they share with birds and many reptiles.

The principles underlying the establishment of uricotely were first formulated by J. Needham, who approached the problem from the standpoint of vertebrate embryology; they were then extended to adult invertebrates by Delaunay. The argument is that ammonia is highly toxic, and must therefore be excreted rapidly from the body. This, as we have already seen, presents no difficulty to aquatic animals with permeable surfaces. Terrestrial animals, however, are no longer surrounded by water, and must reduce their loss of this essential substance to a minimum, perhaps by the development of an impermeable body surface, as commonly occurs in arthropods. This means that nitrogenous excretion can no longer take place through the surface; instead, it is now dependent on the specialized functioning of excretory organs, which thus take on an importance in excretion that they have not achieved in ammonotelic forms. Nevertheless, the urine that they produce cannot be used for the removal of nitrogen in the form of ammonia, for this would require the loss of an extravagant quantity of water if the concentration of the ammonia were to be kept below the level of toxicity.

To deal with this difficulty animals make use of two main devices. One of these is to combine the ammonia with carbon dioxide to form urea, which is much less toxic than ammonia, and can therefore be excreted (ureotely) in more concentrated solution with a consequent reduction in the loss of water. This device is used in mammals (and is employed for a different reason in elasmobranch fish), but it has found little application in invertebrates. Indeed, it has the disadvantage of still demanding a considerable output of water. Urea does, however, play a significant part in the metabolism of oligochaetes, which have a characteristic pattern of nitrogenous excretion that shows them to be essentially fresh-water animals that are exploiting damp terrestrial habitats; and we shall see later that this is true also of their osmotic and ionic regulation. Some nitrogen is lost in the surface mucus, but much is excreted through the nephridia, and has been identified in the urine collected by Bahl from the Indian earthworm, *Pheretina posthuma*, by allowing the fluid to drain into a glass dish). Ammonia is an important component of this, but so also is urea; opinions differ as to which predominates, probably because the pattern varies with the physiological condition and perhaps also with geographical distribution. Uric acid is a minor component, amounting only to at most 1.5% of the excreted nitrogen.

An important factor in the excretion of earthworms is their chloragogenous tissue, a specialized derivative of the coelomic epithelium which lies close against the wall of the intestine. The chloragogen cells store fat and carbohydrate, and they are also the main site of deamination, so that there is some analogy between them and the hepatic cells of vertebrates. Ammonia and urea pass from the cells into the coelomic fluid, and are swept through the nephridial funnel into the tubule. Waste particles (including mineral aggregates of muscovite or mica) accumulate within the chloragogen cells, and are released when these break up. The material is largely carried away by amoebocytes, either for deposit or for passage through the epidermis, but finely divided particles can also be taken up by the narrow ciliated parts of the nephridial tubules.

Leeches, like the oligochaetes to which they are so closely related, have also a fresh-water pattern in their disposal of nitrogenous waste, but they produce a much higher proportion of ammonia. This accounts for 72% of the total excreted nitrogen, urea accounting for only 5–10%. The reason for this is a strange one, being dependent on a capsule which is inserted in leeches between the nephridial funnel and the tubule (Fig. 12–10). This capsule contains two species of bacteria, one floating in the capsular fluid, and the other forming a carpet-like pile which is readily mistaken by the unwary for ciliation. These bacteria convert much of the more complex nitrogenous waste into ammonia, the urinary content of which falls if the animals are treated with antibiotics.

Particulate waste material accumulates in coelomic corpuscles and also in the botryoidal tissue, which in leeches largely represents the chloragogenous tissue of oligochaetes. In the primitive *Theromyzon*, where there is still an open communication between the nephridial funnel and the tubule, this material eventually passes into the latter. In more advanced forms, such as *Glossiphonia* and *Hirudo*, however, the opening has been blocked (Fig. 12–10), and in correlation with this the cilia of the funnel beat outwards into the coelomic cavities that surround the funnels. This is unexpected, but again there is an explanation: the capsule has become a site for the production of coelomic amoebocytes, which are thus swept out of the funnel. Particulate waste must now either be stored permanently in the body, or else broken down into a soluble form for removal by the nephridia. It follows that in these more advanced leeches the nephridia operate entirely by filtration.

The second device used by terrestrial animals is the excretion of nitrogen predominantly as uric acid (uricotely). This is of particular advantage to terrestrial animals. Uric acid is relatively insoluble, so that it can be precipitated from solution and removed in a solid or semifluid form, and this is undoubtedly one reason why the use of uric acid as the chief nitrogenous waste product is particularly characteristic of birds and terrestrial reptiles. For the same reason a dry urate excretion is found in the Onychophora, myriapods, insects, and certain gastropods. In myriapods and insects it is correlated with the development of Malpighian tubules as excretory organs. The Onychophora lack these, but are able to excrete their uric acid from the intestinal epithelium into the intestinal lumen. It is removed from here within a peritrophic membrane, which is expelled at intervals of about 24 hr. Despite the development of this adaptation, the Onychophora still possess

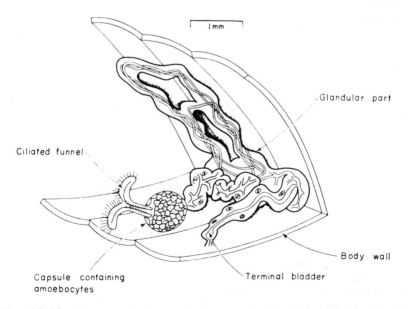

Fig. 12-10. Diagram showing nephridium of *Glossiphonia complanata, in situ*. After Oka, from Mann, 1962. *Leeches*. Pergamon Press, Oxford.

coelomic excretory organs, which retain their primitively segmental arrangement. Water containing ammonia is expelled from them, but only at infrequent intervals. It would seem that these organs must be regarded as a vestige of the aquatic ancestry of the group, with their activity now reduced to a minimum so as to avoid unnecessary waste of water.

Another factor suggested by Needham is that uricotelic animals usually develop from cleidoic eggs, which are eggs that are protected against water loss by being surrounded by a relatively impermeable membrane. Uricotely permits the retention of nitrogenous waste within this membrane in the form of uric acid. Thus, to take one example, the uric acid content of the egg of *Lymnaea* increases from 0.5 mg/100 ml at cleavage to 4.5 mg/100 ml at hatching. *Peripatopsis*, the onychophoran in which nitrogenous excretion has been most closely studied, is viviparous, but here, too, the embryo accumulates uric acid, ridding itself of this after birth. This provides an instructive contrast with mammals, for the mammalian foetus relies upon its mother to remove its nitrogenous waste through her blood stream. It is because of this that it can afford, unlike the chick embryo, to excrete its nitrogen as urea; a circumstance that may well have influenced the establishment of uricotely in mammals.

The incidence of uricotely in gastropods is illustrated in Table 12–1, from data given by Needham. These are animals in which the excretory organs play an important part in the removal of nitrogen, for crystals of uric acid accumulate in the secretory cells of their kidneys, to be discharged from the body at intervals. The values given in the table are thus an index of accumulation, and may not always reflect the true differences in excretory rate of the animals listed. However, very large amounts of uric acid accumulate in some species, and only negligible amounts in others, and it is reasonable to suppose that such differences are indices of different intensities of uricotely.

On this assumption, the adaptive value of uricotely for terrestrial animals is

Table 12–1 Uric acid content of the kidneys of molluscs (mg/g dry weight of kidney. Data compiled from Needham by Potts, 1967. *Biol. Rev.*, **42**, 1–41.

TERRESTRIAL PULMONATES		LITTORAL PROSOBRANCHS	
Helix pomatia	700	*Littorina neritoides*	25
Limax maximus	205	*L. saxatalis*	5
L. flavus	31	*L. obfusata (littoralis)*	2.5
TERRESTRIAL PROSOBRANCH		*L. littorea*	1.5
Pomatias (Cyclostomata) elegans	1,000	MARINE PROSOBRANCHS	
FRESH-WATER PULMONATES		*Buccinum undatum*	4
Lymnaea stagnalis	115	*Gibbula umbilicalis*	2
Planorbis corneus	41	*Nucella lapillus*	4.5
Ancylastrum fluviatilis	4		
FRESH-WATER PROSOBRANCHS			
Viviparus fasciatus	35		
Bithynia tentaculata	150		
Hydrobia (Potomapygus) jenkinsi			

certainly suggested by some of the data. Consider, in particular, the prosobranchs. Four of the species of *Littorina*, adapted for life in different regions of the littoral zone, show a gradation in their uric acid content up to a maximum in *L. neritoides*, which lives an almost terrestrial life in the splash zone. In a closely related form, the terrestrial *Pomatias* (*Cyclostomata*), uric acid constitutes about 100% of the dry weight of the kidney, a striking contrast indeed with the low levels found in marine prosobranchs.

Other data, however, cannot be so easily explained. Needham suggested that the large amounts of uric acid in certain fresh-water pulmonates (e.g. *Lymnaea*, *Planorbis*) were a consequence of these animals having secondarily reverted to an aquatic life after previous adaptation to life on land. It is not certain, however, that this view of their ancestry is correct, and it has been argued that they may well be derived from a stock that was never completely emancipated from water. If this be so, it is more difficult to account for their uricotely. The fresh-water prosobranchs are also troublesome. The negligible uric acid content of *Hydrobia* is to be expected, particularly since it is known to be a recent immigrant into fresh water from brackish waters. But the high content of *Viviparus* is difficult to explain, for there seems no question of it having had a terrestrial phase in its ancestry.

No doubt the enthusiasm with which comparative biochemistry was seen, in the 1930s, as bringing new strength to classical evolutionary theory, has proved to have been over-optimistic. But it is fair to recall that the propositions made at that time were not always lacking in caution. J. Needham himself presented the data of nitrogen excretion as 'an instance of the way in which comparative biochemistry may contribute to the problems of evolution rather than as a conclusion fully worked out for one special field'. What happened was that the further working out revealed unforeseen complications, the prime source of which, as we have already argued, must surely be the widespread distribution in animals of common metabolic pathways. It is the readiness with which analogous but remarkably similar biochemical adaptations can be independently evolved from these pathways that confuses the explorer of phylogenetic history. Comparative biochemistry can still illuminate the adaptive organization of animal adaptation and its evolutionary background. But for firm and reliable phylogenetic conclusions we need to have evidence drawn from the widest possible range of evidence, and here the findings of comparative biochemistry may provide only a single parameter.

The situation is well illustrated by *Onchidella celtica*, which has only a low level of uric acid. The Onchidiidae were formerly regarded as pulmonates, perhaps derived from land snails, and thus constituting a terminal strand in the evolution of the Stylommatophora. On this basis it was suggested that the low level of uric acid in *Onchidella* was an example of reversal of evolution. An alternative view, however, is that the Onchidiidae are more likely to be opisthobranchs, sharing many features with the pulmonates because of the close origin of the two groups and because of the similarity of their diet and their air-breathing habit. On this interpretation, the low level of uric acid could be regarded as the retention of a primitive character. Here, then, the biochemical evidence can readily be incorporated into more than one evolutionary pattern, but it cannot guide us in our initial formulation of that pattern.

It remains to remark that ammonia, urea, and uric acid do not constitute the whole sum of invertebrate nitrogenous excretion. Emphasis upon them may reflect their familiar importance in vertebrate excretion, and the interest of the evolutionary speculations with which they have become involved, but other products also merit consideration. For example, part of the nitrogenous waste of animals comes from the metabolism of the pyrimidines and purines of the nucleic acids. The pyrimidines yield ammonia. The purines may be excreted direct, as adenine or guanine, or they may undergo degradation to ammonia along the uricolytic sequence shown in Fig. 12–11. This sequence may be interrupted at any stage, the excretory product being correlated with the general nature of the excretory adaptation, so that ammonotelic forms, for example, may excrete purines as ammonia, whereas uricotelic forms may not. Because of this potential flexibility, it is an over-simplification to regard the term uricotelism as implying the excretion of nitrogen exclusively or even predominantly as uric acid. As far as insects are concerned, uric acid is an important excretory product in most terrestrial forms investigated, but the uricotelism of the group extends to the excretion of allantoin and allantoic acid, which are degradation products of uric acid. Since these resemble uric acid in their relative insolubility and in their nitrogen content, the argument relating nitrogen excretion to habitat is not thereby affected. A more serious departure from the generalization is that the mode of nitrogen excretion may vary at different stages of the life history, but even here the variations may be correlated with habitat; the moist and richly proteinaceous environment in which the larval blowfly (*Lucilia*) flourishes is a case in point (Table 12–2).

The facts so far outlined indicate a close link between the degradation of proteins and purines, and this is of considerable evolutionary interest. It will be

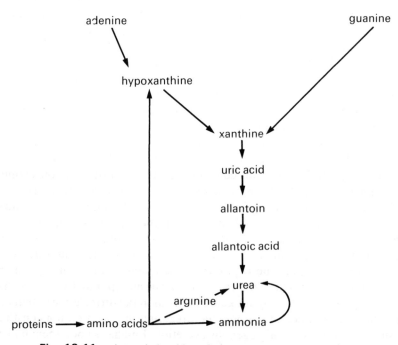

Fig. 12-11. Interrelationships of nitrogenous excretory products.

Table 12-2 Quantity of nitrogen excreted in different nitrogenous end products, expressed as a proportion of the nitrogen in the predominant end product, which is represented as 1.00. Data compiled from various authors by Bursell, 1967. *Adv. Ins. Physiol.* **4**, 33–67.

	Uric acid	Allantoin	Allantoic acid	Urea	Ammonia	Amino acids
DIPTERA						
Lucilia sericata adult	1.00	0.30	—	—	0.30	—
L. sericata pupa	1.00	0.00	—	—	0.15	—
L. sericata larva	0.05	0.02	—	—	1.00	—
HEMIPTERA						
Rhodnius prolixus	1.00	—	—	0.03	—	trace
ORTHOPTERA						
Periplaneta americana	1.00	0.00	0.00	—	—	—
Blatta orientalis	0.64	0.64	1.00	—	—	—

appreciated that the pathways of purine and nucleic acid metabolism are common to all animals. The existence within these pathways of uric acid thus suggests that the evolution of uricotely as an adaptation for terrestrial life may have been based upon the modification of already existing enzyme systems. Similar considerations apply to the ureotely of vertebrates, for urea is formed from arginine through the mediation of arginase, an enzyme that is widely distributed in microorganisms, plants, and animals. Indeed, it may be that this enzyme mediates the formation of urea in molluscs by breaking down the arginine in their diet since these animals lack the ornithine cycle that is responsible for the formation of urea from metabolic nitrogen in vertebrates. In one way and another, we can visualize that specialized modes of excretion might have evolved by the modification of existing metabolic pathways. To this extent aquatic animals may have possessed in their mode of nitrogen metabolism some measure of biochemical pre-adaptation to terrestrial life. We shall find that their mechanisms of ionic regulation also provided them, in marine conditions, with some pre-adaptation towards the exploitation of fresh-water habitats.

But this argument cannot be pressed without reservations. Considering the high degree of elaboration of the crustacean excretory organs (see also Sec. 13–3), it is very remarkable that the terrestrial members, as well as the aquatic ones, have remained ammonotelic. The only fully terrestrial ones are the Oniscoidea, isopods (e.g. *Porcellio*) which live and reproduce on land. Small amounts of uric acid have been identified in them, but their nitrogen is excreted predominantly as ammonia (Table 12–3). There is certainly no justification for attributing the uric acid to the influence of terrestrial life, for a comparable amount is found in the fully aquatic isopod *Asellus*.

It used to be thought that these terrestrial isopods excrete much less nitrogen per unit weight than do their aquatic relatives, and it was thus suggested that their chief adaptation to terrestrial life was a reduction of nitrogen metabolism rather than a modification of it. However, it now appears that they have actually acquired a high degree of tolerance of ammonia, which they release as a gas through their cuticle. The production of the gas was understandably overlooked by earlier investi-

Table 12-3 Partition of nitrogen in excreta of some Crustacea, expressed as percentages of total nitrogen. From Parry, 1960. In *The Physiology of Crustacea*, vol. 1 (T. H. Waterman, ed.). Academic Press, New York.

Species	Habitat	NH_3	Urea	Uric acid	Amino N
Marinogammarus pirloti	Marine	87	0	0	2
Orchestia sp.	Littoral	70	1	0	11
Ligia oceanica	Littoral	83	0	0	6
Porcellio leavis	Terrestrial	57	0	4	1
Gammarus locusta	Brackish water	80	1	0	7
G. pulex	Fresh water	70	9	0	3

gators, and this gave a misleading impression of a reduced nitrogen metabolism. A remarkable feature of this adaptation is a diurnal cycle of release of both water and ammonia through the cuticle, presumably reflecting fluctuations in its permeability. Maximum release occurs when the animals are resting in shelter, which could well be an adaptation to conserve water by reducing its loss when the animals are active and exposed.

It will be appreciated that a change of front has now become necessary, and that we need to replace an explanation of what was thought to be a reduction in nitrogen metabolism with an explanation of why this reduction has apparently not occurred. One suggestion is that by avoiding the need to detoxicate ammonia to urea or uric acid the animals conserve energy which can be more usefully expended in the search for food. But nature may have more surprises in store, and it is evident that there are physiological and ecological issues here which call for further investigation.

We have discussed elsewhere the paucity of terrestrial forms among crustaceans. Whatever explanation of this may be suggested, it is apparent in the present context that not all groups have exploited paths of biochemical evolution that, on theoretical grounds, would seem to have been available to them. Another illustration of unpredictability is seen in the Arachnida. Spiders conserve water by excreting nitrogen predominantly as guanine, which appears as crystals in the Malpighian tubes.

While, therefore, ammonia, urea, and uric acid are the main types of nitrogenous waste material discharged from the animal body, it would be rash to assume that there are no other excretory products to be identified, even though they may be present in only very small quantities. Chromatography makes possible the separation of substances that would never be detectable by classical methods of analysis, and it is sobering to learn that this technique has disclosed the presence of no less than 45 components in the urine of the octopus. It is not surprising, therefore, that published data of the composition of the nitrogenous waste of animals customarily leave a substantial proportion as unidentifiable non-protein nitrogen. The physiological importance, if any, of these unknowns has not yet been determined; as Ramsay remarks, the imagination recoils from contemplating the magnitude of the effort required to attack the problem. In the meantime, however, our knowledge of nitrogen excretion is sufficient to show that this process, particularly in aquatic invertebrates, is not enough in itself to account for the complexity of the various types of excretory organ. It is this that brings us to consider the problems of osmotic and ionic regulation.

13
Osmotic and Ionic Regulation

13-1 IONIC REGULATION IN MARINE ANIMALS

Studies of the osmotic pressure of body fluids are favoured by the comparative ease with which the osmotic pressure of a small volume of fluid can be determined by measuring the depression of its freezing point (expressed as Δ) below that of water. Data obtained for the body fluids of a wide range of organisms show that the great majority of marine invertebrates have an internal osmotic pressure that agrees very closely with that of the sea water in which they live. This is one reason why we believe that life originated in the sea, or at least became associated with it at an early stage, for it suggests that protoplasm may at one time have been adjusted to carrying on its metabolic activities while being directly bathed by sea water.

Osmotic relationships, however, are not the only ones that have to be considered in this context; the ionic composition of the body fluid is no less important. Information regarding this aspect has been slower to accumulate, but much accurate information has now been secured as a result of the development of flame photometry and other techniques adapted for the chemical analysis of the small quantities of fluid available. Earlier work had shown that there was a general resemblance between the ionic composition of the sea and of the body fluids of animals, a resemblance indicating, like the osmotic data, a close relationship between all living organisms and the marine environment. At one time, indeed, it was suggested that the body fluids were essentially sea water. However, marine invertebrates, despite being usually isosmotic with the sea, maintain an ionic composition of their body fluids markedly different from that of normal sea water. This is even true of the coelenterates, for fluid from the mesogloea of *Aurelia* has significantly more potassium and less sulphate than the surrounding sea water. Other groups of the less highly organized invertebrates show a similar state of affairs; echinoderms have high potassium values, as also do polychaetes, while the latter may show a reduced sulphate content, like *Aurelia*.

When these differences were first demonstrated it was argued that the body fluids of present-day animals might reflect the chemical composition of the sea as it was during earlier periods of evolution. This view, however, is not supported by geological evidence, which suggests that there has been very little change in composition, and certainly not enough to account for the observed differences. Consequently the ionic peculiarities of body fluids are now regarded as being of adaptive significance, and there is good reason for believing that they were established at a very early stage of evolution. Nevertheless, there is sufficient general resemblance between the ionic composition of the sea and of body fluids to justify a belief that the composition of the latter was established in a marine environment.

That these differences are the result of active regulation on the part of the animals can be demonstrated by comparing the composition of freshly drawn plasma or coelomic fluid with that of a similar sample that has been dialyzed against sea water, and which is thus in passive equilibrium with the latter. The composition of this dialyzed sample will differ from that of the sea, because body fluids contain large quantities of protein, the indiffusibility of which gives rise to Donnan effects, and calcium ions may form insoluble complexes with them. But its composition differs from that of the freshly drawn fluid—a difference that must result from the active regulation that goes on in the living body (Table 13–1). This phenomenon is known as ionic regulation, which we may define, following Robertson, as the maintenance in a body fluid of concentrations of ions differing from those that would result from a passive equilibrium with the external medium.

Ionic regulation of the body fluids, while always apparent, is comparatively limited in the lower invertebrates, but in the most highly organized ones, such as the decapods and the cephalopods, it may extend to every ion. On the whole, calcium and potassium tend to be more concentrated in the extracellular body fluids than in the external medium, while magnesium and sulphate are less concentrated; an increase in chloride compensates for the reduction in sulphate. We must conclude, therefore, that the tissues of marine invertebrates require a medium of specialized ionic composition. Since this situation is detectable in coelenterates, it was probably established at an early stage of evolution.

From the point of view of cell function the composition of the body fluids is merely a means to an end, for the cells have to maintain their own specialized ionic composition. Thus the excitability of nervous and muscular tissues depends upon the maintenance of a high internal concentration of potassium, some 20–50 times that of the ambient fluid. Sodium and chloride, by contrast, are relatively dilute, their concentrations outside the cells being, respectively, 3–15 and 5–50 times that within. This distribution of ions, which is responsible for the potential drop across the cell membrane known as the resting potential, appears to depend upon the resting membrane being moderately permeable to potassium and chloride, and relatively impermeable to sodium. Such sodium as does pass in is pumped out by an active metabolic process, essentially secretory in nature, and called the sodium pump. When these cells are excited by the application of a stimulus, there is a change in the permeability relationships, which leads to sodium entering the cell and potassium passing out; the immediate result is the establishment of a reversed potential called the action potential, the resting conditions being rapidly restored again during a recovery period. This principle of operation is certainly very widespread in animals,

Table 13-1 Ionic regulation in some marine invertebrates

	Concentrations in plasma or coelomic fluid as percentage of concentration in body fluid dialyzed against sea water					
	Na	K	Ca	Mg	Cl	SO_4
COELENTERATA						
Aurelia aurita	99	106	96	97	104	47
ECHINODERMATA						
Marthasterias glacialis	100	111	101	98	101	100
TUNICATA						
Salpa maxima	100	113	96	95	102	65
ANNELIDA						
Arenicola marina	100	104	100	100	100	92
SIPUNCULOIDEA						
Phascolosoma vulgare	104	110	104	69	99	91
ARTHROPODA						
Maia squinado	100	125	122	81	102	66
Dromia vulgaris	97	120	84	99	103	53
Carcinus maenas*	110	118	108	34	104	61
Pachygrapsus marmoratus†	94	95	92	24	87	46
Nephrops norvegicus	113	77	124	17	99	69
MOLLUSCA						
Pacten maximus	100	130	103	97	100	97
Neptunea antiqua	101	114	102	101	101	98
Sepia officinalis	93	205	91	98	105	22

*Webb (1940). †This grapsoid crab is the only animal in the table which is hypoosmotic (ionic concentration 86% that of sea water). From Robertson, 1957. In *Recent Advances in Animal Physiology* (B. T. Scheer et al., eds.) University of Oregon Publications, Eugene, Oregon.

and may even be universal. It seems possible, therefore, that ionic regulation may have been established very early in evolution as a mechanism for securing the essential requirements of irritability and response. But even if this were so, it would still be difficult to account for the differences in pattern of ionic regulation between one species and another. It has been suggested that the narcotic action of the magnesium ion explains the low concentration of this ion in the blood of some crustaceans, but against this must be set the high level of magnesium in the very active *Loligo*. Probably it is asking too much to seek explanations that can be generalized to cover such widely separated groups.

In one field of adaptation, however, we have an explanation for highly specialized ionic regulation. As we know from our own experience, the freedom of movement of animals in the sea is greatly restricted unless they can achieve neutral buoyancy by bringing the density of their bodies to about that of sea water. The substitution of lighter ions for heavier ones is a way of achieving this, and is probably the explanation of the reduction of sulphate ions in medusae. It has been calculated that the amount of lift given by this is about 1.5 mg/ml of body fluid: this

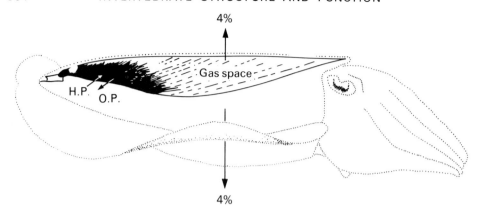

Fig. 13-1. Diagram summarizing our knowledge of the cuttlebone. The one shown here has a density of about 0.6. Liquid within it, which is marked black, almost fills the oldest and most posterior chambers. If they were filled with gas, this would tend to tip the tail of the animal upwards. The newest 10 or so complete chambers, which lie centrally along the length of the animal, are completely filled with gas. These chambers can give buoyancy without disturbing the normal posture of the animal. The hydrostatic pressure (H.P.) of the sea is balanced by an osmotic pressure (O.P.) between cuttlebone liquid and the blood. In sea water the cuttlebone gives a net lift of 4% of the animal's weight in air and thus balances the excess weight of the rest of the animal. From Denton and Gilpin-Brown, 1961. *J. mar. biol. Assoc. U.K.*, **41**, 365-381.

may seem a modest gain, yet, given the low proportion of organic matter in these animals, the effect on their buoyancy must be substantial.

Other groups use other techniques for exploiting osmotic regulation to the same end. For example, cranchid squids, such as *Cranchia scabra*, have a remarkable degree of buoyancy which enables them to remain almost motionless in sea water. This faculty depends on the possession of an enormous coelom, the fluid contents of which constitute about two-thirds of the total weight of the body. The specific gravity of the coelomic fluid (1.011) is lower than that of sea water (1.012), so that the animals gain about 15 mg of buoyancy from each millilitre of the fluid. This gain results from exploitation of the high level of ammonia production of cephalopods, a level which is probably associated with their carnivorous habits. In these particular squids the coelomic fluid is isosmotic with sea water, but the usual cations have been largely replaced by lighter ammonium ions. These accumulate to the extent of providing some four-fifths of the total cation content, and thereby lower the specific gravity of the fluid.

Another example, illustrating a different cephalopod solution of the same problem, is seen in *Sepia*, an animal so buoyant that it can poise without effort in the water as it stalks its prey. Here the chambered cuttlebone serves as a variable-buoyancy tank (Fig 13-1). Its chambers are filled in part by gas (largely nitrogen, always at a pressure of about 0.8 atm) and in part by fluid, and the proportions of these can be varied. Thus, writes Denton, just as a submarine commander controls the buoyancy of his ship by varying the water content of its tanks, so the cuttlefish varies its density by varying the mass of liquid in its cuttlebone, increasing the mass and reducing the gas content at greater depths. But it cannot do this simply by opening and closing a cock. How, then, does it prevent hydrostatic pressure from flooding the cuttlebone with water? The answer is that in animals near the surface the fluid approximates to sea water in its osmotic concentration, but at greater depths the concentration is lowered by ionic movements. Thus the cuttlebone fluid becomes markedly hypotonic to sea water and also to the blood. The effect of this is to create an osmotic force which balances the tendency of the increased hydrostatic pressure to push water into the gas spaces of the bone.

The device of the cuttlebone recalls to some extent the hydrostatic function of the swimbladder of teleost fish, but the mechanism differs in one important respect, apart, of course, from its anatomical basis. Fish rely upon the swallowing of air, or on the secretion of gas into the bladder, for adapting to surface waters, and the response is a rapid one. In *Sepia*, by contrast, the gas plays a passive role, merely diffusing into the spaces left by the movements of fluid. Because of this it is slower to equilibrate than a swimbladder, and consequently tends to conserve energy by avoiding rapid adaptation during brief excursions to the surface. As so often, cephalopod organization compares favourably with that of vertebrates (p. 146).

13-2 THE MAMMALIAN NEPHRON AS A MODEL

We have noted that excretory organs are often closely involved in the flux of water and electrolytes between the organism and its environment, and we must now examine the contributions which they make to ionic regulation. It will become apparent, however, that they are only one element in the regulatory system. Adaptations affecting surface permeability and the active uptake of ions, for example, make essential contributions to regulatory systems which show, in their complexity, the close interrelationship of the component parts of living organisms. So far as the excretory organs of invertebrates are concerned, our understanding has been profoundly influenced by investigations of the mode of functioning of the vertebrate kidney tubule, and by the techniques used for these. Here, of course, we have an illustration of the ways in which the underlying unity of living systems allows studies of one group of animals to shed much light on the organization of quite different groups. It will thus be convenient to recall in very general terms what are the steps involved in the production of urine in mammals (Fig. 13-2).

The process begins in the renal corpuscle, with ultrafiltration from the glomerulus into the lumen of Bowman's capsule. The ultrafiltrate resembles the blood plasma in its composition, except for the virtual absence of protein, most protein molecules being unable to pass through the filter. In addition, the distribution of ions can be affected by Donnan effects, and by binding to plasma proteins. The pressure responsible for ultrafiltration is the force exerted by the blood pressure, which must be sufficient to overcome the osmotic pressure of the plasma proteins, and also the back pressure in the kidney tubules. The filter itself is formed by the wall of the glomerulus, which consists of squamous endothelium, basement membrane, and the lining epithelium of the capsule. Electron microscopy shows that there are perforations in the endothelium of the filter, ranging in size from 50–100 nm. It also shows that the epithelial cells (podocytes) on the other side of the membrane have arms and secondary processes that inter-

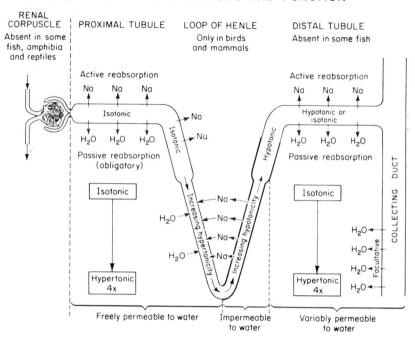

Fig. 13-2. Mechanism of urine formation in the mammal. After Pitts, 1959. *The Physiological Basis of Diuretic Therapy*, Thomas, C. C. Springfield, Ill. From Hoar, 1966, *General and Comparative Physiology*. Prentice-Hall, Englewood Cliffs, New Jersey.

digitate with each other on the basement membrane, and leave tortuous free areas for the passage of the ultrafiltrate.

The composition of the fluid so produced is modified during its passage along the tubule of the nephron, largely by resorption of water and physiologically important solutes such as glucose and sodium chloride; secretion into the fluid also plays an important part. Resorption, which involves the movement of solutes against the concentration (electrochemical) gradient, depends on active transport, and thus demands energy. Active transport is particularly characteristic of the proximal tubule, where it is associated with the presence of epithelial cells possessing a brush border, in which alkaline phosphatase is concentrated; here the movement of water is passive, accompanying the active reabsorption of sodium, which moves with its accompanying chloride. Glucose and other substances are also taken up at this stage, but the fluid remains isosmotic with the plasma. Factors which we cannot here consider, including counter-current multiplication, cause changes in tonicity of the fluid as it passes along Henle's loop and the first part of the distal tubule. Further resorption occurs distally and in the collecting ducts, and here the uptake of water is variable, being regulated by the antidiuretic hormone released from the pituitary gland.

The interrelationship between ultrafiltration, resorption, and secretion can be studied by making use of certain test substances which have clearly defined behaviour. One of these is inulin, a plant polysaccharide with a molecular weight of 5200.

This passes into the ultrafiltrate, but is neither secreted nor resorbed by the tubule. Thus the ratio of its concentration in the urinary fluid to its concentration in the blood (U/B ratio) will be unity in the capsule, and it will remain so throughout the tubule provided that water is neither added to nor removed from the fluid. Passage of water into the fluid will lower the ratio, while removal of water into the blood will increase it.

Many of the substances in the ultrafiltrate are, however, resorbed wholly or in part. Their U/B ratios in the finally formed urine will then depend upon the relative movements of these solutes and of water. In the extreme case of glucose, which is totally resorbed in normal conditions, the ratio falls from unity in the ultrafiltrate to zero in the tubule fluid. If, however, the animal is treated with phlorizin, which abolishes glucose uptake by the tubular epithelium, the U/B ratio will remain the same as for inulin. Finally, substances such as phenol red or para-aminohippuric acid, which are both filtered and secreted, will give U/B ratios greater than that of inulin.

Inulin studies provide a convenient baseline for measurements of kidney activity, provided that the initial plasma concentration of inulin is known. For example, the quantity of inulin excreted in a given time is a measure of the filtration rate. Alternatively, this can be expressed as the inulin clearance rate, which is the volume of blood that must be filtered in unit time to yield that amount of inulin. It follows from what has been said above that the ratio of the clearance rate of other solutes to the clearance rate of inulin shows to what extent tubular activity is involved in the handling of those substances.

13-3 SOME PRINCIPLES OF INVERTEBRATE RENAL FUNCTION

Our present understanding of the functioning of invertebrate excretory organs rests upon elegant studies of a number of invertebrate genera. Favoured techniques, as with vertebrate studies, have been the removal of minute quantities of fluid by micropuncture, the chemical analysis of these samples (by flame photometry, for example), and the use of radioisotopes to determine the patterns of ionic flux. For obvious reasons information is derived mainly from the larger molluscs and arthropods, and even here the number of species investigated is still too few to form a satisfactory basis for generalization. Thus, it is well to beware of confident assertions that 'most' members of a group excrete in a particular way. On the other hand, it is fair to remember that where morphological studies show resemblances in details of organization, it is a reasonable, if provisional, assumption that the organs concerned will function along similar lines. What has emerged very clearly, however, is that the functional organization of invertebrate excretory organs resembles that of the vertebrate kidney in many respects. Understandably enough, even if vexatious to the comparative anatomist, the invertebrate organs are increasingly referred to as kidneys.

A convenient starting point for our analysis is that there is a continuous outflow of water through the excretory organs of crabs and lobsters. This outflow is demonstrable by closing the excretory apertures. The result of this is complicated by the possibility that closure may reduce filtration by increasing pressure within the excretory organ, but it can certainly lead to fluid accumulating in the animal to the extent of some 5% of the body weight in 24 hr. Two main factors are concerned in the maintenance of this outflow: the permeability of the body surfaces (particularly the respiratory ones) and the activity of the excretory organs. These organs, however, are not merely passive ducts. The fluid passing through them is usually isosmotic with the blood, but if its composition is compared with that of the blood plasma it is found that there are important differences. As Table 13–2 shows, sulphate is passed out in excess of its concentration in the blood; sodium, potassium, and calcium, however, are clearly being conserved, for their concentrations in the excretory fluid are less than their concentrations in the blood.

By thus controlling the removal of ions from the body the excretory organs contribute to the regulation of the composition of the body fluids, but in doing so they inevitably bring about a loss of both water and salts, and this loss has to be made good. Because of this there is a continuous entry of water into the body, probably to some extent through the gut epithelium, but mainly through the permeable external surfaces of the body, particularly those of the respiratory organs. How this inflow of water is contrived in animals that are isosmotic with the sea is by no means clear, but in any case it is only part of the problem, for ions must also be taken up. Some of these will be entering against the concentration gradient, for, as we have seen, certain ions are maintained internally at a concentration higher than that in which they are found in the sea. Their passage into the body must therefore be due to active transport (Sec. 2–4). This is evidently an essential part of ionic regulation, and it must therefore have been established early as a fundamental property of living organisms.

Table 13–2   Antennal gland secretion and renal sac fluid compared with blood plasma

	Concentrations as percentages of plasma values (on water-content basis)					
	Na	K	Ca	Mg	Cl	SO$_4$
CRUSTACEA (Antennal gland secretion)						
Maia squinado	100	98	99	109	101	214
Cancer pagurus	97	81	90	125	96	134
*Carcinus maenas**	95	78	94	390	98	224
Palinurus vulgaris	98	65	86	137	101	98
Homarus vulgaris	99	91	64	180	101	159
Nephrops norvegicus	98	83	81	130	101	106
Palaemon serratus†	82	86	95	670	106	380
CEPHALOPODA (Renal sac fluid)						
Eledone cirrosa	102	90	87	89	97	136
Sepia officinalis	79	50	70	68	100	215

*Webb (1940).   †Parry (1954), on ml basis. *Palaemon* shows hypo-osmotic regulation, but 'urine' is iso-osmotic with blood.   Remainder of analyses from Robertson (1939, 1949, 1953). From Robertson, 1957, *op. cit.*

One point of resemblance between vertebrate and invertebrate excretory processes is provided by ultrafiltration, which occurs in the excretory organs of many invertebrates, just as in the nephron of vertebrates. Little is known about nephridia, and present evidence again comes primarily from molluscs and crustaceans. The insects, as we shall see later, are a special case. Part of the evidence derives from studies of the excretion of inulin and glucose. Injection of inulin is followed by its appearance in the urine of several species, including marine forms (*Octopus*, *Homarus*), fresh-water forms (*Anodonta* and the crayfish), and terrestrial forms (the pulmonate snail *Achatina*). On the assumption (known to be justified at least for the lobster) that inulin is not metabolised in invertebrates, this is presumptive evidence for ultrafiltration.

The primitive filtration site in molluscs was probably the heart, as it still is in *Anodonta* (Fig. 13-3). If the pericardial membrane is punctured in this animal, fluid (which would normally pass on into the kidney through the renopericardial opening) can be continuously drained from the cavity. This location of the site favours ultrafiltration, for the fluid passes out through the wall of the heart, exactly where blood pressure is maximal. However, filtration sites in molluscs are not always the same as this. Certain other molluscs, among them the fresh-water prosobranch *Viviparus*, also filter into the pericardial cavity, but in *Octopus* filtration takes place from appendages of the branchial hearts into the small coelomic spaces that surround them (Fig. 13-4). Terrestrial pulmonates are different again, for in these animals filtration takes place from the renal vein (Fig. 13-5). The pericardial cavity is not involved, and in consequence no fluid can be drained from it. Finally, to take an arthropodan example, the site of filtration in the crayfish is the coelomic end-sac of the antennal gland (Fig. 13-17, p. 378). Evidence conformable with this conclusion comes from electron micro-

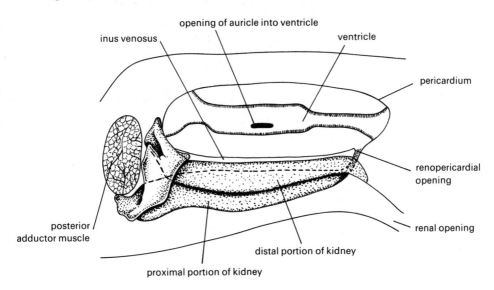

Fig. 13-3. Reno-pericardial relationships in *Anodonta*. From Potts, 1967. *Biol. Rev.*, **42**, 1–41.

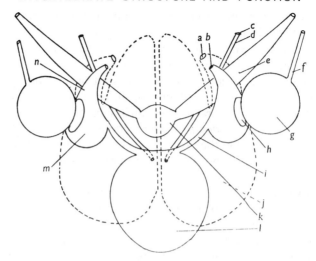

Fig. 13–4. Diagram showing the relationships between the excretory system, circulatory system and derivatives of the coelom in the octopus. *a*, Urinary pore; *b*, reno-pericardial canal opening into the renal sac; *c*, gonopore; *d*, gonoduct; *e*, auricle; *f*, afferent branchial vein; *g*, branchial heart; *h*, branchial heart appendage; *i*, aquiferous canal; *j*, renal sac; *k*, ventricle; *l*, gonadial coelom; *m*, pericardial cavity; *n*, reno-pericardial canal. From Harrison and Martin, 1965. *J. exp. Biol.*, **42**, 71-98.

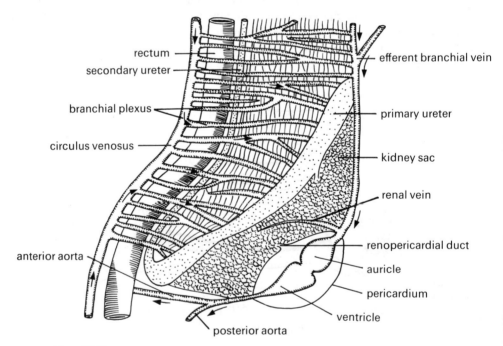

Fig. 13-5. Kidney of the pulmonate *Helix pomatia*. The circulus venosus collects blood from all parts of the body and passes it into the branchial plexus from whence it drains either into the efferent branchial vein or into the kidney. Blood from the kidney is collected by the renal vein which joins the efferent branchial vein as it enters the auricle. From Potts, 1967. *op. cit.*

scopy, which shows that the lining cells are podocytes, resembling those of the vertebrate glomerulus.

Further evidence of the occurrence of ultrafiltration in molluscs is obtained from chemical analyses of the blood and pericardial fluid. In *Viviparus* the two fluids resemble each other very closely (Table 13-3); Donnan effects are negligible, but there is some indication of calcium binding to the blood proteins. Further confirmatory evidence is that urine production is influenced by pressure. In *Achatina* the flow of urine can be abolished by applying a back pressure equivalent to 12 cm of water, while in *Octopus* the flow can be shown to vary with the blood pressure. In general, however, values for the hydrostatic and colloid osmotic pressures of invertebrate blood systems are scanty and unreliable, so that there is little direct evidence that pressure relationships are adequate to effect filtration.

The second respect in which molluscan and crustacean excretory mechanisms resemble those of vertebrates is a necessary corollary of the occurrence of ultrafiltration. Physiologically valuable solutes that pass through the filter must, as far as possible, be returned to the blood, and there is good evidence that resorption does indeed occur. Both glucose and sodium have been shown to be resorbed by active transport in the tubule of the antennal gland of crustaceans, and in the renal sac of *Octopus* and *Haliotis*. Chloride is also resorbed, but this may simply accompany the sodium, although there is some evidence for its independent uptake.

Particularly clear evidence comes from glucose studies. It is significant that this substance, although commonly present in the blood of invertebrates, is normally absent from their urine. The mechanisms responsible for the active transport of glucose has been shown to be energy-dependent, and, as in vertebrates, to be associated (in crayfish, for example) with cells possessing a brush border with alkaline phosphatase. Confirmatory evidence is provided by injecting animals with phlorizin. Again, as in vertebrates, this completely blocks resorption of glucose, so that its clearance becomes similar to that of inulin. In the octopus, for example, most of the glucose resorption occurs in the renal sac (Fig. 13-4), but some does take place in the pericardial cavity; as may be seen from Fig. 13-6, this resorption is abolished by phlorizin, again exactly as in vertebrates.

Furthermore, and paralleling the situation in diabetic mammals, injection of glucose into invertebrates, with consequent elevation of the blood sugar values, results in this substance being lost in the urine as a result of saturation of the active transport mechanism. Estimates of the threshold values (values at which glucose begins to appear in the urine) suggest 100 mg glucose per 100 ml of blood in the

Table 13-3 Composition of medium, blood, and urine in *Viviparus viviparus* from stream water (mM/l ± S.E.). From Potts, 1967, *op. cit.*

Iion	Medium	Blood	Pericardial fluid	Kidney-sac fluid	Final urine
Na	2.5	32.5 ± 0.8	32.0 ± 1.5	13.0 ± 1.7	9.0 ± 1.7
Ca	3.3	5.8 ± 0.6	4.8 ± 0.2	2.5 ± 0.5	1.5 ± 0.02
Cl	8.0	29.0 ± 1.0	29.0 ± 1.2	13.0 ± 1.1	10.0 ± 1.2
Δ as NaCl	5.0	38.5 ± 1.4	37.0 ± 1.5	13.5 ± 2.1	7.0 ± 1.8

362 INVERTEBRATE STRUCTURE AND FUNCTION

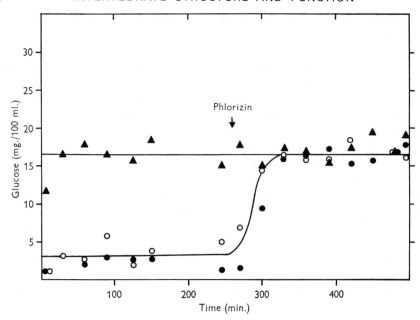

Fig. 13-6. The relationships of the normal concentration of glucose and pericardial fluid in the octopus before and after the administration of phlorizin. ▲, blood; ●, right pericardial fluid; ○, left pericardial fluid. Phlorizin given from 245 to 450 min at a dosage of 10.3 mg/kg. From Harrison and Martin, 1965. *op. cit.*

lobster and 200 in the crayfish. These figures may be compared with the threshold value in man, which is 180 mg.

Evidence for the resorption of ions is given, as already explained, by the U/B ratios. For example, the ratio for sodium may fall to 0.6 in fiddler crabs (*Uca*) kept in concentrated sea water. Theoretically, such values could be obtained through the secretion of water into the kidney tubule, but inulin studies give no evidence for this, since the U/B ratios for this substance do not fall below 1, although they may rise as high as 30 in some species. This indicates water absorption accompanying the absorption of electrolytes, with the water moving along the osmotic gradient. There is no convincing evidence for active uptake of water in these types of excretory organ (but see p. 386).

Finally, there is evidence for tubular secretion, which we have seen to be the third element in mammalian nephron functioning. The evidence is partly inferential, derived from analysis of the composition of normal excretory fluid, but it comes also, and particularly clearly, from studies of the excretion of dyes. For example, the injection of phenolsulphonphthalein into the octopus is followed by the appearance of this substance in the urine, eventually to give a U/B ratio of more than 100, in specimens in which the inulin ratio is 1. However, the ratio for phenolsulphonphthalein depends upon the blood concentration. At high concentrations (exceeding 3 mg/100 ml) the ratio is near to unity, and it is at low concentrations that the higher values are obtained. The inference is that phenolsulphonphthalein is removed by filtration at high blood concentrations, but by secretion at low concentrations.

These examples will serve to show how very closely the functioning of crustacean and molluscan excretory organs resembles that of the mammalian kidney. Of course, anatomical and histological resemblances have long encouraged the belief that functional analogies between these several organs must exist, but it is doubtful whether any morphologist would have hazarded the suggestion that the resemblances would be quite so close. In part this situation reflects the principle that, given the limitations of biological patterns of structure, similar ends must often be attained in similar ways. But we are justified in suspecting also that there may well be great biological advantage in the filtration/resorption mechanism. What is the advantage likely to be?

Kirschner suggests that it lies partly in the flexibility of the system, and partly in its evolutionary potentiality. Ultrafiltration makes possible the removal of unwanted or harmful substances, as is seen in the fate of inulin. Secretion by itself could not compete with this facility, because the secretory epithelium could not be expected to have adapted to the removal of molecules that are not commonly encountered. There is, however, a disadvantage in ultrafiltration: resorption must be established in parallel with it, in order to avoid the loss of valuable material, and this can be metabolically expensive. We shall see later that this becomes a particularly important consideration in fresh-water animals, but it is a much less serious one in marine forms, which live in a medium with no lack of essential ions. It is, of course, significant that it was in the sea that the organs that we are considering must first have evolved.

As for the evolutionary potentiality of the filtration/resorption mechanism, this lies in the possibility of modifying the ultrafiltrate in its passage through the excretory system by applying absorptive and secretory processes to it. This adaptability makes possible the establishment of means of diluting the urine or of concentrating it; devices that can help to resolve the problems of controlling water and electrolyte fluxes in fresh-water and on dry land, and which thus provide some pre-adaptation for the exploitation of these environments. These aspects will be examined later.

At this stage, however, it is worth reflecting that the study of invertebrate excretory organs sheds indirect light on the problem of the origin of vertebrates. The vertebrate glomerulus provides a means of removing the excess water that enters the body along the osmotic gradient in fresh-water species. In marine fish, however, it is a disadvantage, because their hypotonicity results in water moving outwards, so that they need to conserve it, which they do in part by reduction of the glomerulus. Recognition of these facts led at one time to the suggestion that vertebrates arose in fresh water, and that ultrafiltration through the glomerulus was an adaptation to fresh-water habitats, to provide for removal of the water drawn in by osmotic action. But there is now good palaeontological evidence that vertebrates arose in marine or estuarine conditions, and that their fresh-water stage was a later one, achieved after their kidney mechanism had already been established.

The comparative study of invertebrate mechanisms shows us that this argument is plausible, for ultrafiltration in marine invertebrates cannot possibly be a product of fresh-water ancestry. We have already hinted that their kidneys provide a regulatory mechanism that may have facilitated entry into fresh water. So also, no doubt, with the vertebrates. But whereas the major invertebrate groups provide us with abundant evidence of the value of filtration/resorption mechanisms in marine conditions, the evidence from vertebrates is almost non-existent, being confined to the myxinoids (hag-fish). These marine animals, isosmotic with the sea, and regulating their body fluids along lines similar in principle to the mechanisms of marine invertebrates, provide us with evidence of what was probably the primitive mode of functioning of the vertebrate kidney. Here is one respect in which there is no sharp physiological separation of invertebrates from vertebrates.

13-4 PROTOZOA AND FRESH-WATER LIFE

The passage to fresh water has proved to be well within the capacity of protozoans, a fact so familiar that it is easily taken too much for granted. In view of what has been said above regarding the dependence of protoplasm on conditions comparable with those of a marine environment, we might expect the internal osmotic pressure of fresh-water forms to be substantially above that of the external medium. Measurements of their volume changes in solutions of different molarity have given direct evidence that this is so.

For example, if *Amoeba proteus* is placed in various lactose solutions it is found that the weakest concentration in which a decrease in its volume can be detected is 0.005 M, which suggests that this is about the value of its internal osmotic pressure. Hence there must be an osmotic flow of water into its body when it is in fresh water, and presumably this is true of other fresh-water protozoans as well. Even if the frequent presence of a cuticle reduces the permeability of their surface, and we cannot be sure that it does, the animals must still maintain nutritive, respiratory, and excretory relationships with the external medium. We may expect them to accumulate water as a result of their own metabolism, of uptake in food vacuoles, and of their surface permeability. Of these three factors the last must present a particularly acute problem to animals which, because of their minute size, have an enormous ratio of surface area to volume. As for ionic regulation, we know virtually nothing of this in Protozoa, but we have earlier seen that protozoans possess contractile mechanisms that may well be similar in principle to those of the Metazoa. If this is indeed so, we may expect these mechanisms to depend upon the regulation at least of the movements of potassium and sodium ions, although at present this is no more than supposition.

It is evident, even from these limited considerations, that fresh-water Protozoa must certainly be able to control the water content of their bodies. This is probably one of the functions, and possibly the primary one, of the contractile vacuole. This structure, which accumulates fluid and periodically discharges it, is a well-known feature of most fresh-water protozoans. It has often been said that it is restricted to these, but this is not so, for it occurs in many marine flagellates and ciliates, although it is usually absent from parasitic forms. It is found also in the motile stages of some fresh-water algae, and in some of the cells of fresh-water sponges.

The functional importance of this organelle is indicated by its high level of organization. In *Amoeba* (Fig. 13-7) it has a simple spherical form, with a bounding membrane that is about 0.5 μm thick. The presence of many mitochondria around it indicates that it is involved in a high level of metabolic activity. During its period of enlargement (diastole) it is carried forwards in the plasmasol; small vacuoles have

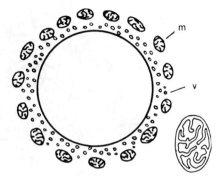

Fig. 13-7. The contractile vacuole and associated structures. The vacuole is enclosed in a membrane having a triplet structure similar to that of the external membrane and is surrounded by mitochondria (*m*). Between the layer of mitochondria and the membrane are found vast numbers of small vacuoles (*v*), which presumably burst into the main vacuole. The small vacuoles appear to form spontaneously in the space between the mitochondria and the vacuolar membrane. The smaller figure shows the arrangement of the internal membranes of a mitochondrion. From Mercer, 1959. *Proc. R. Soc. B*, **150**, 216-237.

been observed to contribute to it at the beginning of this phase. Eventually it becomes enclosed in the plasmagel, and may then be left in a posterior position, but discharge (systole) soon takes place through a temporary pore in the surface, and a new vacuole then starts to grow in the same region. In protozoans with a more defined body form the contractile vacuole has a fixed position and structure. In *Euglena*, for example, it opens into an anterior invagination, the flagellar pocket. Here a group of small vacuoles contributes to form the main one, a procedure not fundamentally different from that in *Amoeba*. In the ciliates, in conformity with their generally advanced organization, the contractile vacuoles (commonly more than one) are more complex. Fibres may be associated with them, and they are fed by a complex and fixed system of secondary channels, which extend considerable distances through the body as minute vacuoles (Figs. 13-8, 13-9).

The underlying mechanism of these contractile systems is little understood. The fluid cannot enter by filtration, as it is often believed to do in the excretory organs of metazoans, because the contractile vacuole is only connected to the outside medium at systole. For this reason, the internal hydrostatic pressure could not be relieved by filtration of fluid into the vacuole, since the total volume within the limiting cell membrane would not be reduced. It is therefore suggested that during diastole there is active secretion of water through the vacuolar wall, and that systole is effected by contraction of the wall. Its thickness suggests that a layer of structural

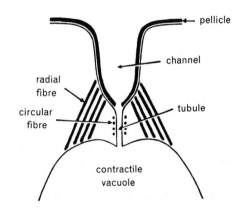

Fig. 13-8. Diagram showing the complex structure of the duct of the contractile vacuole in the suctorian *Tokophrya*. The existence of the circular fibres is highly probable but not yet definitely established. After Rudzinska, from Grimstone, 1961. *Biol. Rev.*, **36**, 97-150.

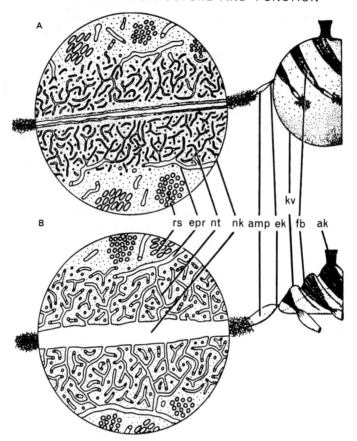

Fig. 13-9. Schematic drawings of the contractile vacuole apparatus in *Paramecium* showing main vacuole to right and one radial canal, with enlarged circular inset, to left. Upper drawing (A) shows radial canal in systole, main vacuole in diastole; lower drawing (B) shows radial canal in diastole, main vacuole in systole. *ak*, discharge canal; *amp*, ampulla of radial canal; *ek*, injector canal; *epr*, endoplasmic reticulum, showing continuity of this system with *nt*, 'nephridial' tubules forming a sponge around *nk*, 'nephridial canal'; *fb*, bundles of fibrils in vacuole wall; *kv*, main contractile vacuole; *rs*, clusters of membranous tubules. From Schneider, 1960, and from Pitelka, 1963. *Electron Microscope Studies of Protozoa*. Pergamon Press, Oxford.

protein might well be present in it, and it has been observed that it is weakly birefringent during diastole but that this property disappears at systole. This behaviour could be interpreted in terms of the folding or displacement of protein molecules such as has been postulated as a basis for amoeboid movement.

Early observers of contractile vacuoles compared them with the kidneys and analogous organs of higher forms, and regarded them as being organelles of nitrogenous excretion, but this interpretation no longer seems plausible, at least as an explanation of the primary function of the vacuoles. There is direct evidence that protozoans excrete their nitrogen as ammonia, and perhaps to some extent as urea, and it seems likely that these substances could be readily lost through the body surface. The favoured view is that contractile vacuoles are primarily osmoregulatory.

This does not destroy, of course, the validity of comparisons drawn between them and excretory organs, for the latter, too, are by no means restricted in function to nitrogenous excretion, even in the most advanced animals. Moreover, we have seen that the first appearance of excretory organs may well have been conditioned primarily by the need for regulating the flux of water and ions.

Belief in the osmoregulatory function of contractile vacuoles has rested partly upon the fact that these organelles are particularly characteristic of fresh-water forms; the conclusion may thus seem somewhat weakened by their presence in many marine ones. The weakness is not, however, serious, for the presence of protonephridia in so many of the lower marine invertebrates sufficiently shows that there is need for such regulatory organs (see below) in the sea as well as in fresh water, and we have seen good reasons why this should be so.

In any case, there is experimental evidence to support the osmoregulatory interpretation. For example, the rate of output of the vacuoles can often be decreased by increasing the osmotic pressure of the medium, and increased by decreasing it. The output from marine peritrichous ciliates increases as much as 70–80 times when these animals are immersed in 10% sea water. Moreover, new contractile vacuoles appear in marine, and also in parasitic, forms if these are subjected to osmotic pressures lower than those to which they are adjusted. Then again, the addition of cyanide to the external medium suppresses the activity of the vacuoles, presumably because of its inhibitory action on respiration; this results in an increase in the volume of the body. There are good grounds, then, for concluding that one of the many functional resemblances of Protozoa to Metazoa is the use of devices for solving the osmotic problems encountered when animal life, adjusted to maintaining itself in a marine environment, began to exploit fresh water.

We have mentioned above that contractile vacuoles occur also in fresh-water sponges. Their presence in these animals was first reported over 100 years ago and was confirmed more recently by Jepps, who found them in the amoebocytes and choanocytes of finely teased fragments of the fresh-water sponges *Ephydatia* and *Spongilla*. Two or three of them exist near the base of the collar of the choanocytes, where they are said to be particularly large in 'that revealing moment just before death'. Significantly, in view of what we have said of their importance in fresh-water Protozoa, they have not been positively identified in any marine sponge. We can only surmise that these vacuoles are fulfilling, in the sponge body, functions similar to those that we have attributed to them in Protozoa. Their presence is a striking indication of the particularly close association that is held to exist between the two groups. They have not been reported in any other metazoan; not even in the coelenterates, where, in the absence of any differentiated excretory or osmoregulatory organs, we might have expected to find them.

This lack is perhaps associated with the almost exclusively marine distribution of coelenterates. Penetration into fresh water has been achieved by only a very few species; some medusae are included among these, but the best-known examples are *Hydra* and its relatives. *Hydra* is a good example of the way in which familiarity can blind us to the unexpected, for, with no obvious sign of surface protection or of specialized osmoregulatory structures, it is yet well able to maintain hypertonicity in its fresh-water habitat. The ectoderm and endoderm cells, which are highly permeable to water, probably maintain an internal osmotic pressure equivalent to about 0.04–0.05 M sucrose, for they begin to shrink in concentrations of 0.04 M. Moreover, studies with radioisotopes have shown that sodium, potassium, and bromine are concentrated in *Pelmatohydra oligactis* at levels above those of the external medium, so that these animals, like *Aurelia* in the sea, are evidently able to carry out ionic regulation. Unfortunately, the mechanisms by which they achieve these results remain unknown; presumably they depend upon properties of the epithelia that form the internal and external surfaces of the diploblastic body.

13–5 METAZOA AND FRESH-WATER LIFE

The problems involved in the responses of marine invertebrates to changes in salinity are of great interest because of their bearing on the colonization of fresh water. As we have seen, these animals are highly permeable, and live in dynamic equilibrium with sea water as far as their water and ion relationships are concerned. Usually, and particularly when they live below the tide marks, they do not encounter changes of salinity in the external medium, and they are not adapted to deal with them. In consequence, when they are confronted with such changes under experimental conditions, the osmotic pressure of their body fluids varies with that of the water in which they are placed, and is virtually isosmotic with it; they are said to be poikilosmotic, or to be osmo-conformers. An example is *Hyas araneus* (Fig. 13–10).

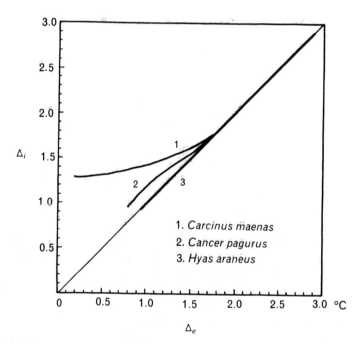

Fig. 13-10. Blood concentrations as freezing points (∇_i) of three marine crustaceans as a function of the external medium (∇_e); *1, Carcinus maenas* (∇_e 2·88–0·19°C); *2, Cancer pagurus* (Δ_e 2·41–0·82°C); *3, Hyas araneus* (Δ_e 1·80–0·96°C). From Robertson, 1960. In *The Physiology of Crustacea*, vol. 1 (T. H. Waterman, ed.), 317–339. Academic Press, New York.

When osmo-conformers are placed in dilute sea water, water enters the body as a result of osmotic uptake, while salts are lost along the gradient of concentration through the excretory organs and through the permeable body surface. Soft-bodied forms therefore swell in dilute sea water, and shrink in concentrated sea water. Often, however, they show some capacity for regulating their volume. This necessarily involves some adjustment of the salt content of the body, since they continue to remain isosmotic with the external medium. For example, if *Aplysia* is placed in 75% sea water, it at first gains weight for several hours as a result of the osmotic uptake of water. It then enters on a recovery phase of active regulation, during which it loses weight, as a result of loss of water, and at the same time makes a corresponding reduction in its total salt content. If, during this recovery phase, it is transferred back to normal sea water, it continues to lose weight; this is because it now suffers an osmotic loss of water, consequent upon its loss of salts during the recovery phase.

The cells of these animals will be affected in the same way, for a reduction in the salinity of the blood will cause swelling of the cells and tissues, as well as disturbance of their ionic balance. The extent to which poikilosmotic forms can survive changes in the salinity of the external medium therefore depends upon the extent to which their tissues can continue to function at different salinities. Those that can withstand only a very narrow range of salinity are termed stenohaline; those with a wide range of tolerance are termed euryhaline. The range is sometimes surprisingly wide. The osmotic pressure of the blood of *Mytilus edulis*, for example, closely follows that of the surrounding water, whether it is living in the North Sea, at a salinity of 30‰, or in the Baltic, at 15‰. It can even survive in dilutions of 4–6‰, which must clearly involve considerable adaptation on the part of its tissues. Other examples of this are seen in *Aurelia aurita*, which tolerates a salinity of 6‰ in the Baltic, *Membranipora pilosa*, which tolerates 4‰, and *Mya arenaria*, which tolerates 5‰.

The capacity of poikilosmotic animals to survive changes in the salinity of the medium, so strikingly evident in experiments, is of great importance for survival in the littoral zone, where salinity changes are frequent and rapid. This is well seen in molluscs, which are particularly characteristic of this habitat. One factor in their success is their ability to isolate themselves from the environment; *Mytilus* can close its valves, *Littorina* can retract into its shell, and *Patella* can adhere to rock surfaces. These devices, however, are of only limited application, and it may not always be possible to rely upon them. Prolonged exposure to low salinities, for example, might cause irreversible damage to the tissues through flooding of the cells by the inward movement of water. An important safeguard here is that the animals can regulate intracellular osmotic pressure. This process, which is found also in *Arenicola* and crustaceans, as we shall see, is termed isosmotic intracellular regulation, as contrasted with anisosomotic extracellular regulation of the composition of the body fluids themselves. It accounts for about 99% of the osmotic pressure of the blood of molluscs being due to inorganic ions, whereas only about 50% of that of the intracellular osmotic pressure is due to these. The remaining 50% is made up by amino acids and various other organic compounds such as taurine and glycine-betaine. The intracellular concentration of these compounds, and hence the intracellular osmotic pressure, can be regulated in response to changing osmotic concentration in the body fluids, presumably as a result of some modification of the balance between degradation and synthesis. As an illustration of this, various euryhaline molluscs have a lower level of amino acids in the tissues when in dilute sea water than when in more concentrated media, and a concomitant increase in ammonia excretion may be demonstrable in the dilute media, as illustrated by *Macoma* (Fig. 13–11). It will be noted that ammonia excretion in this animal returns, after 16 days in the dilute medium, to the lower level associated with 100% sea water. This is presumably because a new metabolic steady state has been reached, but the underlying mechanism for this is not known.

Exposure to low salinity, whatever the adaptive response may be, probably leads to considerable stress, which must reduce the metabolic efficiency. For example, the oxygen consumption of *Mytilus* at a salinity of 15‰ is lower than in normal sea water, and there is also a reduction in its ciliary activity and rate of heart beat. One would not expect, therefore, that tolerance of very low salinities could provide a basis for the exploitation of fresh-water habitats, yet the swan-mussel, *Anodonta cygnea*, succeeds in defying this expectation. How does it do this? The amino acid content of its muscles is actually much lower than that of *Mytilus*, the concentration of α-amino nitrogen in its fast adductor muscle being 10.5 mM/kg water, the corresponding value for *Mytilus* being 155–389. This accords with the view that organic compounds are used in isosomotic regulation in *Mytilus*, but not in *Anodonta*. In fact, the latter survives in fresh water because its tissues are adapted to a blood plasma with a freezing point depression Δ of only $-0.08\,°C$. This indicates an extraordinarily low value for the osmotic pressure of an internal medium, and it must surely make the animal's body fluid one of the most dilute among living organisms.

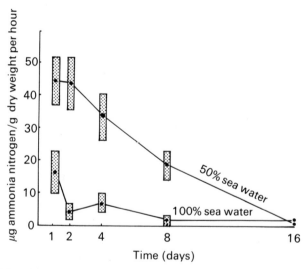

Fig. 13-11. Ammonia excretion of *Macoma inconspicua* as modified by salinity and time. Each mean value shown by a solid dot represents twelve determinations. The bars indicate the confidence limits at the 95% level. Values are based on dry weights of soft parts of animals. From Emmerson, 1969. From Schoffeniels and Gilles, 1972. In *Chemical Zoology*, vol. 7, 393–420 (Florkin and Scheer, eds.). Academic Press, New York.

Contrary to the more usual situation in fresh-water invertebrates, *Anodonta* is highly permeable to water. Considerable quantities of fluid are passed out through its pericardial coelom and kidneys, the amount in intact animals being of the order of 1.9 ml/hr for a 100 g animal at 15°C. The fluid within the pericardial cavity is relatively free of protein and is isotonic with the blood. Probably it is formed by ultrafiltration from the blood through the wall of the pericardial coelom, for the hydrostatic pressure of the blood is high enough to provide for this. From here the fluid passes into the kidneys, which are active in ionic uptake, for they discharge a hypotonic urine.

It may at first seem surprising that *Anodonta* is not able to make use of this active uptake in its kidney to maintain a much higher internal osmotic pressure, comparable, for example, with that achieved by fresh-water fish. Potts has suggested that part, at least, of the explanation of this is found in an analysis of the kinetics of the situation. A factor of great importance is the surface permeability. We have seen that this is high, and it is necessarily so in *Anodonta* because of its ciliary feeding habit and the large surface area of its ctenidia (p. 266). This must result in a prohibitively heavy influx of water. It has been calculated that if *Anodonta* had to exist in fresh water with an internal osmotic pressure equal to that of sea water it would have to carry on osmotic work demanding an energy consumption of 62 cal/per 100 g body weight in order to maintain its salt content. A low internal osmotic pressure reduces the amount of metabolic work required, and the exceptionally low value in *Anodonta* probably reduces it by a factor of several thousand. In one sense, then, this solution of the osmotic problem of fresh-water life is an efficient one. Yet, as we have suggested earlier, so dilute an internal medium must be far from ideal for the functioning of highly specialized cells, and this may well account for the Bivalvia remaining predominantly marine animals. In any case, *Anodonta* still has its problems. The heavy influx of water leads to a filtration rate so high that the loss of sodium in the filtrate per hour amounts to about 5% of the total extracellular sodium. It has been calculated that the replacement of this sodium, by uptake through the body wall and the kidney, could require energy amounting to as much as 20% of the animal's total metabolism.

Anodonta, by maintaining an internal osmotic pressure higher than that of fresh water, clearly has some capacity for osmoregulation, which we may define, following Robertson, as the regulation of the total particle concentration of body fluids at levels different from those of the external medium. Other osmoregulators are able to regulate to a higher level of internal osmotic pressure, and it is these that provide a more promising basis for the exploitation of fresh-water habitats. *Nereis diversicolor*, for example, is a euryhaline form that may be found in natural conditions at a salinity of 4‰. Here it can regulate so that it is hyperosmotic to the external medium, although its internal osmotic pressure is lower than that of normal sea water. In this it differs from *Perinereis cultrifera*, which swells in salinities as high as 20‰, and from *Arenicola*, which also swells at reduced salinities and cannot survive at salinities lower than 12‰. Both of these species are osmo-conformers. The concentration of their body fluids follows that of the external medium, and they show little capacity for regulation of the consequential changes in body volume in response to changes of salinity. Nevertheless, *Arenicola marina*, being able to survive at reduced salinities in an isosmotic state, is clearly euryhaline. (We have mentioned the damaging effects that salinity changes can have upon cells.) Survival of the tissues of *Arenicola* at lowered salinities is aided, as in molluscs, by a decline in the concentration of intracellular amino acids, and particularly of alanine and glycine. In a medium of 50% sea water these show a decline to 180 nmole/l as compared with the normal level of 427 nmole/l.

The success of *Nereis diversicolor* as an osmoregulator is related in part to ionic permeability. The point is well brought out in comparisons of *Nereis limnicola*, which is a fresh-water form, *N. diversicolor*, and *N. (Neanthes) succinea*, which is less tolerant of low salinities than is *N. diversicolor*. *N. limnicola* shows the lowest rate of salt loss in fresh water, and *N. succinea* the highest, while *N. diversicolor* is intermediate. *N. diversicolor* is in one respect less flexible than might be expected, for its permeability remains constant despite changes in external salinity, but it can afford this limitation because it is able to absorb sodium by active uptake at low salinities. This can best be understood by considering the sequence of events when the worm is transferred from normal sea water, with which it will be in isosmotic equilibrium, to dilute sea water.

It first gains weight by the osmotic uptake of water, but later it accommodates (Fig. 13–12) and becomes equilibrated at a new level of water balance, with a weight higher than its starting weight but lower than the maximum first attained. During the earlier gain in weight it loses salt, but at the final equilibrium position it is still hypertonic to the medium, and this condition it maintains by active regulation. It can be shown, by adding tracer doses of ^{24}Na to the medium, that in normal sea water sodium passes into the body by diffusion at rates of 240–275 μg hr^{-1} g^{-1} wet weight. This can be ascribed to passive exchange with the medium. In small dilutions this sodium influx falls in proportion to the degree of dilution, so that the animal is behaving as a simple osmometer. But at greater dilutions the uptake is higher than can be accounted for in this way. For example, uptake at a salinity of

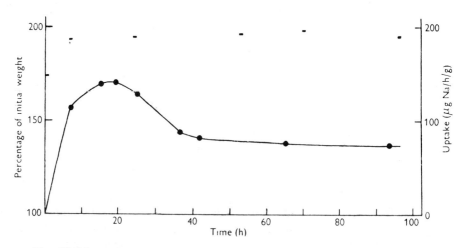

Fig. 13-12. Uptake of sodium per hour (per gram wet weight) by *Nereis diversicolor* during the period of weight regulation when transferred from sea water of 35‰ salinity to that of 9‰ salinity. Continuous line indicates weight regulation. From Fretter, 1955, *J. mar. biol. Assoc. U.K.*, **34**, 151–160.

9‰ is actually higher than at 18‰ (Fig. 13–13), and these values remain more or less the same after equilibration. (Fig. 13–12). The conclusion is that active uptake of sodium must be taking place against the concentration gradient. Whether chloride is replaced by active uptake as well, or merely by diffusion, aided by feeding, is uncertain. It has been thought that the movement of this ion is wholly passive in this instance, loss of chloride at low salinities being reduced by the production of a urine that is hypo-osmotic to the blood, as happens in earthworms (p. 380). The greater anatomical complexity of the nephridium of *Nereis diversicolor*, as compared with that of osmo-conforming nereids, has been thought to support this possibility.

Osmoregulation in dilute sea water is better understood in certain decapod crustaceans. We have seen that *Hyas* is poikilosmotic; so also are many other crabs and lobsters that normally live in full-strength sea water. Some decapods, however, have penetrated into brackish water, which may be defined as ranging in salinity from 30‰–0.5‰, normal sea water being of about 35‰. An example of these is the common shore crab, *Carcinus*, which ranges from below low tide level to estuarine water of less than half the concentration of sea water. At high salinities *Carcinus* is iso-osmotic with the medium, but at lower ones it can maintain, by active uptake (mainly through its gills), a hyperosmotic state (Fig. 13–10). This enables the animal to survive the fluctations of salinity to which it is exposed in its littoral and estuarine habitats; yet the osmoregulation is only partial, and the cells are subjected to considerable changes in the osmotic pressure of the fluids that surround them. This osmoregulatory capacity is shared by other euryhaline crustaceans, although there is much variation in the osmotic level which they can maintain. It is made possible by the interaction of several factors: reduction in surface permeability, active transport, and the intracellular osmotic adjustment already discussed. At best, however, the osmoregulation of these animals is incomplete, for, as we have seen, it would require excessive metabolic work to maintain high ion levels in the blood at very low salinities.

The intracellular adjustments involve in part the inorganic ions. Potassium remains relatively constant in *Carcinus*, but there is a marked fall in the intracellular concentrations of sodium and chloride as the osmotic pressure of the ambient

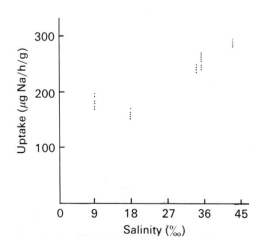

Fig. 13-13. Rate of uptake of sodium by *Nereis diversicolor* in sea water of different salinities. (Rate expressed as micrograms of sodium per hour per gram wet weight.) From Fretter, *op. cit.*

medium declines. Amino acids, however, are also involved; in particular, intracellular levels of glycine, proline, glutamic acid, and alanine fall, whereas arginine is unaffected, perhaps because of the importance of arginine phosphate in cell energetics. The result of these changes is that when *Carcinus* is transferred from normal sea water to a diluted medium there is at first an increase in its rate of nitrogen excretion owing to the breakdown of amino acids. Conversely, when it is placed back in a more concentrated medium it responds with a decreased rate of nitrogen excretion which persists while amino acids are being accumulated. A similar regulation of amino acid metabolism has been recorded in *Eriocheir* (see below).

A limiting factor in the maintenance of the ion and water balance of *Carcinus* is that its antennal glands, unlike the kidneys of *Anodonta*, can do no better than produce a urine that is isosmotic with the blood. In other words, when the surrounding medium is dilute, and ions are consequently scarce, the kidneys do nothing at all to help to conserve these ions by active uptake. They are simply maintaining ionic regulation, as they do in osmo-conforming crustaceans in normal sea water. In brackish water, therefore, *Carcinus* loses ions in its urine as well as by diffusion outwards through the surface of the body, and it is to the handling of this problem that active uptake and reduced permeability make their contributions.

The osmolality of the body fluids of *Carcinus* in relation to the surrounding medium is illustrated in Table 13–4. The data are expressed in terms of sodium, the influx and efflux of which can be followed, as in *Nereis*, by the use of the radioactive isotope as tracer (Fig. 13–14). In normal sea water the animal is isosmotic with the medium, but with some accumulation of sodium. Its urine is isotonic with its body fluids, so that it removes some sodium; this, however, represents only 2.7% of the total sodium loss in normal sea water, the rest taking place by diffusion through the permeable external surfaces (Table 13–4). In these conditions the sodium is largely replaced by passive influx, active uptake being very low at this level of blood sodium.

According to Shaw's analysis, the active transport mechanism becomes fully activated when the blood sodium falls slightly below 400 mM/l (Fig. 13–15), so that the animal can now maintain the sodium concentration of its blood against a fall in the concentration in the medium. There is, however, a limit to this capacity. Since the active transport system has now been fully activated, continued decline in the external medium cannot be indefinitely countered. Soon, therefore, the blood sodium

Table 13–4 The partition of sodium loss in *Carcinus* for animals in normal, 50%, or 40% sea water. From Shaw, 1961. *J. exp. Biol.*, **38**, 135–152.

External medium (% normal sea water)	Mean blood sodium conc. (mM/l)	Total outflux for 50 g animal (μM/hr)	Mean urine production (% body wt/day)	Sodium loss through urine for 50 g animal (μM/hr)	Loss through urine as % total outflux
100	460	1,267	3.6	34.5	2.7
50	340	1,009	23	163	16.1
40	300	891	30	188	21.1

Fig. 13-14. Apparatus for measuring ^{24}Na outflux in *Carcinus*. From Shaw, 1961. *J. exp. Biol.*, **38**, 135–152.

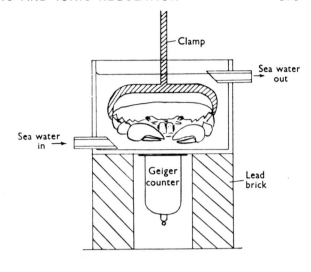

Fig. 13-15. Relation between the rate of active uptake of sodium and the blood sodium concentration in *Carcinus*. From Shaw, 1961. *op. cit.*

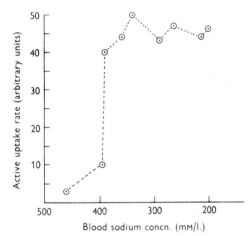

concentration starts to fall, and it thereafter follows the concentration level of the ambient medium but at some distance from it (Fig. 13–16). The consequential changes in the pattern of sodium flux are seen in Table 13–4. In 40% sea water the renal loss of sodium now accounts for over 20% of the total loss, while active uptake, already fully employed in 50% sea water, shows no further increase.

The use of active uptake as an adaptation to brackish conditions does not involve any new physiological principle, for, as we have seen, many marine organisms concentrate potassium and calcium in this way (Table 13–1), and we have emphasized the significance of this as a pre-adaptation facilitating movement into dilute media. *Carcinus*, however, goes further than these animals in its capacity to concentrate both sodium and chloride by active uptake. All of this requires the expenditure of energy, which must bear heavily upon brackish-water animals, and it is precisely here that a reduced surface permeability is helpful. If this can be reduced to a level below that of osmo-conforming marine species, there is a corresponding economy in energy out-

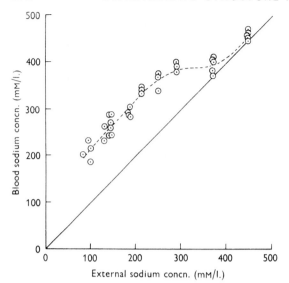

Fig. 13-16. Relation between the sodium concentration of the blood of *Carcinus* and that of the external medium. From Shaw, 1961. *op. cit.*

put. Our reference to nereids has already shown the value of reduced permeability, and it is not surprising, therefore, to find that this is common in brackish-water forms as well as in fresh-water ones. Its effect, as far as water movement is concerned, is seen on comparing the responses of the marine *Cancer* and the brackish-water *Carcinus* when their excretory pores are blocked. The resulting accumulation of urine leads to an increase in weight which is greater in *Cancer* than in *Carcinus*, indicating that the former is the more permeable to water. Other types of experiment have shown comparable differences in permeability to calcium and iodide ions.

Even with the advantage gained by reduced permeability, however, the type of adaptation seen in *Carcinus* would be a metabolically extravagant way of exploiting fresh-water habitats. The limits to its possibilities are probably reached by the grapsoid crab *Eriocheir sinensis*, which has penetrated into fresh water and which succeeds in maintaining there a relatively high internal osmotic pressure ($\Delta = -1.18°C$). It does so, however, by no more than the mechanism that we have just described, succeeding in this because of its rapid active uptake of ions. The gills are probably the seat of this uptake, for they can absorb sodium chloride from concentrations of no more than one-fortieth that of the blood. This animal, therefore, must work very hard to maintain its fresh-water life, and is compelled to do so because, like *Carcinus*, it receives no help at all from its antennal glands; these continue to exercise their old function of producing a urine isotonic with the blood.

Eriocheir has to migrate to the sea to breed, although young animals can penetrate hundreds of miles up river, and its osmoregulatory capacity is adapted to facilitate this change of medium. Not only can it remain hyperosmotic in fresh water; it can also remain hypo-osmotic in concentrated saline, differing in this respect from *Carcinus*. The regulatory capacity of the latter is confined to maintaining hypertonicity in dilute sea water. Like *Eriocheir*, however, it depends upon a high salinity for breeding, its larvae being unable to develop in salinities of less than 28‰.

Potamon niloticus is another fresh-water crab which, like *Carcinus* and *Eriocheir*, secretes an isosmotic urine, although in this animal permeability has been so far reduced that the urine output, which may be as little as 0.6–0.05% body weight per day, is less than that of marine forms. This, however, is an exceptional condition. In general, reliance upon an isosmotic urine, with the consequential heavy demands for active uptake of ions, has not provided a fertile solution to the problem of freshwater adaptation. There is, however, an alternative solution: the production of hyposmotic urine. This adaptation, already encountered in *Anodonta*, is found also in gastropods and crayfish, while its establishment in vertebrates (lampreys, teleost fish, amphibians) has made an important contribution to the passage of higher vertebrates from water to dry land.

The internal osmotic pressure of *Astacus*, at about $\Delta = -0.80°$, is notably higher than that of *Anodonta*, and this will enhance the metabolic efficiency of the tissues. It is made possible in part by low surface permeability, which is well shown in comparison with that of *Carcinus*; the extrarenal sodium loss in a 50 g *Carcinus* is 891 μE/hr, whereas in *Astacus* of similar weight it is only 7.5 μE/hr. Even so, ions are lost in the urine, the outflow of which in *Astacus* may amount to 8.2% of the body weight per day. The loss is compensated for by active uptake, in which the movements of sodium and chloride ions are secured by two distinct mechanisms, but here, of course, the crayfish does gain some real advantage from its secretion of hyposmotic urine, since ions can be conserved while the surplus water is removed. This capacity, which is found also in the fresh-water *Gammarus pulex* and the brackish-water species *G. duebeni* and *G. zaddachi*, constitutes an important contribution of the coxal glands to the exploitation of fresh and brackish water, and is reflected in the structure of these organs.

The significant element in this structure is the presence in *Astacus* of a long secretory tubule following the labyrinth (Figs. 13–17 and 13–18). This tubule, which is absent from the excretory organs of the marine *Homarus*, is known to be the region in which the urine is diluted. The study of samples of urine removed from the excretory organ of the crayfish shows that the fluid in the end sac is isosmotic with the blood, and that it becomes less concentrated during its passage through the tubule. The situation in *Astacus* is closely analogous to the mode of life of teleost fish in fresh water, for they, too, absorb ions through their gills and excrete a hypotonic urine. By analogy with the vertebrate kidney tubule, filtration in the excretory organ of *Astacus* might be expected to occur in the end-sac, and re-absorption of ions in the tubule, but we have no right to assume that natural selection will have produced identical results in crustaceans and vertebrates, nor does the available information support this assumption. Certainly Peter's data (Fig. 13–18) indicate resorption of chloride in the tubule distal to the labyrinth, but Riegel's micropuncture studies (Table 13–5), using the crayfish *Oronectes* and *Austropotamobius* (*Astacus*), have shown that sodium, chloride, and potassium are actually resorbed throughout the whole of the antennal gland, including the labyrinth and even the bladder. Despite the hypotonicity of the urine, however, the U/B inulin ratio indicates that water absorption occurs in the coelomosac and labyrinth. Furthermore, the ratio in the bladder is lower than that in the preceding part of the tubule, which means that the bladder must contribute to the dilution of the urine, and is apparently more important than the distal tubule in this respect.

Gammarus shows a modification of the antennal gland similar to that seen in

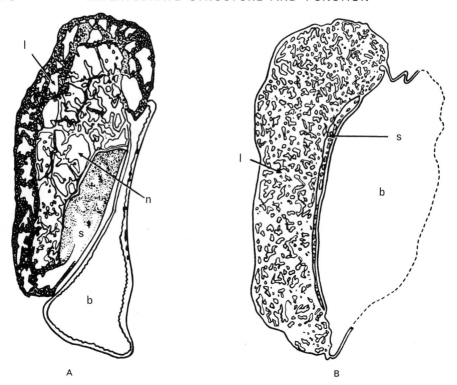

Fig. 13-17. Transverse sections of the antennal glands of *Astacus astacus* (A) and *Homarus gammarus* (B). *s*, end-sac; *l*, excretory tubule or labyrinth; *n*, additional tubule; *b*, bladder. From Parry, 1960. In *The Physiology of Crustacea*, vol. 1 (T. H. Waterman, ed.), 341-366.

Astacus, and to a degree that is well correlated with the habitats of the several species mentioned. The marine *G. locusta* has a short excretory tubule, the fresh-water *G. pulex* a long one, and the brackish-water *G. duebeni* one of intermediate length. In correlation with these differences the two latter species can both produce a hypotonic urine. *G. duebeni* is notable for the speed with which it can vary the concentration of its urine in response to changes in the salinity of the external medium; this capacity is probably to be correlated with its frequent occurrence in salt marshes, which are particularly liable to sudden fluctuations in salinity.

Table 13–5 Mean values ($\pm$S.D.) for the U/B ratios of urine removed by micropuncture from various of the antennary glands of crayfish. The difference between the mean U/B ratio for the coelomosac and for the labyrinth, proximal tubule, and distal tubule are statistically significant ($P = 0.10$), but the difference between the coelomosac and bladder is not. From Riegel, 1965. *J. exp. Biol.*, **42**, 379–384.

Coelomosac	Labyrinth	Proximal tubule	Distal tubule		Bladder
			proximal	distal	
1.16 ± 0.16	1.35 ± 0.26	1.49 ± 0.28	2.12 ± 0.96	1.81 ± 0.56	1.60 ± 0.92
P	$\cong 0.02$	$\cong 0.03$	$\cong 0.08$	$\cong 0.06$	> 0.20

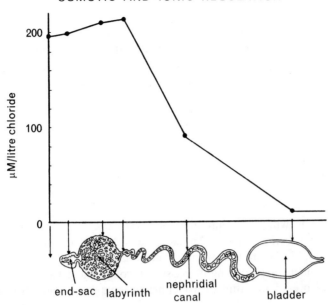

Fig. 13-18. Diagram of the antennal gland of *Astacus astacus* (after Marchal) with a graph of Peters' data of the chloride content of the excretory fluid. From Parry, 1960. *op. cit.*

From all of this we can conclude that the structure and mode of functioning of the coxal glands of crustaceans have enabled these organs to play a significant part in the adaptation of the group to fresh-water life. In this habitat, however, and also in the sea, the capacities of the glands are insufficient by themselves to meet the animals' needs; as already emphasized, adaptations affecting permeability and active uptake of ions at the body surfaces are also of the utmost importance.

13-6 FROM WATER TO DRY LAND

The water relationships of terrestrial animals, and the mode of functioning of their excretory organs, depend very much upon the route by which they entered onto dry land, and their degree of specialization to that habitat. It must be remembered, too, that a terrestrial environment is not necessarily a dry one for all of the animals that exploit it. For those dwelling in soil and litter it may often be a very wet one, presenting problems that are exemplified in some of the adaptations of earthworms. These animals can be regarded as having evolved from fresh-water forms, their nearest relations being the freshwater oligochaetes. This relationship is indicated by their habits, for earthworms commonly require moist soil for their surroundings, and can readily survive immersion in fresh water. Their body surface is highly permeable, as it must be for them to be able to respire through it; thus transference from damp soil to fresh water results in an uptake of fluid sufficient to produce an increase in weight of some 15% in 5 hr. This is a passive entry, the water following the osmotic gradient between the external medium and the inside of the body.

The dampness of the soil accounts for earthworms typically having metanephridia that are specialized for the removal of water. The osmotic pressure of the coelomic fluid is equivalent to $\Delta = -0.31°C$, while the blood is slightly less concentrated ($\Delta = -0.29°C$). The fluid from the metanephridia, however, is markedly hypotonic ($\Delta = -0.06°C$), which shows that these organs, like other excretory organs, can exert some regulatory function. Ramsay, by collecting fluid from different parts of the tubule, showed that in the first part of the metanephridium (the narrow tube) the urine remains more or less isotonic with the blood. The fall in osmotic pressure begins to occur in the middle tube, but is mainly a function of the wide type, although whether this is affected by influx of fluid, or efflux of ions, or both, remains unknown (Figs. 13–19, 13–20).

The production of this dilute urine, which is a measure of the essentially freshwater character of this group of animals (their excretion of ammonia is in conformity with it), must be of value on land when the soil is very wet. Some earthworms, however, have become adapted for life in drier soils by the development of complex nephridial systems which discharge in part into the intestine (enteronephridia) instead of directly to the outside. An example is the Indian earthworm, *Pheretima posthuma* (cf. Fig. 13–21). In this species, as in *Lumbricus terrestris*, the nephridial fluid becomes less concentrated during its passage through the organ.

There is also evidence that protonephridia have some regulatory function, at

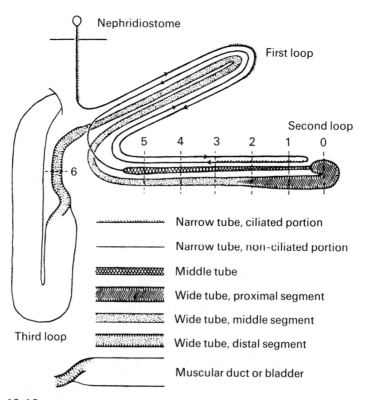

Fig. 13-19. Diagram of the nephridium of *Lumbricus* (from the account by Meisenheimer). For reasons of clarity the loops are shown displaced from their normal positions. The numbers indicate certain positions at which samples were collected. From Ramsay, 1949. *J. exp. Biol.*, **26**, 65–75.

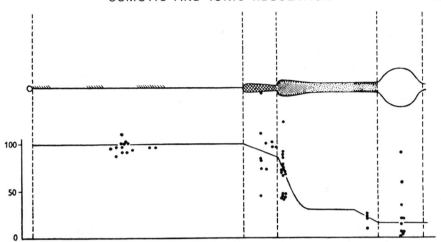

Fig. 13-20. Osmotic pressure of the urine at different levels in the nephridium. The osmotic pressure of the Ringer's fluid surrounding the nephridium has been equated to 100. Individual observations are shown as points; the line drawn through the points represents the interpretation placed upon them. From Ramsay, 1949. *op. cit.*

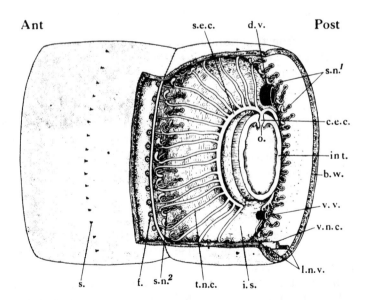

Fig. 13-21. Diagrammatic representation of the enteronephric nephridial system and its relative position as found behind the 74th segment in *Megascolex cochinensis*. *bw*, body-wall; *c.e.c.*, common excretory canal; *d.v.*, dorsal vessel; *f.*, funnel of a septal nephridium; *int.*, intestine; *i.s.*, intersegmental septum; *l.n.v.*, lateral neural vessels; *o.*, opening of the common excretory canal into the lumen of the intestine; *s.*, seta; *s.e.c.*, septal excretory canal; $s.n.^2$, septal nephridia of the right and left sides; *r.n.c.*, ventral nerve-cord; *v.v.*, ventral vessel; *t.n.c.*, terminal nephridial canal.

least as far as water relationships are concerned. In trematodes, for example, the protonephridia bear contractile vesicles, which can accumulate fluid and discharge it periodically, while in the cercaria larvae the rate of discharge decreases with increase in the osmotic pressure of the external medium. We have already mentioned that the protonephridial system is particularly well developed in the terrestrial nemertine, *Geonemertes*. It is certainly more extensive than in the marine nemertines from which this animal has presumably been derived, and this suggests that, like the metanephridia of earthworms, it may be producing a copious flow of water. Pantin has shown that the protonephridia of *Geonemertes* certainly do excrete fluid, and that the activity of the flame cells is related to the water contents of the animal; partial drying results in a cessation of their activity. Undoubtedly this animal may lose water at a great rate, not only by evaporation from its body surfaces, but also in laying eggs, and in secretion of the mucus which it uses in locomotion. It reduces the loss by living in sheltered and damp situations (under logs and stones, for example), and it buries itself when it begins to become dry. As in the earthworm, then, the functioning of the excretory system of *Geonemertes*, and its capacity for removing water, has to be evaluated in the light of the types of habitat into which the normal reactions of the animal will take it. Pantin points out that its preference for damp logs may result in it being sometimes surrounded by pure dew or rain water. It will then be in danger of excessive hydration rather than of desiccation, and in this circumstance the action of the protonephridia will be highly advantageous.

Instructive illustrations of some of the problems involved in terrestrial adaptation are provided by crustaceans. As we have seen (p. 197), some terrestrial forms are found in the amphipods; these, which all belong to the Talitridae (a group familiar as the sand-hoppers of the drift zone of sandy shores, have doubtless evolved by movement to land from the littoral zone. This is true also of the better-known anomuran and brachyuran crabs which have developed semi-terrestrial habits. These crabs are of particular interest because of their reliance upon behavioural adaptations, combined with ability to resist osmotic changes in the body fluids, and with a varying capacity to regulate the composition of the blood.

An example of a species which has not progressed very far along this line of adaptation is the mole-crab, *Emerita talpoida*. Living on sandy shores, and moving up and down with the tides, it swims and burrows and thus avoids exposure to desiccation. Even so, it is reported to be able to survive for up to 24 hr in either 125% or 75% sea water.

The anomuran coconut crab, *Birgus*, is more advanced. It, too, avoids desiccation by remaining in moist burrows, and it can withstand changes in blood concentration up to a level equivalent to about 118% sea water, but it is also adapted to drink water, and to keep its gills moist by placing water in its gill chambers. Another example, still more advanced, is the brachyuran land crab *Gecarcinus*, which rarely enters water, and can tolerate blood sodium concentrations ranging from 329 mM/l to 600 mM/l. It can live for several months in sand moistened with fresh water, apparently because it can take up water from the sand by means of pericardial sacs. These structures, which are diverticula of the pericardium, project into the branchial chambers and have at their tip a tuft a setae which collect and transmit the water by capillary action. This water is then absorbed, or perhaps passed to the gills.

Of all crustaceans, however, it is an isopodan group, the Oniscoidea, which have made the best progress towards life on land. These animals might be said to have been preadapted for terrestrial life with their isopodan characters, which include a ventrally flattened body and a brood pouch, yet they have developed little in the way of specific adaptations for such a life, and they remain dependent to varying degrees upon a damp environment. They have not waterproofed their exoskeleton, and they continue, like their marine ancestors, to rely upon their pleopods for respiration. The modifications of their pleopods are, however, modest terrestrial adaptations; those isopods possessing pseudotracheae (Fig. 13–22) can respire better in dry air than can those without them.

To judge from the habits of their immediate relatives, these isopods must have evolved from marine ancestors, their path of entry onto land being through the littoral zone. This is suggested by *Ligia oceanica*, a familiar isopod of the splash zone, which has not quite emancipated itself from the sea shore, and which has unmodified pleopods. What little is known about the water relationships of isopods indicates that these animals have remained as much marine forms as the earthworms have remained fresh-water ones. The osmotic pressure of their blood is high, as would be expected if they are of marine origin, but they seem to have little power of regulating it. They appear to survive by being able to tolerate the internal osmotic fluctuations that result from their surface permeability. Thus *Ligia* can tolerate fluctuations in the composition of its blood reanging from $\Delta = -1.44°C$ to $\Delta = -3.48°C$. All of these isopods are further aided by behavioural responses which are evoked by loss of water. *Oniscus*, *Porcellio*, and *Armadillidium* drink water (as also, incidentally, does *Birgus latro*), while *Ligia* takes up water by the anus.

Edney has pointed out that a possible explanation of the limited extent of the terrestrial adaptations of isopods may be found in the problems presented by passage through the littoral zone, as compared with passage through the water of estuaries and swamps. In the latter, and particularly in swamps, aerial respiration can become established while the animals concerned are still protected by an aquatic habitat from the large temperature fluctuations that are encountered on dry land. This was, in fact, the background of the evolutionary origin of terrestrial vertebrates. In the littoral zone, however, aerial respiration has to be established at the same time as temperature fluctuations are being encountered; and in these conditions the retention of a permeable body surface could be highly advantageous. The reason for this is that transpiration through the surface is an important means of lowering the body

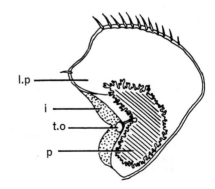

Fig. 13-22. Exopodite of the first pleopod of *Porcellio scaber*, to show the extent of the pseudotracheae. *i*, interior border of the pleopod; *l.p*, lamina of the pleopod; *p*, pseudotracheal area; *t.o*, tracheal opening shown by presence of air bubble. After Verhoeff, from Edney, 1960. In *The Physiology of Crustacea*, vol. 1 (T. H. Waterman, ed.), 367-393.

temperature during periods of potential heat stress, for, just as in the temperature regulation of birds and mammals, it removes the latent heat of vaporization of the water. The significance of transpiration is illustrated in Fig. 13–23. A specimen of *Ligia* resting in sunlight has a temperature several degrees lower than that of a cockroach (*Blatta*) placed side by side with it. This is because the cockroach does not transpire as effectively, although it does so sufficiently well to keep its temperature a little lower than that of a dead *Ligia*. Edney suggests that it is because of the value of transpiration that terrestrial isopods sometimes expose themselves to direct sunlight. As Fig. 13–24 shows, the body temperature of a *Ligia* on a sunny day can be lower when the animal is exposed than when it is sheltered beneath stones. The reason is that in the latter position transpiration is more difficult, and is not aided by air convection.

We have seen in various contexts that arthropod structure has great potentialities for adaptation to terrestrial life, but these are only fully realized in the insects. In this respect the achievements of these animals present the most striking contrast possible with the meagre progress of crustaceans, which by analogy with vertebrates, have really not proceeded beyond the amphibian stage of adaptation to dry land. As with vertebrates, the primary adaptation in insects is the waterproofing of the surface, here by the laying-down of the wax layer near the surface of the cuticle; a device that is not, of course, restricted to insects, but that is found also in other terrestrial arthropods, including mites, ticks, and probably spiders. Because of this, insects are outstanding in their ability to survive in really dry conditions; they are further aided by their invaginated respiratory surface, and their spiracular closing mechanisms. Asso-

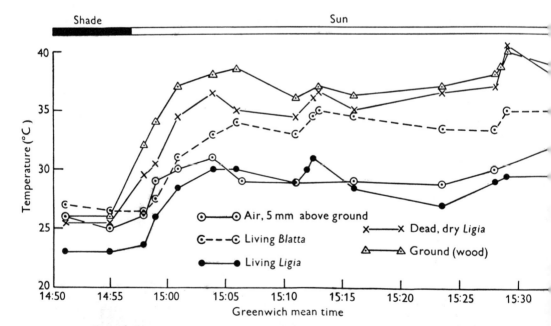

Fig. 13-23. The internal temperature of a living cockroach (*Blatta orientalis*), compared with that of a living and of a dead dry *Ligia*, during insolation (24 July) upon wood. Relative humidity 39-45%; wind speed c. 500 cm/sec. From Edney, 1953. *J. exp. Biol.*, **30**, 331-349.

OSMOTIC AND IONIC REGULATION 385

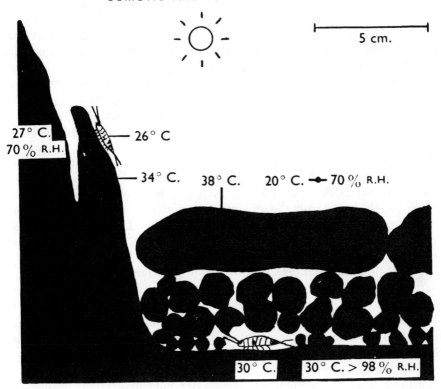

Fig. 13-24. Diagrammatic vertical section of base of red sandstone cliff and shingle inhabited by *Ligia*, to show the microclimatic conditions and internal temperatures of the animals, at *c.* 14.00 Greenwich Mean Time in August. From Edney, 1953. *op. cit.*

ciated with these features is the development of the Malpighian tubules, which replace the coxal glands as excretory organs. Our discussion of arthropod phylogeny has already shown that this adaptation must have evolved at least twice: in the Arachnida on the one hand, and in the myriapod-insect assemblage on the other.

Even within the insects Malphighian tubules vary in detail, but in principle they have walls composed of a single-layered epithelium, and they are bathed in the blood of the haemocoel. Water passes into the lumen of the tube, together with nitrogenous waste and dissolved salts, but how these substances enter is not entirely clear. There is some evidence for the occurrence of filtration in the end sac of the coxal glands, although even in that instance the evidence is not conclusive. In insects the position is a different one, for the blood pressure is very low, and the Malpighian tubules are surrounded on all sides by the blood, so that it is difficult to see that external pressure could have any other effect than to cause the collapse of the tubes. It seems likely, then, that the initial entry of the fluid and of its solutes must involve active secretion. In any case this certainly takes place in the reverse direction, either in the lower parts of the tubes, or in the rectum, for the material finally voided by the animal is a concentrate in which urates have been precipitated. This, as we have suggested earlier, can be regarded as a contribution of the excretory system to the conservation of water.

The mode of functioning of Malpighian tubules may well vary in detail as much as does their structure, but the observations of Wigglesworth have provided valuable information regarding their operation in *Rhodnius prolixus*. Here the upper (distal) two thirds of the tubule has a granular epithelium, and fluid is present in the lumen; the lower (proximal) one third has a relatively non-granular epithelium, but its lumen is filled with crystals of urates. When the animal feeds it ingests a large volume of blood. Within 3–4 hr the water contained in this has been absorbed and excreted through the Malpighian tubules, the result being that the proximal parts of the tubes are flushed free of their urates. After this phase is over, crystals appear again in the proximal parts of the tubes; they are derived, it is believed, from sodium or potassium urate that is secreted with the water into the distal end of the tube. The contents increase in acidity as they pass down, presumably because water and bases are re-absorbed into the blood from the proximal region, so that they become available for combination with more nitrogenous waste. The uric acid crystals, precipitated in the more acid mediums continue to be washed out into the gut for many days after the meal.

Investigations on other insects, showing the importance of active transport, and the participation of the rectum, can be illustrated here by only two examples. In the stick insect, *Dixippus morosus*, the urine in the Malpighian tubule is isotonic with the haemolymph, but has a high ratio of potassium to sodium, resulting from the abundance of potassium in the plant food, and its active transport into the tubule. Within the rectum, however, there is further modification, with extensive re-absorption of water and salt by rectal glands. The other example is provided by the larvae of *Aëdes aegypti*, which normally live in fresh water, but can survive in distilled water and also in nearly isotonic solutions of potassium or sodium chloride. Regardless of the medium, however, the haemolymph of these larvae continues more or less constant in composition, while the tubule fluid remains isotonic with it, and continues to show a high potassium/sodium ratio. Clearly these larvae are capable of ionic regulation, but it cannot be the tubules that are carrying out the regulation. If they were, their fluid could not retain this constant composition when the animal was being subjected to such varying regulatory demands. The explanation of the apparent paradox again rests on the functioning of the rectal glands, as is seen when the composition of the urine is determined after it has passed out of the tubules and traversed the rectum. It is then found that the fluid finally excreted from larvae kept in distilled water has a very low concentration of sodium and potassium ions, and is hypotonic to the haemolymph. In larvae kept in sodium chloride solution, however, it is nearly isotonic, but with a high/potassium ratio, while in those kept in potassium chloride solution it is still nearly isotonic, but now has a low sodium/potassium ratio. These results show that the rectum is ultimately responsible for regulating the composition of the haemolymph; this it is able to do because it possesses capacities for active transport of sodium, potassium, and chloride.

In addition, however, the rectum can actively transport water. This is remarkable, for in vertebrates, and also, it is believed, in many invertebrates, the movement of water is a passive process, along the gradients established by the active transport of ions. The evidence for its active transport in insects, however, is very convincing. It rests essentially on demonstrations that in certain conditions water can be transferred out of the rectum when its contents are markedly hypertonic. This is why the mealworm larva, *Tenebrio molitor*, can produce a solid pellet of uric acid and faeces. In this animal, living in flour with virtually no free water available, the conservation of water is aided by a close functional association of Malpighian tubules and rectum, the whole complex being enclosed in a perirectal membrane. Some other insects, and also certain ticks, can take up water by active transport from air which is low level of relative humidity. One insect (the firebrat, *Thermobia*) can do this against a gradient equivalent to an osmotic pressure difference of some hundreds of atmospheres.

A fully terrestrial animal does not have to deal with osmotic problems in the way that aquatic and semi-terrestrial ones do, but ionic regulation remains important. Little is known of this aspect of insect excretion, although there is some evidence that sodium and potassium can be regulated, as one would expect. On the other hand, many insects have returned to water for part or the whole of their life cycle, and their problems now become the same as those that we have already discussed for primitively aquatic forms. One might expect that they would rely upon their impermeable surface for reducing salt loss and water gain, and this seems to be so in the larva of the alder fly, *Sialis*; the plasma chloride of this animal ranges from 0.15–0.34% sodium chloride, and no less than 6 weeks starvation in distilled or tap water is needed to reduce this value to 0.06%. Re-absorption in the Malpighian tubules doubtless contributes to this retention, for the urine contains little sodium or potassium, and is devoid of chloride. The success of the protective mechanisms may be judged from the fact that the animal has no capacity for active uptake of ions, and presumably obtains all that it requires from its food.

It has been mentioned earlier that aquatic insects may develop structures that have been called tracheal gills. These are present in the *Sialis* larva, and in this animal they seem to be truly respiratory. In some other aquatic forms, however, their appearance is deceptive, and their name unjustified. Examples are the larvae of *Chironomus*, *Culex*, and *Aedes*, which have outgrowths that are now called anal papillae. These structures were formerly called gills, but it is now known that they are organs through which water can enter or leave the body, according to the osmotic pressure of the medium; they are also capable of active uptake of ions. Among the evidence for this may be mentioned that larvae of *Aedes aegypti* shrink in hypertonic media, but do not do so if their hind end is tied off, while larvae of *Chironomus* and *Culex* can take up chloride from very dilute solutions, but are unable to do this if the anal papillae are destroyed or ligated. Moreover, it can be demonstrated, in both experimental and natural conditions, that there is a good correlation between the state of development of the anal papillae and the salinity of the medium in which the larvae live. In general, the papillae are large in fresh-water forms and reduced, or even absent, in those living in saline waters. Anal papillae, then, like the other regulatory organs that we have discussed, illustrate the danger of naming organs, and of inferring their function and mode of operation, without adequate experimental evidence.

14
Sources of Information

14-1 CODED SIGNALS

The maintenance of life demands continuous adjustment of the organism to changing conditions in its environment. Some of these changes will have no significant effect upon an animal—to these it may be insensitive. Others may influence its capacity to survive or to reproduce, or at least to carry out with efficiency its normal range of activities—to these it must be adapted, which means that it must be able to respond to them in such a way as to promote its survival. An animal must therefore possess sensitive structures, called receptors, which can be excited by an appropriate range of signals or stimuli from the environment. They provide it with the information which enables it to formulate its adaptive responses.

This pattern of stimulus and response must have existed from the very beginning of life. It probably depended in the first instance upon the disturbance by the environment of some fundamental property of protoplasmic organization. This property is perhaps to be found in the structure of the living membrane that forms the surface of the cell (Sec. 2–4). The significance of the membrane in this connection is well seen in two types of cell that are intimately concerned in animal behaviour: the nerve cell (neurone, neuron) and the muscle cell. We have seen that they differ in composition from that of the surrounding medium in having a higher concentration of potassium and lower concentration of sodium. In association with this the membrane carries an electric charge; it is said, therefore, to be polarized, the charge being termed the resting potential. An active control of potassium concentration is a widespread feature of living systems (p. 352), and it is likely, as we have already suggested, that polarization of the surface membrane is a common, and perhaps universal, property of living matter. This suggests that from an early stage of evolution the primary effect of environmental disturbance was to create localized states of instability in surface membranes, involving changes in their ionic permeability. This would have resulted in a flow of ions which could bring about some measure of depolarization.

Such membrane disturbances can be conducted over cell surfaces, but the effect diminishes with distance from the point of initial disturbance, and the conduction is said to be decremental. It provides a rudimentary mode of conveying information, but clearly it can only be of significance over very short distances. Theoretically it could provide for transmission through the body of a protozoan, particularly one like *Amoeba* in which there are no visibly differentiated organelles that might serve for conduction. Unfortunately, the small size of these animals makes it difficult to secure data sufficiently precise to confirm this, although there certainly are differences of potential across their surface membranes.

In the giant amoeba *Pelomyxa carolinensis* the potential difference at the surface is about 80–100 mV, with the positive charge external; this is associated with an internal potassium concentration greater than the external, and if the external concentration is increased the potential differences is correspondingly lowered. It is impossible to judge for certain whether or not such changes in surface charges are involved in the responses of Protozoa, although there is no evidence against this. Increased excitation of *Opalina* changes the direction and wave pattern of the ciliary beat, and theoretically, as already suggested (Sec. 3–2), we could account for this by supposing that there are corresponding changes in the polarization of the body surface.

There is, indeed, evidence that the coordination of the beat of the compound cilia (membranelles) of *Stentor* depends on the propagation of a cytoplasmic change which is induced by the beat of one membranelle, and which then promotes the beat of the adjacent one. The occurrence of such cytoplasmic transmission (neuroid transmission) is obviously difficult to verify, nor is it always necessary to appeal to it. It is thought that the coordination of ciliary beating is often due to hydrodynamic linkage of the beat of adjacent cilia, these being able to influence each other through the water layers which they transport during their movement (Sec. 3–2).

Communication in sponges is clearly at a primitive level. Mechanical stimulation of these animals produces contraction of the body, so that some form of signal can evidently be transmitted from one cell to another. The contractions remain localized, however, so that all that sponges seem to have achieved is some decremental spread of response over short distances. In this respect, as in others, they are essentially assemblages of cells with little coordination between them (Sec. 4–4). Something much more is needed to provide for the long distance signalling that is essential for the organized responses of the highly differentiated bodies of metazoans, and it is this that is the special province of the nervous system. It is believed that this system is absent from sponges, but it is well established in the coelenterates, and is a familiar and characteristic feature of all metazoan animals.

The nervous system is the result of the exploitation of an important invention, the specialized cell called the neurone. The significance of this cell lies in its ability to translate excitation into coded signals termed nerve impulses, which it can distribute (or propagate) over long distances without decrement. Neurones are highly diversified in form, but we may follow Bodian in recognizing that the common principle of their organization is the possession of three components: a cell body, a dendritic zone, and an axon (Fig. 14–1).

Like other types of cell, the neurone must carry out trophic or vegetative functions. These are mainly located in the cell body, which is responsible for the maintenance of the dendritic zone and axon, and for the production of certain secretions, called the neurohumours and the neurohormones. The dendritic zone is the receptor

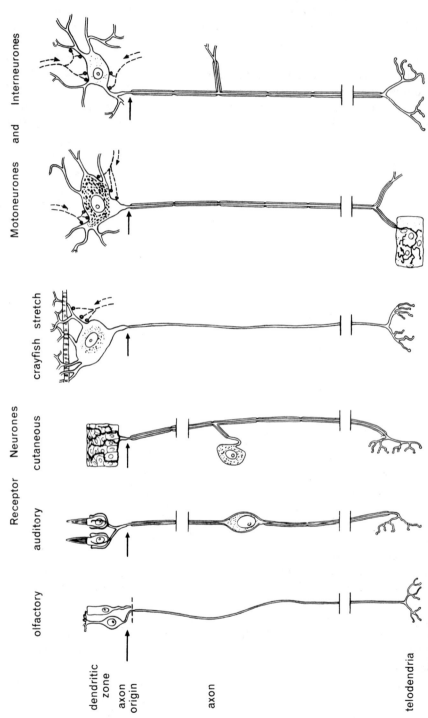

Fig. 14-1. Diagram of a variety of receptor and effector neurones. Except for the stretch receptor neurone of the crayfish, the neurones shown are those of vertebrates. From Bodian, 1962. *Science, N.Y.*, **137**, 323–326. Copyright 1962 by the American Association for the Advancement of Science.

region of the cell. It may be specialized as a receptor structure, sensitive to environmental stimuli; or it may be part of a motor neurone or internuncial neurone, in which cases it will receive the endings of the axons of other neurones, and be excited by their activity. It will then often consist of a series of branched processes called dendrites.

The axon, or nerve fibre, is a specialized outgrowth which conducts nerve impulses away from the cell body. It branches at its distal end to form the axon telodendria, a region that differs from the rest of the axon, and resembles the dendritic zone in that it is not concerned with the propagation of the nerve impulse itself. Instead, it is typically specialized so that when a nerve impulse reaches it, it transmits a signal (usually chemical, but sometimes electrical) across a barrier called the synapse; these signals bring about the excitation of other neurons, or they may excite the effector structures (usually muscle or gland cells) that are the agents of response. Because of spatial differences in the properties of neuronal cell membranes, the location of the synapses that end on them is an important element in the complexity of functioning of the nervous system. A further possibility is that the axon ending may be associated with blood vessels to form neurohaemal organs, specialized for the release of neurohormones into the circulation (Sec. 17–2).

14–2 PROPERTIES OF RECEPTORS

We may begin our analysis of the nervous system by considering some aspects of the organization and mode of functioning of receptors. These structures are differentiated in all multicellular animals above the level of the sponges. Their characteristic properties are well seen in the stretch receptors of the lobster and crayfish (Fig. 14–2), as studied by Eyzaguirre and Kuffler. In this instance the receptor is the dendritic zone of a neuron that sends dendrites into a strand of muscle (Fig. 14–3) and that gives off an axon that runs into the central nervous system. The dendrites are excited when the muscle is stretched or contracted, and it is supposed that the excitation normally begins in the peripheral ends of the dendrites. These, it is thought, are specifically sensitive to stretch, which causes their surfaces to become depolarized, with consequent changes in their membrane potentials. By inserting a microelectrode into the interior of these neurones (Fig. 14–4), we can take direct readings of the initial or resting membrane potential, and show that the permeability changes resulting from excitation evoke the appearance of a new potential, called the receptor potential.

This change in potential is a graded one, varying according to the intensity of the stimulus (Fig. 14–5). If the change is sufficiently large, it will evoke a localized depolarization in the non-receptive region of the cell membrane, probably at the axon hillock, where the axon takes origin from the cell body (Fig. 14–1). for the threshold is usually lowest at this point. This depolarization is not a graded one; it is an all-or-nothing response, which evokes a similar depolarization in its immediate neighbourhood. So, by repetition of this process, a wave of depolarization is propagated without decrement along the whole length of the axon. This propagated disturbance is the nerve impulse. The potential associated with it, called the action potential, is the electrical expression of its passage. The electric excitability of the axon, and its consequent capacity to produce and propagate potentials, is a funda-

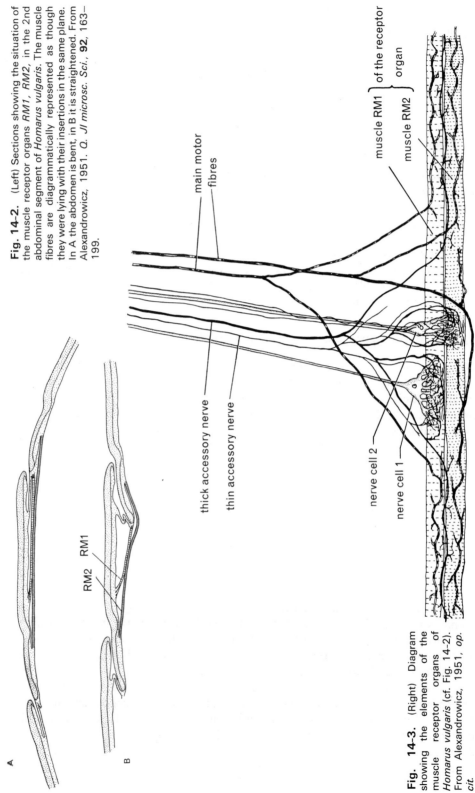

Fig. 14-2. (Left) Sections showing the situation of the muscle receptor organs RM1, RM2, in the 2nd abdominal segment of *Homarus vulgaris*. The muscle fibres are diagrammatically represented as though they were lying with their insertions in the same plane. In A the abdomen is bent, in B it is straightened. From Alexandrowicz, 1951. *Q. Jl microsc. Sci.*, **92**, 163–199.

Fig. 14-3. (Right) Diagram showing the elements of the muscle receptor organs of *Homarus vulgaris* (cf. Fig. 14-2). From Alexandrowicz, 1951, *op. cit.*

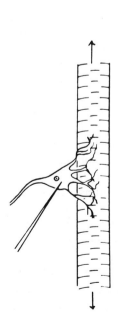

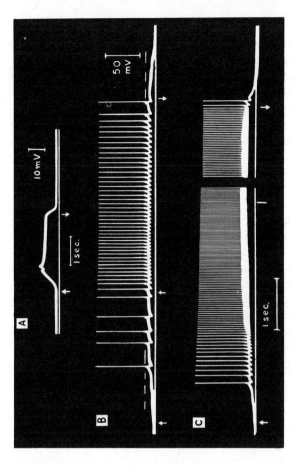

Fig. 14-4. (Above) Diagram showing a nerve cell making contact with a receptor muscle strand, seen through the dissecting microscope, with a capillary electrode in position. From Eyzaguirre and Kuffler, 1955. *J. gen. Physiol.*, **39**, 87–119. Reprinted by permission of the Rockefeller University Press.

Fig. 14-5. (Right) Records from slowly adapting sensory nerve cells of the receptor muscle. A, Subthreshold stretch, applied for the period indicated by the arrows, evokes only a membrane potential change. B, Stretching at just above the threshold evokes irregular discharge of nerve impulses; additional stretch (second arrow) increases the discharge rate. C, Discharge rate increases as stretch is increased (to straight line) and remains constant during maintained stretch (to second arrow). From Eyzaguirre and Kuffler, 1955. *op. cit.* Reprinted by permission of the Rockefeller University Press.

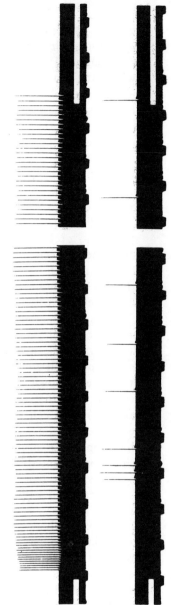

Fig. 14-6. (Right) Oscillograms of action potentials of a single optic nerve fibre of *Limulus* in response to prolonged illumination of the eye. For the top record, the intensity of stimulating light was 10,000 times that used for the bottom record. Eye partially light-adapted. Signal of exposure to light blackens out the white line above time marks. Each record interrupted for approximately 7 sec. Time marked in 1/5 sec. From Hartline et al., 1952. *Cold Spring Harb. Symp. quant. Biol.*, **17**, 125–141.

mentally important invention which permits information to be transmitted in the body through long distances without loss.

Continued stimulation of a receptor cell will produce continued excitation (within the limits mentioned below), but any one response is followed by a brief period of absolute refractoriness during which no response can be shown. Continuous stimulation therefore produces a series of discrete responses; continuous illumination of the eye of *Limulus* (Fig. 14–6), for example, results in the passage of a train of responses, each involving a characteristic sequence of ionic fluxes and changes in polarization in the nerve fibres concerned. These changes can be electrically recorded as spikes which are our visible evidence of the transmission of information along the axon. The rate of discharge of these impulses varies with the level of the graded generator potential, and hence with the intensity of the stimulus, as may be observed in Fig. 14–5, B and C. Commonly, however, the continued application of a stimulus to a receptor eventually stops the discharge of impulses. The receptor is no longer being excited, and has adapted to the stimulus. The discharge of impulses from some of the stretch receptors of the lobster can persist for several hours if the stimulus is maintained; these are slow-adapting receptors. Others are fast-adapting, for in them the discharge ceases in less than a minute.

This course of events is, we believe, generally applicable to the action of receptors which we think of as initiating a train of nerve impulses by establishing a local depolarization at the proximal end of the nerve fibre. This fibre may be a product of the receptor cell concerned, as in the example of the crustacean stretch receptors. Such a cell, known as a primary sense cell (Fig. 14–1), may be a primitive form of receptor. Alternatively, the conducting nerve fibre may belong to a separate (afferent) nerve cell, which is then said to innervate the receptor cell, the latter being termed a secondary sense cell (e.g. the auditory receptor in Fig. 14–1). In this case, the receptor potential evokes in the afferent nerve cell a depolarization (called the generator potential) which generates action potentials if its peak is sufficiently high, this again depending upon the intensity of the stimulus.

The sequence of events is a complex one, but it is yielding to experimental analysis. The primary function of the receptor element is to capture from the stimulus some energy which causes the organism to respond to the stimulus. The amount of energy involved may sometimes by infinitesimally small, as when a male moth detects the scent of a female at a great distance. Even in man, where the sense of smell is relatively feeble, the evil smelling substance, skatol, can be smelt when its concentration is so low that there may be little more than one molecule of it for each olfactory cell of the nasal epithelium. We must suppose, then, that the receptor includes some amplifying device, although we can only speculate as to what this may be.

By some means or other the energy initially provided by the stimulus is transformed into a series of electrical events in the axon—the train of action potentials that constitute the nerve impulses. This transformation is called transduction. One result is that a continuous stimulus is coded into a series of discrete nerve impulses. It is in this coded form, therefore, that information obtained from the environment is transmitted through the nervous system. The code is a very simple one, the only variable in it being the frequency of the nerve impulses, which increases with increase in strength of the stimulus. Yet, for all its simplicity, it is adequate to provide the organism with a very wide range of information. This is because each type of receptor, sensitive—as we shall see—to one particular type of stimulus, transmits impulses along a particular set of nerve fibres to its own particular region of the central nervous system. The information can thus be interpreted and acted upon according to the source from which it has been received. It is important that this should be so. An animal needs a great variety of information if it is to have an adequate basis for adaptive response.

14-3 MECHANORECEPTION

The maintenance of a rich input of information demands a high level of specialization among receptors, for each type is organized to respond to one particular mode of stimulation. At one time it was customary to classify receptors as exteroceptors, sensitive to external stimuli, interoceptors, sensitive to internal ones, and proprioceptors, sensitive to the tension and movement of muscles, tendons, and joints. It is often more useful, however, to relate them to the physical character of the stimuli to which they are sensitive. On this basis we can classify invertebrate receptors into mechanoreceptors, responding to mechanical deformation; chemoreceptors, responding to changes in the chemical composition of the environment; and electromagnetic receptors, responding to quanta of light and radiant heat over a spectrum that varies in different species, but that extends from the infrared to the ultraviolet. In considering these three types, we shall refer to a few of the more closely studied examples. It is probable that the principles involved in their operation will be shown by future work to be of wide general applicability.

The primary source of mechanoreception was presumably the sensitivity of cell surfaces to contact with stationary or moving objects. Changes in membrane potentials and ionic composition may have ensued, and any of these that evoked responses leading the organism to move away from harmful material would from the beginning have been of adaptive value. It is likely that free nerve endings in the body surface often serve for mechanoreception in metazoa, but more complex structures have also been evolved, particularly in the arthropods. In these animals, as in the vertebrates, information regarding the position of the joints of the skeleton and the tension of the muscles is essential for the rapid and effective execution of responses. One example from the crustaceans is the muscle stretch receptor that we have already considered; it may be compared with the muscle proprioreceptor of vertebrates. Another is the statocyst, particularly striking in showing that the arthropods and the vertebrates have independently evolved two very similar types of receptor, sensitive both to gravity and to acceleration. No less striking is the resemblance in these respects between vertebrates and *Octopus*.

The statocyst (Fig. 14–7) occurs among the Crustacea in the Malacostraca, more particularly in the Mysidacea and the Decapoda, situated either at the anterior or posterior ends of the body. Morphologically it is an invagination of the body surface which forms a sac filled with fluid. The wall of the organ bears sensory setae which may be of two types, as they are, for example, in crabs. One of these types is the statolith hair, stimulated by movements of particles called statoliths. These particles, which are of higher density than the fluid in which they are suspended, consist either of a secretion of the animal itself, or of sand grains cemented by a secretory product. The statoliths, together with the sensory hairs and fluid, are shed at the moult. If

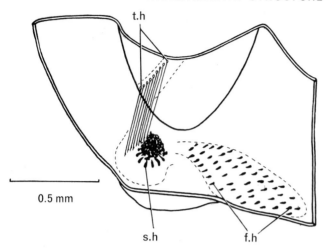

Fig. 14-7. Inside of isolated ventro-caudal corner of left statocyst of *Carcinus maenas*, seen from dorsofrontal aspect. The hair fields surrounded by interrupted lines are clearly transparent in contrast to rest of statocyst wall. *t.h*, caudal end of thread hair row; *s.h*, statolith (hook) hairs touching statoliths with their tips; *f.h*, free hook hairs with tips directed laterally. From Cohen and Dijkgraaf, 1961. *The Physiology of Crustacea*, vol. 2 (T. H. Waterman, ed.), 65-108. Academic Press, New York.

iron filings are substituted for the statolith grains—as they can be, for example, in *Palaemon*—the position of the animal can be controlled by a magnet. This component of the statocyst therefore functions as a position receptor, the orientation of the body being maintained in response to excitation of the statolith hairs produced by gravitational force acting on the statoliths. The mode of action is remarkably similar in principle to that of the maculae acusticae of the membranous labyrinth of vertebrates. Just as in the latter animals, angular displacement of crustaceans brings about compensatory reflexes which affect the limbs and eye stalks. Moreover, the position receptors of both groups resemble each other in that, irrespective of stimulation, they spontaneously generate a continuous discharge of nerve impulses. What they respond to is displacement of their sensory hairs in either of two directions, this bringing about an increase or decrease in rate of discharge. In both groups, the spontaneous discharge also contributes to the maintenance of muscular tone.

No less remarkable is the parallelism between the functioning of the second type of sensory setae found in the more specialized statocysts. These hairs, called the thread hairs, have no contact with the statoliths, but instead are stimulated by movements of the fluid within the statocyst, being highly sensitive to changes in angular rotation about any of the three main axes of the body. Stimulation is followed by compensatory reflex movements of the eye stalks (Fig. 14–8), which turn contrary to the direction of rotation when the rate of rotation is increasing, and resume a normal position when the speed becomes constant. The results are similar to those evoked by changes in angular rotation in vertebrates, with the difference that in crustaceans there is no structural elaboration comparable with the semicircular canals. In both groups this particular reflex movement of the eyes contributes to maintaining a uniform field of vision during acceleration.

Considering the degree of convergence between crustaceans and vertebrates in this form of mechanoreception, it is surprising that position receptors of this type are almost wholly absent from insects. The Diptera are an exception, for they possess halteres which evoke reflex reactions to rotational movement during flight. Cohen and Dijkgraaf suggest that this difference may be related to differences between the

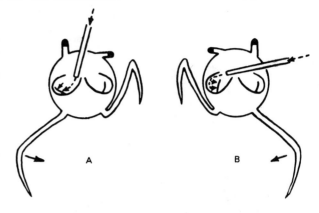

Fig. 14-8. Compensatory eyestalk and limb reflexes evoked in *Astacus astacus* by bending statocyst hairs in opposite directions with a fine water jet (statoliths removed). A, bending the hairs laterally. B, bending the hairs medially. From Cohen and Dijkgraaf, 1961. *op. cit.*

state of equilibrium of crustaceans and vertebrates on the one hand, and insects on the other. The equilibrium of the two former groups is unstable. Insects differ from them in having a lower centre of gravity, and a greater surface area relative to their weight, factors that make them mechanically more stable even when flying.

Nevertheless, other forms of mechanoreceptors are well developed in insects, as they must be in active animals with an articulated skeleton. Connective tissue is conspicuous in this connection, and neurones have become associated with it to form organs that are specialized for signalling the stresses that result from movements of the body wall. In the simple form found in *Periplaneta* these receptors each consist of a multipolar neurone embedded in a strand of connective tissue. In grasshoppers and locusts the neurone has become encapsulated by connective tissue, and the whole is attached to a muscle. In Lepidoptera there is still further specialization: the muscle fibre that bears the receptor complex has become separated from the remainder of the muscle, has developed a separate motor innervation, and contains two giant nuclei. At this stage it has something in common with the vertebrate muscle spindle; its functions are similar in principle to those of the spindle, and also to those of the crustacean stretch receptor, but the latter differs structurally in being a compound organ with as many as four neurones on one muscle fibre. We see in these examples how similar needs condition the evolution of similar receptor mechanisms, with sufficient differences in detail to indicate the independence of their origins.

Mechanoreceptors are also associated with the exoskeleton of insects, occurring as hair sensilla and campaniform sensilla. The former (Fig. 14-9) are often arranged as the sensory cells of hair plates which are stimulated by the movements of the inter-segmental folds of the cuticle. They serve, therefore, to signal the position of the parts of the body, without being dependent, like the muscle receptor organs, upon the development of muscular tension. The campaniform sensilla are essentially domes of cuticle, their associated receptor cells being stimulated by the strains produced by the bending of the skeleton. These sensilla are often found on the legs, in positions suggesting that they signal the forces acting on the skeleton of the stationary animal.

Finally, insects possess chordotonal sensilla, lying on elastic strands extending between two regions of the body wall. These receptors, too, can signal changes in

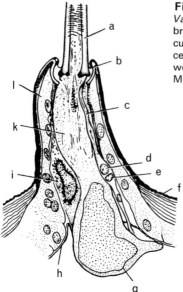

Fig. 14-9. Section through base of tactile bristle in larva of *Vanessa urticae* (after Hsü). *a*, base of hair; *b*, articular membrane; *c*, scolopoid body; *d*, sense cell; *e*, neurilemma cell; *f*, cuticle; *g*, trichogen cell; *h*, basement membrane; *i*, tormogen cell; *k*, vacuoles in trichogen cell; *l*, epidermis. From Wigglesworth, 1947. *The Principles of Insect Physiology* (3rd ed.). Methuen, London.

tension, but they also give rise to highly evolved auditory organs. In this form the receptors, arranged in groups which differ in their sensitivities, are stimulated by vibrations of a membranous tympanum that overlies them, and which presents an obvious analogy with the ear drum of higher vertebrates. They are highly sensitive, sometimes to frequencies that are far above the limit of perception of the human ear, an example being the ability of some insects to sense the ultrasonic pulses generated by the echo-locating bats which prey upon them. Audioreception, indeed, is an important factor in the life of insects, not only because of their terrestrial life, but because the exoskeleton lends itself to the production of sounds, as is well seen in the bowing of the tegmina of male grasshoppers by their hind legs. Experimental study has clearly demonstrated that insects, although limited in auditory function, can recognize the sounds which are characteristic of their own species, and which are proving to be of taxonomic value. Moreover, the grasshopper *Chorthippus* produces a number of distinctive sounds which are associated with equally distinctive situations that, from a human standpoint, would have emotional overtones. Faber has been able to distinguish, among no less than 12 such sounds, the song of the male in the absence of the female, the serenading of the female by the male (ending, if uninterrupted, in a shout of triumph), the duet of rival males, and the song of the copulating male.

Just as insects lack the statocysts of crustaceans, so the latter seem to lack hearing organs. It is true that crustaceans are sensitive to vibration, but this, as Cohen and Dijkgraaf point out, is not the same as possessing specialized sound receptors which are stimulated by pressure waves of characteristic velocity. Hearing, defined in this formal way, has not yet been clearly demonstrated in any crustacean. Perhaps the predominantly aquatic habit of the group is a contributory factor, but it cannot be the whole explanation, for teleost fish often have an acute sense of hearing, aided in many of them by the development of Weber's ossicles. We can only remind ourselves that groups and species have receptors that are adapted to respond to particular ranges of sense modalities, and that it is upon these that their behaviour depends. The consequence for us is that the world is full of noises and other sources of information that are quite unavailable to human receptor systems, unless, of course, we supplement them with electronic equipment.

14–4 CHEMORECEPTION

The limitation of our own receptors is evident in chemoreception. It is difficult for us to appreciate the significance of this sense modality in animal life, for in man it is greatly reduced in importance. In many other species the situation is wholly different, for they possess an acute sensitivity to extraordinarily small chemical changes in the environment. Later we shall consider how products of metabolism, for example, provide a basis for animal communication and responses. We have, in fact, already noted this in the feeding processes of coelenterates. It is not clear which of the receptor cells of these particular animals are specifically concerned with chemoreception, but some information is available for certain other groups.

In turbellarians chemoreceptive cells occur in the cephalic pits or grooves. Ablation experiments demonstrate that these receptors are responsible for increasing the movement of the animals in the presence of food and for guiding them towards it. This is essentially a 'seeking' reaction; other chemoreceptors on the head operate when the animal is sufficiently close to the food, and initiate feeding. The chemoreceptors of flatworms are primary sense cells, elongated and slender, with stiff cilia-like processes projecting from their distal ends. Similar cells in the surface epithelia of annelids may be chemosensitive, but this assumption is based only upon observations of the intact animal, and experimental evidence at the cellular level is commonly lacking. It is known, however, that the whole surface of earthworms is sensitive to saline solutions, whereas sensitivity to sugar solutions is restricted to the prostomial region. This indicates that chemoreceptors must be widely distributed over the body, and that they are differentiated into more than one type.

Much more precise information is available for certain arthropods, for in these animals the association of receptor cells with differentiated outgrowths of the cuticle greatly aids their study by electronic techniques. In certain flies the legs and labellum bear chemoreceptors that are excited by contact with chemical substances, and that initiate such feeding responses as the protrusion of the proboscis. As with so many arthropod receptors, these structures are innervated setae. In the blowfly, *Phormia*, to take one example that has been extensively studied, the chemoreceptors of the labellum are setae, each with a sac of cells suspended beneath it. Within the sac are the trichogen and tormogen cells that are concerned respectively with the secretion of the seta and with the formation of the socket. Associated with these are chemoreceptive neurones, which may be as many as five in number. Of these, several have distal processes that extend up to the tip of the seta. This is the permeable region of the seta, and it is at this point that the processes can be chemically stimulated. One process, however, ends at the base of the seta. This is believed to be a mechanoreceptor, for the seta remains sensitive to touch after its tip has been removed, but will no longer respond to chemical stimulation.

These and other observations on sensitivity depend upon applying fluid-filled micro-pipettes to the setae and using them as recording electrodes. They can be inserted either in the side walls of the setae or in their tips, the advantage of the latter position being that the fluid in the micro-pipette can then serve as a source of chemical stimulation. By inserting one electrode into a seta and another into the head of the animal we can record action potentials from a single seta. As with the stretch receptor considered earlier, the stimulus is coded into nerve impulses, the frequency of which varies with the strength of the stimulus. The responses of the individual receptor cells can be distinguished by analysing the action potentials.

These studies have shown that the chemoreceptors are sensitive to at least 18 sugars, and that stimulation of a single hair will elicit the characteristic feeding response of proboscis protrusion. The sensitivity is extraordinarily high, so much so that the complete behavioral response can be produced by stimulating a single hair with sucrose in a concentration of only 0.00001 M. A variety of other substances, including salts, acids, and alcohols, will also stimulate the hair, but with them the response is rejection, shown by a withdrawal of the proboscis, if this is already extended.

Are these two contrasted responses, feeding and rejection, mediated by the same receptor cell, or are these cells differentiated in their sensitivity? An answer to this is suggested by the electrical responses resulting from the two forms of stimulation. The rejection response is accompanied by a discharge of large spikes from the sensilla fibres, whereas the feeding response is accompanied by a discharge of small ones. This suggests that the two responses arise from two different cells, one sugar-sensitive, the other a non-sugar receptor. But this is not all. Rejection proves to be not a single modality of response, but one that can be evoked in two different ways. One of these involves stimulation of the rejection receptor, which is believed to be specifically sensitive to monovalent salts. The other depends on inhibition of the acceptance receptor by substances that do not stimulate the rejection receptor. Furthermore, there is evidence of the existence of a specific water-sensitive receptor, distinct from the carbohydrate receptor, while female flies, at certain stages of their reproductive cycle, seem able to distinguish between protein and carbohydrate acceptance.

The mechanism of this differential sensitivity is by no means understood. It appears to depend in part upon the structural configuration of the molecules concerned, but there is evidence also that different molecules may act at different sites on the receptor cells. Many suggestions have been made as to how the action is exerted. Presumably it must involve some disturbance of the surface membrane; this, according to the general concept outlined earlier, would result in ionic movements that could establish a generator potential. Davies has argued that the disturbance may result from the odorant molecules becoming temporarily incorporated into the membrane, the specificity of the association being determined by the characteristic shapes of the molecules.

Other factors influencing the feeding responses of *Phormia* are the general state of excitation of the nervous system, and the adaptation of the receptors. Thus a tactile stimulus, the bending of a seta, will not normally evoke extension of the proboscis, but it will do so if the fly has been without water and food for a day, when the animal is apparently hypersensitive. After further tactile stimulation the response disappears; the seta has thus adapted to this particular stimulus, but neighbouring setae will be found to be unaffected. Similarly, any one seta can be successively adapted to water and to increasing strengths of sucrose solution. Finally, the extension of the proboscis is only the initiation of the feeding process. Once it has occurred, stimulation of chemoreceptors on the labellae is necessary to bring about sucking, and, indeed, to regulate the extension, while further stimulation of papillae between the labellae is needed if the feeding process is to be continued.

Thus, starting with a sensory seta, which itself contains several functionally distinct receptor elements, our analysis leads us to see even a simple behaviour pattern as dependent upon coded signals from a large number of diverse receptors. Under normal conditions, of course, many sensory setae will be involved in the feeding response of *Phormia*, and many receptors additional to the chemical ones. This is a measure of the complexity of organization needed in the nervous system, which is responsible for the handling of these signals. We shall later consider some of the steps by which this has been achieved.

14–5 PHOTORECEPTION

The capture of energy from the environment by photoreceptors involves the absorption of quanta of light by photosensitive pigments, the resultant chemical change in these pigments being the source of the generator potential. In the vertebrate eye there are four such pigments. All of these are carotenoid derivatives, consisting of a substance called retinene (vitamin A aldehyde) combined with a specific type of protein called an opsin. One or other of two main types of opsin may be joined with one or other of two types of retinene, retinene$_1$ being derived from vitamin A_1, and retinene$_2$ from vitamin A_2. Thus are formed rhodoposin, iodopsin, porphyropsin, and cyanopsin.

It is a striking demonstration of the unity underlying organic diversity that carotenoids are widely, and perhaps universally, used in photoreceptor systems, not only in animals, but also in the systems that are concerned with the phototropisms of moulds and higher plants. Moreover, pigments similar to those of vertebrates, although not containing either vitamin A_2 or retinene$_2$, have been demonstrated in the complex eyes of both arthropods and cephalopods. It is not clear whether these similarities of biochemical organization have been independently evolved, or whether they are derivatives of a fundamental pattern achieved very early in the history of life. Such, however, is the importance of sensitivity to radiant energy that the latter alternative would seem the more probable.

An explanation of the wide use of carotenoids in visual receptor systems has been put forward by Wald. He ascribes it to the peculiar capacity of these compounds for geometrical isomerization. Other types of natural pigments lack this capacity, for their atoms are held firmly in position in ring structures. Carotenoids, however, possess straight-chain conjugated systems, the stereochemical configurations of which can be readily modified. What is particularly remarkable is that the pigments involved in photoreception seem always to have a certain configuration known as 11-*cis*-, meaning that there is a *cis*-configuration at the eleventh carbon aton. Isomerization involves the transformation of this into the more stable all-*trans*-configuration, and it is this change that occurs when a quantum of light (a photon) is absorbed. The selection of this particular configuration is a consequence of its possessing a unique degree of photosensitivity; in other words, it can be isomerized by light with the maximum efficiency of energy transfer.

The most primitive manifestation of photosensitivity is the so-called dermal light sense, a diffuse sensitivity that does not depend upon any obviously recognizable receptor system. Indeed, since the nature of the receptor structures is unknown, it is not even certain that they are dermal, so that the designation of this phenomenon may be something of a misnomer. The property is widely spread through all the major phyla, more particularly in aquatic forms, and often co-exists with morphologically differentiated eyes. It is apparent in *Amoeba*, which changes its direction and speed of movement in response to stimulation by light. It is seen, too, in the isolated mesenteries of anemones, which contract in response to local illumination. They continue to do so even after the nervous system has been immobilized by anaesthesia, so that local reflexes cannot be involved. This suggests that certain coelenterate cells may be able to act in response to illumination as independent effectors, preserving a type of response similar to that of *Amoeba*.

In more complex animals the dermal light sense may play an important part in orientation, either through general or local reactions. Thus local illumination of the body of *Ciona* brings about a bending of the animal, produced by local contraction of the musculature, whereas general illumination causes closure of the siphons. We might suppose this latter effect to be caused by stimulation of the pigment spots that are conspicuous around the rims of the siphons, yet this is not so. They can be removed without affecting this reaction, so that the name of eye spots, frequently applied to them, is misconceived. Other examples of such reactions are to be seen in *Holothuria*, which retracts its tentacles when these are locally illuminated, but reacts to general illumination by moving and turning. They also occur in tubicolous polychaetes, where the orientation of the tubes may be influenced by illumination. With regard to the evolution of photosensitivity, it is significant that the maximum sensitivity of the dermal light sense lies at wavelengths between 470 and 580 mμ. This agrees with the known properties of photosensitive pigments, and certainly suggests that the dermal light sense, and primitive photosensitivity in general, depend upon the presence of small amounts of pigments similar to those of differentiated receptor systems. This could well have been the starting point of the evolution of such systems.

Two main types of highly differentiated photoreceptor system have appeared in the invertebrates: the compound eyes of arthropods and the camera-type eyes of cephalopods. Enough is known of the mode of functioning of these, and of their probable past history, to show that they represent the evolution, along two very different lines, of organs that have some striking points of similarity with the vertebrate eye, not only in their pigments but also in certain details of their structural organization. Indeed, this is an aspect of animal organization to which we have more than once referred—a convergence resulting from the widespread distribution of a common biochemical ground plan. In this instance the common feature is, of course, the nature of the photosensitive pigments.

Simple types of eyes are seen in the free-living Platyhelminthes and in the Annelida, where they are often composed of sensory cells associated with screening pigment cells. In their simplest form they may be no more than pigment spots, forming part of the general epithelium, but more usually they sink inwards to form cups. In the Turbellaria the pigment cells are often arranged to form the wall of an open bowl, the bipolar receptor cells projecting into this through its aperture. In such an eye there can be no possibility of forming an image when there is no refractive structure. Sometimes a lens is present, but in general these organs are doubtless restricted to the differentiation of light and darkness, and in this way they make it possible for the animal to orientate itself with respect both to the intensity and to the source of the illumination. The distal ends of the receptor cells are differentiated to form a rod border, similar in ultrastructure to the rod-like differentiation in the receptor cells of more complex eyes (see later).

Cup-like arrangements of pigment cells are common in the eyes of polychaetes, but a higher level of differentiation is reached in this group. The epithelium of the cup may produce secretions that fuse to form one or more lenses, and groups of sensory cells may be closely collected together to form ommatidia, recalling the unit structures of the compound eye of arthropods. Indeed, in sabellids (*Branchiomma*, for example) the ommatidia themselves may be grouped together to form a rudimentary type of compound eye. No doubt a similar tendency played an important part in the ancestors of arthropods, contributing to the establishment of their characteristic compound eyes. The oldest known examples of these are found in the trilobites of the Lower Cambrian, where they were present in a fully developed form that could well have been as efficient functionally as those of modern arthropods. Unfortunately, this tells us nothing of the origins of these eyes, but the uncertain affinities of trilobites, the evidence of convergence in the evolution of arthropodization, and the grouping of ommatidia in certain polychaetes, go some way to make acceptable the possibility of the independent evolution of compound eyes in the ancestry of modern arthropods.

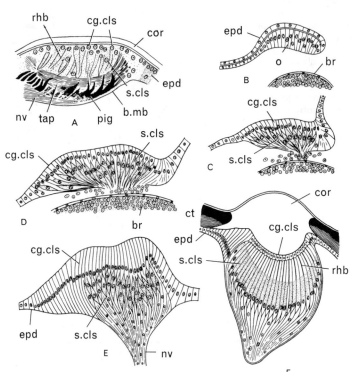

Fig. 14-10. Examples of the structure and development of dorsal ocelli of insects. A, dorsal ocellus of *Machilis*. B–F, stages in the development of a dorsal ocellus of male of *Formica pratensis*, and mature median ocellus of same. *b.mb*, basement membrane; *br*, brain; *cg.cls*, corneagenous cells; *cor*, cornea; *ct*, cuticula; *epd*, epidermis; *nv*, nerve; *o*, ocellar rudiment in epidermis; *pig*, pigment; *rhb*, rhabdome; *s.cls*, sense cells; *tap*, tapetum. Adapted from Snodgrass, 1935. *Principles of Insect Morphology.* McGraw-Hill, New York. Used by permission.

404 INVERTEBRATE STRUCTURE AND FUNCTION

Evidently the optic cup is the foundation upon which the varied types of arthropod eye have been formed. It is seen in the Onychophora, in which the eyes are a pair of closed vesicles into which a lens is secreted. It is seen also in the median eye of the nauplius larva, and in the dorsal ocelli of insects (Fig. 14–10). From a purely morphological point of view it is not difficult to visualize the evolution of such structures into the vertically elongated groups of cells that make up the ommatidia of the compound eye. This type of eye is seen at its simplest in the lateral eye of *Limulus*, which has provided material for experimental analysis of much elegance. It consists of about 1,000 ommatidia, each of which is formed of twelve retinula cells, which are the receptor cells, together with an eccentrically placed nerve cell (Fig. 14–11). The retinula cells are arranged like the segments of an orange, packed radially around the denderite of the nerve cell. Within the ommatidia is a photosensitive pigment based upon vitamin A_1 and retinene$_1$; these, as we have seen, occur in the vertebrate eye, but the analogy with the latter is even closer than this.

Electron microscopy reveals that the rods and cones of the vertebrate retinula cells contain membranes arranged as closely packed discs. It is supposed that the

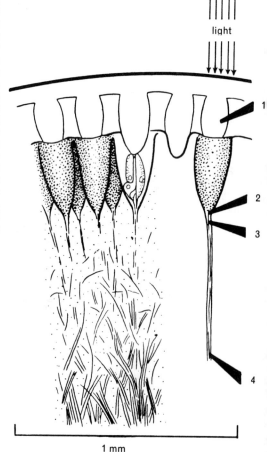

Fig. 14-11. Schematic drawing, representing a section of lateral eye of *Limulus*, in a plane perpendicular to surface of cornea, as seen in fresh preparations. Transparent cornea at top, showing crystalline cones of the ommatidia; the heavily melanin-pigmented conical bodies of these form a layer on the inner surface of the cornea. On the left, a group of ommatidia is represented, with indications of bundles of nerve fibres traversing the plexus behind the ommatidia, collecting in larger bundles that become the optic nerve still farther back. One of these ommatidia has been represented as if the section had passed through it, revealing the sensory component, also as if sectioned. On the right an ommatidium with its nerve fibre bundle is represented as it appears after having been isolated by dissection and suspended, in air, on electrodes (moist cotton wicks, from chlorided silver tubes filled with sea water) represented by the solid black triangles. From Hartline et al., 1952. op. cit.

visual pigments are arranged on or in these membranes, and that isomerization causes permeability changes, resulting in ionic movements and changes in membrane potential. Within the retinula cells of *Limulus* there are arrays of parallel tubules that suggest an obvious analogy with the ultrastructure of the vertebrate rods and cones. A similar ultrastructure is found in the rod-like elements which are found in photoreceptors throughout the animal kingdom, and this, taken in conjunction with the similarity of the visual pigments already mentioned, makes it highly probable that the mechanisms of energy capture and trausduction are always fundamentally the same. This is certainly well shown in the ommatidium of *Limulus*, where, by using microelectrodes, we can demonstrate generator potentials, and show that the eccentric nerve cell (the corresponding elements are, of course, arranged quite differently in the vertebrate retina) gives rise to the nerve impulses already mentioned.

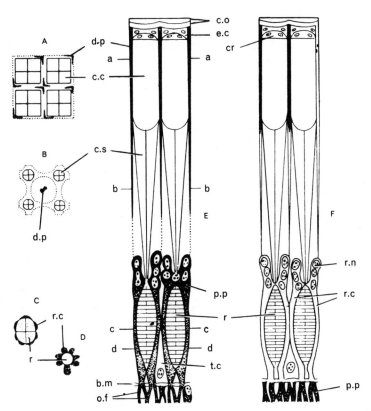

Fig. 14-12. Structure of the ommatidia of *Astacus astacus* (redrawn from Bernhards). A, B, C, D, tangential cross-sections of four (A, B) ommatidia or one (C, D) ommatidium at the levels indicated in E by *aa*, *bb*, *cc*, and *dd*, respectively. E, radial (axial) section of two ommatidia showing the pigment in the light-adapted condition providing maximum shielding of each single unit. F, similar view in the dark-adapted condition with a condition of minimum light shielding by pigment. The facet diameter of such an eye is about 60 μm. *b.m*, basal membrane; *c.c*, crystalline cone; *c.o*, cornea (lens); *c.s*, stalk of crystalline cone; *cr*, crystalline cone cells; *d.p*, distal pigment; *e.c*, corneagenous cells; *o.f*, optic nerve fibres of retinal cells; *p.p*, proximal pigment; *r*, rhabdome; *r.c*, retinular cells; *r.n*, retinular cell nucleus; *t.c*, tapetal cell. From Waterman, 1961. *The Physiology of Crustacea*, vol. 2 (T. H. Waterman, ed.), 1–64.

Compound eyes of a more advanced type are found in crustaceans and insects; while varying greatly in detail, they are essentially of the plan in Fig. 14–12. Two refractive bodies are present: the lens, which is a biconvex thickening of the general cuticle, and the crystalline cone, secreted by a group of vitrellar cells. These two determine the fixed focal length of the ommatidium. Below them is a group of seven to eight receptor (retinula) cells, each of which forms a fibrillar rhabdomere along its length, the several rhabdomeres fusing to form the rhabdome, a structure of high refractive index. The visual pigment is probably located within the rhabdomeres, which may therefore be compared with the rods and cones of the vertebrate retina. Photostimulation is presumably initiated within the rhabdome, its result being the propagation of nerve impulses into the optic nerve along the nerve fibres that arise from the base of each retinula cell.

We can distinguish two main morphological types of compound eye, the apposition eye and the superposition eye (Figs. 14–13, 14–14). In both types the ommatidium is structurally a separate and self-contained unit, separated from its neighbours by groups of distal and proximal pigment and reflecting cells. Where the types differ is in the degree of functional independence of the ommatidia, a factor that has an important influence on the mode of functioning of the eye.

Fig. 14-13. Diagram of an ommatidium of an apposition eye. *c.l,* cornea lens; *c.c,* crystalline cone; *s.c,* sense cells. In the cross-section the individual rhabdomeres can be recognized in the rhabdome (*r.m*). From Kuiper, 1961. *Symp. Soc. exp. Biol.*, **16**, 58–71.

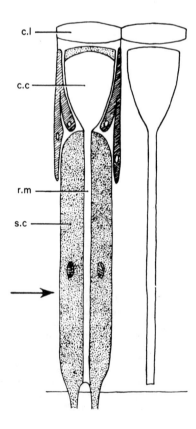

The apposition type is particularly characteristic of terrestrial, littoral, and diurnal species, including the Hymenoptera and Diptera amongst the insects. Its receptor cells are greatly elongated, extending distally as far as the crystalline cone. According to the classical mosaic theory of vision, each ommatidium receives light from the very small area that corresponds to its geometrical projection. This means that the image formed by the eye as a whole must be a mosaic of areas corresponding in number to the ommatidia. In the superposition eye—found, for example, in lobsters, crabs, and nocturnal insects—the retinal cells are widely separated from the crystalline cone, and it is supposed that each receptor cell is stimulated by light that has entered through a number of ommatidial lenses. This type of eye is suited for vision in dim light, which might not be an adequate stimulus if each receptor cell received only the rays that had entered through its own ommatidium. The resulting overlapping of images, however, will yield poor resolution, whereas much better resolution is to be expected from the isolated images formed by the apposition eye.

This interpretation would indicate that the pigment cells cooperate in visual function by modifying the degree of isolation of the ommatidia. Their behaviour varies greatly from species to species, but in general they seem to be of greater importance in the superposition eye. Here, the pigment moves in strong light to screen the retinal cells, so that they can only be stimulated by light entering along the ommatidial axis. The eye is now said to be light-adapted, a condition which probably protects it from high light intensities. In dim light the pigment moves to expose the retinal cells, which can now be reached by light from a wider source, a condition in which the eye is said to be dark-adapted (Fig. 14-12).

Evidently, many arthropods are unable to secure precise discrimination of form in their visual field. Often they will only distinguish degrees of shade and

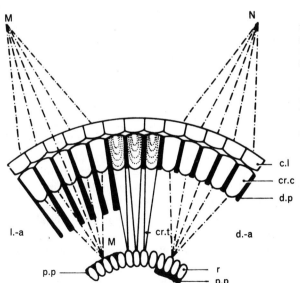

Fig. 14-14. Diagram of the light rays in the superposition eye. The position of the distal pigment (d.p) and the proximal pigment (p.p) is shown in the light-adapted state (l.-a) and the dark-adapted state (d.-a). In the three central ommatidia the hypothetical laminated structure of the cones is indicated; here also the crystalline tracts (cr.t) are drawn. The corneal lenses (c.l) are usually not drawn in such diagrams though they often have a greater refractive power than the cones. m and n are two light sources and r is the retinula. From Kuiper, 1961. op. cit.

illumination, but even so the compound structure of the eye probably aids detection of movement, because of the continuously changing pattern of ommatidial illumination. These matters are still very imperfectly understood, but new experimental approaches are leading to re-assessments of old problems. For example, it has been widely supposed that the apposition eye favours the discrimination of form through acuity of resolution. Electronic studies, however, have shown that the field of vision of a single ommatidium is much larger than was formerly believed. Thus a single ommatidium of the eye of *Locusta* receives light over an angle of 20°, which means that a single point source must stimulate many ommatidia. Yet this same eye can detect movement, and discriminate pattern, when the angle subtended by the movement or source is only of the order of 0.3°. The sensitivity of the compound eye in this instance is certainly much higher than might have been predicted—perhaps, according to one suggestion, because additional images are formed deeper in the eye by groups of ommatidia. It is at least quite clear from behaviour studies, to which we shall be referring later, that some insects are well able to discriminate form and, by virtue of this capacity, to learn the landmarks in their neighbourhood.

Some arthropods, both crustaceans and insects, can discriminate the plane of polarized light, and can also discriminate colour, with (at least in insects) a marked ultraviolet sensitivity. It has been supposed that the basis of colour vision in vertebrates is the presence of three pigments, differing from each other in their spectral sensitivity, and there are now some reasons for believing that a similar explanation may apply to the compound eye. Electronic techniques have made it possible to explore the responses of single retinal cells in the compound eyes of the blowfly, *Calliphora erythrocephala*, and to show that the degree of depolarization depends upon the wavelength of the light, as well as upon its intensity. This has led to the recognition of three types of receptor cell in this eye. All three show a peak response at a wavelength of about 350 mμ, in the ultraviolet range, but they differ from each other in respect of a second peak. In one of the types, the most abundant one, this peak lies at about 490 mμ (green-type receptor), while in the other two it lies respectively at below 470 mμ (blue-type receptor) and at above 520 mμ (yellow-green type). Obviously the differences between these several pigments are very small, and so the central nervous system must be highly specialized to differentiate between the varying patterns of excitation that are transmitted to it from the eyes. In this there is nothing either surprising or improbable. It is much easier to imagine the evolution of colour vision by the accumulation of small changes in pigment properties rather than by the sudden emergence of entirely new types of photosensitive substance.

Whatever details of its operation may be disclosed by future research, the capabilities of the compound eye already demand our respect. Waterman points out that *Daphnia*, with only 22 ommatidia in its median compound eye, and with less than 200 neurosensory cells, can perceive colour, analyse polarized light, and perform visual avoidance responses. How it does so is obscure, and some involvement of a dermal colour sense cannot be excluded, but even so, this aspect of arthropod organization must surely appeal to those for whom economy of material is a prime characteristic of elegant design.

We have suggested that the history of the arthropod eye is to be traced through the increasing elaboration of the simple vesicular eyes found in lower invertebrates; this also seems to be the case in the molluscan eye. In the more primitive forms, as, for example, in *Patella*, the eyes are no more than open invaginations of the epidermis, lined with pigment and receptor cells, but with no lens. In more advanced forms the opening of the vesicle is narrowed and closed, the vesicle then containing either a fluid secretion or a lens. Such organs are presumably used for orientation, since they can have no capacity for form discrimination. More specialized arrangements are seen in the more active molluscs, such as *Pecten*, where the pallial tentacles bear highly differentiated eyes, each with a lens and an inverted retina. Even here, however, these organs can do no more than detect movements and shadows, capacities that are none the less of the first importance in animals that are capable of such rapid swimming responses.

As might be expected, it is in the cephalopods, with their generally high level of activity and response, that the molluscan eye reaches its peak of differentiation. Superficially, the eye of dibranchiates such as *Octopus* greatly resembles that of vertebrates, and presents in this respect a classical example of convergence (Fig. 14–15). Yet its mode of development is different, while in some ways its retinal

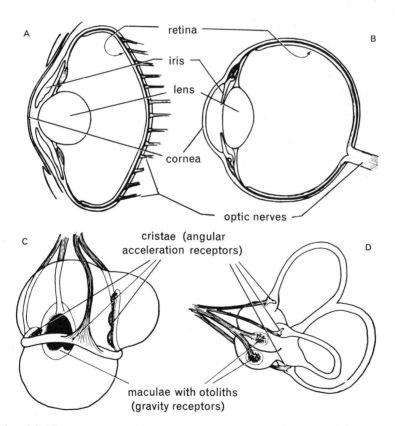

Fig. 14-15. Parallel development of sensory receptors in man and the octopus. A and B are longitudinal vertical sections through the eyes of an octopus and of a man respectively. C and D show the statocysts of *Octopus* and the labyrinth of the inner ear of man. (C is adapted from Young, 1960, *Proc. R. Soc. B*, **152**.) From Wells, 1961. *Advmt Sci., Lond.*, **17**, 461–471.

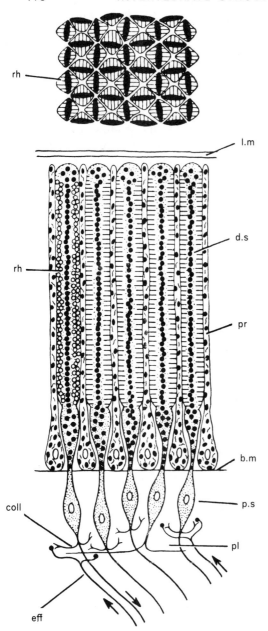

Fig. 14-16. Diagram of probable arrangement of the elements of the retina of *Octopus*. *Above*, as seen in tangential section; *below*, in radial section. *b.m*, basal membrane; *coll*, collateral of retinal fibre; *eff*, efferent axon to the retina; *d.s*, distal segment of retinal cell; *l.m*, limiting membrane; *p.s*, proximal segment of retinal cell; *pl*, plexus beneath retina; *pr*, process of supporting cell (probably thinner); *rh*, rhabdomes. The second retinal cell from the right has been wrongly shown; it should carry a fibre to the optic nerve. From Young, 1961. *Biol. Rev.*, **36**, 32–96. Used by courtesy of the Cambridge Philosophical Society.

organization is more akin to that of the compound eye of arthropods. In *Nautilus* the eye is still simple, being a cup-shaped invagination with a small opening to the outside, and lacking any lens. The dibranchiate eye still develops in this way, so that it is formed in its entirety direct from the epidermis; the vertebrate eye, by contrast, develops centrifugally as an outgrowth of the central nervous system. These differences in mode of development account for the cephalopod retina being the direct type, with the sensory ends of the receptor cells direct towards the source of light, whereas the retina of vertebrates is inverted, with the sensory ends turned away from the source of light. The eventual resemblances between the eyes, however, are all the more remarkable. In the cephalopod the invagination closes over, the lining forming the retina while a lens develops at the point of fusion. Later an iris is formed, so that a pupillary opening is delimited, and the whole becomes covered by a transparent cornea associated with eyelids. Focusing is effected through the ciliary muscle which contributes to the suspension of the lens, relaxation of the muscle leaving the lens in position for distant viewing.

The optic potentialities of this eye, by analogy with the vertebrate eye, would seem to be excellent, yet it is limited in its capacities by the organization of the retinal elements. These are arranged (Fig. 14-16) in groups that recall to some extent the ommatidia of the arthropod eye. The retinula cells, which are pigmented, are arranged

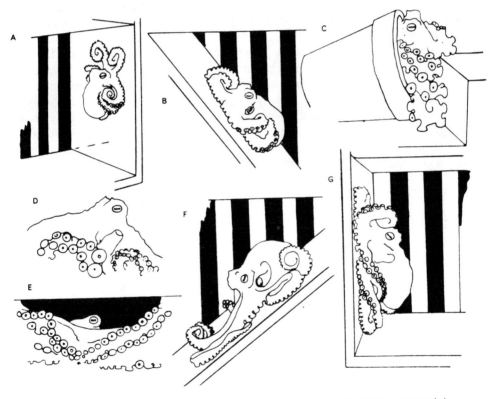

Fig. 14-17. Orientation of the eyes before and after bilateral statocyst removal. In unoperated animals the slit-like pupil normally remains horizontal or very nearly so (A–E), whatever the position of the octopus. After removal of both statocysts this ceases to be true, and the orientation of the retina, as indicated by the position of the pupil, thereafter depends upon the position in which the animal is sitting (F–G). Pictures traced from projections of Kodachrome transparencies: C, D, and E are of comparatively large (500 g) octopuses, the rest of small (15–25 g) animals in an aquarium set up in front of a vertically striped background; in B and F the aquarium, with the animals sitting on the bottom, has been tipped through 45°. From Wells, 1960. *J. exp. Biol.*, **37**, 489–499.

in groups surrounding a rhabdome, which is composed of four rhabdomeres. Like the rods and cones of vertebrates, and the rhabdomes of arthropods, these rhabdomeres have a complex fine structure, each consisting of piles of tubules. The important feature of these piles, and one that probably has a fundamental effect upon the mode of behaviour of these animals, is that the piles are so arranged that two of them lie in the horizontal plane and two in the vertical plane. Each pair (vertical or horizontal, as the case may be) is related to a distinct nerve fibre.

Behaviour studies of the octopus show that this animal has marked powers of form discrimination, but that these depend largely upon the estimation and comparison of horizontal and vertical extents. Thus it can readily discriminate between horizontal and vertical rectangles but cannot discriminate between oblique ones. It has been suggested that this limitation is a consequence of the horizontal and vertical orientation within its receptor elements, an orientation that is also detectable in the dendritic fields of the optic lobes into which the impulses from the eyes are discharged. The importance of this orientation is further suggested by the octopus maintaining a constant orientation of its eyes, irrespective of the position in which the body may be held (Fig. 14–17). This orientation depends upon information received from the statocysts. If these organs are removed the orientation of the eyes is disturbed; for example, they may be held at an angle of 90° from their normal position, and in these circumstances the animal may interpret vertical rectangles as horizontal ones, and vice versa. Clearly the apparent convergence of the cephalopod and the vertebrate eye is superficial and deceptive, and is applicable only to the more general features of the structure of these organs. Closer analysis shows that the mode of functioning of the dibranchiate eye, and its relationships with the statocyst, attain a very high level of specialization in details that have no parallel in the vertebrates.

15
Primitive Nervous Systems

15–1 COMPONENTS OF BEHAVIOUR

We have so far viewed animal behaviour in the simplest possible terms of a pattern of stimulus and response, dependent upon the transmission of coded information from receptors to effectors. In exceptional cases a receptor cell may also be responsible for the final reaction to the stimulus. Such a cell (the cnidoblast of coelentrates is an example) is called an independent effector. Much more often, however, the excitation is propagated from receptor to effector through pathways formed by chains of two or more neurones. Commonly this includes passage through a central nervous system, which interconnects different pathways, but this is not an essential feature of neural organization; there are primitive types of nervous system without a differentiated central nervous system.

Where a central nervous system does exist the neural pathways constitute the familiar reflex arcs of classical physiology. The fundamental features of these are afferent limbs, conveying impulses centrally, efferent limbs conveying impulses peripherally to the effectors, and a central nervous system providing central connections. These pathways may be laid down during development, in which case they are genetically determined just as are other structural features. They are thus independent of the experience of the individual, and they form the neural basis of inborn reflex actions. These are characterized in general by the simplicity of their neural pathways, by the association of a sharply defined stimulus with a correspondingly restricted response, and by the termination of the response when the stimulus ceases to act. Such simple and inborn reflex responses are by no means restricted to animals with a central nervous system, as we shall particularly see in coelenterates. Hence we usually extend the concept of reflex response beyond its classical limits to include the behaviour of animals with primitive and non-centralized nervous systems.

Reflex responses, as defined above, appear as unitary components of behaviour. The observed behaviour of animals, however, is commonly of much greater com-

414   INVERTEBRATE STRUCTURE AND FUNCTION

plexity. When inborn it is expressed as instinctive behaviour, which typically consists of reflex actions that are unified or integrated into patterns of activity involving the whole organism. In contrast to reflex responses, instinctive behaviour is usually excited by a complex pattern of stimulation, termed a releaser. Moreover, it is associated with an internal drive which is manifested in seeking, or appetitive, behaviour, in which the animal actively explores the potentialities of its environment instead of passively waiting, as it were, to be stimulated. Although the effect of drive is often clearly evident, the concept is somewhat abstract, and its physiological basis is not well understood. Perhaps it results from characteristic patterns of activity in the nervous system, and more particularly in the central nervous system.

How adaptive behaviour can be effectively organized out of a relatively small equipment of reflex responses is illustrated in the behaviour of the polyplacophoran mollusc, *Lepidochiton cinereus*, as analysed by Evans. His analysis starts with the observation, readily made in the intertidal zone, that when the tide is out the animal is usually on the under-surfaces of stones. If one of these stones is turned upside down in bright sunlight, the animals creep over its surface until they are again on the surface that is now lowermost (Fig. 15–1). Clearly this is adaptive behaviour, protecting the animals from the adverse effects of exposure. Evans has shown that it depends upon responses to three factors in the environment: light, gravity, and humidity.

When exposed to bright light the animals move at random, the rate of movement being to some extent dependent upon the intensity of illumination. They do not orientate their movements by the direction of illumination, at least under experimental conditions, although possibly they can do this if they are exposed to strong sunlight. Nevertheless, despite their lack of orientation, they collect in the shadowed area of a Petri dish, simply because they arrive there by chance, and then, having done so, slow down and stop (Fig. 15–2A,B). A locomotory reaction of this type, in which the stimulus brings about a variation in linear velocity, is called an orthokinesis (*orthos*, straight); *Lepidochiton* shows a negative orthokinesis, a response which is one factor determining its position below stones.

Another factor is humidity, together with the degree of immersion. Chitons will not move across a dry surface. But on a moist surface they are active, much more so

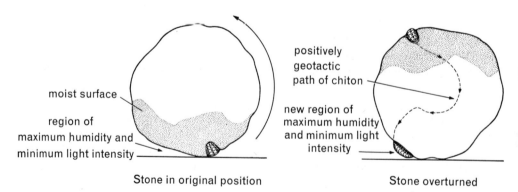

Fig. 15-1. Movements of a chiton when the stone bearing it is overturned. Adapted from Evans, 1951. *J. Anim. Ecol.*, **20**, 1–10.

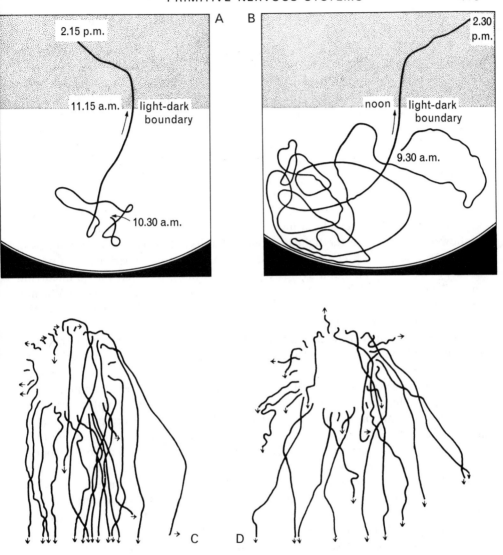

Fig. 15-2. A, B, movements of two chitons placed in the illuminated halves of Petri dishes. C, D, paths of chitons on vertical plates out of water, C, on glass, D, on roughened Perspex. Arrows show in C and D orientation of the head after $2\frac{1}{2}$ hr. From Evans, 1951, *op. cit.*

than when completely submerged in water. Consequently they are stimulated to move when they become exposed as the tide falls, so that they are likely to find protection before the substratum dries. In this they are aided by their response to gravity, which varies according to whether they are exposed or immersed. If they are placed on a glass plate below the surface of the water they move in all directions, showing no response to gravity. When, however, the plate is removed from the water they move downwards (Fig. 15–2C,D), a response to gravity that, in the intertidal zone, will lead them to the underside of stones when the tide falls. A response like this, in which movement is orientated with reference to the source of the stimulus, is termed a taxis; *Lepidochiton* is negatively geotactic.

Two other examples illustrate the same principles. The location of *Littorina neritoides* on the seashore, which is in crevices towards high-water mark, is determined primarily by its responses to light and to gravity. In addition to being geonegative, it is photonegative when it is under water and the right way up, but photopositive when it is upside down. If, however, it is out of water, it is uniformly photonegative. These reactions result in it moving upwards, and in and out of crevices, when it is under water. When, however, the tide falls and it is left exposed, its photonegative response leads to it remaining within the crevices where we find it.

The burrowing amphipod of mud-flats, *Corophium volutator*, is photonegative when moving on the surface, so that it tends to pass down towards the water-line. It is photopositive when swimming, so that it tends to move up towards the waterline. It burrows more readily in light than in the dark, so that it tends to burrow in shallower waters where the light is stronger. But its behaviour is modulated by still other responses, which can be demonstrated in laboratory experiments. It prefers deoxygenated sediments and avoids very shallow ones, it prefers fine-grained sand to coarse-grained, and it is sensitive to the microbial life of the substratum. We shall see in chapter 20 that these reactions, which circumscribe the distribution of this animal on the shore, parallel the reactions of many marine larvae.

Effective though these patterns of adaptive behaviour clearly are in ensuring optimum exploitation of the habitat, they are based upon a few comparatively simple reflex responses. Perhaps they would not usually be described as instinctive behaviour, for this term could seem more appropriate to the more elaborate behaviour patterns in, for example, the lives of social insects. Often, however, complex instinctive behaviour is found, on analysis, to be made up of a series of reflex responses to separate releasers. An example of this is the courtship behaviour of the Queen butterfly (*Danaus gilippus*). Here copulation is preceded by a sequence of actions by one or other of the two partners, the response of one partner serving as a releaser for the next response of the other partner (Fig. 15–3). An important element in this behavior is the production by the male of a scent which is disseminated by the structures called hair pencils that are erected during the courtship. This scent, which is an example of a pheromone (Sec. 17–8), arrests the female's non-specific escape flight, and probably keeps her quiescent after the male has induced her to alight.

This, therefore, is a field of analysis in which we cannot draw rigid boundaries. Behaviour that is apparently complex may yet depend upon very simple neural mechanisms, as, for example, when the feeding response of *Phormia* is evoked by the stimulation of a single receptor. To some extent, then, we can look in animal behaviour patterns for a continuous spectrum of organization, ranging from simple reflex responses to the grouping of these into highly elaborate behaviour patterns. Underlying all of this is the phenomenon so very characteristic of living organisms, and so very difficult to analyse: the fact that they behave as wholes rather than as the sum of their constituent parts. Their behaviour shows integration, which we have already mentioned as a process unifying the actions of an organism into patterns that involve the whole individual. We can formally define this as a phenomenon in which the generation of a single nerve impulse by a receptor may evoke in the effector system a response greater or less than could be produced by the arrival of one

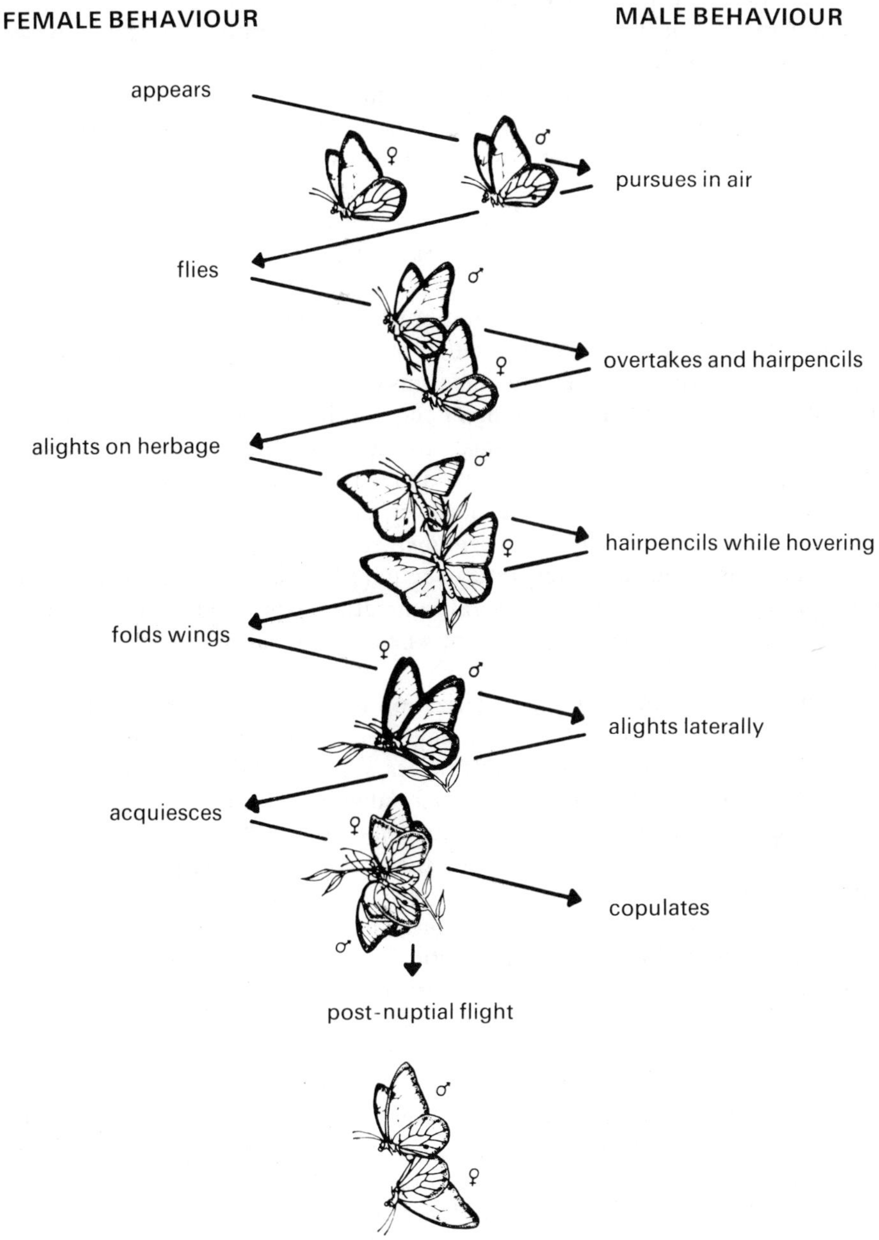

Fig. 15-3. Summary of the stimulus–response reaction chain in the courtship of the Queen butterfly. The male behaviour is shown on the right and the female behaviour on the left. From Brower *et al.*, 1965, *Zoologica*, **50**, 1–39.

impulse. What this means for the animal is that it can give a unified response to a diversity of signals. We shall see that integration is provided for in even the simplest of nervous systems. But one of the advantages gained by the increasing elaboration of neural organization is undoubtedly improvement in integration, with all the possibilities of behavioural elaboration that flow from this.

Integration depends in part upon inerconnections within the nervous system; these permit efferent pathways to be linked with a wide range of incoming information. But no less important are the physiological properties of nerve cells, and particularly of the synapses which make the connections between them. A brief outline must serve to illustrate this aspect of neural organization.

The result of the arrival of a nerve impulse at a synapse depends in part upon the properties of that synapse. If it is an excitatory chemical one, each impulse will release a minute quantity of the transmitter which reacts with receptors on the synaptic membrane of the post-synaptic cell to produce a brief excitatory post-snyaptic potential (EPSP) which results from the transient depolarization. Unlike the action potential, this is not an all-or-nothing effect, and it cannot be propagated without decrement. It may, however, be adequate to depolarize the non-synaptic membrane of the post-synaptic cell; this will evoke immediate spike formation. But a single impulse may be inadequate for this; however, the arrival of a train of impulses can evoke a series of EPSP's which add arithmetically (more or less) until the resultant depolarization is sufficient to generate an action potential. This additive effect constitutes summation. It is called temporal summation, when many impulses arrive at the same synapse, or spatial summation, when they arrive at different synapses on the same cell (Fig. 14–1, p. 390).

Another factor influencing depolarization is that repeated impulses can cause the EPSP to grow to a point at which it may evoke an action potential even without the additive effect of summation. This process, which permits the total effect of the impulses to be greater than the sum of their individual effects, is called facilitation. Other possibilities include synaptic fatigue, which is a decline of synaptic potential with repetition, or, at inhibitory synapses, the evoking of a hyperpolarization which results in inhibition of action potentials.

Inhibition commonly occurs in the central nervous system, but this is not always so; peripheral inhibition is a characteristic feature of crustaceans, in which both excitatory and inhibitory nerves may innervate a single muscle fibre. The result of this, in conjunction with other factors such as summation and facilitation, enables these animals to secure fine ccntrol over muscular responses with a relatively small number of motor neurons. This is another example of economy in arthropodan organization, here permitting adaptive responses to be mediated by the small central nervous system of these animals. Vertebrates, by contrast, rely upon the subdivision of their muscles into large numbers of groups of fibres (motor units), each innervated by a single motor neuron. Control of the force exerted by the muscle is here contrived by varying the number of motor units that are brought into action.

These, then, are some of the features of neural organization that promote integration. But there is another aspect of behaviour to be considered. Survival in a constantly fluctuating environment would be difficult, if not impossible, if it depended solely upon rigidly determined patterns of activity, established during individual development under the influence of genetic control. Some degree of flexibility of response is therefore required. Many factors contribute to this. For example, responses depend very much on the internal state of the animal; thus, the reactions of a hungry animal will differ from those of a well-fed one. *Phormia* has already given us one illustration of this. Another is provided by the sheep tick which is found at the tips of blades of grass when hungry, but which moves downwards when it has had a meal. Reactions will also vary with age and state of development; in particular, the state of sexual maturity will be of paramount importance. Here, as in other conditions, the secretion of the chemical regulators known as hormones is often important in influencing the excitability of the nervous system.

Perhaps most important of all, however, is the modification of behaviour in the light of the previous experience of the animal. This is the phenomenon called learning, which we may define, following Thorpe, as that process which manifests itself by adaptive changes in individual behaviour as a result of experience. Learning takes many forms, which will be mentioned here only briefly, leaving examples to be dealt with in later contexts. Probably its simplest form is habituation, or the cessation of response to a repeated stimulus when the stimulus is without significance in the mode of life of the animal concerned. The capacity for habituation is clearly a fundamental need of animals. Without it there must be a waste of energy which could be more usefully employed to promote survival. Not surprisingly, therefore, it is widespread in the animal kingdom; it is seen in Protozoa (p. 420), and it may very well be a universal property of excitable living material.

Another type of learning is called associative learning. One example of this is known in classical physiology as the development of a conditioned reflex. This is observed when two different stimuli are applied simultaneously and repeatedly to the animal. In these circumstances the reflex response that is normally given to one of the stimuli (this stimulus is called the reinforcement) becomes transferred to the other stimulus. This can now evoke a reflex response that is not an inborn one, and the animal is said to be conditioned to this second stimulus, or to have developed a conditioned reflex.

The other type of associative learning is called trial-and-error learning. In this the animal develops an association, during appetitive behaviour, between a stimulus and a response which is an element of that behaviour. Both the stimulus and response, however, precede the reinforcement, which, in contrast to classical conditioning, exerts a retroactive influence in linking them. Further, the motor response is not an inevitable inherited response to that reinforcement. Such learning is seen when an animal, moving in a simple T-shaped or V-shaped maze or choice chamber, learns to take one path as a result of receiving a distasteful stimulus after entering the other path. Learning to run more complex mazes is another example. In experiments of this type the reinforcement (often called the reward) may be food, or perhaps some form of 'preferred' environment.

A reward is not, however, an essential feature of learning, for there is also latent learning, in which associations are formed without any obvious immediate reward. We shall see illustrations of this in the lives of social insects; worker bees, for example, learn landmarks in the neighbourhood of the hive preparatory to undertaking long foraging flights. Finally, there is insight learning. We can recognize this in ourselves, but find it difficult to define what it involves in terms of neural structure and functioning. We can only remark that it appears to be expressed as the sudden solution of a problem, perhaps through the sudden reorganization of the past experience of the animal. It is the foundation of scientific method, for it is out of this capacity that hypotheses are born.

Enough has now been said to show that there is much variation in the behaviour patterns shown by animals, and in the flexibility manifested in their responses. To a large extent these variations reflect differences in the level of organization of the nervous system. We shall now examine these differences to see something of how the requirements of adaptive behaviour are met in the main groups of invertebrates. In doing this, however, it is worth remembering that studies of the nervous system and behaviour, more than most fields of biological investigation, confront us with the problem of evaluating the nature of scientific evidence. Thorpe has warned of the inadequacy of information theory and computer design principles as a complete guide to the workings of the human brain. It is wholly proper for biologists to seek, by the reductionist approach, for an explanation of complex neural mechanisms in the relatively simple terms of physics and chemistry. But the holistic approach has a part to play as well, and Thorpe has insisted that eventually we may have to depend for further progress upon some quite new understanding of the joint relationships of bodily and mental activity. There is no justification, he suggests, for dismissing what we call 'mind' as a by-product of the operations of a stupendous computer. He was arguing on behalf of the human brain, but the argument, if acceptable for man, must be no less applicable in principle to the lower types of nervous system from which our own has ultimately evolved.

We are dealing here with issues fundamental to the nature of living material, which we cannot hope to penetrate at our present level of knowledge. In the evolution of the nervous system, as of other systems, there has been an exploitation of fundamental properties of living material that antedate a differentiated nervous system, and do not depend upon it. The situation is well illustrated by *Paramecium*. We have referred earlier to fibre systems in ciliate protozoans which have been thought to function in coordination or integration (Sec. 3–2. But it has never been proved conclusively that they do this, and it is difficult to see them as evolutionary precursors of nervous conduction, remembering how dependent this is in metazoans on surface membrane phenomena (Sec. 14–2). Nevertheless, the behaviour of *Paramecium* is not completely predictable. According to Jennings, it is variable in a way that shows at times the slight beginnings of the modification of behaviour through previous experience. In *Stentor* there are even signs of habituation. Jennings found that after it has responded a few times to some very weak stimulus it begins to react differently; the change is not due to fatigue, but seems rather to have some regulatory character. When a current of water from a capillary tube is directed against its disc the animal at first contracts. But soon it expands again, and next time it does not react to the current, but continues its normal activities unmoved.

Attempts to demonstrate associative learning in protozoans have been less successful, but they have certainly served to illustrate problems of experimental design and interpretation. In one example a single paramecium was placed in a drop of culture medium which was arranged so that one half of the drop was at 15°C and the other half at 42°C; the cooler half was darkened, while the warmer half was illuminated. In these circumstances the animal remained predominantly in the cooler half, giving the characteristic avoiding reaction at the boundary. After 90 min both halves were brought to 15°C. For a time the paramecium still gave the avoiding reaction and so remained in the darker half, but the reaction slowly weakened until after about 15 min the animal was swimming freely throughout both the dark and illuminated zones. This behaviour was at first attributed to the temporary establishment of a conditioned response to illumination, the stimulus of high temperature having acted as the reinforcement.

Later, however, a simple control experiment carried out by another investigator showed that this was a misinterpretation. In this new experiment the illuminated half of the drop was heated before the animal was placed in it, and then after 90 min the temperature was equalized to 15°C. The animal was now introduced for the first time. It immediately gave an avoiding reaction at the boundary, although it had had no experience of encountering a temperature barrier there, nor did one now exist. The explanation proved to be that heating of the water modifies its chemical composition by driving off dissolved gases, resulting in a difference in chemical composition between the two halves. The animal in the first experiment was responding to this chemical difference after the temperature had been equalized, while the disappearance of the response after 15 min was a consequence of the drop acquiring uniformity of composition through diffusion. There had, in short, been no conditioning at all.

Not the least interesting aspect of this episode is that Jennings had long before shown that *Paramecium* was sensitive to just such changes in chemical composition of the medium. As Jensen points out, there are two modes of approach to behaviour studies. An investigator may study one topic (such as responses to light or to temperature) throughout a range of species, or he may elect to study the whole range of reactions throughout one species alone. Both modes of approach are required, and they really need to be coordinated. In the present instance much confusion would have been avoided from the beginning if the results obtained by Jennings from his comprehensive studies of *Paramecium* had been applied to the interpretation of an experimental situation that was not as simple as it seemed.

15–2 COELENTERATES AND THE NERVE NET

A fundamental and primitive feature of neural organization in the coelenterates is the arrangement of nerve cells in a continuous layer to form an irregular nerve net or plexus. Neural pathways become differentiated within this net, foreshadowing the association of fibres into the macroscopically visible tracts that constitute the nerves of more advanced nervous systems. As this association develops in higher groups, the primitive nerve net diminishes in importance, although it may still have some significance.

From previous discussion we know that the responses of the coelenterates show a high level of adaptive efficiency. How the requirements of this adaptive behaviour are met within the structural limitations of what appears to be a very simple plan of neural organization is of particular interest. Our knowledge of the structure of the coelenterate nerve net derives from the nineteenth-century studies of Schäfer and of O. and R. Hertwig. Within the epidermis and gastrodermis of coelenterates there exists a plexus of nerve cells that are usually bipolar or tripolar, although occasional multipolar ones may be present. These cells lie immediately above the muscle fibres, and apparently never within the mesogloea.

This localization was clearly emphasized in the descriptions given by the early authors, although, as Pantin has shown, a debasement of text figures has led to the belief that the nerve net extends into the mesogloea. This error has been encouraged by later accounts based upon the use of silver impregnation. The results of this technique are difficult to interpret, for it also impregnates connective tissue fibres; it is then possible to confuse them with nerve fibres. It is important, therefore, that descriptions of the distribution of these fibres should be checked against observations made with other techniques as well. For this purpose *intra vitam* staining with methylene blue is particularly valuable, although the action of this dye is erratic, and not all species respond well to it. Its use by Pantin and his colleagues, however, confirmed the main conclusions of nineteenth-century workers, and, we are told, led Pantin to formulate the Hertwig law: 'the Hertwigs were always right'.

The properties of the coelenterate nerve net are well exemplified in the responses of the anemone *Calliactis* to mechanical or electrical stimulation of the column. These responses can be related to the several components of the muscular system (cf. p. 73), the contractions of which can be recorded separately (Fig. 15-4). A weak stimulation may result in a local contraction of the circular musculature, leading to the initiation of a peristaltic wave, or the animal may shorten. During this shortening the disc may remain expanded, the parietal muscles being the only effectors involved. A stronger stimulation will also cause withdrawal of the disc through contraction of the mesenteries, while a very strong mechanical stimulation will evoke an immediate protective contraction of the marginal sphincter. These varied responses occur irrespective of where the stimulus is applied, which shows that there is a diffuse conduction of excitation in all directions through the net. The form of the responses varies, however, with the strength of the stimulus: as the strength increases so also does the distance over which its effect spreads.

At one time it seemed that these properties were irreconcilable with the fundamental principle that nerve impulses are propagated without decrement of excitation. Pantin's analysis of the situation was the first to show clearly that the diffuse and

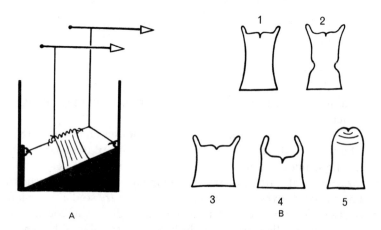

Fig. 15-4 A, method of recording sphincter and mesenteric movements in the anemone *Calliactis parasitica*. B, diagrammatic representation of symmetrical responses of the animal. From Pantin, 1935. *J. exp. Biol.*, **12**, 139–155.

apparently decremental conduction was simply a result of the morphological organization of the nerve net, and that its physiological properties were not fundamentally different from those of typical nerve cells. The explanation depends firstly upon the fact that the net is composed of separate cells that are not fused together, but that connect with each other at synapses. This was, indeed, the view of the early workers, and later studies have confirmed it, although it does not follow that all nerve nets are of this kind. It is believed, for example, that the atrial (visceral) nerve net of amphioxus is without synaptic barriers, and these are probably also absent in *Hydra* as a result of secondary fusion of the processes of the initially separate nerve cells. In anemones, however, the cells remain separate, and it is this, in conjunction with the properties of the synapses, that accounts for the observed peculiarities of response in these animals.

In the oral disc, to take one example, a single electrical stimulus will produce either no response at all, or at best only a slight one. If the stimulus is repeated, muscular contraction begins, and then, as the frequency of stimulus is increased, so the size of the response increases and also the distance over which it is propagated. Now if this were a result of decremental conduction of nerve impulses we might expect a single wave of muscular contraction that would diminish in extent as it passed farther away from the point of stimulation. What we actually find, however, is something quite different. At first the only muscles to respond are those close to the point of stimulation; those farther away do not do so until a number of stimuli have been given, after which these muscles, too, are drawn into the response. The explanation is that the responses of the anemone are normally evoked not by a single stimulus but by a series of stimuli, arising, perhaps, through the touching of a number of mechanoreceptors. The early nerve impulses travel only short distances, being arrested at the synaptic barriers, but they leave the system in a state of excitation. This excitatory state aids the passage of subsequent impulses, provided that they follow immediately; thus the impulses resulting from prolonged stimulation can travel progressively farther. This is the phenomenon earlier referred to as facilitation; in the form described here it is called interneural facilitation.

The properties of the receptors also influence the response, as is shown when the pedal edge of *Calliactis* is stimulated by dropping a weight upon it. This mechanical stimulus is followed by a discharge of impulses leading to contractions of the sphincter muscle. Fig. 15–5 shows a falling-off in the number of impulses, and an increase in the interval of time between them, even though the mechanical stimulus is maintained. This is due to sensory adaptation; the receptors adapt to the stimulus and may eventually cease altogether to respond. Facilitation is shown in the way that the contraction increases when the impulses follow close together (Fig. 15–5). As they decline in number and frequency there is a corresponding decline in facilitation, and so the response diminishes and finally vanishes.

A further factor that influences the responses of *Calliactis* is the specialization of its muscles; this accounts for the varied character of the responses mentioned earlier. Each muscle gives a facilitated response at a characteristic frequency range of stimulation, this range being related to the natural speed of contraction of each muscle. For example, contractions of the circular muscle of the column are called forth by an extraordinarily low frequency of stimulation, ranging from one every 10 sec to one every 6 sec; this muscle takes many minutes to reach its maximum

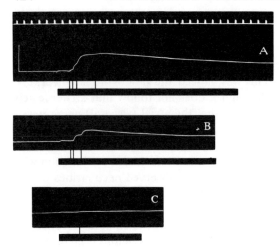

Fig. 15-5. Graded responses of the sphincter of *Calliactis parasitica* to mechanical stimulation of pedal edge. Black band shows duration of mechanical stimulus. Vertical lines show arrival of impulses at sphincter. First line corresponds to arrival of second impulse; first impulse causes no response. Note occurrence of adaptation. From Pantin, 1935. *op. cit.*, **12**, 119–138.

contraction. When the frequency of stimulation is one every 3 sec (Fig. 15–6) there is a very slow contraction of the parietal muscles. At one every 2 sec the longitudinal mesenteric muscles contract in advance of the parietals. At one every 1.1 sec there is a response from the sphincter, longitudinal mesenteric, and parietal muscles. At one every 0.6 sec the sphincter response predominates, and there is no response from the parietals. Thus, although the nerve net of the column provides, from the morphological point of view, a simple unit of diffuse conduction, interneural facilitation and specialization in the effector system allow varied and graded responses.

Even this by no means exhausts all the specialization involved in the action of the primitive nerve net. Examination of the rates at which nerve impulses are propagated through the net shows that there are certain areas in which propagation is very rapid. These areas of through-conduction are especially pronounced in the mesenteries, where they can conduct vertically at a rate of 120 cm/sec as compared with the much lower rates in the column. These vertical mesenteric pathways connect with a ring of rapid conduction near the sphincter, the system as a whole providing for the rapid protective contraction referred to earlier. The existence of these through-conduction tracts was inferred initially from physiological studies; it has since been confirmed by direct observation of the structure of the nerve net. For example, the retractor muscles of the mesenteries mediate the retraction reflex, and it is precisely on the retractor (endocoelic) faces of the mesenteries of *Metridium* that there are well-developed lattices of relatively thick nerve fibres derived from large bipolar cells, predominantly oriented along the oral–aboral axis (Fig. 15–7). The input to this nerve net is from endodermal sense cells; the output is rapidly distributed throughout the effector muscles. On the opposite (exocoelic) face of each mesentery is a sheet of radial muscle. This is a weak muscle, functionally distinct from the retractors, and its associated nerve cells form a much sparser lattice of thinner fibres. Probably this plexus is quite separate from the endocoelic one, which clearly provides the through-conduction pathways that are indicated by the physiological evidence.

We have now seen enough of the actinian nervous system to understand something of the principles upon which it is built. Its fitness to serve the needs of these

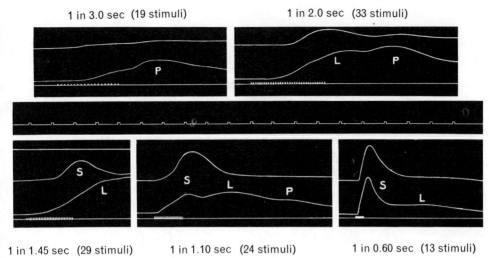

Fig. 15-6. Successive facilitation of parietal (P), longitudinal mesenteric (L), and sphincter muscles (S) in *Calliactis parasitica* with increasing frequency of stimulation. Time in half-minutes. From Pantin, 1935. *op. cit.*

animals, in conjunction with the muscular and hydrostatic skeleton (chapter 4), is again well exemplified by *Calliactis*. This anemone lives on stones on the sea bottom, but is often found also in a mutualistic relationship (p. 676) on shells of *Buccinum* that are inhabited by hermit crabs, and its nervous system is particularly well adapted to this mode of life. Following Pantin's analysis, we can regard the column as a machine for bearing the disc, which is itself concerned with feeding. As the anemone is carried around by the crab the column must necessarily be subjected to a great deal of mechanical shock, but it is to the animal's advantage if much of this is ignored, for it involves no danger, and response to it would interfere with feeding. On the other hand, there will sometimes be more severe disturbance, to which the anemone must give some protective reaction. Facilitated responses, leading under severe stimulation to contraction of the sphincter and retractor muscles, are ideally fitted for these requirements. It is equally advantageous for the feeding reactions of the disc to be separated from the protective reactions of the column. Moreover, we have seen earlier that the feeding of anemones involves asymmetrical behaviour of the tentacles and mouth. Local stimulation of a tentacle causes it to bend over towards the centre of the disc, while prolonged stimulation brings into action the mouth and the other tentacles. This sequence, in itself highly adaptive, is an expression of interneural facilitation. The emphasis in the disc upon this, rather than upon through-conduction pathways, allows the individual responses of the tentacles and the efficient manipulation of the food.

The division of the actinian nervous system into partially distinct functional components shows the possibility of elaboration that is latent in the superficially simple plan of the nerve net, but the elaboration does not stop at this point. Further possibilities are illustrated by the situation in more active coelenterates, of which the ephyra larva of *Aurelia* will serve as an example. Sexual reproduction of the

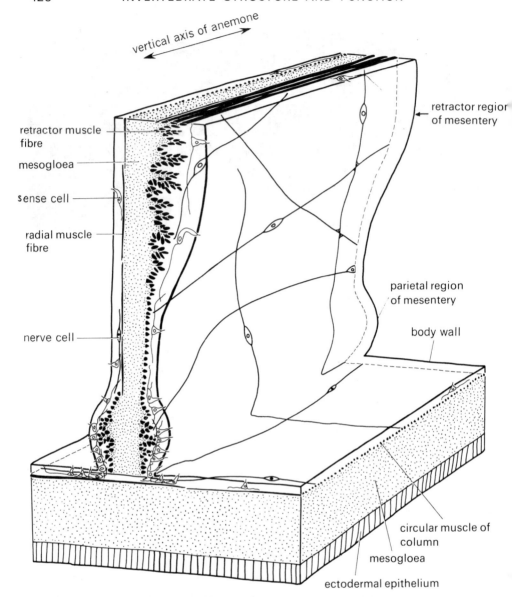

Fig. 15-7. Diagrammatic summary of the histological elements of the mesentery and body wall of an anemone. The endocoelic face of the mesentery is shown. From Batham, Pantin, and Robson, 1960. *Q. Jl microsc Sci.* **101**, 487–510.

mature medusa leads to the formation of the planula larva which settles and develops into a small polyp called the scyphistoma. During winter and spring this reproduces asexually (Fig. 15–8) by transverse fission (strobilation) to form ephyra larvae which swim away as immature medusae. The ephyra has eight arms, separated by the initially undeveloped interradial areas, each arm ending in two lappets (Fig. 15–9).

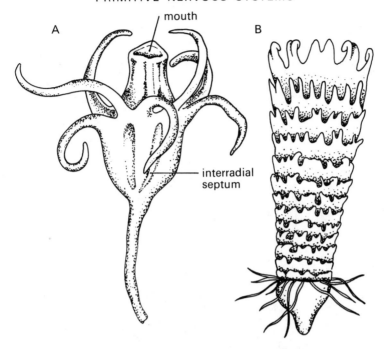

Fig. 15-8. A, scyphistoma of *Aurelia*. B, strobilation of scyphistoma. After Schechter, from Meglitsch, 1972. *Invertebrate Zoology* (2nd ed). University Press, Oxford.

The locomotor movements consist of rhythmic and symmetrical contractions called the swimming beat. These are sharply differentiated from the specialized feeding reactions (Fig. 15-9). Food material (initially protozoans, but later including small copepods) is caught on the lappets, the arm involved bending over to the mouth, which itself moves to meet it. Each arm can act independently in this respect, so that several arms can each catch copepods in rapid succession. On the other hand, a strong stimulus may cause a sustained contraction called the spasm, in which all of the arms fold tightly over the oral surface.

The functional dichotomy of locomotion and feeding is reflected in the nervous system. This includes two distinct nerve nets, one of which is thought to be associated with the swimming beat and the other with the feeding reactions. Indeed, it would be difficult to account for the behaviour of the ephyra larva in terms of only one nerve net, for the feeding response can occur in the absence of any swimming beat, while regular locomotor contractions can occur without any trace of the maintained contraction of an arm. What is needed is a further development of functional isolation within the nerve net, which we have seen to be partially attained as between the disc and the column of the anemone; the differentiation of two nets within the same epithelium provides for exactly this (Fig. 15-9).

One of these nerve nets is a synaptic system consisting of large bipolar cells, 6-10 μm long, with fibres that are usually less than 1 μm thick. This system, called by Horridge the giant fibre system (by analogy with the situation in other inverte-

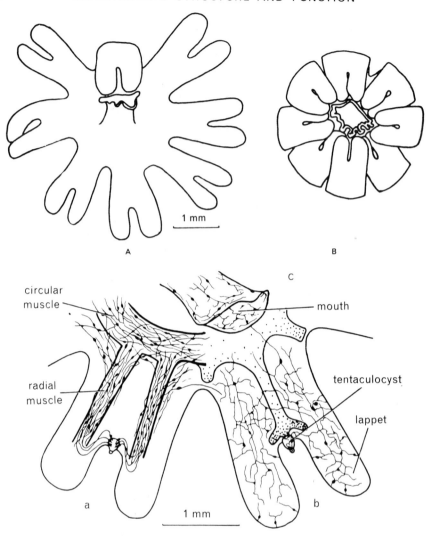

Fig. 15-9. A, the coordination of the mouth and one arm in the feeding response of the ephyra larva of *Aurelia*. B, the long-lasting contraction of all the arms in the spasm. C, two arms of the bell of the ephyra larva, showing *a*, the muscle-strips and the giant fibre system, and *b*, the cells of the diffuse net and the underlying gastric cavity. From Horridge, 1956. *Q. Jl microsc. Sci.*, **97**, 59–74.

brates, Sec. 16–4, is mainly restricted to the sheets of circular and radial muscle fibres. It is believed to be responsible for the coordination of the swimming beat. The second nerve net is histologically quite distinct, and it is distinct also in its distribution, for it is dispersed over the whole of the epithelium. It is composed of bipolar sensory cells, with bipolar and multipolar cells in the main body of the net, these being smaller and their fibres more delicate, than those of the other system. This more delicate net is believed to coordinate the feeding response and the spasm.

This functional interpretation of the two nets is based partly upon deductions from the anatomical structure and from the observed behaviour of intact animals, but it is corroborated also by experiments. For example, if the band of circular muscle and its giant fibre net are cut through at the base of each arm, these arms will now beat each at its own rate, independently of the others, whereas the whole animal can still give a properly coordinated spasm. Other evidence for functional independence of the two nets involves the tentaculocysts. These are hollow structures, with a mass of crystals at their tip and including at their base a group of nerve cells that forms a rudimentary ganglion, which is probably connected to both nerve nets. The function of these cells is not fully understood, but it is believed that the impulses which evoke the swimming beat are distributed from them, and are at least to some extent initiated in them. Whatever their precise role, however, it is found that if an arm is removed it will continue to beat in isolation, provided that it still contains its tentaculocyst. If a cut is then made at one side of the latter, in such a position that the nerve fibres passing from the ganglion to the radial muscle are interrupted on that side, only the strip of muscle on the other side, where the nerve fibres are intact, will continue to beat. The whole arm, however, can still be induced to give a feeding response of spasm, indicating that this reaction is mediated independently of the locomotor response.

Evidently the deceptively simple morphological organization of the nerve net of coelenterates permits considerable elaboration of behaviour. We have noted the existence of independent effectors as part of their behaviour mechanism, and we have now seen something of the contribution made by functional specialization within the nerve net. This type of nervous system is defined by Bullock and Horridge as an anatomically dispersed system of neurons, permitting the spread of excitation through a number of neurones diffusely and in any direction. They list the factors that control the spatial extent and decrement in intensity of response mediated by such a system. Including in these factors are those that we have now encountered: through-conduction, polarization, facilitation, regional differentiation, and the existence of two systems, slow and fast, in the same epithelium. Another factor is the presence of spontaneous patterns of activity, differing in different parts of the body, and reflected in the spontaneous movements that we noted earlier (p. 73). Then also, in lower animals as in higher ones, fluctuations of the internal state play an important part, influencing the interrelationships of the spontaneous rhythms and determining the pattern of response at any given moment. Changes in state provoked by hunger or satiety are only one obvious example of this.

Reviewing this list, and remembering that it is not exhaustive, we can appreciate more easily the integrative capacity of the nerve net, and how complex behaviour can be founded upon it. One example of its possibilities is exemplified in D. M. Ross's description of the way in which *Calliactis* becomes attached to shells (Fig. 15–10). The process involves first the attachment of the tentacles and oral disc to the shell, then the freeing of the pedal disc from its previous attachment, followed by its adhesion to the shell, and finally the release of the oral attachment and the establishment of normal posture. The sequence of movements shows a remarkable degree of coordination, during which the animal receives a flow of information about the chemical and physical properties of the shell. It will not, for example, attach to shells that have

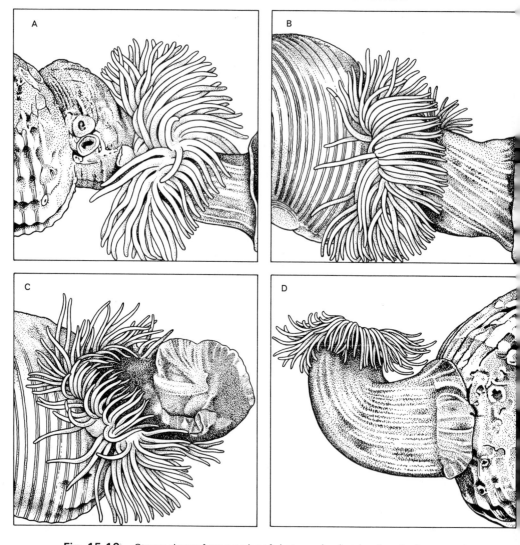

Fig. 15-10. Stages drawn from a series of photographs showing the attachment and settling of *Calliactis parasitica* on a shell occupied by *Pagurus bernhardus*. A, establishment of contact between anemone attached to wall of tank and shell (small area of pedal disc not securely attached). B, tentacles establishing firm attachment to shell; column of anemone twisting; pedal disc loosening its hold on glass. C, oral disc expanded to provide large area adhering to shell; pedal disc freely suspended; column bending to bring pedal disc into position for attachment to shell. D, most of pedal disc now attached, assisted by a constriction of the column immediately above it; oral disc detached and column beginning to straighten out. Approximate times involved from A: B, 5 min; C, 9 min; D, 22 min. After photographs by Ross, 1960. *Proc. zool. Soc. Lond.*, **134**, 43–57.

been boiled in alkali. This information is then translated into a series of responses, some of them asymmetrical, which are adapted to the achievement of a precisely defined goal.

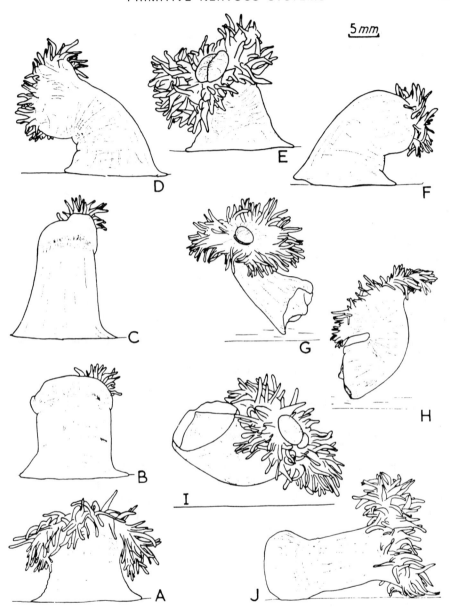

Fig. 15-11. *Stomphia coccinea.* A, normal appearance. B, responding to stimulation by starfish; oral disc partly retracted and sphincter contracting. C, extension of column after its contraction; oral disc beginning to expand. Both actions due to contraction of circular muscle of column. D, Lateral bending caused by rapid contraction of limited area of parieto-basilar muscles. E, lateral bending towards observer. Note folds of stomodaeum protruding through mouth as a result of increased internal pressure. F, lateral bending. Note raised edge of pedal disc. G, swimming. The animal is free from substratum and is actively swimming. G, H, and I illustrate the side-to-side bending movements. H, swimming. I, swimming. Note conical shape of pedal disc. J, period of inactivity after swimming. Note that the oral end of the animal is supported by the turgid oral disc and tentacles. From Sund, 1958. *Q. Jl microsc. Sci.*, **99**, 401–420.

Another impressive example is the sequence of movements evoked in the anemone *Stomphia* by contact with a starfish (Fig. 15–11). The anemone first contracts, but soon elongates and becomes very turgid. It then begins a writhing movement in which the oral disc is whipped from side to side. Releasing itself from the substratum, it propels itself through the water by means of the same writhing action, settling down on its side after a few minutes in an exhausted and virtually inexcitable condition. Within 1–2 min it has recovered, and assumes its initial sessile posture.

Horridge, in his discussion of this behaviour pattern, points out how astonishing it is that so complex a reaction can occur in animals with a nerve net. It is especially elicited when the tentacles touch one or other of two particular species of starfish. At least 15 other genera are ineffective, although the reaction can also be evoked by electrical stimulation, or when the base of the column is touched by the nudibranch *Aeolidia papillosa*. Remarkable sensory discrimination must therefore be involved. Analysis of the events shows that a complex sequence of muscular responses is brought into action; release, for example, is accomplished by the basilar muscles of the pedal disc, while the writhing movements are accomplished by sequential contractions of parts of the parieto-basilar system. Probably there is an innate fixed action pattern, but where and how is the necessary information stored to activate these sequential contractions, and how is it called forth? We do not know, nor, indeed, can we even be sure that the reaction has any adaptive significance. It is not obvious that it provides an essential protection from predation, for starfish are not adventurous. A single individual may take more than a week to come into contact with the anemone, even when both animals are confined together in only $1\frac{1}{2}$ cu. ft of water, and it is not even clear that they normally occupy the same habitat. Perhaps, suggests Sund, the reaction is a form of protection against the risk of being buried by mud and silt, but again we cannot be sure.

It is sobering to reflect that there is no easy understanding of the ways of even the so-called elementary nervous system. We may claim, as reductionists, to know something of the component parts of these responses, but we can hardly pretend to understand how they are integrated into the behaviour of the whole animal. Here, as so often, we are describing aspects both of the life and of the organization of an invertebrate, while still remaining a very long way indeed from understanding what the animal is really doing.

15–3 ECHINODERMS

The study of the echinoderm nervous system, which will be considered here mainly with reference to *Asterias*, presents considerable technical difficulties, some of which will be mentioned later; in consequence, a very great deal remains to be learned about it. Clearly, however, it is structurally a relatively simple system, much of it retaining a primitive position in close relationship with the epidermis. Brain and ganglia are lacking, but some concentration of nervous elements has occurred, for circular and longitudinal nerve cords are present. This is an important advance, for it makes possible the development of separate conducting systems and the beginnings of a central nervous system.

Much of the surface epithelium of echinoderms is sensory, with ciliated cells that bear microvilli. Some of these cells bear a single cilium surrounded by a collar of microvilli, and are examples of the choanocyte-like cell type mentioned in an earlier context (Sec. 4–4). Below the epidermis, which is sensitive to touch, light, and chemical stimulation, are the cell bodies of multipolar neurons, together with an array of miscellaneous cells, some of which secrete mucus, while others are pigment and glial cells. Axon-like processes of the epidermal sensory cells extend through this region, forming a plexus, but whether this is a true nerve net is doubtful; probably not in echinoids, but it may be so in asteroids. The plexus is thickened along the edges of the ambulacral grooves in starfish to form marginal nerves (see Fig. 15–13) that extends the length of the arms. The epidermis along their course is said to have as many as 70,000 sensory cells per square millimetre.

Closely associated with this superficial plexus in *Asterias*, and immediately overlain by the epidermis, is the ectoneural nervous system (Figs. 15–12, 15–13), the main features of which are a circumoral nerve ring and five radial nerve cords, one of which extends from the ring down each arm. Each cord is composed of tracts of fibres, small in diameter (0.3–1.0 μm), and with little visible internal structure. The sensory epithelial cells send fibres into the radial cords. Probably, therefore, many of the fibres in the cords originate in this way, for few cell bodies can be seen in them, and there are no ganglia. Other fibres must originate from the multipolar neurones which lie beneath the epidermis, many of

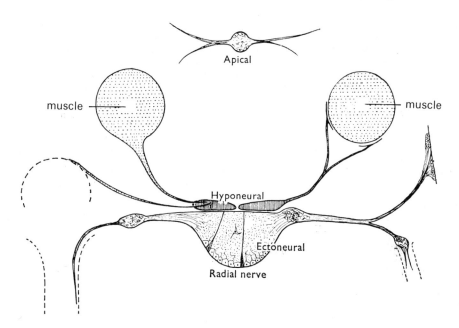

Fig. 15-12. Diagrammatic cross-section of the nervous tissue of the arm of a starfish, to show the arrangement of the nervous system. Central pathways are within the radial nerves and the circumoral ring. These central tracts give rise to a series of pseudo-metamerically repeated series of nerves, some of which run to the musculature and are purely motor, while others are of mixed composition. From Pentreath and Cobb, 1972. *Biol. Rev.*, **47**, 363–392.

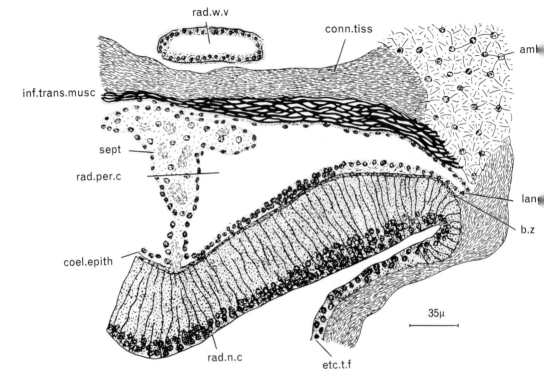

Fig. 15-13. Transverse section of half of the radial cord of *Marthasterias glacialis* showing the innervation of the inferior transverse muscle by Lange's nerve. Fixed corrosive-acetic, stained Mallory's triple stain. *amb.oss*, ambulacral ossicle; *b.z*, boundary zone; *coel.epith*, coelomic epithelium; *conn.tiss*, connective tissue; *ect.t.f*, ectoderm of the tube-foot; *inf.trans.musc*, inferior transverse muscle; *lang.n*, Lange's nerve; *rad.n.c*, radial nerve cord; *rad.per.c*, radial perihaemal canal; *rad.w.v*, radial water vessel; *sept*, septum. From Smith, 1937. *op. cit.*

which are likely to be interneurones. Small fibre tracts, similar in appearance to the main cords, run out from the latter to the tube-feet, ampullae and pedicellariae; two types of axon are thought to occur in these, one type being sensory and the other motor. The ectoneural system thus seems to be mainly sensory, but with some motor function as well.

One feature that is probably of functional importance is the presence in the main cords of small areas of neuropile, in which the nerve fibres show close interweaving instead of the regular parallel alignment seen elsewhere. Similar areas are found locally in the subepidermal plexus. The fibres within the neuropile contain numerous vesicles, which range from 35 nm to over 100 nm in diameter, and which may reasonably be supposed to contain chemical transmitter substances. (Acetylcholine is thought to be present in sensory and motor fibres in echinoderms, and either dopamine or noradrenaline in the fibres of the interneurones.) It is supposed that the organization of the neuropile areas provides for the transmission of signals from the sensory fibres to those of the interneurons and motor cells. Typical synapses, however, are rarely if ever detectable, so that passage of the chemical signals must be relatively unorganized and non-specific, as is suggested also by the varicose appearance of many of the supposedly pre-synaptic fibres.

The ectoneural system is the most constant feature of the echinoderm nervous system, being present in all except the crinoids, which lack a complete circumoral nerve ring. Two other components, however, may be present in one or other of the main groups. One of these is the hyponeural system, an exclusively motor component which lies more deeply that the ectoneural system, but is closely related to it in function. It is best developed in ophiuroids, where it innervates the muscles responsible for the characteristic flexure of the arms. In asteroids it is represented by Lange's nerve (Fig. 15–13). The third component is the apical system, a motor system which is present only in asteroids and crinoids. In the latter it forms an important nerve centre, but in asteroids it is inconspicuous.

Since the hyponeural system is exclusively motor, it is necessary to explain how sensory information is conveyed into it. At one time it was thought, on the evidence of methylene-blue staining, that there were nervous connections between the ectoneural and the hyponeural systems, but it is not always possible to distinguish between nervous and non-nervous fibres in such preparations, and electron microscopy has failed to show these connections. On the other hand, the connective tissue barrier between the two systems is very thin at certain places, sometimes no more than 10 nm thick, and it contains in these areas granular material like that found in synaptic clefts in higher animals. It has thus been suggested (but it remains to be proved) that chemical transmission from the ectoneural to the hyponeural system takes place across these boundaries. If this indeed be so, the communication must again be relatively non-specific, as we have already seen for the neuropile areas of the ectoneural system. This, however, could be adequate for the requirements of echinoderm behaviour, which often seems to involve responses that are not very precise. They could conceivably be mediated by the excitation of small groups of nerve cells serving the same general areas of effector organs.

We can now conveniently sum up this interpretation of echinoderm neural organization in terms of a model formulated by Cobb, bearing in mind that there is still much in it that is hypothetical, and (as we shall shortly see) that there is still little direct physiological evidence for its mode of functioning. Sensory signals from a wide area of epidermis (Fig. 15–14, **1**) are passed to the ectoneural neuropile areas (**2**). Motor fibres run from these areas to local effectors (**3**), while the areas are interconnected by interneurones (**4**). There are indirect connections across connective tissue boundaries, between ectoneural interneurones (**5**) and the hyponeural system (**6**), perhaps involving primitive forms of synaptic endings; connections are thus effected with the muscles that are innervated by the hyponeural system. There is some evidence that this innervation is further mediated by neuropile areas in the hyponeural nerves. It will be apparent from Fig. 15–14 that such a system could have been highly dispersed when it first evolved, and presumably it still continues to be so in part. The radial and circumoral cords can be visualized as having evolved through the concentration and compression of parts of the system into clearly defined tracts, primarily to facilitate communication between those neuropile areas which have also been included within the cords. It will be noted that the model, which is intended simply

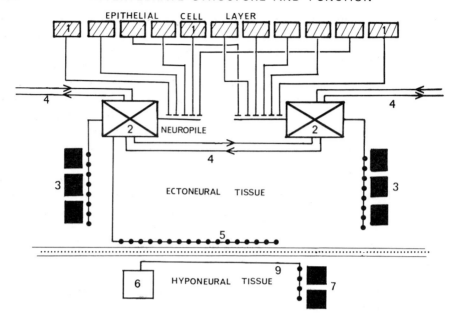

Fig. 15-14. Diagrammatic representation of a hypothetical model proposed for the echinoderm nervous system, based mainly on asteroids. Sensory cells (1) connect to neuropile (2). Motor neurons connect directly with local effectors (3). There are direct interneuronal links (4) between areas of neuropile. There is an indirect link via interneurons and synapses (5) between the neuropile and the hyponeural nerves (6) and the muscles (7) they serve. These is some evidence that neuropile (9) occurs also among the hyponeural nerves. From Cobb, 1970. *Z. Zellforsch.*, **108**, 457–474.

to aid our current understanding, and to provide a basis for further investigation, carries no implication of any phylogenetic relationship with the simpler nervous system of coelenterates, nor does it suggest any link with the chordates. The echinoderm nervous system, like so many other features of the group, is very much *sui generis*, and its origins remain a matter of speculation.

In the meantime, it would be helpful if this model could be tested against the results of physiological experiments or studies of echinoderm behaviour. Unfortunately, this can only be done at present in the most general terms. Electric potentials have been recorded from the nerve cords of starfish and have been shown to be conducted decrementally. Consistent results, however, are hard to obtain, and it is not even possible to be sure that electrical stimulation of the nervous system will always set up electrical activity within it. In any case, these are only records of mass conduction; it has not been possible to obtain repeatable recordings of potentials from single neurones, and this frustrates attempts to identify neural pathways. However, recordings from isolated radial nerve cords of sea urchins show two components of the potentials which perhaps indicate two distinct classes of fibres.

Behavioural studies also present difficulties, both in planning and interpretation. Consistent results are difficult to obtain, partly because responses are influenced by overall levels of excitation evoked by local sensory fields, and also because experiments involving cutting of the nervous system, which might be expected to clarify the course of nerve tracts, have an adverse effect on responses by reducing the general tone of the effectors. Investigators have tended to concentrate on the spines, pedicellariae and tube-feet, and have disclosed complex behaviour patterns for all of these, seen, for example, in the movements with which the spines of sea-urchins respond to changes in light intensity. The activities of the tube-feet of starfish are probably the easiest to observe, however and studies of these have given results that are in line with what we know of neural structure, while also showing how limited this knowledge is.

The locomotor steps of the tube-feet of the starfish involve a sequence of three phases: retraction, pointing in the line of advance, and protraction. When the animal is walking (Fig. 15–15), most of its thousand or so feet are so engaged, and the impression that they give is one of considerable disarray, for there is no discernible phase relationship coordinating the movements of these tube-feet into a common rhythmic pattern.

Nevertheless, there is evidence of neural control. The feet show a common direction of movement, for all of the stepping feet, irrespective of the arm to which they belong, point and step in the animal's line of advance (Fig. 15–15). The direction of movement varies from time to time, but at any given moment one arm will be anterior; this is regarded as the leading arm, and its feet will be stepping along the ling of its longitudinal axis. Variations in this general pattern of movement are brought about in response to stimulation. These variations may result in changes in

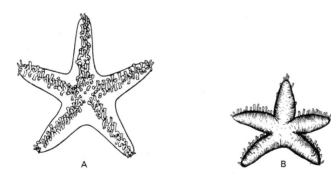

Fig. 15-15. Starfish in ventral view, showing (A) the absence of phase relationship between the movement of the tube-feet, but (B) the common direction of movement of the stepping feet. From Smith, 1950. *Symp. Soc. exp. Biol.*, **4**, 196–220.

the speed of stepping (from 3 to 10 steps per minute in *Asterias rubens*), in changes in behaviour of the individual feet (in connection with feeding, for example), or in changes in the direction of movement.

The responses of the tube-feet can be further studied by inverting a starfish over the open end of a broad glass tube (Fig. 15-16). Stimulation of the dorsal surface by pressure with a probe for varying times evokes several patterns of response, which can be attributed to the progressive spread of impulses through the nervous system. The first response consists of movements of the pedicellariae and spines, the former opening and closing their valves. The maximum response is shown only in the immediate neighbourhood of a brief stimulus; even within a distance of 5 mm the movements are slight, and farther away they are altogether absent. The implication is that we are here seeing the action of local pathways, provided for in the model, with propagation limited by the need for facilitation. This is apparent also in the response of the tube-feet, which consists initially of extension and bending of those feet that lie immediately beneath the site of the stimulus (Fig. 15-16). In addition, there is lateral protraction of the feet that lie distally and proximally to the point of stimulation. This must depend upon the propagation of impulses through the radial nerve cord, for if part of them is removed the response is abolished (Fig. 15-16). The conduction is decremental, the response diminishing with distance from the stimulus. This, of course, is to be expected from the result of electrical stimulation, already mentioned.

The responses so far described are limited to the neighbourhood of the site of stimulation, but there is a further response which is not limited in this way. Within a few seconds of applying a more prolonged stimulus, all the stepping feet through-

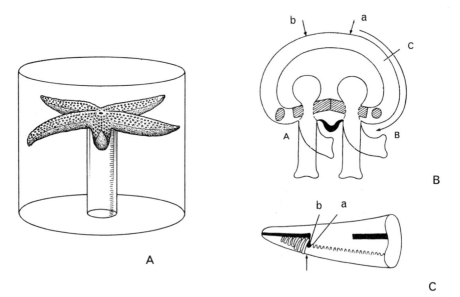

Fig. 15-16. A, position of starfish to show responses of tube-feet. B, effect of tactile stimulus upon underlying tube-feet. C, responses of tube-feet to decremental conduction through the nerve cord; responses are eliminated by removal of a section of the cord. For further explanation see text. Adapted from Smith, 1950. *op. cit.*

out all the arms may show an increased rate of stepping, often accompanied by a change in the direction of their movement. This undoubtedly depends upon propagation of impulses through the radial nerve cords and the nerve ring with which they are connected, for if the cord is cut in any one arm the feet distal to the cut no longer respond.

One conclusion drawn from observations of the movements of starfish is that there must be some degree of central control. We have seen that when the animal is moving normally, one of the arms is temporarily anterior, and all the tube-feet of all the arms are stepping in the line of movement. The source of the control has been thought to be indicated by the behaviour of a single arm that has been isolated from the rest of the body. If this arm has no connection with any part of the nerve ring it will move predominantly with the base foremost. If, however, it retains even a small piece of the ring, then it will usually move with its tip foremost. This, in conjunction with other observations, has been interpreted to mean that there are controlling centres in the ring, perhaps one for each arm. Unfortunately, it is impossible as yet to relate this hypothesis to any feature of the structure of the nervous system; a difficulty which, taken in conjunction with others mentioned earlier, leaves the problem of control to await further analysis.

15-4 HEMICHORDATES

There are good grounds for regarding the Phylum Hemichordata as related to echinoderms within the Deuterostomia (Sec. 18-5), and the nervous system gives some support to this view, for its structure in enteropneusts shows some resemblance to the ectoneural system of echinoderms, although at a more primitive level of organization. Unfortunately, our understanding of it is still very imperfect.

A plexus of nerve fibres lies at the base of the epidermis throughout the enteropneust body (Figs. 15-17, 15-18), interpenetrated by fibrous extensions of the epidermal cells like those in the starfish. It extends also along the anterior region of the alimentary tract. Out of the epidermal system, and still remaining within the epidermis, there are differentiated through-conduction pathways, formed of fibres that are grouped into longitudinally directed nerve cords (Fig. 15-17). Many of these cords are very small, but two large ones are prominent in the trunk, one mid-dorsal and the other mid-ventral, while there is a prominent mid-dorsal one in the proboscis (Fig. 15-18). Completing this mid-dorsal differentiation there is found in the collar a neurocord, formed by the rolling-up of the epidermal nervous system in the mid-dorsal line. This structure, which separates from the overlying epidermis, has often been compared with the hollow spinal cord of the vertebrates, and has been regarded in consequence as one of the truly chordate features of the hemichordates. In fact, however, there is no evidence that it exerts any of the specialized integrative activities of a true central nervous system. Like the other cords, it has only a small amount of neuropile, and it is consequently difficult to homologize any of them with the chordate central nervous system.

A peculiarity of the enteropneust nervous system is that the cell bodies of the neurones lie entirely outside and above the plexus, with their nuclei at a lower level than those of the epidermal cells. These cell bodies are sometimes clearly identifiable as bipolar neurones, and presumably these correspond to the interneurones which,

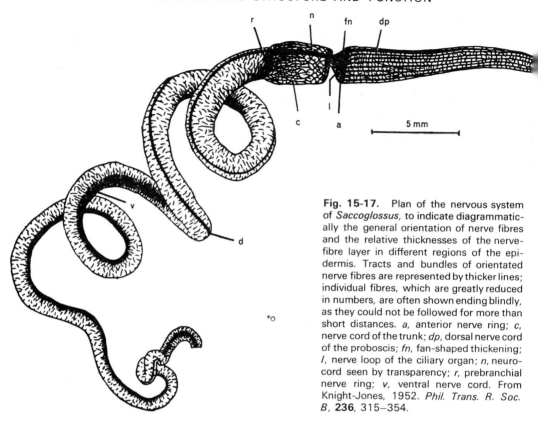

Fig. 15-17. Plan of the nervous system of *Saccoglossus*, to indicate diagrammatically the general orientation of nerve fibres and the relative thicknesses of the nerve-fibre layer in different regions of the epidermis. Tracts and bundles of orientated nerve fibres are represented by thicker lines; individual fibres, which are greatly reduced in numbers, are often shown ending blindly, as they could not be followed for more than short distances. *a*, anterior nerve ring; *c*, nerve cord of the trunk; *dp*, dorsal nerve cord of the proboscis; *fn*, fan-shaped thickening; *l*, nerve loop of the ciliary organ; *n*, neurocord seen by transparency; *r*, prebranchial nerve ring; *v*, ventral nerve cord. From Knight-Jones, 1952. *Phil. Trans. R. Soc. B*, **236**, 315–354.

in the starfish, are enclosed within the plexus. The receptor cells are primary sense cells, with distal fibres passing into the plexus. They are said to be more numerous than the interneurones; a feature which, in conjunction with the absence of sense organs, contributes to the simplicity of behaviour of these animals.

A puzzling feature of the enteropneusts, and one that contributes to the view that their nervous system is more primitive than that of echinoderms, is the absence of any clearly defined motor system corresponding with the hyponeural system of the starfish. However, the motor system varies in its degree of development in echinoderms in accordance with the degree of muscular activity. The condition in enteropneusts may therefore be correlated with the relatively inactive life of these animals. Nevertheless, their reactions certainly demand some innervation of the muscles of the body wall, and it is supposed that this is provided by the passage of individual fibres through the basement membrane, without them being anywhere collected together into nerve tracts.

Just as the nervous system of the starfish is adapted for the needs of an animal that moves over the sea bottom, so that of the enteropneusts is adapted for the life of a burrowing animal. For example, burrowing depends upon the passage of peristaltic waves backwards over the body surface, and particularly over the proboscis; these waves can be shown to depend upon the propagation of excitation down the dorsal nerve cord, from the extreme anterior end. If the proboscis is cut through

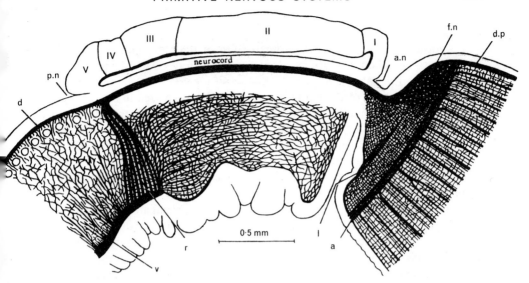

Fig. 15-18. Plan of the nervous system of the collar and adjacent regions of *Saccoglossus*. The dorsal half of the collar is seen in sagittal section, to show the neurocord and the thickness of the epidermal nerve-fibre layer in successive zones of the collar epidermis. The remainder is in surface view to show the arrangement of fibres in the epidermal nerve-fibre layer. Bundles of orientated nerve fibres, represented by thicker lines, are greatly reduced in number for the sake of clarity. a, anterior nerve ring; a.n, anterior neuropore; d, dorsal nerve cord of trunk; d.p, dorsal nerve cord of proboscis; f.n, fan-shaped thickening of the nerve-fibre layer on the dorsal part of the posterior surface of the proboscis; l, nerve loop underlying groove of ciliary organ; p.n, posterior neuropore; r, prebranchial nerve ring; v, ventral nerve cord of trunk. I–V, zones of collar epidermis. From Knight-Jones, 1952. *op. cit.*

transversely, in such a way that the two portions are connected only by a narrow bridge of tissue containing the dorsal nerve cord, the peristaltic waves will continue to pass without interruption (Fig. 15–19). Conversely, if the dorsal nerve cord is cut out at one point, but the whole of the rest of the proboscis is left undamaged, the waves from the front end will stop at the cut while an entirely independent series will start immediately behind that point. Thus the cord is seen to be a polarized through-conduction pathway, impulses from which are doubtless propagated laterally

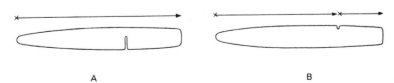

Fig. 15-19. The propagation of burrowing waves along the proboscis of *Saccoglossus ruber*. Bulges are initiated opposite the points marked with a cross, and travel posteriorly as indicated by the arrows. A, peristalsis is not interrupted by a cut through the ventral half of the proboscis. B, peristalsis is interrupted by a lesion of the dorsal nerve cord; an independent series of waves starts behind the lesion. Adapted from Knight-Jones, 1952. *op. cit.*

through the epidermal nerve net. The structure of the proboscis nervous system is thus well adapted for the orderly propagation of impulses evoking peristalsis.

Another characteristic behaviour pattern of the enteropneusts is a retreating movement, in which the animal contracts and draws backwards in its burrow. This involves the passage of peristaltic waves forwards from the extreme hind end of the trunk, instead of the backwards-directed passage seen in burrowing. Experiments similar in principle to those already mentioned show that in this case it is the ventral cord of the trunk, but not the dorsal one, which is responsible for the maintenance of the trunk movements. Direct recordings from the ventral nerve cord have confirmed these observations by showing that it has through-conduction tracts with rapid conduction and short latent period. In these respects, it differs from the radial nerve cords of echinoderms. Excitation in the collar is propagated through the neurocord, but in the proboscis it is propagated through the whole nerve net and not, as might perhaps be expected, through the dorsal nerve cord. In contrast to what is found when the burrowing movements are analysed, cuts through the proboscis will not interrupt the forward progress of the waves, provided that a bridge of epidermal tissue is left. Moreover, this bridge need not in this case include the dorsal nerve cord.

One feature of the neurocord is that it contains giant neurons, giving off giant axons that pass forwards to the anterior end of the proboscis, and backwards through the branchiogenital region. By analogy with the known functions of such giant neurones in other invertebrates (Sec. 16–4), we might expect them here to mediate rapid responses of substantial parts of the body. It may indeed be that they are involved in the abrupt contraction of the anterior end of the trunk that results when the proboscis is touched.

In any case, it is clear that the regular arrangement of the fibres of the plexus, and the concentration of some of them into nerve cords, has provided for the establishment of definitely orientated pathways, as was suggested earlier in this chapter. Moreover, this system despite its apparent simplicity, has powers of integration which ensure that the animal reacts with the behaviour pattern of a whole organism. For example, if an individual is stimulated mechanically at a point halfway along the body, it may respond with a burrowing reaction, or it may show the retreat movement. But whatever the response it will be a total and integrated one; provided the nervous system is complete, burrowing and retreat behaviour will never occur simultaneously. There is nothing to suggest that the neurocord is particularly active in this regard, or that it functions like the central nervous system of the vertebrates. Probably, however, the ventral nerve makes some contribution, for recordings show that it contains, in the hepatic region of the trunk, an integrative centre where impulses are modified in number and pulse rate. But to say this is to say little. The source of the influence that welds separate reflex patterns into an integrated whole remains fascinatingly elusive, in these animals and in many others as well.

The nervous system of the pterobranchs, less well known than that of the enteropneusts, seems to be even more primitive, although neoteny associated with the minute size of these animals may contribute to this appearance. There is a ganglion in the collar (Fig. 10–17, p. 278) which is supposedly homologous with the neurocord of enteropneusts, but it is solid, remains intraepidermal and lacks giant cells, in which respects it differs from the neurocord. One peculiarity of the nervous system is that the synaptic vesicles range widely in size and contents. It has been suggested that this diversity exemplifies a primitive condition, out of which have evolved the more uniform populations of the nervous systems of higher forms. The same argument might, perhaps, be applied to the vesicles of the echinoderm neuropile. Clearly the pterobranch nervous system merits much more study, for it may well be closest to the type of system that was present in the supposedly common ancestors of echinoderms and chordates.

16
Advanced Nervous Systems

16–1 TRENDS IN NEURAL EVOLUTION

As we trace out the history of invertebrate life we see animals developing increasingly complex behaviour patterns, that enable them to exploit their environment with ever-improving efficiency and with an endless variety of means. The nervous system plays a key role in this history, with certain trends in its organization clearly apparent. The primitive nerve net rapidly diminishes in importance, and becomes very difficult indeed to identify, although in some groups at least it still retains significance. In its place there is established a system of nerves, formed of tracts of fibres that convey impulses into and out of a central nervous system, a situation foreshadowed in the development of functionally differentiated tracts and centres within primitive nerve nets.

The particular importance of the central nervous system is that cell bodies, apart from those of the receptor cells, become largely localized within it. Many of these cells are connecting cells: the association cells, internuncial neurones, or interneurones as they are variously called. These form links in the reflex pathways, and make possible the complex of central pathways and junctions that are the structural basis of advances in integration. Another important feature of the central nervous system, although one not entirely confined to it, is the formation of systems of giant fibres, which we shall consider later. These carry further another tendency that we have already encountered: the formation of through-conduction pathways that improve the efficiency of reactions by increasing the speed of propagation of nerve impulses.

Finally, since freely moving animals usually move in one direction, and consequently are bilaterally symmetrical with the major receptor systems at the anterior end, there is marked specialization at that end of the central nervous system. This, which is part of the process of cephalization that we have already encountered, leads to an increasing domination of the cephalic end of the central nervous system over the remainder of the organism, and thus to the appearance of that morphologically and physiologically complex structure called the brain.

These trends (in the analysis of which we shall follow the lines laid down in the comprehensive review of Bullock and Horridge) are already present in the Phylum Platyhelminthes. Here a central nervous system with cephalic dominance is well established, together with some capacity for short-lived habituation. Trial-and-error learning has also been claimed for these animals. In one set of experiments the planarian *Phagocata gracilis* was forced by electric shocks to turn in a particular direction, and was supposedly trained to associate this direction with increased illumination or with a vibratory stimulus. Experiments of this type, however, are very difficult to interpret, for reasons which will be explained later (p. 457), and this field of platyhelminth biology remains a controversial one.

In free-living platyhelminths there are primitively three to five pairs of longitudinal nerve cords, connected by circular commissures which themselves connect with a submuscular nerve plexus. The commissures may be arranged in a regular series or in an irregular one, the former constituting a pattern termed orthogonal (Fig. 16–1). In triclads (Fig. 16–2) the pair of ventral cords tend to predominate, and may be the only cords remaining. In polyclads (Fig. 16–3) the cords tend to branch and to merge into the submuscular plexus. Regardless of these details, however, the cell bodies of afferent neurones are distributed peripherally, while those of the interneurones and motor neurones are distributed throughout the cords, being also grouped at the anterior end of these to form a pair of cephalic ganglia which constitute a rudimentary brain. Here, as also in the rest of the nerve cords, the nerve fibres form a central neuropile, with the cell bodies arranged peripherally around them. This pattern is very common in the central nervous systems of invertebrates.

The nerve cords of platyhelminths are evidently essential for the maintenance of spontaneous movement and for the coordination of responses, since isolated pieces of the body only show these features if they retain some part of the cords. Accounts of the influence of the brain vary, but on the whole, behaviour seems to be incomplete in its absence. In the polyclad *Yungia*, for example, spontaneity of movement is largely lost after removal of the brain (decerebration), yet creeping and even swimming can still be evoked by strong stimulation. Triclads, by contrast, can move spontaneously, if more slowly, after decerebration, but they may be unable to recognize food, although they can swallow it if their pharynx touches it. Already, then, we can see the brain beginning to exert a dominating influence on behaviour.

Little is left in platyhelminths of the primitive elementary nerve net, for the peripheral submuscular plexus does not conduct diffusely, and is not a nerve net in the sense earlier defined. In the pharynx, however, there is a true nerve net which can coordinate feeding movements in the isolated organ; it shows in this a spontaneity which is inhibited in the intact animal by the central nervous system.

16–2 METAMERIC NERVOUS SYSTEM AND LOCOMOTION IN ANNELIDS

The metameric nervous system of annelids, which is derivable from the orthogonal system of platyhelminths, is based upon a ground plan which is also discernible in

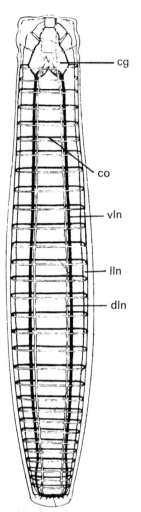

Fig. 16-1. Nervous system of *Bothrioplana semperi* (Order Alloecoela) cg. cerebral ganglion, co, commissure dln, lln, vln, dorsal, lateral and ventral longitudinal nerve cords, After Reisinger, from Hanström, 1928. *Vergleichende Anatomie des Nervensystems der wirbellosen Tiere.* Springer, Berlin.

arthropods. This plan comprises a pair of ventral cords, bearing segmental ganglia and giving off several pairs of nerves in each segment; anteriorly, the cords continue as circumoesophageal connectives which end dorsally in the cerebral ganglia (brain). Bipolar receptor cells lie peripherally, while within the nerve cords are found the motor neurones (motoneurones) and the interneurones. These neurones are concentrated in the ganglia, where, as in platyhelminths, their cell bodies are arranged peripherally around a central neuropile.

Motor neurones tend to innervate only the muscles of their own segments, so that it is possible for segments to act independently. The axon endings have not been easy to find, at least in polychaetes, but it has been suggested that the motor innervation in these animals may be multiterminal, and perhaps also polyneuronal. In theory, this could provide for separate slow and fast motor fibres evoking slow and fast motor responses and could go some way to account for the complexity of locomotor movements. Integration is, of course, essential for the carrying out of these

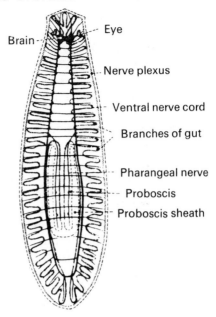

Fig. 16-2. Nervous system and alimentary tract of *Procerodes segmentata* (Order Tricladida). After Lang and Bütschli, from Hanström, 1928. *op. cit.*

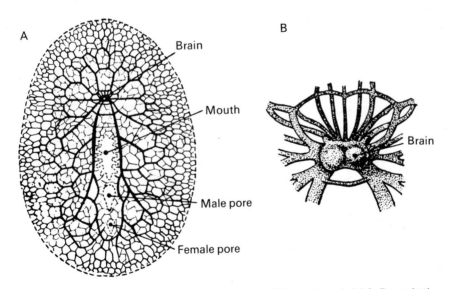

Fig. 16-3. A, nervous system of *Planocera graffii* (Order Polycladida), B, cerebral ganglion and associated nerves of the same. After Lang and Bütschli, from Hanström, 1928. *op. cit.*

movements, and we shall see more than one way in which this is ensured. In general, the afferent fibres of the sensory cells run through several segments within the nerve cord, and this is true also of the fibres of the association neurons. In consequence, adjacent segments can interact in their reception and integration of signals.

448 INVERTEBRATE STRUCTURE AND FUNCTION

The two ventral nerve cords become closely associated. Primitively, however, they remain widely separated, and are then connected transversely to form the ladder-like system that is found not only in certain annelids but even in primitive crustaceans such as *Artemia*. Generally speaking, the more primitive the nervous system the closer remains its relationship with the ectoderm in the adult, although the nerve cords are never intra-epidermal (as they are in the echinoderms and hemichordates). Such primitive appearances, however, may sometimes be a consequence of secondary simplification, associated with small size and neoteny, as we have already suggested for pterobranchs. This may partly explain the apparently primitive form of system in the archiannelids, for despite their name these animals probably owe much of their simplicity to secondary reduction in size.

The finer structure of the annelid nervous system can be illustrated in nereids from the studies of J. E. Smith (Fig. 16-4, Fig. 6-6, p. 161), based largely upon the use of methylene blue staining. Bipolar sensory cells are well developed, being particularly numerous on the parapodia (especially on the cirri) and in the ventral body wall. Information from these receptors is conveyed into the central nervous system by afferent fibres running in the segmental nerves, four pairs of which arise from each of the segmental ganglia. Of these nerves, I and IV (Fig. 6-6, p. 161)

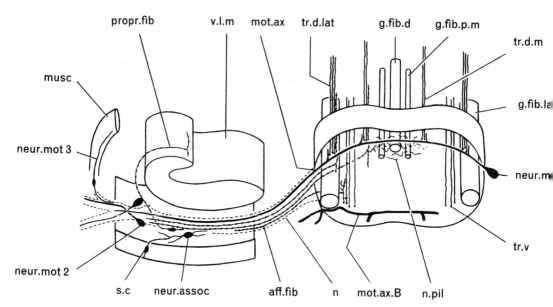

Fig. 16-4. Diagram to show in a generalized form the central and peripheral interneural connections of the component neurones of a segmental nerve. *aff.fib*, afferent fibre; *g.fib.d*, dorsal giant fibre; *g.fib.lat*, lateral giant fibre; *g.fib.p.m*, paramedial giant fibre; *mot.ax*, motor axon of dorsal emergence; *mot.ax.B*, motor axon of ventral emergence; *musc*, muscle; *n*, segmental nerve; *n.pil*, neuropile; *neur.assoc*, association neurone of the subepithelial plexus; *neur.mot, neur.mot(2 and 3)*, motor neurones of the first, second, and third order; *propr. fib*, proprioceptor fibre; *s.c*, sensory cell; *tr. d. lat*, dorso-lateral fine-fibre longitudinal internuncial tract; *tr.d.m*, dorso-medial tract; *tr.v*, ventral tract; *v.l.m*, ventral longitudinal muscle. From Smith, 1950. *op. cit.*

carry fibres from receptors on most of the body surface, II from the parapodia, and III from the mechanoreceptors (proprioceptors) of the dorsal and ventral longitudinal muscles.

A remarkable feature of this aspect of neural organization is the disparity between the number of receptor cells in each segment and the number of fibres in each segmental nerve. Smith finds that the number of afferent fibres in each of the four segmental nerves is only 3–4, 6–8, 2, and 6, respectively, yet there are probably not less than 1,000 sensory cells per square millimetre of the body surface alone. This disparity is partly accounted for by the presence of a nerve plexus which lies close below the basement membrane of the epidermis, and which comprises the nerve fibres of multipolar association cells (Fig. 16–4). Many of the sensory cells discharge into this plexus. From it there arise nerve tracts which pass into the segmental nerves, but the constituent fibres join together so that as the nerves pass inwards there is a continuous reduction in the number of fibres. Not all of the afferent fibres follow exactly this course, however. The proprioceptor fibres, and some of those from the epidermal receptors, pass into the segmental nerves without communicating directly with the plexus at all; nevertheless, the principle of reduction in the number of fibres is unaffected.

The motor fibres of the segmental nerves are even fewer than the sensory fibres, the four nerves of each segment having respectively 1, 3, 1–2, and 4. These fibres, which arise from cell bodies in the segmental ganglia, are more numerous towards the peripheral end of the nerves. Their arrangement thus parallels that of the sensory fibres, and with essentially the same pattern of organization, for the multiplication is effected by the introduction into the motor pathways of intercalary or relay neurones. The more central of these are termed second-order motor neurones, while those that finally supply the muscles are the third-order neurones (Fig. 16–4).

Two conclusions are suggested by this anatomical arrangement. The great preponderance of peripheral fibres in both the sensory and motor pathways may provide for the short-circuiting of the passage of the nerve impulses, so that these do not have to traverse the central nervous system. If this is so (and it is only a suggestion), responses to stimulation can be mediated by local reflexes at the periphery, without involving the central nervous system at all. In addition, the limited connections between the latter and the periphery through the segmental nerves will permit some central integration. Since so few fibres pass to the central nervous system, however, it seems likely that it can only receive broad patterns of information rather than precise detail. The peripheral receptors show much variation in form. This suggests that there must be a good deal of peripheral sensory discrimination, yet from the point of view of the central nervous system much of this information must be wasted. Nevertheless, the importance of central integration is shown by the number of cells that are set aside for this purpose. The sensory cells are situated peripherally, while each segmental ganglion contains only a few large motoneurones; most of the cell bodies in the ganglia must therefore belong to association neurones.

The mode of action of the polychaete nervous system is conveniently illustrated by reference to locomotion. We have seen that annelid movement involves a delicately programmed interaction of the muscles of the body wall with the hydrostatic skeleton, and with the segmental musculature of the parapodia and the chaetae. The execution of this programme must depend upon the integrative action of the nervous system, for although the segments can act independently, yet their individual activities are moulded into the unified behaviour of the whole individual. This is seen when a polychaete begins to move. The first movement forwards is effected by a stepping action of parapodia situated at about four-segment intervals from each other. This action begins at the front end of the body, and spreads rapidly backwards; it is then linked by the movements of the parapodia of the intervening segments, so that the whole of the body becomes involved in the peristaltic cycle (Fig. 16–5). Other evidence of local action is that headless worms and even short pieces of the body can exhibit normal locomotor patterns.

One factor that contributes to the independent action of the segments is spontaneous rhythmic activity in the ganglia, which is revealed by study of the action potentials in the annelid nerve cord. In the earthworm, for example, a rhythmic activity can be detected in the central nervous system during normal peristaltic movement, the rhythm being identical in frequency with that of the muscular contractions. Theoretically, this neural rhythm might be a causal factor in locomotion, arising spontaneously as in the cyclical respiratory and feeding behaviour of *Arenicola* (pp. 166, 305), and evoking muscular responses in the form of reciprocal excitation and inhibition. In fact, however, the situation is not as simple as this. Certainly the isolated nerve cord shows a spontaneous rhythmic activity, but the pattern is often quite different from that of the normal peristaltic rhythm. This lack of correspondence is even clearer in the leech. The nerve cord of this animal shows an electrical rhythm identical with the swimming frequency, but only while it is in normal physiological connection with a body that is displaying swimming movements. It shows no such rhythm when it is removed from the body.

Central automaticity, therefore, is not enough to account for integrated movement. Another factor is required, and this is provided by peripheral excitation evoked by the stimulation of segmental exteroceptors or proprioceptors. Reflex responses to this stimulation are readily demonstrable in intact earthworms. For example, a decapitated worm suspended by cotton threads in water shows no peristaltic rhythm, but this rhythm appears if the worm is removed from the support of the water into the air (Fig. 16–6). The same result ensues if the animal is subjected to tension while it is in water. It appears that in both instances the movement is evoked

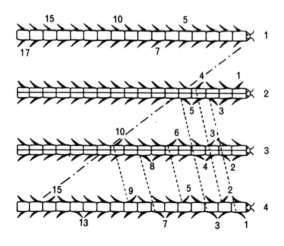

Fig. 16-5. Diagram showing the start of slow ambulation in *Nereis*. Note the rapid spread of the ambulatory pattern (–.–.–) over the whole body from head to tail, and the movement of this pattern from tail to head at a much slower rate (––––––). From Gray, 1939. *J. exp. Biol.*, **16**, 9–17.

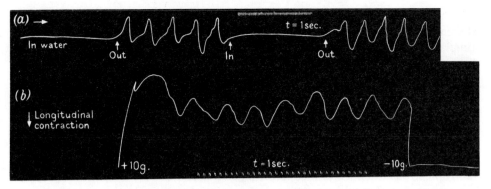

Fig. 16-6. Dependence of the rhythmicity of decapitated earthworm on longitudinal tension. A, a preparation suspended horizontally in water shows no rhythm; the rhythm emerges when the preparation is exposed to tension from its own weight by removal from water to air. B, a preparation freely suspended in water shows no rhythm but quickly exhibits a rhythm on applying longitudinal tension; it becomes inactive again as soon as the tension is removed. The records read from left to right. From Gray and Lissmann, 1938. *J. exp. Biol.*, **15**, 506–517.

by the stimulus of stretching. Another important factor is tactile stimulation applied to exteroceptors on the ventral surface. Thus, a suspended earthworm will show peristalsis while its body is in contact with the substratum, but may cease to do so when it is removed from that contact.

A situation similar in principle, but complicated by a more specialized body form, is seen in the medicinal leech, *Hirudo*. As in the earthworm, terrestrial locomotion depends upon the passage of waves of muscular contraction, which involve successively the longitudinal and circular muscle layers, with the dorso-ventral muscles remaining inactive. In the intact leech these movements are regulated by excitation arising in the suckers. Fixation of the anterior sucker to the substratum is followed by a wave of contraction of the longitudinal muscles, while activity of the circular muscles follows fixation of the posterior suckers. Yet the suckers are not the only factors, as can be seen if the posterior one is removed, and the anterior one denervated. The resulting preparation can still show normal locomotor movements, but only as long as it is in contact with the ground. The movements cease if the animal is lifted off the ground by passing threads underneath it. Thus the preparation is essentially similar in this respect to an intact earthworm; both require tactile stimulation of the ventral surface if they are to move. The intact leech differs from the earthworm, however, in that its specialized suckers provide time signals for the initiation of the waves of muscular activity.

The leech also differs from the earthworm in being adapted for aquatic as well as terrestrial movement. These two types of movement depend on different patterns of muscular activity, and are evoked by different patterns of stimulation. Aquatic movement involves characteristic up-and-down undulations. During these the dorso-ventral muscles are held in contraction, and the circular muscles are relaxed, while waves of contraction pass rapidly backwards down the longitudinal muscles, alternately dorsally and ventrally. The slow rhythms of terrestrial locomotion are only shown when there is ventral tactile stimulation of the suckers or ventral surface, whereas the rapid rhythms of aquatic locomotion are only shown when that tactile stimulation is removed. Thus an intact leech suspended in water shows long-sustained swimming movements, and so also will the preparation already mentioned, in which the suckers are no longer functional.

Finally, the integration of endogenous rhythms and the segmental reflex responses to stimulation must be incorporated into the total behaviour pattern of the whole animal. This depends upon central conduction in the nerve cords, as can be demonstrated in polychaetes by cutting through the nerve cord but leaving the animal otherwise intact. The continuity of movement of the whole body is now interrupted, but the normal locomotor pattern is still shown independently by the two regions anterior and posterior to the point of section. In this simple way we learn that the integration of the segments depends upon the propagation of impulses from segment to segment along the length of the central nervous system.

Central conduction is also readily shown in earthworms. We can divide one of these animals so that the two portions are left connected by the ventral nerve cord alone. Peristaltic movements occur in both portions, but these are completely coordinated with each other, even when there is no possibility of one portion stimulating the other by pulling upon it. This coordination can only be attributed to the conduction of excitation through the nerve cord. The importance of the latter is further shown if isolated pieces of the body, 20–40 segments long, are suspended in a saline bath. The application of gentle tension or touch to such pieces evokes peristalsis, but only if the nerve cord has been left intact. If it is removed from these pieces, the response is no longer obtained.

The interaction of reflex responses and central conduction makes it possible for the contraction of the muscles in any one segment of the earthworm to be evoked in one or other of two ways: by excitation transmitted centrally from another segment, or by the stretch receptors of one segment being stimulated by tension exerted by the next adjacent segment. The first possibility is shown in the experiment, just mentioned, in which two pieces of an earthworm are left connected only by the ventral nerve cord. The second possibility is shown by the fact that peristaltic movement can continue down the length of an earthworm even after complete severance of the body into two portions, provided only that these portions are tied together by threads.

As far as polychaetes are concerned, the integration of movement, dependent in part upon the organization of the segmental ganglia, is aided (in *Nereis virens*) by two other factors: the linking of adjacent segments by the oblique and diagonal musculature, and the overlapping of segmental sensory fields in the dorsal epidermis. Records of unit responses obtained from a segmental nerve while the surface is being explored with a probe (Fig. 16–7) show that the sensory field of one segment overlaps that of the segment on either side. Excitation in all three segments can thus be transmitted through one nerve. It is supposed that this allows the activity of an individual segmental ganglion to be influenced by the approach and direction of a locomotory wave, for excitation evoked in sensory units by the contraction of muscles on their own side can be transmitted to the muscles of the opposite side of the body, and of an adjacent segment. This and other possibilities are indicated by the hypothetical model in Fig. 16–8.

ADVANCED NERVOUS SYSTEMS 453

We may conclude that the rhythmic and integrated actions of annelidan locomotor muscles, to judge from the few species so far studied, depend upon spontaneous endogenous rhythms of activity in the nerve ganglia, but that these must be maintained and modulated by reflex responses to peripheral stimulation, which are themselves facilitated by central integration and by the organization of the musculature. This combination of factors is of great adaptive value, for it makes possible the suiting of the rhythmic pattern to local conditions in the environment. At the same time, central conduction ensures that when the appropriate pattern has been determined, it will be distributed uniformly down the length of the body. Gray and Lissmann's analysis, for example, suggests that the stretch reflexes of the earth-

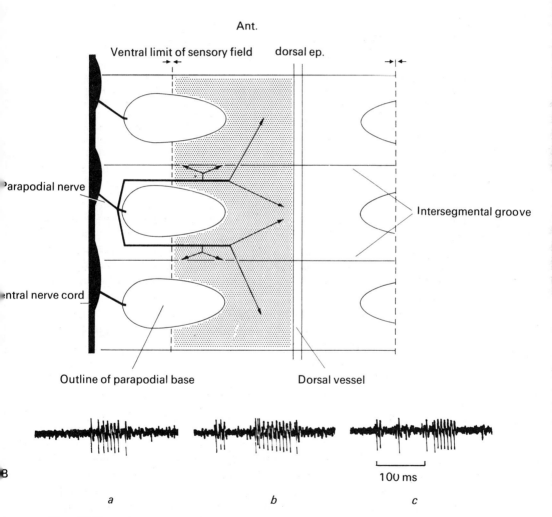

Fig. 16-7. A, diagram of the sensory field of the dorsal epithelial bipolar cells of one half segment of *Nereis virens*. Sensory field indicated by stippled area. B, responses from the dorsal epithelial receptors, recorded in the parapodial nerve, to stimulation in *a*, the segment in front, *b*, the same segment, and *c*, the segment behind. From Dorsett, 1966. *Proc. R. Soc. Lond., B*, **164**, 615–623.

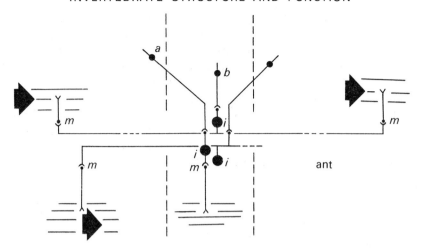

Fig. 16-8. Hypothetical model, based on anatomical evidence, showing how information received from the epithelial bipolar cells of *Nereis virens* might coordinate the propagation of a longitudinal wave, passing from left to right. For example, sensory units (a) might transmit excitation to the longitudinal muscle of the contralateral side. Sensory units (b), stimulated later, might then transmit excitation to muscle of the ipsilateral side over several segments in an anterior or posterior direction! *i*, interneurones! *M*, motor neurones. From Dorsett, 1966. *op. cit.*

worm do not operate when the worm is moving over a smooth surface, but become important when the surface is rough, and thus offers variable resistance to the body surface. In these circumstances, therefore, both the tactile and the stretch reflexes could cooperate with the conducting role of the central nervous system to produce muscular efforts adjusted to the external conditions. Whether local reflexes, suggested earlier as a possibility in polychaetes (p. 449), play any part at all in this is not clear, but there is no decisive experimental evidence that they do so.

In view of the results achieved by local activity and central conduction, one may ask what influence is exerted upon the locomotion of an annelid by its brain. This structure can be thought of as a further development (and a very considerable one) of the type of brain seen in platyhelminths. We have seen that it is the concentration of sensory structures at the anterior end of bilaterally symmetrical animals that determines the appearance of cerebral ganglia, and the point is well shown within the annelids themselves. Oligochaetes and leeches, with a virtual lack of specialized cephalic sense organs, have a simply organized brain. In contrast, polychaetes possess a diversity of cephalic sensory structures (palps, tentacles, eyes), and, in correlation with this, they have a relatively complex brain, distinguishable into regions that are functionally related to the several types of receptor.

In the most advanced polychaete brains it is possible to identify many cell masses and fibre tracts. Some of these cells are neurosecretory ones, as they are in all annelids (Sec. 17–5); the functions of other can be deduced from their peripheral connections. Some of the cell masses are thought to be motor or sensory centres, although the localization of function has been little explored. Others probably have integrative functions, an example being the corpora pedunculata of the mid-region of the brain.

These centres are so called because of their mushroom-like appearance, which results from a mass of cell bodies being associated with a stalk of nerve fibres. These fibres give off many short collateral branches, which implies that the cells concerned are involved in associative functions. What is of particular interest here, apart from the high level of neural organization, is that very similar structures are found in the arthropod brain, where (in bees, for example) they certainly are important association centres. Such a close resemblance is suggestive of homology, although the relationship between present-day annelids and arthropods is so remote that we cannot discount the possibility of convergence, which we have already seen to play such a prominent part in arthropod history. Bullock and Horridge remark that if the corpora pedunculata of these two phyla are not truly homologous, then the forces making for parallel development are indeed remarkable; which, of course, they are.

Although the brain of oligochaetes and leeches is simpler than that of polychaetes, the difference is probably largely conditioned by differences in sensory input, for the effect of decerebration is similar in principle in all three groups. A decerebrate earthworm can crawl normally, and can also eat and copulate, but it is restless and unduly active, its sensory efficiency is diminished, and its burrowing is less effective. Decerebrate nereids are also over-active in general, although they do not feed or burrow. Further, they are insensitive to light, and they show reduced chemosensitivity. Such results show that the annelid brain, in addition to providing differentiated centres, exerts also a generalized inhibitory control of movement. In this respect it has close functional links with the suboesophageal ganglion, for if this is removed from either earthworms or nereids the animals become inactive. A similar effect is seen in leeches. Removal of the brain leaves these animals hyperexcitable and restless; if, however, the suboesophageal ganglion is removed as well, there is a loss of muscular tone, while terrestrial movement only takes place in response to strong stimulation.

It seems, then, that the suboesophageal ganglion exerts some excitatory influence. We have seen that endogenous rhythms in the segmental ganglia play an important part in locomotion. The suboesophageal ganglion, by virtue of its anterior position and its association, through the circumpharyngeal commissures, with the specialized receptor systems, may well play a leading part in the origin and maintenance of this excitatory state in the ventral nerve cord. The brain clearly exerts an antagonistic inhibitory action on these lower centres, while no doubt it modulates their activity by transmitting to them a flow of information from the anterior receptors. This, of course, is part of the significance of cephalization. These functions do not, however, exhaust the importance of the anterior end of the annelid nervous system. It can influence the reactions through which the animals adjust to their environments, and we shall see later that the annelid brain influences growth, regeneration and reproduction through its secretion of neurohormones. Thus our interpretation of the results of decerebration must at present be very provisional, and cannot be formulated solely in terms of classical neurophysiology.

Some of the implications of this discussion are relevant to the results of experiments on learning in annelids. Habituation is readily demonstrated, its adaptive value being expressed in responsiveness to likely sources of harm, economy of effort, and some capacity to discriminate between signals. Responsiveness to danger signals does not preclude habituation to them. A simple stimulus (a shadow, for example) may, for a polychaete worm, be a signal of a predator, but repeated stimulation is likely not to be. Habituation to it will thus make for economy of effort, but it is a delicately balanced adaptation, and its course depends very much on circumstances. A spider will take a long time to habituate to vibrations from its web, for these are always likely to signal food, and it is said that even when the animal has habituated it will readily respond to vibrations coming from another part of the web.

Economy of effort in annelid habituation is well shown by earthworms in their manoeuvering of leaves into their burrows, for leaves which are too tough or too large are left alone after some 10–12 trials. Discrimination is illustrated by the tube-dwelling polychaete *Branchiomma vesiculosum*, which rapidly withdraws into its tube when stimulated by a sudden decrease in illumination or by a moving shadow, either of which might signal a predator. Habituation in response to the decreased illumination is very rapid, but is slower for the shadow, perhaps because this is a more common index of danger (Fig. 16–9).

The complexities inherent in habituation are well shown by an errant polychaete, *Nereis pelagica*, which lives in the brackish mud of estuaries, and feeds by protruding its head. Mechanical shock, a moving shadow, or variation in light intensity cause the animal to retract into its burrow, but it can habituate to all of these stimuli, and can also discriminate between them. Thus the worms soon cease to respond to a moving shadow, but many of such habituated animals will still retract if tested with a mechanical shock (Fig. 16–10). Similar specificity of response, and the consequent saving of effort, is found in animals much more advanced than worms. Nestling chaffinches are an example, quoted by Manning. These give a food-begging reaction in response either to a dark shape appearing over the nest, or to jolting of it, both of which are signals that may indicate the arrival of a parent. The nestlings

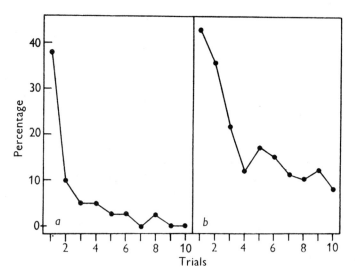

Fig. 16-9. Habituation in tube-dwelling *Branchiomma vesiculosum*, tested when all the worms were expanded. Elimination of the response to a sudden decrease in light intensity (a) is more rapid than the extinction of the response to a moving shadow (b). From Nicol, 1950. *J. mar. biol. Assoc. U.K.*, **29**, 303–320.

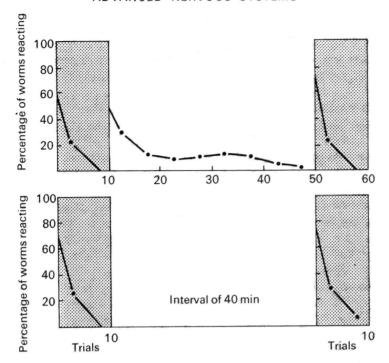

Fig. 16-10. Habituation in *Nereis*. Trials were given at 1 min intervals, the response being the sudden withdrawal of the worm into its tube. *Shaded areas:* responses of a group of 20 worms to a moving shadow. *Unshaded areas:* responses to a mechanical shock with trials at one-minute intervals (upper graph) or a 40-minute period with no trials (lower graph). The upper graph shows that half of the worms respond to a mechanical shock after habituating to a moving shadow, and that habituation to the shock is slower than to the shadow. Comparison of the two graphs shows that habituation to the shadow is lost after 40 min, whether or not mechanical shocks intervene, so that habituation for the two types of stimuli is quite distinct. From Clark, 1961. *Anim. Behav.*, **8**, 82–91.

readily habituate to one of these signals, but can still respond immediately to the other.

Simple T-shaped or V-shaped mazes have been used by many experimenters to establish that trial-and-error learning occurs in annelids. For example, an earthworm, moving along a T-shaped tube from the light into the dark, may be trained to develop the habit of entering one arm if, on entering the other arm, it encounters a lining of sandpaper and receives an electric shock. Reinforcement is provided by allowing it to enter a dark, moist chamber after making the correct choice. Similar results have been obtained with snails.

However, supposed demonstrations of associative learning in the lower invertebrates (all invertebrates below the level of arthropods and cephalopods, let us say) must be evaluated with great caution, for there are many difficulties in their planning and interpretation. For example, *Nereis virens* can learn quite quickly to turn into one arm of a T-maze if, like the earthworm just mentioned, it receives an electric shock on turning into the other arm (Fig. 16–11). This seems good evidence

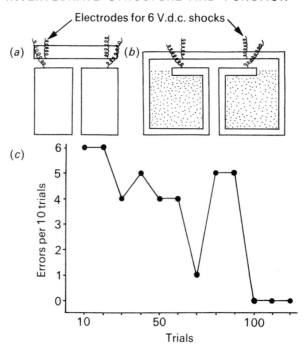

Fig. 16-11. Learning in *Nereis virens*. Diagram (a) shows a T-maze used in preliminary experiments by Evans; performance was unsatisfactory (see text). Diagram (b) slows a later modification, in which the worm was rewarded for a correct choice by being allowed to remain in a dark compartment for 5 min or longer. There were about 25 trials per day. (c) shows the performance of a typical individual under these conditions. From Wells, 1965. Adv. mar. Biol. **3**, 1–62.

for associative learning, but it was not easily secured, for Evans found that the performance was only satisfactory when the worms were rewarded for their correct choice by being allowed to remain in the dark for some minutes (Fig. 16–11). Earlier experiments, in which they were ushered back into the apparatus with a paint-brush, gave less good results, for this treatment was itself a form of punishment, and was discouraging.

Among other difficulties that have to be considered in experimental design is the effect of prior handling. Thus, an earthworm, stimulated on one side of its body, will tend to turn towards the other side when it is later tested in a maze. And then there are the influences which environmental or internal cycles may exert on behaviour. Not surprisingly, as it now seems, the learning ability of earthworms varies with the phase of the diurnal cycle at which the tests are carried out. Then, too, the development of fatigue or of sensory adaptation may give false appearances of learning, although these can be excluded by careful experimental design. It is because of sensory adaptation that we rapidly become unaware of our clothing and eventually, perhaps, of a hair-shirt. In theory, this phenomenon might seem to account for *Nereis*, in some of the experiments just mentioned, ceasing to withdraw its head after repeated stimulation with a probe. In fact, however, this could not be the explanation, for the worm remained aware of the probe, and still tried to seize it. Habituation is here a reasonable explanation of the cessation of head retraction.

One complicating phenomenon is sensitization. This is the term applied when an animal increases its response to a stimulus after receiving either a reward or a punishment. For example, *N. diversicolor* will crawl to the end of a glass tube to obtain food, but withdraws if subjected to a sudden increase in illumination. In one experiment, however, 4 out of a group of 19 worms responded to the light by crawling on. All the group were then fed, and were tested again after a 30 min interval; 12 of the 19 now crawled on. In some way the presentation of food had sensitized the response of the worms to light; a result that could easily give a false impression of succesful conditioning in experiments aimed at bringing about an association of food with illumination.

It has been suggested that sensitization might have provided an intermediate stage in the evolution of associative learning, but whatever the truth of this, it is dangerous, as Manning points out, to try to formulate a phylogenetic sequence of learning ability. Animal evolution must have depended heavily upon the ability of nervous systems to store the effects of experience. Nevertheless, learning is only one of the ways in which animals are adapted for survival, and we should not be led by our own prejudices to assume that it is always the best one. M. J. Wells argues that it must be evaluated in relation to the physiological background, and also in relation to what the animal is built to do. The conditions of life of some animals, and particularly marine ones, are so constant that selection may well have favoured the programming of a stable and unchanging behaviour pattern. In these circumstances a limited capacity for learning should not be regarded as indicating a low level of total physiological capacity. In other circumstances, however, an environment providing complex and fluctuating stimulation may well have promoted the establishment of much flexibility in behaviour.

Granted, however, that forms of learning do occur in annelids, what part in them is played by the supra-oesophageal ganglion? There is evidence that it does not dominate the process to the same extent that the brain does in animals with more advanced nervous systems. For example, behaviour acquired by *Nereis* in the type of learning experiment just mentioned may still persist even after the supra-oesophageal ganglion has been disconnected from the rest of the central nervous system, so that it can have no effect on the worm's choice.

The difficulty in conducting such experiments is that disconnection or removal of the ganglion must often affect the behaviour of the animals by interfering with the inflow of essential information from the cephalic sense organs. Much of this information passes through the ganglion, and is lost to the animal when that ganglion is disconnected. Fortunately the peristomial cirri of nereids provide an exception to this, for they are connected directly with the ventral nerve cord. The retention of the behaviour pattern in the experiment just mentioned depends upon this connection being preserved, so that the animal retains part of its sensory input despite the disconnection of its supra-oesophageal ganglion. Here at least, then, we have evidence that retention of a new behaviour pattern does not require retention of that ganglion. However, it would be wrong to assume that the ganglion plays no part in the learning processes of nereid worms, for training in some types of experiment is more difficult in its absence. Clark summarizes the position by concluding that the supra-oesophageal ganglion is not a unique memory storage centre in polychaetes, although it is involved in some way in learning.

16–3 ASPECTS OF ARTHROPOD BEHAVIOUR

The locomotion of annelids depends upon a segmented nervous system functioning at a comparatively simple level of organization. The arthropod system is more complex, but its structure and functioning illustrate the same interaction of centrally driven rhythms, sensory excitation, and cephalic modulation—all deployed for a more wide-ranging exploitation of the environment.

This is well illustrated by the studies of Hughes and Wiersma on the neural basis of the movements of the abdominal appendages (swimmerets) of the crayfish, *Procambarus clarkii*. We have seen how important is the metachronal beat of crustacean appendages in feeding and respiration. This metachrony is probably a fundamental property of the group, persisting in the crayfish primarily for respiratory purposes. Study of this particular animal, therefore, sheds light on the role of the nervous system in crustacean locomotion generally.

Three pairs of nerve roots arise from each of the abdominal ganglia (which are, of course, closely fused pairs of ganglia). Of these roots, the second pair (Fig. 16–12) carry sensory fibres from stretch receptors in the abdomen, while the third pair are motor to the flexor muscles of the abdomen. They are not, therefore, directly involved in swimmeret movement, and can be severed without affecting it. The important roots in this context are the first pair in each segment, for they carry afferent impulses from the interoceptors of the swimmerets and motor impulses to their muscles. The contribution of these roots can be investigated in an abdomen that has been completely isolated from the cephalothorax. The swimmerets in this preparation show normal rhythmic movement, and the associated nervous activity can be studied by leading off and recording the action potentials in the first pair of roots, or in units obtained by splitting them.

In this way it can be shown that bursts of activity correspond in phase with the rhythmic limb movements. Moreover, these bursts continue even when the nerves have been cut distally, so as to prevent any transmission of peripheral sensory information. It follows that limb movements are associated with the firing of impulses that originate within the abdominal ganglia. Nevertheless, impulses from the receptors also play a part in these movements, as the following evidence shows. Fibres run into the nerve cord from proprioceptors that detect movement of the basal joints of the swimmerets. These fibres fire when the appendages are moved, and, because they have central connections that extend over at least two segments, they provide a basis for integrating the beat of adjacent limbs. Much can be achieved, however, even when most of the sensory input is lost.

For example, the rhythmic discharges recorded from an isolated abdomen are similar to those obtained from the cord of an intact animal, but only as long as at least one of the first nerve roots retains its peripheral connections. If, however, these are severed, so that the cord is wholly deprived of any peripheral sensory input, the rhythm no longer corresponds with the normal one. Probably, therefore, sensory

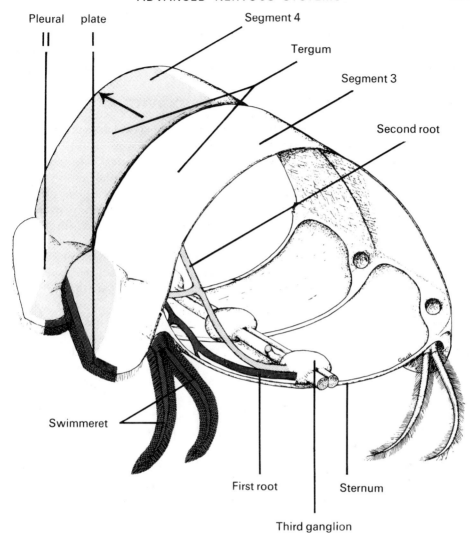

Fig. 16-12. Stereogram of the third and fourth abdominal segments of a crayfish showing the sensory innervation of the right half of the third ganglion. First root's fields: dark shade. Second root's fields: light shade. Arrow indicates the position of the right fast and slow abdominal stretch receptors for the joint between the fifth and fourth segments. From Hughes and Wiersma, 1960. *J. exp. Biol.*, 37, 291–307.

input determines the integration of the central firing of the individual segments into the pattern of normal metachronal rhythm. No doubt it also contributes information that makes it possible for the swimmeret movement to conform fully to the requirements imposed by the environment. Normally, of course, all of the abdominal segments are interconnected both by these sensory fibres and also by interneurones. The forward passage of impulses through these connecting tracts probably accounts for the delay that marks the metachronal beat, with the anterior segments beating last.

Studies of the influence of cephalic centres on locomotion in crustaceans are reminiscent of the results of comparable experiments on annelids. The hermit carb, *Pagurus*, can still walk after decerebration, but it does not give its usual response to an empty mollusc shell, which it would normally enter for its home. Loss of sensory imput is presumably a major factor in this instance.

The influence of the crayfish brain on swimmeret movement is shown by the fact that rhythmic movement of the appendages in an abdomen isolated from the rest of the body may be even more active than in the intact animal, which suggests that the brain may normally exert some inhibitory influence. This conclusion is strengthened by the fact that stimulation of certain fibre bundles in specific areas of the circumoesophageal commissures inhibits discharge of impulses through the first nerve pairs in the abdominal segments. Stimulation of other bundles, however, can stimulate such discharge (Fig. 16–13). Evidently, then, the brain has the means to exert considerable influence on swimmeret movement by increasing or decreasing the activity of the abdominal ganglia.

As might be expected, principles similar to these are equally apparent in insects, in the movement of both wings and limbs. Commands to action, which can be shown to originate centrally in the thoracic ganglia, are modified by the input of information from proprioceptors, and by excitatory and inhibitory impulses from the cephalic centres, these latter being themselves influenced by the input from the anterior sense

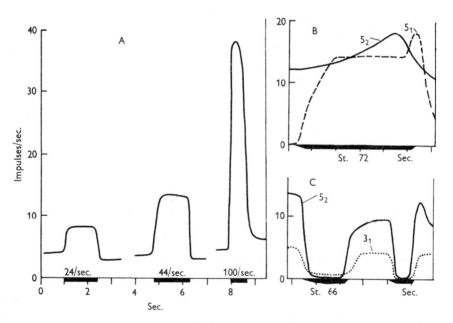

Fig. 16-13. *Procambarus clarkii.* Abdominal nerves intact except those from which recordings are shown. A, effect of stimulating a small bundle of fibres in area 72/74 of the circumoesophageal commissure at different frequencies on the discharge of impulses in a single unit of the contralateral first root of the third ganglion. B, impulse frequency in two units of the ipsilateral first root of the fifth ganglion following stimulation (50/sec) of a bundle in area 72. C, same preparation as B, stimulation in area 66 produces inhibition of units in the first root of the third and fifth ganglia. From Hughes and Wiersma, 1960. *op. cit.*, 657–670.

organs. In this way the independent capacities of the individual segments are integrated into the sequential movements of flight and ambulation described earlier, and which are readily adaptable to conditions prevailing in the environment.

Flight in insects is initiated by various reflex mechanisms, one being the tarsal reflex, which evokes flight when the tarsi lose contact with the ground. Other examples are reflex flight response to wind pressure on the antennae of Diptera or wind-sensitive sensillae on the head of the locust *Schistocerca*. Similar sensory excitation is necessary in Diptera and the locust for the maintenance of flight once it has been initiated. The effects of removal of the cephalic centres recall the results of comparable experiments on annelids. Flight in the locust is maintained after destruction of the brain, but tends to vanish when the suboesophageal ganglion is also destroyed. It seems likely, therefore, that the latter centre normally supplies some form of excitation to the thoracic ganglia. Central firing can occur in these ganglia even in the absence of proprioceptive feedback. Evidence for this is that bursts of activity can be recorded in the thoracic nerves to the wings even if the flight muscles are completely removed, provided that the preparation is appropriately stimulated (as, for example, by the pressure of wind acting on the head). Normally, however, proprioceptive feedback, arising from sensillae at the bases of the wings, is an essential part of the locomotor mechanism, and leads to appropriate adjustments of the form of the wing movements. For example, the wings of the locust normally exert an average and constant lift force; this force can be maintained under experimental conditions when the inclination to the wind is changed artificially, the control being effected by nervous reflexes that bring about a twisting of the wing.

The walking of insects is governed by similar principles of nervous organization, the capacity of the ganglia for independent action being readily demonstrable in cockroaches. An isolated leg of *Periplaneta*, if still connected with its ganglion, will execute a stepping movement when the tarsus is stimulated by traction. The extension of the leg is brought about by the depressor muscle. If, in a suitable preparation, the experimenter arranges for a sudden increase in resistance to the extension, a sudden burst of high-frequency impulses discharges into the muscle, to be followed by a resumption of a steady discharge at a higher rate than before. Normally, this response would serve to overcome the resistance. A similar effect is produced by pressing on the trochanter of the leg, because this stimulates the proprioceptive campaniform sensilla. This response to peripheral stimulation is one means by which coordination of the limbs is effected during walking; movement of an anterior leg will exert a tractive force on a posterior one, which will then respond reflexly to this stimulus. Because of this effect of peripheral excitation, a cockroach will continue to show coordinated walking movements even after its nerve cord has been cut right through in the thorax.

Coordinated movement in this animal is thus possible, at least to some extent, without the participation of the central nervous system. Nevertheless, the transmission of excitation along the nerve cord through the internuncial neurones must also be important in normal conditions. Circumstances doubtless vary in different species. For example, the walking movements of the stick-insect, *Dixippus*, are abolished if the thoracic nerve cord is completely transected. If, however, only one of the two longitudinal commissures is cut, and the other left intact, normal walking is still possible. It follows from this that the central nervous system includes some provision for alternative pathways which can be brought into use when the normal ones are blocked.

Although alternative pathways are valuable as a contribution to flexibility in behaviour, their use is not itself a learning process, for it does not involve the establishment of new associations; yet the existence of multiple pathways may clearly be a factor favourable to the development of learning. Relevant to this is the demonstration by Horridge of a simple form of learning in the thoracic ganglia of the cockroach. The method (Fig. 16-14) involves the use of a headless cockroach, so arranged that a particular leg receives an electric shock whenever it falls below a certain level. After about 30 min there is evidence of a lesson having been learned, for the behaviour of the preparation changes: the stimulated leg is raised so that it receives fewer shocks. Particularly striking is the result of associating a second headless animal in series with the first one. The second, because it is in series, receives shocks at varying leg positions, so that it is unable to associate the shocks with any one position. Thus no lesson can be learned. The resulting difference between the two headless animals is seen if they are disconnected and then retested after being reconnected in parallel. Each preparation now receives the shock when the stimulated legs fall to the same level. The first animal, however, has already learned its lesson, and shows this by raising its leg and thus receiving fewer shocks than the second, which has had no previous opportunity to learn (Fig. 16-15).

So far as intact animals are concerned, there is abundant evidence, from observation and experiment, of the learning capacity of arthropods. Habituation is wide-

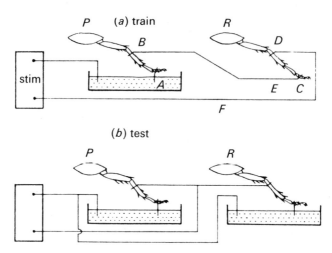

Fig. 16-14. The arrangement of the stimulating leads for the 'training of legs of headless cockroaches. (a) During the initial training the two animals are connected in series. Preparation P receives a regularly repeated shock, the stimulus being led in through wire A and out through wire B. The stimulus passes to the foot of the second preparation R at C, the circuit being completed by wire D. Since the two preparations are connected in series, R receives a shock irrespective of where it places its leg. After 40–45 min training, the wire is cut at E, and the preparations are connected in parallel (b) for testing. Either now receives a shock separately if it lowers its leg beyond the critical point. From Horridge, 1962. *Proc. R. Soc. Lond.*, B, **157**, 33–52.

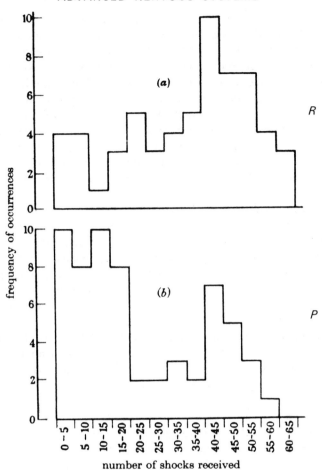

Fig. 16-15. Histograms of the frequency of occurrence of different numbers of shocks in 60 intervals of 2 min (a) for R cockroaches, (b) for P animals when placed on retest. With the interval between stimulator shocks used in this series, the maximum number which could be received was 80 per 2 min, and with this arrangement the R animals, which had been unable to 'learn' and which therefore had a poor leg-raising response, gave rise to the peak at 40–45 shocks per 2 min. P animals, which had been able to 'learn', gave rise to the left-hand peak in (b). From Horridge, 1962. op. cit.

spread. To give only one example, active locust hoppers belonging to the migratory phase *gregaria* of *Locusta migratoria*, become quiescent when isolated from the swarm. Attempts to promote sustained activity in the isolated insects are frustrated by their ready habituation, within a day or so, to a range of ingeniously diversified stimuli, including tapping the floor of the cage, tickling them with a brush, bombarding them with sand, and providing them with mirrors so that they can see themselves jumping. Why, then, do they not habituate to the movement of their companions? Thorpe suggests that inactive individuals in a swarm might fail to secure food; activity may therefore be maintained by food reinforcement.

Associative learning by classical conditioning has been convincingly demonstrated in the blowfly *Cynomyia*. This shows a typical unconditioned response in extending its proboscis when its tarsal sense organs are stimulated by contact with solutions of sugar. In one set of experiments the fly was exposed for 1 sec to the odour of coumarin, a non-toxic substance which smells (to us) like new-mown hay. This stimulated olfactory organs which can be shown to be restricted to the antennae and labella. The fly was then placed in contact with a sugar solution, which evoked the proboscis response, and it was then allowed to feed for 2 sec. After three conditioning periods, of six exposures each, and spaced at 12-hr intervals, an association had developed between the two stimuli, the proboscis now being evoked in 90% of the tests by coumarin alone. Untrained flies showed only a very small degree of response to coumarin.

Conditioning during the normal life of an insect is illustrated by the silver Y moth (*Plusia gamma*), which visits flowers for food. The newly emerged moth finds flowers by their scent, and is not attracted by those without scent, as can be shown by concealing the flowers in a container (Fig. 16–16) with only a small opening. Once it has fed, however, it begins to seek by sight as well as scent, having developed an association between the two. Eventually it develops strong association in a three-part complex of food, scent, and colour, and because of this seeks only one species of plant.

Trial-and-error learning has been demonstrated by the use of simple T-mazes; isopods, malacostracans, myriapods, and insects have been among those tested. There is much individual variation, and even allowing for this, success is not always achieved. Failure may be due to inadequate attention being given to the design of the apparatus. *Lithobius*, for example, can be trained in a T-maze, but only if the two walls differ in texture. It is also possible that in the less successful learning experiments there has been insufficient respect for the animal's sensory equipment, or for its normal motivation and drive. (The same, of course, might be said of ourselves.) To give only one example of this, larvae of the mealworm beetle (*Tenebrio molitor*) can be trained to avoid a rough surface if this is associated with an electric shock. So also can the adults, but these learn better than their larvae. One possible explanation of the difference is that the tarsi of the adults are better adapted to detect differences of surface texture.

Success in experiments involving more complicated mazes demand great patience in the experimenter; such experiments seem to have been largely restricted to social insects and cockroaches. The remarkable behaviour of the former will be referred to later (Chapter 21). As regards cockroaches, *Periplaneta* has been successfully trained in an open maze of sheet copper (Fig. 16–17), supported over water, the reward being access to the insect's cage. On the first trial, we are told, the cockroach 'almost invariably rushes off into the water'. Sooner or later it begins to search, relying largely upon its sense of touch, and carefully examining the corners and edges of the maze, and the adjacent spaces. It 'enters blind alleys, occasionally falls into the water, makes its toilet one or more times, perhaps engages in a few acrobatic stunts'. Finally, by accident, it finds its home. Graded improvement both in time and in the number of errors (Fig. 16–18) suggests trial-and-error learning. There is much individual variation, and memory markedly lapses after some 12 hr.

Finally, arthropods, and more particularly insects, show remarkable capacities

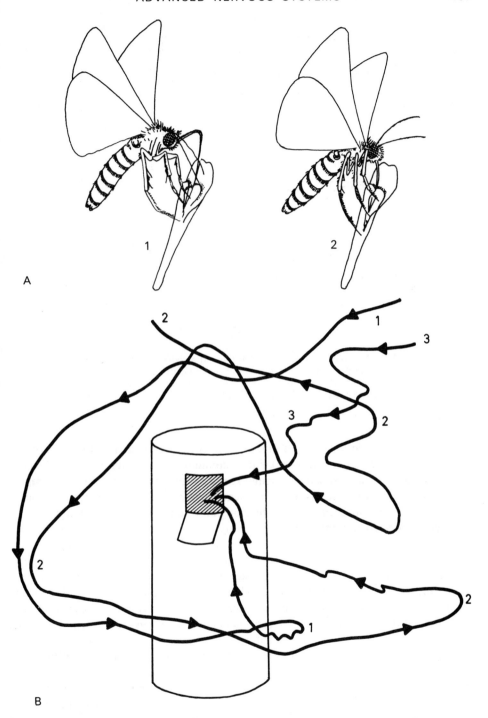

Fig. 16-16. A, feeding positions of the Silver Y moth, *Plusia gamma*. *1*, introduction of proboscis; *2*, sucking. B, three successive flight paths of a newly emerged moth to a concealed flower. From Schrammer, 1941. *Zool. Jahrb. Syst. Ök.*, **74**, 375–434.

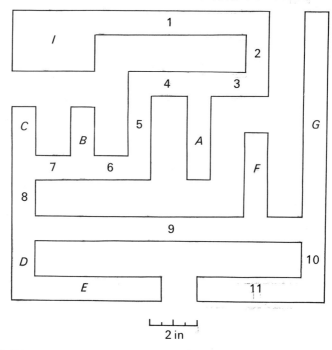

Fig. 16-17. Diagram of a maze used in training a cockroach. *I*, platform on which the insect is placed: *C*, the point from an incline leads to the cage. Numbers 1–7 indicate direct path to *C*. Other numbers and letters indicate blind alleys. From Turner, 1913. *Biol. Bull. Woods Hole*, **25**, 348–365.

for returning to a predetermined point, often in association with the care of the young, or with some other aspect of social life, although the phenomenon is by no means restricted to social and subsocial forms. This may depend in part on light–compass reactions (orientation with reference to the position of the sun), but it certainly depends in some instances upon the learning of visual cues in the landscape, as can be demonstrated by moving objects in the area around the home base. Not the least impressive result of such experiments is the finding that insects can battle with the experimenter by very quickly making use of entirely new landmarks. The Digger-wasp *Philanthus* has been shown to be able to learn such a new situation in as little as 9 sec during a single orientation flight. This is an example of the type of learning earlier referred to as latent learning. It is strikingly demonstrated in the provisioning behaviour of the solitary Hunting wasp *Ammophila pubescens*, as studied by Thorpe.

Ammophila feeds its larvae on caterpillars, which it has to hunt and paralyse; then, because of their weight, it has to drag them back to its burrow, sometimes for a 100 m or more, and past a variety of obstacles. It can orientate itself on the ground even after being moved in a closed box (Fig. 16–19), presumably in part because of prior aerial reconnaissance, and certainly with only the minimum of supplementary survey flights during its return. Probably it is orientating itself in large measure with respect to the sun, but even so its behaviour suggests some capacity to use a 'world-model' of the region around its burrow. Thorpe, who used metal screens as obstacles,

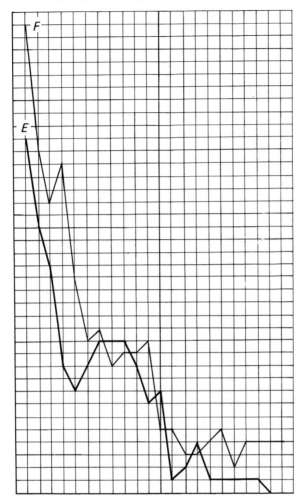

Fig. 16-18. Learning curves constructed from the average reactions of 10 cockroaches. The spaces from left to right represent successive trials. *E*, total number of errors. *F*, total minutes needed to run the maze. From Turner, 1913. *op. cit.*

records how the insect diverges around them without investigation, without waste of time, and with the maximum economy of effort. On one occasion, having done this twice, it solved the problem on the third test by climbing to the top of the screen and flying down on the other side, always without letting go of its caterpillar. This remarkable performance is suggestive of insight, but Thorpe is unwilling to regard detour behaviour, however perfectly executed, as conclusive evidence of this, for continuous optic feed-back is often important in such cases. But he concludes that the behaviour of *Ammophila* shows almost uncanny knowledge of the details of the terrain.

Arthropod nervous systems, like those of annelids, contain many sensory cells but relatively few motor cells and interneurones. The flexibility of arthropod behaviour, and their impressive powers of association learning, may thus seem sur-

470 INVERTEBRATE STRUCTURE AND FUNCTION

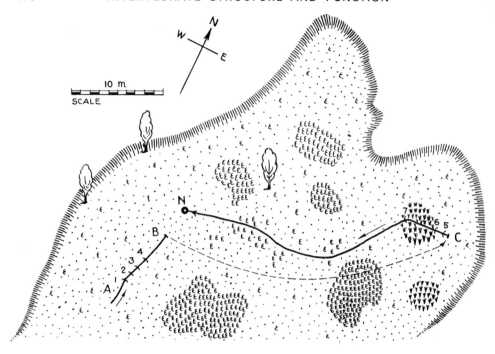

Fig. 16-19. Diagram to show the route taken during detour and displacement experiments by an individual of the Hunting wasp, *Ammophila pubescens*, engaged in dragging its prey towards the nest. Heavy line indicates course of insect. A, point at which first observed. Numbers 1–4 indicate points at which a metal screen was placed in its path for detour tests. B, point at which insect captured. Broken line indicates transfer of insect in box to release point C. numbers 5–7 indicate further detour experiments. N, nest. The shading indicates a light depression in an area of gravelly heathland with small patches and scattered plants of *Erica* and *Calluna* indicated by the symbol *E* and with small birch trees (*Betula*) about 4–6 ft high indicated by the conventional tree symbol. The remaining symbol indicates tussocks of *Juncus*, etc. Conditions: bright sunshine, July. Time taken by insect approximately 15 min. From Thorpe, 1950. *Behaviour*, **2**, 257–264.

prising. Part at least of the explanation, however, is doubtless to be found in the extensive branching of their nerve processes, and the variety of connections thereby made possible, so that the nervous system is very much more flexible in action than is the rigid exoskeleton. Quite complex responses are certainly mediated reflexly by very few neurones. For example, a crayfish defends itself in a threatening situation by a characteristic reaction in which the claws and the anterior end of the body are raised; behaviour that would seem to be complex, but which can be evoked in its essentials by stimulation of a single fibre in the circumoesophageal commissure. Then again, the escape reaction, in which the tail is violently flapped, involves many muscles, but these can be brought into action by a single impulse in one of the giant fibres (p. 475). However, the neural basis of these responses is not quite as simple as these observations suggest. During the defence reaction, for example, the animal may turn towards the source of the threat; this will involve the excitation of additional neurones. The escape reaction, too, involves repeated muscular contraction, producing a series of flaps, and it is likely that this prolongation is aided by the muscle receptors. These will be stimulated by the initial flapping, and the impulses that they send into the nerve cord may help to maintain the excitation of the reflex paths.

What these responses do suggest is that there need be no hard and fast line between simple reflexes and more complex patterns of instinctive behaviour. These latter may depend upon a chain of reflex pathways, in which each phase is stimulated by the immediately preceding one. For example, the feeding of a decapod crustacean may involve first the seizure of food by the cheliped, then its transfer to the mouth parts, and finally its swallowing if it proves to be acceptable. This sequence of reflex actions can be completed even by a crayfish from which the brain has been removed. Nevertheless, the brain must normally exert some regulatory action, for a brainless animal may burst its stomach by continuing to feed even after that organ has been completely filled.

The pattern of neural pathways involved in such a complex response must be preformed in the ventral nerve cord, with the brain exerting some degree of integration. However, the integrative capacity of the arthropod brain needs to be evaluated in the light of its small size, and free of anthropomorphic preconceptions. For example, the brain of a blowfly has a wet weight of only 0.84 mg, and probably does not contain more than about 100,000 cells. Furthermore, the cell bodies are concentrated around the periphery, where, unlike those of vertebrate brains, they are not well placed to enter into a variety of synaptic connections. These factors, together with the small size of the cells, must limit integration, and thus the perceptual world of the animal is probably less rich than that of a vertebrate.

But arthropods deserve to be measured also in the more relevant context of invertebrate evolution. Pantin, contrasting *Ammophila* with an actinian, remarks that both animals probably have much the same number of nerve cells, perhaps about 10^5; what they can respectively do with their predictor machinery, however, depends upon how these nerve cells are organized. The anemone must depend upon limited behaviour patterns initiated by tactile, mechanical, and other information secured by receptor cells at the body surface. A crucially important step in the evolution of higher forms of behaviour was initiated when such simple patterns of stimulation were superseded by world-models, abstracted from a complex of sensory information. This involved not only improvement of receptor machinery, but also the evolution of the three-dimensional nervous systems which are apparently essential for making the necessary correlations that we have been considering. Both ants and men, argues Pantin, react to stones rather than to the two very different sensory inputs by which they actually detect these objects.

The neurological basis of these arthropod achievements remains difficult to define with any precision, but the cerebral centres presumably play an important part, as we have seen them to do in other aspects of behaviour. That honey-bees can be trained to associate food with specific colours is certainly related in some way to the mushroom bodies of the brain, for it is lost when these bodies are removed.

But at least regulating centres in arthropods need not always be cerebral ones. An illustration of this is seen in the behaviour of the male mantis. The last abdominal ganglion of this animal has a copulatory centre that is inhibited by the suboesophageal ganglion until the inhibition is specifically removed. Removal of the brain produces no sexual activity, for it is not responsible for the inhibition. But removal of the suboesophageal ganglion, and hence of its inhibitory centre, results in copulation being carried to completion. This is why, when the female sometimes eats the male, anterior end first, during copulation, the copulatory act itself continues without interruption; she may be 'more deadly than the male', but her action, so far from being inimical to propagation, may actually be said to further it! The world of such animals is a strange one, and difficult for us to penetrate, either by experiment or by intuition. 'Looking at their rigidly armoured bodies', writes Dethier, 'their staring eyes, and their mute performances, one cannot help at times wondering if there is anyone inside.'

Whether there is or not, from such patterns of reflex activity is built up the complex adaptational behaviour that is observed in animals under natural conditions. Current interpretations suggest that these patterns are only manifested when the animal is in a specific state, indicated by the occurrence of the restless activity which is called appetitive behaviour. If, for example, it is hungry, it will search for food; this is appetitive searching, so organized that it tends to lead the animal to those conditions in which food is likely to be found. The essential requirement here is that the animal should encounter specific stimuli (the releasers, mentioned earlier) which are essentially patterns of stimulation to which the animal is adapted to respond. The response is presumably possible because the connections in the central nervous system are so arranged that they are excited by a particular pattern of sensory input. In more abstract terms, the central nervous system is said to possess a releasing mechanism, which, as a result of the stimulation of the releaser, initiates the final step in the behaviour pattern, which is called the consummatory act. When this has occurred, the initial appetitive behaviour disappears. Once the animal has fed, it will no longer search for food. Although, however, the consummatory act is final as regards one particular segment of behaviour, it may yet in its turn constitute the appetitive behaviour for another complex of reactions.

Appetitive behaviour may be very varied in its origins. In one sense it must always be a consequence of some specific physiological state, but this may be determined by external (exogenous) or internal (endogenous) factors. The appetitive behaviour of hunger, for example, may perhaps be directly stimulated by the receptors of the alimentary canal, particularly of those parts that are concerned with the storage of food. Frequently, as with reproductive behaviour, hormones must play a part, and this aspect we shall consider later. In other instances, however, behaviour patterns may arise spontaneously; without, that is, being evoked by specific receptor activity. We have seen examples of this in the respiratory and feeding cycles of *Arenicola* (Sec. 11–3), which appear to be evoked by spontaneous innate rhythms within the nervous system. These are said to be endogenous, since they arise internally, and quite independently of peripheral stimulation. Such behaviour is an exception to the principle that we have been developing in this discussion, which implies that inborn behaviour depends on adaptive reflexes initiated by stimulation of peripheral receptors. This principle, which expresses what may be regarded as the classical view of reflex behaviour, remains of wide application. Endogenous rhythms, however, are certainly common in animals, operating at many frequencies and at all levels of life, and often so regular that they have been referred to as 'animal clocks'. Presumably they depend upon oscillation in biochemical or physiological systems, but the nature of the oscillations, and their mechanisms, remain obscure. Perhaps they often originate within the nervous system, but this cannot always be so, for rhythms have been demonstrated in unicellular as well as in multicellular organisms, and even in metazoan cells maintained *in vitro* (human erythrocytes are an example of this). Whether they occur in prokaryotic organisms is less certain. One suggested explanation of them is that the oscillations may originate in membrane properties, but this is no more than a possibility.

The relative importance of endogenous and exogenous (external) factors in the initiation, driving and regulation of rhythms has been a matter for much controversy, but it is now widely held that endogenous origins predominate. Evidence that rhythms are endogenous depends on evidence that they continue under uniform conditions, free of the environmental fluctuations to which they are normally related. Rhythms that continue in this way are said to be free-running. But endogenous rhythms are commonly influenced by external factors, for their significance to animals is that they are adaptively related to the regular and cyclic fluctuations which characterise the environment provided by our planet. External influences are therefore needed to synchronize the internal oscillations with the external fluctuations to which the rhythms are adapted. In other words, the internal clock must be set to the correct time. The environmental agent responsible for time-setting is called a Zeitgeber (time-giver), and it is said to entrain the rhythm. The possibility that some rhythms may be exogenous cannot, of course, be excluded *a priori*. If they were, then the Zeitgeber might also initiate and drive the rhythm or it could be that a combination of external stimuli might be involved.

These principles are well exemplified by the cockroach, which has a daily activity cycle reaching a maximum shortly after the onset of darkness. If it is kept in constant light or darkness, the rhythm continues free-running for at least some days; it now lacks a Zeitgeber, and, as is common in such cases, it ceases to coincide exactly with the 24-hour environmental cycle. For this reason, such cycles are termed circadian (*circa*, about; *dies*, day). When the cockroach is reintroduced to a regular alternation of light and darkness, the rhythm conforms once again, with maximum activity occurring soon after the beginning of the dark phase. Here the stimulus of the change from light to dark is the Zeitgeber, acting through the ocelli. If the artificial cycle is so arranged that it reverses the normal 24-hr cycle of light and dark, the animal will conform to the reversal, becoming maximally active soon after what would normally have been its dawn, but which is now its dusk. The device, widely used in laboratories and zoological gardens, is invaluable for the human observer, who need not disturb his own circadian cycles (which include the ability to perform accurate observations) in order to observe those of his fellow animals.

No doubt the most familiar cases of rhythmic activity in animal life are the annual reproductive cycles of vertebrates, often, but by no means always, dependent upon the annual fluctuation of photoperiod, but we shall later see other illustrations of the importance of rhythms in invertebrates (they are no less common in plants, which provided some of the earliest experimental material for this field of study). Examples include the involvement of the lunar photoperiod in the reproductive rhythms of nereid worms (Sec. 17–5), and the ability of insects to allow in their directional homing for the movement of the sun, presumably through the operation of some form of internal 'clock' (Sec. 21–4).

16–4 GIANT NERVE FIBRES

The activity cycles of *Arenicola* are not the only example of a departure by invertebrates from the classical picture of reflex behaviour as being dependent upon internuncial neurones linking afferent and efferent pathways. Giant nerve fibres are another example. These were first described in the central nervous system of crustaceans in 1836, but it was another 25 years before they were seen in annelids. Their nature was at first obscure, for giant fibres, as seen in transverse sections of polychaetes and annelids, look very unlike the expected appearance of nerve fibres. Nevertheless, their nervous character had been well established by 1900, for it became evident that they contained neurofibrillae, that they belonged to giant neurones, and that they possessed myelin sheaths which could be blackened by osmic acid. They are now known to be of widespread occurrence, being found, for example, in cestodes, nemerteans, archiannelids, polychaetes, oligochaetes, arthropods, molluscs, and hemichordates. In the squid their diameter may reach 700 μm, as compared with 20 μm for a representative vertebrate nerve fibre.

The pattern of organization of these giant cells shown much variation. To take one familiar example, there are three dorsal giant fibres in the nerve cord of earthworms, the lateral ones reaching a maximum diameter of 50 μm in *Lumbricus*, and the median one 75 μm. There are also two smaller ventral ones, but these are not well understood. The dorsal ones, which extent throughout the body, are divided by septa into segmental units, each of which is in effect an interneurone, with a giant soma and nucleus, and differing from a more typical motoneurone only in its large size. In some parts of the body of earthworms the septa are missing, so that here the fibres are syncitial, while in polychaetes, where giant fibres are also present, this condition obtains throughout the body. These and other variations suggest that giant fibres have arisen independently many times. In hemichordates, for example, they form a system peculiar to the group, composed of cells that are variable in size and that are largely restricted to the neurocord. From here they give off giant fibres, some of which run forwards in the dorsal cord of the proboscis, while others run backwards in the ventral cord of the trunk.

If giant neurones have in fact arisen independently on a number of occasions, it is to be expected that they must have a considerable adaptive value. This is indeed, so, for they permit exceptionally rapid propagation of nerve impulses, and are particularly concerned in reactions that result in sudden retreat movements in response to potentially harmful disturbance. Thus *Nereis* responds to touch by giving its 'startle' reaction, which is a rapid jerk caused by contraction of the longitudinal musculature, usually accompanied by a parapodial reaction as well. This involves the giant fibres (Fig. 16–4), which provide for three pathways of rapid conduction. A single median fibre conducts from sensory afferent fibres in the anterior region, paired paramedial ones conduct from afferents in the posterior region, while lateral giant fibres, which are the largest ones, conduct after there has been stronger stimulation at any point in the body. Transmission takes place in both directions from the point of excitation. Certain structural features contribute to the highly adaptive value of the system. Sensitivity to warning signals is ensured by having many sensory fibres converging on the giant fibres. It is, however, important that these fibres should not come to dominate the central nervous system through their continual activity. This is provided for by having rapid accommodation at the junctions of the peripheral sensory fibres with the giant fibres, and at the junctions of the latter and the fast motor axons. Significantly, giant fibres are absent from polychate species that do not show these 'startle' reactions.

The action of giant fibres is well shown also in earthworms. In *Lumbricus*, for example, it has been shown by cutting the giant fibres, by studying their development and regeneration, and by recording their action potentials, that these axons conduct the impulses that bring about the rapid end-to-end contraction that form the well-known protective response of this animal. The connections of the giant fibres in earthworms are complex. They have afferent connections with the neuropile, and efferent ones with small motoneurones and also with giant ones, the axons of which pass out in the segmental nerves. Because of this complexity, the exact form of the responses depends upon the position of stimulation. When the anterior end of the earthworm is touched, the head is withdrawn and the tail is anchored, the anchoring being aided by the flattened form of the hind end and by its chaetae being directed forwards. This response, which is an adaptation providing for retreat when the animal is exploring with its anterior end protruded from the burrow, depends on impulses passing backwards along the median giant fibre, which is connected with anterior receptors. If the posterior end is touched when the worm is out of its burrow, the response is different. The head now remains stationary and the tail is pulled forwards, this response being mediated by the lateral giant fibres, which are connected with posterior receptors. Thus the median fibre normally conducts backwards and the lateral ones forwards, but this is a result of their pattern of connections, and not of any limitation in their inherent capacities; in experimental conditions they can be made to conduct in either direction. Repeated prodding of an earthworm leads to rapid fatigue of the 'startle' response, the fatigue occurring, just as in *Nereis*, at the junctions of sensory fibres with the giant fibres, and at the junctions of the giant fibres with the motoneurones. Peristaltic movements can, however, continue despite this fatigue, which is therefore thought to be a safety device ensuring that the muscles required for peristalsis are not fatigued by the time that normal locomotion is required. Evidently, there is much in common between polychaetes and oligochaetes in the functioning of their giant-fibre systems.

Many other examples of the dependence of escape reactions upon giant fibres might be given, but a few must suffice. In the crayfish *Cambarus* there are two pairs of these fibres (Fig. 16–20). Two median ones arise from cell bodies in the brain and extend back to the telson, while two lateral ones, which do not reach the brain, arise from segmental components and are connected with each other by segmental branches. Impulses in both pairs of giant fibres excite large axons which run out in the third pair of segmental nerve roots (cf. Fig. 16–12, p. 461) and innervate the longitudinal flexor muscles of the abdomen. Stimulation of this system evokes the characteristic flapping of the abdomen (p. 470) which draws the animal rapidly backwards, and it also causes movements of the antennae and of the thoracic and abdominal appendages.

Finally, giant fibres occur in molluscs, where two groups, the nudibranchs and the cephalopods, illustrate their functioning in quite different circumstances. The nudibranch *Archidoris* has seven giant neurones in its cerebro-pleuro-pedal complex,

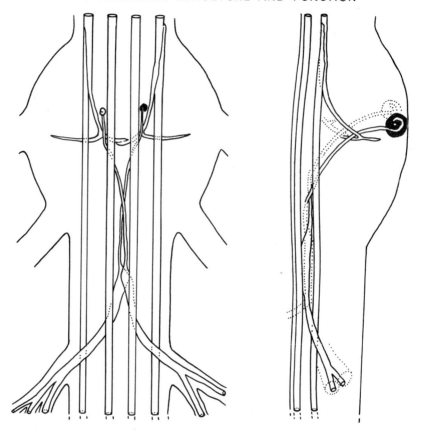

Fig. 16-20. Giant nerve fibres of the crayfish *Cambarus*. From Johnson, 1924. *J. comp. Neurol.*, **36**, 323–373.

one being present in each of the two cerebral ganglia. Axon branches from these are distributed symmetrically through the cerebral, pleural and pedal nerves to supply a number of effectors (Fig. 16–21), including those involved in the animal's escape reaction. This reaction (Fig. 16–22) necessarily differs in nature from that of a worm or crayfish, for *Archidoris* is a slow-moving animal and does not swim. The complete reaction, which can be evoked by repeated tactile stimulation of the mantle, involves withdrawal of the rhinophores (chemoreceptive tentacles), the branchial ring, and the edge of the mantle, together with closure of the mantle over the gill, and the production of mucus from glands in the mantle and foot. It clearly involves several different motor pathways, and it is thought that the giant cerebral neurones serve to coordinate the various components. Evidence comes from anatomical studies, and from intracellular recordings, which show that the neurones fire in response to various stimuli, including shadows and touch.

Giant fibres are also well developed in certain cephalopods, where they are associated with the high level of activity, and particularly with jet propulsion. In the squid, *Loligo*, a pair of first-order giant cells in the pedal ganglion region of the

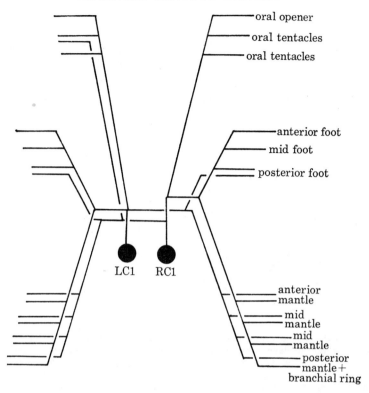

Fig. 16-21. Distribution of axon branches of the left (*LCl*) and right (*RCl*) cerebral giant neurons of *Archidoris pseudocarpus* to different effector systems. The diagram, based on electrophysiological mapping of RCl and LCl, summarizes the symmetrical output relations of the two neurons. The branching pattern has been simplified for the purposes of illustration and the diagram is not intended to reflect the actual sequence of branching of the axons within the neuropile. From Blackshaw and Dorsett, 1976. *op. cit.*

brain make synaptic connections with second-order cells that have cell bodies in the pallio-visceral lobe of the brain. Giant fibres from the latter cells pass out along the mantle connectives to the stellate ganglion. There they make proximal and distal synapses with third-order giant cells (Fig. 16–23), the axons of these leaving the ganglion by the stellar nerves to innervate the mantle after extensive branching. We have already seen (Sec. 5–5) that the adaptive value of this system lies in its contribution to the efficiency of jet propulsion. Many fibres are caused to contract at the same time, so that a single impulse can bring about a widely distributed response, while it can also evoke powerful twitches of very high frequency. The ability of squid to move at several knots and to maintain station with ships owes much to their giant fibres. Nor should the value of the system to the biologist be forgotten. It has provided material for the quite exceptional series of studies in nerve chemistry and physiology which established the foundations of our understanding of the action potential.

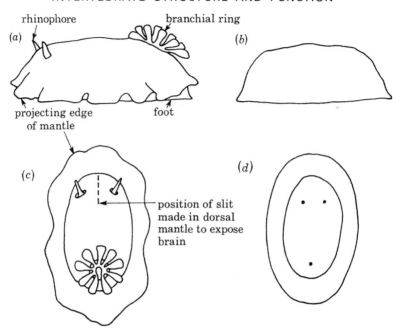

Fig. 16-22. The escape behaviour of *Archidoris pseudocarpus*. (a) and (c) show lateral and dorsal views with the branchial ring and rhinophores fully extended. (b) and (d) show the same views, with the projecting edge of the mantle, the branchial ring and the rhinophores withdrawn, and the mantle edges closed over the retracted branchial ring and rhinophores. From Blackshaw and Dorsett, 1976. *Proc. R. Soc. Lond.*, B., **192**, 421–437.

All of these giant fibre systems are adaptations ensuring maximum speed of conduction in the mediation of responses that are of particular importance in preserving the life of the animal. They are, in fact, an extreme expression of the selection pressure that has favoured increased efficiency of nerve conduction. We first saw this illustrated in the development of through-conduction pathways in primitive nerve nets. The improvement secured by the development of giant fibres is sufficiently indicated by measurements made in earthworms. These show that the fine fibres conduct at about 0.025 m/sec, the lateral giant fibres at 7–12 m/sec, and the median giant fibre at 17–25 m/sec. The difference is partially due to the reduction or elimination of synaptic barriers in the giant fibre pathway, and partially to the fact that rates of conduction increase with increase in the diameters of the axons, because of a concomitant decrease in electrical resistance of the axoplasm. Conduction speed is also increased by the presence of a myelin sheath around the fibre. This imposes a fast rate of conduction because its high electrical resistance prevents current from passing through it. The principle has been widely used in the nervous system of vertebrates, which is perhaps why they have not developed the use of giant fibres. Finally, the peripheral connections of the giant fibres often facilitate speed of response. The incoming pathways from the receptors may be short, for example, and in some instances, as in the crayfish, there may be a continuous and uninterrupted pathway between the central cell body and the peripheral musculature.

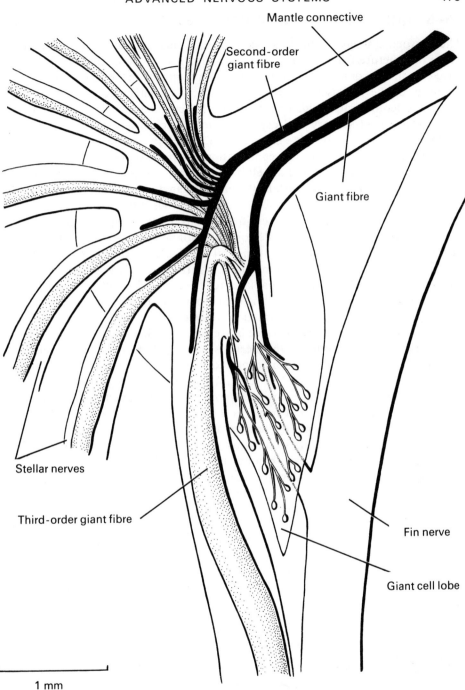

Fig. 16-23. The giant fibres and their synapses and cell bodies in the stellate ganglion of *Loligo pealii*. Dimensions are to scale from a small specimen. Modified after Young, from Bullock and Horridge, 1965. *Structure and Function in the Nervous System of Invertebrates*, vol 2. W. H. Freeman & Co., San Francisco and London.

16–5 MOLLUSCAN NERVOUS SYSTEM

The nervous systems of the Platyhelminthes, Annelida, and Arthropoda show a range of structure from which much can be learned of the factors that have influenced the history of neural organization. It is a comment on the scale of the adaptive radiation of the molluscs that they show a very similar range. Within this single phylum is demonstrated a historical process that culminates, in the cephalopods, in the production of sensory and central nervous mechanisms that demand serious comparison with those of vertebrates as well as with those of arthropods.

The absence of metameric segmentation is a point of obvious difference from the neural organization of the annelids and arthropods, but apart from this there is a very similar trend in the establishment of a ganglionic system controlling local reflexes. The primitive form of the molluscan nervous system (Fig. 16–24) can be judged from its appearance in present-day chitons. Here, as in platyhelminths, there is some centralization, but very little concentration of nerve cells; less, indeed, than in those animals, for there are no cerebral ganglia. Longitudinal nerve cords are present, linked by transverse commissures to form a ladder-like system, with pedal cords running along the foot and pallial cords lying more laterally. These cords link with a ring of nervous tissue encircling the oesophagus, but the only ganglia present are a pair of buccal ganglia, innervated from the ring and concerned with the action of the radula. Elsewhere the nerve cells are scattered, in the ring and in the cords.

The subsequent history of this nervous system reflects the modes of life of the main molluscan groups. It is to be expected that a well-organized central nervous system, capable of a high level of integration and conferring adaptive flexibility upon responses, will only appear when its evolution is promoted by its high selective value. This condition does not occur in the gastropods and bivalves, which, with certain exceptions, are essentially inactive or even sedentary animals, living without a wide range of environmental stimulation. Even when such stimulation does occur, the invention of the molluscan shell has meant that the response is retreat rather than exploration. It is doubtless because of this that the ganglia of these animals are centres of reflexes that involve comparatively restricted sensory and motor areas, and that show only a limited degree of interaction.

In the gastropods a pair of pleural ganglia appears at the anterior end of the pallial cords, while the cords themselves form the visceral loop, bearing a pair of parietal ganglia and a visceral ganglion, which may also be paired. These features are probably the more primitive ones. Later developments involve further concentra-

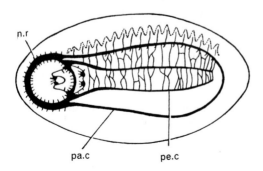

Fig. 16-24. The primitive molluscan nervous system. *n.r*, nerve ring; *pa.c*, pallial nerve cord; *pe.c*, pedal nerve cord. From Morton and Yonge, 1964. *The Physiology of Mollusca*, vol. 1 (Wilbur and Yonge, eds.). Academic Press, New York.

tion of the nerve cells, with a pair of cerebral ganglia appearing in the nerve ring, and a pair of pedal ganglia at the anterior end of the pedal cords; these cords then disappear. Thus arises the typical gastropod ganglionated nerve ring. The torsion of gastropods leads to a twisting of the visceral loop, the original left and right parietal ganglia now forming respectively the subintestinal and supra-intestinal ganglia. The subsequent changes in this visceral loop reflect the complex evolutionary history of the gastropods. The common trend throughout all the higher forms is an increasing concentration of the ganglia, the sub- and supra-intestinal ganglia being drawn into the main nerve ring, with a shortening of the visceral loop. In the Opisthobranchia this process is associated with the reversal of torsion and the untwisting of the visceral loop; in the Pulmonata this does not occur, although little indication of the original twisting of the loop can be seen.

At the structural level of analysis the influence of cephalization upon the gastropod nervous system is sufficiently obvious; no less striking is its virtual absence from the bivalves. These animals have no head, and their mode of life is sedentary and withdrawn. In correlation with this, the nervous system is of a very simple character. It is derivable from the primitive plan that we have outlined above, with cerebral ganglia above the oesophagus, pedal connectives leading from them to the pedal ganglia in the foot, and visceral connectives running back to the visceral ganglia lying under the posterior adductor muscle. Pleural ganglia are associated with the cerebral ones, being completely fused with them except in the protobranchs.

Both in gastropods and bivalves the fields of action of the ganglia are restricted ones, as is demonstrable by conventional experimental procedures. For example, the pedal ganglia of gastropods innervate the foot. If they are removed from *Aplysia* there is an increase of tonus in the foot, and also increased contraction of that region. If, on the other hand, the cerebral ganglia are removed but the pedal ganglia left undisturbed, there is an increase in locomotor activity and increased excitability of the lateral parapodia of the foot. These results suggest that the pedal ganglia inhibit tonic contractions, and that the cerebral ganglia, like the brain of annelids and arthropods, inhibit locomotor activity, in this case by action through the pedal ganglia. The local actions of the latter have been demonstrated by stimulating nerves running inwards to the ganglia from the parapodia. The result is to evoke contractions on the other side, indicating that the ganglia are relay points for reflex responses to stimulation of the foot.

The evidence available from bivalves also indicates localized action of the ganglia. Thus the cerebro-pleural ganglia innervate anterior structures, including the palps, otocysts, osphradia, and the anterior adductor muscle. If these ganglia are removed from *Mytilus*, the foot is still able to creep and also to spin the byssus, being apparently controlled in these activities by the pedal ganglia. The visceral ganglion innervates a large part of the body, including the gills, siphons, pallial sense organs, much of the mantle, and also the posterior adductor muscle. Its removal from *Mytilus* abolishes the opening and closing of the valves which are the animal's response to changes in the conditions of the water.

The simplicity of response of the bivalve is correlated with its ability to secure complete protection within the shell, and because of this the regulation of the action of the adductor muscles is of crucial importance. These muscles are of compound structure, being composed of two types of fibre; one of these is striated, and is capable of rapid or phasic contractions, while the other is unstriated, and is responsible for the sustained tonic contractions that maintain closure of the valves. Barnes's description of the regulation of these contractions in *Anodonta* gives a good illustration of the mode of operation of bivalve ganglia.

If the animal is maintained in a fixed position, with one valve of its shell attached to a recording lever, it exhibits two rhythms. One of these is a 'slow' rhythm, in which a period of quiescence, with the valves closed, alternates (with frequencies of 3–30 per week) with a period of activity, when the valves gape open. The other rhythm is a 'rapid' one, the adductor muscles showing rapid contractions, followed by slow relaxations, with a frequency of up to 20 per hour. Study of animals from which the cerebro-pleural ganglia have been removed, or of preparations consisting of the posterior adductor muscle and the visceral ganglia, show that the 'slow' rhythm is a function of the unstriated muscle fibres, and that it is jointly controlled by the visceral and cerebro-pleural ganglia. The visceral ganglia produce a tonic contraction, while the cerebro-pleural ones inhibit this at intervals. The 'rapid' rhythm, on the other hand, is a function of the striated muscle fibres, and is controlled entirely by the visceral ganglia, without the participation of the cerebro-pleural ones. Both rhythms are independent of external stimulation, and seem to result from inherent properties of the ganglia concerned, yet they can be modified by stimulation. Thus vibration or rotation of the animal evokes a reflex response in which the tonus of the adductor muscles is relaxed and a new period of activity initiated.

The regulation of the rhythmic contractions of the anterior adductor of *Anodonta* is similar in principle, but this muscle is remote from the visceral ganglia, and probably the cerebro-pleural ganglia alone control both rhythms. This doubtless explains the form of the nervous system of *Pecten*. In this animal, as we have seen, there is a wide departure from the mode of life typical of bivalves. The rapid contractions of the striated fibres of the posterior adductor muscle are in this case responsible for jet propulsion, while the anterior adductor muscle has been lost. In correlation with these changes the visceral ganglia have come to form the largest single component of the nervous system. In an animal which, like all bivalves, is essentially headless, cephalization of the nervous system has no meaning: the cerebro-pleural ganglia are small, and have moved some way backwards.

16–6 NERVOUS SYSTEM AND LEARNING IN CEPHALOPODS

As we have already seen in other contexts, the history of the cephalopods has followed lines very different from those of other molluscs, the fundamental difference lying in their pelagic and predatory mode of life. No doubt they must at one stage have shared a common ancestry with the more sluggish, and more typical, members of the phylum, but of this we know nothing: cephalopod shells occur in Cambrian deposits, so that the independent history of the group must extend over some 500 million years. The only link with past ancestry that we have with us today is *Nautilus*. This animal is described by Anna Bidder as 'odd . . . dumb, and impersonal . . . a *long way* from other living cephalopods'. To judge from its shell, it has probably changed little since Cretaceous times. Its eyes are of simple structure, lacking lens and iris, and it is unlikely that the animal recognizes objects by sight; chemoreception must be more important. Like other cephalopods, its central nervous system shows marked cephalization, but it is still comparatively simple, with only three pairs of lobes connected by commissures to form a ring-shaped complex. With the characteristic specializations of cephalopods setting in so early, there is little profit in seeking in these lobes for precise homologues with the parts of the nervous system of other molluscs. A functional analysis is altogether more revealing, for to this the cephalopods lend themselves very well.

Nautilus has been little studied from this point of view, but the intelligence of the other modern forms, and the ease with which some of them can be maintained in the laboratory, has attracted much attention to them. The sacrifice of their external shell is an index of their commitment to a pelagic life (although the retiring habits of *Octopus* make it an exception in this regard). Associated with this is the speed and freedom of movement of these animals, their well-developed sense of vision, and their remarkable capacity for colour change, to name only some of their outstanding adaptations. No less significant is the complex organization of the central nervous system. This forms a substantial concentration around the oesophagus, and is commonly referred to as the brain.

We may distinguish in the brain of *Octopus* (Fig. 16–25) as many as thirty distinct lobes, each composed of a layer of nerve cell bodies, enclosing a central mass of fibres. Ventral to the oesophagus is a group of suboesophageal lobes which are concerned in motor responses, and which in a very general way recall the ganglia of other molluscs. Direct stimulation with electrodes reveals something of their function. Stimulation of the posterior chromatophore lobe expands the chromatophores of the mantle and funnel; in this instance there seems to be a direct link between the central nervous system and a specific motor response. Stimulation of other parts of the suboesophageal lobes produces motor responses in various parts of the body, such as the arms, funnel, and ink sac, and there is some regional localization of the relevant centres, as indicated in the nomenclature of the lobes. This localization is not, however, as sharply defined as with the chromatophore response.

Local stimulation of these centres in the cuttlefish (*Sepia*) evokes isolated sets of muscular contractions, abstracted, as it were, from the complete behaviour patterns, and showing no sign of the integrations that normally characterize those patterns. The explanation of this is that the supra-oesophageal portion of the brain (Fig. 16–26) contains lobes which integrate or organize motor responses, and can therefore be termed higher motor centres. Direct stimulation reveals that the anterior and posterior basal lobes are concerned. Stimulation of the anterior basal lobe evokes movements of the eyes, head, and tentacles, such as are used in feeding; these are sufficiently coordinated to be clearly recognizable as elements of normal behaviour patterns. Similarly, stimulation of the posterior basal lobes produces coordinated respiratory and swimming movements of the fins and funnel, together with the escape reactions mediated by the giant fibre system, and also a variety of visceral responses. In no case are the muscles concerned directly innervated by the basal lobes. The connections of the latter are with the motor centres of the suboesophageal lobes, and it is through these that the integrating action of the basal lobes is exerted.

Naturally, direct electrical stimulation is a crude way of testing central nervous function. Not only is the stimulus itself quite abnormal, but the central nervous system is deprived of the sensory inflow from other parts of the body that normally

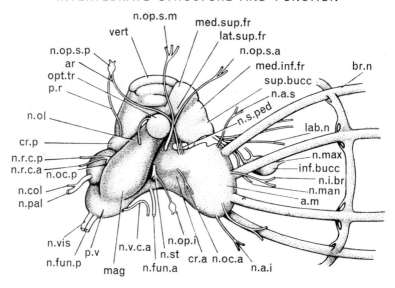
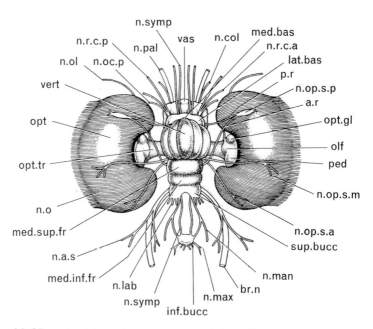

Fig. 16-25 Lateral (*above*) and dorsal (*below*) views of the brain of *Octopus vulgaris*. *a.m*, anterior suboesophageal mass; *br.n*, brachial nerve; *cr.a*, anterior chromatophore lobe; *cr.p*, posterior chromatophore lobe; *inf.bucc*, inferior buccal lobe; *lat.bas*, lateral basal lobe; *lat.sup.fr*, lateral superior frontal lobe; *mag*, magnocellular lobe; *med.bas*, median basal lobe; *med.inf.fr*, median inferior frontal lobe; *med. sup.fr*, median superior frontal lobe; *olf*, olfactory lobe; *opt*, optic lobe; *opt. gl*, optic gland; *p.v*, pallioviseral lobe; *ped*, peduncle lobe; *sup.bucc*, superior buccal lobe; *vert*, vertical lobe; *vas*, vasomotor lobe. Nerves not listed. From Young, 1964. *A Model of the Brain*. Clarendon Press, Oxford.

accompanies environmental stimulation, an inflow that includes the feedback of the proprioceptive and other information upon which depends the smooth execution of movements. These limitations can in part be overcome by a study of the effects of brain lesions; results thus obtained are largely in agreement with the conclusions suggested by the stimulation experiments. Removal of the basal lobes seriously interferes with the organization of integrated movements. For example, loss of the anterior basal lobe of one side results in the animal circling continuously with the undamaged side of the brain on the inside of the turn, while loss of the complete supra-oesophageal region of the brain results in the animal swimming backwards. Other experiments have shown that the optic lobes are involved in the maintenance of muscular tone, and that the suboesophageal region of the brain can organize simple reactions, but not the more complex ones that demand widespread integration. In short, the suboesophageal region can be said to include the lower motor centres, while complex integration is carried on in the higher motor centres of the supra-oesophageal region.

This, however, leaves unexplained certain areas of the supra-oesophageal region that are called 'silent areas'. If these are removed there is no disturbance of the motor behaviour of the animal, so that they are evidently not concerned with the organization of integrated movements. By analogy with the vertebrate brain we might expect

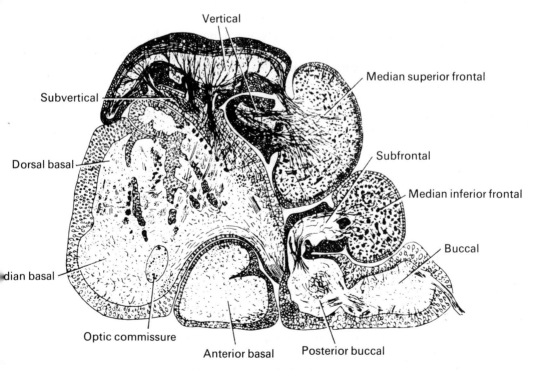

Fig. 16-26. A longitudinal section through the supra-oesophageal lobes of the brain of *Octopus*. The section is cut slightly to one side of the mid-line. After Young, from Wells, M. J., 1965. *Adv. mar. Biol.*, **3**, 1–62.

the 'silent areas' to be concerned with memory and learning; extensive experimentation has shown this to be so, and has suggested important and far-reaching hypotheses regarding the nature of these phenomena in animals in general.

These experiments have exploited the capacity of *Octopus* for the acute visual and tactile discrimination mentioned earlier. The experimental principle, as regards visual discrimination, depends on the use of moving sheets of plastic, cut into well-defined shapes such as circles and rectangles. The animals can be taught to attack one such shape by rewarding them with food, and to avoid others by punishing them with an electric shock. This is the procedure of the classical conditioning experiment. In this instance, to follow Young's analysis, the results show that the nervous system must retain a coded representation of a situation, expressed within some particular set of neural processes, that is stored in the memory of the animal. It may be helpful to draw an analogy with a computer, in which the coded data correspond to the representation of past situations. From this point of view the process of learning is the setting-up of new representations. No animal, however, can afford to depend solely upon this process, for it must be able to survive during the period when the new representations are being established. Accordingly it will also possess to a greater or lesser degree the capacity for making responses that do not have to be learned. This aspect of behaviour, which we have termed instinctive, must also depend upon the presence of representations within the nervous system, but these representations will have been established during embryonic life as a result of the action of the genetic code inherited from the parents.

For all its acuity, the visual discrimination of *Octopus* has marked limitations as regards the shapes that can be distinguished. For example, vertical and horizontal rectangles can be distinguished from each other, but not oblique ones, a limitation that we have seen to be associated with a structural peculiarity of the eye. This and other observations have led to the suggestion that the nervous system operates by computing the ratios between the horizontal and vertical projections of the figures being observed. That some such process is involved is certainly suggested by the marked tendency for horizontal and vertical orientation in the structure of the nervous system. As we have already noted, the retina is organized in this way, and so also are the dendrites of the optic lobes into which the optic nerve fibres discharge. Moreover, the responses depend upon the maintenance of a horizontal orientation of the pupil, through the mediation of the statocysts (p. 412).

These are examples of how modes of response and learning capacities in animals are limited by features of neural organization; a principle that is as significant for the comparative psychologist as it is for the particular species concerned, and that reinforces the lesson that has already been taught us by *Paramecium*. Another illustration is provided by the tacile sense of *Octopus*, as analysed by M. J. Wells. In this instance association experiments have shown that the animal can distinguish the surface texture of objects seized by its tentacles, but that it differs from man in being unable to distinguish the actual mode of distribution of surface irregularities. All that it can apparently do is to estimate the proportion of the surface that is elevated, or, as we might say, its general roughness.

This is probably connected with another limitation of *Octopus*: it cannot learn to distinguish objects by their weights. This is not because it lacks the proprioceptors upon which we rely for weight estimates; their existence is shown by the animal being able to compensate for the weights of objects when it handles them with its tentacles. The limitation arises because the movements of the tentacles are regulated in local motor centres (axial ganglia) in the arms (Fig. 16–27). It is there that the proprioceptive information is used, so that it does not become available to the brain as a basis for learning. Not only, then, is the animal unable to learn to discriminate by weight; lacking proprioceptive information, it cannot judge the position of its mechanoreceptors, and because of this it cannot estimate the surface patterning of an object. Indeed, the consequences of this limitation go even further, for, despite the undoubted intelligence of the animal as measured by its capacity for learning to deal with new environmental situations, it is quite unable to learn to use its tentacles in new manipulative skills.

Wells suggests that this peripheral restriction of proprioceptive information is a consequence of the cephalopods having flexible bodies. Animals with articulated skeletons can readily provide for central monitoring of the positions of the parts of their bodies, since they require information only from relatively few stations. Large animals with flexible bodies would need information from more stations, and the computing of this information would seem to demand brains very much bigger than they actually possess. The decentralization of reflex motor control avoids the need to develop a disproportionately large brain, but at the same time it limits their learning capacities. The activity of the local ganglia in *Octopus* is seen in the way in which the arms of completely brainless animals can carry out movements of seizure and rejection. These presumably depend upon chain reflexes, such as we have encountered in arthropods, and that are here evoked within the individual arm by stimulation of its receptors. Thus the only control that has to be exerted by the computer mechanism of the brain is the signalling to the lower centres of instructions either to attack or to retreat.

The analytical approach to the study of the behaviour of the octopus, designed

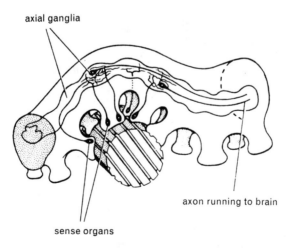

Fig. 16-27. Diagram of part of the arm of an octopus touching a plastic cylinder. The pathways from the receptors are inferred from physiological and degeneration experiments. After Wells and Wells, from Young. 1964. *A Model of the Brain.* Clarendon Press, Oxford.

to determine its powers of discrimination, has been carried out on animals sitting in a home of bricks in an aquarium tank, from which they emerge to attack suitable objects. But their total behaviour pattern is, of course, much more complex than this, as is shown by the results of detour experiments, based upon their intense interest in crabs. Figs. 16–28, 16–29 illustrate a type of experiment in which the octopus could see the crab but, in order to obtain it, had to move into a corridor where it no longer had sight of its prey. It had then to move along the corridor and make an appropriate choice of direction (left or right) at the end of it. The first response of the animal is to try to attack the prey through the glass, so that it is usually necessary first to lead it through the maze. Soon, however, it learns to complete the detour spontaneously. The learning is shown in the decline, during a sequence of trials, of failures to secure the prey (Fig. 16–30). These failures are either attempted attacks through the glass, or abortive entries into the corridor which end in it returning to the attack instead of completing the detour to the crab.

As regards the learning process in *Octopus*, there seem to be two partially

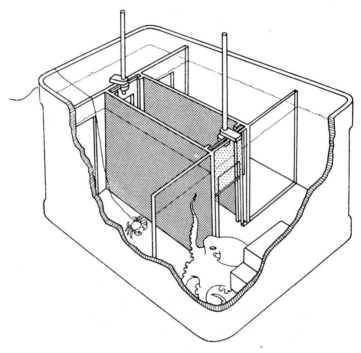

Fig. 16-28. The apparatus used for detour experiments with the octopus. The animal lived in a pile of bricks in a 'home' compartment, and was usually confined to this compartment between trials by lowering a shutter. There was a second shutter at the far end of the central corridor. On leaving the corridor at this end the octopus entered a choice compartment, like the home compartment but without bricks and with doorways leading into the two feeding compartments, on either side of the passage. The time taken to make a detour was measured from the moment of entry into the corridor until the octopus passed into one or other of the feeding compartments. If the octopus went into the correct feeding compartment it was rewarded. The crab was removed as soon as the octopus entered the corridor. From Wells, M. J., 1967. *J. exp. Biol.* **47**, 393–408.

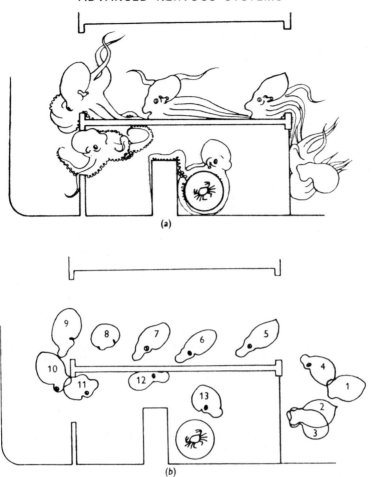

Fig. 16-29. Postural changes during a typical detour by an octopus, drawn by tracing projections from a cine film taken from above the tank; 45 cm maze. (a) shows the position of arms, head, and body at various stages in the detour, and (b) the position of the head and body, at more frequent intervals. Position 1, during the animal's approach from the home, is followed by 2, one sec later; 3 is three sec after 2, and was followed by a period spent struggling against the glass, position 4 being filmed 13 sec after 3 as the octopus began to move around the corner into the passage. Positions 5–13 follow at 2 sec intervals. Details of the arms are given for positions 2, 5, 7, 9, 11, and 13. From Wells, 1964. *op. cit.*

distinct mechanisms in the brain (Fig. 16–25). One of these, comprising the optic, superior frontal, and vertical lobes, is concerned with visual learning. Information is first transmitted to the optic lobes, and is distributed from there to the superior frontal lobes, and on to the vertical lobes. Anatomical studies show that fibres from the latter pass back to the optic lobes, so that there is a circuit of information through the vertical lobes. A fact of great potential significance is that this circuit makes use of a very large number of cells, the vertical lobes containing some 25 million cells in comparison with the 1–2 million fibres that connect them with the superior frontal

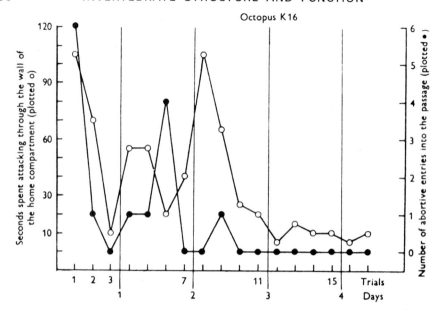

Fig. 16-30. A summary of the performance of an octopus which completed 17 trials before it climbed out of its tank and died. O, time spent attacking through the trasparent wall of the home compartment before going into the passage for the first time; ●, the number of abortive entries into the passage, at each trial. This octopus failed to make a detour within 10 min at the first trial, but succeeded at all trials thereafter. From Wells, 1964. *op. cit.*

lobes. The optic lobes are even richer in cells, each containing some 60 million cells, as compared with 30 million for the whole of the remainder of the nervous system. Undoubtedly the vertical lobes are of great importance in visual learning. If they are removed, the animal substantially loses a previously learned lesson. There are signs, however, if the lesson is retaught, that traces of the old lesson remain in its memory. Such an animal can still learn fresh lessons, but it does so more slowly and less accurately than an intact animal, and only does so at all if the discriminations involved are simple ones.

A suggested interpretation of these facts, argued in detail by Young, is that cells in the optic lobes are pre-set to respond to particular dimensional ratios, and that during the learning process they become conditioned to signal either attack or retreat, according to the information supplied to them. At this stage the conditioning would affect only those cells that are under the direct influence of the receptor system. If the process developed no further the lesson could only be a limited one, for the animal would be unable to generalize from one shape to another closely similar one, or from stimulation of one part of the visual field to another. In fact, however, *Octopus* shows marked powers of generalization, and this, it is argued, could depend upon the activity of the vertical lobes. These, by storing representations, and by distributing excitation back to wider fields of cells in the optic lobes, could extend the area over which conditioning is established.

Force is given to this argument by consideration of the process of tactile learning. The inferior frontal and subfrontal lobes are concerned here. Removal of these regions completely abolishes the capacity to learn by touch. They are separate from the centres concerned with visual learning, yet they are not entirely independent of them, for the two systems overlap in the vertical lobes. Removal of the latter affects tactile learning in essentially the same way as it affects visual learning. This is in accord with the suggestion that the vertical lobes facilitate generalization, yet it must be emphasized that this whole concept of their function is still highly theoretical. In fact, the organization of the subfrontal lobe is much like that of the vertical lobe, and this has led to an alternative suggestion that the latter has no unique function at all, but merely serves as a pool of cells available for the extension of memory. Or again, it may be concerned particularly with representations involving the combination of tactile stimuli with visual ones.

As Young rightly emphasizes, such problems demonstrate our almost total ignorance of the nature of learning, and of the changes that are involved in the conditioning process. It is a measure of this ignorance that the process has tended to be discussed over the years in terms of those particular aspects of neural functioning which were fashionable subjects of research at the time—structural, chemical, or electrical, as the case might be. The current use of the terminology of electronic engineering should be viewed with this in mind. A comparison of neural organization with the organization of computers is immensely helpful in facilitating discussion through a formal and intelligible terminology, but it by no means follows that living organisms actually function in this way (p. 420).

17
Chemical Coordination

17-1 NEUROHUMOURS

In our review of the general properties of the neurone we saw that it is a cell specialized to carry out two distinct yet interrelated functions: the propagation of nerve impulses and the manufacture and discharge of secretions. We have so far been considering primarily the first of these, and it is now necessary to consider the second, which is no less important.

The secretory activity of the neurone plays more than one part in the physiological organization of animals. The first to be demonstrated was the production of the chemical transmitter substances (neurohumours) which provide for the chemical transmission of excitation at synapses (Sec. 14–1). In general, these are formed in the cell body of the neurone, probably by the mitrochondria and Golgi material. They are then passed down the axon and are stored at the axon terminal; from here they are released on the arrival of nerve impulses, exciting the next cell in the chain of action, neurone or effector cell as the case may be.

The action of chemical transmitters, and the criteria for establishing their function, can best be illustrated by brief reference to acetylcholine, which acts as a chemical transmitter at the nerve endings of the parasympathetic nervous system of vertebrates. These endings, which are termed cholinergic, are shown by electron microscopy to contain small bodies called synaptic vesicles, each with a mean diameter in the range 30–40 nm. If the nerve fibre is examined following stimulation some of these vesicles appear empty, as though their contents have been discharged. It is thus supposed that they contain units or 'quanta' of acetylcholine, which are discharged in proportion to the number of nerve impulses arriving at any particular ending.

To maintain a precise relationship between the impulses and their effect the chemical transmitter must be quickly destroyed, otherwise the response could be prolonged out of all proportion to the instruction transmitted through the nerve fibre. With acetylcholine this is ensured by the existence of cholinesterase, an enzyme which rapidly inactivates the transmitter substance. Another essential is that the responding cell shall be highly sensitive to minute quantities of the transmitter substance that excites it. From all of this it is apparent that several conditions must be met before any particular substance can be safely interpreted as a chemical transmitter substance. There must be evidence of its storage in the nerve fibres concerned, and of its release under physiological conditions; there must also be a very high level of sensitivity to it in the cells that they excite, and there must be some means of rapid inactivation of the substance. It is important to keep these criteria in mind; the presence, and even the wide distribution of certain substances in nervous tissues, cannot in itself be accepted as establishing that these substances are neurohumours.

Acetylcholine is present in certain protozoans and in all of the main invertebrate groups, with the exception of sponges, coelenterates and urochordates. Whether it is physiologically active in this wide range of animals is far from clear, but good evidence of its neurotransmitter function comes from bivalves. It exerts an inhibitory effect on the heart of *Venus mercenaria*, and it is reasonable to suppose that this effect is a physiological one, and that it is controlled by enzyme action. The inhibition can also be evoked by stimulating the visceral ganglion, and this effect can be enhanced and prolonged by the addition of eserine, which inhibits the action of cholinesterase.

Acetylcholine is present in considerable amounts in the nervous system of decapod crustaceans, and exerts an excitatory action on the heart of many crustaceans. Yet the concentrations required to affect the heart are sometimes too high to meet the essential requirement of high sensitivity on the part of the reacting tissue; moreover, it has not yet been clearly established that acetylcholine mediates the transmission of excitation from nerves to skeletal muscle in these animals. The situation in other groups of invertebrates is, in general, no less obscure. The central nervous system of the cockroach, however, has a high concentration of acetylcholine and also a high level of cholinesterase activity, from which it is supposed that acetylcholine may be a chemical transmitter in insects in the central nervous system (although not at the neuromuscular junctions, where, in contrast to vertebrates, cholinesterase in absent). This is a matter of toxicological importance, for certain insecticides are inhibitors of cholinesterase, their action resulting in the accumulation of excess acetylcholine in the animals. The importance of precisely understanding the physiological significance of acetylcholine, and, indeed, of other neurohumours, is thus self-evident. It has been suggested that injudicious use of insecticides of this type could possibly have a ruinous effect upon aquatic crustaceans, and thereby weaken the food resources of other and commercially important forms.

Noradrenaline, a catechol amine, acts as a chemical transmitter in vertebrates at the post-ganglionic sympathetic nerve endings, which are in consequence termed adrenergic endings. Together with the closely related adrenaline it is also released from the chromaffin tissue (or medulla) of the mammalian adrenal gland (or its homologue in lower vertebrates) when this is stimulated through its sympathetic nerve supply. These two compounds function as hormones, that can be said, in very general terms, to enhance and prolong the effect of sympathetic stimulation. They are stored in large amounts in the cells of the chromaffin tissue, in the form of granules that react with potassium dichromate or iodate to give a brown oxidation product. This is the so-called chromaffin reaction, from which the medullary tissue of the adrenal gland takes its name. Information regarding the distribution of these two catechol amines in the invertebrates is unfortunately so fragmentary that it is impossible to judge of their possible physiological importance in these animals. The most that can be usefully said here is that cells giving a chromaffin reaction are found in annelidan nerve cords, and that both adrenaline and noradrenaline can be extracted from these. It has thus been suspected that adrenergic nerve fibres may exist in annelids, but even this is uncertain.

An indole alkylamine, 5-hydroxytryptamine, is widely distributed in a variety of tissues, including the nervous systems of vertebrates as well as of certain invertebrates. There are reasonable grounds for regarding it as a neurohumour, although the evidence is still very scattered and incomplete. The most convincing evidence is obtained from molluscs, in which it probably functions as an excitatory transmitter. Other substances are also suspected of functioning as chemical transmitters in the invertebrates. These include γ-aminobutyric acid, thought to be an inhibitory transmitter in crustaceans, and glutamic acid, thought to be a motor transmitter in these animals and in insects.

Evidently, chemical transmission is an aspect of invertebrate neural organization that needs much further study, but, on the assumption that the mode of functioning of the neurone will be fundamentally uniform throughout the animal kingdom, we may properly suspect the wide occurrence of neurohumoral substances. Fortunately there is another aspect of neural secretory activity in which the evidence has accumulated much more satisfactorily, and which provides a most striking demonstration of a unified mode of functioning throughout a very wide range of animals, both vertebrate and invertebrate.

17-2 HORMONES AND NEUROHORMONES

On purely etymological grounds it would be reasonable to regard the production of chemical transmitter substances by nerve cells as a process of neurosecretion, and to term the cells neurosecretory cells. It has, indeed, been suggested that all neurones should be so designated, but there is a more widely accepted convention, which we will follow here, that restricts the term neurosecretion to a different type of neural activity. This convention regards the 'ordinary' or 'typical' neurone (the *neurone banale* of French endocrinologists) as concerned with the propagation of nerve impulses and with the secretion of the chemical transmitter substances that we have mentioned. In addition to these 'ordinary' neurons there is a second type (Fig. 17-1) that differs from them in containing a secretory product that is readily stainable and that can be seen with the light microscope. It is referred to as a neurosecretion. This material, like the neurohumours, arises in the cell body, passes down the axon, and is released from the nerve endings. The fundamental difference is that it is not restricted to local and transitory action. Instead it passes into the blood stream and circulates round the body, to produce specific physiological effects at points that may be remote from the region of its release. The cells producing this substance are customarily referred to as neurosecretory cells; their function is the secretion of neurohormones, which form part of the endocrine secretion of the body.

So important is this concept that we must examine some of its implications further before considering specific illustrations of it. First, it must be emphasized that

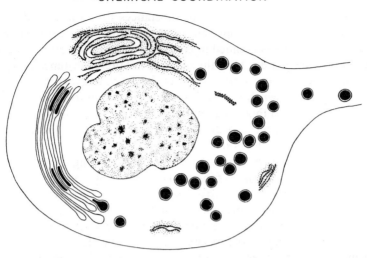

Fig. 17-1. Diagram based on electron micrographs of neurosecretory cells of the supra-oesophageal ganglion of *Lumbricus terrestris* illustrating formation of granules. Synthesis is presumed to begin in the ergastoplasm (= rough endoplasmic reticulum, above the nucleus) from where they are transferred to the Golgi apparatus. Lying between the lamellae, the product becomes visible as an electron-dense substance which eventually fills Golgi vesicles (to the left of the nucleus). Next, the granules leave the Golgi lamellae which furnish the membranes of the neurosecretory granules (below the nucleus). They accumulate in the intercisternal space of the ergastoplasm which becomes disorganized. Eventually the granules are carried off along the axon of the cell (to the right of the nucleus). From Scharrer and Brown, 1962. *Gen. & compar. Endocr.*, **2**, 1–3.

the mere existence of stainable material within a neurone does not justify regarding it as a neurosecretory cell. Such material may consist of inert metabolic products, such as lipofuscin granules, that would certainly be unable to exert specific effects upon other tissues. The function of the product is, therefore, an essential criterion of its status as a true neurosecretion. To put the matter in a formal way, we can say that the term is applicable only when it can be shown that the products are acting in a manner characteristic of the particular group of secretory products that we call hormones. What exactly do we understand by these products?

We have so far considered coordination and integration as being the field of action of the nervous system, and as being mediated by the propagation of nerve impulses. Yet we have seen also that the body fluids, and more particularly the blood systems, provide communication systems by which the products of digestion and metabolism are transmitted. It is perhaps not surprising, then, that these fluids are also used for the transmission of chemical substances that act in conjunction with the nervous system to convey information and to promote integration. Sometimes these substances may themselves be the by-products of metabolism. A well-known example of this is the way in which carbon dioxide acts in vertebrates to regulate the respiratory centre; we have seen that it also acts in a similar way in certain insects. Many of these substances, however, are specialized secretory products, and it is these that are termed hormones. According to the classical definition, based initially upon studies of vertebrate physiology, hormones are substances that are formed in particular regions of the body, usually in specialized glands, and that are discharged into the blood stream. This process is called internal or endocrine secretion, and the glands are termed endocrine glands. After the discharge of these substances, they are transmitted in the blood stream to other parts of the body, where they produce specific regulatory effects. The term hormone is derived from the Greek *hormaein* (to impel or arouse to activity), and thus refers only to the regulatory effect. In practice, however, the concept of action at a distance, with the substance forming part of the communication system of the body, is an essential part of the full definition.

Endocrine glands were originally conceived as being formed of epithelial secretory tissue. The importance of the concept of neurosecretion, as defined above, is that it extends endocrine activity to the nervous system, thereby bringing into close association two modes of coordination, chemical and neural, that were earlier regarded as distinct. It is thus all the more important that the hormonal character of the products of supposed neurosecretory cells should be clearly established; yet this may present great difficulties. In vertebrate studies it has often been possible to achieve chemical characterization and even synthesis of hormones, and, in consequence, to define very precisely their physiological effects. The small size of many invertebrates is a serious obstacle to such success, although the obstacle is being overcome. At present, however, the evidence has often to be secured by indirect methods. These include the extirpation of supposed endocrine tissue, the injection of extracts of it, and histological and cytological study designed to establish changes in the appearance of the animal's cells that can be correlated with physiological changes in the body. Here the electron microscope has proved of the greatest value, particularly with neurosecretory cells. In these it has been shown that the material visible with the light microscope is composed of elementary granules having a diameter from about 100 to 300 nm.

Neurosecretory cells, then, are neurones that are carrying out a type of secretory process distinct from the neurohumoral secretion of the 'ordinary' neurone, but the distinction must not be allowed to obscure the fact that both are nevertheless engaged in secretion as a basis for the use of a chemical language. Finlayson and Osborne, in emphasizing this point, suggest the term 'neurocrine' to embrace all of the secretory functions of neurons, whatever their type. Neurocrine activity thus emerges as the central feature of animal communication, with the neurone as the command cell. Adaptive diversity of command is achieved by regulation of the amounts of secretion produced and the rates at which they are discharged, these being related to the visible differences in the nature and sizes of the granules in which the secretions are stored.

The evolutionary relationship between 'ordinary' and neurosecretory neurones is obscure. One view is that neurosecretory cells could have been derived by specialization from 'ordinary' ones. Another view is that neurosecretory cells were the first to appear, perhaps derived from epithelial cells functioning in the regulation of growth, and perhaps already capable of electrical conduction, for there is clear evidence, from invertebrates as well vertebrates, that neurosecretory cells can transmit nerve impulses. Later, these cells might have developed processes which, with the further elaboration of electrical conduction, made possible the evolution of rapidly and locally acting control mechanisms. We shall see that this second view accords with the fact that the lower invertebrates rely largely, if not entirely, upon neurosecretory cells, rather than upon epithelial glands, for endocrine coordination, for this implies that the epithelial type of endocrine gland evolved later than the neurosecretory one.

It remains to add that the release of the neurosecretory product is an important and characteristic aspect of the functioning of the neurosecretory cell. Unlike the 'ordinary' neurones, these cells do not usually end in synaptic connection with neurones or effector cells, although there are exceptions to this generaliztion. Their axon endings are typically in close physical relationship with blood vessels, the result being the formation of compound structures called neurhaemal organs. These organs are the storage and release centres of the neurosecretory system. It is probable that the neurosecretory product is always stored there within the axon bulbs, for there is no convincing evidence for the existence of granular neursecretory material within the blood vessels. Presumably the secretory product is release from the granules as hormonal molecules that diffuse through the membranes into the blood stream.

Here again, however, there is variation. Discharge of the neurosecretion at neurohaemal organs, although part of the original or classical concept of neurosecretion, is, as we have indicated, no longer an essential criterion of the process. The fibres may instead make direct contact with effector cells, which may be neural or endocrine or neither; they may, for example, be muscle fibres. Another possibility is that the neurosecretory axons may end close to the effector cells, but be separated from them by narrow spaces across which the released secretion must diffuse. In these instances the neurosecretions are not fulfilling the classical requirement of hormones, i.e. that they must be transmitted through the blood stream. However, definitions are our servants, not our masters, and the fact is that the classical definition of a hormone applies to only one part of a much wider spectrum of chemical communication processes. We shall be considering further illustrations of this later.

17-3 ENDOCRINE REGULATION IN CRUSTACEANS

Crustaceans and cephalopods share with many fish, amphibians, and reptiles a remarkable capacity for changing the pattern and intensity of their colouring. In considering this phenomenon we must first distinguish between morphological colour change, which depends upon alterations in the total amount of pigment, and physiological colour change, which depends upon the movement of pigment granules within specialized cells called chromatophores. We are not concerned here with morphological change, nor shall we deal with the physiological change in cephalopods, for in these particular animals the effector cells are unique in being controlled by extrinsic muscle fibres. The chromatophores of crustaceans and vertebrates are cells with branched processes; pigment granules are present in them, and these can either be concentrated in the centre of the cell, giving a minimum display of colour, or dispersed through the cell body and its processes, giving a maximum display. The pigments are of more than one colour; black, red, or yellow, for example. This creates possibilities of elaborate patterns, but at the same time requires correspondingly elaborate regulation. Usually any one chromatophore will contain only one type of pigment, but several cells with different pigments may often be united in crustaceans to form syncytial complexes called chromatosomes.

It is usually assumed that physiological colour change is of adaptive value to the animals concerned, and this is probably so, although the nature of the adaptation takes many forms. Colour change may, for example, be a response to the colour of the background, or to the intensity of illumination. There may be a diurnal rhythm of change, which is possibly sometimes of thermoregulatory advantage, as in the fiddler crab, *Uca*, in which black pigment becomes concentrated when the temperature rises.

Whatever may be the adaptive significance of colour change in crustaceans, there is a fundamental uniformity in the means by which it is controlled, for it is under the influence of neurohormones secreted by a complex neurosecretory system. The vertebrates differ in this respect. In cyclostomes, fish, Amphibia, and reptiles hormones (melanocyte-stimulating hormones, MSH) certainly play a part, but they are not neurosecretory in origin; they arise in the epithelial secretory tissue of the pars intermedia of the pituitary gland. In teleost fish other hormones may also be involved; their origin, if they exist at all, is uncertain, but they are not neurosecretory products. Further, in teleosts and reptiles, and perhaps also in selachian fish, the nervous system directly innervates the chromatophores, so that these are under combined nervous and endocrine control. This is certainly not so in crustaceans, which have no innervation of their chromatophores. There are thus differences between crustaceans and vertebrates in their mechanisms of colour control. The differences, which reflect the independent origin of the phenomenon in the two groups, are discussed further below.

The key to understanding the crustacean mechanism was contained in a discovery made in the 1870s by Pouchet, who found that the colour responses of shrimps to background conditions were eliminated if their eye stalks were removed. The correct explanation of this was not apparent at the time. The observation suggested that colour change in these animals was controlled by the nervous system after this had been activated by the visual receptors of the eye stalk, and not until 1928 was it shown that chromatophores of blinded shrimps would respond to injections of eye-stalk extracts. This indicated that there might be endocrine tissue in the eye stalk, regulating colour change through its hormonal secretion. This interpretation has since been amply confirmed by a wide range of evidence.

Within the eye stalk (Fig. 17–2) is a brain centre, the medulla terminalis, distally to which lie three optic centres, the medulla interna, the medulla externa, and the lamina ganglionaris. Clusters of neurosecretory cells exist in this nervous tissue, the principal cluster (or at least the one to which attention was first directed) being called the medulla terminalis X-organ. There is also a sensory pore or sensory papilla X-organ. This is so called because it is separate from the other groups of cells, and is typically associated with a sensory pore or papilla, which seems from its histological structure to be in part sensory and in part neurosecretory. Other X-organs have also been described. Undoubtedly this neurosecretory system is very complex, both in its structure and its function, and we know it is concerned in the regulation of much more than colour change. Irrespective of the functions of the various types of neurosecretory cells, however, many of their fibres share a final common path in that their fibres end within the eye stalk in a neurohaemal organ called the sinus gland. The axon terminals within this organ contain electron-dense granules, 100–300 nm in size, of the general type that we have already mentioned; they are the product of the neurosecretory cells, transmitted to the gland along their fibres. The discharge of the secretion into the blood stream is probably brought about by nerve impulses initially

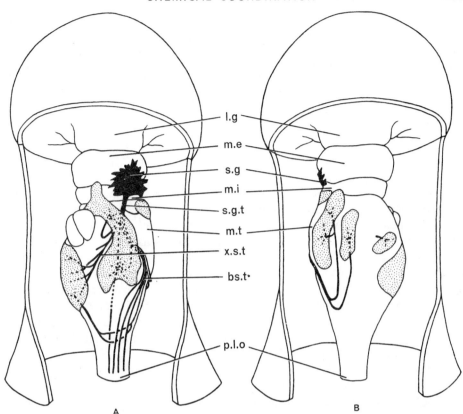

Fig. 17-2. The eye stalk neurosecretory system of the crayfish *Orconectes virilis*. A, dorsal view of the right eye stalk; B, ventral view. Stippled areas indicate regions in which neurosecretory cells occur. The principal neurosecretory fibre tracts to the sinus glands are shown. *bs.t*, brain-sinus gland tract; *l.g*, lamina ganglionaris, first optic ganglion of the eye stalk; *m.e*, medulla externa; *m.i*, medulla interna; *m.t*, medulla terminalis; *p.l.o*, optic lobe peduncle; *s.g*, sinus gland; *s.g.t*, sinus gland tract; *x.s.t*, tract from medulla terminalis X-organ to sinus gland. After Bliss, from Charniaux-Cotton and Kleinholz, 1964. *The Hormones*, vol. IV (Pincus et al., eds.). Academic Press, New York.

evoked by stimulation of the appropriate receptors. This discharge is facilitated by the close association of the sinus gland with a blood sinus, an association from which it derives its name. The term 'gland', however, is a misnomer; it was given to the organ before the phenomenon of neurosecretion had been defined, and when the secretion visible within the organ was presumed to arise there.

The recognition, from 1951 onwards, of the importance of neurosecretion in the control of crustacean colour change led at first to emphasis being placed upon the role of the X-organ and sinus gland complex. But we have already noted that crustaceans possess several types of chromatophore. Not only do these differ in the pigment that they contain, but there is also evidence that some at least of these types can exhibit independent responses. Out of these observations has grown what may be termed the multiple-hormone hypothesis of colour control. According to this the responses depend upon a number of hormones, acting to some extent independently of each other and capable of evoking individual responses from particular types of chromatophore. Associated with this has been the demonstration that neurosecretory cells exist in parts of the central nervous system other than the centres of the eye stalk. One region of particular importance is the tritocerebrum, and the tritocerebral commissure which is anatomically derived from it. Neurosecretory fibres, arising in the tritocerebrum, run to the commissure, which they leave in the post-commissure nerves. At one point on these nerves the surrounding membrane, the epineurium, is enlarged to form an epineural plate, or post-commissure organ. The structure, which contains many neurosecretory droplets, is thought to be a neurohaemal organ, similar in function to the sinus gland. The injection of extracts of this region evokes responses in particular types of chromatophore. It is believed that several hormones are concerned, and that these differ, at least in part, from the hormones released from the sinus gland. We are thus led to the view that crustaceans must secrete a number of chromactivating hormones. Support for this has come from purification studies, using gel filtration and ion exchangers, for these have made it possible to separate factors with distinctive actions, and to proceed some way with their purification and chemical characterization. Thus a hormone which has the precise effect of concentrating the red pigment of the chromatophores of the shrimp *Pandalus borealis* has been shown, by structural analysis and by synthesis, to be an octapeptide with the structure pyroglutamic acid—leucine—aspartic acid—phenylalanine—serine—proline—glycine—tryptophan—NH_2.

Crustacean colour change is thus a well-established demonstration of the importance of neurosecretion in regulating adaptive responses. But what is particularly striking is the wide distribution of neurosecretory cells in the crustacean nervous system. They are by no means restricted to the sites already mentioned. Moreover, such diverse types exist that it would seem improbable that they should only regulate the behaviour of chromatophores. There is, in fact, ample evidence that crustacean neurosecretions regulate a wide range of physiological activities.

We have already mentioned the adaptive significance of the movements of the pigments of the compound eye. In the eye of *Palaemonetes vulgaris* (Fig. 17–3) these pigments comprise a distal pigment, situated in distal pigment cells; a proximal pigment, situated in the retinular cells; and a white reflecting pigment, situated in tapetal cells that lie between the proximal ends of the ommatidia. In bright light the distal and proximal pigments screen the sides of the rhabdome so that the ommatidium is stimulated mainly by light entering along its axis. This is the light-adapted condition (p. 406). In darkness the pigments move into the dark-adapted condition; the rhabdome is unscreened so that the ommatidum is stimulated by light entering from a larger source, while the reflecting pigment is so situated that it scatters incident light and still further increases the stimulation of the rhabdome. Visual acuity is less in this condition, but maximum use is made of the available light rays.

These pigment movements are in part under hormonal control, as can be shown by injecting extracts of eye stalks into dark-adapted *Palaemonetes*. The result of this is that the distal and the reflecting pigments move into the light-adapted condition. Further and convincing evidence is that the distal pigment in *Palaemon* (*Leander*) will pass permanently into the dark-adapted condition if the sinus glands are extirpated, even though the eyes are left quite undamaged. Curiously enough, the reflecting

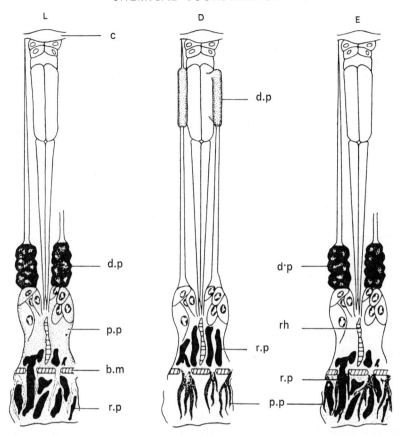

Fig. 17-3. Ommatidia from the eyes of *Palaemonetes vulgaris*, showing the general structure and the position of the retinal pigments under various conditions. L, from an eye in the light condition; D, from a dark-adapted eye; E, from the eye of an animal which, after adaptation to darkness, was injected with eye stalk extract prepared from light-adapted prawns. *c*, cornea; *d.p*, distal pigment; *p.p*, proximal pigment; *b.m*, basement membrane; *r.p*, reflecting pigment; *rh*, rhabdome. From Kleinholz, 1936. *Biol. Bull. mar. biol. Lab., Woods Hole*, **70**, 159-184.

pigment is unaffected by this exirpation. However, we may conclude that some of these pigment movements (certainly those of the distal pigment, and probably those of the reflecting pigment) are regulated by a neurohormone released from the sinus gland. The hormone is known to be secreted within the eye stalk, exactly like some of the chromactivating secretions, for active extracts can be obtained from various parts of the nervous tissue of the stalk, as well as from the sinus gland itself. It proves to be an octadecapeptide, distinct, therefore, from the red-pigment concentrating hormone just mentioned. Nevertheless, there are aspects of retinal pigment migration that still remain unexplained. It is not clear how the position of the proximal pigment is regulated, for example, nor is it certain that dark adaptation is effected hormonally, although there is some evidence suggesting that a separate dark-adapting hormone may exist.

The regulation of pigment movements is a very specialized and limited aspect of the physiology of crustaceans. Emphasis upon colour change in accounts of crustacean endocrinology is a result of the historical accident that, being a conspicuous feature of their behaviour, it has for a long time attracted attention, and has lent itself well to experimental analysis. However, considering the wide range of activities that are controlled by hormonal action in the vertebrates, and considering the complexity of crustacean organization, we might reasonably expect to find hormonal mechanisms fulfilling a variety of functions in crustaceans. Information here is still very limited, but at least it is certain that moulting is regulated by hormones (Fig. 17–4).

The moulting cycle of crustaceans comprises four stages: premoult (proecdysis), in which occur preparatory events, including partial reabsorption of the old cuticle and partial synthesis of the new one; moult (ecdysis), when the old exoskeleton is shed; postmoult (metecdysis), in which the formation of the new exoskeleton is completed; and intermoult, in which tissues grow and metabolic reserves are laid down. The length of the intermoult varies, but two main patterns can be distinguished. In one of these (diecdysic cycle) the intermoult is short and the premoult long, while in the other (anecdysic cycle) the intermoult is long and the premoult short.

Evidence for hormonal control of the cycle in decapods is obtained by removing both eye stalks during the intermoult. This accelerates the onset of the next moult, but the acceleration can be abolished by implanting sinus glands into these eyestalkless animals or by injecting extracts of the eye stalks. The effect is most clearly seen in anecdysic forms (e.g. crayfish and the fiddler crab, *Uca pugilator*), but it is demonstrable also in diecdysic ones (Fig. 17–5). The inference from these and other experiments is that the eye stalks contain a moult-inhibiting neurohormone, supposedly a peptide, which is secreted by the medulla terminalis X-organ, and released by the sinus gland. The relationship between the sinus glands and the X-organ in this context has been well established by experiments involving their removal and implantation.

Exactly how or where this moulting hormone acts is still uncertain, but in any case it is not the only one involved in the process. In crustaceans, as in vertebrates, the hormones are not exclusively neurosecretory ones. Some arise from epithelial glands. These seem to play only a minor part in arthropods as compared with their wide range of functions in vertebrates, but they certainly make an essential contribution to the control of moulting in crustaceans. The glands concerned are the Y-organs, a pair of structures lying either in the antennary or the second maxillary segment. They are the source of the true moulting hormone, for which reason they are also termed the ecdysial glands. The evidence for this rests upon various extirpation and implantation experiments. Thus removal of the Y-organ from *Carcinus* during the intermoult or during the preparatory stage of moulting will prevent the moult from taking place. The inhibition is a permanent one, provided that the animals receive no further treatment. If, however, Y-organs are implanted into them the moulting cycle is resumed. Physiological relationships between the X-organ and sinus gland complex and the Y-organ are suggested by the results of experiments in which both the Y-organ and the sinus gland were extirpated: this treatment prevents moulting, whereas removal of the sinus gland alone, with the Y-organ remaining in position, produces the acceleration of moult to which we have already referred.

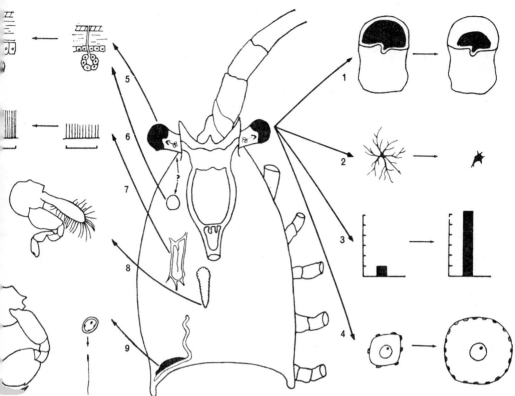

Fig. 17-4. Hormonal functions in crustaceans. On the right side *1–4* represent hormones from the eye stalk; on the left, *5–9* represent eye stalk and other endocrine effects. *1*, light-adapting distal retinal pigment hormone; *2*, chromatophorotropins; *3*, hyperglycaemic hormone; *4*, eye stalk ablation results in ovarian growth, through precocious vitellogenesis. *5*, moult-inhibiting hormone of the eye stalk, probably acting normally on the Y gland; *6*, the Y gland, from which is secreted a moult hormone; *7*, the pericardial organ, extracts of which accelerate the heart rate; *8*, ovarian hormones regulating female secondary sexual characters; *9*, androgenic gland of male, regulating spermatogenesis and secondary sex characters in male. From Charniaux-Cotton and Kleinholz, 1964. *op. cit.*

A substance believed to be the moulting hormone has been extracted from crustaceans and characterized and synthesized as 20-hydroxyecdysone (crustecdysone; β-ecdysone). Full confirmation that it is indeed the moulting hormone is still awaited, but certainly it evokes precocious moulting when injected into *Uca* and the crayfish *Procambarus*, and it stimulates the synthesis of RNA and protein in the hepatopancreas of the latter. It is presumed that this substance, which is a steroid (Fig. 17–11, p. 511), is secreted by the Y-organs, and it has been suggested that these are inhibited by the moult-inhibiting hormone. However, another view is that the two hormones are antagonists, the moult-inhibiting one restricting in some way the action of 20-hydroxyecdysone upon the epidermal cells. The latter view implies the possibility that both are present together in the haemolymph, their relative levels varying from species to species, and from stage to stage in the moult cycle of any one species. This

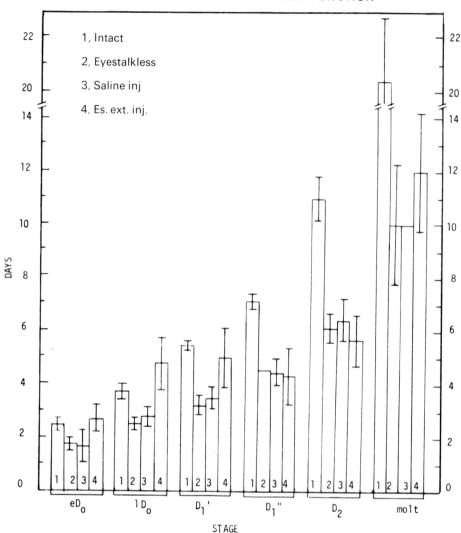

Fig. 17-5. Comparison of period (ordinates) from ecdysis to successive stages in premoult in intact (1), eyestalkless (2), saline-injected eyestalkless (3) and eyestalk-extract-injected eyestalkless (4) shrimps (*Palaemonetes pugio*). The numbers of animals per group in groups 1–4 were respectively 24, 20, 10, and 12. The vertical line in each bar represents 95% confidence limits. The animals ranged in size from 25 to 40 mm. Stage eD_0, primordia of seta-forming areas not visible; ID_0, areas present as thin lines; D_1', seta-forming areas widen; D_1'', new setae become visible; D_2, new setae clearly seen. From Freeman and Bartell, 1975. *Gen. & compar. Endocr.*, **25**, 517–528.

could account for certain contradictions in some of the experimental findings, particularly with regard to differences between anecdysic and diecdysic forms. Overriding these uncertainties, however, is the indication of close resemblances between the control of moulting in crustaceans and insects. We shall return to this later (p. 511).

Other aspects of crustacean metabolism may well be under hormonal control (cf. Fig. 17-4), but the evidence is sometimes very incomplete. Among the processes to which hormonal regulation has been attributed are the uptake of water, which is an important feature of moulting, calcium metabolism, also involved in moulting, and the metabolism of nitrogen and carbohydrate. Hydromineral metabolism, the importance of which is by no means restricted to the moult, is almost certainly under hormonal control. For example, the crayfish *Procambarus* has a brain factor which increases the chloride level in the blood and also the influx of sodium into it. Removal of the eye stalks decreases the concentration of chloride, and increases the influx of water and the flow of urine, while injections of brain homogenates into normal crayfish increase the level of blood sodium. Generalization, however, is difficult, for in marine and estuarine forms the hydromineral balance is differently regulated, perhaps by different factors.

A fundamental difficulty in interpreting the evidence in this field is simply that too little is known at present of crustacean physiology, of the biochemical pathways in these animals, and of the interaction of the various hormones (established and supposed) with each other. Those who have some knowledge of the immense complexity of vertebrate endocrine systems, and of the array of experimental evidence upon which their interpretation is based, will be content to wait for comparable advances in our knowledge of crustaceans before attempting to generalize about the endocrine systems of these animals. It must be sufficient now to accept the fundamental importance of hormones in their regulatory mechanisms, and to avoid supposing that these will always be organized on vertebrate lines.

A good illustration of the danger of extrapolating from vertebrate hormonal systems to invertebrate ones is given by the hormonal regulation of sexual differentiation in crustaceans. There is some evidence that in females of *Orchestia* a hormone secreted by the ovary controls a permanent secondary sexual character, the presence of oostegites, and a temporary one, the presence of ovigerous hairs on the oostegites. Removal of the ovaries leads to loss of the hairs, but these reappear in the castrated females if ovarian tissue is implanted into them. The production of sex hormones by the gonads is a familiar feature of vertebrate sexual organization, but the situation in the males of higher crustaceans is quite different from that in male vertebrates, where the male hormone is secreted by interstitial tissue in the testes. The male hormone of crustaceans is secreted by the androgenic glands (Fig. 17-6), which are entirely independent of the gonads, being attached to the hinder end of the vasa deferentia (except in isopods, where they lie more anteriorly, and may even be attached to the testes).

The hormone of the androgenic gland, which, it is surmised, may be a protein or polypeptide, is responsible for the differentiation and maintenance of the testes and also of the secondary sexual characters. If the androgenic glands are removed from males, and the appendages also removed, the animal regenerates sexually undifferentiated limbs, while the testes may transform into ovaries. Conversely, implantation of androgenic glands into females results in the ovaries becoming transformed into testes, while the appendages become masculinized. Transformation of the appendages takes place even if the ovaries are removed from the females before the implantation of the androgenic glands, which shows that the male hormone acts directly on the limbs and not through the transforming gonad.

The sexual endocrinology of crustaceans is very remarkable, and could not have

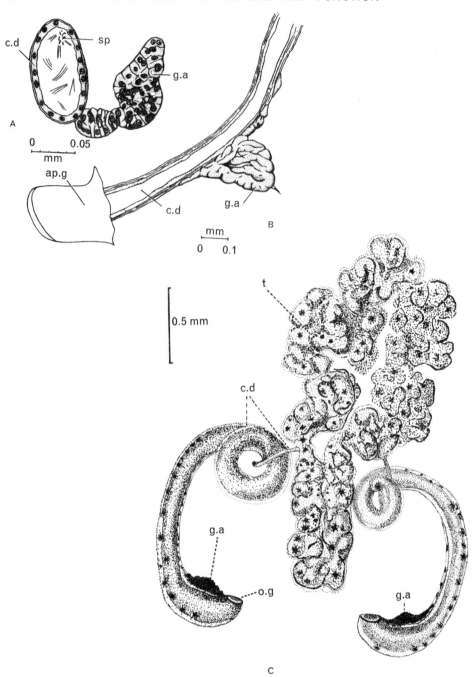

Fig. 17-6. Androgenic gland of *Orchestia gammarella*. A, transverse section; B, appearance of fresh gland. *ap.g*, genital apophysis; *g.a*, androgenic gland; *c.d*, genital duct; *sp*, spermatozoids. From Charniaux-Cotton, 1957. *Annls Sci. nat., Zool.*, 11 sér., **19**, 411–560. C, reproductive system and androgenic gland of a male *Palaemon serratus*. *cd*, genital duct; *g.a*, androgenic gland; *o.g*, genital opening; *t*, testis. Charniaux-Cotton and Huguet, unpublished.

been predicted by arguing from analogy with the vertebrates. Indeed, it is precisely because of this that the phenomenon of parasitic castration, known in crustaceans since the nineteenth century, proved impossible to interpret satisfactorily until this endocrinological mechanism had been unravelled. Parasitic castration is a form of intersexuality in which males become feminized when they are parasitized by rhizocephalan cirripedes or by epicaridian isopods (Fig. 17-7). The widespread nature of the effect which in crabs, for example, involves changes in the shape of the abdomen and in the development and form of the appendages, is clearly suggestive of some endocrine disturbance. For a long time it was supposed that this disturbance was associated with the destructive action that the parasites were known to have on the gonads. By analogy with vertebrates, it seemed reasonable to assume that destruction of the testes resulted in a reduced output of male sex hormone, and a consequent reduction of male differentiation. There was always difficulty in accepting this view, however, for there was no close correlation between the degree of gonadal destruction and the degree of masculinization. The discovery in 1954 of the androgenic gland resolved this particular problem, for it was then apparent that parasitic castration resulted from the destruction of that gland by the parasites. There could be no better illustration of the danger of pressing too far an argument by analogy, when the groups concerned are as widely separated as are the crustaceans and the vertebrates.

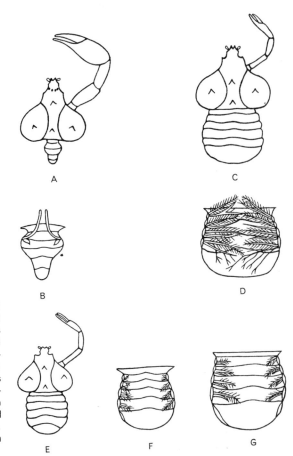

Fig. 17-7. The crab *Inachus*. A, normal male with short small abdomen and broad claw; B, male abdomen without pleopods from beneath; C, normal female with broad abdomen and small claw; D, female abdomen from beneath with hairy pleopods; E, sacculinized male with female characters (small claw and broad abdomen); F, abdomen of sacculinized male from beneath with pleopods; G, abdomen of sacculinized female. After Smith. From Hanstrøm, 1939. *Hormones in Invertebrates*. Clarendon Press, Oxford.

508 INVERTEBRATE STRUCTURE AND FUNCTION

The action of the androgenic glands is equally well shown in those crustaceans which undergo natural sex reversal. An example is the protandric shrimp, *Pandalus platyceros*, which spends one to three years as a male, and then transforms into a female. The change coincides with the disappearance of the androgenic glands, which are thought to be under neuroendocrine control, for removal of the eyestalks delays the transformation and prevents atrophy of the glands. Measurement of RNA synthesis in the gland, using material from specimens that have been injected with uridine-5-^{3}H, shows that removal of the eyestalks increases the synthesis, while reimplantation of the sinus gland decreases it (Fig. 17–8). Evidently, then, an eyestalk factor inhibits the gland, and thus, presumably, controls in some way the sex reversal. The control may not, however, be a direct one, for 20-hydroecdysone treatment also increases RNA synthesis in the androgenic gland (Fig. 17–8). Perhaps, then, sex reversal in this shrimp is mediated by 20-hydroxyecdysone which, as we have seen, is supposed also to be the moulting hormone. As usual, much remains to be learned.

17–4 ENDOCRINE REGULATION IN INSECTS

Studies of hormonal coordination in insects have largely centred around moulting and growth, which are regulated by neurosecretory mechanisms bearing remarkable resemblances to those of crustaceans. Here again we are concerned with neurosec-

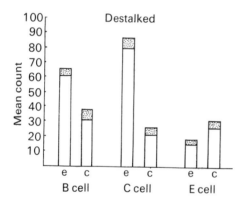
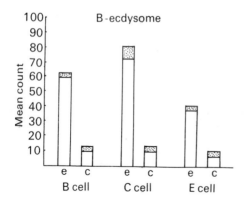

Fig. 17-8. Mean autoradiographic silver grain counts of three types of gland cell in the androgenic gland of the protandric shrimp, *Pandalus platyceros*. Above, destalked shrimps and intact controls; below, β-ecdysone-treated shrimps and intact controls. Open bar represents nuclear counts; dotted bar represents cytoplasmic counts. Each mean is based on a 2500 μm^2 area of the androgenic gland. e, experimental group; c, control group. From Foulks and Hoffman, 1974. *Gen. compar. Endocr.*, **22**, 439–447.

retory cells in the central nervous system. An important assemblage of these is found in the median region, or pars intercerebralis, of the protocerebrum, where they commonly form four groups, two medial and two lateral. We may compare these cells with those of the ganglionic X-organ of the crustacean eye stalk. Like the latter, they are connected with a neurohaemal organ, which in this instance is formed by the corpora cardiaca, a pair of bodies lying behind the brain and close to the aorta (Figs. 17–9 and 17–10). The corpora cardiaca are comparable, therefore, with the sinus glands, but they are probably more complex in function than the latter organ. They are composed in part of neurosecretory nerve endings, but in addition they contain cells, some of which are probably themselves secretory. However, it is certain that neurosecretion formed in the pars intercerebralis passes down axons into the corpora cardiaca; there it is stored and eventually released, perhaps after further processing by the secretory activity of the cells of the corpora.

It is probable that different brain hormones with diverse effects are secreted in the neurosecretory cells of the pars intercerebralis. The most closely studied is the thoracotropic effect, which is attributed to a thoracotropic hormone. This is so called because it activates a pair of glands called the thoracic glands (*trope*, turn, i.e. towards). The relationship is best understood by considering the sequence of events that is believed to determine moulting in the larva of the hemimetabolous bug, *Rhodnius*, the suitability of which for experimental study has been brilliantly exploited by Wigglesworth. (Strictly, the young stages of hemimetabolous insects are nymphs, not larvae, but the distinction is commonly ignored in endocriological literature.)

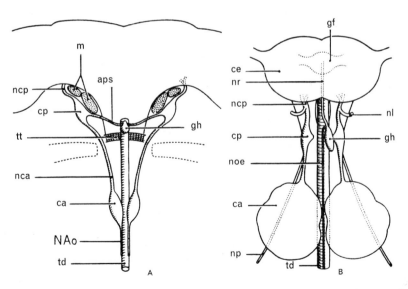

Fig. 17-9. (Above) Retro-cerebral complex in A, the psyllid *Homotoma ficus*, and B, the cocid *Pulvinaria mesembryanthemi*. *aps*, paracardio-sympathetic anastomosis; *ca*, corpora allata; *ce*, brain; *cp*, corpora cardiaca; *gf*, frontal ganglion; *gh*, hypocerebral ganglion; *nao*, aortic nerves; *nca*, *ncp*, nerves to corpora allata and cardiaca; *noe*, oesophageal nerve; *np* posterior nerve; *nr*, recurrent nerve; *td*, gut; *tt*, tentorium. From Grassé, 1951. *Traité de Zoologie*, vol. 10 fasc. 2. Masson, Paris.

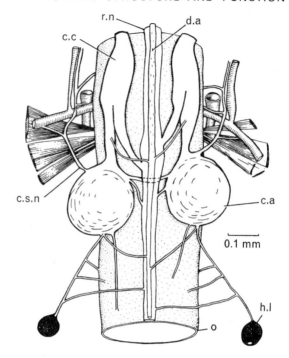

Fig. 17-10. (Right) The corpora cardiaca, corpora allata, and their nervous connections in the cockroach, *Periplaneta americana*. *c.a*, corpus allatum; *c.c*, corpus cardiacum; *c.s.n*, corpus allatum-suboesophageal ganglion nerve (enlarged); *d.a*, dorsal aorta; *h.l*, head lobes; *o*, oesophagus; *r.n*, recurrent nerve. From Harker, 1960. *Cold Spring Harb. Symp. quant. Biol.*, **25**, 279–287.

The initial stimulus to moult in this animal is the stretching of the wall of the alimentary canal by the single large meal which the larva takes in during each instar. As a result of this stimulation, nerve impulses are transmitted to the brain and promote the secretion of the thoracotropic hormone and its release from the corpora cardiaca. This hormone then stimulates the thoracic glands, changing the appearance of the nuclei and cytoplasm of their cells. These changes culminate in the release from the thoracic gland of another hormone, which is called ecdysone because it initiates the complex of metabolic processes that constitute ecdysis or moulting. These processes include induction of the synthesis of dopa decarboxylase, which is required for the tanning of the cuticle. A convenient basis for bioassay of the hormone is thus the *Calliphora* (blowfly) puparium reaction, which is the hardening of the cuticle of the larva (pupariation) just prior to the formation of the pupal stage (pupation). Ecdysone has been isolated, and chemically characterized and synthesized, thanks initially to the preparation of 250 mg of it from 4000 kg wet weight of *Bombyx* (silkworm) pupae. It is a steroid (Fig. 17–11) which is secreted as α-ecdysone and then rapidly transformed into β-ecdysone (20-hydroxyecdysone, crustecdysone). This is presumed to be its active form within the animal, although both forms are biologically active, so that it is often convenient to refer simply to ecdysone as the

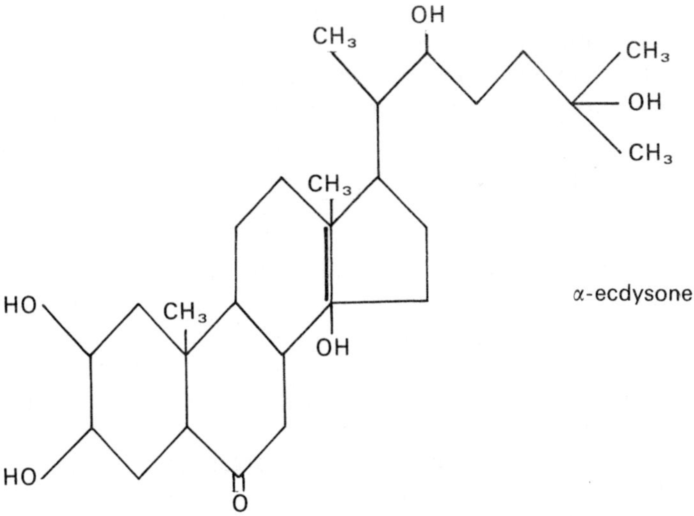

α-ecdysone

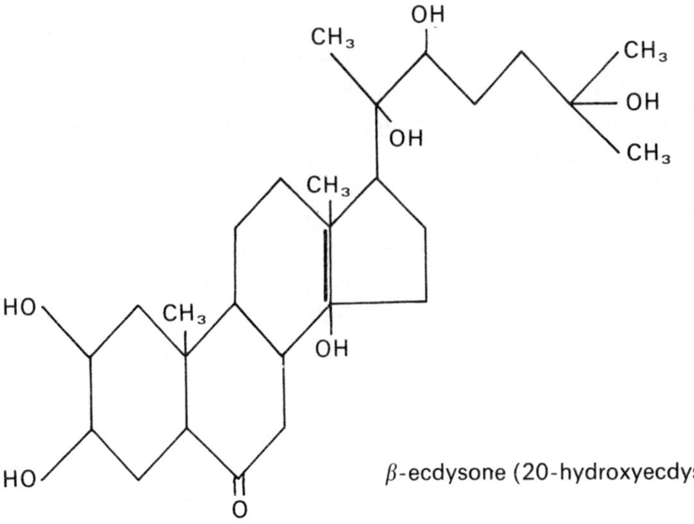

β-ecdysone (20-hydroxyecdysone; crustecdysone)

Fig. 17-11. Ecdysones.

hormone. It is synthesized by insects from cholesterol which they have to obtain from their food, since they cannot synthesize it for themselves.

It can now be seen why the presence of 20-hydroxyecdysone in crustaceans is of such interest, for this reveals a very close resemblance between the hormonal mechanisms that regulate moulting in those animals and in insects. It is obviously attractive to suppose that the thoracic glands and the Y-organs are homologues, possessing identical biochemical pathways, but the complexities of arthropodan phylogeny and the possibility of convergence, discussed in chapter 7, impose caution. It does seem likely, however, this is an example of latent homology (p. 28), some common genetic endowment having been turned to very similar function in the two groups. That steroids should appear as hormones in arthropods is also of interest in view of the important contribution that they make to the vertebrate endocrine system. The gonadial sex hormones of vertebrates are steroids, and so also are the hormones of the adrenal cortex, which regulate ion and water exchanges and certain other aspects of metabolism. Here, then, are examples of the biochemical potentialities of the sterol ring structure being exploited in arthropods and vertebrates along similar lines, yet quite independently.

Rhodnius passes through five larval (nymphal) instars during its growth and development. The moult at the end of the fifth instar also involves metamorphosis into the adult stage, with the formation of wings and the complete reproductive equipment. This does not mean, however, that the larval moults involve no more than simple increase in size, with all morphological transformation confined to the metamorphic moult. On the contrary, the wing lobes of *Rhodnius* show some relative increase in size during the earlier larval instars, and the reproductive organs undergo some differentiation. Nevertheless, so much transformation takes place at metamorphosis that there is clearly some major difference in the factors that influence this moult as compared with those influencing the earlier non-metamorphic ones. This difference lies in the functioning of an epithelial (non-neurosecretory) endocrine gland, the corpus allatum. This gland, which develops from the ectoderm, lies close to the corpora cardiaca. It is innervated by the nerve supply of the corpora cardiaca, and it appears to receive neurosecretory products from them. Possibly, then, the corpus allatum can be stimulated by neurosecretory fibres from the pars intercerebralis.

The corpus allatum secretes a hormone, called the juvenile hormone, which is believed to arise initially within the cell nuclei and to be processed further in the cytoplasm of the gland cells. In simple terms this hormone can be thought of as inhibiting metamorphic change. While it is circulating, the moult evoked by ecdysone will be a larval moult. It is not, however, present during the last larval moult, and its absence makes possible the drastic morphological changes of metamorphosis. We have seen, however, that some advance towards the adult form can be detected at the earlier moults, and this in itself shows that we are stating the role of the juvenile hormone in over-simple terms. To obtain a closer approximation to the truth it is helpful to adopt an analysis that has been developed by Wigglesworth.

According to this the developing insect is thought of as having the capacity for existing in either larval or adult form. The particular form that it assumes depends upon whether at any given stage it is under the influence of the genes that determine larval structure, or of those that determine the adult structure. On this view the function of the juvenile hormone is to maintain the activity of the genes that determine larval form. In the absence of this hormone at the end of larval life these genes can no longer be active, and those determining adult form can exert their effect. There would probably be general agreement on this aspect of the interpretation. What is less clear is what determines the small degree of differentiation that does actually occur at the larval moults. One view ascribes this to a continuing fall in the amount of juvenile hormone released, which consequently allows the expression of some degree of adult form. Another suggests that it is determined by the pattern of release of thoracotropic hormone, variations in the amount of juvenile hormone having no influence upon the result. For our present purposes, however, we can disregard this particular problem, for it does not affect the validity of the fundamental principles of moulting and metamorphosis that we have briefly outlined.

These principles could not have been established without prolonged experimentation, involving the application of exceedingly elegant techniques to animals that fortunately have the capacity for surviving drastic surgical treatment. Examples of these are parabiosis experiments in which two decapitated *Rhodnius* are joined together by their cut surfaces. If an animal is decapitated one or two days after its meal it will not moult, but it will do so if decapitation is delayed until a critical period of several days has elapsed. This is because the brain hormone has to exert its action upon the thoracic glands for several days before the glands can come into full secretory activity, and release adequate amounts of ecdysone. The presence of the latter in the blood can be demonstrated by uniting in parabiosis two decapitated larvae, one of which has been decapitated before the critical period and the other after that period. Both will moult, although the former would not have done so had it been left in isolation. The result is ascribable to the fact that united animals have a common circulation and a common share of ecdysone derived from the larva that was decapitated after the critical period.

The influence of the juvenile hormone can be similarly demonstrated by parabiotic union of a first-stage larva, decapitated before the critical stage, with a moulting fifth-stage one (Fig. 17–12). The latter undergoes metamorphosis, as is to be expected, but so also does the first-stage larva, so that this becomes a small and precocious adult. This is because it is under the influence of the ecdysone of the moulting animal, and has no source of juvenile hormone to permit the expression of larval structures. Conversely, a fourth-stage larva, decapitated after the critical stage, can be united with a fifth-stage one that has been decapitated before the critical stage. The fifth-stage larva should metamorphose when it moults, but does not, in fact, do so, because both juvenile hormone and ecdysone from the fourth-stage larva are circulating in its body.

These and other experiments have provided an extensive and wholly convincing demonstration of the hormonal regulation of growth, moulting, and metamorphosis in *Rhodnius*, but this is only one species in a class that include more species than all the remainder of the animal kingdom (Fig. 1–2, p. 6). The number of other species that have been critically investigated from this point of view is small indeed; certainly less than 0.01% of the total number of insect species. Yet we can be confident that the fundamental principles are of wide applicability in the group, for they depend upon the secretory activity of organs that are common features of insect organization. At the same time, much variation is likely in the precise way in which these principles operate in any particular species.

One aspect of this variation concerns the relationship between the thoracotropic hormone and the thoracic gland. In *Rhodnius* this is determined by the output of the hormone that is evoked by the stimulus of ingestion. But the locust, for example, feeds continuously during its larval instars, yet the thoracic gland only produces ecdysone at the time appropriate for evoking moulting. One possible explanation of the difference is that in this animal the brain secretes its hormone continuously, and that it is continuously released from the corpora cardiaca. The stimulus for production and release is thought to be distension of the pharyngeal wall by feeding, or, at ecdysis, by the swallowing of air. During the larval instar

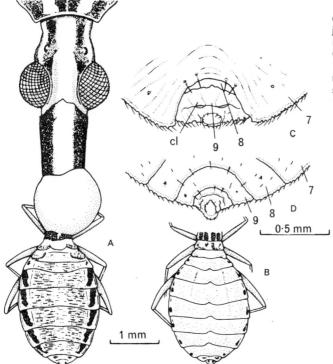

Fig. 17-12. A, precocious 'adult' produced from first nymph of *Rhodnius* by joining it to head of a moulting fifth nymph. B, normal second nymph for comparison. C, terminal segments of the precocious 'adult' (male); D. the same in normal second nymph. Figures indicate homologous sterna; *cl*; claspers. From Wigglesworth 1934. *Q. Jl microsc. Sci.*, **77**, 191–222

the released neurosecretion is used in normal growth and metabolism, and its concentration does not rise sufficiently high to stimulate the thoracic glands. Near the time of ecdysis the animal ceases to feed, so that these metabolic processes decline. In consequence there is a diminished demand for the brain hormone, which now reaches a concentration high enough to stimulate the thoracic glands, thus initiating the moult. This hypothesis awaits further experimental exploration, but in the meantime it provides a good example of the flexible way in which the principles of insect endocrinology can be applied.

Another illustration is given by the giant American silkworm moth, *Hyalophora* (*Platysamia*) *cecropia*. This passes during its holometabolous life history through a diapause, a specialized condition of arrested development. This is a well-known phenomenon among insects, usually interpreted as an adaptation favouring survival during adverse environmental conditions. Diapause in this species sets in at the beginning of the pupal instar and persists during the winter. It results from the brain being inactive; it does not secrete thoracotropic hormone, and if microelectrodes are inserted into it it is found to be electrically silent, showing a lack of spontaneous activity. The end of diapause, and the initiation of adult development, require the renewal of activity in the brain, and the resumption of neurosecretion. This will only occur if the pupa is chilled for some 10 weeks at a temperature of 3–5°C. Without this chilling (if, for example, the pupa is maintained at normal room temperature) diapause cannot end. This analysis, like that of moulting and metamorphosis in *Rhodnius*, is well substantiated by experimental analysis. For example, pupae from which the brains have been removed cannot metamorphose. But a long chain of such pupae, united in parabiosis, can be induced to metamorphose if a chilled (activated) brain is implanted into the most anterior member of the chain. This is because their thoracic glands can now be activated by the thoracotropic hormone that diffuses through the chain from the activated implant. The action of these glands is well seen in the responses of isolated abdomens. These lack the glands, and could never normally metamorphose, although they can survive for long periods if they are attached by their cut surface to cover-slips. If thoracic glands from activated pupae are implanted into such abdomens they are able to metamorphose under the influence of the ecdysone released from the inplants; indeed, they may even survive to lay eggs.

The control of moulting and metamorphosis is the best documented aspect of insect endocrinology, but as with colour change in crustaceans, the concentration of research upon one particular facet is a consequence of historical accident and of experimental convenience, and we can expect insect hormones, like crustacean ones, to have a diversity of functions. To some extent this could be provided for by diversified specialization of the targets, permitting one hormone to have multiple effects (there is some evidence of this in crustaceans), but there are good grounds for supposing that both crustaceans and insects do indeed secrete a number of hormones. Gel filtration of insect brain extracts permits the separation of peptide factors with varying actions on heartbeat, lipid metabolism, and nitrogen metabolism, but the nature and number of these factors remains uncertain. This is partly because they still await chemical characterization, and partly because of the presence in the extracts of substances that are only of pharmacological significance. Nevertheless, suggestive evidence is certainly accumulating.

Thus hormonal regulation of carbohydrate metabolism seems likely in locusts, where two hyperglycaemic factors have been detected. One of these is secreted by the brain and released from the corpora cardiaca, while the other is secreted in intrinsic cells of the latter. Particularly well established, however, is the hormonal regulation of excretion in insects. In *Periplaneta* a diuretic hormone is released from the last abdominal ganglion, while in locusts the brain secretes two factors, one of them being antidiuretic and the other diuretic. The diuretic one of locusts, which is released by the stimulus of feeding, or which can be introduced by implanting corpora cardiaca, increases fluid secretion by the Malpighian tubules (Table 17–1) and reduces rectal

Table 17–1 Rates of fluid secretion by Malpighian tubules in intact locusts (*Locusta*). From Mordue, 1972. *Gen. & compar. Endocr.*, supplement 3, 289–298.

	Number of animals	Fluid secretion (μl/hr $\pm$ standard error)
Starved	6	7.5 ± 0.7
Starved, then fed for 2.5 hr	5	27.5 ± 1.7
Starved + 0.5 pair of corpora cardiaca	5	14.2 ± 0.8
Starved + 1 pair of corpora cardiaca	5	23.0 ± 1.8
Starved	9	7.7 ± 1.2

reabsorption. The antidiuretic one increases rectal reabsorption but has little if any effect on the Malpighian tubules; the stimulus for its release is not known.

Two hormones may not seem essential for regulating excretion, although hormone interaction and antagonism is, in fact, a common feature of endocrine systems, in vertebrates as well as invertebrates. Probably, in some insects, a reduction of tubule secretion is effected simply by reduction of the amount of circulating diuretic hormone. This seems to be so in *Rhodnius*, where a diuretic neurohormone is released (in this case from the fused thoracic and first abdominal ganglia) within 30 sec of feeding. (It will be observed that insects are vigorous exponents of interspecific variation.) The antidiuretic hormone of locusts is perhaps most important during periods of water conservation, when, as we have seen, it would increase rectal reabsorption without having much effect on the Malpighian tubules.

Similar considerations apply in principle to the corpora allata, which are certainly not restricted in their action to regulating the expression of larval characters. In some insects (probably in many) they resume activity in the adult; this is so, for example, in *Rhodnius*, where they are essential for the production of ripe eggs, as they are in other insects as well. Yolk deposition depends upon the presence of sex-specific proteins in the haemolymph. Synthesis of these substances, which are taken up intact into the oocytes, depends upon the secretion of juvenile hormone. Removal of the corpora allata (allatectomy) results, therefore, in the cessation of this synthesis (certain moths are an exception). In addition, it is likely that cerebral neurosecretion is also involved, because it has a more general effect on protein synthesis.

In some insects the corpora allata also influence the activity of the accessory reproductive glands, but wherther these and other effects attributed to the corpora allata are mediated by one hormone or by several is still unknown. Clearly, however, this gland has some influence upon sexual activity. Yet there is here an important difference between crustaceans and insects. In the former group, as in vertebrates, sex hormones have a far-reaching effect upon sexual differentiation, but in insects this is not so. In these animals the sex chromosome mechanism determines the sexual pattern of all parts of the body, and cannot be overridden by hormones from implanted tissues as it can in crustaceans and vertebrates. That the two latter groups should resemble each other closely in this principle of organization, and that crustaceans should differ so markedly from insects, is one of several aspects of arthropodan endocrinology that raises interesting questions of homology and analogy in the organization of endocrine systems. These, however, will be more easily discussed when we have considered the situation in other invertebrate groups.

17-5 ENDOCRINE REGULATION IN ANNELIDS

Annelids were drawn later into the field of endocrine research than were crustaceans and insects, but there is now ample morphological and experimental evidence that neurosecretory regulation is well established in them, and that probably non-neurosecretory hormones are also involved.

The cerebral ganglia of polychaetes, which are the best understood of these animals, contain several types of neurosecretory cells with associated fibre tracts, each of which itself contains different types of fibres, as indicated by the sizes of the elementary granules. These fibres contribute to a glandular complex, exemplified by that of *Eulalia viridis* (Figs. 17–13, 17–14), which is similar in principle to that found in certain other polychaetes, although with differences in detail. Three groups of neurosecretory cells, differing in the size of their granules, give off fibre tracts which end against the pericapsular membrane surrounding the cerebral ganglion; their endings form there a typical neurosecretory neuropile. In some polychaetes (*Nereis* and *Nephtys*, for example) certain of these fibres pass on into the infracerebral gland (see below), but this does not happen in *Eulalia*. Two other groups of cells give off fibres which also end against the membrane, this time as swollen endings called secretory end-feet, which contain clusters of mitochondria as well as neurosecretory granules.

The pericapsular membrane, together with the perivascular membrane of the associated dorsal blood vessel, forms a structure, supposedly an epithelial gland, which is termed the infracerebral gland because of the position of the corresponding structure in other polychaetes, although it is actually medio-posterior in *Eulalia*. It contains two types of presumed secretory cell. One type has processes that extend through the gland. The other type gives off a process (proximal process, Fig. 17–14) that runs through the pericapsular membrane into the cerebral ganglion. In some polychaetes (*Nereis* and *Nephtys*, for example) some neurosecretory fibres from the ganglion pass on into the infracerebral gland, but this does not happen in *Eulalia*. Much remains to be learned about this remarkable glandular complex of polychaetes; but it clearly provides a good basis for the production of neurosecretory and non-neurosecretory hormones, and there is already satisfactory evidence that this does occur.

Experimental evidence for hormonal regulation in polychaetes comes in part from studies of growth and regeneration in nereids. These worms grow rapidly when they are young, with proliferation of new segments. Later the growth rate declines,

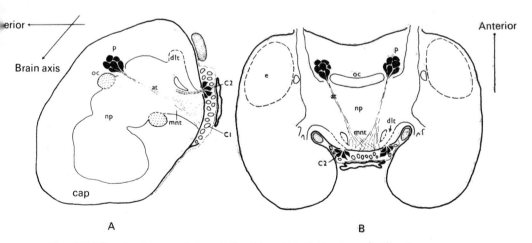

Fig. 17-13. A, diagram of the relationships of cerebral neuroendocrine structures in *Eulalia*, as seen in a thick sagittal section. The median (*mnt*) and the dorsolateral (*dlt*) neurosecretory tracts, shown in outline, abut the brain capsule (*cap*). Axons from the secretory end-feet cells (*p*) form a tract (*at*) directed to the neurosecretory neuropile. The infracerebral gland is composed of two cell types (C1 and C2); *np*, Neuropile; *oc*, optic commissure. B, the same, as seen in a thick frontal section; *e*, compound eye. From Whittle and Golding, 1974. *Gen. & compar. Endocr.*, **24**, 87–98.

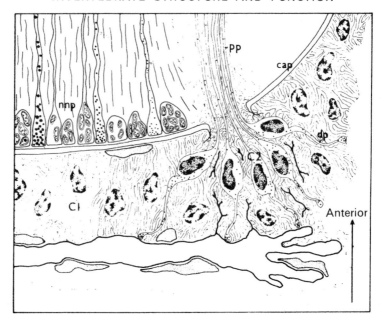

Fig. 17-14. Diagram of the components of the cerebrovascular complex in *Eulalia*, as seen in frontal sections. Processes of the C2 cells of the infracerebral gland pass through the capsule into the cerebral ganglion. *dp*, Distal process; *pp*, proximal process; *nnp* neurosecretory neuropile; otherwise, as for Fig. 17-13. From Whittle and Golding, 1974. *op. cit.*

until increase in length comes to depend mainly on the enlargement of already existing segment. This reduction in growth rate is accompanied by a parallel decline in regenerative capacity, which suggests that both processes may be controlled by the same regulating factor. Theoretically this factor might be no more than the ageing of the tissues, but in fact this is not the correct explanation. On the contrary, there is good evidence that the occurrence of both growth and regeneration depends upon a neurohormone produced by the cerebral ganglion, and that their decline results from a reduction in this secretion.

If the worm is decerebrated (i.e. the cerebral ganglion removed) at the time of posterior amputation, regeneration will not take place, except that local inductive action of the ventral nerve cord can evoke the formation of a pygidium and anal cirri (cf. Eulalia, Fig. 17-20, p. 524). The decerebrate worm can, however, be made to regenerate by implanting into it a cerebral glanglion taken from another worm, provided (for a reason to be seen later) that this is an immature animal. This action of the ganglion must be a humoral one, for it occurs even when the implanted ganglion is lying free in the coelom. Further, it is exerted over a long period of time, for decerebration of a worm which has already been regenerating for some days results in a slowing down of the rate of regeneration (Fig. 17-15). Transplantation of the ganglion from one amputated worm to another shows, indeed, that the ganglion can continue to function for as long as seven months, which is far longer than would be needed to ensure normal regeneration of one individual.

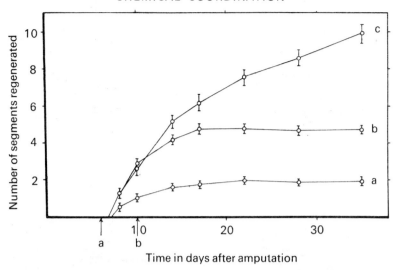

Fig. 17-15. The effect on regeneration of decerebration carried out at different time intervals after segment loss. a, Decerebration after 6 days; b, after 10 days; c, control. From Golding, 1967. Gen. & Compar. Endocr., 8, 356–367.

Growth and regeneration are not the only aspects of nereid development that are subject to hormonal regulation. This extends also to sexual maturation, as is particularly well illustrated by the phenomenon of epitoky. In some species of nereids sexual maturity is accompanied by various somatic changes (Fig. 18-6) which include enlargement of the parapodia, development of a different type of chaeta, and histolytic changes in the body musculature (Figs. 17-16, 17-17, 17-18). The changes are confined to a particular part of the body, termed the epitoke; as a result of them, the worm undergoes a transformation or metamorphosis into the heteronereid phase, which is adapted for sexual swarming. The morphological changes are correlated with a change in the mode of locomotion, from the type of swimming described earlier (Sec. 6-3), in which undulations of the body are important, to a faster type which depends mainly on parapodial action. Hence, the histolysis of the longitudinal musculature. As a result, the heteronereid of *Nereis vexillosa* can swim at about ten times the speed of the non-epitokous *N. diversicolor*.

If the cerebral ganglion is removed from individuals when they are at an appropriate stage of development, but are not yet sexually mature, they will undergo precocious metamorphosis into the heteronereid. The conclusion is that the cerebral ganglion of sexually immature worms secretes a hormone that inhibits epitoky. This is confirmed by the demonstration that the premature epitoky of decerebrate nereids can be prevented by the implantation of cerebral ganglia from immature worms. On the other hand, implanted ganglia are unable to prevent such precocious epitoky if they have been taken from nereids that are themselves in the heteronereid phase. It follows from all this evidence that the attainment of full sexual maturity in nereids requires the withdrawal of an inhibitory cerebral hormone. The hormone is neither sex-specific nor species-specific, and is present even in nereids that do not have a

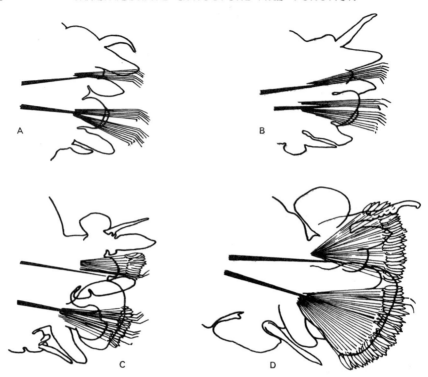

Fig. 17-16. Modification of the parapodia of *Perinereis cultrifera* at metamorphosis (Bauchot-Boutin and Bobin, 1954). A, parapodium of the immature worm. B, enlargement of the parapodial lamellae and ventral cirrus at an early stage in metamorphosis; some nereid chaetae have been shed, but the heteronereid chaetae have not yet erupted at the surface. C, appearance of the heteronereid chaetae and continued shedding of nereid chaetae; further enlargement of parapodial lamellae. D, parapodium of male heteronereid, a few nereid chaetae remain to be shed. From Clark, 1961. *Biol. Rev.*, **36**, 199–236. Used by courtesy of the Cambridge Philosophical Society.

heteronereid phase in their reproductive cycle. Ganglia from immature individuals of *N. diversicolor*, a non-epitokous form, can thus inhibit precocious epitoky in decerebrate individuals of epitokous species presumably because the hormone inhibits sexual maturation as a whole, regardless of whether or not this involves heteronereid transformation.

It will be apparent that withdrawal of the inhibitory hormone parallels exactly the decline in regenerative capacity of nereids. One hormone would therefore be sufficient to account for the facts, and it may well be that only one is indeed involved, serving both as a promoter of growth and an inhibitor of sexual development. If this is so, we have a remarkably economical way of preventing competition between the demands of the two processes. It is an arrangement particularly well suited to worms, such as nereids, which reproduce only once after gaining sexual maturity, and then die. Other polychaetes have different types of life cycle, and it may be because of this that in them, as we shall see, the hormonal regulation is also different.

Fig. 17-17. A, atokous parapodium of *Nereis pelagica*, from the middle region of the body. B, parapodium of an experimentally produced heteronereis of *N. pelagica*, from the same region. *c.d*, dorsal cirrus; *c.v*, ventral cirrus; *l*, parapodial lamellae; *s.h*, heteronereid chaetae; *s.n*, nereid chaetae. From Durchon, 1960. *Bull. Soc. zool. Fr.*, **85**, 275–301.

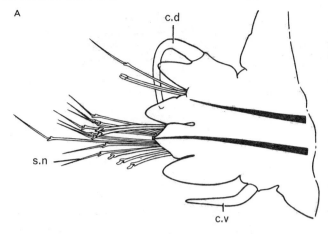

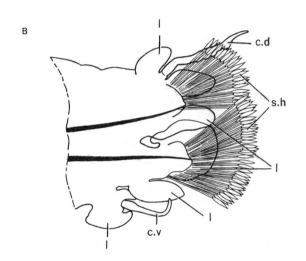

Nevertheless, the effect of the hormone (if one hormone it be) on nereid reproduction is not wholly inhibitory; it does have some positive gonadotropic effect, as is shown by study of oocyte maturation, which consists in polychaetes of successive stages of protein, polysaccharide, and mucopolysaccharide synthesis. In *N. diversicolor* the oocytes at first grow slowly, after their proliferation into the coelom; later they pass into a phase of rapid enlargement, and then complete their maturation in a final phase of slow growth. Removal of the cerebral ganglion during the initial phase of slow growth evokes a precocious assumption of the rapid growth phase (Fig. 17–19). Presumably, therefore, the slow growth is an expression of the gonadotropic action of cerebral secretion. This is apparent, too, from the fact that a continued, although small, amount of secretion is necessary for the completion of normal maturation of the eggs; after complete decerebration they do not reach full size, despite their accelerated growth, while their yolk deposition is abnormal. The eventual

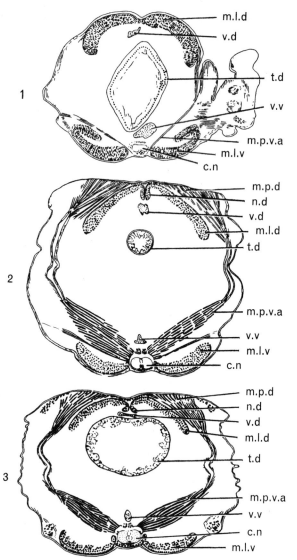

Fig. 17-18. Transverse sections of *Platynereis dumerilii*. *1*, atokous form; *2*, natural heteronereis; *3*, experimental heteronereis. *c.n*, nerve cord; *m.l.d*, dorsal longitudinal muscle; *m.l.v*, ventral longitudinal muscle; *m.p.d*, dorsal parapodial muscle; *m.p.v.a*, anterior ventral parapodial muscle; *n.d*, dorsal neoformation; *t.d*, intestine; *v.d*, dorsal vessel; *v.v*, ventral vessel. From Durchon, 1960. *op. cit.*

normal onset of the rapid growth phase can be ascribed to a decline in secretory activity in the cerebral ganglion, and a consequent fall in the concentration of the circulating hormone. This, of course, corresponds exactly with the suggested mode of regulation of epitoky, which thus appears as one special aspect of sexual maturation. Throughout the sexual maturation of *Nereis*, normal development of the oocytes depends upon a delicate balance between the amount of circulating hormone and the responses of its ovarian target tissue. A high concentration inhibits the development of oocytes, while too low a concentration results in abnormal development. What is demanded is a progressive decrease in the titre of circulating hormone.

The special significance of this gonadotropic effect of cerebral secretion becomes apparent when nereids are compared with *Eulalia*, for this worm, a

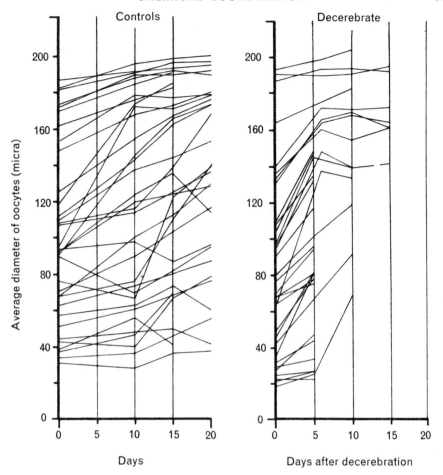

Fig. 17-19. Growth of oocytes in intact and decerebrate worms. *Left*, intact animals; *right*, animals decerebrated immediately after the initial determination of the average oocyte diameter. From Clark and Ruston, 1963. *Gen. & compar. Endocr.*, **3**, 529–541.

member of another family, can continue to regenerate even after decerebration, and can even improve upon the performance of intact worms (Figs. 17–20, 17–21). The cerebral ganglion is still needed, however, for normal egg maturation, for vitellogenesis cannot take place after the prostomium has been removed. It remains to be proved that this effect is actually a hormonal one, although all the probabilities favour this. Indeed, much more experimentation on other species is needed before generalizations can safely be made, for there is clearly much variation between one polychaete family and another with respect to the hormonal control of reproduction.

In syllid worms, for example, reproductive maturation is hormonally regulated, but the secretion is believed to originate in the proventriculus (Fig. 17–22), without involvement of the cerebral ganglion. *Arenicola marina* is another illustration of this

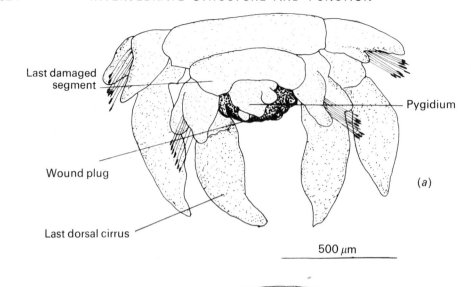
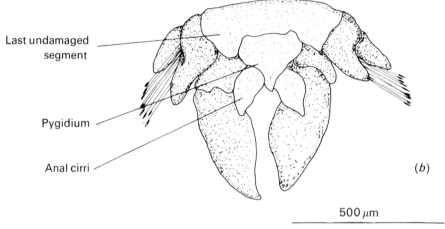
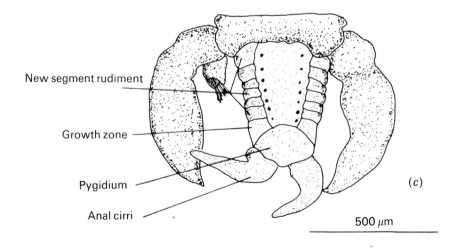

Fig. 17-21. The extent of caudal regeneration in brain-intact and decerebrate *Eulalia* after the loss of caudal segments. Vertical bar = standard error of the mean; * differences between paired means are significant. Note that in this particular experiment the decerebrate worms regenerated significantly more segments than did the brain-intact controls. From Olive and Moore, 1975. *op. cit.*

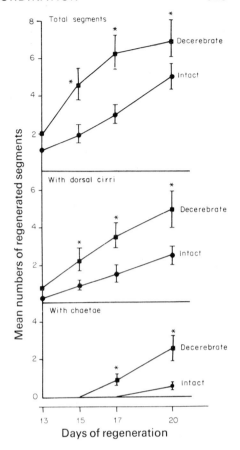

variation. Both the final ripening and the spawning of the eggs of this lugworm (but not vitellogenesis) are controlled by a hormone that is probably secreted by the cerebral ganglion. But in contrast to the situation in *Nereis*, this hormone is an excitatory one, present when breeding is imminent, and absent at other times.

These and other observations have now shown that the nereids, favoured types of laboratory instruction, and the first to be subjected to thorough endocrinological analysis, are, in fact, atypical; in which respect they resemble many other laboratory types in declining to conform to our convenience. It could be that the ability of some polychaetes to regenerate after decerebration is simply due to their dependence upon neurosecretory cells located elsewhere than in the cerebral ganglion. But it could also be that the hormonal programming in nereids may be a specialized one, correlated (as suggested earlier) with a monotelic life cycle, in which

Fig. 17-20. Drawings of regenerating *Eulalia viridis* after the loss of caudal segments. (*a*), Ventral view during wound healing and formation of the pygidium; (*b*), ventral view, showing formation of pygidium and anal cirri; (*c*), dorsal view, showing rudiments of six segments but no dorsal cirri; these, and the chaetae, form later. Note that in this species decerebration has no consistent effect on regeneration (cf. Fig. 17–21). From Olive and Moore, 1975. *Gen. & compar. Endocr.*, **26**, 259–265.

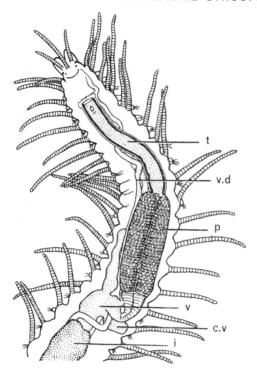

Fig. 17-22. Pharyngeal region of *Syllis amica*. *c.v*, ventricular caecum; *i*, intestine; *p*, proventriculus; *t*, proboscis; *v*, ventriculus; *v.d*, dorsal vessel. From Durchon, 1960. *op. cit.*

growth leads to a single phase of sexual reproduction, and this in its turn to death. Other polychaetes, such as *Eulalia*, have polytelic life cycles, in which survival may extend over many years with regularly recurrent reproductive phases. These species may, perhaps, show a more primitve type of control, in which the neurosecretory action of the cerebral ganglion is primarily gonadotropic, and growth and regeneration are not dependent upon its secretion.

Successful fertilization obviously demands that individuals of both sexes must come together when they are sexually mature. This may be effected very simply, as it is in the non-epitokous *N. diversicolor* (See Fig. 18–6), where the male enters the burrow of the female. Alternatively, this species may spawn on the surface of the substratum. The epitoky of other nereids (and also of many syllids) is a more sophisticated device for securing the same end, for it enables the two sexes to swim freely to the surface, the males often arriving first and emitting their sperm when joined by the females.

A further condition for success is that adequately large numbers of individuals shall become sexually mature at the same time. Adaptations to secure this, in conjunction with swarming, have been evolved independently in eunicid polychaetes as well as in the nereids and syllids. The Eunicidae include the palolo worms, well known as examples of the correlation of spawning with phases of the lunar cycle. *Eunice viridis*, of Fijian and Samoan waters, and *E. fucata*, of the Dry Tortugas, are illustrations of this. The hind parts of these animals contain the ripe germ cells, and these parts are released, while the anterior ends remain behind and presumably regenerate the lost segments. In principle, however, the process is an epitokous one, similar to that found in swarming nereids and syllids (see also p. 553). The three groups thus show very clearly how natural selection, acting on common potentialities, can lead to the parallel yet independent evolution of similar adaptive mechnisms. In this particular instance, however, an important unifying influence is the capacity of these worms for the neurosecretory regulation of sexual maturation.

This regulation operates in conjunction with an environmental factor that brings about a more or less simultaneous reduction of neurosecretory activity in the cerebral ganglia of separate individuals of a population. Not surprisingly, in view of the lunar periodicity that is a feature of the process, this factor is the cyclical fluctuation in the intensity of moonlight. Evidence that the worms can respond to the weak light of the moon comes from studies of *E. fucata*. Immature individuals are photonegative to light of greater intensity than 0.01 foot-candles, which means that they are confined to burrows at the full moon. The epitokal regions of mature worms, by contrast, are photopositive at intensities above 0.005 foot-candles, which means that they are stimulated to swarm at the quarter moon.

The way in which the lunar cycle becomes related to the maturation process of polychaetes is best understood in *Platynereis dumerilii*, a worm that has a spawning maximum around the phase of the new moon. Sexual maturation can be artificially induced in this species by exposing individuals to varying photoperiods corresponding to those of the lunar cycle, always provided that the worms have reached a certain critical stage of development. They will swarm over a number of days, but with a peak at 17 days after the start of the reduction of the photoperiod. It is thus supposed that this reduction initiates the decline in production of the cerebral hormone which is a necessary condition for the attainment of sexual maturity. Confirmatory evidence for this has been obtained by ultraviolet irradiation of the cerebral ganglion. The radiation damage results in a reduced production of the hormone, and this, too, is followed by precocious sexual maturation with a peak at around 17 days after the onset of the treatment. The agreement in the timespan in the two types of experiment provides good grounds for believing that in natural conditions the decline in lunar photoperiod can evoke sexual maturation and so bring about the phenomenon of lunar periodicity of spawning.

There is also evidence that cycles of lunar photoperiodicity can be imprinted on worms, so that the effect of the cycles is manifested after an interval of time. For example, individuals of *P. dumerilii* can be induced to spawn synchronously by exposure in the laboratory to appropriate cycles of photoperiod. This response may be shown even if the treatment is terminated before sexual maturity, and constant illumination substituted for the changing photoperiod. Synchronous spawning may then occur after exposure to continuous illumination for up to three months following the end of the treatment.

Hormonal regulation has been less thoroughly studied in oligochaetes and leeches than in polychaetes, but some common principles can be detected. Neurosecretory cells occur in the cerebral ganglia of earthworms, and in the brain and nerve cord of leeches. Earthworms (*Lumbricus, Allolobophora*) can regenerate missing posterior segments, and, as in nereids, this regeneration is prevented by removal of the cerebral ganglia (although this is said not to be true of *Eisenia foetida*). But it seems that the presence of the suboesophageal ganglion is also essential, although this may be a consequence of the vascular arrangement referred to below.

As regards reproduction, there is an obvious difference between the organization of the two groups in that oligochaetes are hermaphrodite and possess, in the ventral glands and clitellum, important secondary sex characters. It has long been known from castration experiments that the development of these characters is independent of the gonads, and studies of *Eisenia foetida* suggest that this development is regulated by a neurosecretory product of the brain. Removal of this organ, together with the circumoesophageal connectives and the suboesophageal ganglion, results in a reduction in size, a disappearance of the secondary sex characters, and cessation of egg laying, together with interference with the maturation of both ova and sperm. Within 4 to 7 weeks the normal reproductive condition is, however, restored, correlated with a regeneration of the nervous system and a restoration to cerebral neurosecretory cells. The product of these is discharged into blood vessels that run along the circumpharyngeal connectives to the suboesophageal ganglion; it is because of the interference with these vessels that removal of those regions also inhibits reproduction.

A feature of the life cycle of some earthworms is the occurrence of diapause, usually induced by dry conditions in summer, and differing from the effect of simple water lack in that the associated inactivity is not accompanied by partial dehydration. It is possible that cerebral neurosecretion may inhibit this diapause, and it has also been thought to be involved in the regulation of ion content and integumental water permeability of the earthworm, *Lumbricus*. We have already noted aquatic features in the organization of this animal (Sec. 13–6), which can survive in aerated pond water as well as in moist soil. Animals transferred from pond water to moist filter paper undergo dehydration with consequent loss of weight, the rate of loss being similar in intact and in decerebrate worms. When the animals are returned to the water they undergo rehydration, but the rate of this is 70% higher in decerebrate than in intact ones. Moreover, the coelomic fluid of the decerebrate ones has a significantly lower osmotic pressure. These results suggest that the brain may secrete some factor which decreases the permeability of the skin to water, and which affects also the movement of ions. Unfortunately, however, decerebrate worms provided with implanted brains show no difference from the decerebrate ones without implants. The suggestion thus remains unconfirmed, for successful replacement therapy is an essential requirement for the complete verification of hormonal control. Nevertheless, the proposition remains attractively plausible, in view of the undoubted involvement of the endocrine system in water and ion metabolism in other groups of invertebrates.

17–6 ENDOCRINE REGULATION IN MOLLUSCS

That endocrine regulation occurs in molluscs is certain, but relevant information is often incomplete and confusing. The presence of much stainable material in the neurons has complicated the interpretation of supposed neurosecretion, while the great morphological diversity of the nervous system adds to the difficulties. Moreover, extirpation and implantation studies are hard to achieve without serious physiological disturbance to the animals. *In vitro* studies have proved helpful, but even these are not easy to interpret, for it is not always safe to extrapolate from the behaviour of an organ in culture fluid to its likely behaviour within the body. All of this accounts for the contradictory results that have appeared in this field of study, and for the consequent difficulties of generalization.

Striking results have, however, been obtained from studies of the control of sexual reproduction in *Octopus*, which is clearly shown to be under the control of the optic glands. These structures, which are epithelial glands situated on the optic stalks (Fig. 17–23), enlarge at the onset of sexual maturity and secrete a gonadotropic hormone. This promotes maturation of the germ cells and development of the genital ducts in both sexes, and also the synthesis of yolk protein by the ovarian follicle cells (Fig. 17–24). Organ culture experiments have shown that the hormone is the same in both sexes.

There is little information relating to neurosecretory cells in cephalopods, the importance of the nervous system in the present context being that the activity of the optic glands is repressed in the immature animal by inhibitory nerve fibres, the source of this inhibition being in the subpedunculate lobes of the supra-oesophageal region of the brain. This can be shown by appropriate lesion experiments (Fig. 17–23). For example, removal of the source of the inhibition leads to enlargement of the optic gland on that side; so also does optic nerve section, optic lobe removal, and optic tract section. Here, then, control of the maturation of the gonads is operating through the optic receptors and neural pathways, with no participation of neurosecretory mechanisms. Whether light is always the crucial factor in other cephalopods is uncertain; their range of habitat and life cycle leaves open the possibility that other factors may also be involved.

Information is now accumulating regarding endocrine regulation in other mol-

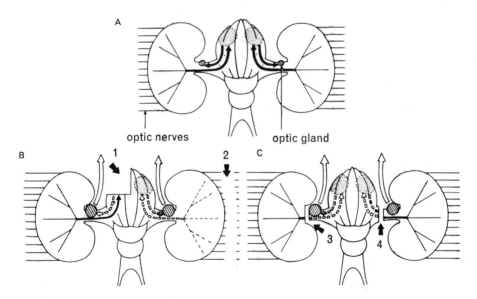

Fig. 17-23. The mechanism of hormonal control of gonad maturation in *Octopus*. A, situation in an immature, unoperated *Octopus*, where secretion by the optic glands is held in check by an inhibitory nerve supply. B, two operations that cause the optic glands to secrete a product causing the gonad to enlarge, being (*1*) removal of the source of the inhibitory nerve supply, and (*2*) optic nerve section. C, further operations having the same effect upon the gonads, thus eliminating the possibility that there is also an excitatory innervation, being (*3*) optic lobe removal and (*4*) optic tract section. From Wells and Wells, 1959. *J. exp. Biol.*, **36**, 1-33.

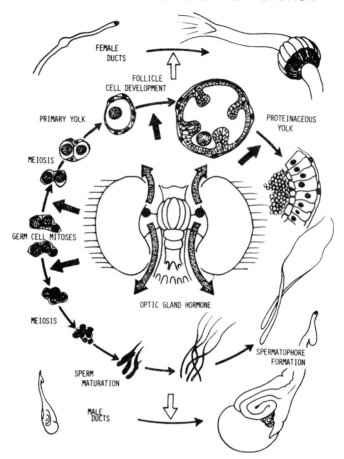

Fig. 17-24. Outline of major steps in sexual maturation of male and female *Octopus*. Arrows indicate processes requiring optic gland hormone. Open arrows indicate conclusions based only on *in vivo* experiments. Shaded arrow is based on *in vitro* experiments in *Sepia* by Richard (1970. Année Biol., **9**). Solid arrows indicate that both *in vivo* and *in vitro* experiments confirm regulation. From O'Dor and Wells, 1975. Gen. & compar. Endocr., **27**, 129–135.

luscs, but the results are not always easy to evaluate, and generalization would be premature. A number of hormones have already been postulated, but it should be remembered that these have not yet been isolated or chemically characterized. It must suffice now to refer briefly to some of the results obtained from gastropods by means of organ culture, in conjunction with experiments involving the removal or destruction of selected regions of the nervous system, and their subsequent reimplantation.

Neurosecretory cells are well defined in the various ganglia of these animals, their secretion being released in diffuse terminal areas which are neurohaemal centres. There are, however, some complications. One is the presence in many nerve cells of stainable inclusions which are not neurosecretions; care is therefore needed in histological assessments. Another is the presence in fresh-water pulmonates

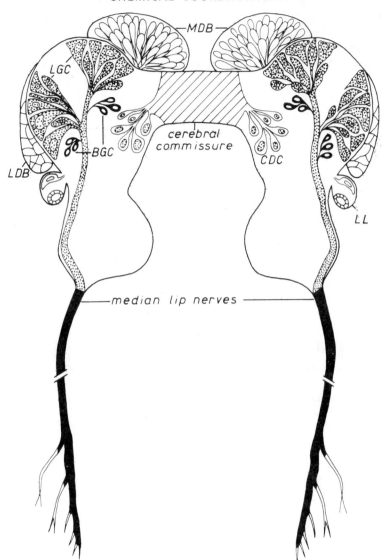

Fig. 17-25. Diagram of the cerebral ganglia of fresh-water gastropod *Lymnaea stagnalis*. BGC, bright green cells; CDC, caudodorsal cells; LGC, light green cells; LDB, laterodorsal bodies; LL, lateral lobes; MDB, mediodorsal bodies. The periphery of the median lip nerves is the neurohaemal area of the light green and bright green cells. The periphery of the intercerebral commissure is the neurohaemal area of the caudodorsal cells. From Geraerts, 1976. *Gen. & compar. Endocrin.* **29**, 61–71.

(basommatophorans) of paired structures called the dorsal bodies (mediodorsal bodies, and sometimes laterodorsal ones as well), attached to the perineurium of the cerebral ganglia (Fig. 17–25). Their likely importance is indicated by the existence of what seem to be equivalent bodies in other gastropods and in the polyplacophoran amphineurans.

In the fresh-water snail *Lymnaea stagnalis*, which is a well-studied example, the nervous system contains at least nine types of neurosecretory cell (Figs. 17–25, 17–26), named by reference to their staining reactions. The cerebral ganglia contain three types of these cells: light green, bright green, and caudodorsal (Fig. 17–25). Cauterization and replacement studies have shown that the light green cells, arranged in two paired groups, produce a growth hormone which, like the growth hormone of vertebrates, has a stimulatory action on growth throughout the body. Attached to the cerebral ganglia are paired lateral lobes (Fig. 17–25), containing three types of neurosecretory cell, together with an epithelial structure called the follicle gland. Removal of these lobes results in excessive growth, indicating that they produce a growth-retarding hormone. The effect of the removal, however, is only seen if the growth hormone is still being secreted. Removal of these lobes leads also to a decline in egg production. Body growth and reproduction (particularly the production of metabolically expensive egg masses) are probably antagonistic in *Lymnaea*, as they are in hydra and nereids, and it is thought that perhaps the lateral lobes may be of special importance in regulating the balance between these two activities.

The dorsal bodies produce one or more hormones which promote both the male and female phases of reproductive development, recalling, therefore, the

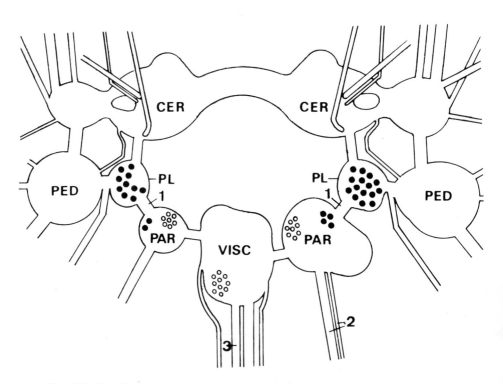

Fig. 17-26. Diagram of the central nervous system of the fresh-water gastropod *Lymnaea stagnalis*. The location of the 'dark-green cells' is indicated by dots, and of the 'yellow cells' by circles. CER, cerebral ganglia; PAR, parietal ganglia; PED, pedal ganglia; PL, pleural ganglia: VISC, visceral ganglia. 1, Pleuro-parietal connectives; 2, right pallial nerves; 3, anal nerve. From Bonga, 1972. *Gen. & compar. Endocr.*, supplement, 3, 308–316.

function of the optic glands in cephalopods. In the female phase they accelerate maturation, vitellogenesis, ovulation and oviposition. They also promote the growth and cellular differentiation of the female accessory sex organs, in which twelve different types of secretory cell have been distinguished, responsible for the production of the perivitelline fluid used in the nutrition of the embryo, and also for the materials used in cementing the egg masses. The dorsal bodies do not affect the differentiation of the male accessory sex organs, nor does the growth hormone affect them, although it does have a general stimulatory action on their growth. Whether there is a hormone in *Lymnaea* that specifically promotes the development of these organs is not clear, but the evidence suggests that such a hormone does function in at least some terrestrial pulmonates (stylommatophorans). In these, the immature duct system develops its male parts if it is implanted into an animal in the male phase, but its female parts if it is implanted into one in the female phase. The implication is that separate male and female hormones are involved, but evidence as to their sources is conflicting; the gonads and the cerebral ganglia have both been implicated.

Gastropods are commonly protandric hermaphrodites. It seems likely that the transition from male to female phases is hormonally controlled, but the details are not yet clear. In *Lymnaea*, the lateral lobes shorten the male maturation period by accelerating spermatogenesis, but complete removal of the lobes only retards this, without entirely preventing it, so that there is room for another hormone to be functioning. In *Helix aspersa*, spermatogenesis is wholly dependent on an androgenic factor secreted by the cerebral ganglia. Perhaps, then a similar factor, additional to that of the lateral lobes, is present in *Lymnaea*. For the present, a possible working hypothesis is that the male phase in pulmonates is initiated by the release of an androgenic hormone from the cerebral ganglia, the later withdrawal of this hormone leading to the onset of the female phase. The argument is complicated by some evidence that the optic tentacles of gastropods may contain endocrine tissue which regulates spermatogenesis. This, however, remains uncertain, and it is far from clear that the tentacles are, in fact, endocrine structures.

Other evidence for hormonal regulation in *Lymnaea* comes from studies of osmoregulation. Two cell types are particularly involved, these being termed, again with reference to their staining reactions, the dark green cells and the yellow cells. Exposure of the snails to deionized water, which activates water elimination and ion uptake, evokes signs of release of secretion from both of these cell types and their axons, while exposure to saline, which suppresses diuresis and ion uptake, leads to their regression. Some results of extirpating the pleural ganglia, where the dark green cells are conspicuous (Fig. 17–26), suggest that these cells stimulate diuresis, and that the yellow cells stimulate active uptake. There is much of interest in these and similar observations, which serve most of all to emphasize the need for much more investigation of hormonal regulation in these puzzling and difficult animals.

17–7 EVOLUTION OF ENDOCRINE SYSTEMS

Evidence of hormonal regulation in other groups than those so far mentioned is very fragmentary. Supposed neurosecretory cells have been described in a wide range of invertebrates, but the evidence for their hormonal function is often incomplete. In any case it is clear that invertebrate hormones are not always neurosecretions, and it is to be expected that more epithelial endocrine glands remain to be discovered. Nevertheless, present evidence indicates that neurohormones are of predominant importance in invertebrate endocrine systems, and it seems likely that the early evolution of these depended largely upon the secretory capacity of the nerve cell, and that exploitation of this capacity antedated the appearance of the epithelial type of gland.

Certainly, conditions in coelenterates suggest that neither organs nor a blood system need have been required for the appearance of neuroendocrine activity. Presumed neurosecretory cells, containing stainable droplets, occur in the nerve net of hydra, mainly at the bases of the tentacles and in the hypostomial region. These droplets move to the ends of the axons, where they are released. Probably the cells concerned are of a primitive type, able to function as ordinary nerve cells as well as neurosecretory ones, which is in line with the view that the two types found in higher forms have had a common ancestry.

That neurosection in hydra is involved in the regulation of growth and budding is indicated by the fact that the droplets appear in the apical zone during these phases, and also during regeneration of the trunk region by an isolated apical end. Further, organic substances that can promote budding and regeneration have been extracted from hydra, and are maximally concentrated at the apical end. However, the relation of these substances to the visible neurosecretion, and also to the axial metabolic gradients referred to later (Sec. 21–2), remains uncertain.

The supposed neurosecretory droplets disappear during the sexual reproductive phase; moreover, extracts of the apical ends of sexually mature individuals cannot induce growth in other individuals. Conversely, the apical end of an immature hydra will inhibit sexual development if it is grafted onto a sexually mature individual. It appears, therefore, that neurosecretory control in hydra provides for the dissociation of the growth and asexually reproducing phase from the sexual one; a device similar in principle to that seen in nereid worms. One consequence of this dissociation is seen in the fate of the interstitial cells. During the growth phase these develop into cnidoblasts and nerve cells, but during the sexual phase they develop into germ cells.

Evidence from other groups of the lower invertebrates is no less fragmentary, although it at least supports the view that endocrine regulation evolved very early, if only as some form of diffusion of regulatory substances. For example, grafting the anterior third of an individual of the sexual strain of the planarian *Dugesia tigrina* onto the posterior two-thirds of an individual of a non-sexual strain, induces the appearance of testes and copulatory organs in the latter. It has been said, too, that regeneration of ocelli in *Polycelis nigra* normally depends upon the cerebral ganglion being present, but can also be induced by extracts of it; however, it is not clear that this effect is a specific one, for homogenates of chick embryos can also do this. Supposed neurosecretory cells have been described in the brain of planarians, and it may be that they are related to these effects, but the exact nature and function of these cells, which are present also in nemertines, remains to be clarified.

More clear-cut information comes from echinoderms, for spawning in sea stars is regulated by the interaction of two hormone-like substances. One of these, a polypeptide with some 23 amino acids, was initially found in extracts of the radial nerves, but has since been found also in extracts of the tube-feet, body wall, and cardiac stomach. In all of these regions there are nerve fibres which contain stainable granules. There is thus some presumption, but certainly not proof, that this substance may be a neurosecretion, especially since it is absent from extracts of the pyloric caeca, where the nervous tissue lacks these granules. This polypeptide is a gonad-stimulating substance, which induces the synthesis by the gonads of the second substance, l-methyladenine. This induces maturation and release of the eggs in the female, and release of the sperm in the male.

The organization of the endocrine systems of molluscs remains very obscure, but the predominance of neurosecretion in the regulatory mechanisms of annelids and arthropods is clear enough; not only in the groups already mentioned, but probably in others also, for there are plausible accounts of neurosecretory pathways in myriapods and arachnids. It is of great interest to find that in this respect the annelids and arthropods resemble the vertebrates. A feature of the latter animals is the existence of neurosecretory cells in the hypothalamus, some of which discharge their products along nerve fibres into the pars nervosa of the pituitary gland, which serves as their neurohaemal organ. Other products are discharged into another neurohaemal organ, the median eminence. From this they pass through local vessels to the pars distalis of the pituitary gland, serving to regulate its activity and, through this, the activity of certain peripheral endocrine glands.

Much has naturally been made of the remarkable similarity of organization shown by the X-organ and sinus gland complex, the pars intercerebralis and corpora cardiaca complex, and the hypothalamus and pituitary complex. Yet this similarity is to be expected. What is common to all of these systems is the neurosecretory cell itself. Given this, the anatomical relationships necessarily follow, for they become essential for efficient organization of secretion, storage, and release. We are dealing, therefore, with convergent evolution, resulting from the existence in unrelated groups of a common principle of neurosecretion. That convergence accounts for the degree of resemblance between vertebrates and arthropods is clear, but how far this also applies within the arthropods themselves is more difficult to judge, although we have seen that convergence must have played an important part in the evolution of these animals. Certainly, the X-organ and sinus gland, and the pars intercerebralis and corpora cardiaca complexes, are sufficiently dissimilar in their anatomical relationships to suggest that they, too, could be products of convergent evolution.

The possible relationship between the thoracic gland of insects and the Y-organ of crustaceans is equally uncertain. The mode of control exerted by the thoracic gland differs from that exerted by the Y-organ, for the former evokes moulting while the latter inhibits it. This difference, however, would not in itself provide an argument against the identity of their products. More we cannot say.

One of the unresolved problems of endocrinology, vertebrate as well as invertebrate, is the mode of action of hormones. In general, they must react with specific receptors on their target cells. Thereafter, however, their effect within the cell often depends upon the release of cyclic AMP (adenosine $3',5'$-monophosphate) and of calcium, which act jointly as second messengers that mediate the action of the hormone. It is, however, also possible that a hormone may act directly upon the nucleus of its target cell. This seems to be true of the ecdysones, which evoke swellings ('puffs') on the giant chromosomes of the insectan salivary gland (in the larva of *Chironomus*, for example). Each of these puffs is the site of production of a specific messenger RNA, individual puffs being characteristic of particular tissues and of particular stages of development. It is supposed that this is the visible expression of the hormone directing the metabolism of the cell so as to favour specific biosynthetic pathways. This aspect of chemical regulation, however, has ramifications which cannot be pursued here.

17-8 PHEROMONES AND ALLELOCHEMICALS

Chemical communication systems not only provide for regulation within the body of an individual animal. They also coordinate the activities of individuals of the same species, while (Chap. 20) they play important parts in maintaining and regulating relationships between more than one species. Such uses may well have provided the earliest forms of chemical communication, for they are arguably an inevitable result of the release of metabolic products from the animal body. These must to some extent affect neighbouring individuals, and, if the relationship proved beneficial, it is to be expected that it would be improved by the agency of natural selection; the products would be further elaborated into specialized secretions, and sensitivity to them would be enhanced.

Substances acting in this way as specialized agents in external chemical transmission extend the classical principles of endocrinology, which were elaborated in relation to internal communication. The novelty in the type of communication that we are now considering is primarily the external transmission; the adaptation of chemical substances to the stimulation of specific receptors is no different in principle from the adaptation of hormones to the stimulation of specific internal tissues. Indeed, this situation shows very clearly that endocrine systems, in the classical sense of the term, are part of a much wider system of chemical relationships; a system that is concerned in a diversity of ways with the regulation of living processes. One very important group of these externally distributed chemical agents comprises the substances called pheromones (derived, by mutilation, from the Greek roots *pherein*, to bear, and *hormaein*, to excite). These are defined as external secretions which convey information from one individual to another of the same species, and which evoke (often through olfactory receptors) specific reactions in the recipient.

An example of this type of communication at the protozoan level is seen in the aggregation of the amoeboid individuals of the slime mould *Dictyostelium*. This aggregation, which occurs when food is scarce, is promoted by the action of a substance which is either cyclic AMP or which evokes the release of cyclic AMP. This involvement in external communication of a substance which, as explained in the previous section, is usually a second messenger, is a striking example of a close link between hormonal and pheromonal mechanisms.

Examples of pheromonal communication in metazoans are to be seen in the regulation of spawning. The interaction of lunar periodicity and the neurosecretory activity of the cerebral ganglion of certain polychaetes is one means of securing this, but it does not exclude the possibility that chemical signals may be exchanged between the swarming animals. An example of such signalling in polychaetes is seen in *Nereis succinea*; a secretion from the mature eggs and the gravid females induces the males to spawn, and the presence of sperm then induces the females to spawn. An example from another group of animals is the spawning of female oysters, which is evoked by some substance present in the sperm and testes of the males. In this instance a chain reaction develops, for spawning of the females then induces spawning of other males, which in their turn stimulate other females.

The use of pheromones as sex attractants, drawing together the sexes for copulation, is an important field for their action, and one possible example of this is seen in the mate-seeking behaviour of the male copepod *Eurytemora affinis*. Tests with tethered females show that males can detect and locate them when up to 22 mm away, their responses involving turns, loops, and somersaults which are carried out in a way suggesting that they are reacting to a gradient of some specific chemical diffusing from the female.

Our knowledge of this type of chemical communication system is still often fragmentary and inferential, but there are many other illustration which are very well documented. This is particularly true of the social life of insects, the path to which, it has been remarked, was surely paved with pheromones. We shall refer to this aspect later (Sec. 21–4). Pheromones, however, are widely used by other insects as well, often as sex attractants, and some of these have been studied in great detail, not least because of the potential economic importance of the subject. The readiness with which a single female moth can attract males over very large distances is an indication of the extraordinary sensitivity of these communicating systems; a sensitivity that has been amply confirmed in certain instances by the isolation of the compound concerned. This isolation makes possible the use of biological assays, which may involve observing the behaviour of a male insect, or recording the activity of its antennal nerve, when its receptors are stimulated.

One substance studied with particular detail is the sex attractant (bombykol) of the silkworm moth, *Bombyx mori*. Secreted by the female from abdominal glands, it both attracts the male and arouses it to erratic dancing and to attempts to copulate with the female or with test objects. It has been chemically characterized as a 16-carbon doubly unsaturated alcohol, *trans*-10, *cis*-12-hexadecadien-1-ol (Fig. 17–27), the result of 30 years' work by Butenandt and his colleagues, and the extraction of no less than 500,000 scent glands from virgin females. The airborne molecules enter the olfactory sensilla through pores which are estimated to total up to 4.5×10^7 in number on each of the antennae, and which provide access to some 25,000 receptor cells. It has been shown, by the use of tritium-labelled bombykol, that a response can be induced in the receptor by only one molecule of the pheromone, while the threshold for the behavioural response of the insect is 200 impulses per second, induced by 200 bombykol molecules. There is noise in the system, resulting from the spontaneous activity of the receptor, amounting to 1,600 impulses per second. Calculation shows that the signal to noise ratio (i.e. 200 to 1,600) is just above the

$$CH_3(CH_2)_2C\overset{H}{=}C-\overset{H}{C}=\overset{H}{C}(CH_2)_9OH$$

Fig. 17-27. Bombykol.

minimum for the transmission of information. This means, in terms of adaptive organization, that the male silkworm moth is very well equipped to respond to low concentrations of the female's pheromone. The same is likely to be true for other insects, for the sensitivity of the silkworm moth is certainly not unique. The female cockroach, *Periplaneta americana*, secretes a sex attractant that has been identified as 2,2-dimethyl-3-isopropylidene cyclopropyl propionate; quantities of less than 10^{-14} μg can evoke a response from the male. And this response is a powerful one. Cockroaches starved in the laboratory for four weeks, and then presented with a choice of pheromone or food, chose the pheromone until they were very near to death.

The sensitivity may be associated with extreme specificity of receptor response, permitting the reproductive isolation of species that occur together. Thus two New York populations of the noctuid moth *Amathes c-nigrum*, thought at one time to be conspecific, are effectively isolated because one is attracted by *cis*-7-tetradecenyl acetate and the other by its geometrical *trans*-isomer. There are obvious practical implications here for the control of insect pests. It is realistic to envisage the blocking of the reproductive cycle by distributing synthetic pheromones in quantities sufficient to overload the atmosphere and thus to interfere with the odour gradients which the male would normally follow. Trials with the cabbage looper moth (*Trichopulsia ni*) suggest that less than 0.2 g per acre per night of synthetic pheromone could be sufficient for this purpose. Other possibilities are to use the pheromones as bait to trap the males (plastic traps are said to have created a litter problem), or to block the action of the pheromones by releasing formaldehyde vapour or one of the artificial pheromones of our own species, Chanel No 5.

We have seen that bombykol, in addition to being a sex attractant, also stimulates the male to copulate. This is an example of the use of a pheromone as an aphrodisiac. This term, which has met some criticism because of its anthropormorphic overtones, needs careful and objective definition, which can be satisfactorily encompassed by requiring that the pheromone, to be regarded as an aphrodisiac, should be released, or active, after the two sexes have been brought together, and

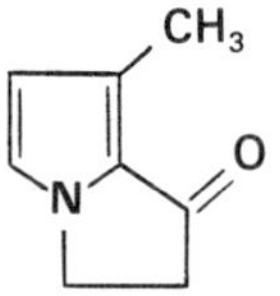

2,3-dihydro-7-methyl-1*H*-pyrrolizin-1-one

$$HO(CH_2)_3\underset{H}{C}=C(CH_3)(CH_2)_2\underset{H}{C}=C(CH_3)CH_2OH$$

3,7-dimethyl-*trans*-2, *trans*-6-decadien-1, 10-diol

Fig. 17-28. Compounds identified from the hair-pencils of the queen butterfly, *Danaus gilippus berenice*. Above, the pheromone. Below, the viscous adhesive. From Birch, 1974. In *Pheromones* (M. C. Birch, ed.). North-Holland Publishing Company, Amsterdam and London.

that it must be seen to facilitate courtship and/or copulation. We have seen an example of this in the courtship behaviour of the Florida queen butterfly, *Danaus gilippus berenice* (Fig. 15–3). The hair-pencilling here transfers to the body of the female cuticular particles. These carry a terpene alcohol (Fig. 17–28) that sticks the particles to the female, and a ketone (crystalline pyrrolizidinone) which is the actual pheromone, and which can only stimulate one of the several types of sensilla present on the female's antenna. About 1–5,000 of these sensillae must be exposed to ensure the success of the courtship, but as these amount to only about 5% of the total, this requirement is not too exacting. The hair pencils, however, are an obligatory requirement. Without them, the male can court but cannot seduce.

It remains to mention in the present context one other aspect of chemical communication. This is the release by individuals of one species of substances that influence the responses of individuals of another species. The term 'allelochemicals' has been suggested for such substances, which differ from pheromones in that the communication subserved is interspecific instead of intraspecific. Allelochemicals are demonstrably important in bringing together the partners in the various types of symbiotic relationship that we shall be considering in later chapters. No doubt this must often result from one species having 'captured' and turned to its own end some metabolic product of the partner species. One very remarkable example of this is provided by the flagellate protozoans that are symbiotic in the intestine of the wood-eating roach *Cryptocercus*. The sexual stages of these protozoans, which are only found when the insect moults, are believed to be produced in response to the ecdysone secreted by their host, which means that there is here an adaptation to what is itself a highly specialized hormonal secretion. It has been suggested that the response of the symbiotes may be to chemical changes set up in the intestine by the hormone, rather than directly to the hormone itself; even so this would still remain a striking demonstration of the possibility of the reproduction of one species being chemically regulated by another.

Another illustration is provided by the mite *Proctolaelaps nauphoetae*, which parasitizes the gregarious cockroach, *Nauphoeta cinerea*, this being its only host. The mite locates the cockroach, which lives in aggregations of several hundreds, by means of three chemical cues of increasing discrimination (Fig. 17–29). First, it responds to the odour from cockroach faeces and general debris, without regard to the species of cockroach concerned. This cue initiates random searching (foraging), which eventually brings it near to potential hosts. The next clue is given by the saliva which the cockroach distributes over its body. At this level of discrimination the mite responds to the saliva of the host and of other members of the family (Blaberidae), but not (as far as tests have gone) to the saliva of members of other families. This leads to direct orientation. The final clue is a glycosaminoglycan (a poly-amino sugar), which is a host-specific attractant. On detecting it, the mite moves to the host's abdomen, grooms itself, and looks for an unoccupied breeding site. If it fails to find a site, it leaves the host and resumes foraging behaviour.

This behaviour pattern is sufficiently impressive, but the high level of specialization that can be achieved in allelochemical relationships is even better shown by the polyphemus moth, *Antherea polyphemus*, the larvae of which will feed only on the leaves of the red oak, *Quercus rubra*. Attention was drawn to this moth when it was observed that females in outdoor cages attracted males over large distances, whereas

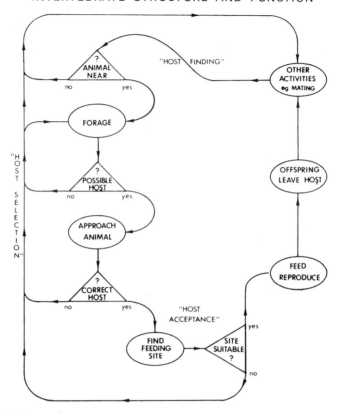

Fig. 17-29. A flow chart showing host discrimination by the mite *Proctolaelaps nauphoetae*. Triangles represent behaviour requiring chemosensory discrimination; ovals indicate other activities. From Barth and Hanson, 1975. *Nature*, **257**, 788–790.

those reared and maintained in the laboratory consistently failed to mate successfully, even though males were present with them. The explanation is that the female releases a sex pheromone which attracts the male and activates it sexually, much as with *Bombyx*. This pheromone, however, can only be released by the female in response to a volatile *trans*-2-hexanal which is produced by the oak leaves. This acts on antennal receptors of the female to evoke, through the brain, the release from the corpora cardiaca of a hormone which then evokes the protrusion of the female genitalia and thus the release of the pheromone. An odd consequence of this is that females from which the antennae have been severed can attract males for up to 6 hours after the operation. This is because nerve impulses are propagated from the injured ends of the antennae, and provide an adequate stimulus for release of the pheromone, comparable to the impulses which would normally be generated by the plant allelochemical. A subtle feature is that many other plants also produce this allelochemical, but they also release masking substances which prevent it from stimulating the female. Here, then, is an elegantly adaptive interaction between an allelochemical and a pheromone, which ensures that reproduction and oviposition shall take place only in the presence of the leaves which are obligatory food for the larvae.

The examples given here are merely a few highly specialized aspects of chemical communication; they need to be seen as part of a much wider and sometimes less specialized network of communications and signals. It could well be, as suggested by Lucas, that the excretion of metabolites may provide pervasive and subtle means of regulating the interspecific relationships within plant and animal communities. This is particularly likely in aquatic communities, where water allows the ready exchange of materials to regulate behaviour and reproduction, and hence the population structure. The potentialities of chemical communication are thus as diverse as are the habits of the organisms that share in it, and are certainly not restricted to various modes of attraction. Allelochemical interactions may influence the growth, behaviour, and well-being of the individuals concerned. They may be primarily of advantage to the recipient (these have been called kaironomes), as, for example, when they release feeding reactions. Or they may be primarily of advantage to the transmitting organism (these have been called allomones), as, for example, in the use of repellents. These and other aspects of chemical communication are woven deeply into the organization of living systems, and surely take us back to the earliest stages of biological evolution.

And now, at the centre of the web, stands our own species. Whether it has its natural pheromones is uncertain, but its capacity for manufacturing attractants, repellents, depressants and toxicants is sufficiently obvious; a complex of allelochemical artifacts which may yet eliminate the whole of the subject matter of this book.

18
Patterns of Reproduction

18–1 PROTOZOAN LIFE CYCLES

The replication that is a fundamental property of living organisms is expressed in two forms in the life cycles of animals: asexual reproduction and sexual reproduction. Of these, the asexual type is in a sense the simpler; it involves no more than the equal partitioning of the genetic material between two daughter organisms, with a correlated division of the associated soma, and it normally depends upon mitotic division of the chromosomes. The essential feature of sexual reproduction is the formation of gametes, which fuse in pairs during syngamy, the fusion products being zygotes. A reduction of chromosome number is needed at some stage of the life cycle to compensate for the doubling that occurs at syngamy. This reduction is provided for by meiosis, which can be interpreted as a specialized derivative of mitosis.

In the Metazoa meiosis takes place during the maturation of the gametes. It does not necessarily occur at the corresponding stage of the life cycle of the Protozoa, but in some of these animals it certainly does so. It is then known as gametic meiosis, which is found in the Heliozoia and the Ciliophora. Alternatively, meiosis may occur during one of the early divisions of the zygote; examples of this, which is known as zygotic meiosis, are seen in certain flagellates (e.g. *Chlamydomonas*) and in the Sporozoea. All such forms are necessarily haploid during their adult, or trophic, stage. It has been suggested that zygotic meiosis may be primitive in the Protozoa, and that gametic meiosis evolved gradually through extension of the growth phase of the zygote leading to a corresponding extension of the diploid phase of the life cycle.

Clearly there are differences between the nuclear process of Protozoa and Metazoa, as we have already noted in another context (p. 30). They are not, however, fundamental, and there is no doubt of the essential similarity of the sexual processes of the two subkingdoms. This similarity is not surprising, for the sexual processes of animals as a whole are fundamentally similar to those of plants. It seems likely, therefore, that sexual reproduction appeared at a very early stage of evolution, and that animals and plants inherited their characteristic features from a remote common ancestry. There is good reason why this should be so. The sexual method of reproduction confers an important advantage in ensuring, through meiosis and syngamy, the recombination of genes and the variation that follows from it. Without it, the only source of genetic variability would lie in mutation, and the amount of this would be insufficient to provide a basis for the establishment of new and more effective patterns of organization through the action of natural selection.

Such peculiarities as we find in the sexual processes of Protozoa are in part an inevitable consequence of the small size of these organisms. Because of this, syngamy may take place between vegetative individuals which transform directly into reproductive ones (gamonts) without undergoing division; they are then known as hologametes. Some species of *Chlamydomonas* are examples of this. Exceptionally, the syngamy may involve nuclei alone, these forming and fusing within single undivided organisms in the process called autogamy (p. 544). Gametes more akin to those of Metazoa may be formed by division of the gamonts. These gametes, which are known as merogametes, may be similar in general appearance (isogametes) or markedly dissimilar (anisogametes), but even when they look similar there may be some physiological differentiation between them. This differentiation becomes apparent when they are unable to unite indiscriminately. They may then be divisible into two groups such that members of one group will fuse only with members of the other group; these two groups may then be thought of as representing two 'sexes'. It is, in fact, likely that the origin of sexual differentiation lay in the establishment of divergent physiological properties between two categories of gametes. These properties, which could well have increased the efficiency of syngamy, may later have been expressed in morphologically visible terms, leading to the division of labour between two types of gamete that is so familiar in the Metazoa. This in its turn would have led to the correlated differentiation of the adults into two sexes, divergently specialized for the production of the two types of gamete.

Although sexual reproduction may thus be thought of as a fundamental property of the Protozoa, these animals make extensive use of the asexual method, and many of them probably reproduce exclusively in this way. This is doubtless because asexual fission provides for a rapid increase in number, and for the efficient exploitation of localized sources of food. An illustration of this is seen in the rapid build-up of a parasitic infection that often follows when an infective stage has gained entry into a host. For the reasons given above, however, we must regard the absence of sexual reproduction in certain protozoan life cycles as a secondary specialization, analogous to the biochemical regression that we have noted elsewhere.

Asexual reproduction (typically by longitudinal binary fission) is common in the Mastigophora, but there has been a widespread loss of sexual reproduction. The only euglenid known to reproduce sexually is *Scytomonas* (*Copromonas*), while *Noctiluca* is the only dinoflagellate known to do so. *Noctiluca* forms uniflagellate isogametes, the parent dying after they have been released (Fig. 18–1).

Where sexual reproduction does persist in the Mastigophora, there may be much variation of detail. Examples of this are found amongst the complex polymastigote and hypermastigote flagellates which live symbiotically in the alimentary tract of termites and the wood-eating roach *Cryptocercus*, where, as we have already

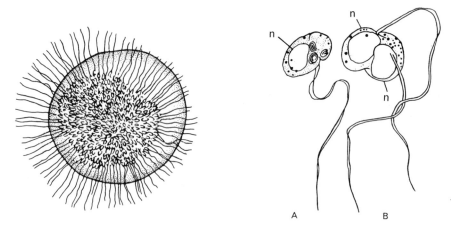

Fig. 18-1. Left, *Noctiluca:* gametogenesis, × 75. The relative size of the 'swarmers' at the oral pole is exaggerated. *Right,* living 'swarmers'. A, single gamete; B, syngamy. *n*, nucleus. From Mackinnon and Hawes, 1961. *An Introduction to the Study of Protozoa.* Clarendon Press, Oxford.

seen, the life-cycle is correlated with the moulting cycles of the hosts. Only a few examples can be mentioned here. The hypermastigote *Trichonympha* is haploid, with zygotic meiosis. The adults become gamonts which encyst and then divide to form two gametes of opposite 'sex'; these leave the cyst and fuse in pairs. The polymastigote *Macrospironympha* is diploid, with gametic meiosis. Encystment here takes place after the first meiotic division, the products encysting separately; each undergoes a second meiotic division, and the two products then leave the cyst and swim away. Another diploid polymastigote, *Urinympha*, undergoes a single meiotic division without any division of the body, the two pronuclei then fusing. Here, therefore, the sexual process is autogamy. These variations, among closely related forms living in similar habitats, are difficult to explain, but perhaps, as Sleigh speculates, the sheltered environment encourages the diversification.

Sexual processes remain well established in the Phytomonadida (Chlorophyceae), which includes many familiar holophytic forms such as *Chlamydomonas*. Some species of this genus form hologametes, as already mentioned, while others give rise to small merogametes which are either isogametes or anisogametes; they fuse in pairs to form zygotes which pass into resting stages within a thick wall. Particularly notable in this group are the colonial forms (p. 639), with sexual reproduction that ranges from isogamous hologamy to an extreme of anisogamy resembling the differentiation of ova and sperm in Metazoa. *Gonium* illustrates the simplest situation, the members of the colony becoming free and acting as isogametes. *Eudorina, Pleodorina,* and *Volvox* illustrate the other extreme, with a marked division of labour in which some individuals enlarge without division to form macrogametes (or ova) while others produce biflagellate microgametes (or sperm) by repeated division. The group has long been used to illustrate the principle that we have mentioned above, of the evolution of sex by the progressive modification of gametes, although this does not mean, of course, that the particular genera concerned represent a continuous evolutionary series. It has also been surmized that phytomonads could have been the forerunners of the higher green plants.

The Mastigophora are not alone among Protozoa in showing extensive loss of sexual reproduction. This is found also in the Sarcodina, where it is particularly characteristic of the Amoebida. Thus the life history of *Amoeba proteus* consists of a sequence of growth and binary fission. In constant conditions this follows a very regular pattern, with a division at perhaps every 24 hr at a temperature of 23°C. This species has been said to form protective cysts, but it is far from certain that it actually does so; it may be that the capacity for encystment, like the capacity for sexual reproduction, has been lost in *Amoeba proteus*, and that growth and fission are the only events of its life history.

This is not necessarily true of all the free-living Amoebida, and it certainly does not apply to the entozoic Entamoebidae. Most of these produce cysts that are of high adaptive value in that they serve for the transmission from host to host that is essential for such forms. An example is *Entamoeba histolytica*, found in the intestine of man and other primates, and also in the rat; it is frequently harmless, but sometimes it invades the gut wall and sets up the symptoms of amoebic dysentery. This organism is transmitted in the form of spherical cysts, containing four nuclei, and entering another host as a result of ingestion. Within the intestine a quadrinucleate amoeba emerges and divides to form eventually eight uninucleate organisms. *Entamoeba gingivalis* is an exception to this type of cycle, in that, like *Amoeba proteus*, it has lost the capacity for encystment. This is presumably correlated with its habitat, which is the human mouth, and particularly the gums; this mode of life readily allows transmission by kissing.

The loss of sexual reproduction in the Amoebida is by no means characteristic of all the Sarcodina. It is retained in the Foraminiferida and the Heliozoia, the foraminiferan *Elphidium* (*Polystomella*) being an example of a sarcodinan in which there is a regular alternation of asexual and sexual generations. The small zygote gives rise to a diploid asexual stage (agamont), often referred to as the microspheric stage because the first chamber of the spirally coiled shell is a small one. Its nuclear division begins early, so that the microspheric form is multinucleate. Eventually it divides asexually to form a number of diploid amoebulae, each of which secretes a shell of which the first chamber is large; the resulting organism (gamont) is often called the megalospheric stage for this reason. In due course it divides to form a number of biflagellate gametes; the fusion of these completes the life cycle.

It will be apparent that protozoan life cycles provide a basis for close adaptation to changes in the environment. Growth is a reflection of the nutrient value of the surrounding medium, while encystment clearly provides a means of withstanding adverse conditions, although the factors evoking it are often not understood. This relationship between modes of reproduction and environmental conditions is particularly well shown in the complex life cycles of parasitic Protozoa. As we have suggested above, an asexual phase, following invasion of a new host, makes it possible to exploit, through multiple fission, the nutrient qualities of the environment, and to establish a firm hold upon the host. Not all parasites, however, take advantage of this possibility, and in some life cycles greater emphasis is placed upon the other essential requirement: the ensuring of transmission.

In *Entamoeba* the procedure is a simple one, and is dependent only upon encystment and a small degree of asexual reproduction. In many other forms, however, and notably in the Sporozoea, an alternation of sexual and asexual generations is found. This alternation operates in two main patterns, characteristic respectively of the

Subclasses Gregarinea and Coccidia. These, as already indicated, are haploid forms, their gamonts dividing mitotically (if microgametes are produced), and meiosis occuring at the first zygotic division.

Gregarines, typically parasites of the body cavities and guts of worms and insects, are exemplified by *Monocystis*. Here the adult haploid trophozoites become gamonts, which encyst in pairs and give rise to numerous gametes. Those produced by one individual fuse with those from the other to form zygotes, each of which develops a spore coat that provides protection during transmission. Within this coat the diploid zygote divides, with meoisis, to form eight sporozites. As in many other gregarines, this is the only asexual stage. The sporozoites grow into adult trophozoites which pass into the gamont stage without dividing; such gregarine infections thus tend to be light ones.

Coccidians, which probably evolved from gregarines (p. 709), differ from the latter in being typically either intracellular or blood parasites, and in undergoing extensive asexual multiplication before the adult becomes a gamont. This multiplication is termed schizogony, the adult being called at this stage a schizont. They differ also in that macrogametes are formed without division of the gamonts (macrogamonts), while microgametes, usually biflagellate and free, are formed from microgamonts after repeated divisions. A cyst is laid down around each separate zygote. In *Eimeria steidae*, parasitic in the epithelium of the bile ductules of the rabbit, the schizogony increases the intensity of infection (see also p. 709); the merozoites enter other epithelial cells and repeat the schizogony in several cycles before giving rise to gamonts. Sexual reproduction occurs in the rabbit, the infection being transmitted through the zygote, which, protected in a cyst (oocyst), divides with meiosis to form eight sporozoites and is passed out in the faeces.

In *Plasmodium* (the malaria parasite of vertebrate erythrocytes), schizogony prior to gametogenesis has become part of the adaptation to transmission through an invertebrate (arthropod) vector from one vertebrate host to another. Sexual individuals (gamonts, often termed gametocytes) are formed in numbers large enough to ensure transmission, but the production of gametes, together with fertilization, is now specifically evoked by the change of environment resulting from passage into the vector. This change can be artificially simulated by drawing off some blood from the infected host. In this type of cycle the zygotes are unprotected and motile, because the products of their fission are transmitted directly into the vertebrate host from the invertebrate one, and do not, therefore, have to resist adverse conditions in the external environment.

Schizogony is thus a valuable feature of parasitic life cycles, but its absence in many gregarines is not necessarily to be regarded as primitive. Certain gregarines that live in marine invertebrates do, in fact, show schizogony in what is now thought to be a primitive form. An example is *Selenidium mesnili*, from the intestine of the tubicolous polychaete *Myxicola infundibulum*. This gregarine reproduces by schizogony during an intracellular phase, the eventual association of the gamonts taking place in the gut lumen. Apparently most gregarines have secondarily lost this schizogonic stage. Certain specialized forms that live in insects, however, seem to have reacquired it.

The protozoan cycles so far mentioned, based upon various combinations of asexual reproduction, sexual reproduction, and the formation of protective cysts, can be related without difficulty to the principles familiar in metazoan reproduction, and they thus show the fundamental unity of the reproductive processes of the two subkingdoms. However, the diversification of protozoan cycles has certain genetic consequences. For example, mutant genes are expressed more readily in haploid organisms than in diploid ones, since they are not masked by dominant alleles. This could facilitate rapid adaptation, but against this is the disadvantage that haploid organisms lack a hidden reserve of variability. As for sexual reproduction and meiosis, the gene recombination that results from this favours the testing of mutants in new gene complexes, and their spread through populations. Moreover, sexual differentiation of the gametes may promote variability by bringing together gametes from gamonts with different genetic constitutions. Asexual reproduction, by contrast, has a conservative tendency, for all the offspring (clone) of one asexually reproducing individual have the same genetic composition. Only rarely, in these circumstances, will mutant genes be placed in complexes favourable for their expression. These and other genetic implications of protozoan cycles can readily be appreciated, yet it is not always easy to account for the reproductive characteristics of particular groups. This is especially true of the one remaining major group, the Subphylum Ciliophora, the members of which stand apart from other protozoans in so many aspects of their highly specialized organization.

The reproductive processes of ciliates have been most closely studied in *Paramecium*, to which we shall here mainly refer. Asexual reproduction is the only means of multiplication possessed by ciliates. It takes the form of a highly characteristic binary fission, differing from that of flagellates in that the division is transverse, separating an anterior half from a posterior one. Since the macronucleus divides by amitosis, and the micronucleus by mitosis, each half starts at once with a full nuclear equipment. In other respects the two daughter forms are at first different from each other, the feeding structures, for example, passing to one of them so that the other has to develop new ones; these, it is said, are formed at least in part by budding from the old structures. The process of fission, which may last for about 30 min, takes place regularly under healthy culture conditions, perhaps as often as three times in 24 hr. So far, then, the life cycle is straightforward; growth and fission alone are involved, for cysts are believed not to be formed in this genus. This does not, however, constitute the whole of the life cycle, which is greatly complicated by the occurrence of sexual processes that seem to be essential if the stock is to remain viable. These processes, which are highly characteristic of the group, are of three kinds, called conjugation, autogamy, and cytogamy; they have been intensively studied in recent years, particularly by Sonneborn, in whose hands they have provided material for far reaching genetic investigations.

Conjugation in *Paramecium aurelia* (Fig. 18–2) will take place when cultures are subjected to certain defined conditions, including some degree of starvation, and maintenance at certain levels of temperature and illumination. It involves the association of the animals in pairs, but this association is not a random one; it can take place only within limits which themselves indicate the high level of specialization involved. Within the species there is much physiological variation, the basis of which, according to Sonneborn, can be interpreted as being the existence of separate varieties and mating types. Each variety (nine of these have been identified in *P. aurelia*) includes two mating types, and conjugation can only take place between

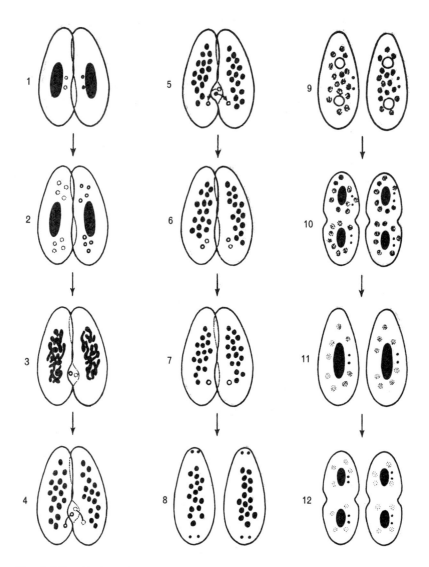

Fig. 18-2. Nuclear changes at conjugation in *Paramecium aurelia*. *1*, two parental animals, each with one macronucleus and two diploid micronuclei. *2*, formation of 8 haploid nuclei from the micronuclei in each conjugant. *3*, 7 nuclei in each conjugant disappear; the remaining haploid nuclei pass into the paroral cones; macronucleus breaks up. *4*, the nuclei in the paroral cones divide mitotically, forming 'male' and 'female' gamete nuclei. The 'female' gamete nuclei pass back into the interior of the parental animals. *5–7*, fusion of 'male' and 'female' gamete nuclei from opposite mates. *8*, each fusion nucleus divides twice, mitotically. *9*, two products of each fusion nucleus differentiate into macronuclear rudiments (white circles); the other two into new micronuclei. *10–12*, return to normal state of one macronucleus and two micronuclei per animal; fragments of old macronucleus gradually lost. From Beales, 1954. *The Genetics of Paramecium aurelia*. Cambridge University Press, London.

individuals belonging to the same variety but to opposite types. If the two different types of the same variety are mixed under suitable conditions, the animals first clump together in large groups, but within an hour they begin to separate and then conjugate in pairs, first by attachment at their anterior ends, and later in the region of their oral surfaces.

There follows a series of nuclear changes, which involve the fragmentation of the macronuclei and the division of each of the micronuclei (two of which are present in each individual of this species) into four. Seven of these daughter nuclei disappear, while the eighth, lying now in a protuberance of the oral surface called the paroral cone, divides again. It is believed that meiosis occurs during the first two micronuclear divisions; thus the nucleus in the paroral cone is already haploid, and its division is mitotic. The haploid phase, however, is only a brief one. One of the two nuclei in the paroral cone in each individual passes back into the general cytoplasm; this nucleus now remains stationary, and is regarded as the female nucleus. The other is regarded as the male nucleus, for it crosses over into the other conjugant and fuses there with the female one. The process is reciprocal, the two partners exchanging their male nuclei. The conjugants now separate as ex-conjugants and undergo fission so that each gives rise to four daughters; during this the old macronuclei finally disappear. Meanwhile the fusion nuclei divide and their products diverge in size so that each daughter comes to contain the normal complement of two micronuclei and one macronucleus. All the surviving nuclei are thus derived from the original fusion nucleus of the ex-conjugant parent.

The significance of conjugation, which has been much debated, cannot be evaluated without considering autogamy and cytogamy. Autogamy includes a nuclear reorganization. This is similar in principle to that resulting from conjugation, but differs in that it is not accompanied by conjugation, so that only one individual is involved. Cultures of *P. aurelia* undergo autogamy when they have been well fed for a time and then starved. Nuclear changes take place identical to those occurring in a conjugant, down to the movement of a nucleus into the paroral cone and its division. Then, however, these two daughter nuclei fuse to form the fusion nucleus; from this a new nuclear equipment is formed, the old macronucleus having broken down. Cytogamy, less easy to demonstrate and perhaps less frequent in its occurrence, involves a pairing of animals as in conjugation. It differs from the latter process, however, in that no exchange of nuclei takes place; instead, the fusion nucleus is formed from two nuclei within the same animal, essentially as in autogamy.

The sexual nature of these processes is clear, since it involves the fusion of two haploid nuclei that can be compared with male and female gametes. Whether the individuals are sexually differentiated, however, is less easy to assess. The varieties into which they are divisible are possibly comparable with physiological species; not only are they unable to effect cross-conjugation, but they differ also in certain characters, such as the duration of conjugation and the conditions in which it will be initiated. The existence of mating types within each variety cannot, however, be regarded as a sexual differentiation into male and female groups, for each type provides both male and female nuclei. It thus seems more acceptable to regard the individuals as hermaphrodites that are capable both of self-fertilization and cross-fertilization. This means that the differentiation of mating types must be regarded as one of the many specialized characteristics of the ciliates.

Quite apart from this difficulty of relating the sexual processes of ciliates to those of other animals, there remains the even greater problem of the significance of the processes for the ciliates themselves. Two views have been much discussed. One is that if cultures are prevented from conjugating they gradually decrease in vitality, this being shown in a decline in their rate of binary fission. Eventually such cultures are thought to die out, a result that has been compared with the senescence of the body (soma) of a metazoan. A contrasting view is that vitality does not decrease if the culture conditions are maintained with sufficient care. According to this interpretation, cultures can divide and survive indefinitely, even without conjugation. This view, however, was developed before the occurrence of autogamy had been established. It now seems likely that autogamy is taking place when cultures survive in the absence of conjugation.

Probably, therefore, we can accept that conjugation or autogamy do provide some rejuvenating effect, but it remains uncertain what that effect may be. If conjugation were the only process involved we could ascribe its benefits to the gene recombination that must result from the syngamy of the haploid male and female nuclei. In autogamy, however, this recombination takes place only once, at the first act of autogamy. Thereafter the nuclei of any given individual must necessarily be homozygotic; no further recombination can therefore take place, except for an occasional mutation. Similar considerations apply to the suggestion that benefits may result from an exchange of cytoplasm. Probably this exchange does sometimes occur during conjugation, but it clearly plays no part in autogamy.

One possible explanation of the rejuvenation is that the old macronucleus becomes in some way degenerate or ineffective, and that the nuclear reorganization provides for its replacement. Here it is of great interest that individuals emerging from either conjugation or autogamy may sometimes replace their macronuclei from one of the small fragments of the old macronucleus instead of from one of the micronuclear products. Studies of such animals have revealed that the macronucleus determines the phenotype of *Paramecium*; this follows from the fact that individuals in which a macronucleus of one genotype is combined with micronuclei of another genotype will show the phenotype of the macronucleus. Further, since a fully functional macronucleus can regenerate from a macronuclear fragment representing no more than one-fortieth of the complete nucleus, it follows that the complete macronucleus must contain 40 or more sets of genes. This suggests that during the normal repeated binary fission of *Paramecium* the macronuclei may eventually accumulate sets of genes that are unbalanced because of inequalities in the amitotic divisions of these nuclei. A situation could thus arise in which replacement of the effete macronucleus by a new and properly balanced one might be the only way of ensuring adequate maintenance of the phenotype. This, however, is supposition. Sonneborn concludes that there is no wholly adequate explanation of the rejuvenating effect of sexual processes in *Paramecium*.

18–2 ASEXUAL REPRODUCTION AND POLYCHAETE LIFE CYCLES

The value of asexual reproduction in providing for rapid increase in numbers, and for efficient exploitation of localized sources of food, is evident enough in the Protozoa. Rotifers and aphids behave in an analogous way, by producing entirely female generations with parthenogenetic eggs, but rapid and truly asexual reproduction is scarcely found in the Metazoa, especially when they have gained some degree of size and complexity of differentiation. Asexual reproduction, when it does occur in these animals, takes the form of budding or fragmentation; processes that involve both growth and the regeneration of missing parts, and that draw heavily upon the nutritive resources of the organisms. It is particularly characteristic of the coelenterates, and fresh-water rhabdocoels and triclads, certain polychaetes and oligochaetes, and the Urochordata. This might seem to suggest that regeneration is particularly associated with the lower invertebrates, but evolutionary status is not the only factor concerned. A. E. Needham has pointed out that while both regeneration and asexual reproduction can be regarded as primitive properties of organisms, their distribution in the animal kingdom as a whole is consistent with the view that they have been selected in certain groups because of their high adaptive value.

It is easier to appreciate the essentially primitive nature of regeneration in view of what we now know of the rapid and continuous exchange of materials between the organism and its environment. Not only is there replacement of the molecules lost in the wear and tear of physiological processes; radioisotope studies show that even those components of the body that seem superficially to be stable are in fact being constantly renewed. Because of this we cannot clearly distinguish between the replacement of lost parts, that we call regeneration, and the processes of normal metabolism. Rather must we suppose that regeneration has simply been brought into special prominence under the influence of natural selection, where conditions make it advantageous and the level of organization of the animal allows it to be practicable.

Similarly, we can view asexual reproduction in the Metazoa as being, in its simplest form, an extension of normal growth. This is most evident in the coelenterates, where it is manifested in the budding that is the basis of colony formation. This, as we shall see later, carries a substantial adaptive advantage. In worms, with their higher level of differentiation, asexual reproduction is more complex, usually involving fragmentation of the body and subsequent regeneration of the missing parts. The close correlation between regeneration and asexual reproduction is, indeed, particularly well shown in these animals, for they display a wide spectrum of phenomena that range from the straightforward replacement of a damaged part to specialized acts of fission that are closely correlated with sexual reproductive processes. The damaging by a predator of part of an elongated body must be a common hazard for them; considerable survival value must consequently attach to the ability to shed the damaged part (a process called autotomy) and then to replace it. This ability is found, for example, in *Chaetopterus* (Fig. 18–3). If the anterior end of this animal is pulled, as it might well be by a predator, a contraction of the circular muscle occurs between segments 12 and 13, and the body breaks into two pieces at that point. Here the autotomy, which is followed by regeneration, results in the reproductive and current-producing regions being preserved at the expense of the possibly less valuable head region. Such are the regenerative powers of this animal that a complete worm can be regenerated from a single isolated segment (Fig. 18–4). The anterior and posterior surfaces of this will grow and differentiate until the normal number of segments has been replaced, the initial segment thus coming to occupy once again its correct position in the body. The only limit in this process is that regeneration of the anterior part of the animal can only take place from segment 14 forwards; posterior regeneration, however, can be successfully completed by any segment.

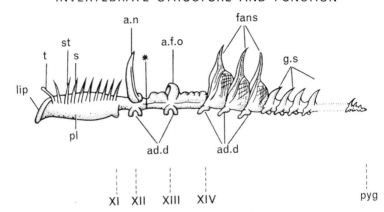

Fig. 18-3. Diagram of *Chaetopterus*. *a.f.o*, accessory feeding organ; *a.n*, aliform notopodia; *, autotomy level; *g.s*, genital segments; *pl*, plastron; *pyg*, pygidium; *s*, stout setae; *st*, setigerous segment; *t*, tentacle; *xi-xiv*, segment numbers. From Berrill, 1928. *J. mar. biol. Ass. U.K.*, **15**, 151-158.

Autotomy and regeneration could clearly provide, in suitable circumstances, for reproduction as well as replacement, and well-known examples of this exist. The syllid polychaetes are particularly notable for this, fragmentation (stolonization) of their bodies being associated with a highly specialized organization of their segments. Autotomy takes place in *Autolytus*, for example, if it is placed in dilute sea water, the fragmentation being brought about by very strong contractions of the longitudinal musculature. The breaking points are predetermined by the position of specialized septa, the positions of which are indicated by white transverse lines.

In natural conditions the anterior and posterior pieces of the syllid may regenerate respectively the missing tail and head (Fig. 18-5a,b); this regeneration may actually precede the fragmentation (Fig. 18-5c), and may be accompanied by pygidial budding (Fig. 18-5d). In some instances the level at which the head is differentiated is predetermined. Thus, in *Autolytus edwardsi* a new head forms in a segment situated within 16-22 segments from the hind end; more heads then appear in front of it, so that a chain of individuals of zooids is formed. As many as 29 such individuals have been observed in *Myrianida*, all for a time attached to the parent stock. In *Autolytus pictus* the new head is said always to appear on the anterior half of the fourteenth segment from the anterior end. If the worm is cut into two through the anterior half of its thirteenth segment it will regenerate a new head at the anterior cut surface, but another will also differentiate on the anterior part of segment 14. Here it would seem that segment 14 is inherently specialized for head production; in classical embryological terminology, this segment is a presumptive head. One other example of the highly specialized budding of syllids may be mentioned, from among many. In *Syllis ramosa*, which is commensal within a siliceous sponge, lateral buds grow out tail-first from various segments. Some of these buds then produce secondary lateral buds before they separate from the parent.

Budding carried to this level of specialization provides scope for a significant degree of multiplication, and even more striking results are possible on theoretical gounds. Thus there are species of the nemertine *Lineus* with a capacity for regenera-

Fig. 18-4. Anterior and posterior regeneration in *Chaetopterus* from (fan) segment 14. From Berrill, 1928, *op. cit.*

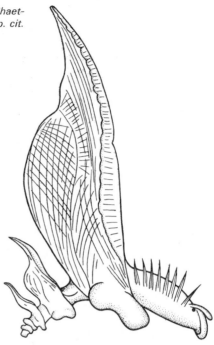

tion so well developed that, according to one calculation, as many as 200,000 individuals could be formed out of the fragments of one parent. Even at the higher level of organization of oligochaetes, a species of *Nais* (*N. paraguayensis*) is theoretically capable of producing 15,000 individuals in two months. We need not suppose that this potentiality is ever realized in practice. As we have seen, there are limits to the efficacy of asexual reproduction for increasing numbers, and it is thus significant that in one of its most specialized forms it is closely associated with sexual reproduction. This is the situation that we have already encountered in certain polychaetes, where fission, in conjunction with epitoky, ensures a successful completion of external fertilization.

The epitokes (Fig. 18-6) of nereids (heteronereids) are simply the intact bottom-dwelling individuals that have become transformed into free-swimming swarmers, and this is true also of certain syllids, such as *Odontosyllis*. Events in this latter family, however, take a variety of courses, sexual reproduction often being associated with the budding and fragmentation processes mentioned above. Sometimes, as in *Syllis hyalina*, it is the hind region which becomes separate as a group of sexually mature segments. Here the separated piece does not develop a head, doubtless because its independent life is a brief one, concerned only with swarming. In other instances a head may be formed, but it remains incomplete, lacking pharynx and jaws. In *Myrianida* and *Autolytus* chains of zooids are formed by asexual reproduction, the zooids breaking off as sexually mature individuals. The zooids of *Myrianida* do not produce their germ cells until after they have attained some size, so that there is here a true alternation of generations, the founding parent of the chain functioning as the asexual phase.

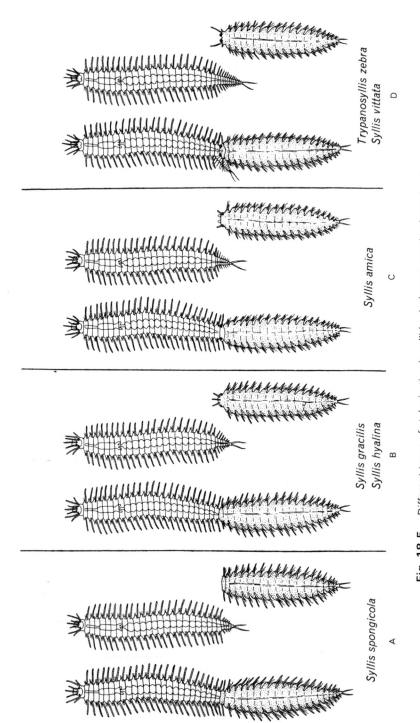

Fig. 18-5. Different types of stolonization in syllid polychaetes, ranging from simple fission on the extreme left to stolonization with pygidial budding on the extreme right. Note the different stages at which the head may differentiate. From Durchon, 1960. *op. cit.*

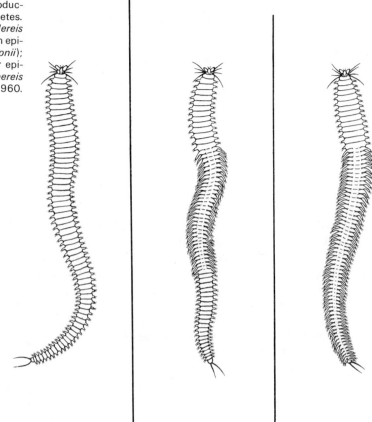

Fig. 18-6. Modes of reproduction in the nereid polychaetes. *Left,* without epitoky (e.g. *Nereis diversicolor*); *middle,* median epitoky (e.g. *Perinereis marionii*); *right,* median and posterior epitoky (e.g. *N. pelagica*; *Platynereis dumerilii*). From Durchon, 1960. *op. cit.*

18–3 UNITY IN THE EARLY DEVELOPMENT OF METAZOA

Sexual reproduction in the Metazoa usually involves the fusion of two highly specialized gametes, the sperm and the ovum. These show a division of labour that, as we have seen, is closely paralleled in the sexual reproduction of many protozoans. The specialization of the sperm is sufficiently obvious from the way in which it is adapted for locomotion and for the conveying of genetic material in a concentrated form. The specialization of the ovum is less obvious, apart from the fact that it is often an unusually large cell, heavily loaded with food reserves. Its true nature was a matter for controversy long before the occurrence of syngamy had been observed, or the events of sexual reproduction understood. During the seventeenth and eighteenth centuries it was supposed by some that the future organism was preformed in the egg, and that its development involved the unfolding of a structure that was already established within it. This was the preformation theory of animal development. Opposed to it was the epigenesis theory; according to this there was no preformed structure within the egg, development consisting instead of the progressive laying-down of new organization.

There was an inherent absurdity in the classical preformation theory, for it implied that the first females must have contained the rudiments of all subsequent generations. Apart from this, however, the advances in the techniques of microscopy during the nineteenth century seemed in any case to favour the epigenesis theory, for they failed to reveal any signs of preformed embryos in eggs. Yet towards the close of the century the controversy was re-opened, or developed in new terms, as a consequence of the application of the experimental method to embryological problems. This led to the demonstration that in certain circumstances a complete embryo can develop either from an isolated part of an egg, or from parts of the whole of two eggs that have been artificially fused together. For example, it is possible to separate the blastomeres of the 2- and 4-cell stage of the cleaving egg of sea-urchins, and to show that any one of these blastomeres can continue to cleave on its own and to develop into a larva that is unusually small but that is in other respects normal. Conversely, if two sea-urchin eggs are removed from their membranes and brought carefully together they may sometimes fuse and give rise to a single larva of abnormally large size. Such results suggested on first analysis that the egg in these early stages must be free of any regional differentiation, and that its subsequent development must be truly epigenetic. Capacities for future differentiation seemed to be distributed equally throughout the egg, which could thus be described as an equipotential system, any one part of which contained a share of all the properties of the whole. During normal development, one such part would form only part of the embryo, but under abnormal conditions it could give rise to the whole embryo. Eggs behaving in this way were referred to as regulation eggs.

Other eggs, however, proved to have different properties. As an example, the unfertilized egg of an ascidian contains in its cytoplasm three zones, each with a characteristic pigmentation. This egg has a polar axis, at one of which, the animal pole, is a zone of clear cytoplasm, the ectoplasm. Around the surface of the egg is a thin layer of mesoplasm, while the rest of the cell, including the other (vegetative) pole, consists of endoplasm.

Fertilization of this egg is followed by streaming movements in the cytoplasm that redistribute these zones. The ectoplasm and endoplasm come to lie respectively in the animal and vegetative regions; meanwhile the mesoplasm forms a yellow crescent in the future ventral region, and a new zone, the grey crescent, forms opposite to it in the future dorsal region. The fertilized egg now has a bilateral symmetry, and also a regionally differentiated organization, for the four zones already differ in their developmental potencies. These zones become confined to particular groups of cells during subsequent cleavage and gastrulation, and the fates of these cells are determined in particular ways. Those containing the ectoplasm produce the ectoderm, those containing the endoplasm produce the alimentary tract, those containing the yellow crescent produce mesoderm, and those containing the grey crescent produce the central nervous system and the notochord. As a result, isolated blastomeres will give rise to little more than those structures which they would normally have produced had they remained in their correct positions. In other words such an egg behaves like a mosaic of parts, each of which has its own independent properties. Regulation appears impossible, so that eggs behaving in this way have been termed mosaic eggs.

We now know that this distinction between regulation and mosaic eggs is a false one as far as the fundamental mechanisms of development are concerned, although

PATTERNS OF REPRODUCTION 557

it retains a certain descriptive value. The falsity can be demonstrated in experiments involving the separation of the blastomeres of the sea-urchin. If these are separated later than the 4-cell stage, after the first transverse cleavage furrow has formed, it becomes evident that the developmental potencies are already distributed differentially along a polar axis (Fig. 18-7). For example, one half (the animal half) of the cleaving egg will give rise to a larva with a large apical tuft but no alimentary tract,

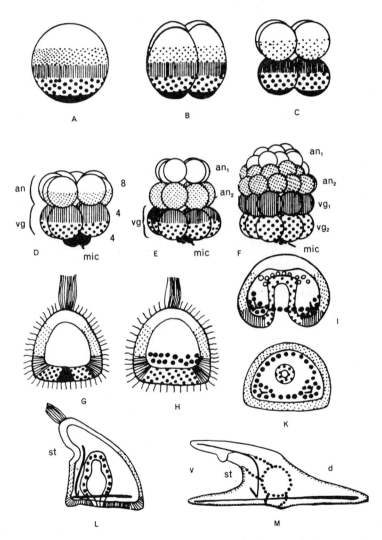

Fig. 18-7. Schematic representation of the larval development in the sea-urchin, to show the primitive position of the materials destined to each region. A–F, cleavage; G–M, gastrulation and establishment of some larval characteristics. *Black*, micromeres; *coarse dotting*, entoblast, secondary mesenchyme, and coelomic material localized in vg_2 during the 64-cell stage (F); *vertical hatching, light stippling*, and *white*, ectoblastic areas respectively located in vg_1, an_2, and an_1 during the 64-cell stage (F); *d*, dorsal; *v*, ventral; *st*, stomodaeum. Modified after Hörstadius, from Dalcq, 1938. *Form and Causality in Early Development.* Cambridge University Press, London.

while the opposite (vegetative) half produces a larva with an alimentary tract and skeletal structures, but with no apical tuft or ciliary band. Some form of polar differentiation is, in fact, common in eggs; it is determined in part, perhaps, by their position of attachment within the ovary and by the way in which the nutritive reserves are deposited. It is believed that in the particular case of the sea-urchin there are two interlinked patterns of differentiation in the unfertilized egg. One is the gradient of potencies distributed along the polar axis. The other is a pattern of differentiation lying in the superficial layer, or cortex; this, it is supposed, forms with the polar axis a system of coordinates that controls the progressive further differentiation of the cleaving egg.

The results of such experimental analysis can be expressed in terms of the cellular information theory discussed in earlier chapters. The fertilized egg contains within itself all the genetic information needed to programme its development, this information constituting what may be called the total genome. We may thus think of differentiation as the process in which the cells of the developing embryo are brought to select between the diverse potentialities of this genetic endowment. The way in which the selection is achieved can be visualized as being similar in principle to hormonal regulation of gene action (p. 535), for we can suppose that differentiation is determined by chemical messengers which repress some genes but permit the expression of others. The important difference is that hormonal regulation in the fully developed animal provides for continuous and flexible adjustment, whereas the results of differentiation in the embryo are much more stable. In many animals, and particularly in the higher ones, genes that are not to be used in a particular cell lineage may be switched off so firmly that their activity cannot be re-opened. Indeed, if this were not so, stability of organization of the metazoan body could never be achieved.

These chemical messengers that we are postulating as factors in differentiation correspond to the inducing substances, or evocators, that have been postulated by experimental embryologists to account for the phenomenon of induction: the process by which one embryonic tissue influences another tissue to develop in a particular direction. The classical example of this is found in amphibian embryos, in the action of the mesoderm of the archenteron roof upon the overlying ectoderm, which results in the latter giving rise to the central nervous system. This primary induction, and the chain of secondary inductions that follow it, must depend upon some chemical transmission from the inducing tissue to the responding one. The concept recalls that of the relationship between effectors and repressors postulated in the Jacob–Monod theory of gene regulation, for it seems reasonable to suppose that embryonic inducers also act by repressing or activating operons, and thereby influencing the course of differentiation. This, however, is largely supposition, for it is in the nature of the case difficult to provide direct demonstration of the action of these inducing substances upon the genome.

Nevertheless, some indication of it has been available since 1899, when Boveri gave a classical description of the behaviour of the nuclei of the nematode *Ascaris megalocephala* during early development. He showed that the germ cells are traceable to a 'stem-cell' that is distinguishable in the 2-cell stage, and that gives rise to a germ-line distinguishable in each succeeding cleavage because it retains its full complement of chromatin. The somatic nuclei differ in losing part of their chromatin at mitosis (a process called chromatin reduction) so that they are smaller and paler. This can be taken as an illustration of an irrevocable switching-off of genes, which is here achieved by some cytoplasmic inducing substance causing the genes in question to be actually discarded from the cells.

In general, the nature of the postulated inducing substances remains obscure. Nevertheless, it is reasonable to suppose that it is the production and spatial localization of such substances that are the essential characteristics of chemodifferentiation. As already suggested, they must to a large extent be laid down while the egg is maturing in the ovary, in patterns that are perhaps determined by the position of the egg within that organ, but which, even so, are ultimately attributable to instructions from the parental genome. The patterns may be invisible, and if, as in the ascidian, they are expressed in colour, this is fortuitous, and a bonus for the experimenter.

In any case, the difference between mosaic and regulation eggs does not result from the presence or absence of chemodifferentiation, for, on the argument outlined here, this is an essential feature of all development. The difference must depend on various factors, including the rate at which chemodifferentiation proceeds, and the relationship of its developing pattern to the main axes of the embryo. The experiments described earlier show that the regulative capacities of the first four blastomeres of the sea-urchin are a consequence of chemodifferentiation being distributed in such a way that each of those blastomeres has a share of all the developmental potencies that have been defined at that stage. If the first cleavage furrow of the sea-urchin were transverse instead of meridional, the egg would be regarded as mosaic in type. To the organism, however, this counts for nothing. What is important is that the cleavage pattern, however it be programmed, should be firmly defined, for only thus can an orderly pattern of differentiation be achieved. Another factor that may be relevant to this analysis of early development is the permeability of the cell walls of the early embryo. Bullough, in a through discussion of these matters, points out that the chemical messengers in mosaic development tend to be confined to the cells as they are cut off, whereas in regulation development they escape and create local environments. They can then determine the fate of cells that may be experimentally transplanted into those regions.

The analysis set out above, for all its speculative content, is strongly suggestive of an underlying unity in the modes of development of metazoan eggs. Yet it is none the less possible to distinguish, at a different level of analysis, certain patterns that are of the greatest value in helping us to understand the history and relationships of some of the main invertebrate groups. To judge the validity of this proposition, however, we must first examine how far embryological data are in general a reliable guide to phylogenetic history.

18–4 EMBRYOLOGY AND PHYLOGENY: THE ORIGIN OF METAZOA

Belief in the phylogenetic value of embryological evidence may be said to stem from the enunciation by von Baer, in 1828, of four fundamental principles, which have been restated and discussed in contemporary terms by de Beer. In essence, these principles state that the general characters of an animal (those that it shares with members of other groups) develop before the special ones that are peculiar to it. As a corollary of this, we find that animals resemble others more closely in the earlier stages of their development than they do in the later ones. In the early nineteenth century these principles carried no evolutionary implications. Following the publication of the *Origin of Species*, however, it was appreciated that embryological resemblances could be interpreted as evidence of common ancestry, indicative of the common ground plan of organization from which the specialized forms of today had diverged.

This point of view was elaborated by Haeckel into the biogenetic law, or the theory of recapitulation, according to which an individual passed through its ancestral history during its early development. This interpretation of embryology no longer commands assent. The reasons for this cannot be fully discussed here, but essentially they rest upon the recognition that adaptive evolution is grounded upon mutations and their recombination in sexual reproduction, and that new characters are therefore products of the complete ontogeny of individuals, and are not merely added on to the end of ancestral ontogenies. Early stages of embryology, therefore, are not a summary of past history, compressed by the addition of new characters in later stages of embryology. What they show us are the foundations upon which later specialization has been based; foundations that have often changed little because they have not been subject to adaptational modification.

Nevertheless, the interpretation of embryological data is not so straightforward as this may suggest. The major difficulty is that adaptive modification is by no means confined to the adult. The early stages of development, too, can undergo adaptive change, giving rise to characters that have been evolved in relation to special requirements of embryonic or larval life. These are called caenogenetic characters, in contrast to palingenetic ones, which are adaptations of the adult. Caenogenesis may thus seriously disturb the phylogenetic value of embryological evidence, and this is another reason for rejecting the over-simple concept of the recapitulation theory. Because of this, the comparative embryologist, just as much as the comparative anatomist, has to base his judgments upon a careful assessment of what is primitive and what is specialized in any particular situation. Moreover, for the same reason, it is unwise to interpret embryological data in isolation. Estimates of phylogenetic relationships need to consider all the evidence available from the form and function of the adult as well as of the young stages. Yet, given this cautious approach, the evidence of embryology will often be the surest guide to phylogeny, since the earliest stages of development seem to be generally more stable than the later ones.

Consider the problem of the evolutionary origin of the Metazoa, and the bearing upon this of the mode of development of the Porifera and the Cnidaria. The sexual reproduction of the sponges involves the formation of sperm and ova (Fig. 18–8). These resemble the corresponding gametes of Metazoa so closely that the principle of division of labour that they embody was probably established before these groups originated from their protozoan ancestors. Indeed, we have seen already that this principle operates in certain Protozoa at the present day. In other respects, however, the sponges behave in ways peculiar to themselves. For example, the developing oocyte moves like the amoebocytes which are common in the sponge body, and it grows by engulfing other cells, or by receiving nutriment from them. It is a sign of the low level of organization in the sponge body that these oocytes may sometimes arise by transformation of choanocytes, as also may the sperm; indeed, entire flagellated chambers are said to transform into nests of sperm. Yet another peculiarity is the way

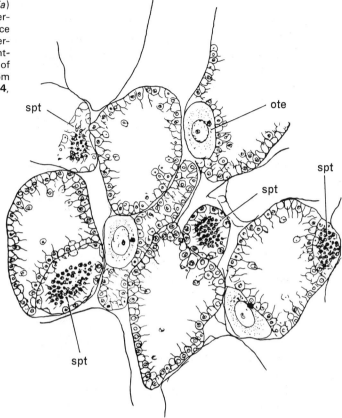

Fig. 18-8. Part of sponge (*Grantia*) showing oocytes (*ote*) and nests of spermatids (*spt*) with tails. The entire substance of this sponge contained patches of spermatids and isolated oocytes; it was a slightly protogynous hermaphrodite, as some of the oocytes were being fertilized. From Gatenby, 1920. *J. Linn. Soc. (Zool.)*, **34**, 261–296. Reproduced with permission.

in which the sperm first enter choanocytes (Fig. 18–9) or wandering amoebocytes, which then transmit them to the ova.

This is so individual a picture that we may reasonably suspect that sponges diverged very early from the main metazoan stem, or, as we have suggested earlier, had an origin entirely independent of the Metazoa. This conclusion is borne out by the course of later development. In some Calcarea the fertilized egg develops first into a stomoblastula. This is composed in part of non-flagellated cells surrounding a mouth, and in part of flagellated cells which have their flagella directed inwards towards the blastocoel. Later the stomoblastula undergoes a process called inversion, in which it turns itself out so that the flagella are directed towards the outside. Thus it becomes the amphiblastula, which is the characteristic larva of the Calcarea (Fig. 18–10). This larva, having completed its development within the maternal mesenchyme, passes to the outside through the canal system as a hollow structure, one half being composed of slender flagellated cells and the other of larger unflagellated ones. The flagellated hemisphere is anterior during swimming, and would thus seem to correspond with the animal hemisphere of a metazoan gastrula, but this interpretation is belied by subsequent events. Soon the flagellated cells become covered over by the non-flagellated ones. The latter then produce the dermal epithelium, together with the porocytes and scleroblasts (which secrete the major part of the spicules),

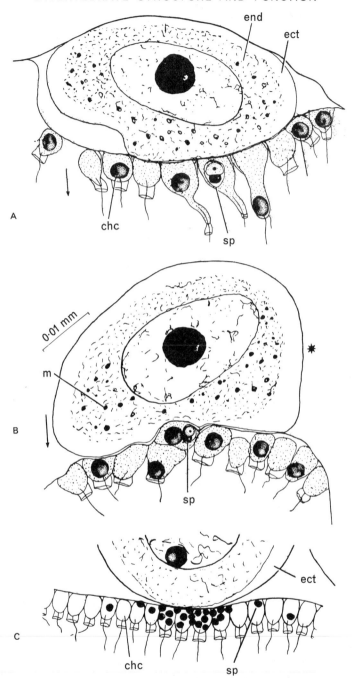

Fig. 18-9. A, B, two stages in the fertilization of *Grantia*; in A the spermatozoon (*sp*) has just entered a collar cell; in B the collar cell has lost its collar and is sinking below its fellows. C, part of an oocyte and the covering wall of collar cells (*chc*). The black dots represent the comparative positions of 25 spermatozoa in that number of cases examined in the stage drawn in B. *ect*, ectoplasmic, and *end*, endoplasmic regions of oocyte; *m*, mitochondria. From Gatenby, 1920. *op. cit.* Reproduced with permission.

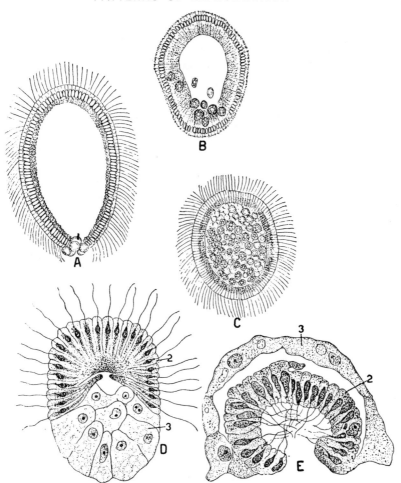

Fig. 18-10. A–C, stages in the formation of parenchymula of *Leucosolenia* by unipolar ingression; D, amphiblastula of a calcareous sponge (*Sycon*); E, the same after gastrulation and attachment. 1, So-called archaeocytes; 2, flagellated cells; 3, granular cells. From Hyman, 1940. *The Invertebrates: Protozoa through Ctenophora.* McGraw-Hill, New York and London.

while the flagellated cells become the choanocytes and produce also the amoebocytes. In this way there is formed a simple asconoid body (olynthus stage, Sec. 4–4).

This transformation of the amphiblastula, which takes place within a few hours or days of the initiation of swimming, is a simple example of metamorphosis (p. 587) which in this instance may be regarded as reversing the inversion that took place at the stomoblastula stage. Many other sponges, including some Calcarea, have a solid parenchymula larva (Fig. 18–10), which has an almost complete covering of flagellated cells enveloping the future dermal and mesenchymal cells. At metamorphosis the external and internal layers must therefore exchange layers, and so the flagellated cells pass inwards and become covered over by the others. As with the amphiblastula,

564   INVERTEBRATE STRUCTURE AND FUNCTION

the fates of the two main types of cell have no exact parallel in the Metazoa, but this is partly because their arrangement is reversed. The choanocytes, which could be said to correspond to a precociously developed metazoan endoderm, are anterior in the amphiblastula, instead of posterior, as they would be in a metazoan. However, the differences are so considerable that it is probably wise to confine the use of the terms ectoderm and endoderm to metazoan embryology.

A characteristic adaptive feature of poriferan development is the formation of asexual reproductive bodies called gemmules, which are found in all fresh-water sponges (Spongillidae) and also in some marine forms (although their ecological significance in the latter is obscure). Fresh-water life, as we shall see, involves special hazards that call for appropriate adaptations, and gemmule formation is evidently a device for withstanding the marked seasonal changes that characterize it. A gemmule is formed by an accumulation of amoebocytes (archaeocytes) which feed on other nutritive amoebocytes (trophocytes) and lay down yolk at their expense (Fig. 18–11). Other cells gather to form an epithelium over the surface of the developing gemmule and secrete a protective shell of spongin, while still more cells (scleroblasts) bring a contribution of spicules to reinforce the shell. Finally, the archaeocytes become binu-

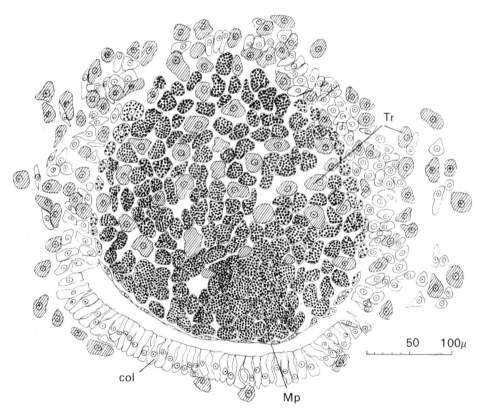

Fig. 18-11. First stage in the development of a gemmule in *Ephydatia*. *col*, columellar cells; *Mp*, primordial membrane; *Tr*, trophocytes. From Brien, 1968. In *Chemical Zoology*, vol. 2. (Florkin and Scheer, eds.). Academic Press, New York.

cleate, by undergoing division without division of the body. This marks the beginning of what is really a form of hibernation, for the gemmule can now survive the winter in an inert state, while the original sponge body dies down. When conditions are favourable, in the following spring, germination takes place, the archaeocytes emerging through a micropyle in the shell and giving rise to a new sponge body around it (Fig. 18–12).

The mechanism of this adaptation is characteristically poriferan, in the way in which different members of the cell republic come together for a specific purpose, yet in one respect it shows a remarkable parallelism with insect diapause. In some species (e.g. *Ephydatia fluviatilis*) the gemmules can germinate as soon as they are brought into water at room temperature, regardless of the conditions to which they have previously been exposed. Normally, their development is simply inhibited by low temperature. In other species (e.g. *E. mulleri*) the gemmules cannot so germinate unless they are first exposed for several weeks to low temperatures of the order of 0°C. This is true diapause, in the sense in which we have earlier used this term (p. 514), and it has the obvious advantage of ensuring that the gemmules will not germinate prematurely, and in advance of the imminence of suitable conditions for the life of the new sponge body.

But how is premature germination prevented in *E. fluviatilis*, which does not undergo true diapause? It is believed that security is ensured in this species by the parent sponge body secreting a substance that inhibits gemmule germination. The parent survives until late in the winter; during its survival the gemmules are unable to germinate, although they will do so as soon as they are removed from the parent. This control may be thought of as an example of chemical regulation, equivalent in some sense to hormonal regulation, but operating by diffusion at a level of evolution that antedates the differentiation of secretory organs and a vascular system.

With the Cnidaria and Ctenophora we pass to animals that are truly metazoan (or eumetazoan, as they are sometimes called). However, we have seen that their

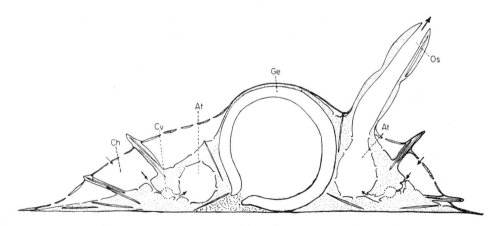

Fig. 18-12. Diagram of optical section of a young leucon, hatched from a gemmule of *Ephydatia*. *At*, Atrial cavity of excurrent system; *Cv*, flagellated chamber; *Ch*, hypodermal cavity of incurrent system; *Ge*, empty wall of gemmule; *Os*, osculum. From Brien, 1968. *op. cit.*

mutual relationships are decidedly uncertain. They have much the same grade of diploblastic organization, but they differ in many characters, and not least in their sexual development. Cleavage in the coelenterates is indeterminate and of a generally unspecialized nature, whereas that of ctenophores is very highly specialized, and is, indeed, one of the classical illustrations of determinate and mosaic development. Perhaps the ctenophores are best regarded as an early offshoot of the coelenterate stem, and one which subsequently pursued an entirely independent line of evolution (Sec. 4–3). The coelenterates themselves can be more easily related with higher metazoans, since they have preserved more of what is usually visualized as the primitive equipment of the Metazoa (but see below, p. 573). For this reason the course of their development has a wider significance.

Cleavage in coelenterates leads to the formation of a hollow blastula (coeloblastula), which is set free as a ciliated larva in those species in which the ova develop outside the body. Endoderm forms in the Hydrozoa by unipolar or multipolar ingression or delamination, giving a solid gastrula (stereogastrula). This constitutes the ciliated planula larva, which is the earliest free-swimming stage in those many hydroid species in which the ova develop within the reproductive individuals called gonophores. The planula, despite its lack of mouth and enteron, has something of the characteristic cellular differentiation of the phylum, with nematocysts, muscle processes, and sensory, nervous, and secretory elements. After a brief free-swimming existence, ranging from some hours to a few days, the endoderm cells become arranged to form the lining of a coelenteron. The organism now settles on a suitable surface, develops a mouth and tentacles, and gives rise to the first hydranth of a new colony, the oral end of the polyp arising at the posterior end of the larva. In some hydroids (*Tubularia* and *Myriothela* are examples), the planula stage remains within the gonophore (Fig. 4–3) and develops there into the tentaculate actinula larva. This has the form of a polyp, but lacks the stalk of the fully formed individual. It leaves the parent, but soon attaches to the substratum after a period of creeping; it then completes its growth into a typical polyp.

Gastrulation in the other two classes of the coelenterates may take place by invagination of the coeloblastula. It does so in some Scyphozoa, and typically also in the Zoantharia, which contrast in this respect with the Octocorallia, in which the endoderm forms by delamination. These two types of gastrulation have been associated with two contrasting views of the origin of metazoans. However, the fact that members of two subclasses of the one Class Anthozoa differ in this respect suggests that the difference can only be of minor phylogenetic significance. We shall return to this matter later.

The gastrula of anemones develops into a ciliated larva in which the mouth persists at the site of the blastopore, so that the oral pole of the future polyp is posterior, as it is in hydrozoans. The larva eventually transforms into a sessile polyp by the development of septa.

Development in the Scyphozoa varies. The planula of *Aurelia* settles and develops into a polyp-like scyphistoma larva, which later gives off minute young medusae (ephyrae) by strobilation (p. 427). The ephyrae, usually with eight bifurcated lobes, then grow into the mature medusa. In the related genus *Pelagia*, however, the planula transforms directly into an ephyra without passing through a sessile scyphistoma stage. We shall see later the possible significance of this.

The fundamental problem in interpreting the life histories of coelenterates, and one that is also highly relevant to theories of the origin of Metazoa, is whether the polyp or the medusa is the more primitive and thus closer to the common coelenterate ancestor. Either point of view can be argued with some plausibility. On the first interpretation, the ancestral coelenterate was a sessile polyp, sexual reproduction becoming secondarily associated with a motile medusa to facilitate dispersal. We shall adopt here, however, as a basis for argument, the second view, derived from one formulated in the nineteenth century by Brooks. This supposes that the ancestral coelenterate was a motile proto-actinula, which was planktonic or bentho-planktonic in its habit. Emphasis upon planktonic life could have led to this proto-actinula evolving into a medusa. Having regard to the differences between the medusae of the Hydrozoa and the Scyphozoa, this would probably have occurred independently in the two classes. Settling of the actinula, after the evolution of the medusa, could then have occurred in both hydrozoans and scyphozoans. In the Hydrozoa it led to its evolution into the polyp, which became thereafter increasingly important in the life cycle, giving rise to polymorphic colonies as a result of budding. On this interpretation, *Pelagia* might be regarded as a model of the primitive scyphozoan life cycle, while the Anthozoa must have arisen through settling of the proto-actinula prior to the evolution of the medusa, for there is no indication that is was ever present in the anthozoan life cycle.

The Trachylina, a hydrozoan group with the polyp reduced or absent, provide a good model of this supposed ancestral hydrozoan life cycle. In many trachylines the planula larva does not settle, but develops into an actinula larva, which then transforms into the adult medusa by flattening of the body and invagination of a subumbrella (Fig. 18-13). This could well be a starting point for coelenterate life histories, although this does not mean that the trachyline medusae of today are themselves pri-

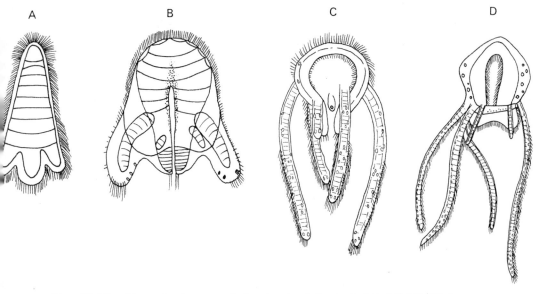

Fig. 18-13. Direct development of the planula of the trachyline *Aglaura* into a medusa. The swimming planula (A) forms mouth and tentacles (B), and develops into an actinula (C), which then metamorphoses into the medusa (D). From Hyman, 1940. *op. cit.*

mitive forms. The Trachylina also show us what can readily be interpreted as stages in the evolution of metagenesis (Fig. 18–14). One example is *Cunina proboscidea*, which has a complicated life cycle, in the course of which planula larvae parasitize another trachyline, *Geryonia*, giving rise to a stolon from which medusae are budded. Another example is *Pegantha*, which begins its development in the maternal medusa and eventually forms actinulae which bud off other actinulae. If, as is said, the primary actinula remains in being, it may be regarded as the founder polyp of an asexually reproducing generation, giving what is essentially a typical hydrozoan life cycle. Another instructive trachyline example is the fresh-water medusa, *Craspedacusta sowerbyi*, which was introduced into Europe with an Amazonian water lily. It has a minute polyp stage, formerly known as *Microhydra ryderi*, which forms small colonies from which small medusae are eventually budded off.

By some such paths, or by others of which we know nothing, there could have evolved the life cycles of the colonial Hydrozoa. Here the polymorphism (more will be said of this later, p. 643) has led to the production by the colony of feeding gastrozooids and reproductive gonozooids, the latter being often modified to blastostyles

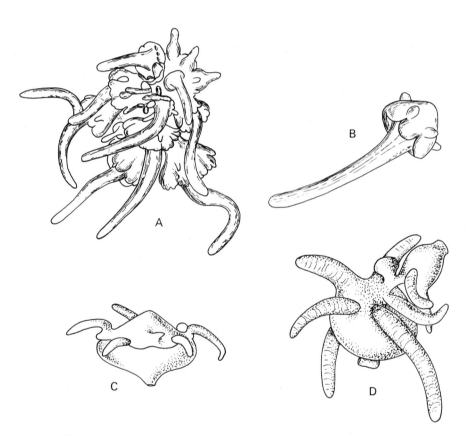

Fig. 18-14. A, stolon of Cunina, found attached to another trachyline (*Rhopalonema*), in process of budding medusae; B, one of the young medusae released from the stolon; C, young actinula of the trachyline *Pegantha*, in gastric cavity of mother; D, later actinula budding other actinulae. From Hyman, 1940. *op. cit.*

(e.g. *Obelia*), in which polypide features have been lost. The colony may also bear individuals called dactylozooids, which serve for protection. These are derived by simplification of the structural plan of the gastrozooid. Mouth, tentacles, and enteron are reduced or lost, so that in its extreme specialization the dactylozooid may have the form of a tentacle, bearing many nematocysts; less highly modified types may still bear terminal tentacles, as do the protective spiral zooids of *Hydractinia*.

The reproductive structures bud off motile medusae which, by sexual reproduction, give rise to planulae that settle and form new colonies. On the interpretation being adopted here, the further history of this type of life cycle has involved competition between polyp and medusa for the favours of natural selection, and the evidence indicates that the balance has shifted towards the polyp at the expense of the medusa. A possible explanation is that dispersion is sufficiently secured by the planulae, and that the responsibility for survival is better met by the sessile colony than by the relatively small and vulnerable medusae. However this may be, there are many illustrations of the reduction of the medusa in the Hydroida.

In some of the Calyptoblastea, such as *Obelia*, the blastostyle, enclosed within its peridermal case, the gonotheca, buds off gonophores which develop into medusae that are then released from the colony. These medusae (called Leptomedusae) typically have shallow bells, numerous tentacles, statocysts, and gonads on the radial canals. A characteristic feature of the developing bud is the invagination of ectoderm cells to form a hollow structure, the entocodon, which is the rudiment of the subumbrella, and around which the radial canals are pushed out from the gastrodermis. Often, however, the bud never completes its development into a medusa, but remains in position as a gonophore. Its true nature is revealed in some cases by the presence of the entocodon and canal rudiments, but different degrees of arrest of development are found, with loss of the entocodon, and, finally, the complete elimination of the bud, with the germ cells developing directly on the wall of the blastostyle. It would be difficult to read this series in the opposite direction (gradual evolution of the medusa), and it is this that justifies the view that the life cycles in the Order Hydroida have been simplified by reduction of the medusa.

Comparable reduction of the medusa has taken place in the Gymnoblastea, although on a different basis of polymorphism. In this group the gonophores are formed on various parts of the colony, including the stolon, the stalks of the hydranths, or the hydranths themselves. An example of the last is *Tubularia* (Fig. 4–3), in which the gonophores are borne on a circlet of long stems that are situated just above the proximal tentacles. In some instances, however, the hydranths are modified, by loss of mouth and tentacles and reduction of the coelenteron, to blastostyles. From the gonophores are released the medusae (Anthomedusae), typically with deep bells, few tentacles, gonads on the manubrium, and ocelli, but lacking statocysts. *Tubularia* is an example of the loss of this stage; here the planula develops into an actinula while still in the gonophore, and then develops directly into the polyp founder of a colony after a brief period of independent creeping.

The life history of *Hydra*, judged from its nematocysts to be related to gymnoblasteans, is an extreme example of this reduction. Total loss of the free-swimming stage in this genus has probably evolved in adaptation to the conditions of freshwater life. Another example, comparable with *Hydra* in some respects, but established in totally different ecological conditions, is the Order Actinulida, created in 1959 for two genera, *Halammohydra* (Fig. 18–15) and *Othydra*. These are minute hydrozoans, 1 mm long, free-living and solitary, completely ciliated, and with essentially an actinula-like organization. They are adapted for life in the interstitial spaces of marine sands, and, in correlation with this, have developed statocysts, which are common in this habitat. There is no evidence of relationship to any stock in which medusae had evolved, except in so far as characteristics of their nematocysts suggest a possible derivation from marine Gymnoblastea. Perhaps they evolved from some primitive stock, in which an actinula colonized the intertidal zone at an early stage of coelenterate history, prior to the evolution of the medusa.

Some Anthomedusae can reproduce asexually by budding off daughter medusae; an example is *Sarsia*, which may carry daughter medusae on the long manubrium. This habit could well have been the origin of a quite different resolution of the

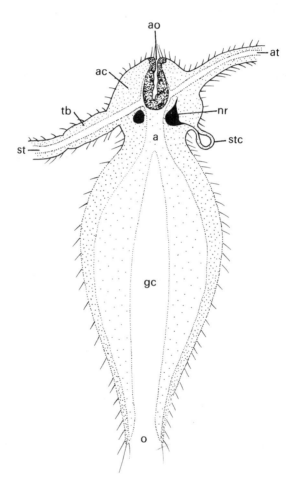

Fig. 18-15. Organization of *Halammohydra*. Key: *a*, axial endoderm; *ac*, aboral cone; *ao*, aboral adhesive organ; *at*, tentacle belonging to aboral girdle; *gc*, gastric cavity; *nr*, nerve ring; *o*, mouth. *st*, tentacle belonging to subaboral girdle; *stc*, statocyst; *tb*, tentacle bulb. From Swedmark & Teissier, 1966. *Symp. zool. Soc. Lond.*, **16**, 119–133.

interaction between medusa and polyp, which is characteristic of the Order Siphonophorida. This comprises remarkable swimming or floating colonies in which both hydroid and medusoid forms are associated together, with much differentiation of labour (see also p. 645). They form two main divisions, the Suborders Calicophora and Physophorida. The Calicophora are typically in the form of long strings of individuals (Fig. 18–16), bearing at the apex one or more swimming individuals, called nectophores. These individuals, which are modified medusae, propel the colony by medusoid jet propulsion. The Physophora are more variable, but typically have some form of float, also derived from a medusoid individual, in which gas is secreted. An example is *Physalia* (Portuguese man-of-war), with a blue float from which are suspended groups of individuals (each group called a cormidium) comprising gastrozooids for feeding, a blastostyle bearing gonophores, and a protective dactylozooid abundantly provided with the large and virulent nematocysts to which we have earlier referred. The development of these animals is complicated, but essentially it involves the production of planula larvae from which colonies arise by budding and differentiation of the colony members.

At this stage we may conveniently consider the bearing of coelenterate life histories upon theories of the origin of the Metazoa. One line of thought derives from Haeckel's blastaea–gastraea theory. Arguing from his recapitulation theory, and from the frequent existence of a blastula stage in metazoan development, he believed that the earliest Metazoa had the form of a hollow blastula. This, it was supposed, could probably have originated as colonies of protozoans, such as are seen today in the Volvocinae. Haeckel further suggested that these hollow, single-layered organisms later developed a two-layered structure by invagination. He thus regarded the formation of a gastrula by invagination in many metazoans as a persistent ancestral stage, corresponding to a coelenterate-like level of evolution. On this view, one would expect to find invaginate gastrulae in coelenterates, for these are envisaged as comparatively primitive forms in which development has been little modified. The fact that invagination does not occur in hydrozoan development is thus an obstacle to the acceptance of Haeckel's theory, unless one is to suppose that invagination has been secondarily discarded in this group. More serious, however, is the general abandonment of the fundamental principle of recapitulation upon which the blastaea-gastraea theory is based.

Nevertheless, Haeckel's concept of the origin of Metazoa has been revived in a modified form by Jägersten. Starting, as did Haeckel, with a blastula stage, he suggests that this began to move over the substratum instead of swimming at the surface, and that invagination into a diploblastic gastrula stage was promoted as a result of the body arching over food material to engulf it. Haeckel's theory supposed that the gastrula evolved while the organism retained its free-swimming habit, and it is admittedly difficult to see the advantage that would have been gained by this development. Jägersten's theory avoids this difficulty, although, like Haeckel's, it makes it necessary to assume that gastrulation by invagination has been secondarily lost in many coelenterates, and particularly in the Hydrozoa.

An alternative theory, and one that avoids this latter difficulty, is the planula–acoeloid theory, first suggested by Metschnikoff. This view, which has been cogently supported by Hyman, among others, supposes that the diploblastic stage of evolution was reached from a hollow blastula by the immigration of cells, just as occurs today

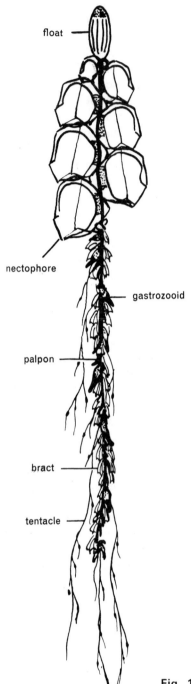

Fig. 18-16. A young specimen of a siphonophore, *Nanomia cara*. From Mackie, 1963. In *The Lower Metazoa* (Dougherty *et al.*, eds.). University of California.

in hydrozoan development. The resulting planula-like stage is visualized as feeding by phagocytosis, so that the development of mouth and enteron was at first unnecessary. The Coelenterata are visualized as developing from this stage as a consequence of the planula evolving, through a transitional actinula-like organism, into a medusa, specialized for locomotion and for the distribution of the germ cells. Colonial organization, with its alternation of generations, could have resulted from the actinula settling on the substratum and budding before the medusa stage. Further, a planula-like stage, creeping over the substratum and feeding by phagocytosis, could conceivably have been the origin of the acoelous platyhelminths, which, as we have seen, still rely largely upon phagocytosis for feeding. From these would then have evolved the higher platyhelminths, together with the vast range of forms that, as we shall argue below, constitute the assemblage called the Protostomia.

The planula–acoeloid theory may fit the facts in a comparatively simple way, but this does not necessarily make it correct. There remains, indeed, an entirely different way of interpreting them. This, which may be called the ciliate–acoeloid theory, has been particularly developed by Hadži. The full implications of this theory, and the wealth of argument with which he supports it, go beyond the scope of our present discussion. It must suffice to say that it is based on the belief that metazoans did not evolve from colonial aggregations of protozoans; instead, they are supposed to have arisen from ciliated protozoans by multiplication of their nuclei in the manner seen in the polymastigote and hypermastigote flagellates (cf. Fig. 23–1, p. 686). This would have been followed by gradual cellularization of the body.

The feature of Hadži's theory that has attracted most attention and controversy, however, is his further supposition that this cellularization would have given rise not to coelenterates but to acoelous platyhelminths. The triploblastic structure of these latter animals is thus regarded as primitive, arising not by invagination and cell immigration, but by the subdivision of a continuous body. A further and fundamental corollary is that the coelenterates are not primitive metazoans, but are secondarily derived from platyhelminths. Thus the subdivision of the animal kingdom into two levels of organization, the diploblastic and the triploblastic, breaks down. Indeed, Hadži goes further than this. He regards the Anthozoa as the earliest of the coelenterates, deriving them from rhabdocoel platyhelminths. The latter have cephalic tentacles, a simple nerve net, and an alimentary tract with some tendency towards folding, and in these respects they show coelenterate features. Moreover, the bilateral symmetry of the Anthozoa can be held to mark their derivation from freely moving ancestors, although we have seen earlier (p. 78) that an alternative explanation of this is possible.

A final feature of Hadži's theory is that the whole of the Metazoa are monophyletic. This view, however, is not easy to maintain unless embryological data are ignored or dismissed as relatively unimportant. We shall suggest below that if we view the life history of individual groups as a whole, and take their embryological development and larval organization into full account, there are good grounds for recognizing a fundamental division of at least the greater part of the invertebrates into two main assemblages. The coelenterates admittedly remain outside these, but whether as a basal stock, or as a secondary derivative of platyhelminths, is not easy to decide. Undoubtedly the latter interpretation raises serious difficulties. In our discussion of the hydrostatic skeleton we showed that the organization of the platyhelminth body was more complex and better suited for movement than is that of coelenterates. It is not easy to see why these advantages should have been abandoned. Nor is it easy to see why coelenterates should have also abandoned the advantage of a central nervous system in favour of exclusive reliance upon a nerve net. Other difficulties could also be mentioned. The complete lack of any trace of spiral cleavage in coelenterates, for example, is understandable if they preceded the platyhelminths, but surprising if they were derived from them; although it might perhaps be held that this derivation took place at a very early phylogenetic stage, before spiral cleavage had become established. A discussion of these and other aspects of Hadži's views must, however, be looked for elsewhere. Rees focuses a vigorous attack upon them, asserting that they have received an unwarranted amount of publicity, that they require too many assumptions, and that many of the arguments are reversible.

It is fair to add that the theory of the derivation of metazoans from a volvocine type of colonial organization also has its difficulties. For example, the coelenterates are carnivorous and are predominantly marine, whereas the volvocids are freshwater phototrophs. The derivation of coelenterates by cellularization of multinucleate ciliates has its attractions from this point of view, and could presumably be accepted without necessarily adopting the view that coelenterates are secondary derivatives of platyhelminths. There is a point, however, at which discussions along these lines lose touch with reality. We are considering events that may have occurred in the remote past, in conditions that no longer obtain, and affecting organisms that no longer exist. The volvocine series of flagellates affords no more than a suggestion of one possible route of origin of metazoans. Even if we accept that a colonial origin of the Metazoa is likely it does not follow that the ancestral types resembled *Volvox*, or, indeed, that they were necessarily phototrophic. But Hardy has argued that they may well have been so, for that would have solved the very real difficulty of how nutrition was carried on while the gap between protozoan and metazoan organization was being bridged. Metazoa may have evolved from early metaphyta, which developed pocket-like invaginations in which small organisms might collect and perish, to provide nutritive products within the cavity. The water plant *Utricularia* catches small crustaceans in such pockets, and the sundew puts tentacles to similar use. Admittedly these are specialized organisms, yet analogous adaptations might conceivably have developed at earlier stages of evolution.

There is surely much truth in Hyman's observation that 'anything said on these questions lies in the realm of fantasy', if only because no intermediate forms survive (which perhaps is hardly surprising). At one time it was thought that the strange animals belonging to the phylum Mesozoa might be intermediate between protozoans and metazoans, which is why they were given this name by van Beneden. They constitute two orders, the Dicyemida, comprising common parasites in the excretory organs of cephalopods, and the Orthonectida, much rarer endoparasites of flatworms and other invertebrates. Dicyemids, which are ciliated vermiform organisms, 6–7 mm long, have a few cells forming an outer layer that surrounds an elongated axial cell. This constitutes the nematogen stage, in which the axial cell divides to form asexual reproductive cells (agametes) that develop directly into daughter nematogens. Eventually the nematogen becomes a sexually mature rhombogen stage, which liberates ciliated larvae. The orthonectids have an asexually reproducing stage which is a multinucleate plasmodium; this eventually produces male and female organisms, 0.1–0.3 mm long, resembling the nematogen stage of dicyemids. The structural simplicity of these puzzling organisms need not mean, of course, that they are primitive forms, nor would their complex life histories be in line with this. Their simplicity may well be a secondary condition, associated with their small size; one widely-supported view is that they are perhaps degenerate platyhelminths.

But the lack of evidence relating to the origin of metazoans is not a reason for declining to discuss a problem that must always be of compelling interest to biologists, and that provides a valuable exercise in the application of biological principles. It is, however, a reason for not pursuing in too much detail an analysis of questions to which we cannot hope to obtain a final answer because essential data are forever lost to us. We may pass instead to a more decisive field of analysis, and consider certain embryological evidence that bears on the problem of the history and relationships of some of the higher invertebrate groups.

18-5 PROTOSTOMIA AND DEUTEROSTOMIA

One fact of fundamental importance is that the polyclad platyhelminths, the nemertines, the annelids, and the molluscs share a common characteristic in the possession of highly mosaic eggs that undergo a specialized form of cleavage called spiral cleavage (Figs. 18-17, 18-18, 18-19). The distinguishing feature of this is that during

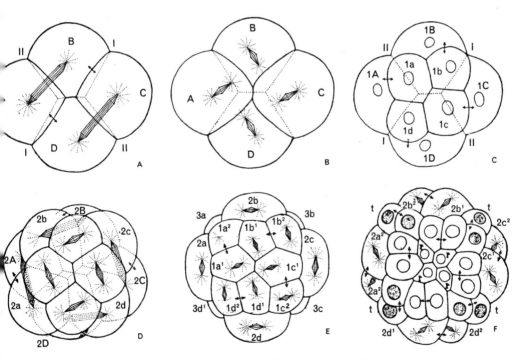

Fig. 18-17. Early stages in the spiral cleavage of the gastropod *Trochus*. A, B, 4-cell stages from upper pole; C, 8-cell stage, first micromere quartet; D, 12-cell stage, transitional to 16-cell, second quartet just formed; E, 20-cell stage (transitional to 32-cell), from upper pole, third quartet formed; F, 36-cell stage from upper pole; showing apical rosette (r), trochoblasts (t), and all the blastomeres of the first two quartets (28 cells). From Wilson, 1928. *The Cell* (3rd ed.). Macmillan, New York.

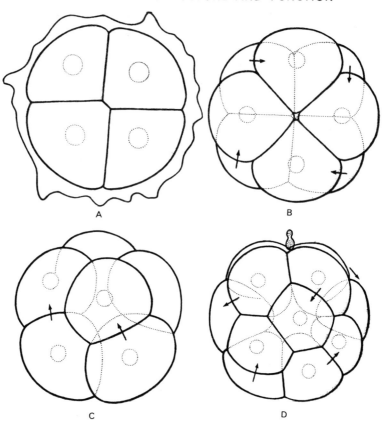

Fig. 18-18. Cleavage of *Polygordius*, from life. A, 4-cell stage, from above; B, corresponding view of 8-cell stage; C, side view of the same; D, 16-cell stage from the side. From Wilson, 1928. *op. cit.*

the third and subsequent cleavages the mitotic spindles are orientated nearly vertically, but at an angle to the polar axis, the displacement from the vertical being alternately to one side of the axis or to the other. The first two cleavage furrows are meridional, and produce four large blastomeres (macromeres), distinguished by convention as the A, B, C, and D blastomeres. The third furrow is latitudinal, and separates four small blastomeres (micromeres) from four macromeres. These first micromeres (1a, 1b, 1c, 1d) are called the first quartet of micromeres. The orientation of the spindles results in each of the four cells of this quartet being displaced to one side of its corresponding macromere (1A, 1B, 1C, 1D), the displacement being clockwise as viewed from the animal pole, in all four of the phyla concerned. (A well-known exception to this, however, is a variety of the pond-snail *Lymnea*. In this variety the first quartet is budded off in an anticlockwise direction and the symmetry of the whole animal is correspondingly reversed, so that the shell coils sinistrally, instead of dextrally as in most gastropods.)

At the next cleavage, in which the displacement is anticlockwise, the macro-

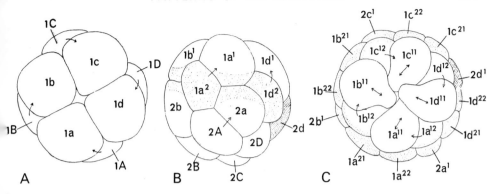

Fig. 18-19. Spiral cleavage of the polychaete *Podarke*, viewed from the anterior pole with the dorsal surface on the right. A, 8-cell stage, B, 16-cell stage, C, 32-cell stage. From Anderson, 1966. *Acta Zoologica*, **47**, 1–42.

meres give off a second quartet of micromeres (2a, 2b, 2c, 2d), and are then termed the 2A, 2B, 2C, and 2D macromeres. At the same time the first quartet of micromeres divides to form two daughter quartets. The component cells of these are designated as shown; for example, the cells formed by the division of 1a are lettered $1a^1$ and $1a^2$, the former being the one lying nearer to the animal pole. The same pattern of cleavage is continued, with successive divisions of macromeres and micromeres, to build up a blastula in which the division products have been alternately displaced clockwise and anticlockwise. As a result, the shortest line that can be drawn from one pole to the other without cutting through the sides of any cell is a spiral, and it is this that gives spiral cleavage its name.

The classical cell lineage studies of spiral cleavage, by Wilson and others, showed that the pattern of blastomeres produced by it, up to the 64-cell stage, is very much the same in principle from group to group and from species to species, regardless of the size of the egg (Fig. 18–20). They showed also that the fates of the blastomeres at this stage are fixed, and predictable with certainty. This, then, is mosaic development in the sense discussed earlier in this chapter. But it follows from the discussion that this is not a result of the spiral cleavage process itself. We must therefore distinguish two separate aspects of the development of these groups: the existence and distribution of chemodifferentiated areas, and the spiral cleavage which segregates them. Both aspects are of great phylogenetic interest.

As regards the cleavage pattern, this is so very specialized and characteristic that it is most unlikely to have been evolved independently in unrelated groups. Provisionally, therefore, we can take this as evidence that the groups showing spiral cleavage are phylogenetically related, and we shall see later that there is further evidence to support this conclusion. As regards the chemodifferentiated areas, which are believed to be present initially in the cortex of the egg, these are not visibly distinguishable, but the cell lineage studies permit the preparation of maps of presumptive areas which seem to follow the same general pattern from species to species (Fig. 18–21). Their segregation between the blastomeres also follows the same general plan, yet with some interspecific variation.

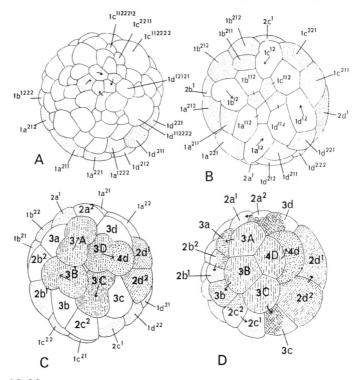

Fig. 18-20. Spiral cleavage in polychaetes. A, *Podarke*, 64-cell stage, from anterior pole; B, *Amphitrite*, the same; C, *Podarke*, 32-cell stage from posterior pole; D, *Arenicola*, the same. From Anderson, 1966. op. cit..

One example of this variation is that in the larger and more yolky polychaete eggs the 2d and 4d blastomeres are conspicuously larger than the corresponding cells of the other quadrants (Fig. 18–20). This is because they make a major contribution to the formation of the trunk segments prior to the beginning of feeding. The 2d blastomere forms much of the ectoderm behind the prototroch, while the 4d one gives off a pair of ventro-lateral mesoderm bands in which the coelom later appears. A more extreme example of variation is that certain cells which give rise to temporary larval structures may, in those species that undergo direct development, give rise instead to permanent adult ones.

The presence of spiral cleavage in certain groups of the animal kingdom, and its absence from others, is an important element in the argument that most invertebrates, together also with the Chordata, but excluding the radially symmetrical Cnidaria and Ctenophora, can be assigned to one or other of two large assemblages, the Protostomia and the Deuterostomia. The major groups included within the Protostomia are the Platyhelminthes, Nemertinea, Annelida, Arthropoda, and Mollusca. The Pogonophora probably belong here as well, although the necessary information about their development is lacking, and so also perhaps do some minor phyla of uncertain affinities. Characteristic of protostomes are spiral cleavage (the arthropods are a special case, which will be discussed below in terms of presumptive

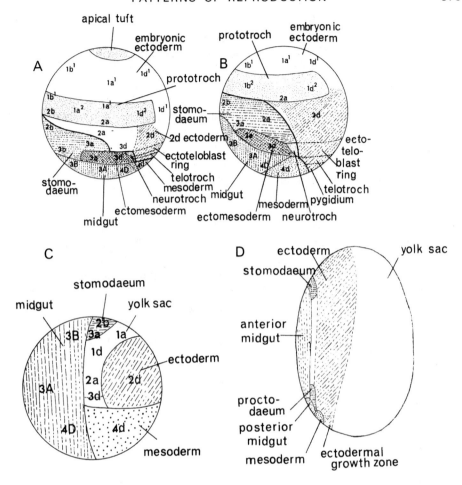

Fig. 18-21. Presumptive areas of the blastulae of some protostomes. A, *Podarke*; B, *Scoloplos*; C, *Tubifex*; D, yolky Onychophora. From Anderson, 1966. op. cit.

areas), development of the mesoderm by cell proliferation, and of the coelom as a schizocoel, by splitting of the mesoderm layer. Further, the mouth forms directly from the blastopore, or in the region where the blastopore closes; hence the term 'protostome'. Finally, the larval stage, when present, is often of the type called the trochophore, or can be interpreted as having a structure closely related to it. A major exception in this respect is the Phylum Arthropoda, in which group there is no sign at all of a trochophore larva, despite the indirect development of many members, and notably of the crustaceans. There can be no doubt, however, of the close relationship of the annelid and arthropod plans of structure, even though, as we have seen (Sec. 7-4), the Arthropoda may have evolved from worm-like forms quite distinct from annelids as known today.

As regards arthropodan embryology, the process that we have termed arthropodization seems to have involved a modification of development so extreme that

580   INVERTEBRATE STRUCTURE AND FUNCTION

embryological links with other groups have been largely lost. Yet the loss has not been total, for it can be demonstrated, by an extension of the analysis just outlined, that the onychophoran blastoderm has an arrangement of presumptive areas fundamentally similar to that of the oligochaete and hirudinean blastula. Certain oligochaetes and leeches (tubificids, lumbriculids, glossiphonids) have primitively large and yolky eggs. These undergo a modified spiral cleavage (Fig. 18–22) which yields a blastula with a pattern of presumptive areas that can be derived from that of polychaetes (Fig. 18–21), modified, however, in association with increased yolk. The secondarily microlecithal eggs of lumbricids, gnathobdellids, and pharyngobdellids show further modifications of cleavage, together with embryonic adaptations for feeding on albumen, but the same pattern of presumptive areas is retained.

Viviparous onychophorans, which develop within the parent and derive nutriment from it, show a secondary loss of yolk, but primitively large and yolky eggs are found in the ovoviviparous forms, which develop within the parent but take little if any nutriment directly from it. In these, the cells, in contrast to those of oligochaetes and leeches, are small and lie at the surface of a central mass of yolk, having become adapted for its extracellular digestion. Nevertheless, the basic pattern of development of the yolky type of onychophoran egg is derivable from that of oligochaetes, even though its spiral nature is not immediately obvious. The derivation is particularly evident when the patterns of presumptive areas are compared (Fig. 18–21). We are thus justified in concluding that the Onychophora evolved

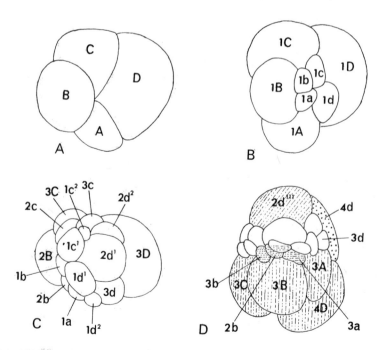

Fig. 18-22. Cleavage in *Tubifex*. A, 4-cell stage; B, 8-cell stage; C, 17-cell stage; D, 22-cell stage. From Anderson, 1966. *Proc. Linn. Soc. N.S.W.*, **91**, 10–43.

from ancestors with spiral cleavage, and this is therefore likely to be true of arthropods as a whole, although this extension of the argument is complicated by the possibility that the onychophoran–insect line has had an independent history (Sec. 7–4).

For reasons previously stated, these ancestors are unlikely to have been annelids in the accepted sense of that term. It follows that the assemblage of Protostomia must be regarded as including, in addition to the surviving groups named above, a wholly extinct group of metameric coelomates that also developed with spiral cleav-

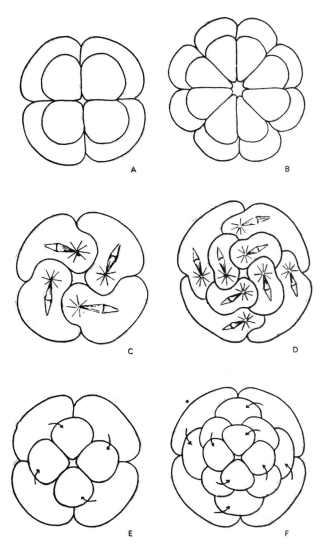

Fig. 18-23. Comparison of spiral and radial cleavage. A, B, 8- and 16-cell stages, radial type; C, D, 3rd and 4th cleavages, spiral type; E, F, 8- and 16-cell stages, spiral type. From Richards, 1931. *Outline of Comparative Embryology.* Wiley, New York.

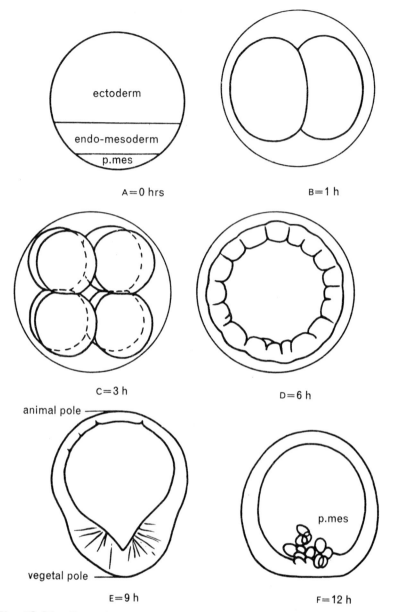

Fig. 18-24. Stages in the development of the egg of *Psammechinus miliaris* into a pluteus larva. A, eggs with the main presumptive regions of the larva indicated. B, 2-cell stage. C, 8-cell stage. D, blastula before hatching. E, blastula shortly after hatching. F, release of mesenchyme. G, mesenchyme blastula at the outset of gastrulation. H, profile of early gastrula, at the end of primary invagination, showing the beginning of dorsoventrality, the ring of primary mesenchyme, and one of the two ventral branches from this ring. I, gastrula at the onset of the second phase of invagination, showing pseudopodial activity at the tip of the archenteron rudiment, further flattening of the ectoderm, and appearance of thickening at the animal pole. J, gastrula at the end of invagination, showing release of secondary mesenchyme from the archenteron tip, early skeletal formation, and further flattening of the ventral side. K, gastrula after oral contact has formed, and the beginning of subdivision of the gut; note the oral invagination (stomodaeum rudiment). L, early pluteus in ventral view, showing the paired coelom formed from the archenteron tip, stronger subdivision of the gut, and further growth of the skeleton. M, later stage of the pluteus development where the mouth has opened, and one of the coelomic sacs has been extended to form the primary pore canal which later opens at the primary madreporic pore (in this case it is the right coelom, but normally the left); the skeleton has extended the ectoderm. From Gustafson and Wolpert, 1963. *Int. Rev. Cytol.*, **15**, 139–214.

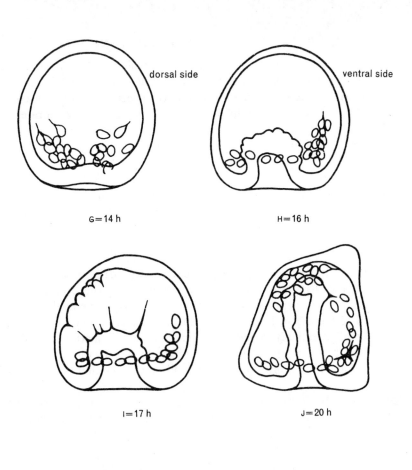

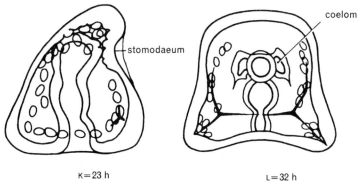

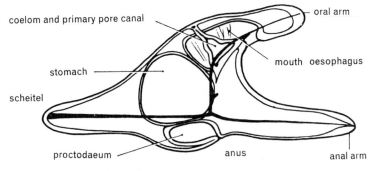

age. The argument is obviously a wide-ranging one, which just because of this, merits close consideration, particularly with reference to the evidence that has been set out so cogently by Manton and Anderson. Our ability to postulate with some confidence the one-time existence of a totally hypothetical group of invertebrates, and to attribute spiral cleavage to it, among other characteristics, demonstrates how powerful can be a rigorous analysis of the data of comparative enbryology and functional morphology.

The Deuterostomia are envisaged as comprising, in addition to certain minor phyla, the Echinodermata, Hemichordata, and Chordata (including the two protochordate groups, the Urochordata and the Cephalochordata). In echinoderms where deuterostome characters are easiest to appreciate, there is no sign of spiral cleavage; instead, a simple pattern of meridional and latitudinal furrows (radial cleavage, Fig. 18–23) leads to the formation of a hollow blastula. After gastrulation the anus typically forms from the blastopore, or at the region where the blastopore closes, but the mouth is a new formation, from which feature is derived the term Deuterostomia (*deuteros*, second). After the proliferation of mesenchyme, the coelom forms from outgrowths of the enteron (Fig. 18–24), so that it is an enterocoel instead of a schizocoel. Finally, indirect development involves the formation of a dipleurula stage, from which several types of highly characteristic larvae differentiate. The organization of these larvae is fundamentally different from that of the trochophore (Fig. 19–12).

These, it must be emphasized, are the fundamental and typical characteristics of echinoderm development, from which individual species may secondarily diverge, for echinoderm embryos, like their larvae (Sec. 19-4), are highly plastic. This plasticity need not affect our general interpretation. Greater difficulties, however, admittedly arise when we consider the other deuterostomes, but these, too, can be attributed to extreme and divergent specialization, in this case affecting all the members of particular groups.

It is at least certain that none of them shows spiral cleavage, but the development of the mesoderm is greatly diversified. In the Cephalochordata the coelom is clearly enterocoelic. In the Hemichordata its mode of origin varies, while in the Urochordata it is reduced, and even its existence is a matter of opinion. In the vertebrates it arises as a schizocoel; in the highly specialized development of the chick, for example, the future mesoderm moves in from the primitive streak without having any direct relationship with the endoderm. Much importance has long been attached to the presence in certain hemichordates of the tornaria larva that closely resembles the dipleurula, from which it differs in having a terminal anus and an additional perianal band of cilia. The tornaria has been widely held to give good evidence for the close relationship of hemichordates and echinoderms. However, there are few certainties in these phylogenetic exercises, and it has also been suggested that the extreme plasticity of echinoderm larvae, and their consequent proneness to convergent and divergent evolution, leaves the status of the tornaria larva, if considered in isolation, open to some discussion.

Evidently, then the patterns of development within the Deuterostomia are far from homogeneous, and, exactly as with the arthropods, embryology here provides us with only a partial picture of phylogenetic relationships. Nevertheless, it can be argued that the concept of the Protostomia and the Deuterostomia is a sound one, for it takes its support from other considerations as well as from the purely developmental ones with which we are here concerned (p. 277). What is particularly convincing is that within each of the two assemblages we can distinguish characteristic and coherent evolutionary trends. These suggest that the members of each assemblage have inherited from a common ancestry a common stock of genetic potentialities. The trends associated in the Protostomia with the history of the coelom, of metamerism, and of the nervous system are one illustration of this. Another is seen in the history of the ciliary feeding mechanisms of the Deuterostomia.

19
Larval Forms

19–1 MODES OF LARVAL DEVELOPMENT

The study of the eggs and embryos produced by sexual reproduction reveals many different patterns of development and life cycle. One variable factor is the degree of protection given to these stages. Here useful distinctions can be made between three modes of development. One is oviparity, in which the parent releases the eggs at an early stage of their development, or even before cleavage has begun. The second is ovoviviparity, in which the eggs are retained within the parent, but are still surrounded by egg membranes for at least part of their development, nutrition being derived mainly or entirely from the yolk. The third mode of development is viviparity, in which the eggs are retained within the parent, with some loss of their membranes, nutriment being derived mainly from the parental tissues. Viviparity (the influence of which we have already encountered in our discussion of the development of onychophorans) can lead in its most complete form to the release of young stages as juveniles that closely resemble adults except in size and sexual immaturity. They will thus have benefited from a high degree of protection. However, variable degrees of protection can be secured even within the limits of oviparity. The eggs may be shed singly, for example, or grouped in gelatinous masses, or protective capsules. Those of aquatic forms may be left to float freely in the upper layers of the water or at the sea bottom, or they may be attached to the substratum. We might suppose that each mode of dealing with the eggs would have its own advantages and disadvantages, and would somehow be correlated with the particular way of life of the species concerned, but it is often difficult to find any clear-cut evidence of this.

    J. E. Smith has mentioned a number of examples illustrating this difficulty. Of three common shore-dwelling nemerteans, which might be expected to show similar life histories, *Amphiporus lactifloreus* hatches its young in a crawling stage, *Lineus ruber* has a short free-swimming phase, while *Cephalothrix* has larvae that may live in the plankton for two or more weeks. Shore-dwelling gastropods, to take only one other set of examples, show a variation of pattern that has no obvious explanation. *Littorina littorea* has larvae that live for four weeks or more in the plankton. *L. littoralis* and *Nucella* have a direct development in egg capsules, while *L. neglecta, L. patula* and *L. rudis*, from the upper half of the shore are viviparous. There is no indication that the maintenance of their populations is influenced by the differences in their modes of development. It might be thought that species lying highest on the shore would stand in most need of the protective advantage of viviparity, but there is no evidence even of this. *L. rudis*, which is covered by most tides, is viviparous, whereas *L. neritoides*, occupying the splash zone and only rarely wetted, relies on a planktonic larva for its survival, releasing its spawn into the sea from September to April.

The difference between direct and indirect development is a familiar feature of life histories. In the former the adult stage is attained by progressive growth and differentiation, whereas in the latter there is a larval stage, differing both in its structure and its habit from the adult, and acquiring adult form through a radical and sometimes sudden metamorphosis. Three main types of these larvae can be distinguished. There is firstly the lecithotrophic larva, so called because it feeds exclusively upon the yolk originally laid down in the egg, and does not, therefore, take any food from external sources. Nevertheless, such larvae may have a long period of life in the plankton, being carried along largely by currents which can distribute them over wide areas. With their independent food supply, they probably have a better chance of reaching metamorphosis than have planktotrophic larvae, but the rate at which they can be produced is restricted by the large size of the yolky eggs from which they develop. Thorson estimates that they form about 10% of the species of temperate and tropical seas that have pelagic larvae. They appear to be absent from arctic waters.

A far commoner type of larva is the planktotrophic larva, so called because the yolk reserves of the egg are soon exhausted, and the organism has to feed on plankton. Such larvae are usually well adapted for prolonged movement, swimming actively for 2–4 weeks in summer months and perhaps for as long as 3 months during the winter. They are cheap to produce, and are well suited for invasion, but against this must be set their almost total dependence upon planktonic food and their prolonged exposure to predators. They are estimated to occur in the life cycle of 55–65% of marine species in boreal seas, and 80–85% of those living in tropical waters. As with the lecithotrophic type, they are hardly represented in arctic species, nor in those that inhabit abyssal waters.

Finally, there are other planktotrophic larvae with only a short period of free-swimming life, amounting to a matter of hours or at most to a very few days. Their chief functions are to spread the larval stock and to find suitable substrata. These are estimated to be produced by about 5% of marine invertebrate species, from the arctic to the tropics.

We shall consider the significance of larval life in the next chapter, but we can distinguish three fundamentally important aspects of it now. There is firstly the need for the delicate young organism to grow in conditions which satisfy its special requirements, and which avoid unnecessary competition with the adult. Secondly, there is the need to provide for dispersal of the species, and thus to avoid overcrowding. Finally, there is the need to select a habitat that is suited to the requirements of the adult. Because of these varied functions larvae often become very highly specialized, and all of them undergo some degree of metamorphosis. It is impossible here to attempt a comprehensive review of the rich diversity of ways in which larvae fulfil the requirements of their species. It must be sufficient to illustrate the general principles by reference to some of the more familiar types.

19-2 SOME PROTOSTOME LARVAE

The trochophore larva (Fig. 19-1), formerly known as Loven's larva, after the Swedish priest and naturalist who discovered it in 1840, is exemplified in the larva of certain species of the archiannelid *Polygordius*. It is approximately biconical in shape, with a ciliated band extending around its equator. This band, which is one of the distinguishing features of the trochophore, is mainly responsible for locomotion. Two other ciliated bands, the metatroch and paratroch, may be present below it, the paratroch surrounding the anus. Another characteristic feature is the arrangement of the alimentary canal. This opens at the mouth, which lies below the equator on the future ventral surface; the anus is at the lower pole, or future hind end. The alimentary canal lies within the body cavity, which is a blastocoel. This cavity contains also a pair of mesoderm bands, developed from the teloblasts, and a pair of protonephridia, together with muscle fibres and mesenchyme cells which represent the ectomesoderm.

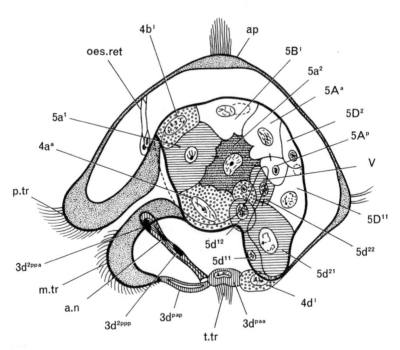

Fig. 19-1. Optical section of young trochophore of *Polygordius* after gastrulation is complete. *a.n*, protonephridium; *ap*, apical plate; *oes.ret*, oesophageal retractors; *p.tr*, prototroch; *v*, valve. For other lettering, see Sec. 18-5. Adapted from MacBride, 1914. *op. cit.*

The receptors and nervous system are of some complexity; a significant fact in view of what we shall later see of the complex behaviour of this and similar larvae in general. At the apical pole is an area of thickened ectoderm, the apical plate, which bears a tuft of supposedly sensory cilia. Beneath the plate is a ganglion, from which extend radial nerves. These are united by one or more delicate nerve rings, the chief one being associated with the prototroch. Statocysts and eye spots are often present towards the apical pole, while other parts of the ectoderm may well be sensory.

The need for metamorphosis is evident in the trochophore larvae of archiannelids and polychaetes, for these larvae, unlike the adult worms, are unsegmented organisms. In its simplest form the metamorphosis of the trochophore involves the backward growth of the lower pole, which thus becomes recognizable as the hind end. Immediately in front of the anus is a growth zone from which segments arise, as they continue to do in the adult. The lower half of the larva becomes incorporated into this segmented region, while the upper half becomes the prostomium. Such is the course of events in certain species of *Polygordius* and in many polychaetes (Fig. 19–2). Comparison with a range of polychaete species, however, immediately reveals the plasticity of larval forms and of their metamorphoses. Larvae, in fact, have their own history of specialization, which may run quite independently of that of the corresponding adults; this, as we shall see, is a principle of fundamental evolutionary importance.

There is much variation in the arrangement of the ciliary bands. Thus the prototroch may be lost, and the ciliated bands confined to the mesotroch (mesotrochous larva) or paratroch (telotrochous), or represented by a series of bands (polytrochous). Alternatively, the larva may be released in a stage of development less advanced than that of the trochophore; this happens in *Nereis diversicolor*, the larva of which lacks mouth and anus, and has been called a protrochophore larva. Conversely, the larva may persist for some time at a stage more advanced than that of the trochophore. An example is the nectochaetous larva, in which the posterior (lower) half of the larva is metamerically segmented, with parapodia and chaetae. Such a larva is well able to swim (as its name implies), and these stages of development are common in the plankton; in some circumstances, however, they creep over the substratum, a mode of life for which the trochophore is obviously unsuited.

The nectochaetous stage shows the larva developing appreciably towards the adult stage without necessarily losing its capacity for planktonic life and hence for dispersal of the species. Its increased size and weight must reduce the efficiency of its cilia as locomotor structures, but the muscular parapodia to some extent assume this function, aided by the often considerable length of the very slender chaetae, which help to support the organism in the water. A possible advantage of this combination of greater size and increased freedom of movement is suggested by Garstang's verses. Having written of the trochophore of *Phyllodoce*, he goes on:

> In this way fares *Phyllodoce*, but *Nereis* can beat her:
> She gives each egg some extra yolk to hatch as *Nectochaeta*.
> The simple stage with prototroch is by-passed in the eggs,
> And each when hatched has three good pairs of parapodial legs.

The circumstances of metamorphosis vary with the species, and, as we shall see, are related to the habits of the adult. Sometimes, as in *Polynoë*, the nectochaetous stage may continue its growth and differentiation until it has formed a small but well-

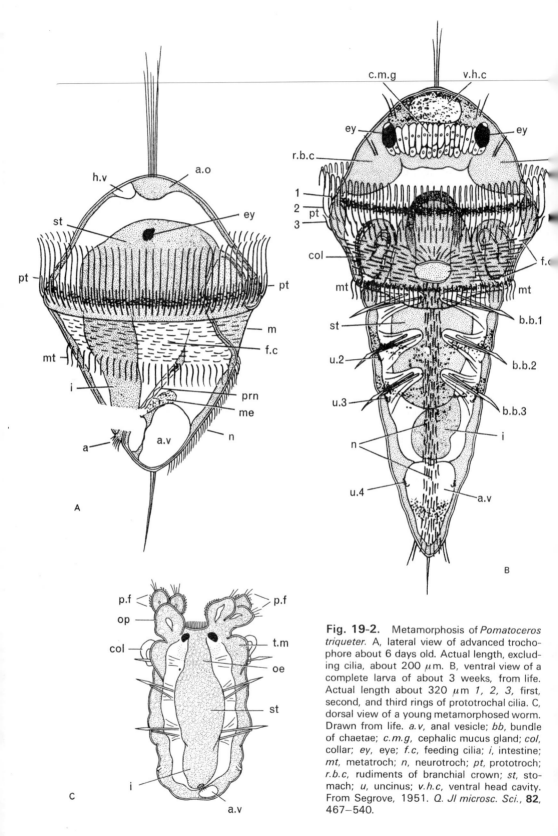

Fig. 19-2. Metamorphosis of *Pomatoceros triqueter*. A, lateral view of advanced trochophore about 6 days old. Actual length, excluding cilia, about 200 μm. B, ventral view of a complete larva of about 3 weeks, from life. Actual length about 320 μm *1, 2, 3*, first, second, and third rings of prototrochal cilia. C, dorsal view of a young metamorphosed worm. Drawn from life. *a.v*, anal vesicle; *bb*, bundle of chaetae; *c.m.g*, cephalic mucus gland; *col*, collar; *ey*, eye; *f.c*, feeding cilia; *i*, intestine; *mt*, metatroch; *n*, neurotroch; *pt*, prototroch; *r.b.c*, rudiments of branchial crown; *st*, stomach; *u*, uncinus; *v.h.c*, ventral head cavity. From Segrove, 1951. *Q. Jl microsc. Sci.*, **82**, 467–540.

developed juvenile worm, and only then does pelagic life cease. Others retain the form of trochophores externally, but develop the body of the future worm, enclosed within the larval body in an invaginated stage. This is seen in the larva of a North Sea species of *Polygordius*, which for some time carries a segmented body enclosed within a cavity that develops by invagination around the anus, but otherwise has something in common with the amniotic cavity of a vertebrate embryo. At a given moment the wall of the trochophore starts to shrink, so that the segmented trunk is evaginated posteriorly. The apical region of the larva is drawn onto the trunk anteriorly, while the equatorial protrochal region shrivels and is lost.

A much more specialized version of this device is seen in the mitraria larva (Fig. 19–3). In *Owenia fusiformis*, described in detail by Wilson, a trochophore stage is reached about 2 days after fertilization, but growth in size continues thereafter, associated with a great elaboration of the prototroch, which becomes folded into a large and sinuous band. This is presumably a locomotor adaptation, correlated with the continued growth and differentiation of the larva. Large provisional larval chaetae, developed from the lower end of the larva, presumably aid in its support, and may also serve to provide some protection from predators. Feeding is carried out, according to Wilson, by the cilia of the metatroch; these sweep food

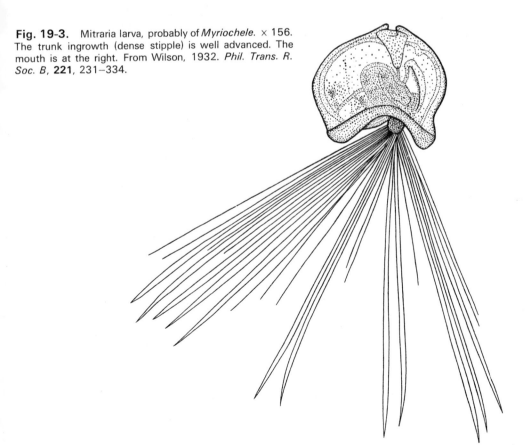

Fig. 19-3. Mitraria larva, probably of *Myriochele*. × 156. The trunk ingrowth (dense stipple) is well advanced. The mouth is at the right. From Wilson, 1932. *Phil. Trans. R. Soc. B*, **221**, 231–334.

particles into the mouth, a procedure that involves them passing around several folds of the prototroch. The rudiment of the future trunk forms in this larva as an invagination between the anus and the mouth, its growth involving a process of invagination and folding (Fig. 19–4). The most anterior segments, in Wilson's words, are 'turned inside out and drawn back over the succeeding segments much as the top of a stocking can be turned inside out and drawn back over the foot.'

The metamorphosis of this larva is a sudden (cataclysmic) process, which occurs about 4 weeks after fertilization, and which is thought to be initiated by contact with fine sand similar to that in which the tube-dwelling adults live. We shall discuss this aspect of habitat selection further below. The trunk is everted into its correct position, the gut of the larva is pulled into it, the remaining larval tissues, including the ciliated bands and nephridia, disintegrate, while the larval chaetae drop away. The whole process is extraordinarily rapid. Most of the main changes are completed within 30 sec of their initiation, and little material is wasted in the process, for the young worm swallows the discarded larval structures at the same time that it begins to secrete its protective tube. The swallowing is completed in about 15 min, the tissues being digested intracellularly during the next 2 days. The whole sequence of events in this larval history, as in that of the *Polygordius* species mentioned above, is a striking solution of the problem of integrating development and dispersal. No less significant, however, is its demonstration of the way in which the specializations of larval life can diverge from those of the adult. The importance of this will be appreciated later, when we consider the larval organization of the higher Protostomia.

Spiral cleavage occurs only in the polyclads among the platyhelminths, which suggests that the course of their development must be more primitive than that of other members of the phylum. It is significant from this point of view that in polyclads and in the Acoela, unlike the other turbellarians, yolk glands are absent, so that yolk is deposited within the ova, in the primitive fashion, instead of being laid down in separate yolk cells. Polyclads are classified into two suborders, the Acotylea and the Cotylea, which are respectively without and with an adhesive disc behind the female gonopore, and which differ in the position of the tentacles. In most of the Acotylea the development is direct, and there emerges from the egg capsule a small flatworm, but in some polyclads, and typically in the Cotylea, there is a larva known as Müller's larva. This has an apical ganglion, overlain by a tuft of sensory cilia, and a typically platyhelminth alimentary tract, with a mouth and a blind intestine. The feature that chiefly relates it to the trochophore is the locomotor mechanism, which consists of a ciliated band lying above the mouth, and corresponding in its position to the prototroch. The characteristic feature of the larva is that this band is situated on the edges of eight posteriorly directed lobes. These disappear after a few days, when a flattening of the body converts the larva into a young worm. The metamorphosis is thus a simple one, and it is easy to see that the larva is essentially an early developmental stage of the flatworm, adapted for a pelagic life by its specialized ciliation.

Development in the nemertines may be either direct or indirect, the latter mode being particularly characteristic of the heteronemertines. The typical larva of these forms is the pilidium larva (Fig. 19–5), so called because of its helmet-like shape (*pilidion*, a small cap). There is an apical plate and tuft (but no other sign of nervous

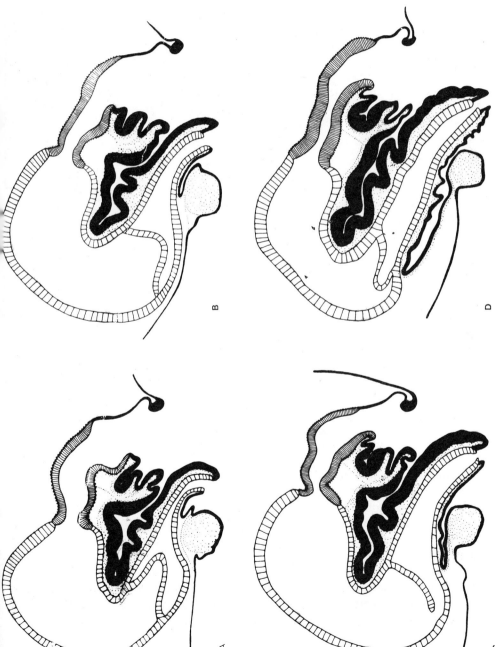

Fig. 19-4. Diagrams of median sagittal sections of the lower part of the mitraria larva of *Owenia fusiformis*, to illustrate the growth of the invaginating ectoderm of the worm rudiment. The larval mouth is above and the anus below. Ectoderm, black; mesoderm, stippled; larval oesophagus, closely shaded; larval stomach and intestine, lightly shaded. A, from a larva 22 days old, before the mesoderm has surrounded the intestine. B, the same age, but more advanced, the mesoderm having met around the intestine. C, the same age, but still more advanced. D, from a larva 27 days old, and ready to metamorphose. From Wilson, 1932. *op. cit.*

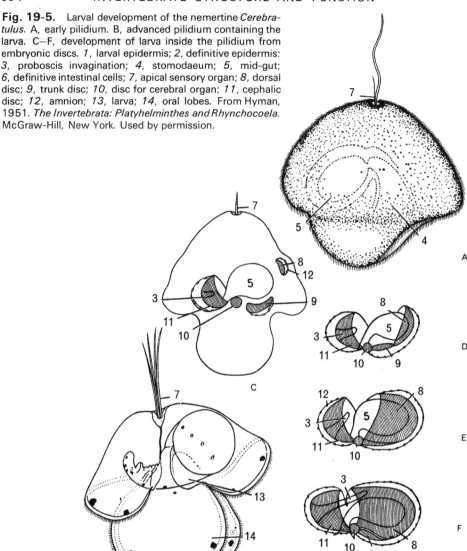

Fig. 19-5. Larval development of the nemertine *Cerebratulus*. A, early pilidium. B, advanced pilidium containing the larva. C–F, development of larva inside the pilidium from embryonic discs. *1*, larval epidermis; *2*, definitive epidermis: *3*, proboscis invagination; *4*, stomodaeum; *5*, mid-gut; *6*, definitive intestinal cells; *7*, apical sensory organ; *8*, dorsal disc; *9*, trunk disc; *10*, disc for cerebral organ; *11*, cephalic disc; *12*, amnion; *13*, larva; *14*, oral lobes. From Hyman, 1951. *The Invertebrata: Platyhelminthes and Rhynchocoela*. McGraw-Hill, New York. Used by permission.

structures), and a mouth that leads into a blind alimentary tract, the organization in this latter respect being platyhelminth-like. As with Müller's larva, locomotion depends upon a lobed ciliated girdle, the cilia in this instance being carried on the edges of a pair of oral lobes that grow down, one on either side of the mouth, giving the larva its characteristic shape. A variant of the pilidium larva is found in the development of *Lineus*. This larva, known as Desor's larva, lacks the apical plate and tuft, and the oral lobes and specialized ciliation, and it remains within the egg membranes. It has been supposed that this is a secondary modification of the more typical indirect development, enabling the whole of the life cycle of this littoral form to be confined to the limits of the shore.

It might be expected that the pilidium larva would undergo a simple and straightforward metamorphosis into the adult form, much as does Müller's larva, but this is not so; a fact that illustrates the plasticity of larval life histories and the consequent difficulty in generalizing from a few examples. Both in the pilidium larva and in Desor's larva metamorphosis is preceded by the development of seven or eight ectodermal invaginations of the larval ectoderm. These separate off to produce a series of sacs with thin outer walls and thick inner ones formed of columnar epithelium. The sacs eventually join together, and the continuous inner wall gives rise to the epidermis of the adult worm. The larval alimentary tract and mesoderm, enclosed by this wall, undergo a good deal of reorganization and differentiation during the metamorphosis, and it is at this stage that the anus develops. At the end of these changes the larval ectoderm and the outer wall of the fused invaginations are shed, together with the apical plate, and the young nemertine emerges from its larval container. The metamorphosis is thus considerably more drastic than might have been expected at this level of organization, and it is not easy to see why this should be so. Functionally, however, as with the mitraria larva, the process provides for laying the foundations of adult structure without interfering with the distributive activity of the larva during its pelagic life. It may well be that these dual requirements, always fundamental to larval life, are here provided for more efficiently than in the simpler developmental history of the polyclads.

Another method of securing this provision is found in the molluscs. As is to be expected in such a large group, there is much variation in the life history, correlated in part with the varied habitats of the adults. Close relationship with the annelids is seen, however, in the occurrence of a trochophore stage in many forms, particularly in the archaeogastropods (e.g. *Patella*), and in the Bivalvia; it is found also in the development of chitons (Amphineura) and *Dentalium* (Scaphopoda). This larva is rapidly transformed into a more complex stage, the veliger larva (Fig. 19-6), which is particularly characteristic of the gastropods and bivalves. In this the prototroch is drawn out into a pair of ciliated lobes, an arrangement that considerably increases the support given to the larva and makes for a more vigorous and controlled locomotion. This development is explained by the advanced stage of differentiation reached by the veliger. It has something of the form of a mollusc, with a shell, a mantle cavity, and the beginnings of a foot, the latter sometimes bearing an oper-

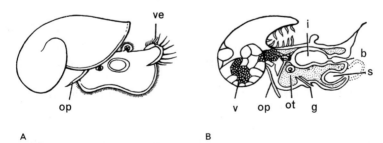

Fig. 19-6. Veliger larva of *Entoconcha*. A, external aspect. B, optical longitudinal section (after Baur). *b*, mouth; *g*, pedal gland; *i*, gut; *op*, operculum; *ot*, otocyst; *s*, sac-like invagination; *v*, residual yolk; *ve*, velum. From Caullery, 1952. *Parasitism and Symbiosis*. Sidgwick and Jackson, London.

culum that can close the opening of the shell. Clearly the improved ciliation is necessary for the support and movement of this heavier and more complex body.

In the early larval life of gastropods there is another complication that is more difficult to explain. This is the torsion, which twists the viscera through 180° relative to the rest of the body, brings the originally posterior mantle cavity to the anterior end, and leaves the originally left side of the pallial complex on the right. We have already seen that this transformation, which is effected very quickly—in some species in a few minutes—is brought about through the contraction of an asymmetrically arranged retractor muscle. This runs from the right side of the shell to be inserted on the left side of the head and foot (Fig. 11–10, p. 311).

We have already discussed the torsion of gastropods from the standpoint of adult anatomy, and have seen that its advantages to the adult are by no means clear. Some have argued that it carries positive disadvantages in that it leads to the contamination of the mantle cavity, and they support this view by emphasizing that these effects are eliminated independently in the opisthobranchs and pulmonates. Against this it can be argued that there is some advantage in the anterior position of the gills and osphradium, since it enables the latter to sample the water into which the animal is moving. However this may be, a widely favoured theory of the origin of torsion is one originally put forward by Garstang. He suggested that it arose as a larval adaptation that was carried over into the adult stage, the adults being then compelled, so to say, to make the best of the resulting situation. Garstang's argument was that prior to torsion the protective value of the larval shell was limited because the delicate head and velum could not be withdrawn into the posterior mantle cavity until the foot was withdrawn; in effect, it was impossible to give them adequate protection. Torsion, which could conceivably have resulted from a genetic mutation promoting asymmetrical muscular development, facilitates the withdrawal of the head and velum, because the mantle cavity is now in a more favourable position. This would have had sufficient survival value to ensure the persistence of this new feature, and thereby to establish the fundamental distinguishing character of the Class Gastropoda. Again in Garstang's words,

> Predaceous foes, still drifting by in numbers unabated,
> Were baffled now by tactics which their dining plans frustrated.
> Their prey upon alarm collapsed, but promptly turned about,
> With tender morsel safe within and the horny foot without!

It can be argued against this explanation that the advantage gained in protection is of little value against larger predators, such as herring, which indulge in mass consumption of these larvae. Yet Garstang's theory remains a favoured one. Indeed, it has been so widely accepted by British zoologists that their students must often feel that it is the only possible explanation of the origin of gastropod torsion. This, however, is not so. Ghiselin, for example, while accepting that torsion confers a functional advantage upon the larva, considers it to be too complex a transformation to have been evolved in a single step. Basing his argument on an earlier view of Lang, he suggests that torsion arose after exogastric coiling had already been established in the larval stage as a by-product of ontogenetic growth processes. This coiling need not have interfered with larval locomotion, but the movements of the adults (and even of the larvae when they began to settle) might well have been handicapped by the difficulty of balancing the shell. Torsion could then have evolved as an adaptation to overcome this difficulty.

Obviously the problem of the origin of this torsion, like various other phylogenetic problems that we consider elsewhere, is likely to remain for ever unresolved. It is just for this reason, however, that it is important to scrutinize the theoretically possible alternatives, and to evaluate them from time to time in the light of developing evolutionary theory. In this context, both of the views that we have outlined stand in need of some form of experimental verification, or at least of a more exhaustive analysis of the developmental factors involved.

Garstang's theory, however, has a significance ranging far beyond the field of molluscan evolution, for it recognizes the immense potential importance of the capacity of larvae for evolving, independently of their adults, in adaptation to their own particular modes of life. Other examples to be mentioned later will show this being carried to a point at which a profound and drastic metamorphosis is needed to bridge the gap between larva and adult—something far in excess of that seen in the mitraria larva. Even more important, we shall be able to consider further the possibility of some of these divergent larval specializations being retained in the adult phase and thereby establishing entirely new types of adult form, much as Garstang's theory visualizes the origin of gastropod form as a result of the retention in the adult of the larval torsion.

19-3 CRUSTACEAN LARVAE

We have seen that the inclusion of arthropods within the Protostomia rests primarily upon inferences based upon adult anatomy, with some support from the early development of the Onychophora (Sec. 18–5). Arthropodan larvae are of little help in this regard, but comparisons of their different forms are instructive in other ways. The Crustacea have a wide variety of larvae, but these are entirely crustacean in form, and show no relationship at all with the ciliated larvae of annelids. In fact, the almost complete loss of cilia by the phylum extends to their larval stages; these, like the adults, rely upon their limbs for locomotion. Whether there ever was a trochophore stage in the remote ancestry of the marine arthropods is unknown, for, if there was, it has certainly been lost without trace. The larvae now hatch in a relatively advanced stage of development, with at least a few functional appendages present and with the accompanying metameric segmentation. In this there is no new principle; we have seen in the nectochaetous larva an example of a metamerically segmented stage that may be achieved at hatching, and crustacean larvae merely illustrate a further extension of this acceleration of development.

The simplest crustacean larva is the nauplius (Fig. 19–7), a three-segmented organism with three pairs of limbs; the first pair are uniramous antennules, the other two being biramous and representing respectively the antennae and mandibles of the adult. They differ in form, however, from the adult appendages, for they are specialized to meet the locomotor needs of the larva, and because of this it is as meaningless to seek in the nauplius for the characteristics of the ancestral crustacean as it would be to interpret the nectochaetous larva as an ancestral polychaete. The phylogenetic significance of larval forms is as difficult to determine as is that of the different modes of cleavage and gastrulation. We shall refer to this prob-

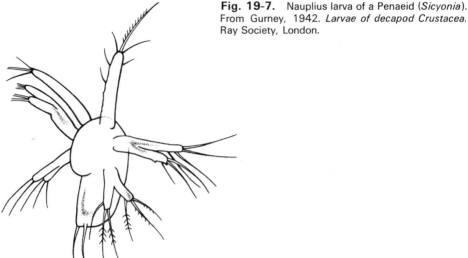

Fig. 19-7. Nauplius larva of a Penaeid (*Sicyonia*). From Gurney, 1942. *Larvae of decapod Crustacea.* Ray Society, London.

lem again later, but it is well to bear constantly in mind that larvae, too, are developmental stages, specialized for functions that are peculiar to them, and that may be widely different from the activities of the adults.

In the branchiopods the further development of the nauplius involves continuous growth, interrupted by periodical moults, at which additional somites and limbs are added; the adult form is thus achieved without drastic metamorphosis. We may reasonably assume that this simple type of life history was the primitive mode of development of Crustacea. In the other groups with free larvae, however, including the Copepoda, Ostracoda, Cirripedia, and Decapoda, we find one or more stages which include a sharply defined metamorphosis. The most obvious reason for this is when larva and adult have entirely different modes of life, as, to take extreme examples, in the barnacles (Cirripedia) and in the crabs (Brachyura). Yet this cannot be the only explanation, for metamorphosis may occur when the modes of life are not greatly dissimilar, and there may even be a sharp transition from one larval stage to another. Such metamorphoses probably reflect marked changes in level of specialization of the stages concerned, or in the pattern of organization of particular functions; such is to be seen, for example, in the transition, during the development of the Penaeidae (see below), from the antennal propulsion of the nauplius larva to the thoracic propulsion of the zoea larva.

The Copepoda typically begin their free-swimming life with a nauplius larva that subsequently undergoes a clearly marked yet simple metamorphosis. There are six nauplius stages, the last one still having three pairs of limbs, but possessing rudiments of five further pairs. The sixth moult gives rise to the first copepodid stage, with an essentially adult form; two pairs of swimming legs are present, and the future regions of the adult body are defined. Subsequent development may involve five copepodid stages, during which the assumption of the final adult form is progressive and direct, with no further metamorphosis, at least as far as free-living species are concerned. In the highly modified parasitic forms, however, there may be only one copepodid stage, and a considerable measure of metamorphic change.

How the adult mode of life influences larval development is well seen in the Cirripedia, sessile forms that are so profoundly modified that on first inspection they are hardly recognizable as crustaceans at all. They begin their free-swimming phase as a nauplius, characterized in this group by a distinctive pair of anterior lateral horns, and by a posterior forked spine. This larva undergoes a series of moults, and then passes at a single moult into the cypris stage, so called because the body and appendages are enclosed within a bivalved shell in a manner that recalls the appearance of an ostracod. Eventually this larva settles, attaching itself to the substratum by its modified antennules and by a secretion of cement glands that are situated at the bases of these appendages. The shell of the cypris is now discarded, and its place is taken by the valved shell of the adult. The behaviour of the cypris larva at settling has been thoroughly studied, for barnacles play a major part in the fouling of ships' bottoms, so that the assessment of the value of anti-fouling paints demands an understanding of the biology of the larvae. Knight-Jones and Crisp have described how, after alighting, the cyprids 'walk about, pulling themselves forward by each antennule alternately, and changing direction only infrequently.... On an unsuitable surface ... they may swim off almost at once'. We shall be examining later the significance of this and similar behaviour patterns in invertebrate larvae.

In the Decapoda the larvae have become more specialized and the life histories correspondingly more complex. One result of this is that hatching is usually at a later stage, the nauplius being no longer recognizable. Yet we should expect that it must at one time have been a normal stage of the life history of the ancestral malacostracans; that this was so is certainly suggested by the life cycle of the penaeid prawns. These are members of the Natantia, but are distinct from the true prawns and shrimps. *Penaeus*, the Mediterranean prawn, hatches as a nauplius with the typical three pairs of limbs. Thereafter, by growth and moulting, it develops rapidly without any abrupt transition through a series of nauplius stages into a protozoea (Fig. 19–8A). This still lacks compound eyes, but the thorax is fully segmented, a small carapace is present (developed as a fold from the maxillary segment), and the telson is forked. The appendages comprise antennules, antennae, and mouth parts, and also the first and second maxillipedes, which are functional biramous limbs; rudiments of the third maxillipedes may also be present. There are altogether three protozoea stages, which may be regarded as a transitional phase in the life history; the third stage has uropods, together with rudiments of all of the legs.

Gurney has emphasized that the organization of the main developmental stages of the higher Crustacea is primarily determined by their mode of locomotion. The nauplius depends upon antennal propulsion, and so, in general, does the protozoea, as is shown by the large natatory exopodites of its antennae. The appearance of the maxillipedes foreshadows the next stage, which depends upon thoracic propulsion. This is initiated by the transformation of the protozoea into the zoea larva (Fig. 19–8B), which then passes through several stages. Initially it is characterized by the possession of stalked eyes and at least three pairs of biramous thoracic limbs which function as swimming appendages. Further, the carapace, which remains free of the tergites in the protozoea, now begins to fuse with them. The remainder of the thoracic limbs are at first only rudiments; the abdominal segments are defined, but their appendages are not yet developed.

In the later zoea stages the remainder of the thoracic limbs develop; they are

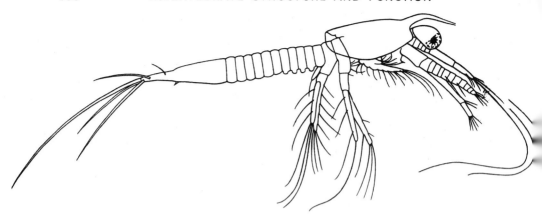

Fig. 19-8A. Protozoea 2 of a Penaeid (*Gennadas*). From Gurney, 1942. *op. cit.*

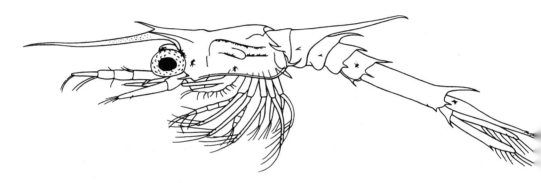

Fig. 19-8B. Zoea of *Gennadas*. From Gurney, 1942. *op. cit.*

all biramous, with large natatory exopodites, so that such larvae are sometimes referred to as the schizopod stage. The abdominal pleopods also appear, but are not yet functional. Finally, with the moult to the post-larval stage, these latter appendages develop setae and become functional swimmerets, while the thoracic limbs lose their exopodites. A curious feature of this stage (which is not quite fully adult) is the temporary loss or reduction of the fourth and fifth thoracic limbs; these reappear later during the course of several moults.

This penaeid life cycle, varying considerably in detail in different species, indicates the primitive form of decapod development. The general trend in other decapods has been towards a shortening of the life cycle, a trend that is probably correlated with the widespread habit of protecting the eggs by carrying them on the body of the parent. Thus the true crabs (Brachyura) hatch as zoea larvae (Fig. 19–9). These have a forked telson, a curved abdomen, and a helmet-shaped carapace which bears two long spines: a median rostral one, extending forwards, and a median dorsal one. Later the rudiments of the swimmerets and of the remaining thoracic limbs appear, and the larva becomes transformed into another clearly defined stage, the megalopa. This is an essentially crab-like form in which the spines become

Fig. 19-9. Larval stages of the common shore crab, *Carcinus maenas*. A, young zoea, shortly after hatching; B, megalopa stage; C, young crab. A × 20, B and C × 10. From Calman, 1911. *The Life of Crustacea*. Methuen, London.

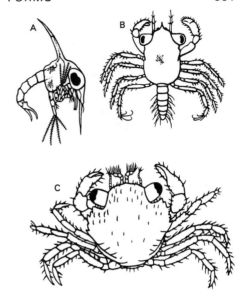

shorter and all of the appendages are present. At first the abdomen is still extended; the megalopa swims at the surface, using the swimmerets, but later it moults on the sea bottom to become a fully formed crab, with its abdomen folded and the abdominal limbs no longer locomotory. The megalopa, often referred to as a larva, can alternatively be regarded as a first post-larval stage, which marks, with its abdominal propulsion, the transition to the specialized form of the adult crab. The Anomura are significant from this point of view. These animals, which include *Galathea*, *Porcellana*, and the hermit crabs, have partly evolved a crab-like organization. They hatch as zoea larvae, but do not pass through a schizopod stage. Hermit crabs have a glaucothoë stage, with symmetrical abdomen and swimming pleopods; this corresponds to a megalopa. The zoea of the Anomura is distinguishable by its long anterior spine, and the two spines that extend backwards from the posterior angles of the carapace; these spines, particularly the anterior one, are enormously and characteristically elongated in the larva of *Porcellana*.

The larvae of lobsters (Astacura) illustrate another abbreviation of the life cycle, for they hatch in the schizopod stage. All of the thoracic appendages are present, with well-developed exopodites, and with the first three pairs of legs chelate. Abdominal appendages are absent or rudimentary in stage 1 larvae, but swimmerets and uropods appear at subsequent moults. During these stages the endopodites of the thoracic appendages lengthen markedly, but the exopodites show only a small increase in size, so that the proportions of the two components change (Fig. 19–10). At the moult from stage 3 to stage 4, after about 32 days of larval life, an essentially adult form is assumed; the functional exopodites are lost from the chelipeds and pereiopods, and an adult behaviour pattern is assumed, with swimming replaced by walking, climbing, and burrowing.

Analysis of the swimming mechanism of the larva of the lobster *Homarus* amply justifies Gurney's argument that larval form is determined by mode of locomotion. The endopodite makes no contribution to this, the significance of the schizopod

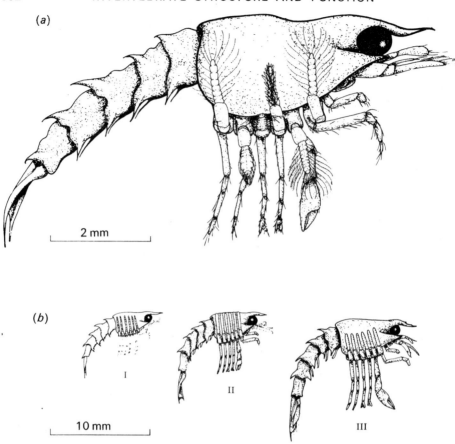

Fig. 19-10. (a), Stage I larva of the lobster *Homarus gammarus* in normal swimming posture, with the exopodites shown in their correct metachronal phase relations to one another. For further explanation, see text. (b), Scale representations of larvae in stages I, II, and III to show the increase in body size during development, and the changes in the relative dimensions of the exopodite and endopodite branches of the thoracic appendages. Note the appearance of swimmeret buds in stage II, and the uropods in stage III. From Neil, MacMillan, Robertson and Laverack, 1976. *Phil Trans. R. Soc. B.*, **274**, 53–68.

limb being that the exopodite is the active component. Its power stroke, which is usually directed ventro-laterally, is effected with the flagellum straight and the field of setae fully expanded, the limbs beating in metachronal rhythm so that they do not interfere with each other's action (Fig. 19–10). On the return stroke the surface area of the exopodite is reduced by over 90% through the setae being drawn inwards, while drag is further reduced by active bending of the flagellum towards the body. An individual which at the end of the four larval stages weighs perhaps only 60–70 mg may in due course achieve a weight of at least 1200 g. Thus larval life takes up only a minuscule part of the total life history, yet it represents a heavy investment in muscular machinery and nervous pathways, including a specialized larval musculature which degenerates at metamorphosis. This is surely an impressive tribute to the importance of the distributive function of larval stages.

A remarkable variant of the abbreviated type of life cycle of the Astacura is that of the spiny lobster, *Palinurus*, and its allies. The larva hatches as the phyllosoma larva (Fig. 19–11), which is so modified that it is immediately distinguishable from any other larval type. In some respects it is a modified schizopod stage, with two pairs of large maxillipedes (the first pair being rudimentary) and three pairs of legs, the fourth and fifth pairs of legs being only small buds. The most obvious external characteristics, however, are the flattening of its thorax, and its glassy transparency; features that give it, in the words of Hardy, the appearance of being fashioned from a coverslip. The phyllosoma larva is an extreme example of planktonic adaptation; it is not easily comparable with any other crustacean larva, and illustrates very well the possibilities of independent larval evolution. For all its schizopod appearance, it is protozoean-like in that the carapace is mostly free from the thorax; yet the development of three pairs of legs in its first stage is a degree of precocity unexpected in a protozoea.

Finally, the true shrimps and prawns (e.g. *Crangon, Palaemon, Hippolyte*) hatch as zoea larvae with three pairs of functional maxillipedes; the remaining legs appear in succession or, sometimes, out of order, with the fifth forming in advance of the third and fourth. These larvae have their own distinguishing characteristics; they lack both the conspicuous median spine of the brachyuran zoea and the backwards-directed ones of the anomuran larva.

19–4 SOME DEUTEROSTOME LARVAE: THE ORIGIN OF VERTEBRATES

The fundamental characteristics of deuterostome larvae are to be seen in the array of larval forms produced by the Phylum Echinodermata (Fig. 19–12). The range of modification found in these larvae, their drastic metamorphoses, and the possible bearing of their own organization upon the problem of the origin of vertebrates, have long made them a focus of interest. We have already seen that an early feature of echinoderm development, following gastrulation and the appearance of mesenchyme and an enterocoel, is the formation of the dipleurula stage (Figs. 18–24, 19–12). At this stage the mouth is ventral; the anus, derived from the blastopore and

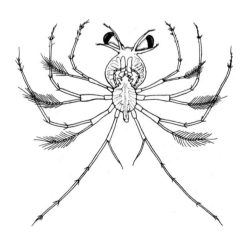

Fig. 19-11. Phyllosoma larva of the spiny lobster, *Palinurus vulgaris*. From Calman, 1911. *op. cit.*

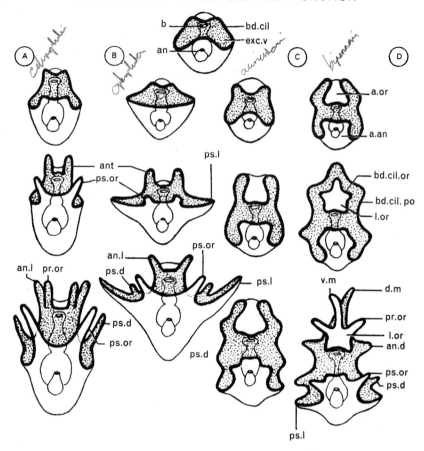

Fig. 19-12. Diagrams illustrating the possible evolution of larvae derived from the dipleurula. Series A represents the transformation of the dipleurula into the echinopluteus, B into the ophiopluteus, C into the auricularia, and D into the bipinnaria. The ciliated band is shown as a thick line. *an*, anus; *a.or*, *a.an*, preoral and anal areas; *b*, mouth; *bd.cil.or*, *bd.cil.pr*, preoral and circumoral (postoral) ciliated bands; *exc.v*, ventral depression. Larval arms: *an.d*, antero-dorsal; *an.l* (and *ant*), antero-lateral; *d.m*, dorsomedian; *pr.or*, preoral; *ps.d*, postero-dorsal; *ps.l*, postero-lateral; *ps.or*, postero-oral; *v.m*, ventro-median. From Grassé, 1948. *Traité de Zoologie*, vol. 11. Masson, Paris.

originally posterior (Fig. 18–24), also becomes ventral, while the digestive canal differentiates into an oesophagus, stomach, and intestine. The general surface ciliation of the embryo becomes reduced to a ciliary band which runs round the margin of a saddle-shaped depression that includes the mouth but excludes the anus. The region of the body anterior to the mouth is the pre-oral lobe, at the apex of which an apical sensory plate and tuft of cilia may be formed. Externally, the dipleurula stage is distinguished from the trochophore larva by the circumoral course of the ciliary band. Internally, it is distinguishable by the paired coelomic enterocoelic sacs and by the absence of protonephridia. Other developmental characteristics of the echinoderms have already been considered (Sec. 18–5).

In the Echinoidea and the Ophiuroidea the dipleurula stage develops into the pluteus larva, the organization and further development of which provides for improvement of locomotion and of suspension in the water (Fig. 19-12). This is a result of the outgrowth of long paired arms which carry the ciliated band with them, and which are supported by a skeletal system of slender calcareous rods. Further improvement results, in some genera, from the specialization of parts of the cilated band between the bases of certain of the arms; these parts, which have specially well-developed cilia, separate from the main band to form structures called epaulettes. The echinoids and ophiuroids differ in the mode of outgrowth of the arms, the resultant larvae being distinguished in consequence as the echinopluteus and the ophiopluteus. The four principal pairs of arms of the echinopluteus (antero-lateral, pre-oral, post-oral, and postero-dorsal) do not correspond exactly with those of the ophiopluteus (antero-lateral, postero-lateral, post-oral, and postero-dorsal), the pre-oral arms of the echinopluteus being absent from the ophiopluteus. The larvae differ, too, in general appearance, the arms being more widely opened out in the ophiopluteus.

These larvae may live for weeks or months in the plankton before undergoing their metamorphosis. This is necessarily a drastic process, so different are the larvae from the adults, yet the metamorphosis of the echinopluteus may be completed within an hour. The speed is made possible by what may be termed careful preliminary preparation, the effect of which is both to prolong the dispersal phase and at the time time to minimise the highly vulnerable phase of transformation. During the metamorphosis (Fig. 19-13) a new mouth appears on the left side of the larva, while the anus shifts to the right, the definitive gut forming a loop between them. There is thus virtually a 90° displacement of axis, clearly shown in the orientation of the body of the future adult, which appears within the larva as the echinus rudiment. Meanwhile, the continued development of the coelomic sacs gives rise, among other things, to the development of the left and right somatocoels (later forming respectively the adoral and aboral body cavities) and of the water-vascular system, which is derived from the left hydrocoel, the right one disappearing. The hydrocoel becomes closely apposed to an ectodermal invagination; this closes off, clearly corresponding at this stage with the vestibule of crinoids (see later), although it is here on the left side of the larva. Soon the hydrocoel begins to grow around the gut to form the water-vascular ring, and outgrowths from it establish its pentamerous symmetry.

The pre-oral lobe region, which corresponds to the stalk of the crinoid larva, makes no contribution to the adult echinoids. At the culmination of metamorphosis, which occurs without any fixation, the vestibule and its associated structures (termed the echinus rudiment) form the oral region of the sea-urchin, which thus represents the left side of the larva, while the right side of the larva gives rise to the aboral region. The larval arms are meanwhile absorbed and their skeletal supports discarded. With the hydrocoel and primary podia already established, the young urchin, probably less than 1 mm in diameter, is immediately capable of independent movement.

The metamorphosis of the ophiopluteus is similar in general principle to that of the echinopluteus, the radially symmetrical organization of the adult being laid down within the larva until this sinks to the bottom under the influence of the in-

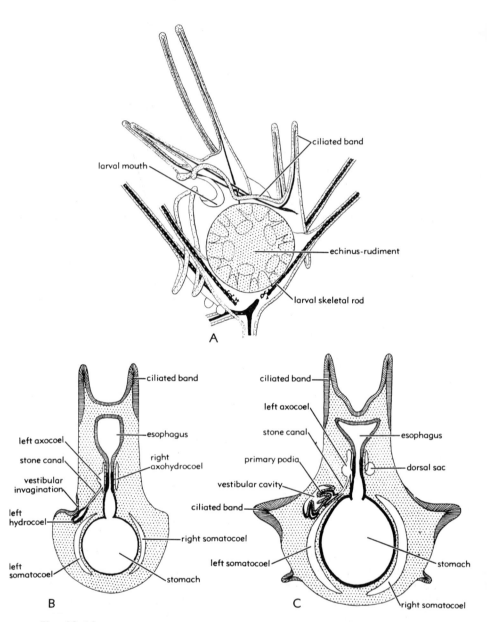

Fig. 19-13. Metamorphosis of echinoids. A, fully formed echinopluteus of *Arbacia* (× 100), seen from left side, showing the echinus rudiment (stippled); B, C, diagrammatic frontal section through echinopluteus showing development of the echinus rudiment on the left side of the larva. From Ubaghs, 1967. In *Treatise on Invertebrate Palaeontology* (R. C. Moore, ed.), Part S, Echinodermata 1. University of Kansas and the Geological Society of America, Lawrence, Kansas.

creasing weight of the developing skeletal material. The arms and their skeleton, together with the anterior end of the larva, are either absorbed or discarded. Among the many differences in detail we need only mention here that the region corresponding to the echinus rudiment develops on the ventral surface of the larva instead of on its left side. Moreover, it does so without the appearance of a vestibule, although the stomodaeum has been interpreted as a rudiment of this.

The characteristic larva of the Asteroidea is the bipinnaria. The ciliated band of the dipleurula stage extends into anterior and posterior folds and then subdivides to form two ventral loops; a smaller pre-oral one and a larger circumoral one. Of these, the pre-oral loop separates off from the main band to give rise to a separate loop which thus encloses a pre-oral lobe. As with the pluteus larva, the locomotion and support of the growing organism are aided by further extension of the ciliated band into arms, but these are not so long as those of the pluteus; moreover, they lack skeletal supports, and so are much more flexible in use.

The metamorphosis of the asteroid larva (Fig. 19–14) is often, but not always, preceded by temporary fixation. In preparation for this there develop, after some weeks of planktonic life, three brachiolar arms, lying anterior to the pre-oral loop. These arms differ from the others in possessing extensions of the coelom, and in bearing adhesive cells at their tips, while an adhesive glandular area or sucker develops between their bases. The larva is now known as a brachiolaria. It attaches to the substratum, first by the brachiolar arms, and then by the sucker, and begins metamorphosis. During this process the anterior region is absorbed, as in the other

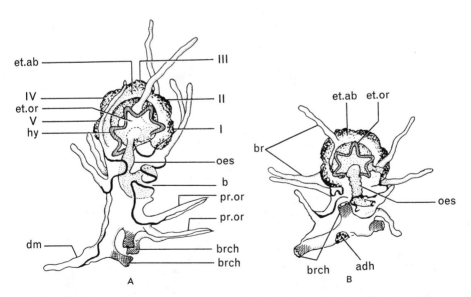

Fig. 19-14. Two stages in the metamorphosis of the brachiolaria of *Asterias pallida*. The attached larva is shown from the left side, so that the oral rudiment of the starfish (*et.or*) is seen in full view. The thick line represents the ciliated bands. I–V, five brachial expansions of the aboral rudiment (*et.ab*); *adh*, adhesive disc; *b*, larval mouth; *br*, larval arms degenerating; *brch*, brachiolar arms; *dm* dorso-median arms; *oes* oesophagus; *pr.or*, preoral arms. From Grassé, 1948. *op. cit.*

608   INVERTEBRATE STRUCTURE AND FUNCTION

examples described above, while the future starfish develops in the posterior region of the brachiolaria. Its formation follows the same general course as that of the young sea-urchin, the left side of the larva becoming the oral part of the disc and the right side the aboral part; there is, however, no vestibule. The metamorphosis may be completed within one day from fixation, the starfish being able by then to pull itself free from the remains of the larva, which Hardy has aptly likened to its perambulator.

A larva closely resembling the early bipinnaria is found in many of the Holothuria, which hatch at around the third day of development as an auricularia larva (*auricula*, a little ear). This has a circumoral ciliated band which is thrown into anterior and posterior folds, but its subsequent history differentiates it from that of the bipinnaria, for it breaks up into sections that become rearranged and extended to form from three to five transverse ciliated bands. This is the doliolaria stage, so called because of its obvious resemblance to the characteristic larva of crinoids. Owing to the elongated and cylindrical form of the holothurian body, the adult stage is attained without such a drastic metamorphosis as is seen in the Asterozoa and echinoids. Such metamorphosis as does occur recalls to some extent the events of crinoid development; the stomodaeal invagination forms a vestibule which, after almost completely closing over, rotates to establish the oral structures at the anterior end.

The development of crinoids is little known. On present information, there is no bilaterally symmetrical larva, development, as exemplified in *Antedon*, leading to the hatching of a vitellaria (doliolaria), with four or five ciliated bands (Fig. 19–15). At

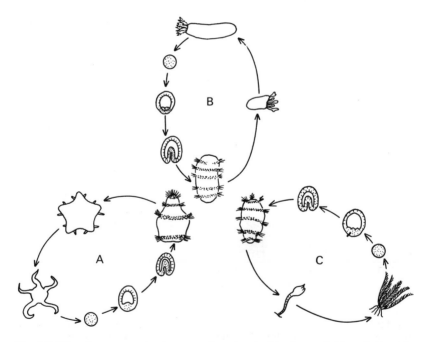

Fig. 19-15. Convergent development of unrelated echinoderms. A, Class Ophiuroidea (*Ophioderma*). B, Class Holothuroidea (*Cucumaria*). C, Class Crinoidea (*Antedon*). In each case the larva is a vitellaria. From Fell, 1967. In *Treatise on Invertebrate Palaeontology, op. cit.*

the anterior or apical end is an apical sensory plate with a tuft of cilia; the opposite end, in accordance with the deuterostome plan, marks the site of the blastopore. The vitellaria swims for a limited period, probably for no more that a few days, and then becomes attached by an adhesive pit which is situated mid-ventrally, and towards the anterior pole. Later the internal organs, which have been differentiating progressively throughout this period, rotate through about 90°; as a result of this the original anterior end of the larva becomes the attachment stalk, while the expansion of the free or oral region makes the larva look like a stalked pelmatozoan. This is the cystidean stage, so called because the organism now resembles one of the extinct cystids. At first it cannot feed, for the stomodaeum is closed over to form the vestibule, but within a few days it opens up and feeding then begins. After remaining in this form for some 6 weeks the larva develops arms, and is now referred to as the pentacrinoid larva. Finally, after some further months of sessile life, it breaks away from the stalk and begins the free-swimming life characteristic of *Antedon* and of most living crinoids.

Before we can judge the phylogenetic significance and the interrelationships of these several types of echinoderm larvae, it must be appreciated that development within each group has been subjected to much secondary modification, leading to abbreviation of the life cycle and to examples of convergence and divergence that are fully comparable with those found in the adults of any group of animals. Only a few examples can be mentioned here.

Abbreviation of the larval life history, and even direct development, are not uncommon in echinoderms. An example of secondary modification is seen in the ophiuroid *Ophioderma brevispina*; this lacks a pluteus stage, having instead a vitellaria larva with four ciliated bands (Fig. 19-15). The appearance of rudimentary skeletal spicules shows that this life history must have evolved from a more typical one involving a pluteus stage. Fell ascribes this type of modification to an increase in the yolk content of the egg, which, he suggests, inhibits metabolism in the bilaterally symmetrical stage so that radial symmetry is established relatively early. It is of particular interest to find that increase in yolk in such forms may modify development in essentially the same way as in the chordates. Invagination is reduced to a solid inpushing of cells; these differentiate to form the mesoderm and the wall of the enteron, the archenteron and the formation of coelomic sacs having disappeared from the ontogeny.

Such shortening of the larval life history is not restricted to the ophiuroids. In asteroids either the brachiolaria or the bipinnaria stages may be omitted, often in association with the production of large yolky eggs. *Astropecten* omits the brachiolaria, and metamorphoses from the bipinnaria stage without undergoing fixation. In *Solaster endeca*, by contrast, brachiolar arms are formed without the larva passing through a clearly defined bipinnaria stage at all. So also in echinoids, where forms with yolky eggs may have their free-swimming stage restricted to no more than a few days, and where the larvae may be simplified to an oval form and uniform ciliation.

Reduction of the larval stage may lead to a fully direct development, which may be associated with brood protection, viviparity and embryonic attachments to the parent, which, however, are not necessarily nutritive. Again a few examples must serve. In the asteroid *Echinaster* the young stages are protected in the ambulacral grooves while in *Pteraster* they are contained in a dorsal pouch or marsupium.

Amphipholis squamata is an example of a truly viviparous ophiuroid. The eggs are shed into the bursal sacs and develop within these into young ophiuroids, which have rudimentary larval skeletons, as in *Ophioderma brevispina*. The bursae, being invaginations of the oral surface that are primarily used for respiration, are presumably well suited for this secondary reproductive function. The embryos become attached to the parent, but this seems not to be a nutritive relationship. It is thought, however, that a nutritive fluid may be secreted into the sacs. Life histories comparable with the above are found in holothurians; for example, eggs may develop in the oviducts or in the coelom, or the young may be protected in marsupia. They are, however, uncommon in echinoids.

Convergence and divergence are marked features of echinoderm life histories, as will already be evident. There has clearly been an independent evolution of cylindrical larvae with ciliated rings (vitellaria larvae) from the yolky eggs of holothurians, crinoids, and ophiuroids, but there are also more specific cases of close resemblances between unrelated species. For example, the larvae of the echinoid genus *Diadema* have an unusual development of the post-oral arms, and a reduction of the others, which gives them a superficial resemblance to an ophiopluteus (Fig. 19–16). As for divergence, this is illustrated in a broad sense in what we have already said of modifications of the life cycles, but it may occur even between closely related forms. Thus the ophiuroid *Ophioderma brevispina* has a vitellaria larva, yet the closely related *Ophiura texturata* has a well-developed and normal ophiopluteus.

Fell has argued that this remarkable plasticity in echinoderm development makes it impossible to attach phylogenetic significance to their larvae. Certainly one cannot possibly justify interpreting them as a strict recapitulation, in the Haeckelian sense, of phylogenetic history. Larvae undergo independent adaptation to their particular modes of life, and to their primary function of ensuring development and dispersal of the species. They cannot, therefore, be precise guides to the organization of adult ancestors; a conclusion that is particularly well demonstrated by the structure of crustacean larvae. This, however, is not a counsel of despair, implying that it is impossible to unravel anything of the history of larval forms.

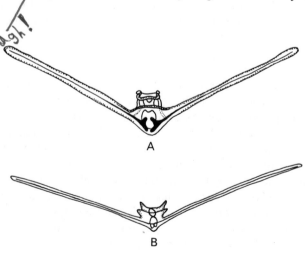

Fig. 19-16. Convergent larval forms in echinoderms. A, an echinoid, *Centrechinus* (*Diadema*). B, an ophiuroid, *Ophiothrix*. From Fell, 1967. *op. cit.*

Rather must they be approached as we would approach adult forms, resting our interpretation of their relationships upon a careful differentiation between the primary features of their organization and the secondary modifications that have been imposed upon them.

Thus it is possible to see in many echinoderm larvae certain common features of organization, including bilateral symmetry, an apical plate and tuft, an antero-ventral mouth, a posterior anus derived from the blastopore, a tripartite and paired enterocoel, and a ciliated band curving to run round the mouth and in front of the anus. These we may regard as primary features, part of the foundation of the concept of the Deuterostomia. They also suggest the possibility that the dipleurula stage (Fig. 19–12) may represent a common ancestor of Echinoidea, Holothuroidea, and the Asterozoa.

For reasons outlined above, it is dangerous to extrapolate from larval organization to adult; yet the drastic character of echinoderm metamorphosis must surely reflect a time in their past history when a bilaterally symmetrical and free-living stage, which may or may not have had the characteristics of the dipleurula stage, settled and in due course developed a pentamerous symmetry. To regard the behaviour of contemporary larvae as an exact recapitulation of ancestral history would be wrong. Events that must have been spread over a vast extent of time are here compressed into hours or minutes; moreover, the execution of the transition has been independently modified in the various groups, as may be seen by comparing the metamorphosis of the crinoids with that of asteroids. Yet here again the evidence available need not be completely ignored. Viewing the metamorphic events of all of the echinoderm groups as a whole, we may reasonably infer that the settling of the ancestral stock took place by the anterior end; this became a stalk of fixation, with the left side of the organism becoming the oral surface and the right side the aboral one. This was presumably followed by the development of radial symmetry, by the rotation of the oral surface upwards and of the aboral one downwards, and by the consequent asymmetrical development of the coeloms of the two sides, the anterior structure of the right side undergoing regression (Fig. 19–17).

If we now extend our analysis to a comparison of the main types of echinoderm larvae, we can readily see the difficulty of applying to them a strictly phylogenetic interpretation. For example, the resemblance between the auricularia of the Holothuria and the early bipinnaria of the Asteroidea might be taken as evidence of a close relationship between these two groups, while a similarly close relationship between the Echinoidea and the Ophiuroidea would be implied by their common possession of a pluteus. We might thus conclude that the ophiuroids, despite their resemblances to starfish in the adult stage, are actually more closely related to the sea-urchins.

However, we have already seen (Sec. 5–3) that the evidence from palaeontology, supported by that from morphology, shows that asteroids and ophiuroids must be closely related, with common ancestors in the early Palaeozoic, and that the echinoids already constituted an entirely separate line of evolution at that time. It follows that the organization of echinoderm larvae can be no guide to the phylogenetic history of the several subphyla. We must regard their special features as independent responses to the demands of planktonic life, influenced, no doubt, by the common genetic potentialities of the phylum as a whole.

The larval history of the other deuterostomes is no less complex than that of the

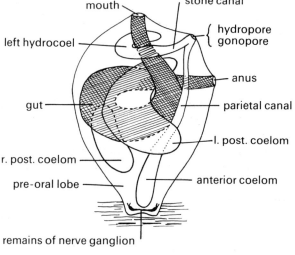

Fig. 19-17. A, diagrammatic reconstruction of a hypothetical Dipleurula, represented crawling on the sea-floor. The ciliated bands are not drawn. B, hypothetical transformation of a Dipleurula into a primitive echinoderm, after settlement by the anterior end, with displacement of the mouth towards the left side and thence to the morphologically posterior pole. After MacBride.

echinoderms. The most straightforward case is that of the Phylum Hemichordata, some members of which have an indirect development with the production of a characteristic tornaria larva (Fig. 19-18). This larva, in its mode of origin and in its organization, differs only in points of detail—notably in the presence of a posterior telotroch, which is a ciliated band additional to the main pluteus-like one—from a dipleurula. On the assumption that this type of larval organization is fundamental to the echinoderms, and an indication of a common inheritance, it is reasonable to regard the tornaria as indicating a close relationship between hemichordates and that phylum. However, the phylogenetic problems presented by the plasticity of echinoderm larvae show that this conclusion needs to rest on firmer ground than that provided by the tornaria; which, indeed, it does (Sec. 18-5).

Fig. 19-18. Two early stages in the development of the tornaria larva. *a*, anus; *a.or*, preoral area; *b*, mouth; *bd.or, bd.po*, preoral and circumoral ciliated bands; *cn*, preanal ciliated band; *coel*, proboscis coelom with pore; *d*, digestive tract; *p*, pore. The ventral depression is stippled. From Grassé, 1948. *op. cit.*

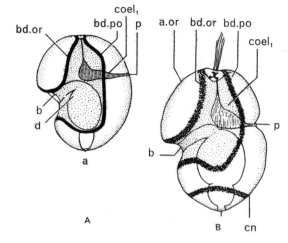

An illuminating feature of the metamorphosis of the tornaria of *Saccoglossus horsti* is the development of a temporary post-anal tail. This is probably a vestige of the attachment stalk of the sessile hemichordates, seen today in *Rhabdopleura* and *Cephalodiscus*. In other respects the metamorphosis of the larva involves little more than simple growth and transformation, without the drastic reorganization that echinoderms undergo. If, then, we accept that hemichordates were derived from the same type of sessile, microphagous ancestor as were the echinoderms, we may infer that they have diverged much less from that ancestral type than have the echinoderms. We may infer, too, that the Enteropneusta have evolved from the Pterobranchia by assuming the limited degree of independent movement involved in their burrowing life. The temporary development of the tail recalls the appearance of the stalk in asteroid metamorphosis. In both instances, although in quite different contexts, the larva carries something of the past with it. The fact that we can confirm our interpretation of the tail of the *Saccoglossus* larva by checking it against the structure of living relatives must surely give us some confidence in our interpretation of the stalk of the asteroid larva.

In some enteropneusts development is direct. As with echinoderms, this is best regarded as a result of secondary loss of the larva, the more so in that an intermediate stage of this is seen in *Saccoglossus horsti*; this has a free-swimming stage of short duration, which does not develop into a tornaria. Undoubtedly the differences in modes of development are to be associated with the yolk content of the egg, as so often in echinoderms. The tornaria is planktotrophic, whereas the larva of *S. horsti* is lecithotrophic, relying entirely upon the yolk laid down in its egg for the whole of its free-swimming activity. This lasts for only 1 or 2 days, at the end of which time the organism is ready to settle and metamorphose. The logical sequel to this is seen in *S. kowalevskii*, which hatches as a worm-like organism, and has no planktonic phase at all.

With the protochordates (the Urochordata and the Cephalochordata) we pass to larvae that show no obvious relationship with any of those that we have so far considered. This is not to say that these groups lack developmental characters justifying their inclusion in the Deuterostomia. In amphioxus, where these characters are easiest to appreciate, we find total and non-spiral cleavage, gastrulation by invagination, the origin of the anus from the blastopore, and an enterocoelic coelom with indications of a tripartite differentiation. The larvae, however, are of more advanced organization than those of echinoderms and enteropneusts, possessing distinctively chordate features that, in the case of the urochordates, provide the main reason for including the adults within the Phylum Chordata.

We are concerned here with only two main larval types, the ascidian tadpole and the larva of amphioxus, for the larval stages of the pelagic urochordates (Thaliacea) are considered to have been modified or lost in correlation with the pelagic life of these animals. The functions of development, dispersal, and habitat selection are all discernible in the ascidian tadpole, but it is the last of the three that predominates. The larva possesses the embryonic rudiment of the adult alimentary canal, but it does not itself feed, for its free-swimming life is restricted to a matter of hours or perhaps to a few days at the most. Its chordate features are particularly well seen in the dorsal nervous system, and also in the locomotor mechanism, which is, however, confined to the tail. This mechanism, like that of the vertebrates, does not employ the principle of the hydrostatic skeleton, but relies instead upon the association of lateral muscle bands with an axial support, the notochord.

Closely associated functionally with locomotion are the ocelli and statocyst in the cerebral vesicle, at the anterior end of the central nervous system. Ascidian tadpoles are at first positively phototropic, and swim to the surface of the water. This behaviour pattern marks the distributive phase of their activity. Soon, however, they become negatively phototropic and positively geotropic, moving downwards into shaded rock crevices and overhanging surfaces that are suitable for the life of the adults. This is the phase of habitat selection, and its importance can be judged, paradoxically enough, in certain species in which the larval stage has been secondarily lost. This has occurred, for example, in certain molgulids that are adapted for life on submerged sand flats. Because of the uniformity of this habitat, and the ease with which it can be thickly populated, a larval stage is no longer as essential as it is in species that require particular types of rock surface. No doubt this explains why these species have secondarily evolved a direct form of development.

The significance of the tadpole in the ascidian life history is thus apparent; but its phylogenetic relationships are puzzling. The most illuminating interpretation is due to Garstang. One line of analysis led him to the suggestion that the tadpole could have been evolved from a larva like the auricularia of echinoderms as a consequence of the further elaboration of locomotor adaptation. This, Garstang suggested, could have led to the approximation of the loops of the ciliary band in the mid-dorsal line, where, with their associated nerve fibres, they could have given rise to the neural folds. The adoral ciliated band, situated inside the mouth and responsible for maintaining a feeding current, might have been the forerunner of the endostyle. It is difficult to evaluate this argument, for there is no direct evidence for it. Although a possibility, it is certainly not essential for our assessment of urochordate affinities. We can visualize the ascidian tadpole evolving as a special product of the urochordate line without weakening the arguments for associating the group with the Deuterostomia.

Where the larval history of urochordates is of outstanding importance is in connection with the second main line of Garstang's thought, in which he attempted to trace the line of evolution from the protochordates to the vertebrates. This, indeed, is a crucial evolutionary issue. Fundamental to the concept of the deuterostomes is the supposition that they all stem from a sessile and microphagous ancestry, a supposition that leaves us with the difficulty of seeing how vertebrate organization could have evolved from such a background. Garstang's answer was analogous to his explanation of the torsion of gastropods as resulting from the persistence of a larval adaptation in the adult. He suggested that the fundamentals of chordate organization, including the nervous system and notochord, were initially established in ancient larval forms like ascidian tadpoles. In due course their metamorphosis was eliminated, so that their larval organization became that of sexually mature adults. This is the process, well known in urodele Amphibia, that we call neoteny (the retention of larval characters into sexual maturity). In this instance it is visualized as leading to a new type of adult organization, one containing the immense potentialities that flowered in the evolution and diversification of vertebrate animals.

The above line of argument does not imply that the vertebrates arose from ascidian tadpoles of the modern type. Indeed, this is rendered virtually impossible by the way in which their free-swimming life has been reduced to a minimum compatible with habitat selection. Garstang's theory demands larvae that have not become specialized in this way, but are rather tending to prolong their independent life. It seems, therefore, that vertebrate origins must be sought in some remote and primitive deuterostome, urochordate-like in fundamental organization, but perhaps not very far from the level of organization of hemichordates, and with a generalized type of tailed larva. Hypothetical though the argument may seem, the contention that new forms of adult organization may arise through neoteny is by no means unsupported by protochordate life histories. It is generally agreed that the Larvacea are neotenous urochordates, preserving their tail into the adult stage, but even more illuminating are conditions within the Cephalochordata.

Amphioxus has a metamerically segmented and planktotrophic larva, highly specialized in its asymmetry; the mouth is relatively enormous, and lies on the left side of the body, probably as an adaptation to the intake of food by a ciliary feeding mechanism (Fig. 19-19). Clearly this larva is much more highly organized for independent life than is the ascidian tadpole; yet some affinity with the latter is seen in the tendency of young amphioxus larvae to attach themselves temporarily to the bottom of aquaria by the secretion of the club-shaped gland. Moreover, older larvae become temporarily attached by three anterior adhesive papillae, which strikingly recall the attachment papillae of the ascidian tadpole.

What is especially significant, however, is that the larval life of amphioxus lasts for as much as 4–5 months, a striking contrast with that of ascidians, and much more reconcilable with the requirements of Garstang's hypothesis. Such behaviour in the earlier stages of deuterostome evolution might very well have led to the further exploitation of the possibilities of pelagic life by the introduction of some degree of neoteny. This, indeed, seems actually to be happening at the present day in the amphioxides larva, a cephalochordate larva, world-wide in distribution, that differs from the typical larva in the prolongation of its pelagic life. It may, as a result, develop as many as 34 pairs of gill slits, in contrast to the 24 which seem to be the normal maximum reached before metamorphosis. But, and more significantly, the gonads

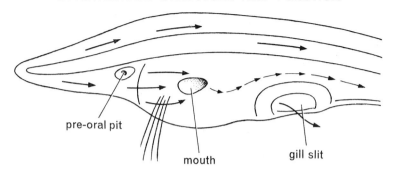

Fig. 19-19. Diagram showing the ciliary currents in the feeding of the young larva of amphioxus. The heavy arrows represent the currents produced by the action of the body cilia; the smaller ones represent the course taken by food particles that have entered the mouth. The mouth is small at this stage, having just pierced; later it rapidly enlarges, and at the same time other gill slits pierce. Adapted from Bone, 1958. *Proc. zool. Soc. Lond.*, **130**, 289–293.

may appear in the amphioxides larva before metamorphosis begins—a clear trend towards neoteny.

As with other problems of larval biology, we must interpret these facts with caution and flexibility. It is not suggested that the vertebrates arose from the amphioxides type of larva, which is probably a specialized product of modern cephalochordates. All that is claimed is that certain processes which operate today may equally well have operated in the remote past, and that, had they done so, they might have led to the origin of the vertebrate line. To this line the Cephalochordata must, on the evidence of their general organization, be closely related. But they have none the less followed a line of evolution that diverges from that of vertebrates. However efficient the exploitation of pelagic life by their larvae, the adult cephalochordates lead a restricted life, sheltered in temporary burrows in sharply defined types of gravelly substrata. It was left for the vertebrates to exploit to the full the possibilities that have always been implicit in the modes of life of pelagic larvae.

The view that larval biology contains the key to chordate ancestry is highly speculative, although, being based upon principles that are sound enough in themselves, it does not lack biological plausibility. It is not, however, the only way of looking at the problem. Another line of argument takes as its starting point the existence of some strange and puzzling fossils that range from the Middle Cambrian to the Middle Devonian. These have an echinoderm-like skeleton of calcite, and have usually been placed in the echinoderms in an assemblage termed the 'carpoids'. Their relationships to other echinoderms are, however, quite obscure, although, since they were bilaterally symmetrical, they have been thought to be remnants of an early pre-radiate echinoderm stock. Jefferies, however, has argued in great detail that two of the 'carpoid' orders, the Cornuta and the Mitrata, should be placed in a Subphylum Calcichordata which, while showing echinoderm affinities, is actually more closely related to the early chordates. His conception is that in the late Pre-Cambrian or early Cambrian a *Cephalodiscus*-like hemichordate began to crawl on the sea floor on its right side, losing in the process its right gill slit and tentacles. Two lines were then derived from this stage. In one, the gill slit of the left side was lost while the remaining tentacles were further elaborated. This gave rise to echinoderms. In the other line it was the remaining tentacles that were lost while the left gill slit was elaborated. From this line evolved the chordates, with the development of a notochord, muscle blocks, and a dorsal nerve cord, their earliest known representative being the carpoid *Cornuta*.

This view leaves the general concept of the Deuterostomia undisturbed, but another view, deriving from an idea first suggested by the Dutch embryologist Hubrecht in 1883, and recently elaborated by Willmer is more radical. According to this, the protochordates could have been derived from nemertines, the proboscis being converted into a notochord and the pharynx into a filter-feeding mechanism, with, of course, many other concomitant changes. Willmer argues further that the process may have developed along more than one line. Amphioxus and the vertebrates, for example, might have started from different nemertine ancestors and have followed parallel courses of evolution.

These unconventional approaches to a perpetually fascinating problem merit more attention than can be given to them here, if only because of their salutary warning that larval life histories may not be the only repositories of the phylogenetic record.

19-5 INSECT LARVAE

Insects, which are primarily terrestrial forms, have necessarily lost any traces of whatever pelagic larvae may have been possessed by their unknown aquatic ancestors. Yet they have developed a new type of larval history, one that presents, both in its origins and its functional significance, an instructive comparison with those of aquatic invertebrates.

It has sometimes been held that insect life histories are composed of a series of stages of progressive differentiation, a view which implies that the larvae are imperfectly developed adults. This is a misleading interpretation. Insects must rather be viewed as carrying to an extreme the tendency, apparent in many other invertebrates, for larvae to evolve a mode of life distinct from that of their adult stages. We have seen that this divergence permits many invertebrates to separate dispersal, in the larval stages, from reproduction, in the adult ones. In insects, however, the division of labour has developed differently, in a way that has immense advantages, and that must have contributed greatly to the success of the group. What has happened is that the larvae have come eventually to concentrate mainly upon growth, sacrificing to this end much of their exoskeleton and much of their mobility as well. The adults have compensated for this by concentrating their energies upon dispersal as well as reproduction. The caterpillar of the giant American silkworm moth, *Hyalophora* (*Platysamia*) *cecropia*, is a striking illustration, being devoted to growth with such concentration that during early summer it increases its weight 5,000 times. It has been aptly remarked that its accumulated assets are then invested in the construction of an essentially new type of organism, a flying-machine devoted to dispersal and reproduction.

As with so many other aspects of insect biology, however, the immense size and range of diversification within the group makes it impossible to do more than consider these matters in the most general terms.

The growth of an insect, like that of other arthropods, is interrupted by periodical moults or ecdyses, the form of the insect between any two ecdyses being called an instar. The primitive type of life cycle is probably that seen today in the primitively wingless insects such as the Thysanura (silver-fish). Here the only major difference between the young stages and the final adult instar is the appearance of the mature reproductive organs in the latter. The change produced at the final moult is therefore so slight that such insects are commonly regarded as undergoing no metamorphosis, and they are termed ametabolous.

In the winged insects the situation is altogether more complex, with two distinct types of life cycle. The more primitive of these is probably that seen, for example, in cockroaches, insects that have apparently persisted from the Carboniferous with very little change. The young stages (nymphs) differ from the adults (imagines) in lacking fully developed wings, although these are growing externally during the nymphal period. Nymphs also lack the sexual armature of the adults, but in other respects they resemble the imagines quite closely; they possess, for example, compound eyes. Thus the metamorphosis, although clearly discernible, is comparatively slight, and these insects are said to be hemimetabolous. Alternatively they are known as Exopterygota, because of the external development of their wings.

The more advanced type of life cycle, termed holometabolous, is characterized by the young stages lacking compound eyes, and differing from the adults very markedly in many other respects, including body form, mouth parts, and often mode of life, which may be adapted to a habitat remote in character from that of the imagines. The young stages are now called larvae, and the metamorphosis is a drastic one, so much so that a pupal stage is interposed between the larva and the final adult instar. The pupa, superficially a quiescent phase, is one in which profound reorganization of the tissues take place. In this type of life cycle the wings develop internally, from buds called imaginal discs, and for this reason the holometabolous insects are alternatively known as the Endopterygota.

It seems probable, to judge from the fossil evidence, that the holometabolous type of life cycle evolved later than the heterometabolous, the earliest examples of it being seen in the Coleoptera, which are known from the Upper Permian. Its adaptive value is clear enough. With the adult stages so freely motile, the functions of dispersal and habitat selection have little significance. The value of the holometabolous life history lies rather in the other aspect of larval life that we have discussed earlier: the opportunity that it affords for divergent evolution of young and adult, and the consequent exploitation of two entirely different types of habitat. The diversified specialization of terrestrial larvae is one demonstration of this. It is illustrated also by those insects that have secondarily adopted aquatic habits during their development, a line of evolution that has been followed by heterometabolous forms as well as by holometabolous ones.

The Plecoptera (stoneflies), Ephemeroptera (may-flies), Odonata (dragonflies), and Hemiptera (bugs) provide examples of the establishment of aquatic habits in the nymph. This has been achieved by the secondary invasion of water from the land, with perhaps an intermediate stage of association with the damp earth at the edges of fresh water. Primarily, the young stages have become so adapted, evidently because of the flying habits of the imagines. In the aquatic Hemiptera (including the water-boatmen and pond-skaters), however, the adults are unharmed by immersion, and in these forms flight has become relatively unimportant. It seems probable that the fundamental advantage gained by the nymphs in this type of cycle is some protection from the extremes of conditions on land, for temperature changes are less drastic in water, and plant food and organic detritus are never lacking. Frequently it is the aquatic nymph that survives the winter. In fact, the adult stage may become of relatively minor importance, as regards its duration; the extreme example of this is seen in the may-flies, where the imago is concerned solely with reproduction, taking no food at all, and often surviving for only a few hours.

No less instructive are the life cycles of holometabolous insects that have taken to the water. Divergence between the habits of larva and adult is here very much more marked, a familiar contrast being between the blood-sucking of the adult mosquito, with its piercing mouth parts, and the microphagy of its aquatic larva, which is dependent upon currents produced by moving its food brushes. In some instances—for example, the Trichoptera (caddis-flies)—the adults are short-lived, just as in some of the Heterometabola. In aquatic beetles, however, the imago may be the predominant phase, but this only emphasizes the importance of water in the life cycle, for in these instances the adult is often aquatic or semi-aquatic in habit. That these highly specialized animals should invade water in this way is a sufficient tribute to its value as a habitat, particularly since many Holometabola with aquatic larvae have had to solve the difficult problem of providing for the escape of the imago from the pupa into the air. Indeed, aquatic life might have played an even more important part in the evolution of insects had it not been for the evolution of flowering plants. The full evolutionary diversification of the Lepidoptera, Hymenoptera, and Diptera broadly coincided with this phase of plant evolution. At the present day the mode of life of the adults of these groups, and the details of their life histories, are closely correlated with the exploitation of the higher plants.

The course of the evolution of insect life histories can only be guessed at, and there is more than one opinion regarding the relationship between the heterometabolan and holometabolan type of development. We shall not discuss these here beyond remarking that there is a divergence between two views: one that the larva is a specialized nymph, and the other that it is a new development inserted into the life cycle in advance of the nymph as a result of earlier hatching. On the former view the pupa is a specialized last-stage nymph, whereas on the latter view it represents a fusion of nymphal stages, the larva being essentially an elaboration of an embryonic phase. Whatever the truth may be, the origin of these life histories is rooted in caenogenetic changes. These have brought about alterations in the relative rates of development of juvenile and adult characters, with consequent divergence of the youthful and adult stages. We have seen that the developmental stages of these animals can be maintained in a juvenile state or switched into the adult one by changes in the hormonal balance. It is likely that a vital factor in establishing these complex and highly adaptive life histories must have been the possession by insects of an endocrine system with a far-reaching capacity for precise regulation and coordination.

20
Larval Lives

20–1 COMPETITION AND COOPERATION

We have been considering the invertebrates largely in terms of individual animals, yet this gives only part of the picture. No animal or plant is an island, complete unto itself. For one thing, it belongs to a particular species, and within that species it is a member of a population, which is the unit within which evolution takes place. At any moment the species consists of one or more such units which may or may not be genetically isolated from each other. But we must press the argument further. The origin and establishment of life involved the interaction of different types of organism, as we have seen, and the maintenance of life at the present day is equally dependent upon such interactions. This is not merely because of the fundamental biological cycles of carbon and nitrogen that link autotrophs and heterotrophs together. At a different level of analysis each individual organism is a member not only of a population within a species, but also of an integrated community (Sec. 1–4) characteristic of particular type of habitat and regulated by processes developed by natural selection. We can here draw an analogy with the evolution of the organization of cells and tissues to form bodies of increasing complexity. This organization depends upon the coordination and integration of the cells; it involves a measure of competition between them for limited resources that are available for their maintenance; and it involves also the cooperation that allows them an influence far greater than any one cell could exert on its own.

The relationships of organisms with their environment are no less complex, and this is true in their developmental stages as well as when they are fully adult. The environment in the broadest sense of the term comprises inorganic or physical factors, and also the biotic factors that arise from the activities of the organisms themselves. All of these are therefore involved in the 'struggle for existence'. This was clearly perceived by Darwin, who emphasized that the term 'struggle' is used in this context in 'a large and metaphorical sense'. The struggle is partly against the physical conditions of the environment, and partly against predators, parasites, and disease. But it also involves competition for essential requirements that are limited in availability: nutrients for maintenance, space in which to function, and a mate to ensure reproduction. Within this general framework it is possible to analyse something of the nature of organic interrelationships, but the analysis is by no means simple. This is not only because of the complexity of organization of natural communities; it is also because the precise relationships of any one species are not constant, but vary with such factors as age, season, and population density.

One useful expression of this has been outlined by Mather (Fig. 20–1). Individuals with some similarity in their needs or activities may come into competition with each other (left side of Fig. 20–1). This is commonly true of members of the same species, although the situation is modified, as we shall see, by polymorphism. The degree of competition will, however, be density-dependent, its intensity in any given population varying with the density of individuals. There will be an upper limit of density beyond which no more individuals can be accommodated, and at this level competition will be maximal. But at lower densities the situation will be otherwise. We may expect that at a certain level, with ample resources, the individuals may be independent of each other, so that their relationships will be neutral. It is also possible to visualize still lower densities, below which the population may be too small to exploit its environment adequately, to deal with the physical stresses that it imposes, or to compete with other species. At this level the relationships between the individuals will be cooperative rather than competitive.

At the other extreme, competition will be minimal or non-existent when the

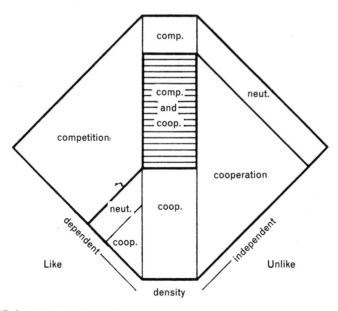

Fig. 20-1. Diagram illustrating types of relationship between individuals. The sizes of the areas are arbitrary. For further explanation, see text. From Mather, 1961. *Symp. Soc. exp. Biol.*, **15**, 264–281.

needs or activities of the individuals concerned are entirely unlike. This situation might exist between two quite unrelated species, although requirements for essential nutrients make it perhaps improbable that two species within the same community will be entirely independent of each other in their demands, unless they are very different indeed in their modes of life. If, however, we assume for the sake of argument that competition may be non-existent in certain circumstances (right side of Fig. 20–1), then the relationship between two such species may be one of neutrality or may give rise to some degree of cooperation. More usually, however, the complexity of the requirements of different species will produce the types of relationship seen in the centre of Fig. 20–1. They are unlikely to be neutral, but may rather be expected to be wholly cooperative, wholly competitive, or a combination of both. Mather points out that this interrelationship of competition and cooperation is fundamental at all levels of biological organization, from the cellular upwards. Competition determines the evolution of adaptation under the influence of natural selection, yet the resultant adaptations may eventually be so complex that they depend on cooperation for their maintenance. It is because of this that control systems are needed, to ensure that the required degree of cooperation is not made unattainable by competition. We shall see that colonial life, with its associated polymorphism, provides excellent examples of this principle in operation.

It follows from these considerations that the action of natural selection is itself necessarily complex, and that it is exerted at more than one level. Within any one group or population it may be acting upon the competitive relationships of the individuals. In respect of the relationships of this group with another, however, natural selection may be acting within the group to favour cooperative relationships of its members and hence the efficiency of the group as a whole, relative to other groups. Nor is this all. We have suggested above that relationships between two different species may also develop a degree of cooperation, and we shall later be considering specific examples of this. But it is possible, as Lucas has argued, that the degree of such cooperation may have been greatly underestimated. We have noted the extent to which organisms are dependent upon the nutrient metabolites produced by other organisms, but there is now much evidence that such dependence extends far beyond the limits of nutritional requirements.

For example, the rate of feeding of barnacles in influenced by the abundance of phytoplankton through a metabolite released by the latter. Again, the rate at which oysters pump water is influenced by the abundance in that water of certain carbohydrates that are released by the phytoplankton. It can be argued that in these examples, the efficiency of feeding of microphagous forms has been increased by the evolution of adaptive responses to metabolites released by the organisms upon which they feed. From this point of view the interactions of the phytoplankton and the animals that depend upon it are subtle ones; more than a blind engulfing of the food that happens to be present, but contributing rather to a greater degree of efficiency in the organization of this particular link in the food chain. The relationship is a cooperative one, in the large and metaphorical sense of Darwin, in so far as the phytoplanktion is contributing to ensure that it shall be efficiently exploited! At this point, however, it will be well to translate these general considerations into an examination of some specific types of interrelationship.

Examples of the combination of competition with cooperation between members of the same species are well seen in gregarious animals. These are animals that grow so closely crowded together that sometimes, as with the pseudo-colonies of *Phoronis*, they may be superficially indistinguishable from true colonies, although in fact the individuals lack the direct organic continuity that is characteristic of colonial forms (see below). Such close crowding may, no doubt, arise by change in particular populations of a species that is not normally gregarious. Truly gregarious species, however, are species in which this is a normal and predictable mode of life; it is part of the adaptive organization with which they confront their environment, and it depends upon specialized patterns of structure and behaviour, such as the settling responses of larvae that we have already discussed.

Knight-Jones and Moyse have pointed out how the pseudo-colonies of tubicolous polychaetes can modify and improve their environment. Associations of *Sabellaria alveolata* increase both the mass and the surface areas of the sandy reefs upon which they grow, while the associations of *Filograna implexa*, increasing by fission of the worms, build up structures in which the individual members enjoy mutual support. The competitive aspect of the relationship is seen in the way in which each worm adds to its tube until it is adequately orientated outwards and has room to expand its branchial crown for feeding.

Crowded populations of barnacles are governed by the same principles. They grow in ridges and hummocks, the distribution of which is perhaps determined initially by the presence of prominences on the substratum which place those that settle there is a more favourable position for feeding. However the pattern may be initiated, the result is a cooperative improvement of the environment, for the effect of the irregularities is to increase both water turbulence and the surface area, and thereby to increase the efficiency of feeding. Individuals in the centre of a hummock must be elongated, but, given this growth response, they are as well placed for feeding as are less crowded individuals that have more space to occupy. The only ill-placed ones are probably the small barnacles in the hollows between the humps; their food supply may be insufficient to ensure breeding, but on the other hand they may provide support for the edges of the humps. In this respect we may think of them as subordinated to the benefit of the aggregation as a whole—a phenomenon of much greater importance in the associations that we call colonies, where it involves a high level of adaptive specialization. These associations are considered later; for the moment we are concerned with the role of larvae in animal relationships with each other and with their environment.

20–2 MARINE LARVAE AND HABITAT SELECTION

Survival demands, amongst other things, the achieving of a compromise between three requirements outlined in the previous chapter: development, dispersal and habitat selection. This compromise is expressed by the establishment of life-history tactics which show much inter-specific diversity; which is not surprising, for they are part of the total adaptation of a species to its habitat, built out of the elements that we have been considering in previous chapters. For example, there may be oviparity or viviparity, indirect or direct development, an extended reproductive season or a restricted one, while these and other choices may be further refined, perhaps by diversification of larval mode of life, or by length of metamorphosis. The justification of such differences, when we come to compare one species with another, is not always easy to see, but this difficulty is of the essence of biological enquiry, and we can hope to enlarge our understanding by extracting some of the principles that underly them. They have been explored in some detail by Stearns. Here we can only suggest a few lines of thought which emerge from his discussion.

Certain conditions must be fulfilled in any life history. For example, parental investment in eggs should yield the maximum effective return after discounting the inevitable mortality of the young stages. Wastage of eggs and larvae, brought about to a large extent by predation, is often enormous, as may be judged from Thorsen's estimate that the females of most marine invertebrate species each produce from several thousands to some millions of ova during a lifetime. The impact of this loss in different species can be estimated by considering rates of egg production, on the assumption that larger rates are correlated with greater wastage. The relevant data show that the higher production rates are found in those species that have planktotrophic larvae with a long pelagic phase. Evidently this method of development is extravagant, but two benefits can be set against it. Firstly, the larva and adult are able to exploit, independently of each other, the resources of two different habitats, so that competition between them is avoided and cooperation can be regarded as a significant factor. Secondly, the prolonged larval phase aids the dispersal of the species, as we have already seen.

The rate of production of young stages must necessarily be influenced also by the availability of food supplies. It can be argued that in a reasonably stable population, many small larvae will be favoured if food resources for them are ample, and if predation pressure is low in relation to their numbers. If, however, food is scarce, fewer larvae will be produced, and there will consequently be a higher risk to the species of loss by predation. Indeed, there will now be some advantage in having a few large progeny which will be at risk for a shorter time. This argument, carried to its extreme, leads us to expect a favouring of viviparity and a non-pelagic life history, for the young will then be still better protected.

Two further examples must suffice to illustrate the different patterns of adaptation which might evolve in particular circumstances. Consider a species which is exposed to an environment in which long periods of stress, fixed in duration, are punctuated by conditions favourable for rapid reproduction and colonisation. Stearns suggests that this could favour the onset of reproduction at an early age; a large brood, augmented by parthenogenesis during the colonising phase; and sexual mating at the onset of stress, producing resistant stages. We shall see such adaptations exemplified in the life histories of cladocerans; aphids provide another illustration. Consider, by contrast, a species living in the intertidal zone, which provides an inherently stable environment, although with regular and predictable cyclic fluctuations. This could favour a concentration of breeding activity at the period of optimum conditions, with little departure from it.

These examples show that the advantages of different types of life-history can be defined in principle, yet it may still prove difficult to justify those that have been adopted by a particular species. Much fuller investigation is clearly needed. To quote another example given by Smith, most shore-dwelling polychaetes have a prolonged larval life, yet the larva of *Arenicola* probably remains in the sand, although it is adapted for independent movement. Very likely this is true also of other larvae, which may well be able to survive without leaving the shore, even though they may appear to be structurally equipped for pelagic life. Many nemertean larvae can cling to the substratum, and many polychaete larvae can probably survive by creeping over the substratum or by living in rock pools. Indeed, one can imagine that a pelagic larval phase might lead to unnecessarily wide dispersal, and that this, combined with the very large production of eggs that it requires, might restrict larval freedom in more specialized life histories. We can say no more of such possibilities than that different species have independently achieved different solutions of the problem, and therefore different equilibria in their life histories.

Irrespective of the length of larval life, habitat selection remains the third important aspect of reproduction for which the larva has to provide, especially in those species that are either sessile in the adult, or that change their position only very little. This is one aspect of metamorphosis, and it is the aspect that is most readily analysed experimentally since it permits a critical determination of the conditions in which metamorphosis occurs. We might expect these conditions to be most flexible in those species that are the freest to move in the adult and that do not, therefore, depend upon larval mobility alone for ensuring survival in a suitable habitat. Little is known of this, although there is some evidence to support the argument. For example, the youngest stages of freely moving shore forms such as *Carcinus maenas* are not normally found on the shore itself, which suggests that their larvae metamorphose in conditions that allow the adults to establish themselves independently in their preferred habitats. The course of events is quite different, however, in species with less mobile adults. Here, as the studies of Wilson and others have shown, there are remarkable adaptive specializations in the receptor systems and behaviour of larvae, even of those which seem, from the morphological point of view, to be of comparatively simple organization.

Wilson has closely studied the development and behaviour of the larva of the polychaete *Ophelia*. This larva (Fig. 20–2) is a simple type of trochophore, which possesses at first a broad prototroch, a small and incomplete telotroch, and groups of neurotrochal cilia extending ventrally between these main bands. It swims freely in the plankton, growing and forming its first few segments, until at the age of about 6–8 days it has three setigerous segments, and is capable of some muscular wriggling of the trunk region. At this stage it is ready to undergo metamorphosis, a process that involves growth of the body, and the loss of the larval ciliation and of the capacity for swimming. The behaviour of the larva changes, and it is now able to crawl among sand particles on the substratum.

This complex of metamorphic changes does not take place abruptly and regardless of the environment in which the larvae find themselves. As soon as a larva is ready for metamorphosis its reactions change in much the same way as those of ascidian larvae. Initially the larvae swim towards the surface and show no reaction to light, but then they become negatively phototropic and positively geotropic, so that they swim towards the sea bottom. At this stage they readily adhere to surfaces, attaching themselves by a secretion, but their subsequent behaviour depends upon the nature of the substratum. They may burrow into it and metamorphose, but if the sand is unsuitable they remain attached for a little time with their body upright, and then swim away. They can repeat this behaviour for several days, so that throughout this period they have the opportunity to test substrata and to reject unsuitable ones.

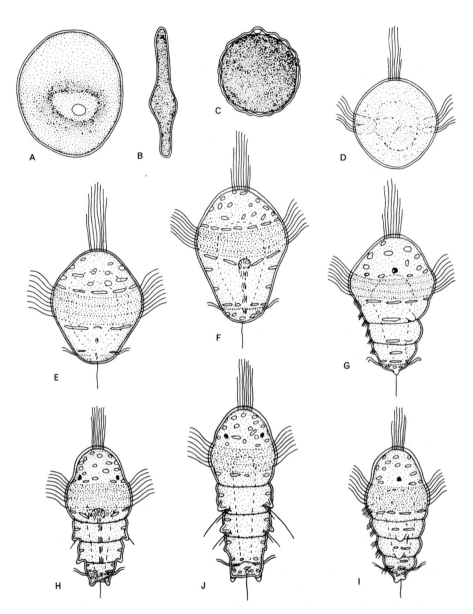

Fig. 20-2. Eggs and larvae of *Ophelia bicornis*. × 225. A, unfertilized egg; B, the same in side view; C, fertilized egg with polar body; D, 1-day-old larva; E, 2-day-old larva, ventral view; F, 3-day-old larva, ventral view; G, 4-day-old larva, view of left side; H, I, 4- to 5-day-old larva, ventral view and view of left side; J, 5-day-old larva, dorsal view; K, L, 7-day-old larva, ventral view and view of left side; M, 11-day-old larva, ventral view; N, young worm, after metamorphosis and 19 days old, dorsal view; O, capillary chaeta; P, winged chaeta. From Wilson, 1948. *J. mar. biol. Ass. U.K.*, **27**, 540–553.

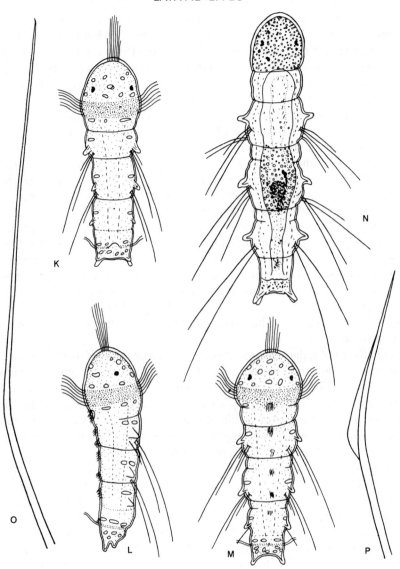

Naturally there is a limit to this, determined by increasing firmness of adhesion and decline in the power of the prototroch. Eventually they must metamorphose or die, but at least their behaviour provides reasonable assurance that some will be able to establish themselves in suitable sand.

What is the nature of the stimuli that determine these responses by such minute organisms? Observations of the settling of larvae on different types of sand show that these types can be classified as attractive, neutral, or repellent. Attractive ones are those on which larvae settle heavily, and then undergo almost immediate metamorphosis. At the opposite extreme are the repellent ones on which few or no larvae settle or metamorphose. The neutral ones are sands on which larvae settle and metamorphose in quite large numbers, if there is no other substratum available. If, however, they are offered a choice by being presented with both an attractive and a neutral substratum in the same vessel, few will select the latter. One factor determining attractiveness seems to be the size and shape of the sand particles, but this is probably of minor importance, for the attractiveness can be removed from the particles by cleaning them with acid, and then washing them thoroughly. Attractiveness can also be removed by heating the sand. This suggests that the stimulus may be an organic one, as does the fact that acid-washed sand grains recover some attractiveness if they are soaked for a long time in filtered sea water.

Extended experimentation along these lines has led to the conclusion that the most important factor in the settling and metamorphosis of *Ophelia* larvae is the presence on the sand grains of bacteria and other micro-organisms. These must be neither too abundant nor too few; dead organic matter is repellent, but so is too heavy a growth of micro-organisms. Probably, then, the relative attractiveness or repellency of substrata depends upon the particular species of micro-organisms that grow on it, and the abundance of the growth. For example, acid-cleaned grains will collect a heavy growth of micro-organisms if they are soaked in sea water containing peptone, but they never become as attractive as do those that are soaked in filtered sea water without peptones or other added nutrients. Presumably the nutrients encourage species of micro-organisms that are not attractive to the larvae, and that do not grow, in untreated sea water.

The patterns of behaviour outlined for the larvae of *Ophelia* are characteristic of many marine larvae. In general, those of benthic forms move towards the bottom, explore the substratum, settle or attach in some way, and then metamorphose. During these processes they show sensitivity to a very wide range of stimuli, including light and currents, the colour and texture of the substratum, its particle size, and its contained microbial fauna and flora, which may be surprisingly complex, even on the surface of a single sand grain (Fig. 20-3). Thus the cypris larvae of *Balanus balanoides* can delay metamorphosis for up to two weeks should they not encounter suitable rock surfaces for settling. During their free-swimming period they are positively phototropic, so that they remain in the surface water, brief periods of negatively phototropic movement enabling them to find rocks on and near the shore. Other larvae are strictly limited in the type of material on which they will settle, the larvae of *Phoronis* selecting limestone rocks, while those of *Spirorbis borealis* settle mainly upon fronds of *Fucus serratus*, being influenced in this by chemical stimuli from the seaweed. (cf. Sec. 17–8).

The performances of these and other minute marine larvae are clearly very remarkable, not least because their very small size, and the consequent close spacing of their receptor cells, must make it difficult for them to detect concentration gradients of chemicals which are themselves likely to be very small. There is some evidence that, because of this, they respond mainly to chemicals which are adsorbed to the surfaces with which they are concerned, and that prior to making the contact they are influenced by other stimuli, and perhaps also by chance movements.

The selection of the substratum by the larvae of sedentary species, especially when these are gregarious, may be facilitated by their recognition of adult members of the same species, often through chemical information. For example, the larvae of the polychaete *Sabellaria alveolata* can recognize the cement which the adults secrete in forming their sandy tubes. Cypris larvae are another illustration of this. They will

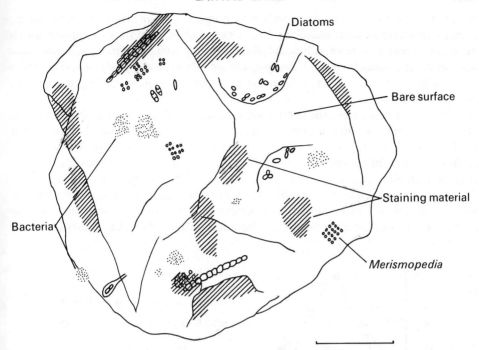

Fig. 20-3. Microbial flora on the surface of a sand grain from an intertidal beach. Note the colonies of bacteria, of diatoms, and of the blue-green alga *Merismopedia* as well as the organic staining material in hollows, and the flat surfaces that are bare. Bar = 100 μm. From Meadows and Anderson, 1968. *J. mar. biol. Ass. U.K.*, **48**, 161–174.

settle on pitted slate panels, for protuberances on surfaces are one of the factors that influence them, but they will only do so if the panels have been soaked in extracts of adult barnacles. In this case the stimulus is probably the arthropodin of the cuticle. There is marked specificity in this reaction, for the larvae of the two barnacle species *Balanus balanoides* and *Elminius modestus* can distinguish between extracts prepared from adults of those two species.

However, gregariousness introduces another problem, and, because of this, selection of the substratum is not the only function of larvae when they are settling. When we consider colonial life, we shall see something of the importance of sessile individuals being sufficiently spaced to avoid over crowding and ensure a balanced exploitation of the habitat. The same consideration applies to all forms of animal life, whether it be birds establishing territorial rights or barnacles distributing themselves over rock faces. Here the larvae of sessile marine invertebrates have an important part to play. Thus larvae of *Spirorbis* explore a *Fucus serratus* frond before finally settling down on it. During this exploratory phase they crawl over the surface and may swim away to another one; the exact behaviour depends upon how long they have previously been swimming, for prolonged activity encourages rapid settling. Contact with another *Spirorbis* induces them to settle, but they metamorphose out of contact with it. Thus, these highly gregarious animals become spaced out in such a way that they have adequate room for subsequent growth. Here the determining factor is apparently their sensitivity to any raised object that they encounter, which need not necessarily be an individual of their own species. The final spacing depends upon the initial density of the larval population, but even at high densities the larvae do no settle as closely to each other as they would do if these limiting factors were not operating.

In *Spirorbis* the relevant stimuli seem to be mainly tactile ones, but the cypris larvae of barnacles show a more complex behaviour. In principle they behave at settling much like *Spirorbis*, undertaking searching movements followed by a final settling. This is influenced by the presence of other individuals of the same species, for these animals also are highly gregarious. According to the analysis of Knight-Jones and Moyse (Figs. 20–4 and 20–5), there are two factors operating in the settling of barnacles, well seen in the slate-settling experiments just mentioned: the response to the species, and the response to protuberances on the surface that is being explored.

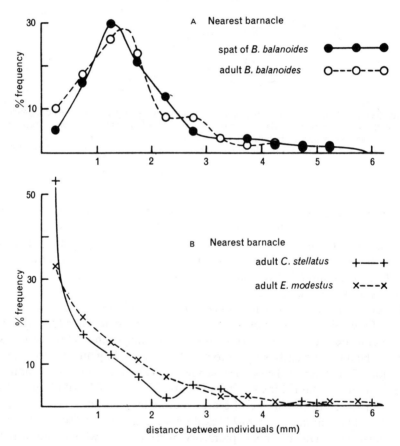

Fig. 20-4. Frequency distributions of distances between cyprids of *Balanus balanoides*, just settled upon limestone, and the nearest previously settled barnacle. The cyprids spaced themselves out from their own species (A), but readily settled alongside other species (B). From Knight-Jones and Moyse, 1961. *Symp. Soc. exp. Biol.*, **15**, 72–95.

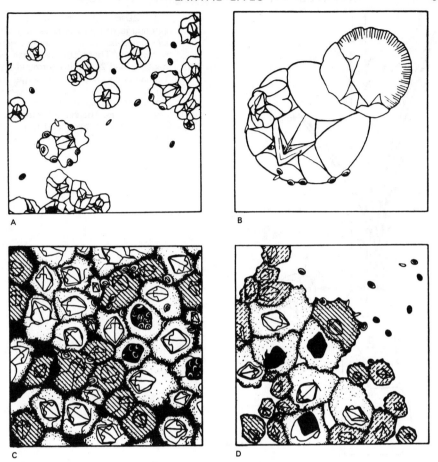

Fig. 20-5. Episcope drawings to show typical distributions of recently settled *Balanus balanoides* in relation to older barnacles. They have spaced themselves out from their own species, but many have settled in contact with other species. Unmetamorphosed cyprids are about 1 mm long and are distinguished by their shape, pointed at each end. Each drawing is of an area 25 mm square. A, part of a smooth slate. There are six *Elminius modestus* in this area, recognizable by their four parietal plates and by the cyprids in contact with them. The remaining adults are *Balanus balanoides*. B, a large *Balanus improvisus* on a shell of *Mytilus*, bearing an adult *B. balanoides* (left of drawing) and several smaller ones. C, dense mixture of adult *B. balanoides* and *Chthamalus stellatus* (the latter cross-hatched). Spat of *B. balanoides* have settled on the *Chthamalus* but not so readily on their own species (though many have settled inside empty plates). D, a piece of limestone rock, with adult *B. balanoides*, *C. stellatus* (cross-hatched), and *E. modestus* (hatched by interrupted lines). From Knight-Jones and Moyse, 1961. *op. cit.*

If both stimuli act together, as when a cypris makes contact with a recently attached member of its own species, the spacing-out reaction follows, and the larva settles at a certain distance from the other individual. If, however, the larva under experimental conditions makes contact with the base of an adult that has recently been detached,

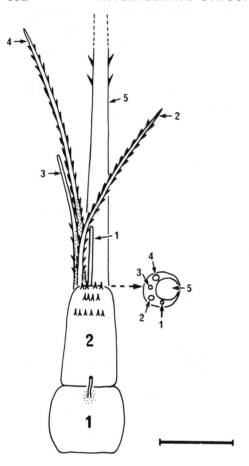

Fig. 20-6. Diagram of a caudal appendage of a cypris larva of *Balanus balanoides* based on scanning electron microscopy. There are two basal segments (1 and 2) and five terminal setae. A small open-ended seta emerges from segment 1. The transverse section shows the arrangement of the five terminal setae. The scale bar represents 50 μm. From Walker and Lee, 1976. *J. Zool., Lond.* **178**, 161–172.

it will settle upon this base because this does not provide the additional stimulus of a protuberance. Then again, if the larva makes contact with an individual of another genus, *Elminius modestus*, it will settle upon this without any prior spacing response; this is because in this instance it does not receive the other necessary stimulus, the presence of an individual of its own species.

The behaviour of cypris larvae implies that they must possess a complex receptor equipment. It is now thought, although on circumstantial evidence, that some of the receptors concerned are located on the antennules. Others are assumed to be borne on the pair of caudal appendages that lie posterior to the thoracic limbs, and that are thought to be sensitive to the contours of the substratum. The apparent simplicity of the caudal appendages, when seen under the light microscope, is deceptive. Scanning electron microscopy shows them to be composed of two basal segments which are provided with spines arranged in a very precise pattern, and which also bear five setae, one large and four small (Fig. 20–6). Clearly, there are potentialities here for sensory discrimination, but much remains to be learned about this.

The effect of the behaviour of these larvae is two-fold. On the one hand it ensures that members of one species, settling gregariously, have adequate room for growth. On the other hand, it makes for acute competition between barnacles of different species, since individuals of one will settle upon those of another, with consequent overcrowding. This may well explain the distribution that is sometimes actually observed upon the shore. As Knight-Jones points out, the barnacle *Chthamalus stellatus* extends well down the shore on the southern coasts of Britain, but on the more northerly ones, where it is approaching the limits of its range, it is restricted much more to the higher levels. The lower levels on the northerly coasts are densely populated by *Balanus balanoides*, which in these regions seems to be the dominating partner in this competition.

The gregarious habit seen in these sessile forms is in one sense an inevitable result of the crowded conditions that develop where marine habitats are highly favourable for growth and reproduction. Yet the crowding must also have a survival value, for otherwise how could mechanisms ensuring gregariousness with optimum spacing have been evolved? The advantages are presumably not different in principle from those that we shall later suggest are gained by colonial organisms. The spacing of the newly settled larvae does not prevent the building-up of crowded populations as the animals grow older. The effect of this is to establish a large surface area for the production of feeding currents, the value of which is evident because crowded barnacles may, given favourable conditions, be as fecund as those that are less crowded. Cross-fertilization is facilitated, and these crowded areas may also be more resistant to wave action, while, by increasing the local turbulence, they may further aid the feeding process. Thus intraspecific cooperation in the exploitation of the environment is possible in solitary species as well as in colonial ones. We shall see later how it may also develop between individuals of more than one species.

It is important, nevertheless, not to be over-facile in attributing supposed selective advantages, for marine larvae, like all of us, are caught up in a complex web of circumstance. The consequential difficulties in interpreting their behaviour are well exemplified by the larvae of *Alcyonidium hirsutum*, a polyzoan which forms greyish fleshy colonies on the fronds of *Fucus serratus*. Choice experiments with pieces of seaweed show that the larvae have a clear preference for this one, with some 35% settlement on it, as compared with about 19% on *F. spiralis*, 8% on *F. vesiculosus*, and 1% on *Laminaria digitata*. There is, however, a marked inequality in the distribution over the intact thallus. Heaviest settlement occurs along grooves and in depressions, with lowest densities at the bases and growing tips of the fronds, and peaks in the middle regions (Fig. 20–7). Why should this be?

The apparent gregariousness, as already mentioned, must favour cross-fertilization and hence will reduce inbreeding, but this is only part of the story, for the primary stimulus to settlement is not the presence of other individuals, but the alga itself. Its attractiveness, which depends upon its chemical composition and secretions, will probably vary from point to point according to the age of different parts of the fronds, and it may perhaps be modulated by the attendant microflora, which will also vary with age. (In this respect, too much reliance on the responses of the larvae to cut pieces of frond can be misleading, for the chemical conditions of these, under experimental conditions, can differ significantly from those of corresponding parts of the intact thallus.)

The resultant pattern of settlement can be seen to have some selective advantage. Thus, the preference for sheltered grooves and depressions, which is a response

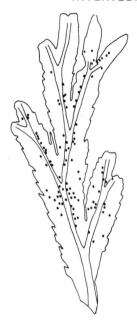

Fig. 20-7. Outline drawing of a frond of *Fucus serratus*, showing (dots) the positions of larvae of *Alcyonidium hirsutum* after natural settlement. From Hayward and Harvay, 1974. *J. mar. biol. Assoc. U.K.*, **54**, 665–676.

of the larvae to surface contours, will protect them from subsequent abrasion. However, even this advantage is not always clear cut, for some of the depressions are due to necrosis, and these areas are likely to be early shed, to the disadvantage of the polyzoan. In any case, life expectancy is not always what might be expected, for the growth, area, and mortality of surviving individual colonies are much the same, whether they are in regions of high or of low density. What, then, is the significance of the gradient in density? Selection may be expected to operate against larvae settling on the extreme base or tip of the thallus, because these are the regions most likely to be destroyed by abrasion and by winter defoliation. Elsewhere on the thallus, where there is also a gradient, it is the most central regions which will be best protected from abrasion. But an additional factor must be predation by the nudibranch *Onchidoris muricata*, which is thought to graze more heavily on the more densely settled areas. Perhaps this counterbalances any advantage gained by the protection of these regions from the abrasion which will most affect the more peripheral and less densely populated areas. Reflection upon these issues must leave one with enhanced respect for the subtlety of the selective process, and with a disinclination to attempt too readily an assessment of the balance of ecological considerations which are involved in the behaviour of marine larvae.

20–3 LARVAL LIFE IN FRESH WATER

We have so far considered only marine larvae, because indirect life histories are particularly characteristic of marine organisms, but they are, of course, by no means confined to them. The small groups of fresh-water sponges and Polyzoa (Ectoprocta) have larvae, and the branchiopod crustaceans hatch as nauplii, but the Cladocera, Ostracoda, and Malacostraca of fresh water all undergo direct development, as also do the fresh-water oligochaetes, leeches, and pulmonate molluscs. The fresh-water mussels incubate their embryos within the gill folds, and eventually liberate the young as the very active glochidium larva, but this is a peculiar secondary adaptation in a highly specialized life history, in which fish serve to distribute the parasitic larva. The only fresh-water mollusc that liberates its young as veliger larvae is *Dressena* (*Dreissensia*) *polymorpha*, but this merely reinforces the general argument: this species, which is found in salinities of 10‰ or less, is a very recent invader of rivers. First recorded from the Volga and the Danube during the eighteenth century, it seems to have spread further west during the nineteenth century.

We have earlier emphasized and discussed the well-known fact that the main animal phyla are far better represented in the sea than in fresh water. Not only are some major groups wholly lacking in fresh-water representatives, but even when they do contain such representatives these are often relatively few and specialized. From this, and from the phylogenetic relationships of the groups mentioned above, we can deduce that much of the fresh-water fauna is of marine origin. The important exceptions to this are the pulmonate molluscs and the insects, together with the fresh-water mites and spiders, all of which must have colonized fresh-water habitats from the land. That terrestrial forms should have lost their primitive indirect life histories is to be expected, so that the absence of larvae from pulmonates is easy to understand. What is less obvious is why the transition from the sea to fresh water should so often have involved the loss of larval stages, yet in fact more than one line of explanation can be suggested.

One aspect that is certainly relevant is the strength and persistence of the currents in rivers and streams, for these endanger the survival of the young by impelling them away from the habitats of the adults. This in itself goes a long way to explain why fragile larval stages have so often been eliminated, and why eggs are often protected by incubation in brood pouches or by deposition in sheltered places. Another factor is the instability of fresh-water habitats, with the recurrent risks of desiccation or freezing. These risks are more prevalent in temperate zones, and this is probably one reason why the number of fresh-water species in any particular group is often found to increase nearer the equator.

But there is another and less obvious consideration that has been emphasized by J. Needham, who has drawn attention to the importance of embryos securing adequate supplies of the various salts that are needed during their growth. Marine invertebrates can readily obtain these from the sea, and the extent to which they draw upon this source of supply is illustrated by data available for the Pacific sand dollar, *Dendraster excentricus*, to take just one example. The total phosphorus content of this organism increases from 760 mg% dry weight at fertilization to 990 mg% at the gastrula stage and 1,230 mg% at the pluteus stage. Some of this phosphorus will have been gained from the combustion of organic material present at the beginning of development, but this accounts for only about 3% of the 62% increase; the remainder must be obtained from the surrounding medium.

Supplies in the sea are certainly adequate for the purpose. In the surface water of the English Channel the phosphorus content may range from 0.0162 mg/l at the winter maximum, down to 0.0032 mg/l at the summer minimum. At times, however, the larvae may have to swim energetically to obtain all of their requirements and Needham suggests that this is one reason for the early hatching and vigorous activity of so many larval types. In fresh water, of course, with its low mineral content, the situation is much less favourable. Here the phosphorus content may range from zero up to perhaps the minimum value quoted above. Thus the young cannot obtain their requirements from external supplies, and they need reserves in the egg. This, then, provides another explanation for the omission of the larval stage. A consequence is that the eggs must be larger, for they must supply all the necessary nutrients, and this in its turn means that far fewer will be produced by fresh-water forms than are produced by marine ones. The common oyster, which is viviparous, spawns 1.8 million eggs at once, while oviparous oysters can produce over 100 million eggs at a spawning. *Anodonta cygnea*, by contrast, spawns about 15,000, although this particular species may sometimes produce as many as 2 million, which is an exceptionally high figure for a fresh-water form. In gastropods the difference is even more noticeable; *Buccinum undatum* may lay 12,000 eggs, and *Nucella lapillus* about 245 capsules that may contain from 400 to 600 eggs apiece, but fresh-water snails lay only 20 to 100, and the fresh-water limpet 5 or 6.

This reduction in number, together with the danger of removal by currents, makes adequate protection of the eggs essential. A need for protection, however, is not peculiar to fresh water, and in fresh-water forms it is frequently met by adaptations similar to those found in marine species. Fresh-water pulmonates fix masses of eggs, coated with jelly, upon stones or vegetation; marine molluscs may also do this. Fresh-water amphipods and isopods protect their eggs within the thoracic brood pouch (Fig. 20–8) formed by their thoracic limbs; so also do their marine peracaridan relatives. The female of the fresh-water crayfish carries her developing eggs attached to the swimmerets, but does not differ in this respect from the lobster and many other crustaceans. It does differ from its marine relatives, however, in that the young do not hatch until they have the form of miniature adults, with a carapace that is globular because of the large food reserves stored within the body (Fig. 20–9). They lack the first pair of swimmerets and the uropods, but have hooked tips to their chelae, so that they are able to cling for some time to the mother. An essentially similar life cycle has been quite independently evolved in the tropical river crabs of the family Potamonidae; in these too the free-swimming zoea has been eliminated and the young hatch as miniature adults.

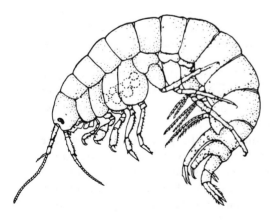

Fig. 20-8. A female *Gammarus*. The position of the eggs in the thoracic brood-pouch, sheltered by the coxal plates, is dotted. From Carpenter, 1928. *Life in Inland Waters.* Sidgwick and Jackson, London.

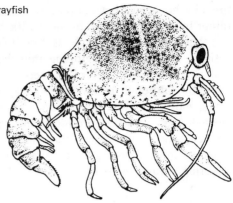

Fig. 20-9. Newly-hatched young of a crayfish (*Astacus fluviatilis*). From Calman, 1911. *op. cit.*

The disadvantages of fresh-water larval stages are so clear that it may seem surprising that they should be so widely retained in the Branchiopoda, a group that is almost entirely fresh-water in distribution. It should be noted, however, that the Cladocera, in which direct development occurs, are the only branchiopods that live in large ponds, lakes, and streams. The remainder (with the exception of *Artemia*, the brine shrimp, which inhabits salt lakes) are restricted to small areas of water and temporary ponds, where there is little danger of the young being swept away. To this extent, then, the production of nauplius larvae by many branchiopods may well limit their distribution. Moreover, these animals, including the Cladocera, show another important adaptation in their life cycles which doubtless improves their fitness for fresh-water life. This adaptation (found also in the Rotifera, another group that is almost exclusively fresh-water in distribution) is the capacity for producing two types of egg; quickly developing ones or resting ones. They are sometimes referred to, although less accurately, as summer and winter eggs, respectively. The production of these two types is closely associated with the capacity for parthenogenesis (development of the egg without fertilization); indeed, in some species males are very uncommon, and are possibly not produced at all.

In Cladocera, to take one example, only females are found at certain times of the year. These lay parthenogenetic eggs that develop very rapidly in the brood pouch, which is a cavity, characteristic of the group, lying under the carapace and dorsal to the body. Multiplication takes place rapidly during this phase. Eventually, it leads to the production of males, the appearance of which initiates normal sexual reproduction. Each of the eggs produced at this phase, normally only one from each ovary, is extruded in a protective covering; this is usually an 'ephippium', formed from the modified wall of the brood chamber and shed at a moult. Within this the segmented egg may remain dormant for a long period, and in this way it can survive the winter. The reproductive cycle is not, however, necessarily as simple as this. There may be several phases of sexual reproduction during the year, or reproduction may be entirely parthenogenetic. The former type of cycle is often found in pond-dwelling Cladocera, liable to frequent desiccation, and the latter in planktonic species, which are free of this hazard. There is evidence, then, of some innate correlation of the pattern of the life cycle with the type of environment. It has been thought, however, that the pattern is not wholly determined by genetic factors, but rather by an inter-action of these with external ones.

This type of life cycle is found in many other branchiopods, and also in rotifers, with quick-developing thin-shelled eggs and resting thick-shelled ones. An alternation of parthenogenetic and normal sexual development is also common, although either type of egg may be produced in either of these two ways; moreover, the alternation of the two phases may be very flexible. Thick-shelled eggs are also produced in copepods, but less commonly than in the Branchiopoda, although some adult copepods can withstand desiccation by entering into a resting stage with a cyst-like secretion that forms a protective covering around the body.

21
Colonial and Social Life

21–1 PROTOZOAN COLONIES

We have referred to colonies as a form of intraspecific association in which individuals are subordinated to the interests of the whole aggregation. True colonies can be defined more precisely as associations in which the constituent individuals are not completely separated from each other, but are organically connected together, either by living extensions of their bodies, or by material that they have secreted. This connection, whatever its precise nature, is more than structural. Commonly the members of the colony are physiologically linked and integrated so that each contributes to the life of the whole colony, but there is much variation in the closeness of the relationship and in its structural and physiological consequences. Colonial life, as thus defined, is necessarily restricted to animals that are of comparatively simple organization, and that can reproduce asexually, for it is fundamentally a result of fission products failing to separate. It is found in the Protozoa, and is particularly characteristic of sponges (although there is a problem of definition in this group, as we shall see), coelenterates, and the Entoprocta. Instructive examples are also found in the Pterobranchia and Urochordata.

A well-known series of colonial organisms is found in the flagellate Protozoa, in the Order Phytomonadida (Volvocida). Although holophytic forms, they have long attracted zoological speculation, for, as we have mentioned earlier, they have been regarded as one path by which multicellular organisms might, in theory, have evolved from unicellular ones. These colonies arise as the result of the continued association of a group of individuals produced by asexual reproduction. In this way there are formed flat plates of 4 to 16 individuals or zooids (*Gonium*), a sphere of 16 zooids (*Pandorina*), or a layer of 32 (*Eudorina*) or 128 zooids (*Pleodorina*) arranged near the surface of a gelatinous sphere. Typically the members of the colonies are biflagellate, with an organization like that of *Chlamydomonas*, and they are held only loosely together in a common membrane or jelly. There is no direct continuity between them, so that they do not strictly fit the definition of a colony given above; but this definition, like others that we shall be introducing later in this discussion, has to be used flexibly, as is well illustrated by *Volvox*. This forms colonies that are essentially further elaborations of the types already mentioned. They are composed of hundreds or thousands of zooids, which are carried on the surface of a gelatinous sphere, but which in this instance retain direct connection with each other by protoplasmic threads.

Irrespective of such connections, the phytomonadidinan colonies exemplify a principle that is of fundamental importance in colonial life: the establishment of an organization that transcends the individuality of the constituent organisms. All the colonies swim with one particular region foremost; they are said, therefore, to be polarized, and to have an anterior pole. In *Pleodorina* and in *Volvox* this functional differentiation is carried further, to a point at which the anterior zooids are unable to reproduce. There is thus a division of labour between the constituent members of the colony, analogous to the division of labour between the cells of a metazoan body and, similarly, leading to a more efficient deployment of the potentialities of the whole. This is emphasized in other ways. For example, in *Pleodorina* the anterior zooids are smaller than the others, and have larger stigmata, implying some differentiation of sensitivity to light. The reproductive processes are another illustration of this differentiation of activity. In the less highly organized colonies each zooid is capable both of asexual and sexual reproduction. Asexual reproduction is effected by the repeated division (multiple fission) of a zooid, with the formation of a daughter colony that then escapes from the parental one. Sexual reproduction in *Gonium* is effected by the escape of individual zooids, which function as isogamous hologametes. In *Volvox*, however, the zooids are differentiated in this respect. Some enlarge without division to form macrogametes, which, as we have already seen, may be compared with ova; others undergo multiple fission to form groups of biflagellate microgametes which are motile, and which may be compared with sperm.

Unquestionably the evolution of colonial life in the Phytomonadida has involved some reduction in the range of activities of the zooids; to some degree they have become subservient to the life of the colony as a whole. Indeed, we may go further; it is precisely this subservience, and the division of labour that goes with it, that is the primary characteristic of colonial life. It is this, rather than purely structural relationships, that defines a colony. Certainly, a colony involves its component members in some loss of individuality, but concurrently a higher order of individuality develops, which is the individuality of the colony as a whole. This, however, remains a highly abstract conception unless we can understand individuality. This understanding eludes us in the phytomonadidan colonies, but we can find a clue in another type of protozoan colony.

This occurs in the Peritrichia, which are ciliate Protozoa with a specialized arrangement of ciliation. Some of these animals are free-swimming, or are free-moving epizoites, but most of them are attached posteriorly by a disc or stalk, and of these some are solitary and some colonial (Fig. 21–1). An unusual association occurs in *Vaginicola* and *Cothurnia*, which sometimes live in pairs in protective cases, but more typical of colonial life are the associations formed by *Carchesium*, *Epistylis*, or *Zoothamnium*, where a number of *Vorticella*-like zooids are united by their stalks. New colonies are formed, from individuals that become

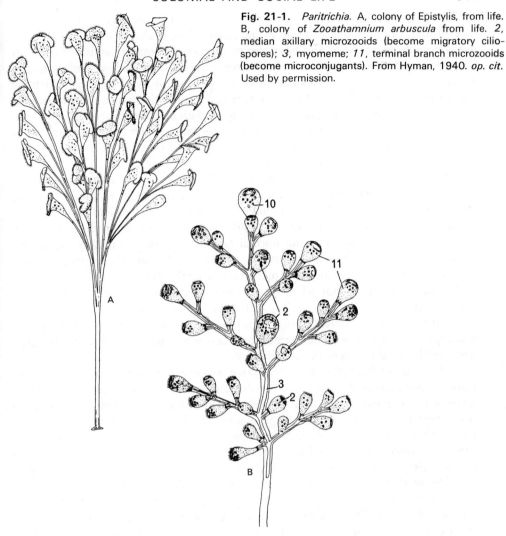

Fig. 21-1. *Paritrichia.* A, colony of Epistylis, from life. B, colony of *Zooathamnium arbuscula* from life. *2*, median axillary microzooids (become migratory ciliospores); *3*, myomeme; *11*, terminal branch microzooids (become microconjugants). From Hyman, 1940. *op. cit.* Used by permission.

free-swimming; they separate from the parent colony, undergoing repeated fission after becoming attached to the substratum. In some genera (e.g. *Carchesium*) the zooids show no differentiation among themselves, yet even so the individuality of the colony is shown by the specificity of its form, which depends upon the precise pattern of branching. In vorticellids in general the contractile myonemes of the stalk are collected into a spiral band, the spasmoneme, the arrangement of which in the colonial genera reflects different degrees of unification of their members (p. 47).

In many colonial peritrichs a new colony may arise from any of the zooids, the one concerned developing a posterior band of cilia and swimming away. Eventually it will attach to the substratum and give rise to a colony by growth and fission. In some species, however, there is some differentiation among the zooids.

In *Zoothamnium*, for example, the end of the axis of the colony bears a terminal macrozooid, the only one of these in the colony. The branches each end in a terminal branch microzooid, and bear also median axillary microzooids and ordinary vegetative microzooids. Of these four types of zooid, the median axillary microzooids and the vegetative ones can swim away as migratory ciliospores, which can then grow into new colonies. The sexual process, as in solitary forms like *Vorticella*, involves the fusion of a motile microconjugant with a sessile macroconjugant. But in *Z. alternans* only the terminal macrozooid can become a macroconjugant, while microconjugants can arise either from the terminal branch microzooids or from certain of the vegetative microzooids.

The basis of this differentiation, and of the influence that coordinates it, has been indicated by experiments in which terminal zooids have been cut off from the colony. Removal of either the terminal macrozooid or the terminal branch microzooids results in the neighbouring zooids assuming both their form and their reproductive functions. Thus the normal fate of a zooid presumably depends in some way upon its position relative to the terminal ones. These perhaps exert some inhibitory influence on the zooids lower down the stem, a supposition supported by the observation that when the terminal macrozooid conjugates, the lateral branches increase their growth; it is as though the influence of the macrozooid is reduced by the physiological changes consequent on conjugation. We shall see that a similar effect exists in coelenterates, and that it may indicate some physiological basis for the establishment of individuality.

The colonial flagellates and ciliates that we have mentioned demonstrate in two ways the advantages that accrue from this cooperative mode of life. The flagellate forms are holophytic; it is thus unlikely that they gain any nutritional advantage from their association, for their requirements are freely available in the water that bathes them. It is conceivable that their association with a gelatinous ground substance allows the transmission of metabolites, and that the differentiation of reproductive zooids depends upon the transmission of nutrients from the vegetative individuals. Probably the main benefit of this type of association, however, is to be found in the locomotor advantage obtained by combining the flagellar activity of a number of zooids. The spherical form of many of the colonies must be significant from this point of view, since for a given volume a sphere exposes the smallest surface area. For these colonies the resistance of the surrounding water is thus reduced to a minimum, and the beat of their flagella acts to the fullest advantage. The gains by sessile and branched ciliate colonies are necessarily different. Apart from the protective advantage of such group associations, which will be especially pronounced where the colony has a continuous system of spasnomemes, the main advantage probably lies in more efficient exploitation of food supplies, with wasteful competition reduced by the orderly spacing of the individuals.

21-2 SPONGE, COELENTERATE, AND ECTOPROCTAN COLONIES

It is customary to regard sponges as being either solitary or colonial in habit. Solitary forms are those with only one osculum, while colonial forms are those with a number of oscula, each osculum with its associated system of canals being considered as corresponding to one individual. This, admittedly, is a somewhat arbitrary point of view, but the fact is that a discussion of individuality in sponges becomes resolved only too easily into a purely verbal issue. Nevertheless, the sponge body has an individuality transcending that of its component cells, however limited may be its powers of coordination. This individuality is revealed in the characteristic form and organization of a particular species, which permits the systematist to handle the group, in principle, in the same way as any of the higher groups of animals. It is shown also in experiments in which sponge bodies have been fragmented by squeezing them through fine gauze, or have been dissociated into suspensions of single cells in calcium- and magnesium-free sea water. The resulting groups or suspensions of cells can associate and build up again the basic organization of a sponge body, with dermal epithelium and choanocyte layer, pores and osculum. We have seen that the relationship of oscular opening to pores and canal system is interpretable in hydraulic terms, and is highly adaptive. Presumably the appearance of additional oscula with continued growth of the body is an expression of these same hydraulic factors. To regard the appearance of a new osculum as marking the differentiation of a new individual is therefore to apply an entirely arbitrary meaning to the concept of individuality. Nothing remotely recalling division of labour between individuals occurs in this type of colony, and we can only conclude that if individuality is indeed associated with a single osculum, it exists there at a lower level of differentiation than anywhere else in the animal kingdom.

We can be more certain in analysing other examples of colonial life. Thus in hydrozoan coelenterates the individuality of the polyps is well defined, as also is their asexual budding; the continued association of their offspring leads to colony formation. The mode of budding varies, establishing characteristic colony patterns in particular species. In fixed forms the colony is attached at its base by strands of living tissue called stolons. In some (e.g. *Hydractinia*) the polyps arise direct from the stolons in an irregular way. More usually a main stem or hydrocaulus grows up from the stolons, and the polyps are borne upon this and upon its branches. The growth of the hydrocaulus largely determines the form of the colony. Growth may, for example, be monopodial; in this a terminal polyp is situated permanently at the end of each branch and of the main stem, and lengthens at a growth zone below the base of each polyp. Alternatively, growth may be sympodial; here the terminal polyp of each branch is temporarily the youngest, and is overgrown by the next one to develop, which arises as a bud at its base. Clearly there is a genetic mechanism, built into each colony, which determinates its form, and to which the individual polyps are subordinated.

Other aspects of the hydrozoan colony reveal this subordination. The organization of a polyp shows a marked polarization, extending from the mouth and tentacles at one end to the closed base at the other end. This is a morphological expression of individuality, but experiment has shown that underlying this there is an invisible physiological polarization. Child demonstrated that different parts of a polyp vary in their susceptibility to lethal agents, and that this variation follows the pattern of a gradient, with maximum susceptibility at the apical (oral) end. Similar gradients are revealed by other criteria. For example, the rate of carbon dioxide production varies, being maximal at the apical end; the rate of reduction of methylene blue also varies, again being maximal at the apical end. Gradients in electric potential have also been observed, not always uniform, but indicating that the apical end is electronegative to more basal regions.

Child used such observations, derived from many groups of animals, and from many stages of their development, to frame a general theory of the nature of individuality and organization. He argued that living material is organized in a spatial pattern, of which the external structure is only one aspect. The pattern is also expressed in differences in the concentration of various substances, and in differences in the rates of their metabolism. The existence of these patterns in the form of gradients of activity is, according to Child's views, an essential element in the development of the axiate organization of polyps, which results in one end of the gradient, the apical end, being able to dominate the rest of the body. The ability to bud, and the pattern of the associated growth, is determined by the capacity of other parts of the body for escaping from the inhibitory influence of this dominant region.

Consider, as an illustration of this concept, the behaviour of the species of *Tubularia* shown in Fig. 21-2. As the unbranched stem of a young individual lengthens, a bud forms at its base and grows out as a stolon. The stolon represents a new gradient system, which has, so to speak, escaped from the inhibitory effect of the polyp's gradient as it has lengthened. For a time the stolon shows no structural differentiation, but as soon as it reaches a certain length (or, according to this interpretation, a certain further degree of physiological isolation) the stolon starts to grow away from the substratum and to differentiate a new polyp.

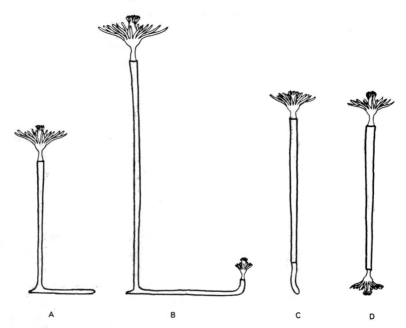

Fig. 21-2. *Tubularia.* A, young individual with developing stolon; B, transformation of stolon tip into hydranth after physiological isolation by increase in distance from dominant hydranth; C, reconstitution of stem piece with hydranth distal and stolon proximal; D, bipolar reconstitution. From Child, 1941. *Patterns and Problems of Development.* University of Chicago Press, Chicago.

We can thus regard differentiation in *Tubularia* as being influenced by the inhibitory action of the dominant end of the polyps. On this interpretation it should be possible to release growth and differentiation by the artificial reduction of the inhibitory influence. This result can, in fact, be achieved, as, for example, in *Bougainvillea*. A colony immersed in an inhibiting agent (M/200 ethyl urethane for 48 hr) shows a disintegration or regression of its polyps and buds, and an outgrowth of stolons. These latter are thought of as the expression of growth gradients, released by the inhibition of the differentiation gradients of the polyps. If the colony is allowed to recover, these growth gradients will themselves become differentiation gradients, and will produce new polyps.

The evidence that hydrozoan polyps are polarized systems is convincing, although interpretation of the action of these gradients remains theoretical, for we can only discuss their nature in general concepts of metabolic activity. Indeed, we cannot be sure whether the gradients are themselves causal factors, or whether they are the expression of other and more subtle features of protoplasmic organization. One important factor is the presence of neurosecretory cells at the apical end of the polyp, at least in *Hydra* (Sec. 17–7). However, it still remains to establish what determines this location, although the regulatory action of the cells on budding is clear enough. But at least the existence of gradients gives us a descriptive basis for visualizing how the individuality of a colony may come to dominate the individuality of its component members. Gradient patterns are detectable in these colonies, as might be expected from what we have just seen of the growth of a *Tubularia* polyp. This is well shown in *Pennaria cavolinii* (Fig. 21–3), a species that has a regularly graded series of lateral branches. If all the polyps are removed, new ones develop in a regular order, beginning at the apical end of the main stalk and of each of the branches. No morphological differentiation is visible in the branches, so that this pattern would seem to be the expression of some physiological aspect of the organization and individuation of the colony.

An important feature of hydrozoan colonies is the polymorphism which we have discussed in earlier chapters. Here we see an illustration of the interplay of

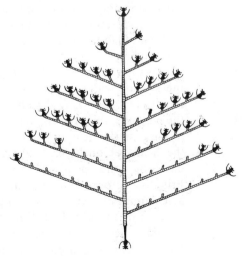

Fig. 21-3. Reconstitution of hydranths in *Pennaria cavolinii*; note that it progresses basipetally in the whole and in each axis. From Child, 1941. *op. cit.*

competition and cooperation, with the competitive element reduced, and with a consequent facilitation of the unified life of the colony. We do not know how this polymorphism is regulated. Genetic factors must certainly be involved, but, if we may draw an analogy from the polymorphism of social insects (Sec. 21–4), we may reasonably suspect that the expression of these is determined by chemical factors, diffusing, perhaps, from the several types of zooid, and ensuring the maintenance of these in suitable proportions. Braverman and Schrandt have shown that the pattern of growth of a hydroid colony can to some extent be simulated by an electronic computer that has been supplied with a set of simple growth rules. This, they point out, is an illustration of the more general proposition that the complexity of living organisms can, in theory, be shaped by a relatively small amount of genetic coding. Whatever its mode of development, however, there can be no doubt of the success of this type of organization. The spectacle of a Portuguese man-of-war seizing and digesting a fish is an impressive demonstration of the power that can be achieved by simple forms of life through the subordination of competition to cooperation. The biggest impact, however, is made by quite a different type of colony, that of the madreporarian corals. These animals, members of the Anthozoa, and essentially anemones that secrete a calcareous skeleton, form colonies of a structurally simple character, lacking any polymorphism, and reproducing without an intermediate medusoid stage. Yet they have played a major part in the building of coral reefs. In doing this they have so profoundly modified their environment that they have made a unique type of marine habitat available for exploitation by other organisms, including man. We shall see later, however, that the coral polyps do not work unaided, their activities being powerfully assisted by a form of association quite different in character from those that we are at present considering.

The principles exemplified in the colonial life of coelenterates operate elsewhere in the animal kingdom where this habit has been developed, but with sufficient differences in results to make comparisons instructive. The Ectoprocta are a group of coelomate animals that are at a higher level of organization than coelenterates. The zooids retain much more of their independence, being connected only by coelomic continuity (Cyclostomata) or by a strand of tissue, the funiculus (Ctenostomata and Cheilostomata). Here the colony is held together in its characteristic form by the external secretions of its members. Even so, some degree of polymorphism and division of labour is found, associated, as in the coelenterates, with feeding, protection, and reproduction.

We have seen that the ectoproctan feeding individual is composed of two parts, the cystid, with a protective zooecium, and the polypide, which includes the mouth, tentacles, and alimentary tract. Protective individuals arise by reduction of the polypide, this process giving rise (in *Bicellaria* and *Bugula*, for example, Fig. 21–4) to the 1 .markable individuals called avicularia. These resemble nothing so much as the heads of birds; each has a movable beak derived from the operculum that closes over the opening of the zooecium in unmodified individuals of many genera. Primitive ectoproctans indicate how this result has come about, for in these the avicularium may still possess a polypide; it then occupies in the colony the normal position of an individual. With further evolution the polypide is lost, and the avicularium becomes merely the appendage of a normal cystid situated near its opening. An even more extreme modification is seen in the vibracula, which are

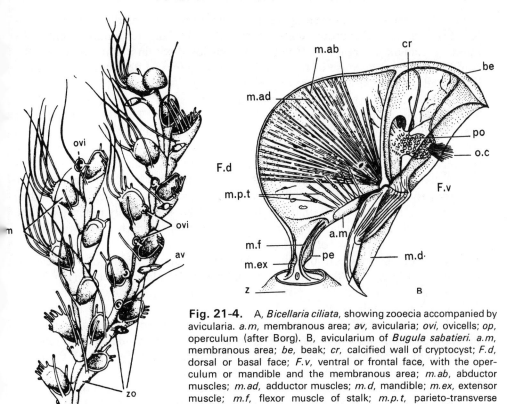

Fig. 21-4. A, *Bicellaria ciliata*, showing zooecia accompanied by avicularia. *a.m*, membranous area; *av*, avicularia; *ovi*, ovicells; *op*, operculum (after Borg). B, avicularium of *Bugula sabatieri*. *a.m*, membranous area; *be*, beak; *cr*, calcified wall of cryptocyst; *F.d*, dorsal or basal face; *F.v*, ventral or frontal face, with the operculum or mandible and the membranous area; *m.ab*, abductor muscles; *m.ad*, adductor muscles; *m.d*, mandible; *m.ex*, extensor muscle; *m.f*, flexor muscle of stalk; *m.p.t*, parieto-transverse muscles of the frontal membrane; *o.c*, ciliated organ; *pe*, stalk; *po*, vestigial polypide; *z*, zooecium forming and carrying the avicularium. (After Calvet.) From Grassé, 1960. *Traité de Zoologie*, vol. 5, fasc. 2. Masson, Paris.

like mobile bristles. These make sweeping movements that keep the colony free of unwanted material, while the avicularia seize animals and other objects that they encounter.

Polymorphism associated with reproduction, structurally simple but with some physiological complexity, is seen in the Order Cyclostomata of the Ectoprocta. In these forms the fertilized egg begins its development as a primary embryo within a modified individual called the ovicell, being nourished from a placenta-like association with nutritive tissue. This nutritional adaptation is probably correlated with the fact that the primary embryo does not itself give rise directly to a new individual, but instead buds off groups of cells that form secondary embryos; these eventually develop into motile larvae.

21-3 ASEXUAL REPRODUCTION AND COLONIAL LIFE IN UROCHORDATES

Finally, and to pass to a very different group, the Urochordata are animals in which colony formation is often an important feature, associated here with a widespread capacity for asexual reproduction and with the secretion of the characteristic tunic. The distinction between aggregated solitary forms and true colonial species is not very sharp in this group; colonies arise when the daughter zooids, produced by budding, remain in organic association and enclosed within a common tunic. Because of this, and because of the high level of structural differentiation, polymorphism and division of labour are commonly absent.

Many ascidians show a regular alternation of asexual and sexual reproduction, corresponding more or less with the sequence of seasons. The sexual phase takes place during the summer months, and the asexual phase during the winter, unless growth ceases during the coldest part of the year. Whether this alternation is a primitive feature of the group is not clear, although it is often assumed to be. This would mean that in the Cionidae and Ascidiidae, families in which the adults are solitary and reproduce entirely by sexual means, a primitive capacity for budding has been lost. It is possible that the botryllids have secondarily redeveloped the budding habit, for they have a peculiar form of budding from the body wall, unlike that of other ascidians in not making use of a stolon.

However, the modes of budding in the urochordates are, in fact, so varied that it is difficult to trace out the history of the habit with complete assurance (Fig. 21–5). A good measure of order can, however, be introduced into this complex field if we follow Berrill's analysis of asexual reproduction in ascidians. The bud may be thought of as essentially a fragment that has become isolated from the control of the parent organism, and which can then grow and differentiate independently. Berrill sees the epidermis as the active agent in urochordates, cutting through the other tissues to form the bud. This contains fragments of one or more tissues, often with nutritive reserves contained in cells called trophocytes, the formation of which may well be a consequence of regression of the parent tissues. This marked tendency of urochordates to regress has probably been a major factor in the evolution of budding within the group, another factor being the remarkable plasticity of the tissues, which makes it possible for a complete organism to be regenerated from only a limited selection of tissue rudiments. Fundamentally, then, we can view budding in ascidians as an adaptation that dissociates the competing nutritive demands of growth and sexual reproduction, and that takes account also of the marked seasonal fluctuations in the food supplies of ciliary feeders. Presumably the process has been elaborated into colony formation because this mode of life confers increased powers of resistance to stress and capacity for exploitation of the environment, as mentioned earlier. In some species the members of the colony are so closely associated that they form a structural and functional unit. In *Botryllus*, for example, the zooids share a common atrium and atrial aperture, and this may perhaps increase hydraulic efficiency. To the extent that the individuals are then interdependent, and are subordinated to the life of the whole assemblage, they may be said to form a true colony. In other instances the association is much looser; often the only connection between the individuals may be that they are embedded in a common tunic, so that there is little to distinguish them from an aggregation of separate individuals.

As regards the evolutionary history of the ascidian colony, there is reason to believe that *Diazona*, which forms large colonies in several fathoms of water, is a very primitive ascidian, and that its mode of budding is also primitive (Fig. 21–6). It provides a good illustration of the epidermis as a primary agent of the process, for at the end of the sexual season, this tissue constricts the posterior part of the

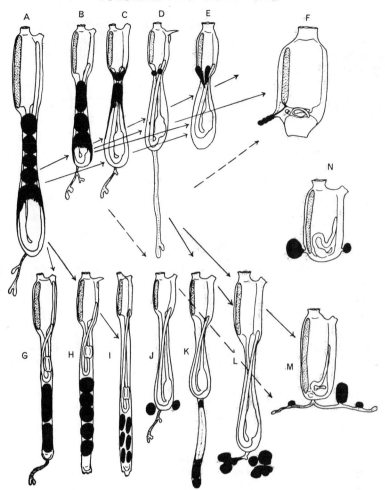

Fig. 21-5. Diagram to show mutual relations of the various methods of budding among the Urochordata. The types represent: A, *Diazona*; B, *Archidistoma*; C, *Eudistoma*; D, *Distaplia*; E, *Diplosoma*; F, a thaliacean; G, *Tylobranchion*; H, *Morchellium* or *Aplidium*; I, *Euherdmania*; J, *Pycnoclavella*; K, *Colella*; L, *Clavelina*; M, *Ecteinascidia* or *Perophora*; N, a Polystelid or Botryllid. The arrows show the probable directions of specialization in the course of evolution, the discontinuous lines possible relationships. From Berrill, 1935. *Phil. Trans. R. Soc. B*, **225**, 327–379.

body into separate buds, while the anterior end is regressing and disappearing. In due course the buds develop into adults by what is essentially a process of regeneration. Here too is a very close correlation with the seasons; asexual reproduction sets in during autumn and early winter, while the regenerative development of the buds, with their contained masses of trophocytes, takes place during late winter.

Clavelina (Fig. 21-7) forms colonies in which the individuals may be more or less free, or may be embedded in a common tunic. It has a cycle very similar to that of *Diazona*, except that asexual reproduction, accompanying the rapid

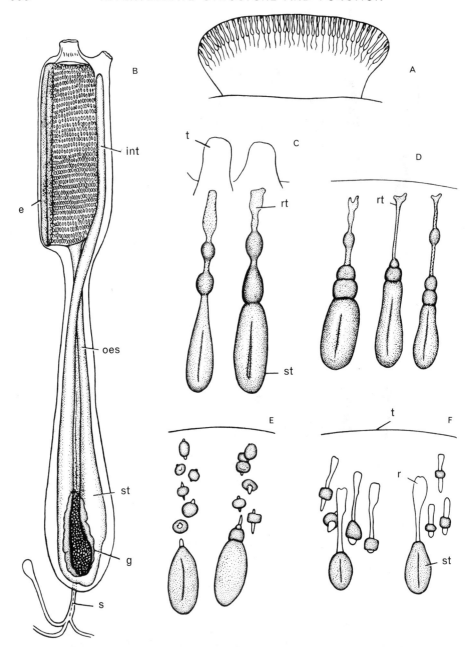

Fig. 21-6. *Diazona violacea.* A, section of whole colony; B, isolated zooid; C, zooids typical of colony taken on November 20; D, zooids of colony taken on January 1, showing regression and constriction; E, F, from colonies taken on January 30 and March 5 respectively, showing isolation and regeneration. *e*, endostyle; *g*, gonads; *int*, intestine; *oes*, oesophagus; *r*, regenerating thorax; *s*, vascular septum; *st*, stomach; *r, t*, regressing thorax; *t*, test at surface of colony. From Berrill, 1935. *op. cit.*

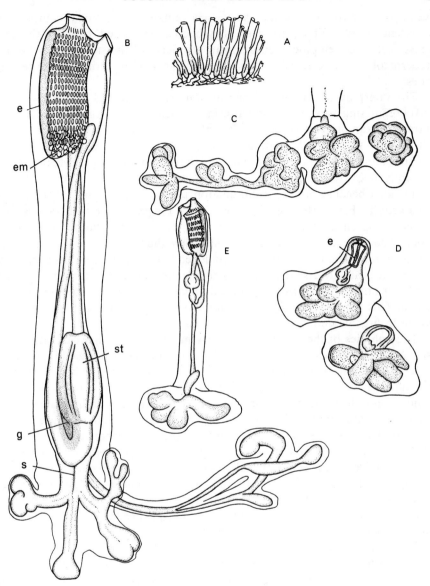

Fig. 21-7. *Clavelina lepadiformis.* A, colony; B, mature zooid showing hypertrophied ventral vessel; C, constriction of ventral vessel following degeneration of zooids to form bud masses; D, development of new zooids from single lobe of each of two isolated parts of ventral vessel; E, later development of bud. *e*, endostyle; *em*, embryos; *g*, gonads; *st*, stomach; *s*, septum of vascular stolon. From Berrill, 1935. *op. cit.*

growth of the newly formed zooids, may also take place during the summer, while the parent zooids are sexually mature. This genus exemplifies the use of a ventral budding stolon, which is here an epidermal outgrowth containing a mesenchymal septum that permits an outflow and return of blood. Trophocytes accumulate in swellings of the stolonic vessel, but these swellings seem unable to develop into active buds while the vessels remain in functional continuity with the parent zooids. When, however, they become separated, or when the parents regress, active development ensues; the ectoderm of the stolon gives rise to the ectoderm of the new zooid, while the vascular septum forms all the remaining tissues.

The interplay of parent and bud in *Clavelina* reminds us of the physiological control regulating asexual reproduction in the coelenterates. Another and somewhat contrasted example is provided by *Perophora* (Fig. 21–8). This also forms stolons with vascular septa, the formation of buds being here initiated by a bulging of the epidermis and an associated proliferation of the septum. No trophocytes accumulate, apparently because the circulation is adequate for the provision of nutrition, and perhaps also because asexual reproduction takes place throughout the period of sexual activity, before there is any regression of the parent forms. In contrast to the stolons of *Clavelina*, those of *Perophora* and *Ecteinascidia*, both members of the Perophoridae, can differentiate new zooids while they are still in organic connection with the adults. The end result is said to depend upon environmental conditions. If circumstances are adverse the daughter zooid may grow at the expense of the parent, whereas in other conditions the parent dominates the situation and may re-absorb the bud. This type of budding can produce true colonies, in which each individual remains in organic connection with the remainder.

The colonies of the Thaliacea are very much more complex than those of the Ascidiacea, largely because of their varied adaptations to a pelagic life. All of them bud by means of stolons, but these differ from the stolons of ascidians in the variety of tissues that they contain (Fig. 21–9). These arise from the pharynx, the atrial chambers, and the pericardium—regions of the body that are well situated to contribute to the stolon contents. In principle, budding is brought about by the constricting activity of the epidermis, as with ascidians, but the relationship between the ascidian and thaliacean type of stolon has been a matter for disagreement. Garstang took the view that the complex stolon of thaliaceans was primitive. Berrill prefers to regard it as a specialization resulting from the increased importance of asexual reproduction in the life histories of these forms; the elaboration of these life histories can be interpreted as correlated with the requirements of pelagic life, which is generally agreed to be a secondary development in thaliaceans.

In the free-swimming colonies of *Pyrosoma* the discharge of the excurrent stream through a single posterior atrial opening, narrowed by a diaphragm, provides for efficient and continuous jet-propulsion. Thus the zooids are here truly subordinated to the life of the colony as a whole. A much more extreme example of this, however, is seen in the life cycle of *Doliolum*, where a remarkable level of polymorphism is achieved during the reproduction of the free-swimming asexual individual called the oozoid. Numerous buds are given off from a stolon, and are moved backwards and upwards onto a dorsal process through the agency of wandering cells, the phorocytes. Those buds that come to lie along the two lateral edges of the process develop into gastrozooids with an enormously enlarged pharynx. They provide for the nutrition of what is now a colony, the parent oozoid losing its own alimentary function and becoming devoted solely to asexual reproduction, so that it is now termed the nurse. Wandering buds also settle along the middle line of the dorsal process; these develop into another type of individual, the phorozooid. Each of these is at first attached by a

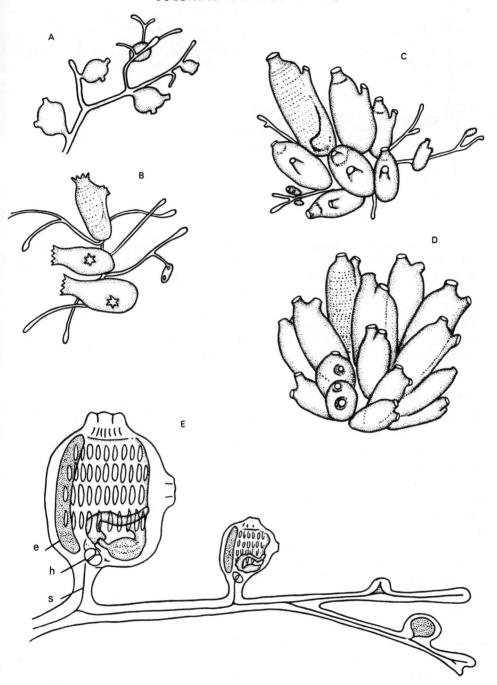

Fig. 21-8. Budding in the Perophoridae. A, small part of colony of *Perophora bermudensis*; B, part of colony of *Ecteinascidia conklini minuta*; C, small colony of *E. conklini typica*; D, part of colony of *E. turbinata*, showing length of ventral stolonic vessels and influence on appearance of colony; E, *Perophora*, showing relation of buds to parent zooid, stolonic vessel, and vascular septum. *e*, endostyle; *h*, heart; *s*, septum of vascular stolon. From Berrill, 1935. *op. cit.*

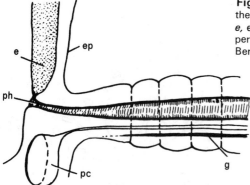

Fig. 21-9. Diagram to illustrate the relations of the budding stolon of the Thaliacea, e.g. *Salpa*. *e*, endostyle; *ep*, epidermis; *g*, gonadial strand; *pc*, pericardium; *ph*, pharyngeal diverticulum. From Berrill, 1935. *op. cit.*

stalk, on which settles another bud. This elongates, and gives rise to a series of buds which remain attached to the phorozooid as a group of developing individuals. The phorozooid eventually breaks free, and leads an independent free-swimming life until the buds that it is carrying have developed into hermaphrodite sexual individuals. These, the gonozooids, are then set free individually to carry out the sexual phase of the life cycle. Unfortunately, the fragility of dolioilids, and their oceanic habit, has so far made it impossible to unravel any of the factors that determine the course of this bizarre and intricate product of the colonial habit.

21-4 SOCIAL LIFE IN INSECTS

The term 'colony' is often applied to the complex societies that are formed by certain species of insects. Since the animals concerned are very highly organized, and retain their individual freedom of movement, the phenomenon clearly does not fall within the strict definition of colonial life that we have so far been using. Yet the term is not inappropriate, for the members of these societies are certainly subordinated to the life of the community as a whole, and they are bound together by mechanisms that are none the less real because they are chemical and physiological rather than structural. In illustration of this we may first review briefly the life of the honey bee, *Apis mellifera*—a species that was known in its domesticated form to the early Egyptians, and that has been carried by man throughout the world.

The bee colony is polymorphic, comprising three varieties or castes. One of these, the queens, consists of fertile females; the other two, workers and drones, are respectively sterile females and fertile males. The queen copulates once only, at the beginning of her reproductive life, when as a virgin queen she is followed out of the hive by a swarm of drones in the nuptial flight. Thereafter she is concerned solely with laying eggs in the cells prepared by the workers; she achieves, in the spring, a maximum rate of about one egg per minute. During the discharge of an egg she is able to release or to withhold sperm, thereby determining whether or not the egg is fertilized; in doing so she effects the primary caste distinction between males and females, for Hymenoptera have a peculiar mode of sex determination in which males can only arise from unfertilized eggs and females from fertilized ones. The factor that determines her action seems to be the size of the cell, for drones are larger than workers and their cells are correspondingly bigger.

A larva hatches from the egg three days after it has been laid, and it has then to be nursed within its cell by workers, which have two important contributions to make in this connection. First, they have to maintain a constant temperature of 35°C. This they do in cold weather by crowding together and imparting their collective heat of metabolism to the hive, while in warm weather they set up a current of air by fanning with their wings. The cooling effect of this current is supplemented by its evaporation of the water which other workers bring into the hive and distribute over the combs. The second function of the workers at this stage is the feeding of the larvae. This completes the determination of the castes, for nutrition is believed to be the factor that determines whether the females develop into queens or workers.

During the first three days of life all the larvae are fed upon royal jelly, which is a secretion of the pharyngeal glands of the workers. After the third day the royal jelly fed to the larvae in normal cells is diluted with honey and pollen from the crops of the workers, and the total amount of food is rationed. These larvae receive, therefore, a restricted diet, and in consequence of this the female ones become workers. The future queens are treated differently. They develop in special cells, large in size and irregular in shape, and are fed exclusively upon royal jelly, which they receive in substantial amounts. On this interpretation then, the diets of queens and workers are qualitatively different, and the determining influence of this is shown by the fact that larvae transferred from worker cells to queen cells will develop at least some of the characters of queens, forming what are called intercastes. The precise nature of the dietary difference remains uncertain. According to one view, royal jelly is rich in nutrients (it contains a large proportion of B vitamins, including pantothenic acid and biotin), and it is supposed that the large supplies of these promote queen development. According to another view, the virtue of royal jelly resides in some specific factor, either ingested as a vitamin-like substance by the workers and stored in their pharyngeal glands, or else actually secreted by them.

There is, however, another interpretation of caste differentiation according to which it is the quantity rather than the quality of the food which is the deciding factor. It is suggested that queen larvae develop surrounded by an excess of food (essentially the mass provisioning to which we refer later), whereas after the third day the worker larvae are fed intermittently, and with less food (progressive provisioning). It seem unlikely that this could be a complete explanation, since artificial rearing of larvae, with mass provisioning throughout, has failed to evoke any sign of queen development. But perhaps the two interpretations are not mutually exclusive; conceivably both quantitative and qualitative factors could be involved.

The production of a sterile worker caste is an example of intraspecific control and cooperation exerted through polymorphism. It solves a problem that is crucial for the maintenance of social organization, in bees as in men: the control of reproduction so that the community remains of manageable size and does not over-exploit its food supply by intragroup competition. The queen is of central importance in this control, as is shown by the behaviour that results if she is removed from the hive. Soon after her disappearance the workers change some of the normal cells, containing young larvae, into emergency queen cells, while the ovaries of these workers, which are normally atrophied, start to develop towards a functional condition. The presence or absence of the queen thus determines both the behaviour and the sexual development of the workers, the loss of the queen leading to changes that tend to increase the production of sexually active females. This control is exerted through the release from the queen of a secretion called queen substance. One component of this is 9-oxodecenoic acid (Fig. 21–10), which is secreted by the mandibular glands of the queen. This component partially inhibits the development of the ovaries of the workers that ingest it, and also inhibits their construction of queen cells. The workers can only obtain this substance by licking it from the queen's body, so that her influence depends upon direct contact with them, and is eliminated if they are separated from her by a double wire screen.

Here, 9-oxodecenoic acid acts in conjunction with a closely related compound, 9-hydroxydecenoic acid (Fig. 21–10). This is a volatile substance, also produced in the mandibular glands, which acts as a primer. The two substances, acting together, completely inhibit ovarian development in the workers, but they have other uses as well: 9-oxodecenoic acid attracts the drones to the queens during the nuptial flight, and also has an aphrodisaic action on the male. Both substances also play a part in swarming; 9-oxodecenoic acid activates the bees, while 9-hydroxydecenoic acid quietens them when they have settled, and thus prevents waste of energy. So the functioning of the bee community depends in many ways upon the linking of its members through chemical messengers that are examples of the pheromones that we discuss elsewhere (Sec. 17–8). In this instance it is remarkable that so much information and regulation can be drawn by the colony from only two compounds.

In theory the bee community can perpetuate itself indefinitely, but for this to be possible an aging queen must from time to time be replaced by a new one, as each queen can only reproduce for four or five years. A single additional queen cell, or at most a very few of them, would be needed for this. In practice, however, the community must do more, for it must propagate itself to compensate for the risk of its destruction by some environmental hazard; this is provided for by the process known as swarming. A number of queen cells are constructed, and eventually the old queen leaves the hive accompanied by many of the older workers. Soon thereafter a new queen emerges from her cell, and within a week or two she flies from the hive on the nuptial flight, accompanied by drones. On her return the remaining workers kill the other young queens, unless, as sometimes happens, the new queen herself leads a swarm of workers from the hive; in this case a second new queen will survive as the reproductive centre of the community. Any remaining drones are eventually discarded, and the hive is then ready for its normal working activity.

We have mentioned that this elaborate control of reproduction is needed to regulate the exploitation of food supplies. Yet this control is not in itself enough,

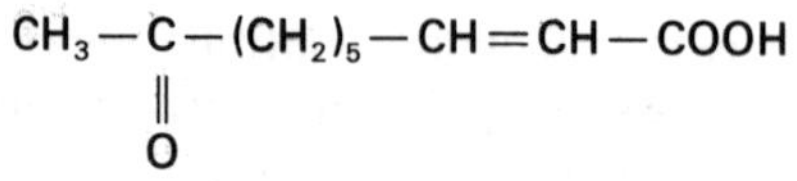

$$CH_3-C-(CH_2)_5-CH=CH-COOH$$
$$\|$$
$$O$$

9-oxodecenoic acid

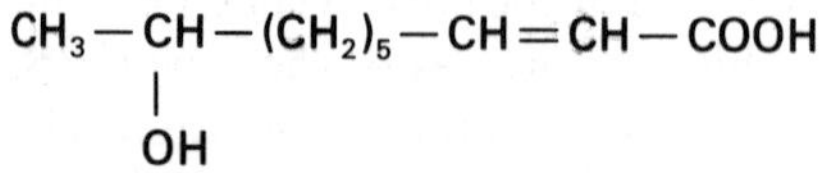

$$CH_3-CH-(CH_2)_5-CH=CH-COOH$$
$$|$$
$$OH$$

9-hydroxydecenoic acid

Fig. 21-10. Pheromones of the hive bee.

for the individual workers must be able to exploit their environment successfully, not only in their own interests but in those of the community as a whole. This result is achieved by a development of genetic and neural mechanisms similar to those found in independently living animals, and remarkable for their precise and complex integration.

One aspect of this is the highly specialized life history of the worker bee.

> Full merrily the humble bee doth sing
> Till he hath lost his honey and his sting
> And being once subdued in armed tail
> Sweet honey and sweet notes together fail.

But of course there is more to it than this, for, although it may only live for five to six weeks, it must carry out during this time all of the essential requirements of the community life, taking these up one by one in a rigidly predetermined order. For the first three days of its life the young worker cleans the brood cells and keeps them warm. Next it becomes a nurse, feeding the older larvae from the third to the sixth day of its life, and the younger larvae and the queen from the sixth to the fourteenth. From the fourteenth to the eighteenth day it is a builder, secreting wax and constructing combs, from the eighteenth to the twentieth day it guards the entrance of the hive, and finally, up to about the fortieth day, it journeys out from the hive to obtain nectar and pollen.

This sequence of behaviour is in part genetically controlled, bringing particular neural and effector systems into action in a predetermined sequence. But the worker bee is more than an inflexible machine. We have seen something of the elaboration of the nervous system and receptor equipment of arthropods. With this as a basis, the worker bee can familiarize itself with the surroundings of its hive, and learn to find its way back to it, memorizing for this purpose the colours, positions, and relative distances of landmarks (Fig. 21–11), and the odours produced by the scent glands of the other bees. Further, bees are able to orientate themselves by the sun, using it as a compass to establish their position, and having also a sense of time which enables them to allow for the movement of the sun during their absence from the hive (Fig. 21–12). The capacity, which is an example of the internal clock mechanism that we have mentioned elsewhere (Sec. 16–3), is maintained even when the dance lasts for many hours, beginning by day and continuing well into the night. The position of the sun changes, but the bee changes her directional information in accordance with the sun's movement, even though she has not observed her position since she entered the hive (Fig. 21–13).

Finally, their eyes are sensitive to the vibrations of polarized light. The direction of these vibrations, in conjunction with the percentage of the light that is polarized, establishes the position of the sun in the sky, and this the bee can detect. Thus they can orientate themselves, even when the sun is obscured, provided that they can see a small patch of blue sky. They are also sensitive to ultraviolet light, and this probably enables them to determine the position of the sun even when the sky is completely overcast, for there will be more ultraviolet light coming direct from the sun than from anywhere else.

These capacities, together with sensitivity to the colour and scent of flowers, ensure success for the individual worker in its round of foraging duties, but they do

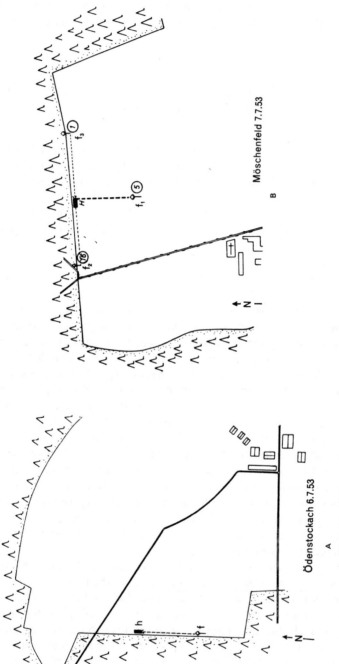

Fig. 21-11. A: On July 6, 1953, in the *afternoon*, a group of bees was trained to the south, along the edge of a forest. B: In the next *morning* the bees saw a similar forest, but this time the edge of the forest ran E–W, i.e. at right angles to the training situation. Sixteen bees were led to the west by the forest's edge; only five bees flew south. (The encircled numbers indicate the number of bees recorded at the feeding places.) From Lindauer, 1960. *Cold Spring Harb. Symp. quant. Biol.*, **25**, 371–377.

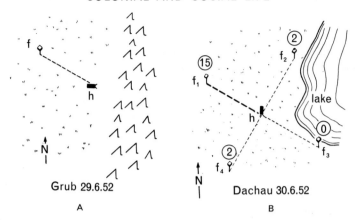

Fig. 21-12. A: A beehive (H) was placed in an unknown region, and a group of bees was fed *in the afternoon* on a feeding table (F) 180 m NW. B: During the night the hive was translocated to another area and, *in the morning*, the bees had to choose one of four feeding tables (F_1–F_4) 180 m NE, NW, SE, and SW. The new landscape did not offer any familiar landmarks; the sun stood at another angle relative to the training line as compared with the previous afternoon. Nevertheless, most bees (encircled numbers) came to the NW table, i.e. the bees had calculated the sun's movement. From Lindauer, 1960. *op. cit.*

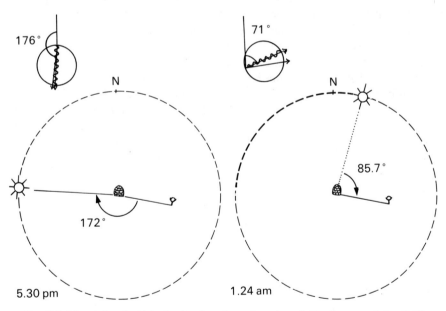

Fig. 21-13. Directional indication in a 'marathon dance'. The bee was fed at 5.30 in the evening, before sunset, at a feeding place in the east. She saw the sun at an angle of 172° to the right of the feeding table. In dancing (*upper left*) she made an error of 4°. After midnight (1.24 a.m.), she was incited to dance in the hive. In dancing, she again referred to the sun in indicating direction. Although she could never have seen the sun's position at this time of night, she made an error of only 14° in calculating its supposed angle (85.7°) with respect to the feeding place. From Lindauer, 1961. *Communication Among Social Bees.* Harvard University Press, Cambridge, Mass.

nothing to ensure the efficient integration of the activities of the foragers as a whole. To provide for this a system of communication has evolved, which has been brilliantly deciphered by von Frisch. It works in this way (Fig. 21–14). When a forager returns to the hive it regurgitates its honey as droplets that are taken up by other workers and distributed to other members of the colony or to the storage cells of the combs. If the source of the honey was near to the hive the forager embarks upon a whirling 'round dance', in which she dances clockwise and anticlockwise in rapid succession. Other workers join in the dance, and, stimulated by the excitement, eventually leave the hive themselves to secure more of the honey. They are aided in their search by their detection of the specific flower scent which the first worker will have brought back with the honey, a good illustration of the close mutual adaptation of bees and the flowers that they frequent.

This, however, is not all that is required to organize the foraging activities. It is important that the workers shall not waste their energies on flying after supplies of nectar that are poor in quality or in quantity. This is ensured by variation in the vigour of the dancing; a limited supply or a low sugar content results in a feebler dance, or in no dance at all, and this will determine how many, if any, workers will search for it: the weaker the dance, the less they are stimulated to fly.

The foragers also need some indication of the distance that they have to fly. This information is given by a second type of dance, the 'tail-wagging dance', in which the bee runs along semicircles, alternately left and right, ending each turn with a straight run back to her starting point. During the straight run she performs a rapid wagging movement of her abdomen. This dance has an effect different from that of the 'round dance', for it indicates that the nectar is at a more distant source. It also signals the distance that must be travelled in order to reach the nectar, for with increasing distance the dance becomes slower, and the time taken to travel along the straight run becomes correspondingly longer. Thus von Frisch found that the straight run was traversed nine to ten times in $\frac{1}{4}$ min if the food source was 100 m away, but only twice if it was 5,000 m away. These figures remain remarkably constant, not only within the same colony at different times of the year, but even in different colonies; evidently the workers have an accurate computer, which seems to depend either on the time that they take to travel a particular distance, or on the energy that they expend in so doing.

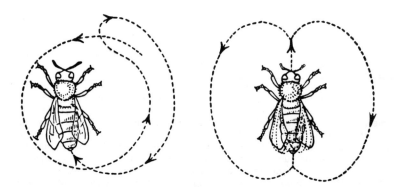

Fig. 21-14. Dances of the honey bee. *Left*, round dance; *right*, tail-wagging dance. From Richards, 1953. *The Social Insects*. Macdonald, London.

Finally, and to complete this very brief survey of truly astonishing adaptations, the 'tail-wagging dance' also indicates the direction of the food supply. In this instance the signal depends upon the bees' use of the sun as a compass, and of polarized light if the sun is obscured. If the 'tail-wagging dance' is on the horizontal platform of the hive, the angle between the axis of the dance and the direction of the sun's rays is the same as the angle between the correct direction of flight and the rays. If, as is more usual, the dance takes place on the vertical comb, the angle of flight is indicated by the angle between the straight wagging run and the vertical. Further, if the wagging run is carried out in an upward direction the feeding place is situated towards the sun, while if it is carried out downwards the feeding place is situated away from the sun. As with all phases of this communication system, remarkable demands are made upon the interpretative powers of the recipients of the information, as well as upon the computing capacity of the successful foragers. The vertical dance takes place in the darkness of the hive, yet the bees are able to translate patterns of vertical orientation into horizontal characteristics of the illuminated world outside. Information relating to sources of pollen is also imparted in the same way, by round and tailwagging dances, the essential difference lying in the characteristic scent of the pollen, which the bees are well able to detect.

The essential features of social life in the honey bee, distinguishing it from a merely gregarious habit, are the abandonment of the uninhibited predatory and reproductive life of the individual, the establishment of division of labour, and the integration of the colony by means of a communication system. Probably the first step in its evolution was the establishment of some degree of maternal care as a result of an association between the mother and her developing young. An often-quoted example of such a situation is the behaviour of the female earwig, which lays her eggs in a hole in the ground and then remains with them, turning them over and licking them, thereby preventing the growth of moulds. When the young hatch they remain with the parent, but only for a few days.

More directly relevant to the evolution of the social habit of bees, however, are the varied types of life history found in many members of this group. Some of these are solitary, each fertilized female hibernating and then raising a brood of males and females by placing eggs and food in holes which she prepares. There is here no community life; the mothers never see their young and the males die after they have mated, while the new generation of females hibernates in isolation. In other species the female lives longer, and is still in the nest when her offspring are emerging, much like the earwig; it is this relationship that contains the potentialities of social life. True social organization, although at a very simple level, is seen in certain species of *Halictus* in which a further important step has been taken: some females work on behalf of the others, without themselves reproducing. The spring females, working alone or in groups of up to three individuals, build a simple nest and produce in it offspring that are all female. These offspring are essentially workers, for their ovaries remain imperfectly developed, the founder females being queens that continue to lay eggs and also guard the nest. It is said (and this is in accordance with the habits of honey bees) that where several founder females are initially associated, only one survives as a queen for the rest of the season. Later in the year males and females arise, mating takes place between them, and the young females hibernate.

Whatever its precise mode of evolution, social life has certainly developed independently along several different lines in insects. In the Hymenoptera it is found among wasps and ants, as well as in bees, and it occurs also in the quite unrelated group of termites (Order Isoptera). Not surprisingly, having regard to the complexity of the adaptations demanded by a fully organized social community, only a relatively small number of species has achieved it. According to Richards, of the 20,200 species of insects in the British fauna, about 88 are social, 37 of these being ants, 44 bees, and 7 wasps.

The characteristic of the hymenopteran colonies is that they are dominated by females, presumably because of the peculiar method of sex determination found in this group, allowing the fertile female direct control over the sex ratio. The mode of origin of the wasp colony can be gauged from the diverse habits of solitary wasps. Some lay their eggs in cells in a nest, and practise what is called mass-provisioning, placing with the egg sufficient food (in the form of paralysed grubs) to serve the young throughout their development. Others employ progressive provisioning, adding more food to the cell after the larva has emerged; this may have been a crucial factor in the origin of the social habit. The colony of the common wasp *Vespula* differs from that of the hive bee in that it is strictly an annual one. During the summer the nest will contain a queen and some thousands of workers, but when colder weather arrives in the autumn the workers and males die, the young queens being left to hibernate and to found new colonies in the spring.

Remarkable as are the specializations of wasps and bees, it is the ants that have most fully explored the possibilities of social life. The group, which comprises over 10,000 species, all of which are social, has exploited terrestrial habitats in every part of the world except the arctic region. We see in ants, more strikingly, perhaps, than in any other insects, the extraordinary potentialities of arthropodization, which in these animals permits the application of social techniques to solve comprehensively a wide range of ecological problems. Elsewhere in the animal kingdom (except in the human species) these problems are dealt with in only a fragmentary way, by species that are specialized for the execution of only a limited range of activities. The comparison of ants with men has often been made. We shall not pursue it here beyond pointing out that one fundamental difference lies in the rigidity of individual behaviour within the hymenopteran colony. The flexible and exploratory behaviour of man, together with his delayed maturation and consequent educability, has made it possible to develop human societies without dependence upon structural and physiological polymorphism. The division of labour is there, however, and so are communication systems; thus comparisons between the human lot and that of the ant are not unprofitable, once the essential differences between the two modes of social organization have been grasped. The comparisons are the more worth while because, as Richards has pointed out, man has probably been social for less than 1 million years, and it is still early to judge his future course. Ants, by contrast, have been organized in societies for much longer even than bees and wasps, certainly for at least 30–40 million years, for they are found preserved in Baltic amber.

It is significant, having regard to the success of ants, that they display more flexibility than other social Hymenoptera. Their nests, built in the soil or out of plant materials, are readily modified and extended, while the animals are less rigidly committed to particular types of food. Moreover, there is within the group as a whole a wide diversification of habit, doubtless reflecting their long evolutionary history.

As regards feeding habits, comparisons can be made with the hunting, food gathering, and cultivation practised by human societies. Some species obtain their food by killing small animals and conveying them to the nest, where they are shared among the members. Others, comparable with human food gatherers, collect seeds or nectar. Such is the association between ants and either aphids or scale insects; this provides the ants with an important source of carbohydrate in the form of liquid excreta (honey dew) of the other insects.

Some ants grade into agriculturists, another human analogy suggested by the leaf-cutting ants of the genera *Acromyrex* and *Atta*. These carry pieces of leaves in procession, 'like Sunday-school children carrying banners', to form in their nests a compost for the growth of the fungus on which they feed. The preparation of the substrate is methodical and thorough, making it difficult, writes Martin, to avoid describing them in anthropomorphic terms as 'efficient and industrious farmers'. Leaf fragments brought into the nest of *Atta* are cleaned, scraped, moistened with saliva, and cut into small pieces, after which drops of liquid faeces are deposited on them. Every step in this process (which they will as readily apply to cornflakes in the absence of leaves) is of elegant adaptive significance, being precisely correlated with certain biochemical limitations of the fungus. This lacks certain enzymes that are needed for the digestion of polypeptides. Fresh leaf material is thus an unsuitable source of nitrogen, and without assistance provided by the ants, the fungus would be at a severe competitive disadvantage; indeed, it cannot survive without them.

The initial preparation of the leaf material removes potentially competitive fungal material and spores, while also liberating cell enzymes and releasing nutrients for immediate use. No less important, the faecal droplets contain nutrients, together with proteolytic enzymes that supplement the limited biochemical resources of the fungus. Comparison of *Atta* with other non-attine species clearly shows the adaptive character of the composition of the faeces in the former, for proteolytic enzymes are absent from the rectal contents of the others (although blowfly larvae do excrete proteolytic enzymes to liquefy the meat in which they live). In short, *Atta* enables the fungus to achieve efficient utilization of protein, while the fungus aids the ant by using its own enzymes to degrade the cellulose which would otherwise not be available for the nutrition of the insect. Martin rightly emphasizes that there is nothing in the biochemical basis of this association which is new, or which departs from fundamental ecological principles. It is, in fact, just one example of the symbiotic relationships which we shall be examining in the following chapters. Nevertheless, we should not underestimate the success with which familiar biological principles and practices are here brought together through the mode of life of these very remarkable insects.

Division of labour and control of reproduction in ants proceeds in principle much as in other hymenopteran colonies. Commonly a colony is founded by a new queen on her return from her nuptial flight, at which time her spermathecae contain enormous numbers of sperm, over 300 million in *Atta sexdens*, according to one estimate. Not surprisingly, some of these queens are believed to survive for as long as 15 years. The workers, which are wingless, and have a small thorax, are variable in form. Sometimes the variation is only a matter of size, but this caste may also be markedly polymorphic, consisting then of smaller workers and larger soldiers, the latter having relatively large heads. The sexual forms, which are winged males and females, are found for only a short season, the males serving merely for the insemination of the new generation of queens. The primary determination of sex depends, as in other Hymenoptera, on whether or not the eggs are fertilized, but apart from this we know little of the basis of caste differentiation, and we cannot assume that the details are necessarily uniform in such a highly diversified group. There is evidence (in *Myrmica rubra*) that the young larvae are plastic, and that their fate depends upon their food supply. It has been suggested that the course of events can be interpreted as a competitive relationship between different sets of organs, with gonads and wing buds developing late, so that sexual females can only arise from larvae that are provided with enough food to enable these organs to complete their development.

Information regarding the modes of integration of ant colonies is no less inadequate, but chemical communication systems are certainly widely exploited. In fact, most of the social behaviour is probably controlled in this way, the agents being exocrine glands discharging a variety of specific secretions which then release innate and stereotyped responses. Odour trails are one illustration of this. They may be exploratory trails, laid by the foraging columns of legionary ants, or the recruitment trails that are laid in many species by workers returning to their nests after discovering food. It seems from studies of the fire ant, *Solenopsis saevissima*, that the strongly attractant qualities of the trail secretion can release all the required behaviour of the followers. The procedure is, in principle, a very simple one, apparently less complex than that of bees. The better the quality and quantity of the food, the more will it be visited. The amount of attractant in the trail will be correspondingly greater, and this will attract still more workers to forage in the same direction. But in *Atta texane* the chemical composition of the trail secretion also plays a part. Rapid recruitment of other workers, once a food source is discovered, is brought about by a volatile component which has little effect after 1 hr. Once the trail is in steady use, however, its persistence is ensured by a non-volatile component. Thus the trail, like that of the fire ant, is reinforced by continued use, until the food has been fully exploited and the ants cease to visit the source.

This exemplifies in a simple way something which a combination of chemical and biological studies is showing to be an important principle of pheromone action: the production of a number of chemical components to evoke the several elements of a complex and integrated response. Another example is given by the bull ant, *Myrmecia gulosa*, which illustrates also how the evolution of social life in ants must have been aided by the addition of complex patterns of aggression to simpler types of behaviour. When this ant is engaged in solitary foraging, it depends on a pattern of aggressive behaviour which is evoked by visual, olfactory and tactile stimuli from the prey. When, however, the nest is under attack, the workers cooperate in a more complex pattern of behaviour, which is evoked by several pheromones. One of these, present in the rectal secretion, is a low-level alerting pheromone which produces only a temporary interruption of activity, but a more vigorous response is evoked by pheromones secreted by Dufour's gland in the abdomen, and by the paired mandibular glands. Dufour's gland secretes a second alarm pheromone which evokes frenzied activity, while the mandibular glands secrete an attack pheromone which evokes grabbing, grasping and stinging.

The slave-making ant, *Formica subintegra*, which captures members of its slave species, *F. subsericea*, and incorporates them into the work of its colony, provides a further illustration. Three esters (decyl acetate, dodecyl acetate, and tetradecyl acetate) are secreted by its enlarged Dufour's gland. These act together to create a penetrating and long-lasting alarm signal within the colony, but they also alarm and disperse members of the slave colony, thereby facilitating their capture. It has been suggested that this is an example of offensive propaganda. Pierre Huber, who discovered slavery in ants in 1810, remarked that fear seemed to prevent the attacked ants from returning to their nest, and he suggested that this was perhaps because they realized that they could no longer remain there in safety. We can now see, however, that no such appraisal is demanded from them, for their caution is adequately accounted for by the persistence of the pheromone of the attackers. It is thought that the abdominal glands of formicine ants were initially used to promote defense by communicating panic alarm within the colony. The elaboration of the glands and their secretions for aggression would then have been a later adaptation. Such adaptations are likely to be widespread in insects, and, indeed, in other groups of animals as well, for the secretion of a blend of chemicals, differing in both their chemical and biological properties (rates of diffusion, for example, and threshold levels for releasing specific responses) offers immense scope for the evolution of signals with a complex information content: and an effective alternative to the slaughter which is the policy of another slave maker, *Polyergus*.

Other examples of chemical communication in ants are the attractants produced by queens, which stimulate workers to tend her and to transport her eggs away. The removal of dead bodies from the nest is stimulated by decomposition products of the corpses, a reaction that may well illustrate, in its use of by-products of metabolism, the way in which some chemical communication systems originated. Chemical control of reproduction, so well authenticated in the hive bee, probably operates also in ants, although the situation is less clear. Certainly there are cases on record in which the presence or the removal of queens influences the fate of the larvae; in *Myrmica*, for example, the presence of a queen is said to result in smaller pupal size and a reduced rate of production of other queens. The effect is not confined to the production of sexual forms. There is evidence that, where the worker caste is polymorphic, the presence of soldiers has an inhibitory effect on the production of other soldiers. Hypotheses based on the supposed distribution of inhibitory pheromones undoubtedly provide the simplest explanation of such phenomena, but not the only one, for it is conceivable that individuals may be influenced by tactile or visual signals (in those species, of course, in which eyes are well developed), or there may be more subtle group effects, dependent upon the composition of the colony and influenced also by external factors. It has been well said of ants that we know in a general way what many species do, yet we do not know how they do it; it is, of course, a comment that is by no means applicable only to this group of animals.

The evolution of social life in the Hymenoptera has clearly been favoured by the peculiar method of sex determination of these animals, and by the high general level of their organization, but there is no reason in principle why it should not also have evolved in other insects. This has, in fact, happened in the termites, the Order Isoptera. If we compare these with the Hymenoptera, we find similar biological results being achieved in very different ways along very different paths of evolution.

We can trace the distinguishing features of termite social organization to two fundamental characteristics of the group. First, their mode of sex determination is of the type more normal in animals—males and females both arise from fertilized eggs, the two sexes differing in the balance between the sex chromosomes and the remaining chromosomes. Probably because of this, both sexes are present in all of the castes, the queen being associated with a king (forming the 'royal couple') instead of relying upon a single act of insemination (Fig. 21–15). The abdomen of the queen becomes enormously enlarged, and she becomes an immensely efficient egg-laying machine, capable of laying at the rate of 36,000 eggs in 24 hr.

The second important characteristic is that termites, being hemimetabolous insects, have a gradual and progressive development through a series of nymphal instars (Fig. 21–16), instead of beginning life as helpless larvae. Thus each member of the colony can soon be useful, while both its form and its function can be modified during its development. This makes for a caste system that is at the same time more flexible and more complex than that of hymenopteran colonies. One caste comprises the soldiers, typically without any sign of wing pads, and to this extent representing an early stage of development. They may be of two different sizes, and they may also be of two distinct types (Fig. 21–17) which do not, however, occur in the same species. One of these types has large mandibles, while the head of the other (the nasute) is produced into a snout from which can be discharged a secretion that is used for attack on other insects. The soldiers do not feed, nor do they carry out the basic work of the community, their function being probably a defensive one. The basic work is performed either by the developing nymphs (in the less specialized colonies) or by a worker caste (in the more specialized ones). The distinguishing feature of a true worker caste is that its individuals are incapable of further growth or development. They, like soldiers, have no wing pads, and may be of two different sizes.

Normally, new termite colonies are formed after the flight of winged sexual forms, which appear in large numbers during a very limited season. The flexibility of the system, however, is shown by the fact that if the royal couple is removed, or dies, certain developing nymphs become capable of reproduction, although they cannot fly, and do not leave the nest. These forms are called substitute sexual forms. Presumably some inhibitory influence exerted by the royal couple normally prevents their development. This effect is well illustrated in some species in which the nest may be enlarged until is spreads far underground. In such conditions the peripheral parts of the nest may be budded off as independent colonies, with substitute sexual forms taking the place of royal couples. This effect suggests that the inhibitory influence of the original royal couple can operate only over a limited distance, a curious parallel with the control of individuality that we have discussed in connection with coelenterate organization.

This is not the only example that can be given of the adaptive regulation of the composition of the termite colony. There is the case of the young colony in which only one nymph normally becomes a soldier; if this is removed, another nymph transforms to take its place, while, on the contrary, the differentiation of the first one can be suppressed if a soldier is introduced into the colony from outside. Then again, if fourth-stage nymphs of *Calotermes* are reared in isolated pairs, both members of the pair will become sexually mature if they are of opposite sex, but only one of the pair if they are of the same sex. Finally, when a number of nymphs of *Prorhinotermes* were placed in an artificial nest many of them died or stayed as nymphs, but the remainder gave rise to an assortment of large workers, soldiers, and substitute sexual

Fig. 21-15. Scene in the royal chamber of the African termite, *Macrotermes bellicosus*, showing the king, queen, and attendant soldiers and workers. After Escherich, from Richards, 1953. *op. cit.*

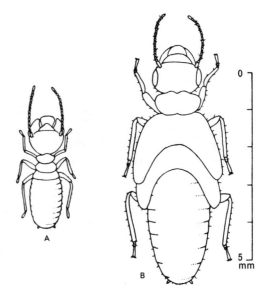

Fig. 21-16. Nymphal stages of *Odontotermes latericius*. A, undifferentiated second-stage nymph; B, fifth-stage nymph with developing wing pads. From Harris, 1961. *Termites: their Recognition and Control.* Longmans, London.

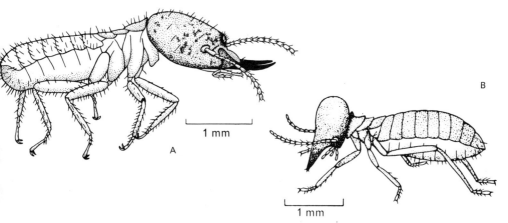

Fig. 21-17. A, soldier of *Coptotermes niger*. B, side-view of soldier of *Nasutitermes infuscatus.* From Harris, 1961, *op. cit.*

forms. The complexity of this transformation can be appreciated when it is realized that some of these nymphs had regressed (the workers and soldiers had lost the wing pads), while the heads of the soldiers and the gonads of the reproductive forms showed progressive development.

We do not know how this developmental integration is effected. It may partly depend upon genetic differences influencing the growth and behaviour of individuals. Other factors might be the signalling of information that influences metabolism, and the transmission of pheromones. Experimental evidence indicating the complexity of the situation has been obtained from *Calotermes flavicollis*. If a colony is divided into two by a double gauze screen, so that one portion contains the royal couple, the other portion then develops substitute sexual forms, all but one pair of these being destroyed. If, however, the gauze screen is single instead of double, substitute sexual forms are again developed, but in this instance they are all destroyed. It is suggested that when there is a single screen the presence of the royal couple is indicated by some signal, perhaps by antennal contact. A chemical inhibitor, which would normally prevent the differentiation of substitute sexual forms, cannot, however, pass, because the gauze acts as a barrier, perhaps by preventing licking. The double gauze partition would prevent direct sensory contact, so that in this circumstance the complete destruction of the substitutes does not occur. The existence of a chemical inhibitor is further indicated by the experiments in which a queen is attached to the screen so that her abdomen belongs to one portion of the colony and her head to the other. Only in the group possessing the abdomen is there complete inhibition of the development of substitute sexual forms; this suggests that an inhibitory pheromone is being released from the abdomen.

Both sexes are mentioned in these experiments, because both are involved in the chemical regulation. Each member of the royal pair inhibits the development of reproductives of its own sex, and stimulates pheromone production by its royal partner. Further, the king produces another pheromone which stimulates the compensatory production of female reproductives if the queen is lost from the colony. Probably other pheromones are involved in other aspects of the community life; in determination of the soldier caste, for example, and in the elimination of any excess production of any caste.

22
Interspecific Associations

22–1 TYPES OF ASSOCIATION

We have seen that the members of natural communities are linked by systems of relationships, partly cooperative and partly competitive, that operate in subtle ways to ensure the efficient exploitation of the environment. Among these relationships there are certain interspecific parterships that are characteristic of particular species, and that are marked by their intimacy, by their permanent or semi-permanent nature, by being obligatory for at least one of the partners, and by the high degree of specialization that they may involve. These relationships have often been described as though they were isolated phenomena, and partly because of this they have sometimes been interpreted in over-imaginative and sentimental terms. In fact, however, they are no more than specialized types of ecological relationship in which the two partners have come to form a functional unit, with its own means of perpetuation, and subject to the action of natural selection.

We have already seen something of the important part played by metabolites in regulating both the multicellular systems of the organisms which produce them, and also the ecosystems in which these individual organisms are adapted to live (Secs. 17–8, 20–1). It is becoming increasingly clear that the intimate interspecific associations with which we have now to deal are, from one point of view, no more than further illustrations of this. In their most extreme form, however, they show how such chemical interrelationships can promote integration to a point at which, it has been aptly said, the concept of organism seems to extend beyond the limits of genetic uniformity that define the individual in our customary use of that term.

But before we can pursue this thought, there are problems of definition to be resolved. These associations are so varied in their nature and in their biological significance that some scheme of nomenclature and classification is needed to describe them. The framing of the appropriate definitions, however, is difficult, for these associations present a continuous spectrum of relationships, so that any classification of them must be somewhat artificial. Because of this difficulty, the terms employed do not always carry the same meanings in different contexts, and can lead only too easily to fruitless semantic controversies. A further handicap is that the terminology came into use at a time when little was known of the physiological basis of the associations. Fortunately, advances in knowledge are now leading to a much more helpful analysis with respect to the metabolic dependence of the partners on each other. For our present purpose we shall define familiar terms in ways that follow some helpful lines of thought set out by Smyth and by Starr, taking heart from the knowledge that a committee responsible for arranging the 29th Symposium of the Society for Experimental Biology, faced with the problem of defining its subject matter, which was *Symbiosis*, 'never quite fully resolved the problem'.

Symbiosis (*symbioun*, to live together) is the living together in close association of (usually two) different species. (This is essentially de Bary's original (1879) definition of the term, but Starr has suggested extending it to include intraspecific relationships as well.) The association must be significant for the well-being, or the unwell-being, or both, of at least one of the partners. Moreover, it must have a time element, being either permanent or at least prolonged. This last proviso relieves us from having to regard predation as a form of symbiosis, although an alternative suggestion (perhaps not to be taken too seriously) is that it might be considered as 'necrotrophic symbiosis with behavioural overtones'. (Necrotrophy is feeding upon dead cells, biotrophy is feeding upon living ones, and saprotrophy, in this context, is feeding upon the non-living environment.) It will be appreciated that symbiosis, as defined here, embraces a number of types of relationship which can be distinguished and severally defined in the following ways.

Commensalism (*cum*, with; *mensa*, table) is a regular and close association in which one partner, often called the commensal, feeds upon surplus food captured by the other. The commensal gains some benefit, but the host suffers no serious disadvantage except in so far as it is robbed of some of its food, although not enough to impede survival or reproduction. It is not always easy to make a distinction between commensalism and parasitism. One that has been suggested is that neither partner in commensalism is metabolically dependent upon the other, but it is not clear that this is necessarily so.

Inquilinism (*inquilinus*, tenant) is a regular and close association in which one partner lives within a host, obtaining shelter thereby, and perhaps also appropriating some of the host's food. It will be obvious that on this definition the only distinction between inquilism and commensalism is that in the former association one partner is within the other; inquilism can therefore be considered as a particular form of commensalism.

Mutualism (*mutuus*, exchanged) is a regular and close association from which both benefit, and from which neither derives any disadvantage. (This type of relationship is sometimes referred to as symbiosis, in a more restricted and, perhaps, more confusing sense than that given above.)

Parasitism is a regular and close association in which one partner (the parasite) lives on or in the body of the other partner (the host), nourishing itself at the expense of the host, but not necessarily destroying it. When it does destroy it, it is doing itself a disservice, and can be regarded as imperfectly adapted to the association. 'It is very well to have friends to lean upon', but Anthony Trollope rightly urged caution in exercising the privilege. A parasite is often said to be 'harmful' to its host, but this introduces an anthropomorphic element into the definition. It is more helpful, and less emotive, to consider a parasite as an organism which is metabolically dependent upon its host, with the degree of dependence varying from species to species. Total parasitism is 100% metabolic dependence. At the other end of the spectrum is zero dependence, which is the condition of a free-living organism. A major factor in the specialization (and, indeed, in the fascination) of parasitism is the need for the parasite to ensure transmission to another host. Often this is effected by an organism called the vector, which serves as a go-between. It is a convention, but not always a convenient one, to regard as the primary host the one in which the parasite attains sexual maturity. The other host (or hosts, for more than one may be involved) is then termed the intermediate or secondary host.

22-2 COMMENSALISM AND MUTUALISM

We can easily visualize some possible origins of commensal relationships. They could be a consequence of the crowded conditions of life in many environments, particularly in the littoral zone, and of the way in which the habits and life histories of some animals bring them into the closest contact with members of other species. So a sessile species might readily become attached to the surface of a free-moving one, perhaps as a result of the accidental settling of larvae upon the host. However, life is unlikely to be as simple as this. It must be remembered, for example, that larvae are highly selective in their choice of substrates, and that the secretions of free-moving animals may often be toxic, and may be specialized to keep the body surface clear of precisely such entanglements. We must certainly expect, therefore, that the maintenance of associations of this kind (whatever their origin may have been) will often involve considerable specialization on the part of one or both partners.

Thus some of the sessile vorticellids that are common on the surface of aquatic crustaceans have come to depend for their survival upon the movement of the host. If they are allowed to become attached to pieces of chitin in artificial conditions they will only continue to live if the fragments are kept in continuous movement. Whether the initial growth of the vorticellid on the host depends upon some specific attraction is not known, but it is clear that the association is far from being simple and unspecialized. This is a case of physiological specialization. The vorticellid *Ellobiophrya donacis* (Fig. 22–1) provides an example of morphological specialization in its attachment to the gills of the bivalve, *Donax vittatus*. The benefits obtained by the protozoan are presumably protection within the shell, and a supply of food from the feeding currents that pass over it. At this level of analysis the association would seem to be a very simple form of commensalism, yet the vorticellid is highly specialized to maintain it, the attachment stalk being bifurcated in a unique way so that it can be virtually padlocked to the meshwork of the gill.

No less specialized are the two-tentacle commensal hydroids that live as colonies upon the tubes of sedentary sabellid polychaetes, and that are sometimes referred to as 'Lar', from the name first given to them by Gosse (Fig. 22—2). Three species of these animals have been described, all belonging to the genus *Proboscidactyla*, and all occupying very similar habitats, for they live respectively on the tubes of *Potamilla torelli*, *P. myriops*, and *Pseudopotamilla ocellata*.

672 INVERTEBRATE STRUCTURE AND FUNCTION

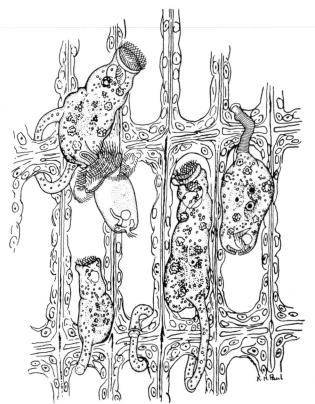

Fig. 22-1. *Ellobiophrya donacis* attached to the gill-grid of its host. Upper left, an individual with a bud that has resulted from a longitudinal fission, showing the posterior circle of cilia. From Baer, 1951. *Ecology of Animal Parasites.* University of Illinois.

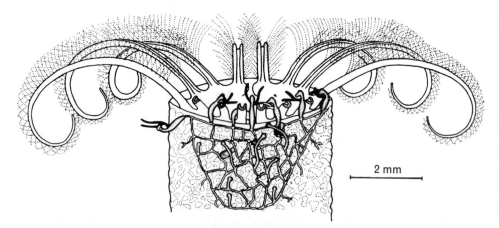

Fig. 22-2. A colony of *Proboscidactyla* sp. in place upon a tube of sabellid, *Pseudopotamilla ocellata*. Several of the worm's tentacles have been cut away, and the number of tentacles shown on it is reduced from the natural condition. From Hand and Hendrickson, 1950. *Biol. Bull. mar. Biol. Lab., Woods Hole,* **99,** 74–93.

The feeding of *Potamilla* is similar to that of *Sabella* (p. 257), depending upon collection and rejection currents upon the tentacular crown, palps, and lips. To these currents the commensals are closely adapted. The colony consists of gastrozooids, orientated around the opening of the worm's tube, and of mouthless gonozooids, situated lower down, along the sides of the tube. The positions are themselves of obvious adaptive significance, and this is emphasized by the behaviour of the gastrozooids. These are in continuous movement, their actions giving them the ludicrously human appearance from which Gosse derived their name (*lar*, household deity). By leaning back they secure food from the incoming current of water, while by leaning forwards, among the tentacles or into the opening of the tube, they can remove particles carried in the ciliary currents. 'The head lobe moved to and fro on the neck' wrote Gosse when he first described them; 'the body swayed from side to side ... while the long arms were widely expanded ... and then waved downwards, as if to mimic the actions of the most tumultuous human passion'.

Hand and Hendrickson point out that the two-tentacled condition of the gastrozooids is well suited for this commensal life. The tentacles are unusually active, and they can be greatly extended to explore much of the host's feeding currents. More tentacles might easily be a disadvantage, for they would tend to become entangled with the branchial crown; their peculiar origin, from a single common area below the mouth, probably helps in avoiding this entanglement. *Pseudopotamilla* reproduces by free-swimming medusae. How these become associated with new hosts for the asexual phase is unknown; we may assume that this involves some response no less specialized than the behaviour of the gastrozooid, for Hand and Hendrickson remark that the species studied by them was never found associated with any animal other than *P. ocellata*. This aspect of animal relationships we shall return to later. In the meantime, we may accept 'Lar' as an example of specialized commensalism, with the benefit accruing solely to the epizoitic coelenterate, as far as we know at present.

Another example of commensalism associated with extensive morphological specialization is seen in the Temnocephalida. These are platyhelminths that are believed to be related to rhabdocoels, and are sometimes classified with them. They live mainly on the body surface or within the branchial chamber of fresh-water crustaceans, and are distinguished from other rhabdocoels by this habit. They are distinguished also by possessing anterior tentacles and a posterior attachment disc which are used for carrying out leech-like movements. It is difficult to believe that this association gives sufficient advantage to the temnocephalids to account for their high level of structural adaptation, particularly since they can feed on diatoms and small animals, and can live for months without their hosts. It seems more likely, especially in view of our earlier discussion of the role of chemical regulation in the organization of animal and plant communities, that this is an example of a relationship involving some form of chemical dependence, operating at a level of subtlety that has not yet been explored. Such relationships (and there are probably many of them) could hardly have arisen ready made. In the case of temnocephalids it is conceivable that in the earlier stages of their evolution some slight advantage in the supply of nutrients, or of oxygen, in the respiratory streams of the hosts, would have provided a foundation upon which a more intimate dependence was later evolved.

Some forms of relationship, which involve little if any modification of the partners, seem to involve protection rather than feeding, but these are instructive in suggesting ways in which more specialized types of commensalism might have been initiated. For example, a variety of species, including creeping ctenophores and crustaceans, inhabit the spiny surface of sea-urchins, primarily obtaining protection but perhaps securing some food as well. Then there is the relationship (cleaner-shrimps are an example) in which many marine and fresh-water organisms clean other organisms, removing ectoparasites and necrotic tissues; presumably feeding is an element here, in which case this is essentially mutualism. Another mutualistic relationship is that between the shrimp *Alpheus djiboutensis* and the gobiid fish *Cryptocentrus cryptocentrus*. The shrimps live in pairs in burrows which are also used by the fish for temporary shelter and rest. It is supposed that the fish benefits by being helped to survive in a shelterless terrain, while the shrimp, which maintains contact with the fish through its tentacles (Fig. 22-3), benefits from this behavioural specialization by having a warning system when it emerges from shelter to feed on sediment.

Coelenterates are involved in many well-known symbiotic relationships, ranging from commensalism to mutualism, which also seem often to involve little adaptive modification in structure, although there may be considerable specialization in behaviour. One example is the association between the coelenterate *Hydractinia echinata* and the hermit crab *Pagurus* (*Eupagurus*) *bernhardus*. Experimental study has shown that the crab has no marked preference for shells that are covered by

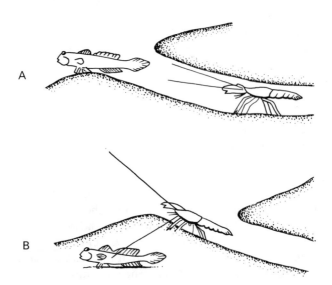

Fig. 22-3. The sequence of behaviour during emergence from the burrow in the mutualism of the shrimp *Alpheus djiboutensis* and the fish *Cryptocentrus cryptocentrus*. (A), the shrimp moves head-first toward the entrance of the burrow; antennae are somewhat raised at an oblique forward angle with the tips 2–3 cm apart. (B), outside the burrow the shrimp keeps one antenna in contact with the fish. After Karplus, from Fricke, 1975. *Symp. Soc. exp. Biol.*, **29**, 581-594 (D. H. Jennings and D. L. Lee, eds.).

the coelenterate. *Hydractinia* is equally independent in habit; it feeds mainly on living plankton, and, although it may share the food of the crab, it does not rely upon this for its regular supply of nutriment. It possesses a form of zooid, called the spiralzooid, that is defensive in function as far as the colony is concerned, but there is no evidence that this zooid is of any value to the crab. In particular, it is not especially developed around the opening of the shell in which the crab lives, where any defensive value would show to the greatest advantage.

There is, however, a significant difference in the distribution of the two partners, and this suggests a possible explanation of the association. *Hydractinia* can live in colder conditions than can *Pagurus*, which is one reason why it is found growing on empty shells in the complete absence of the crab. On the other hand, its freedom for reproduction is limited because its eggs sink to the bottom; thus it cannot reproduce successfully in regions where the substratum is too soft or too poorly aerated to permit survival. Its association with the hermit crab enables it to invade and exploit such regions, to an extent that would be impossible without the support of its host. Consequently, *Hydractinia* is found almost exclusively on hermit crabs in regions where the substratum itself is unsuitable for it. It lives abundantly as independent colonies, however, where suitable rocks and piles are readily available.

A more specialized association of this type has developed between the anemone *Adamsia palliata* and the hermit crab *Pagurus prideauxi* (Fig. 22–4), first described in detail by Gosse. In this association, unlike the previous one, the specializations both of the crab and of the anemone are very evident, involving as they do both structural and behavioural features. Indeed, so intimate is the association that neither partner can survive without the other, once they have grown beyond an early juvenile stage. The crab can recognize the anemone by touch, being aided, perhaps, by chemical stimulation, and it can transfer it from the substratum to its shell. Thereafter the anemone grows in a highly specialized way to enclose the shell, maintaining its

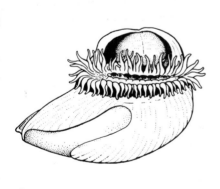
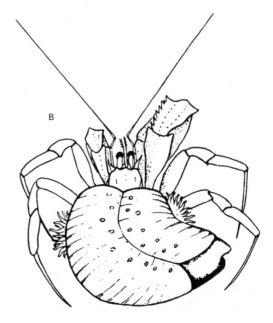

Fig. 22-4. A, shell of *Scaphander*, enveloped by *Adamsia palliata*, seen from below. B, *Pagarus (Eupagurus) prideauxi* in a shell of *Scaphander* which is enveloped by *Adamsia palliata*. From Faurot, 1910. Archs Zool. exp. gén., 5 sér, **5**, 421–486.

mouth immediately below that of the crab so that it is placed in the best position to make use of the feeding processes of the host. As for the crab, it is very active in movement, because of its elongated appendages and because it habitually lives in a shell that is too small to enclose it completely. This is possible because the pedal disc of the anemone secretes a horny membrane that extends over the opening of the shell, and so increases the shelter afforded to the host. No doubt the crab also obtains some protection from the anemone's nematocysts, which are borne on particularly long acontia. Such is the intimacy of this relationship that if the crab is removed from its shell the anemone drops off and soon dies. Here, surely, is mutualism, in the sense of the definition given earlier.

One feature of this association that has attracted particular attention is the mode of transference of the anemone when the crab moves out of one shell to occupy a larger one. This was examined by Gosse, who initially formed the impression that the anemone was an active participant in the removal, and that the crab's behaviour was attuned to that of the coelenterate. During his earlier observations he unfortunately absented himself from his aquarium during the vital hour when the transfer was effected. He concluded from the position of the anemone when he returned that it must have attached itself to the crab and relinquished its hold on the original shell. This drew from him an enthusiastic if incautious paean:

> But what a series of instincts does this series of facts open to us! The knowledge by a crab of the qualities of the new shell; the delay of his own satisfaction till his associate is ready; the power of communicating the fact to her; the power in her apprehending the communication; her immediate obedience to the intimation; her relinquishment of her wanted hold ... all these are wonderful to contemplate, wonderful considered singly, far more wonderful in their cumulation!

Later, however, when he removed an anemone from a shell, he observed that the crab replaced it with its chelipeds; this cast some doubt on his earlier interpretation, and somewhat quenched his wonder, for it suggested to him 'a suspicion that the claws of the crab may have been employed in the transference of the cloaklet from shell to shell'.

This suspicion was well founded; the crab does indeed remove the anemone, as happens also in another well-known association: that between the anemone *Calliactis parasitica* and hermit crabs of the species *Pagurus bernhardus*, *P. striatus*, and *P. arrosor*. The involvement of more than one species of host is an indication of a less extreme specialization. So also is the ability of *Calliactis* to live independently of its host, and the occasional presence upon one shell of a number of anemones which show no regular orientation with reference to the mouth of the host. *Calliactis*, like *Adamsia*, has very long acontia with nematocysts, and the protrusion of these through lateral openings in the body wall doubtless contributes to the defence of the crab as well as to that of the anemone itself. Since the anemones are probably aided in feeding, both by the movement provided by the crab and by the possibility of securing some of its food, the association can be regarded as an example of mutualism, but one that is not as intimately symbiotic as that of *Adamsia* and *Pagurus prideauxi*.

The degree of specialization involved in the *Calliactis* type of relationship should not, however be underestimated. Indeed, it has been clearly demonstrated in Cowles's study of *Pagurus deformis*. This crab, which lives in the shells of *Dolium*, carries two different kinds of anemone, a large greyish one that is usually situated on the sides of the shell, and a smaller and almost colourless one that is attached on the underside below the protruding head of the crab. One shell may bear as many as eight of the former and four of the latter. Cowles observed that after a crab had transferred itself to a new shell it removed the anemones from the old one by clawing, pinching, and pulling them. Its movements were clumsy, and the anemones sometimes slipped off, with the result that the crab temporarily ceased to give them its attention; eventually, however, they became attached, guided into position by the crab's limbs. Two features of this performance illustrate the specialized character of the behaviour of both partners. Firstly, the anemones appear to separate from the old shell more readily and with less disturbance than under the hands of an experimenter. Secondly, the crab's behaviour appears to involve a chain of reflexes that the animal may be unable to complete if it is disturbed. Thus in one instance a crab made no attempt to transfer its anemones when it was removed with them to a new aquarium immediately after it had secured possession of a new shell. Presumably the normal sequence of the behaviour pattern had been interrupted.

We might expect a comparative study of these relationships to reveal various degrees of specialization, and *Calliactis* does, in fact, provide good examples of this. It is associated not only with *Pagurus striatus* and *P. arossor* but also with *P. bernhardus*, and Ross has shown that in this last instance the hermit crab seems to be indifferent to the presence of the anemone, and does not participate actively in establishing the association. The necessary stimulus is given when the anemone comes into contact with a suitable shell, to which it adheres by its tentacles as a result of discharging their nematocysts. Probably some chemical stimulus is derived from the organic deposits on the shell, while the physical features of the latter doubtless play some part also. The situation, in fact, is strongly reminiscent of the relationships between the larvae of *Ophelia* and the characteristics of their preferred sand grains. That the threshold of discharge of the nematocysts can be controlled in some way is shown by the fact that an anemone already attached to one shell will not adhere to another shell when this is presented to it.

In this case, then, the anemone is the active partner in the relationship; we may suppose, therefore, that it was the anemone that initially derived selective advantage from it, perhaps because attachment to a moving eminence helped with the securing of food. The association between *Calliactis* and *P. striatus* and *P. arrosor* are more specialized. In both associations the anemone requires some preliminary stimulation before it will climb onto the shell, with the *P. arrosor* it requires further assistance before it can become firmly attached. As Ross points out, the three relationships are suggestive of an evolving ecological situation, in which there is a shift in emphasis from the anemone to the crab, accompanied by changes in behaviour pattern and presumably in the balance of selective advantages. Already, then, we see some of the problems of definition. Commensalism grades easily into mutualism, as the balance of advantage shifts between the two partners. And our own assessment changes, too, as we analyse these relationships more closely and come to a better understanding of them.

The demonstration that associations between some hermit crabs and anemones depend upon the latter being seized by the host allows us to relate these situations to other examples of crab behaviour, and thus to see them as part of a wider spectrum of adaptations. Certain crabs, of which *Maia* and *Hyas* are familiar examples, rely upon a variety of objects for camouflage and concealment. The surface of their bodies is frequently decorated with algae, sponges, and hydroids in a manner that provides effective disguise in the crowded conditions of the littoral zone. Some at least of these objects are seized by the crabs and placed upon their dorsal surfaces, where they are held in position by hooked setae. In aquaria the animal have been observed to make use of coloured rags; they select appropriate colours, and move on to backgrounds that match their disguise. There is no evidence of such a practice among hermit crabs, but some such behaviour may well have played a part in the evolution of their associations with coelenterates. This is the more likely in view of the remarkable way in which anemones are handled by *Melia*, a crab that is unrelated to those so far mentioned, and that is found on coral reefs.

It is the habit of *Melia* (Fig. 22–5) to carry a small anemone in each of its two chelae, and to brandish these towards a threatening agressor. Observations have shown that this depends upon the crab coming into contact with the anemones during its wanderings. If a crab that has been deprived of its anemones is placed with them in an aquarium it shows no sign of recognizing them until it actually touches them; once it has done this, however, it picks them up by inserting the tip of its first walking leg under the base of the anemone and prising it away from the substratum. The behaviour of the crab is clearly specialized to facilitate the grasping of the anemones, and so also is the structure and use of its appendages. The chelipeds are so slender that they are unsuited for defence or predation, but just because of this they are well adapted for holding the polyps around the middle of their bodies. They do this without injuring the anemones, which soon expand after seizure and maintain their tentacles in full extension. The chelipeds themselves cannot, of course, be used for feeding as they are in other crabs. Instead, the animal relies upon the walking limbs for seizing material from the substratum, the first pair of legs being the most active in this respect. But the most striking adaptation is the way in which the crab exploits the feeding activity of its anemones. If an observer places meat upon the oral disc of one of them, the crab quickly becomes aware of this. Presumably it receives some chemical stimulus from the food, although the possibility of it being sensitive to particular patterns of movement of the anemone is not excluded. Whatever the explanation, however, the crab moves one of its anterior legs forward to the oral disc and removes the food from it, transferring it to its mouth. Should the actinian have begun to swallow the food, the crab declines to be cheated. If necessary it inserts its limb into the stomodaeum and withdraws the food from it.

The remarkable specializations of the crab are not paralleled in the anemone.

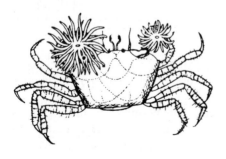

Fig. 22-5. *Melia tessellata* from the Hawaiian Islands, bearing an expanded actinian in each claw. When food is placed on the disc of the actinian, the first ambulatory limbs of the crab reach over and abstract it and pass it to the crab's mouth. From Duerden, 1905. *Proc. zool. Soc. Lond.*, **2**, 494–511.

Both *Sagartia* and *Bunodeopsis* are carried in the chelipeds, and the crab shows no preference for one or the other. Neither seems to be modified for the association, and, although observers of *Melia* have failed to find free-living specimens of the anemones in the same habitat, there seems no reason why they should not live independently. The balance of adaptive advantage is thus the converse of that existing between *Calliactis* and *Pagurus bernhardus*. Even more is this so in a similar association between another crab, *Polydectus cupilifera*, and small specimens of the anemone *Phellia*. Despite the establishment of this association, free-living specimens of the anemone are common around the Hawaiian Islands where the crab is found. Moreover, on Atlantic coasts where *Melia* is absent there are free-living specimens of *Bunodeopsis* that are anatomically indistinguishable from the specimens carried by the crab in other localities. It would appear, then, that this association is not essential to maintain the life of the anemone. Through being carried by the crab it may encounter more food than it otherwise would, but against this must be set its risk of losing this food to its host.

By contrast, the crab probably derives marked benefit from the association, and its adaptations seem to make it actually dependent upon the anemone. The presence of the latter in a cheliped aids both in securing food and warding off enemies; indeed, the chelipeds are of limited value without this aid, for they have little power of grasping food or other objects when these are presented to the animal. The maxillipeds and walking limbs are equally limited in these functions, owing to their lack of chelae. Whether the crabs always carry the anemones is not certain, for they have sometimes been found without them, but this may be only a temporary loss, either from accident or as a result of moulting. The evidence as a whole suggests that the association is a common one, characteristic of the species, and one for which it has become closely adapted. We may regard it as mutualism of a rather one-sided character.

22-3 BEHAVIOURAL ASPECTS

These examples show how far we have now moved from the sentimental and anthropomorphic interpretations of the nineteenth-century naturalists. Of course, we do not fully understand these associations, but at least we can see that they sometimes depend upon structural and behavioural adaptations paralleling those that fit free-living animals to particular niches in their habitat. We must now probe a little more deeply into the nature of some of the adaptations that contribute to these associations.

A crucial problem is to determine how the two partners are brought together. We have already suggested that initially the association must have been an accidental one, promoted, perhaps, by some similarity or complementary feature in the habits of the two species. Such could have been the origin of the associations of anemones and crabs. In some other instances one partner may have provided a suitable settling point for the larvae or juvenile stages of the other—as, for example, in the association between the sipunculid *Aspidosiphon* and the polyp *Heteropsammia*. The latter grows over a gastropod shell in which the sipunculid lives; as a result, the polyp secures some mobility, while the sipunculid is well protected by it, to the point of being almost completely enclosed. In this instance the association is initiated by the larva of the polyp settling on the shell. To what extent this is a selective settling is not known, but even if it is selective rather than a chance distribution there is evidently nothing here that cannot be paralleled in the normal course of larval life histories.

We have seen that the settling of larvae may be profoundly influenced by the physical and chemical properties of the substratum, and similar factors presumably operate in commensalism. An example of this, although demonstrated on adults and not on larvae, is the behaviour of mites that are commensal within the mantle cavities and gills of *Anodonta*. Mites of the species *Unionicola ypsilophorus*, when washed free of any contaminating factor derived from its host, *Anodonta cataracta*, are positively phototactic. If they are placed in an aquarium containing a chemical factor from the host they immediately become negatively phototactic, a reaction that will normally lead them to enter between the lamellibranch's valves. In principle this is similar to the change in the phototactic response of the ascidian tadpole, except that this latter change is initiated within the larva at a particular stage of its free-swimming life. The response of the mites, incidentally, is a highly specific one. Welsh studied three species which were each specifically commensal with three different species of mussels; he showed that reversal of phototaxis occurred in each species, but that it could be produced only by a chemical factor from the particular species of mussel with which the mite was normally associated.

Our review of the importance of allelochemicals in communication (Sec. 17-8), would lead us to expect the type of response shown by these ticks to be widely distributed in commensal relationships. Some convincing evidence for this has been obtained by Davenport in his studies of the responses of polynoid worms. An example is provided by the genus *Arctonoë*, three species of which constitute a commensal complex in the littoral zone of Puget Sound. The principle of Davenport's experiments was to place a worm in a Y-shaped choice tube, the two arms of which were connected to two aquaria (Fig. 22–6). In these were placed various echinoderms. Water from the aquaria was passed through one or other of the two arms, and the worm was then observed in order to see which stream of water it preferred, or, in other words, which arm of the Y it entered.

The species *Arctonoë fragilis* is commensal on the inter-tidal starfish *Evasterias troschelii*; in one type of test this worm distinguished readily between untreated sea water and sea water coming from the aquarium in which a host was contained. Similarly, another species of worm, *A. (pulchra-vittata?)*, was attracted by sea water coming from its own particular host, the holothurian *Stichopus californicus*. The specificity of these responses was demonstrated by two further types of test. In one of these it was shown, that *A. fragilis* is not attracted to sea water coming from the starfish *Pisaster*, which is not its host; similarly, *A. (pulchra-vittata?)* was not attracted to *Cucumaria*, which again is not a host for this species. The other type of test involved cross experiments in which *A. fragilis* (from *Evasterias*) was tested against *Stichopus*, and *A. (pulchra-vittata?)* (from *Stichopus*) was tested against *Evasterias*. Neither of the commensal worms was attracted to the host of its partner species. It may thus be concluded that these polynoids are able to react specifically to a substance or substances released from their own particular host species.

The nature of these chemical attractants is not known; attempts to extract them have proved unsuccessful, possibly because they are very unstable, or because their action is easily obscured by contaminants from the tissues. There is no reason,

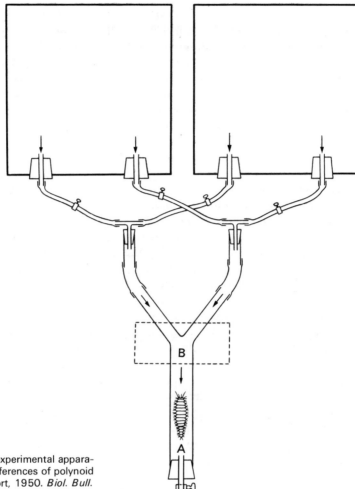

Fig. 22-6. Plan of experimental apparatus for testing the preferences of polynoid worms. From Davenport, 1950. *Biol. Bull. mar. biol. Lab.*, Woods Hole, **98**, 81–93.

however, why they should not be common metabolites, just as the respiratory centre of vertebrates is influenced by that commonest of metabolites, carbon dioxide. In fact, this principle may account for the responses of another polynoid, *Harmothoë*, which is commensal both with a terebellid worm, *Amphitrite gracilis*, and with a eunicid worm, *Lysidice ninetta*. It responds positively to both hosts, but gives no response at all, or at best a very weak one, to other terebrellids which are not its host. If we are correct in assuming that the response is the result of the worm recognizing some simple chemical product, we could assume either that it is adapted to respond to two different substances from the two hosts, or that both hosts are producing one and the same substance. Davenport considers that the latter is more likely, and that the double association results from the chance production by two species of a similar metabolite.

Undoubtedly the responses of commensal species need much further investigation before we can reason with any assurance about the nature. Nevertheless, the general principle of chemical recognition seems to be well established. Davenport mentions other examples in which this appears to be the means by which two species are associated; among these are the commensalism of *Hesperonoë (Harmothoë) adventor* and *Urechis caupo*, to which we have earlier referred. But we cannot suppose that chemical factors are the only means of communication involved. Responses to gravity play a part in the symbiotic relationship between burrowing echinoids and the bivalve *Montacuta* which lives with them. *M. substriata* is geonegative, which tends to keep it near the surface of the sand, where its host, *Spatangus purpurea*, lives. *M. ferriginosa*, however is geopositive, a difference nicely correlated with a different host, *Echinocardium cordatum*, which lives deeper. The polychaete *Nereis fucata*, which lives within the shell occupied by hermit-crabs, comes into association with its host in a different way. First it recognizes the vibrations made by the crab as it approaches, and then it makes contact with the surface of the shell. As a final example, the sheep tick *Ixodes ricinus* depend upon sensitivity to contact, light, gravity, moisture, temperature, and vibration, as well as chemosensitivity. All the sensory perceptions help in finding the host, on which the survival of the ticks depends, and a similarly wide range of perceptions is probably involved in many forms of interspecific relationship.

This must evidently be so in those associations that depend upon sequences of complex behaviour patterns, such as are involved in the relationships of anemones and crabs. In analysing these we can employ the concept that responses are evoked by specific sign stimuli, or releasers. This principle, to which we have already made some reference, was initially formulated by Lorenz to account for the maintenance of intraspecific associations. It operates, for example, in reproductive physiology, where the association of the two sexes, and the initiation of mating behaviour, is evoked by sexual recognition signals displayed by one sex to stimulate the other. An illustration is the enlarged chela of the male fiddler crab. The manipulation of this organ serves as a signal to the female; it is highly specific, and constitutes a barrier to interspecific mating.

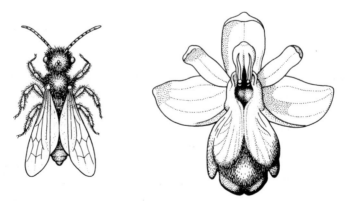

Fig. 22-7. *Left*, the andrenid bee, *Andrena trimmerana*. *Right*, a flower of the orchid *Ophrys fusca*, on the libellum of which the bee will make copulation attempts. From Davenport, 1955. *Q. Rev. Biol.*, **30**, 29–46.

Such a system, providing for animal communication between the members of the same species, could equally well provide for communication between members of two different species. As Baerends points out, the mutual development in two species of a releasing stimulus in the one and a releasing mechanism in the other can only occur if the situation is of benefit to both species, since only then will natural selection favour the necessary adaptation in both of them. There must be many examples of mutual associations in which this condition is met. One particularly striking one, mentioned by Baerends, is the interrelated adaptations of orchids and Hymenoptera that result in pseudo-copulation and sexual exploitation (Fig. 22–7). Certain of these plants are always visited by the male, which fertilizes the flowers while it copulates. The flowers, and particularly the libellum, closely resemble the insect, and this resemblance serves as a sign stimulus that releases the copulation response in the male. It is doubtful whether the insect obtains any benefit from the association, since these flowers do not produce food for it. The flower, on the other hand, is certainly dependent upon the insect, and may be said in a sense to be parasitic upon a clearly defined and essential reflex act of the latter.

23
Further Interspecific Associations

23–1 MUTUALISM AND NUTRITION

We have so far considered examples of commensalism and mutualism in which at least some of the benefits involved can be seen by inspection, although it is likely that there is still much to be learned about them at a deeper level of analysis. There are other relationships, however, in which the associations are so intimate that it is very difficult to evaluate them. Consider, for example, the mycetocytes of insects. Mycetocytes are large cells, probably highly polyploid, that may be scattered in the fat-body, or concentrated into groups, or associate to form multicellular or syncitial structures called mycetomes. These cells harbour yeast-like, rickettsia-like, or bacterium-like bodies, that have been interpreted as symbiotic micro-organisms which supposedly invade the cells when these become embryonically differentiated. The identification of the intracellular bodies as living micro-organisms is, however, far from easy. Some have been successfully cultivated outside the host, and can be accepted unequivocally as true micro-organisms, but in other instances this has not been possible. On the one hand, therefore, it has been suggested that some of the non-cultivable bodies are cell inclusions (mitochondria, for example), while at the other extreme it has been argued that they may actually be symbiotic organisms, but so specialized that they are no longer capable of independent life.

The so-called bacteroids of cockroaches illustrate the problem. These bodies, which are found within mycetocytes, are certainly not mitochondria, for they can be distinguished from these by selective staining. Moreover, they can be eliminated by treatment of the host with antibiotics, which suggests that they are indeed bacteria; it has been difficult, however, to confirm this by cultivating them outside their host, although they multiply within embryonic cells of the host when these are cultivated *in vitro*. Here it is necessary to rely largely upon indirect evidence. During the development of the insect the bacteroids are passed into the oocytes and are eventually distributed only to the mycetocytes, which certainly suggests that they have a specialized role within the body. In short, there is circumstantial evidence of a functional relationship between the micro-organisms and their metazoan host.

One thing that we need to know in such cases is whether benefits are actually obtained by one or both of the partners. It can reasonably be supposed that the micro-organisms would derive some nutritive advantages. Specific suggestions include the possibility that they might metabolize uric acid or ammonia produced by the host, which would thereby benefit in its excretion. Some hosts, however, have been shown to obtain more clear-cut advantages than this. One example is the blood-sucking bug *Rhodnius*, which we have already encountered in several contexts. This has a micro-organism, *Actinomyces rhodnii*, which lives in the lumen of the gut, and which is taken up by newly hatched nymphs from the surface of the egg, or perhaps from dried faeces. If infection is prevented, the insect grows and moults normally up to the fourth or fifth instar, but only a few become adult, and none of these ever reproduces. Normal growth and reproduction, however, are readily restored by allowing the animals to become infected. A clue to the explanation of this is that blood is a restricted diet, poor in vitamins. Probably, therefore, the micro-organisms provide an essential dietary supplement for their hosts.

Confirmatory evidence of this comes from another blood-sucking insect, the human body louse, *Pediculus corporis*, which possesses a mycetome. Growth and reproduction are impaired if this is removed, but can be at least partially restored by injections into the rectum of yeast extract, which is rich in the B vitamins. Given this explanation, it is not surprising to find that symbionts are common in blood-sucking insects. Those that lack them are thought to be those that feed on micro-organisms in their young stages; a habit which presumably enables them to build up an adequate vitamin supply.

Evidently, then, symbiotic micro-organisms can make some contribution to the metabolic processes of the host. Advances in our knowledge of the subtle chemical relationships that may sometimes exist between organisms, and of the complexity of their nutritional requirements, are a warning against expecting the benefits of the relationships to be immediately obvious, or readily open to experimental demonstration. Conceivably the symbiotes may be of value as a source of metabolites, not because the host is unable to manufacture them for itself, but because at certain periods of its life cycle, or under particular stresses, they are needed in greater amounts or in different proportions. This, in theory, could explain why the beetle *Rhyzopertha dominica* suffers no apparent ill effects from the removal of its symbiotes, although it has evolved a mycetome in which to contain them. At least it is evident that associations between insects and micro-organisms must have arisen independently on many occasions. In some instances the micro-organisms are Protozoa, and the analysis of the function of these in aiding termites to digest cellulose provides a well-documented illustration of a truly mutualistic relationship.

The cellulose of plants is a valuable store of energy, but one to which few animals have been able to obtain physiological access. Exceptions include the wood-boring mollusc *Teredo*, the wood-boring isopod *Limnoria lignorum*, and the silver-fish *Ctenolepisma lineata*, which lives on the bark of Eucalyptus trees. These species are said to digest cellulose by their own secretions, but the matter merits further study, for another isopod (*Philoscia muscorum*), which feeds on decaying leaf litter, uses bacteria in its food to break down the cellulose, unaided either by any gut flora or by its own digestive secretions. In general, however, animals have apparently been unable to evolve the cellulase and the β-glucosidases that are needed for the hydrolysis of cellulose (some bivalves, and the gastropod *Helix pomatia*, are amongst the few exceptions), despite the virtually universal distribution of the corresponding enzymes, amylase and maltase, that are used in the hydrolysis of starch.

Termites, nevertheless, feed largely upon cellulose-rich plant material, and do so in such numbers and with such vigour that in warmer countries they are a serious menace to timber, at an estimated annual cost of 1.2 billion dollars. It is only fair to add, however, that less than 10% of the named termite species are actually pests, while the group as a whole makes an important contribution to the carbon cycle by its breaking down of plant litter. Some termites, regarded as the 'higher' ones, feed upon humus, or on leaves and grass, or on wood which has been already partially decomposed by fungi. Some of them (the Macrotermitinae) cultivate fungi within their nests, growing them upon masses of chewed wood, and using them in their diet to provide proteins and, perhaps, vitamins.

The more primitive ('lower') termites, however, feed directly upon the timber within which they excavate their nests. It is the relatively few species of these (belonging to the families Rhinotermitidae, Kalotermitidae, Hodotermitidae, and Mastotermitidae) with which we are here mainly concerned, for the finely fragmented wood particles which they swallow are ingested and broken down by symbiotic zooflagellates (Figs. 23–1, 23–2) which are found elsewhere only in the wood-eating

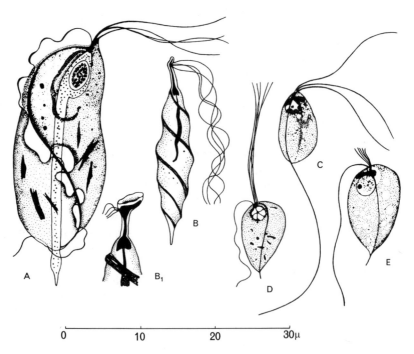

Fig. 23-1. Polymastigote flagellates from *Zootermopsis*. A, *Trichomonas termopsidis*; B, *Streblomastix strix*, and B_1, diagram to show the anterior end of the same; C, *Tricercomitus termopsidis*; D, *Hexamastix termopsidis*; E, *H. laticeps*. From MacKinnon and Hawes, 1961. *An Introduction to the Study of Protozoa*. Clarendon Press, Oxford.

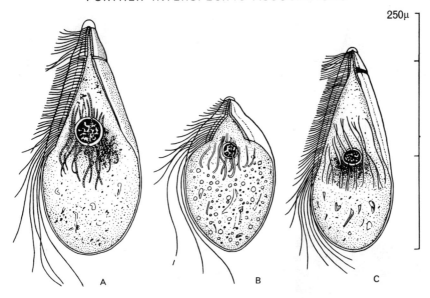

Fig. 23-2. *Trichonympha* from *Zootermopsis angusticollis*. Diagram to show typical forms and average sizes. × 250. A, *Trichonympha collaris*; B, *T. sphaerica*; C, *T. campanula*. From MacKinnon and Hawes, 1961. op. cit.

cockroach *Cryptocercus*. These are scarce or lacking in the higher termites, but they live in enormous numbers in the hind-guts of the lower ones (Fig. 23–3), forming, in the words of one observer, 'a tangled mass of writhing protoplasmic bodies, the movements and appearances of which defy description'. In the nymph of *Zootermopsis* this formidable microfauna has been estimated to make up from one-third to one-seventh of the total weight of the animal.

Among these symbionts are polymastigate and hyhermastigote flagellates of varying complexity, belonging to the Order Metamonadida. *Trichomonas termopsidis*, related species of which are common in the alimentary canal of many vertebrates, is a comparatively simple form, with few flagella, one of which trails and has an undulating membrane. Other members of the fauna, species of the genus *Trichonympha*, are much more complex, with a large number of flagella and many parabasal bodies and axostylar filaments. Cleveland originally demonstrated the role of these Protozoa in the digestion of the wood eaten by their hosts, and later studies have supported his interpretation. Much of the evidence derives from the discovery that these termites can be defaunated (i.e. deprived of their symbionts) by subjecting them to high temperatures or to high oxygen concentrations. Termites treated in this way die within a few weeks if they are supplied with no more than their normal diet of partially decayed wood, but this is not because they have themselves been injured: their lives can be prolonged if they are given cellulose that has been predigested by fungi. Further evidence is that some of the flagellates ingest particles of wood, while a cellulase is present in extracts prepared from cultures of *Trichomonas termopsidis*.

The toxic effect of oxygen on the flagellates results from them being obligate anaerobes, for they cannot release energy by oxidation processes, but must rely instead upon fermentation. The carbohydrates of the wood serve as the substrate for

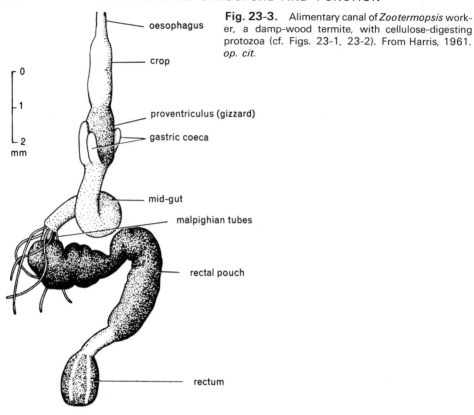

Fig. 23-3. Alimentary canal of *Zootermopsis* worker, a damp-wood termite, with cellulose-digesting protozoa (cf. Figs. 23-1, 23-2). From Harris, 1961. *op. cit.*

this, the products being carbon dioxide and organic acids, with acetic acid as the main component of the latter. These fermentation products are absorbed by the host, and are oxidized by it. Their value is indicated by calculations showing that the oxygen requirements of the termites are about equal to the amount necessary to bring about complete oxidation of the fermentation products of the Protozoa. Since little, if any, of these products are actually excreted, it follows that the insects depend for their energy supplies almost entirely upon the fermentation of cellulose by their symbionts. It might be thought that they would also digest the symbionts and make use of them as a source of nitrogen, but this seems not to be so. More probably they obtain their nitrogen from fungi in their food; these, of course, will have obtained their own nitrogen supply from the wood in which they grow.

Analysis of this aspect of nutrition is complicated, however, by the presence of a vast bacterial population in the gut of termites, higher as well as lower, estimated in one species to reach a minimal concentration of 10^{10} cm^{-3}. Whether they have any cellulolytic function is not clear, but it is certain that in some species, and also in *Cryptocercus*, they can fix atmospheric nitrogen and evolve methane. That the bacteria are the source of this in termites is clear from study of the effects of adding antibiotics to the diet (Table 23–1), for these abolish both the bacterial population and the production of methane, without affecting the protozoans. It is puzzling, however, that they have no effect on methane production in *Cryptocercus*. Methano-

Table 23–1 Methane emission by larvae of *Riticulitermes flavipes* From Breznak, 1975. *Symp. Soc. exp. Biol.*, **29**, 559–580.

Pre-assay diet	nmol Ch_4 emitted (hr^{-1} g^{-1} fresh wt)
Nest wood	27.0 – 73.1
Filter paper: 4 days	149.0
Filter paper: 8 days	1,340.0
Filter paper + antibiotics: 8 days	0.2

genic organisms exist also in the rumen of cattle (see later), and are there known to be a source of protein and vitamins for the host. How far this may be true for termites is uncertain, for there is still a great deal to be learned about the biochemical aspects of this relationship of the dependence of wood-eating termites upon their microfauna, which is present in all of the wood-feeding members of a colony. Where a worker caste is present, the workers feed the old reproductive members; these do not, therefore, have to feed themselves on wood, and it is significant that they do not possess symbionts. The importance of the association is further shown by the speed with which larvae acquire the Protozoa; although they do not possess them on hatching, they secure them within 24 hr, apparently by taking them up from the anus of older members. The Protozoa, however, are equally dependent upon the termites, upon which they rely for their own nutrition. This is readily seen if the insects are starved, for this results in most of the larger Protozoa and many of the smaller ones dying long before the insects do.

The association seems, then, to be a particularly well-balanced symbiosis, with the maximum of mutual benefit and the minimum of exploitation. One reservation must, however, be made in this interpretation. It has been suggested that the symbionts, like most other animals, may not actually produce the cellulase used in their digestive processes, but that they rely upon the production of the enzyme by bacteria living within their bodies. This contention is difficult either to prove or to disprove, and the issue cannot be regarded as entirely settled. Even if this view is well founded, however, it does not effect our general interpretation of the relationship. What it does do is to increase its complexity, for it means that this is an example of hypersymbiosis, in which one symbiotic partner is itself in symbiotic relationship with a third partner of the complex, in this instance the supposed cellulolytic bacteria.

There is a singular and instructive parallel between the digestive processes of termites and those of ruminating mammals, which have also found a way of exploiting cellulose as a source of food. The rumen of these animals, constituting the first, and by far the largest, division of their stomach, contains a rich population of bacteria and ciliate Protozoa, all living in strictly anaerobic conditions. At least 30 species of protozoans have been identified, making up in numbers of individuals as many as 1 million per gram of contents. Included among them are *Isotricha* and *Dasytricha* (belonging to the Order Trichostomida, which includes *Balantidium*, common in the rectum of the frog) and *Entodinium* (Fig. 23–4), *Diplodinium* and *Ophryoscolex* (belonging to the Order Entodiniomorpha), a highly specialized group with anterior membranelles but no general surface ciliation). All can utilize the lower molecular weight compounds made available by bacterial action, but they also engulf and

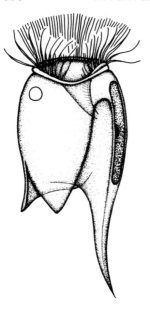

Fig. 23-4. *Entodinium caudatum.* From Sleigh, 1973. *The Biology of Protozoa.* Edward Arnold, London.

digest bacteria and other particulate material. Their relationship with the bacteria is thus complicated. On the one hand, they make sugars and amino acids available to the bacteria, which can survive for some time inside the protozoans as well as outside them. On the other, bacteria in both locations are digested by the protozoans, with a release of amino acids which are then used for synthesizing protozoan protein.

The relation between the micro-organisms and the mammalian host is also complex, and by no means fully understood, although there is clearly mutual benefit. It is certain that fermentation is important in the functioning of the rumen, and that the resulting fatty acids are absorbed by the host and used as a source of energy. To this extent the system is similar in principle to that of the termite's hind-gut; as with the latter, indeed, hypersymbiosis may be a factor, and the apparent cellulolytic activity of *Diplodinium* may actually be due to the presence of intracellular bacteria. From the point of view of the symbionts, as Barnett and Reid point out, the rumen provides a factory of immense capacity, amounting to 27 litres in a cow. In it the symbionts find controlled pH and temperature, mechanical equipment for their use (in the form of churning and rumination), and a very well-developed blood system which functions as a conveyor belt for the removal of their metabolic products. In their absence this specialized vascularization would have little function, so that the organization of the host has been adapted to the needs of the association. There is, however, one difference in comparison with the termites. Unlike the latter, the mammalian host digests its micro-organisms as they pass backwards through the other chambers of the stomach into the abomasum, and it is estimated to obtain as much as 20% of its nitrogen from them. It is arguable, then, that the system is not so neatly balanced as that of the termites, since the latter are thought not to use their symbionts as a primary source of nitrogen. Probably we cannot distinguish sharply between the two associations, but at least the difference indicates how narrow is the boundary between an ideally balanced system and the exploitation of one of the members of it.

23-2 SYMBIOTIC ALGAE

Certain symbiotic relationships between animals and algae have long attracted attention, for they are readily observed and in some instances they lend themselves well to experimental study. They are especially characteristic of protozoans, sponges, and coelenterates; a distribution which suggests that the evolution of these associations has been facilitated by a low level of differentiation and coordination in the multicellular hosts. Perhaps it was easier for the invading organisms to become established in these animals without disturbing the host's organization. We shall see later that the invasion of higher organisms by parasites creates for the invaders the need to override the host's immune responses which are fundamental to the maintenance of its individuality.

These algae have been conventionally named by reference to their colour, but their taxonomic status is now usually well understood. Those in fresh-water sponges (Spongillidae), which give them their characteristic green colour, are called zoochlorellae because of this colour; they are symbiotic Chlorophyceae. Those in marine sponges are blue-green zoocyanellae (symbiotic Cyanophyceae); perhaps the only occurance of blue-green algae in animals, although they are common in marine diatoms. Finally, there are yellow-brown zooxanthellae (symbiotic Cryptomonadida and Dinoflagellida) which are widely distributed in marine animals, and particularly in coelenterates. It is always highly probable that these associations are mutualistic ones, for any association between a phototrophic primary producer and a heterotroph offers a short cut in the cycles of nutrients that we considered in chapter 1. The algae, in addition to being protected, can, in theory, obtain nutrients (nitrates and phosphates, for example, and perhaps carbon dioxide) from the animal, while the latter can obtain oxygen and metabolic products of the algal photosynthesis.

The evidence for fresh-water sponge associations being mutualistic is suggestive, but the case remains to be proved, although certain observations go some way to substantiate it. The population of algae within the sponge is much greater than the population outside it (3×10^{11} l^{-1} of tissue, as compared with 4×10^3 l^{-1} litre of ambient water, have been recorded), while symbiont sponges living in the dark are smaller than those in the light. These facts indicate some metabolic benefit to both partners. Whether or not the algae are digested by the sponge is doubtful, but in any case this is not a necessary requirement, for exchange of material by diffusion is all that is required. Indeed, if the algae were exploited by the animal partner, as they are in certain cases to be mentioned later, the relationship begins to overstep the strict limits of mutualism.

These algal relationships are much better documented in coelenterates, where they are familiar in hydroids, anemones and corals, and where they lend themselves well to experiment. The cosmopolitan green hydra (*Hydra viridis*, or *Chlorohydra viridissima*) owes its colour to a green alga, *Chlorella* sp., which is distinct from the free-living *Chlorella vulgaris*. These live in the endoderm cells, up to about 18 per cell, giving a total population of some 150,000 algae in one individual. Their restriction to the endoderm cells is sharply determined, as is apparent from the behaviour of regenerating fragments of endoderm. These give rise to both ectoderm and endoderm cells, but the former always reject their algae.

This rejection is one indication of the specialized character of the association.

Another is that there is clear mutual recognition between the two partners, as can be shown by the use of aposymbiotic hydra. These are individuals which lack the algae, a condition that can be secured by rearing them from alga-free eggs. Infective and non-infective algae can be fed to these animals, and shown to be treated quite differently. The non-infective ones are engulfed by the endoderm cells into large vacuoles, each with a number of algae, and these vacuoles are soon rejected from the cells. Infective ones, however, are sequestered after engulfment into individual vacuoles which pass to the base of the cell; this remains the characteristic location of the symbionts. Here again much remains to be learned of the mechanisms involved, but one attractive suggestion is that the predetermined movement of the vacuoles depends upon the microtubular architecture of the endoderm cells.

It seems likely that this relationship is a mutualistic one. Hydra reared in the light, and with regular feeding, maintain a constant algal population, but this is reduced if the animals are reared in the dark. For some reason which is not clear, however, this reduction affects only the upper regions of the body (down to and including the budding zone); the algal population in the stalk and pedal disc is not affected. When the hydra are returned to light from darkness, the algae immediately divide and increase in numbers. The algae, then, are affected by the lighting regime, but their persistence during darkness implies that they can live on metabolites produced by the host. Evidence of the passage of nutrients in this way has been shown by the results of feeding ^{14}C-labelled *Artemia nauplii* to hydra kept in the dark. Within 48 hr the algae had taken up 25–34% of the total radioactivity.

The host also benefits from the association, as is shown by growth studies, in which the rate of production of buds is taken as a growth index. Symbiotic and aposymbiotic hydra, maintained in the light, grow at the same rate if they have adequate food, but the green ones grow the more quickly and survive much longer if both types are given a restricted diet (Fig. 23–5). Evidently the hydra are now benefiting from metabolites taken up by the algae. These products are probably complex, but evidence that the uptake does occur comes from the finding that the algae, when freed from the host, release maltose. This can be labelled in the intact animal by incubating green hydra in a medium containing $NaH^{14}CO_3$, the metabolism of which is a convenient index of photosynthesis. The label appears later on in a range of materials, including carbohydrate, lipids, and protein—clear evidence that the host is not only taking up a product of algal photosynthesis but is also metabolizing it. These observations, incidentally, give further evidence of the specialization that underlies the relationships, for free-living (non-symbiotic) *Chlorella* do not release maltose, which is not a normal algal product.

How this specialized relationship arose is, of course, unknown, but the origin can hardly have been a consequence of the algae being ingested by the host, for, as we have seen, hydroids are specialized carnivores. Some observations made on other species suggest that it may have resulted from infection. It is said that algal infection is unknown in *Hydra circumcincta*, but that it sometimes occurs spontaneously in *H. vulgaris*, preceded by enfeeblement of the host, and accompanied by pathological symptoms. The infection can be induced artificially in this species, but is only maintainable for a brief period; in *H. attenuata* it is easier to produce, and the algae are more difficult to dislodge.

The relationship of *Chlorohydra* and *Chlorella*, however it may have arisen, can

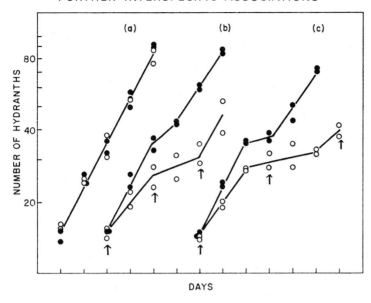

Fig. 23-5. Semi-logarithmic plot of the growth of green (●) and aposymbiotic (○) *Chlorohydra viridissima* (a), fed daily; (b), fed every second day; (c) fed every third day. Arrows indicate time of feeding. From Muscatine and Lenhoff, 1965. *Biol. Bull.*, **129**, 316–328.

reasonably be regarded as a balanced symbiosis. So also, probably, is the association between *Paramecium bursaria* and the zoochlorellae which grow and divide within individual vacuoles in its cytoplasm, as many as several hundred in a single ciliate. The benefit of the association to the ciliate is shown by rearing it in sterile cultures which lack the bacteria upon which it normally feeds. It continues to grow in the light, but in continuous darkness, with photosynthesis impossible, the culture soon dies (Fig. 23–6). Indirect evidence that the algae obtain some nutrients from the association comes from experiments showing that they can continue to divide in the dark, providing that the ciliates are provided with excess food.

It is likely that similar conclusions would hold for the many rhabdocoel platyhelminths, such as *Dalyellia*, which have large numbers of zoochlorellae in their mesenchyme. The young animals acquire these by ingesting them with their food, and the relationship may well have originated in that way. It has been remarked that species of *Dalyellia* possessing symbionts are larger than those without them, and it is supposed that one service rendered by the algae is the utilization of waste products. The oxygen produced by them during photosynthesis may also be of value to species living in stagnant waters, or otherwise subject to oxygen deprivation. The acoelan *Amphiscolops langerhansi* can survive confinement in small bottles of sea water provided that it is kept in the light, but it rapidly succumbs in the dark.

At best, however, mutualistic relationships like these are balanced on a knife-edge, and it is easy for one partner to begin exploiting the other. *Dalyellia* is said to digest senile algal cells, and the boundary between this and the eating of active algae is clearly a very fine one. The delicacy of the balance is well illustrated by the acoelan *Convoluta roscoffensis*, delightfully described by Keeble in a classic account of

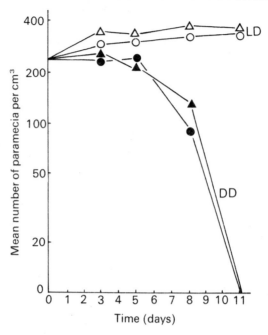

Fig. 23-6. Growth of *Paramecium bursaria* in a bacteria-free inorganic salts medium in 10 hr light followed by 14 hr dark (LD), or continuous darkness (DD). The triangles and circles each represent replicate cultures. From M. Karakashian, 1975. In *Symp. Soc. exp. Biol.*, **29**, 145-173 (D. H. Jennings and D. L. Lee, eds.), after an original figure and data S. Karakashian.

'plant-animals'. These acoelans, bright green in colour, live in the intertidal zone of the western French coast and the Channel Islands, coming to the surface when the tide is out, and appearing then in such numbers that large areas of the shore are coloured green by them. The colour is produced by chlorophyll, a result of infection by symbiotic zoochlorellae (*Platymonas convolutae*), which collect on the egg cases and are ingested after the worms hatch. They perhaps pass through an intracellular stage, but they come to rest between the subepidermal cells, an unusual position for symbiotic algae, which are usually intracellular. They lose their theca, flagella, and eyespot, and become essentially naked protoplasts, which develop complex processes that interdigitate with the host cells and thus secure maximum cellular contact.

If the young worms remain uninfected, which is unlikely to happen except under experimental conditions, they fail to grow. Normally, however, they feed holozoically for a time, and grow rapidly. Later, in a change of habit that seems to occur only in this particular association, the alimentary canal degenerates, and the animal can no longer digest external supplies of food. According to Keeble, it now digests its algae, and eventually dies of starvation after laying its eggs, the supplies of symbionts being then exhausted. This would be frank exploitation, shifting the balance towards the parasitism that we shall be considering later, but unfortunately there is no proof that this digestion actually occurs, although, as Taylor observes, the thought may appeal to one's romantic imagination.

The use of radioactive tracers has, however, done something to clarify the probable metabolic pathways involved in the relationship, and, as far as it has gone, has confirmed the general outline of the classical interpretation (Fig. 23–7). The main products of photosynthesis in the alga are mannitol and starch, the latter of which was identified by Keeble and Gamble, who showed that it disappeared if the animals were kept in the dark, but rapidly reappeared in the light. However, the surface of the algal cell is not very permeable to either of these substances, and there is no evidence that they pass to the animal. What do seem to pass to it are fatty acids, sterols, amides, and amino acids. The wide range of these products is not surprising, for clearly the animal becomes completely dependent upon its algae. The contribution that the animal makes to the relationship results from platyhelminths lacking the enzyme uricase, so that the uric acid formed by the breakdown of their purines has to be excreted in that form instead of being catabolized to urea and ammonia (Sec. 12–2). This uric acid passes from *Convoluta* to its algae, which can then catabolize it to ammonia and utilize the nitrogen. In addition, of course, the algae receive protection and, thanks to the behavioural adaptations of their host, adequate exposure to sunlight. Indeed, the animals might be said to be cultivating their plants.

However we choose to interpret this relationship, cultivation associated with a clear and remarkable example of exploitation is seen in the Tridacnidae, bivalves that are uniquely modified for what we can justly describe as the farming of zooxanthellae. *Tridacna*, the giant clam, which reaches a length of $4\frac{1}{2}$ ft and a weight of 4 cwt, is the largest bivalve known to have evolved in 600 million years. It lives among the corals of the Great Barrier Reef where it rests in a highly unusual position, with the umbo of the shell downwards, but with the visceral mass inverted with reference to the shell, so that the pedal opening is also downwards. This allows the inner lobes of the mantle edges to extend over the free edges of the valves of the shell and to be exposed to the light (Fig. 23–8). The significance of this arrangement is that in these areas there are abundant zooxanthellae, which in this instance are the resting stages of dinoflagellates. They are visible as deep-brown, spherical bodies, carried in amoeboid cells in the blood sinuses of the mantle tissue, and containing starch and oil droplets as food reserves. In effect, their host is exposing them to light, where they can best carry out photosynthesis, and the adaptive significance of this is accentuated by the presence in the upwardly directed mantle surface of many conical protuberances. These contain hyaline structures formed of transparent cells surrounded by a connective tissue capsule. The algae are particularly abundant around these structures, which, according to Yonge's interpretation, seem to concentrate light and thereby facilitate the metabolism of the symbionts.

Confinement of *Tridacna* in sealed containers in light and in darkness has shown no significant differences in the oxygen content and pH of the surrounding water in these two conditions. It follows, then, that the output of oxygen and the removal of carbon dioxide by the zooxanthellae is insufficient to affect the metabolism of the clam. The water, however, is completely depleted of phosphorus, whereas if *Spondylus*, a mollusc without zooxanthellae, is confined in similar conditions the phosphorus content of the water increases owing to the excretory output of the animal. It is suggested, therefore, that the zooxanthellae utilize this material in their own synthetic activities. We shall see later that the same suggestion has been ex-

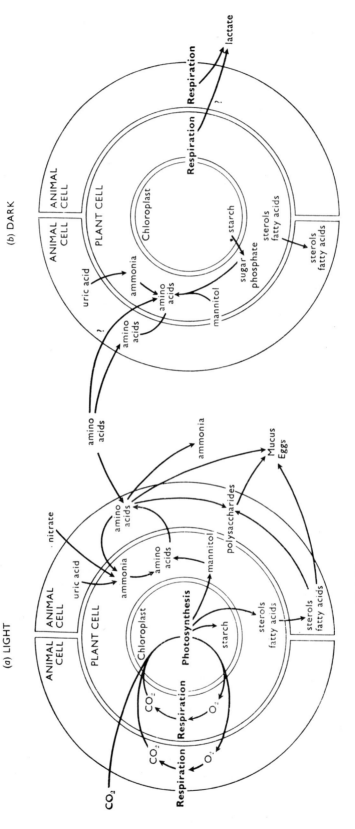

Fig. 23-7. Some probable major metabolic pathways in the association of *Convoluta roscoffensis* and *Platymonas convolutae*. From Holligan and Gooday, 1975. In *Symp. Soc. exp. Biol.*, **29**, 506–523 (D. H. Jennings and D. L. Lee, eds.).

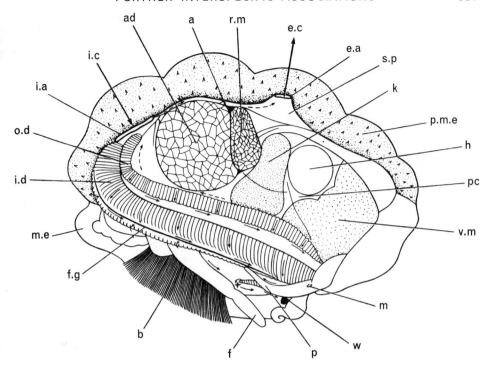

Fig. 23-8. *Tridacna crocea*, drawn from life, to show an individual lying on the left shell valve, right mantle lobe removed. *a*, anus; *ad*, adductor muscle; *b*, byssus; *e.a*, exhalent aperture; *e.c*, exhalent current (represented by arrow); *f*, foot; *f.g.*, food-groove on inner demibranch; *h*, heart; *i.a*, inhalent aperture; *i.c*, inhalent current (represented by arrow); *i.d*, inner demibranch of gill; *k*, kidney; *m*, position of mouth; *m.e*, mantle edge, unpigmented, bordering the pedal gap on the underside; *o.d*, outer demibranch; *p*, labial palps; *p.c*, pericardium; *p.m.e*, pigmented mantle edge of upper, exposed side; *r.m*, retractor muscle of foot; *s.p*, siphonal process of exhalent aperture; *v.m*, visceral mass; *w*, accumulation of waste matter rejected by palps. Small complete arrows show direction of food-collecting currents on the gills; broken arrows show currents in the exhalent chamber. From Yonge, 1931. *Gt. Barrier Reef Exped. Sci. Rep*, **1**, 283–321. Used by courtesy of The Trustees, British Museum (Natural History).

tended to the symbiosis of corals and zooxanthellae, where the biochemical analysis has been carried further.

This, however, is by no means the whole story. *Tridacna*, in addition to being adapted for 'farming' its zooxanthellae by aiding their photosynthesis, is also adapted for exploiting them as a source of food. Evidence for this is obtained by exposing the clams to ^{14}C-labelled carbon dioxide. This is taken up by the algae, and soon thereafter the labelled carbon is found in the clam tissues. The algae are carried from the mantle in phagocytes, which accumulate in large numbers around the digestive diverticula, where they contain zooxanthellae in all stages of digestion. These are apparently used as food, the indigestible remains being carried to the excretory organs, which are much enlarged in order to deal with this material. Unlike *Convoluta roscoffensis*, the giant clam has not carried this relationship to the point of losing its own capacity for securing external food supplies. Nevertheless its alimentary system is certainly modified. The selective mechanisms on the gills and palps are highly developed, while the mouth is small, there are no sorting mechanisms in the stomach, and the digestive diverticula are reduced in number. In Yonge's picturesque phrase, *Tridacna* is specialized to exploit imprisoned phytoplankton as a source of food. This accounts for its great size, for, as we have earlier seen, unaided ciliary feeding mechanisms must limit the size of the animals relying upon them.

Whether the zooxanthellae can be said to derive any advantage from the situation is a matter of opinion and definition. Their perpetuation is guaranteed by the clam, and the fate of the individual algal cell is no worse than it would be in the outside world. Yet the relationship is essentially one of exploitation by one partner. If it does not fit our formal definition of mutualism, this is because our definitions have artificial boundaries. How this particular relationship arose is not easy to see. Tridacnids are obviously an exception to the generalization that algal symbiosis is associated with relatively undifferentiated metazoans, but why this should be so is a matter for speculation. Unlike coelenterates, the lamellibranchs are not specialized carnivores, so that an essential preliminary must have been the ability of the invading organisms to resist digestion. It is conceivable that they first entered the mollusc tissues from coelenterate planulae (see below), already able to withstand digestion by a host.

23-3 CORALS AND SYMBIOSIS

Coral reefs extend over an area of some 68 million square miles in tropical and subtropical seas. Best developed where the mean annual temperature of the water lies within the range of 23–25°C, they do not develop to any significant extent in regions where temperatures fall below 18°C, nor are they found in waters that contain continental sediments. An adequate oxygen supply is important, but this, as we have learned, is not normally a limiting condition for animal life in the sea. Much more important factors for coral reefs are the intensity of surface illumination and radiant energy. This may seem surprising since animals rather than plants are important in constructing reefs, but we shall see that it is a direct consequence of the animals' symbiotic relationships.

A reef, as described by J. W. Wells, mainly consists of the skeletons of reef-building (hermatypic) coelenterates, or corals, and of calcareous red algae. These form an interlocking framework, upon which are deposited sediments derived from the breakdown of the skeletal materials and of the organisms that have secreted them. Thus are provided a range of ecological niches, occupied in part by the commensals, symbionts, and parasites of the primary organisms, and in part by an assemblage of free-moving organisms that are unique in their beauty and variety. This remarkable biological phenomenon is a dramatic demonstration of the results that can flow from the colonial association of lowly animals, fortified in their metabolic activity by the presence within their tissues of symbiotic micro-organisms.

We can conveniently distinguish four types of coral reef (Fig. 23-9): atolls, barrier reefs, fringing reefs, and platform reefs. Atolls are usually oceanic, and without association with land. They consist of low reefs, rising no more than 30 ft above the level of the sea, and enclosing a central area of water called the lagoon. Externally to the reef lies the seaward slope, composed of débris derived from the degradation

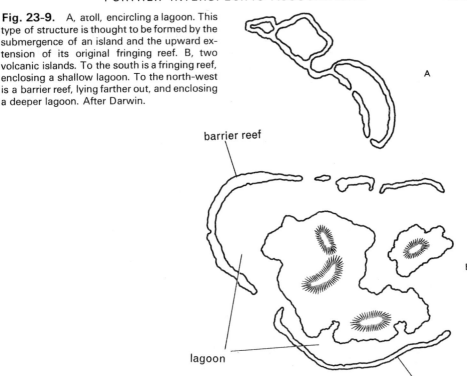

Fig. 23-9. A, atoll, encircling a lagoon. This type of structure is thought to be formed by the submergence of an island and the upward extension of its original fringing reef. B, two volcanic islands. To the south is a fringing reef, enclosing a shallow lagoon. To the north-west is a barrier reef, lying farther out, and enclosing a deeper lagoon. After Darwin.

of the reef, while internally to it are the lagoon slope and floor, of similar origin. Barrier reefs are also low reefs, but they differ from atolls in being associated with land. Between them and the mainland lies a narrow strip of sea, perhaps up to 180 ft deep, which is not much less than the greatest depth attained by lagoons. The Great Barrier Reef of Australia is the best-known example of such a structure, reaching a length of at least 1,200 miles. Fringing reefs are similar in principle to barrier reefs, but differ from them in lying much closer to the mainland; this may be either an island or a stretch of continental coastline, but in either case the intervening water constitutes only a narrow channel. Finally, the platform (table) reefs are, as their names imply, flat structures without lagoons. They rest on the shallower parts of continental shelves; they may appear in the water between the coast and a barrier reef, for example, or they may be associated with atolls.

The recognition of these four categories is a matter of descriptive convenience. From the biological point of view they are essentially similar, arising from the interaction of environmental conditions with the habits of the reef-building organisms, and providing a basis for the establishment of biological communities that are very constant in composition. Nevertheless, the differences in the form of the reefs reflect differences in their modes of origin.

Current views on this derive from a theory originally put forward by Darwin, and based by him not only upon direct study of coral reefs, but also upon the requirements which are known to limit the activity of the coral polyps. These include, as we have mentioned, a dependence upon tropical water temperatures, and also a requirement for light, the polyps being unable to flourish at greater depths than between 20 and 30 fathoms, and rarely below 15 fathoms (90 ft). These limitations create no problem as far as fringing reefs are concerned for they can readily be established on a sloping shore at points where the depth is suitable. Barrier reefs and atolls present an altogether different situation, for these drop sharply down to great depths where it is out of the question for reef construction to take place.

Darwin's solution of the problem of their origin was based upon a comparison of the distribution of reefs with the distribution of areas of present-day or recent volcanic activity. This showed that active volcanoes were associated with the areas in which fringing reefs were distributed, but that areas containing atolls and barrier reefs were free of volcanic activity. He thus suggested that conditions for the development of the latter two types were provided by subsidence, which led to the submergence of peaks of land. Corals would then grow on these when they had sunk to the appropriate depth. Fringing reefs, by contrast, developed in areas where the land surface was either stationary, or was rising with associated outbursts of volcanic activity. It was an explanation which, as Darwin rightly claimed, offered 'a grand and harmonous picture of the movements which the crust of the earth has undergone within a late period'.

This theory, supported by the observations of Dana, and therefore known as the Darwin–Dana theory, has been widely accepted. An additional possibility is that the growth of modern reefs may have been influenced by the locking away of water in the polar ice caps during the ice age. It is argued that the older, Tertiary, reefs would have been cut down during the resulting exposure, and that new growths could then have become established on the margins of the exposed platforms as they were submerged by the later release of the water. This modification of the original theory has the advantage of accounting for the depths of lagoons and lagoon channels being largely constant from reef to reef, for this depth can be explained as determined by the amount of water released when the ice melted. However this may be, it is certain that with Darwin's views as a foundation, we can account for much of the known distribution of coral reefs.

It has long been recognized that this theory could be tested by examining the foundations of reefs, and it has, in fact, been supported by a deep boring made at the Einwetok atoll by the United States Atomic Energy Commission. Previous attempts at boring, by the Royal Society in 1904, had led to inconclusive results, but the later one has shown that the coral sediments extend to a depth of 4/5 mile, where they rest upon the summit of a volcano rising some 2 miles above the sea bottom. Here, then, there must have been prolonged subsidence, the growth of the reef having extended, it is thought, from Eocene times, over a period of some 60 million years. It would be going too far to suggest that this is the only mode of origin of barrier reefs and atolls. Yet little support can now be found for the alternative theory put forward by Murray, according to which these reefs were founded upon the tops of submarine volcanoes after these had been raised to suitable levels by the deposit of sediments.

The complex structure of coral reefs, and their isolation in the sea, make them attractive material for the analysis of the forces that have determined their growth and that permit their continued existence. As we see them today, they are a balance of constructive and destructive agencies. The construction of the reef, and compen-

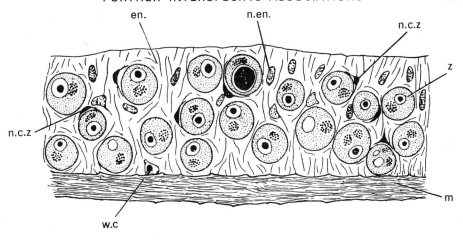

Fig. 23-10. *Goniastrea* sp. Section through endoderm of coenosarc, showing zooxanthellae enclosed within tissue cells. *en*, endoderm; *n.c.z*, necleus of cell containing zooxanthellae; *n.en*, nucleus of endoderm cell; *m*, mesogloea; *z*, zooxanthella. From Yonge and Nicholls, 1931. *Gt. Barrier Reef Exped. Sci. Rep.*, **1**, 135–176. Used by courtesy of The Trustees, British Museum (Natural History).

sation for its continuous erosion, depends upon the continued secretion of the calcareous skeletons of the reef-building algae and coelenterates, and the addition to these of sediments that become cemented by precipitated calcium carbonate. The destructive agencies are in part biological, comprising organisms such as algae, molluscs, echinoids, and fish, that in one way or another erode away the reef material by dissolving it, boring through it, or biting it. No less destructive is the wave action of the surrounding water, which breaks away and redistributes fragments of the reef, and undermines its edge.

This balanced system is founded on the organic productivity of the reef-building organisms, which can be expressed in terms of the production and consumption of oxygen from selected areas of the reef. Such measurements have shown that reefs are more productive than are the open seas around them, and they have shown also that this productivity is not directly related to the abundance of the available plankton. This may be plentiful in lagoons, for example, but much scarcer in the water outside the reefs, yet productivity is no less on the outward slopes than it is on the inner sides. This wide distribution of productivity may be partly a consequence of the large amounts of green algae present on the reefs, but it is certainly to be attributed also to the symbiotic relationships in which the hermatypic coelenterates are involved, and which account for the dependence of these animals upon radiant energy.

All reef-building corals contain zooxanthellae, which are present in the endoderm (Fig. 23-10) in wandering carrier cells, and which are distributed to the next generation in the planula larvae. They are found in the hermatypic Madreporaria, but not in the non-reef-building (ahermatypic) corals such as *Dendrophyllia*. They occur also in the hydrozoan *Millepora*, and the octocorallines *Tubipora* and *Heliopora*, all of which secrete skeletons that contribute to reef structure. They are found, too, in many other coelenterates, including hydrozoans, scyphozoans, and anthozoans, which may live in reef communities without contributing calcareous skeletal material to them. The absence of zooxanthellae from ahermatypic corals is thus all the more significant.

The zooxanthellae of coelenterates, like those of tridacnids, are the resting stages of dinoflagellates (Fig. 23–11), the commonest being *Gymnodinium* (*Symbiodinium*) *microadriaticum*, which is of world-wide distribution. This has suggested the speculation that its initial spread may have been of pandemic dimensions, and (in the light of evidence to be given later) perhaps promoted in the mid-Triassic the rapid evolution of corals at an early stage in the divergence of hermatypic and ahermatypic species. Be this as it may, we are here dealing with yet another symbiotic relationship between an animal and an autotrophic micro-organism, and in this case one that has lent itself particularly well to experimental analysis.

The foundations of our knowledge were laid by Yonge and his colleagues during the Great Barrier Reef Expedition of 1928–29. By keeping corals in sealed glass jars of sea water, it was shown that after 9 hr exposure to light there was a negligible increase of acidity in the medium surrounding a hermatypic coral, amounting to a fall in pH of 0.001. By contrast, the pH fell by 0.219 during the same period of darkness. The inference is that during daylight the zooxanthellae are utilizing the metabolic carbon dioxide output of the animal tissues, but are unable to do so in the dark. This inference was confirmed by the observation that no such difference was seen in the water around an ahermatypic coral (*Dendrophyllia*). Here the medium fell in pH by 0.103 and 0.10 during 9 hr of light and of darkness respectively; the similarity can be attributed to the absence of zooxanthellae, as a result of which CO_2 produced during daylight could not be removed.

Measurements of the phosphorus content of the enclosed sea water gave comparable evidence of the anabolic activity of the zooxanthellae (Fig. 23–12). Using the hermatypic *Favia*, the phosphorus content fell to zero during the first day, in a glass jar that was open to the air, and remained at zero for four days afterwards. With the hermatypic *Psammocera* and *Fungia* there was some increase, but this was very slight for the former. Using *Dendrophyllia*, there was a marked contrast, for the phosphorus content showed a substantial and continuous increase during the same period. Evidenty the zooxanthellae of *Favia* and the other hermatypic species were taking up all or most of the phosphorus excreted by the host. This conclusion was confirmed by the demonstration that the accumulated phosphorus in the *Dendrophyllia* jar was

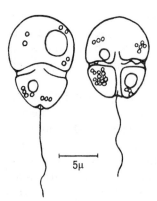

Fig. 23-11. *Gymnodinium* sp., dinoflagellate stage of zooxanthellae from *Acropora corymbosa*. After Kawaguti, from Yonge, 1963. *Adv. Mar. Biol.*, **1**, 209–260.

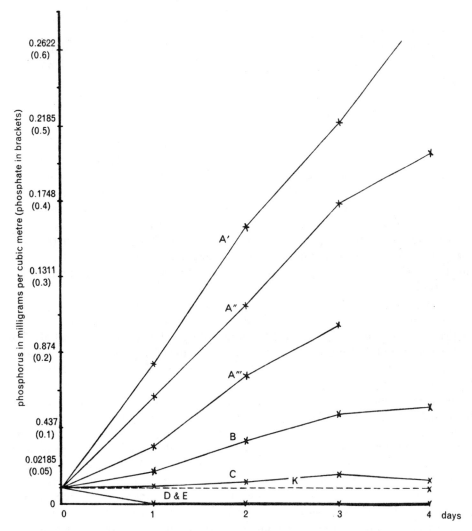

Fig. 23-12. Graph showing exchange of phosphorus between corals and surrounding sea-water. A', A", A, *Dendrophyllia*; B, *Fungia*; C, *Psammocora*; D, *Favia*; E, *Porites*; K, control. From Yonge and Nicholls, 1931. *op. cit.* Used by courtesy of The Trustees, British Museum (Natural History).

taken up in 24 hr when a specimen of *Favia* was placed in the water. Indeed, as much as 2,036 mg of phosphate were removed from the water in 5 days.

These observations led to the suggestion that the symbiotic algae, like those of *Tridacna*, could be thought of as imprisoned phytoplankton, providing nutrients for their hosts and in this instance gaining protection, carbon dioxide, and materials for protein synthesis. Nitrogen as well as phosphates would be needed by them, but this is readily available, for hermatypic corals do, in fact, remove ammonia from the surrounding water. Moreover, *in vitro* studies of *Gymnodinium* show that it can utilize a variety of host excretory products, including amino acids, guanine, adenine, uric acid, and organic phosphates. The metabolic advantages of their situation cannot, therefore, be doubted.

The advantage of the association to the animal, however, is not so immediately apparent. We have earlier noted examples of hosts that eat their algae, and it has been argued that corals may do the same. This possibility, however, was rejected by Yonge, who emphasized that these animals are specialized carnivores; they are believed to feed exclusively on animal prey, using their tentacles, nematocysts, and cilia. The crucial evidence here is the sequence of events that ensues when corals are starved, and when they are kept in the dark. Starvation results not in the digestion of the zooxanthellae but in their extrusion, and many of the extruded organisms, although not all, are dead. So also with the anemone *Anemonia sulcata*, which also contains zooxanthellae. These are gradually ejected if the host is starved in light, while if it is starved in darkness they are all ejected within 2 months. This does not damage the anemone in any way; on the contrary, it continues to live healthily for at least 16 weeks. The same is true of corals. Zooxanthellae cannot flourish in these in the dark, yet a specimen of *Fungia*, fed in the dark, has remained healthy for 164 days; during this time large numbers of the zooxanthellae were extruded, so that its tissues were left almost colourless.

The extruded zooxanthellae pass out through the mesenteric filaments, at the extreme edge of the absorptive zone, and it is this that has particularly led to the suggestion that they are being eaten. Yet there is no good evidence of this at all. The region concerned is also the point where foreign material, such as carmine granules, is excreted; it seems clear, therefore, that the symbionts are simply being extruded from an animal that is no longer able to harbour them, presumably because the experimental conditions have set up some state of metabolic disharmony. It follows, then, that if the zooxanthellae are indeed of advantage to their host, the benefit must lie in some metabolic contribution that they are able to make while they are alive. We cannot, however, generalize from this to include all coelenterates that contain zooxanthellae, as is shown by the instructive example of the Xeniidae, a group of tropical Alcyonacea common on coral reefs. These, too, contain zooxanthellae, but, in contrast to corals, their digestive (ventral) mesenteric filaments are reduced. Moreover, it is said that these animals, despite possessing tentacles armed with nematocysts, do not respond to food. If xeniid colones are kept in the dark, even in the presence of zooplankton, they soon begin to disintegrate and die, yet rapidly regain a normal condition if they are returned to daylight. It would seem, then, that in these animals, again in contrast to corals, the host has become so far adapted to the presence of the symbionts that it depends upon them as a direct supply of food, much as does the giant clam, or the mature *Convoluta roscoffensis*.

But if corals do not feed directly upon their symbionts, what advantages, if any, do they obtain from them? Because of photosynthesis, there is a substantial output of oxygen from coral reefs during daylight, but the animals are unlikely to benefit much from this, for experiments show that they are well adapted to the normal levels of oxygen tension in the sea. More probably, the zooxanthellae make an important contribution to the growth and secretory activity of the coral polyps. Tracer studies of the type already mentioned have shed some light upon this, for they have shown that up to 50–55% of the carbon dioxide fixed by the zooxanthellae may pass to the animal, which can incorporate it into a range of compounds, including lipids and proteins.

It will be noted from Table 23–2 that corals maintained in the dark do show some specific activity after 24 hr, but that this is very much lower than that found in animals incubated in the light. This fixation, which cannot be due to photosynthesis, is ascribable to heterotrophy, which is a supplementary mode of nutrition in the zooxanthellae. These data make it all the more clear that most of the fixed ^{14}C in the tissues of the light-maintained animals is due to algal photosynthesis. This is essentially what we have concluded for the symbiosis of hydra and *Chlorella*, but it is more difficult with corals to be sure that the transfer is of significant benefit to the animal partner, which is well able to feed on the plankton in the surrounding water. However, tropical waters are less rich in nutrients than are those of more temperate seas, so that this could make the transfer a valuable asset.

Yonge emphasized the advantage that corals could gain by the removal of their nitrogenous and phosphate waste, particularly in the crowded conditions of the colonial and reef-dwelling habit. His argument can now be carried further, for the algae have been shown to metabolize nitrogenous compounds to amino acids which can pass back to the animal tissue, permitting an efficient recycling of the nitrogen. It is possible that organic phosphates can be similarly recycled.

But there is also another line of thought. Goreau, using radioactive calcium (^{45}Ca), measured the growth rate of corals in terms of calcium deposition. Comparison of normal corals (*Mancinia areolata*) with those from which zooxanthellae have been removed by keeping them in darkness shows convincingly that the presence of the symbionts greatly increases calcium deposition. Growth is reduced in the dark, although zooxanthellae may still be present; even a cloudy day will reduce cal-

Table 23–2 Estimates of the *in vivo* translocation of ^{14}C from algae to animal tissues in the coral *Pocillopora*, incubated for the times shown in ^{14}C sea water. Radioactivity was estimated as a percentage: cpm animal/cpm animal + algae × 100. Asterisk indicates data not obtained. From Muscatine and Cernichiara, 1969. *Biol. Bull.*, **137** 506–523.

Incubation time (hr)	Radioactivity in animal tissue (%)	Specific activity (cpm/µg protein)	
		Animal	Algae
Laboratory (light)			
1	14.6	86	361
2	19.5	296	681
3	23.2	541	1,180
4	26.9	533	1,451
5	34.5	1,573	1,624
6	34.7	1,218	939
24	41.8	—	—
Reef (Ambient light)			
24	50.3, 36.5	—	—
Laboratory (dark)			
24	49.7	27	116

cification by as much as 50%. One interesting point, suggesting that there is still much to be learned about this situation, is that the rate of calcification in darkness in the presence of the symbionts is greater than it is in light in their absence. In other words, they seem to assist the process even when they are unable to photosynthesize. This could mean that they contribute some growth factor to the animal, an aspect of symbiosis that we have already mentioned in another context.

The contribution of the algae to the calcification process apparently involves their fixing carbon dioxide and bicarbonate. According to Goreau's interpretation (Fig. 23–13), calcium is taken up from the sea water, and passed through the tissues to be absorbed on an extracellular organic matrix where it is incorporated first into bicarbonate and then into carbonate. The reactions are promoted by the removal of H_2CO_3 and its breakdown by carbonic anhydrase to CO_2 and water, the CO_2 being then removed by photosynthesis. Part at least of the contribution made by the zooxanthellae can be attributed, on this view, to their known capcity for absorbing CO_2 and bicarbonate. A general scheme for this phase of coral metabolism is shown in Fig. 23–13. It well illustrates the potential complexity of this mutualistic relationship, even when we confine our attention to only one aspect of metabolism. In so doing it indicates also at what a superficial level so much of our analysis of animal and plant interrelationships has been conducted in the past.

23–4 HOST-PARASITE RELATIONSHIPS

The definition of parasitism as living at the expense of other organisms leads to the reflection that is a fundamental and universal feature of organic communities. The chemical phase of evolution may be said to have culminated in the establishment of organized exploitation. Animals live at the expense of plants, larger animals at the expense of smaller ones, and carnivores at the expense of herbivores. This is not, of course, the whole of the definition. The association of parasite and host is an intimate and continuing one, with the former living on or in the body of the latter, yet the implication that we have suggested is not entirely a verbal one. It emphasizes that parasitism is not a mode of life *sui generis*, unrelated to other types of association, and mainly of significance to us as a source of disease in man and his domestic stock. Like the other types of association that we have been considering, it is one of the main ways in which communities of plants and animals are bound together and regulated, and it needs approaching in part as an aspect of ecological relationships.

But the issues involved are not solely ecological, as becomes apparent when we consider the varying degrees of specificity that are found in the host-parasite relationship. Sometimes there is a very high specificity—that is, the parasite can only live successfully in a single host species, or at best in a small range of related forms. This situation, which reflects a delicate balance of mutual adaptation between parasite and host, is commonly held to show that the relationship is an ancient one, and that the specificity has a phylogenetic significance. The advancing specialization of the host is supposed to have been accompanied by a comparable advance in the specialization of the parasite, while, if the host has become specifically diversified as a result of adaptive radiation, the parasite will show a corresponding diversification. In such cases a taxonomic analysis of the parasitic group may illuminate the taxonomy of the host.

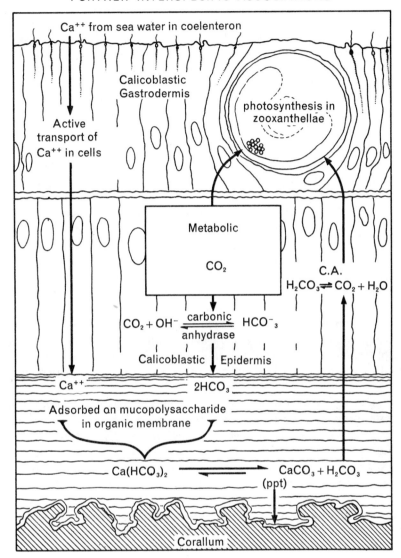

Fig. 23-13. Diagram (after Goreau) showing possible pathways of calcium and carbonate during calcification in a reef-building coral. A diagrammatic cross-section of the calicoblastic body wall at the base of the polyp is shown but the parts are not drawn to scale. The coelenteron and the flagellated gastrodermis containing a zooxanthella are shown at the top of the figure, the calicoblastic epidermis is in the middle, and the organic membrane with crystals of calcareous matter is at the bottom. The direction of growth is upward, i.e. calcium deposition is in a downward direction. From Yonge, 1963. *op. cit.*

Alternatively, a host-parasite relationship may show only a low degree of specificity—that is, the parasite can flourish in a range of host species. In these circumstances the specificity in the relationship may be determined primarily by the overlapping of the ecological distribution of the two partners. This is held to be characteristic of more recently evolved parasitic infections. But ecological relationships are important in even the most extreme examples of high specificity, for they permit the maintenance of the relationships, and must necessarily have contributed also to their origin. No parasitic association could have become established in the first instance had the two partners not been brought together by some common feature of their mode of life.

The ecological importance of host-parasite relationships is not confined to the analysis of their past history; equally significant is the effect of the parasite upon the host. Many parasites seem to have no obvious ill effect upon their hosts. Some do kill them, but these parasites are often regarded as imperfectly adapted forms, since the death of the host must militate against the building-up of large populations of the parasite. Yet to speak of parasites as harmless may well be an over-simplification, for there is obvious difficulty in determining whether or not a wild host is being placed at some disadvantage as a consequence of harbouring other organisms. The presence of these must often create physiological demands arising from the nourishment that they remove from the host, but presumably there is a margin of safety here, and the demands can be met by the ingestion of more food. There are, however, other possibilities. Conceivably a parasite may produce some decrease in the vigour or in the speed of response of its host—effects that may be difficult to overcome, and that may leave the host population more exposed to attack by predators, and less able to achieve a high density of numbers. Parasites can thus be important in regulating the balance of communities. The situation evidently involves a complex and closely integrated adaptation of the host and parasite with each other and with their environment.

As an illustration of this, one of the important biological agents controlling the size of oyster populations is a tissue parasite, *Dermocystidium*. This organism cannot flourish in waters of low salinity, which is one reason why oysters of the southern United States achieve a high density of population in such waters. The oyster, if it were an independent organism, free of parasites, could flourish in waters of high salinity, and could there build up populations of greater density than are actually maintained in lower salinities. That it does not succeed in this is a measure of the important part played by its parasites in determining its survival in particular types of habitat. The effect is seen in the differences in distribution of the oyster along different parts of the Atlantic coast of North America. In the more northerly waters there are larger oyster populations in more saline conditions, because the lower temperatures favour the oyster more than they do its parasites and predators. Farther south the balance of advantage changes, and there the larger oyster populations live in less saline conditions.

Another illustration of the ecological influence of parasites is seen in the history of the eelgrass, *Zostera marina*. *Zostera* beds were at one time the foundation of characteristic communities in areas of high salinity along the Atlantic coast, some of the organisms of these communities being so well suited to the conditions that they lived in densities unequalled elsewhere. During the early 1930s the eelgrass disappeared from all high-salinity waters, and it is believed that one factor in this was a protozoon parasite *Labyrinthula*. As with the oyster, the precise balance between parasite and host is influenced by salinity. The eelgrass, like the oyster, can tolerate lower salinities than can the parasite, and so it survived in the regions of low salinity, and has to some extent recovered elsewhere. One of the ecological and economic effects of the reduction of the *Zostera* beds was the elimination of the United States scallop fishery in those regions, for the bay scallop was an important member of the eelgrass community. Its disappearance, which was presumably a consequence of changes in its biological circumstances rather than in the physical conditions of its environment, meant that some areas of Virginia lost what was formerly a considerable scallop industry. Yet the activity of *Labyrinthula* was not entirely without benefit: some of the flats that were originally populated by the *Zostera* community became occupied by oysters.

The problem of the relationship of parasitism to the other types of association that we have considered raises a number of issues. There is, for example, the question whether it has developed from one or other of those types, or as a direct consequence of the invasion of one organism by another. Then again there are problems relating to the factors that determine the host-parasite relationship; to the degree of mutual adaptation of the two partners; and to the nature of the benefits received by the parasite and the extent to which these are disadvantageous to the host.

As regards origins, we have seen from the beginning of our survey of invertebrate life how wide is the scope for the independent and convergent achievement of similar results in unrelated groups. We must suppose, therefore, that however superficially similar the end results of parasitism may be, the association is likely to have been achieved in many different ways. We have seen that this is in part an ecological issue, for parasitism, like other associations, can only have been evolved when the habits and distribution of the invading organism brought it into frequent contact with its host. Yet the process could certainly have been aided by some preadaptation to facilitate the initial stages of mutual accommodation. Unfortunately, the intense specialization of parasites often conceals their evolutionary history, yet some light can be shed on this if we study non-parasitic members of the same group, while comparative studies of host-parasite relationships can also be very revealing.

Free-living Protozoa are clearly pre-adapted for parasitism by virtue of their habit of forming protective cysts which could readily be ingested by potential hosts. As Baer points out, the Foraminifeida, with their alternation of sexual and asexual generations, show that Protozoa, even when free-living, have the potentiality for evolving the complex life cycles that are such an important and characteristic feature of their parasitic adaptations. These factors enable us to visualize something of the mode of origin of the sporozoean life cycles (Sec. 18–1).

Gregarines, for example, are parasites of the alimentary tract, body cavities, and excretory and reproductive systems of annelids, arthropods, and occasionally urochordates. This, together with their simple life cycle, which usually lacks schizogony, and which depends for its completion upon one host ingesting cysts liberated by another, suggests that these particular parasites first became established in the intestine of invertebrate hosts through chance ingestion of this kind. Gregarines remain typically parasites of the various cavities within their hosts, but they show a tendency to invade cells. This tendency, shown in the entry of the sporozoites of the eugregarine *Monocystis* into the sperm morula of the earthworm host, probably led on to the characteristically intracellular habit of the coccidians. Equally important would have been the greatly increased multiplication made possible by the introduction of schizogony into the life cycle. This is found in only a few gregarines (neogre-

710 INVERTEBRATE STRUCTURE AND FUNCTION

garines), an example being *Ophryocystis*, from the Malpighian tubules of *Tenebrio*. Here, in contrast to *Monocystis*, each gamont produces only one gamete.

Schizogony has an obvious adaptive value, for by increasing the intensity of infection, it increases the chances of successful transmission. Thus it is likely to have been an important factor in the evolution of the coccidian Suborder Haemosporina, which includes the malaria parasites (*Plasmodium*) of reptiles, birds, and mammals. The influence of these upon human history has been profound. One of them, *P. falciparum*, is said to cause more deaths in the tropics than does any other parasite of man, owing to the adhesion of infected blood cells to the capillary walls, with consequent damage to the brain and other organs.

The life cycle of *Plasmodium* (Fig. 23–14) is highly specialized, not least in its adaptation to the habits of the mosquito vectors. The sporozoites, injected with the saliva of the mosquito as it feeds, enter the blood and pass first through an intracellular phase in the tissues of the liver, where they undergo schizogony (pre-

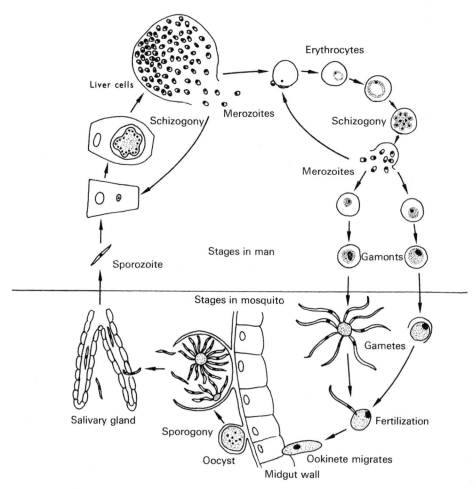

Fig. 23-14. Stages in the life cycle of the malaria parasite *Plasmodium*. From Sleigh, 1973. *The Biology of Protozoa.* Edward Arnold, London.

erythrocytic schizogony). The products (merozoites) may repeat the process, within other liver cells, but most of them invade erythrocytes, where they grow into trophozoites that undergo further schizogony (erythrocytic schizogony), the released merozoites then invading still other erythrocytes. The advantage gained by the parasite from these repeated schizogonies is obvious. The cycle eventually leads in man to the development of certain trophozoites into gamonts. These are taken up by mosquitoes, within which they complete the sexual phase and undergo further multiplication (sporogony).

The possible course of evolution of this remarkable life cycle can be deduced from members of the two other coccidian superorders: the Eimeriina and the Adeleina. *Eimeria steidae* is a convenient starting point. This parasite, which causes extensive liver damage, infests the epithelial cells of the bile ducts of the rabbit, the infection spreading through by schizogony and the movement of the resulting merozoites into other cells. Transmission is by oocysts, formed within the infected cells, and enclosing four sporoblasts, each of which contains two sporozoites. These cysts pass out with the faeces and are taken up by rabbits with contaminated food, and as a result of their coprophagy.

Three other examples must serve. *Schellackia* is an eimeriid parasite of lizards, which undergoes schizogony and sexual reproduction (gamogony) in the cells of the wall of the alimentary tract. From here, however, the sporozoites (in contrast to those of *Eimeria*) pass into the erythrocytes and are then ingested by mites. The life cycle is a simple one; there is no multiplication in the vector, and the primary host is infected by swallowing it, the parasite thus taking advantage of the close association of the mite with the lizard. *Haemogregarina* is an adeleid parasite of vertebrates, with gamonts in the blood cells, and with schizogony occurring there and in other tissues as well. The life cycle is similar to that of *Schellackia*. Again the vectors, which are leeches and ticks, are closely associated with the primary host and infection is usually by ingestion, without prior multiplication. Finally, *Haemoproteus columbae*, a haemosporinan of birds, infects the endothelial cells of the blood vessels, where schizogony occurs. In contrast to *Plasmodium*, there is no erythrocytic schizogony, the gamonts being the only stages found in the blood cells. There are, however, advances upon the two previously mentioned cycles. Multiplication occurs in the vector, which is usually a hippoboscid fly, and transmission is from the fly's salivary glands when it bites the bird.

We may conclude from these three examples (which, of course, are not an evolutionary series) that the evolution of the malaria parasites probably involved the escape of coccidian tissue parasites from the alimentary tract into the blood stream, an extension of multiplication, and an increasingly specialized relationship with the vectors. These must initially have been invertebrates closely and continuously associated with the vertebrate host. The involvement of flying insects, and eventual dependence on the advanced feeding mechanism and behaviour of mosquitoes, would have been later developments, demanding close mutual adaptation between parasite and vector. Having regard to this likely history, it now seems remarkable that the tissue phase of *Plasmodium* was not discovered until nearly 50 years after the discovery of the erythrocytic one. Throughout this period, therefore, the relapses which occurred with *P. vivax* and *P. malariae* remained baffling, for they could not be understood until it was realized that they are due to the persistence of latent stages in the liver. It is easy to be wise after the event. Nevertheless, the existence of these stages might have been deduced much earlier from a knowledge of the life histories of related parasites and the episode certainly illustrates very well the importance of comparative and evolutionary studies.

The verteberate blood stream has also been exploited, independently of the malaria parasites, and in quite a different way, by the flagellate group Kinetoplastida. This order, perhaps derived from euglenids or dinoflagellids, includes the very common free-swimming *Bodo*, with a related parasitic genus *Cryptobia*, and the trypanosomids, a widely distributed group of parasites possessing a characteristic kinetoplast, rich in DNA and associated with the mitochondrion. These parasites occur in four main types distinguished by the position of the main organelles (Fig. 23-15): promastigote (formerly called leptomonad); epimastigote (crithidial); amastigote (leishmanial); trypomastigote (trypanosomal). The promastigote type, with the kinetoplast at the anterior end, is probably the most primitive. From this, the epimastigote can be derived by backward movement of the kinetoplast to a position near the nucleus; the base of the flagellum also moves back, together with its accompanying basal body, and an extension of the flagellar surface forms the so-called undulating membrane connecting with a fold of the body surface. The trypomastigote would be a further stage in this transformation.

The earlier names of these types were derived from genera that conformed to them; trypanosomids, however, are often polymorphic, and more than one of the types can occur as stages in their life cycles. These are seen in their more primitive form in *Leptomonas* and *Crithidia*, which are parasites of the alimentary tract of invertebrates, particularly of insects, and which can complete their life cycles in the one host. Thus *L. jaculum*, which occurs in the water scorpion *Nepa cinerea*, has a promastigote stage which is abundant in the mid-gut, and an amastigote one, rounded and lacking the flagellum, in the rectum. Distribution is by the swallowing of cysts.

Promastigote types occur only in invertebrates, but the other three may occur in them or in vertebrates as well. The trypanosmids of vertebrates, however, require an intermediate host, which is usually a blood-sucking insect, although it may be a leech. The facts suggest that they evolved, in contrast to the haemosporidians, from parasites of the alimentary tract of insects, these parasites probably obtaining entry into vertebrate hosts as a result of the feeding habits of the insects. A similar conclusion is suggested by the life cycle of *Phytomonas*, a trypanosomid which lives in plants and is distributed by plant-sucking insects. The line of argument is well illustrated by *Leishmania*, the cause of kala-azar and oriental sore in man. The parasite lives as the amasti-

Fig. 23-15. The four basic morphological types of haemoflagellate A, promastigote; B, epimastigote; C, trypomastigote; D, amastigote.

gote type within cells of the human reticulo-endothelial system, but transforms into the promastigote type in the alimentary tract of the invertebrate vectors, which are female blood-sucking sand flies (*Phlebotomus*). The flagellate stage is then injected into a new human nost with the fly's saliva. The parasite can also develop in the bed-bug, *Cimex lectularis*, which was therefore thought at one time to be the likely vector. In fact, however, it cannot transmit the parasite to man; an instructive example of the specificity of host-parasite relationships to which we shall return later.

Human sleeping sickness (trypanosomiasis) of tropical Africa is caused by *Trypanosoma brucei gambiense* in the west and *T. brucei rhodesiense* in the east. The two subspecies are morphologically indistinguishable from each other and from *T. brucei brucei*, a non-pathogenic parasite of African wild game, particularly of antelopes, which are natural reservoir hosts for the parasite. Trypomastigote forms occur in human blood, and are taken up by the blood-sucking tsetse fly, *Glossina*, the two subspecies being distributed by different species of the vector. After multiplication by binary fission in the alimentary tract of the insect, the parasites enter the salivary glands, where they give rise to epimastigote forms and, eventually, to infective trypomastigotes which are injected into a new human host. Extensive multiplication by binary fission occurs during the first few weeks, but later the parasites invade the cerebrospinal fluid and brain. This leads to the characteristic depression and eventually (perhaps after as long as several years in the Gambian form) to death, unless the condition is treated.

Trypanosoma cruzi, which belongs to a group of trypanosomes distinct from the *brucei* group, and which is the cause of Chagas's disease of Central and South America, probably shows a primitive pattern of transmission of a trypanosome infection from the invertebrate to the vertebrate host, and thus presents an interesting parallel to the coccidian series just reviewed. The parasites are taken up from the human blood stream by bugs. Infective stages, which develop in the hind-gut of the insects, are deposited on to human skin when the insects defaecate in the course of their meal; these stages are then readily rubbed into the wound as the victim reacts to the irritation. Other mammalian hosts may also become infected by swallowing the bugs, or perhaps by eating other infected hosts. Transmission by inoculation with the saliva through the mouth parts, such as is found in *gambiense*, is likely to have evolved as a later specialization; it clearly affords greater certainty of successful infection.

An interesting byway of parasitological speculation is the thought that Darwin might have been infected by *T. cruzi* when he was bitten by the giant bug *Triatoma infestans* ('most disgusting', he commented), which is a vector of the parasite. If so (and the idea has been disputed), this is one parasite to which we can all feel indebted, for it could well have been the cause of Darwin's chronic ill-health. Thus, by protecting him from the sterile round of committee work, it could have made a powerful contribution to the completion of the *Origin of Species*.

In the groups so far considered there is an absence of close relatives that are also free-living. In many other groups these are still surviving, and our understanding of the host-parasite relationship is correspondingly enlarged. This is notably true of the Nematoda, of which Baer has said that no other single group is so perfectly preadapted to becoming parasitic. We have already noted this aspect of nematode organization. The free-living species of this group, which greatly outnumber the parasitic ones, are animals of peculiar yet generalized structure that are particularly well able, both in the adult stage and during development, to resist environmental stress. In consequence, they have successfully exploited an enormous range of habitats, from polar seas to hot springs, from arid deserts to the ocean depths. Add to this their saprophagous habits and their enormous reproductive potential, and it would be surprising indeed if they had not given rise to successful parasitic forms. The same considerations suggest that this will have happened along many independent lines, and this must surely account for the great variety of relationship between nematode parasites and their hosts, and in the details of their life cycles and modes of transmission. This variety precludes further discussion here, beyond remarking that life in the tissues or body cavities of invertebrates may have been the earliest form of parasitism among these animals. It is significant from this point of view that the parasitic status of their relationships is often inferred, and is unsupported by clear evidence. In some instances the nematodes remain within the host after its death and then feed upon its decomposing tissues. In these circumstances it is difficult to distinguish them from predators, and it is likely that their habit of active searching for such food sources was important in the establishment of their parasitism.

Another group in which free-living and parasitic forms can readily be compared is the Platyhelminthes. The extensive development of parasitism in this group may be ascribable in part to the relatively poor differentiation of the tissues, giving a high capacity for adaptibility to varied environments, and accounting also for their not infrequent lack of absolute host specificity. In addition, there was much pre-adaptation to parasitism, just as in the nematodes. Creeping progression and powers of adhesion, which are ecologically important in free-living forms, and notably in the littoral zone, would have contributed to this. So also would the complex and hermaphrodite reproductive system, facilitating the propagation of isolated individuals, and making possible the production of large numbers of resistant eggs. Here, as in the nematodes, the parasitic forms have evolved along more than one line, but in this instance their history is easier to analyse.

Parasitism is rare in the Turbellaria, one example being a rhabdocoele, *Kronbergia*, found in the haemocoele of amphipods; it presents a remarkable parallel to cestodes, in having lost its mouth and alimentary tract, and helps our understanding of the origins of that group, which, like the trematodes, is exclusively parasitic. In the Class Trematoda the Monogenea are characterized by being mostly ectoparasites with a complicated posterior attachment disc, and with a direct life cycle; the eggs give rise to a larva which is well provided with cilia and with posterior hooks, so that it is adapted for swimming and also for attachment. This larva, called an oncomiracidium, metamorphoses on a new host, so that no vector is required. A few species parasitize amphibians and turtles, but as many as 95% are found on fish, particularly upon their gills; of these species the majority are parasites of elasmobranchs, showing a very high degree of host-specificity. It would appear, therefore, that the Monogenea are ancient parasites that have remained confined to the lower vertebrates. Probably they arose from rhabdocoel-like forms that exploited the skin of early fish, and then, finding protection on the gills, became dependent on bloodsucking for their nutrition.

Particularly interesting from this point of view is *Polystoma integerrimum* (parasitic in the bladder of frogs), for this species, and its polystomid relatives, are the only Monogenea (Fig. 23–16) that have successfully exploited hosts other than fish. This they have done by moving into internal cavities such as the mouth, nostrils,

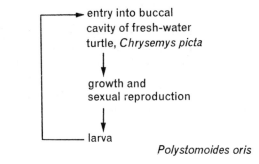
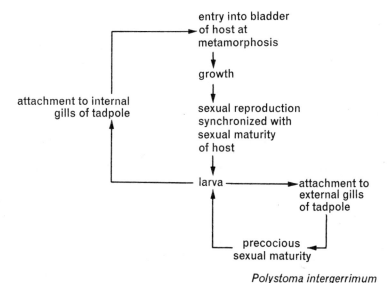

Fig. 23-16. Patterns of life cycles in the Monogenea. In *Polystoma intergerrimum* the normal cycle is shortened by the development of precocious sexual maturity (neoteny) if the larva attaches to external gills instead of to internal ones.

and bladder; never, however, into the digestive regions of the alimentary tract, although the young *Polystoma* passes down the gut of the metamorphosing tadpole to reach the bladder. This limitation is a reminder that a transition from ectoparasitism to endoparasitism, or, in more general terms, from an ectozoic to an entozoic habit, may be attractively plausible in theory, but has not always been easily achieved in practice. In particular, the stresses experienced in the alimentary tract are such that animals must always have found it difficult to establish themselves there unless they possessed from the beginning strongly developed powers of resistance to peristaltic movement and to enzyme action; the Monogenea have perhaps been lacking in these, although there are isolated records of species occurring in the oesophagus or intestine of fish. These tend to support the view that cestodes evolved from early monogeneans.

The Digenea (which include the liver flukes) differ substantially from the Monogenea. Some authorities have nevertheless supposed that they originated from monogeneans, but even if this is so it seems likely that the two groups must have diverged from a rhabdocoel-like ancestry at a very early stage. This follows from the differences in their host relationships. Most digeneans are endoparasites of the vertebrate alimentary tract and its outgrowths, where they exploit the exceptionally rich food resources of these organs. They have a complex and indirect life history, with several developmental stages in which multiplication may occur. Particularly significant is the fact that, although more than one intermediate host may be involved, the first is invariably a mollusc, usually a gastropod but sometimes a lamellibranch or a scaphopod; this is so, irrespective of whether the final host is marine, fresh-water, or terrestrial. When, however, there is a second intermediate host, this may be a member of almost any group of animals. No less significant is the high degree of specificity that digeneans show towards their molluscan hosts, and the low degree that they sometimes show towards their definitive vertebrate ones. This implies that they were first associated with molluscs, and began to parasitize vertebrates at a later stage. (See Fig. 23–17.)

Fig. 23-17. Patterns of life cycles in the Digenea. In the hypothetical ancestral stage there would have been only one host, a mollusc. Later this becomes the first intermediate host. A second intermediate host may (*Bucephalopsis*) or may not (*Fasciola*) be involved. In either case the cercaria commonly encysts to form a tailless metacercaria, which may be partially metamorphosed.

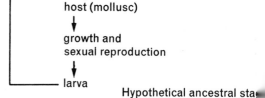

Hypothetical ancestral stage

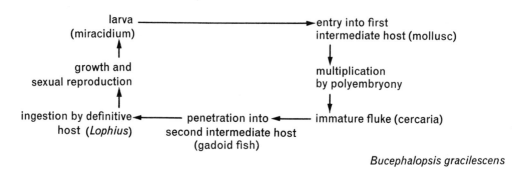

Bucephalopsis gracilescens

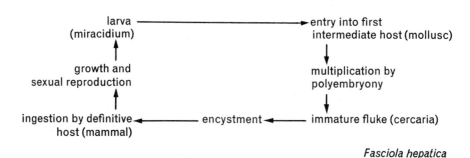

Fasciola hepatica

Thus we can plausibly visualize the Digenea as being derived from rhabdocoel-like ancestors that invaded molluscs, perhaps as a result of ingestion, the invasive stage being the free-swimming larva that is represented today by the miracidium. Whether the worm would have matured within the host is not clear, although it is probably significant that rhabdocoels do now occur in molluscs and echinoderms. An example is *Paravortex*, which is found in the mantle cavity and internal organs of the marine bivalve *Modiolus*. In any case, we may suppose that extension of the life history into a vertebrate host occurred later, when encysted cercariae were ingested by vertebrates and proved able to mature within their alimentary tract. Once this had happened, the group became closely adapted to the ecology of a wide range of vertebrate hosts; this has made possible the proliferation of digenean habits that we find today.

To give only a few examples, *Schistosoma mansoni*, one of the devastating blood flukes of man, is transferred from its molluscan vector to its primary host by the cercaria directly penetrating the skin of human beings that enter the water. *Fasciola hepatica* takes advantage of the amphibious habits of the snail, *Lymnaea truncatula*; the motile cercariae encyst on vegetation as metacercariae, and are eaten by sheep and cattle. *Dicrocoelium dendriticum* enters the bile ducts of as many as 40 different mammalian hosts (including sheep and cattle) by a more elaborate procedure. Groups of the cercariae are enclosed in slime balls, secreted partly by the larvae and partly by the terrestrial snail host, and are then ingested by ants; within the body cavity of this second intermediate host they form metacercariae, and are eventually swallowed by the mammalian host.

Alternative routes into terrestrial vertebrates are seen in *Haematoloechus* spp. and *Lecithodendrium chilostomum*. The former is parasitic in the lungs of amphibians, arriving there by a life cycle that passes first (Fig. 23–18) through molluscan hosts (*Lymnaea* spp.) and then through the nymph of the dragon fly, in which metacercariae are formed; there are then ingested by the primary host when it eats the imago. The life cycle of *Lecithodendrium*, a parasite of bats, depends in a similar way upon the formation of metacercariae in the haemocoel of the larva of a caddis fly.

Exceptionally there may be no motile cercaria stage, as in the life cycle of *Ptychogonimus megastoma*, a parasite of sharks that is elegantly adapted to the habits of two intermediate hosts. The first of these hosts is a scaphopod, from which the cercariae escape still enclosed with the rediae. The latter wriggle slowly on the sea bottom, thereby attracting the attention of crabs, which proceed to eat them. The cercariae escape from the rediae within the crab, encyst in its body cavity, and are readily transmitted to the sharks that prey upon the crabs.

The third major group of platyhelminth parasites is the Class Cestoda, characterized by the anterior attachment organ (scolex), by the loss of the alimentary canal, by the subdivision of the tape-like body into a chain of proglottids, each with a complete set of reproductive organs, and by the development of the egg into a six-hooked embryo (hexacanth embryo, oncosphere) which may hatch as a ciliated coracidium. Oral infection of intermediate hosts leads to the development of various types of metacestodes, which eventually reach their definitive hosts. One view of cestode origins has been that they had a direct line of evolution from monogeneans through digeneans. A possible linking group has been seen in the Class Cestodaria, a group of uncertain affinities, characterized by an undivided body, a single re-

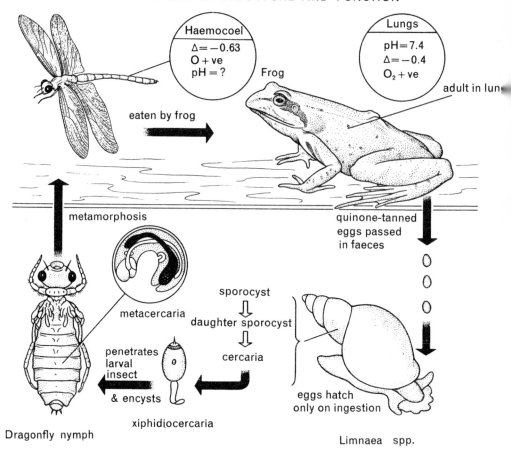

Fig. 23-18. *Haematoloechus variegatus:* the life cycle and some physiological factors relating to it. The eggs are probably embryonated when laid. The details of the cycle are very imperfectly known. From Smyth, 1962. *Introduction to Animal Parasitology.* English Universities Press, London.

productive system, the lack of a scolex, and the development of a ten-hooked embryo. However, this view now receives little support in its original form; apart from what has been said above regarding monogenean and digenean relationships, many other considerations argue against it. The organs of attachment in trematodes and cestodes are very different, and there are important differences also in the reproductive system, particularly as regards the uterus; in trematodes this opens into a genital atrium with the male duct, but it is entirely independent of the atrium in the cestodes. Nevertheless, Llewellyn has suggested that these differences do not necessarily preclude a derivation of cestodes from early monogeneans and has argued that these two groups are more closely related than either of them is to the digeneans.

The characteristic features of tapeworms limit the adults, with few exceptions, to the alimentary tracts of vertebrate hosts, although their larvae use a diversity of intermediate hosts. It is possible, that cestode parasitism, like that of the Digenea, began with the ingestion of rhabdocoel-like organisms. Alternatively, monogeneans may have succeeded in invading the alimentary tract of their hosts, if Llewellyn's arguments are accepted. In any case, the first hosts must have been fish; to this day the more primitive tapeworms survive in the alimentary tract of fish, using other aquatic animals as intermediate hosts.

Subsequently the eucestodes clearly evolved in parallel with the vertebrates, for they are unusually host-specific, with each vertebrate group having its characteristic genera and species. Thus the Orders Tetraphyllidea and Trypanorhyncha are exclusively parasites of elasmobranchs, having successfully adapted to the high urea content of these animals. One trypanorhynchid is able to metabolize this substance, and it has become important in the control of the permeability of tetraphyllideans, whereas in other flatworms it causes irreversible damage. The Order Pseudophyllidea chiefly parasitizes teleosts, birds, and mammals. In this last group of cestodes the transition to land-dwelling hosts has depended upon the fish-eating habits of the latter. An example is *Diphyllobothrium dendriticum*, parasitic in gulls, with a life history (Fig. 23–19) involving copepods as the first intermediate host and fish as the second. *D. latum*, a parasite of man, has a similar cycle. In these instances the parasites have succeeded in exploiting the higher vertebrates by extending their own life cycles.

A more complete adaptation to terrestrial hosts is seen in the Order Cyclophyllidea (Taenioidea), which are mainly parasites of birds and mammals, although found also in reptiles and amphibians. The majority of species are found in birds, a peculiarity that has been attributed to the diversified and almost explosive radiation of that group, as compared with the slower rate of mammalian evolution. In these cestodes the life cycle has come to depend entirely on terrestrial hosts. Examples are *Hymenolepis diminuta* of rodents, transmitted by insects, and *Taenia saginata* of man, distributed by cattle and other mammalian intermediate hosts, who swallow the proglottids that drop from the anus of the primary host. The ultimate limit of success along this line of evolution has been attained by *H. nana nana*, the dwarf tapeworm of man, which can be transmitted directly through the faecal contamination of food; no intermediate host is now required, although arthropod hosts can be employed. (See Fig. 23–20.)

The specificity of host–parasite relationships depends on physiological adaptations that do not differ in principle from those that regulate the relationships and life histories of free-living forms. The precision and delicacy of parasitic adaptations is emphasized by Baer, who points out that the miracidium of the trematode *Opisthorchis felineus* is attracted to a prosobranch snail, *Bithynia leachi*, but not to the closely related *B. tentaculata*, although this occurs in the same habitat and may even be more abundant there. As with the other symbiotic relationships discussed earlier, much more needs to be learned of the basis of these attractions. Physical factors in the external environment certainly play a part in bringing parasites into contact with their hosts. Miracidia larvae respond to light and gravity in ways which tend to bring them into the neighbourhood of their snail hosts. Final contact, however, may be achieved through their sensitivity to the mucus given off by the host. This mucus is of complex constitution; that of *Lymnea truncatula* contains 16 amino acids, glucose, and various lipids, so that there is ample scope for the evolution of highly specific relationships.

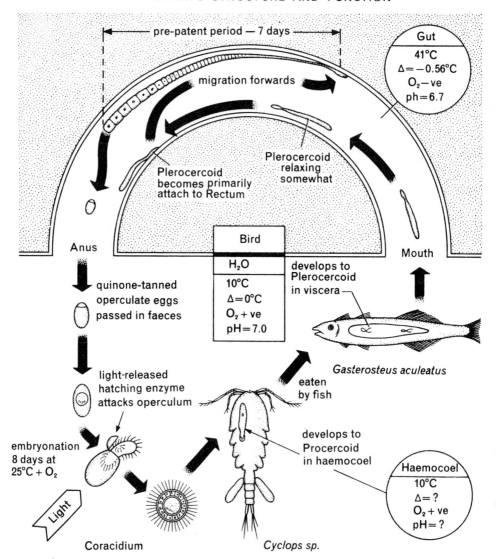

Fig. 23-19. *Diphyllobothrium dendriticum:* the life cycle and some physiological factors relating to it. From Smyth, 1962. *op. cit.*

Cercaria larvae, by contrast, are said not to be so markedly influenced by chemical attraction. Those of *Schistosoma*, released from planorbid snails, are said to locate the skin of their human hosts by chance, although the subsequent penetration is a rapid and highly specialized reaction, involving not only movement but also the secretion of adhesive material together with hyaluronidase and, probably, proteolytic enzymes. The cercariae of some parasites will certainly enter a range of animals, although usually the life cycle can only be completed in the correct host, and misdirected entries lead nowhere. It is well to remember this; specificity is by

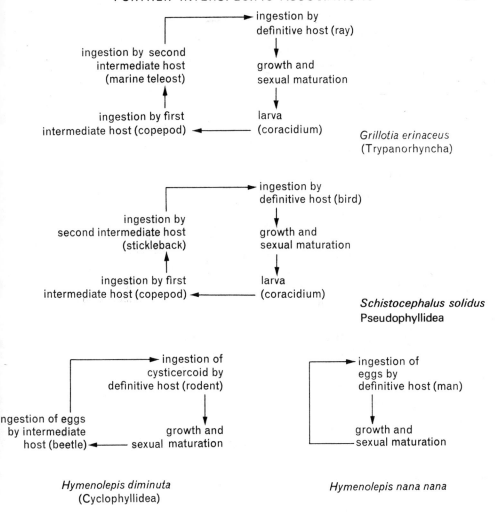

Fig. 23-20. Patterns of life cycles in the Eucestoda. The life cycle of *Diphyllobothrium* (Fig. 23-14) is similar in principle to that of *Schistocephalus*.

no means universal in parasite behaviour, a flaw that has reacted on those investigators who have given erroneous accounts of supposed life histories, based on the mistakes of the parasitic larvae.

The problems involved in the finding of hosts are sometimes formidable, and it is a tribute to the power of natural selection that they have been solved at all. Consider, for example, the difficulty confronting the oncomiracidium of monogenean trematodes (p. 714). The larva, with perhaps only 24 hr of free-swimming life, and able to swim at no more than 5 mm/sec, must intercept fish that can travel perhaps 30 times as fast. Very little is known about the behaviour of these parasites, but, according to Llewellyn, the likely solution to the problem is that the reproductive habits of the parasite are nicely adapted to those of the fish host. The fish tend to concentrate in inshore waters for a limited period for spawning. In some monogeneans the eggs remain attached to the body of the fish, while in others they become free and fall to the bottom. In either case the relative immobility of the fish during their spawning congregation will favour infection by the newly hatched larvae, at least in those localities where most of the eggs have been laid. Infection may, of course, be aided by adaptively organized activity of the oncomiracidia, although here again very little is known.

One example is given by *Diplozoon*, the well-known monogenean parasite of the gills of fish, which consists of two individuals fused together. The oncomiracidium swims at random until it senses the current of water that is being drawn into the mouth of its future host. The movement of its cilia promptly stops, and the larva is drawn into the mouth. *Polystoma intergerrimum* behaves quite differently, attacking its amphibian host by another route, and following a behaviour patterns which, as described by Llewellyn, is engagingly reminiscent of burglarious entry. It glides at random over the skin of a tadpole until it is within about a millimetre of the edge of the operculum. Here its behaviour suddenly changes. 'The larvae then made for the very edge of the operculum, paused for some seconds, and then swiftly entered the opercular cleft during an interval between the exhalent pulses from the gills.' This behaviour clearly implies some perception of water currents.

Periodicity may be an important element of behaviour in the securing of transmission. For example, the cercaria larvae of the trematode *Philophthalmus megalurus*, which lives as an adult in a pocket at the base of the nictitating membrane of birds, encyst on the exoskeleton of crayfish. This vector emerges at dusk, and so do the cercariae. The larvae of the related *Parorchis acanthus* are less restricted in their choice of encystment sites, and they emerge in daylight.

An often-quoted illustration of periodicity, this time from the nematodes, is the filiarian *Loa loa*, which infects the subcutaneous connective tissue of man and monkeys in West and Central Africa, and is distributed by various species of the mangrove fly, *Chrysops*. This eye-worm (so-called because in the course of its movement around the body it passes from time to time across the eye) releases pre-larval stages (microfilariae) which pass into the peripheral blood, from where they are taken up by the vectors. The microfilariae of the human strain are diurnal, almost disappearing from the peripheral blood at night (Fig. 23-21). This accords with the feeding habits of the two vectors, *C. silacea* and *C. dimidiata*, which are also diurnal, their feeding ending at around 17.30 hrs (Fig. 23-22). The simian strain of *Loa loa* is quite distinct from the human one, and behaves differently, its microfilariae appearing in the peripheral blood at night (Fig. 23-21). This accords with the feeding habits of their vectors, *C. langi* and *C. centurionis*, which live in the forest canopy, and which bite from about 17.00 to 21.00 hrs. (Fig. 23-22). This is the time during which monkeys sleep on the branches of the forest trees, and the flies, approaching them from below, can readily bite their projecting hairless buttocks.

Interchange of the two strains of parasite between man and monkey is readily achieved experimentally, the microfilariae continuing to show their innate periodicity, regardless of the host in which they are living. To what extent can such interchange occur under natural conditions? Probably very little, if at all. The fly vectors for the human strain occur in the forest canopy, but are unlikely to have much success in biting the monkeys, for the hairless parts of these are little exposed during the day,

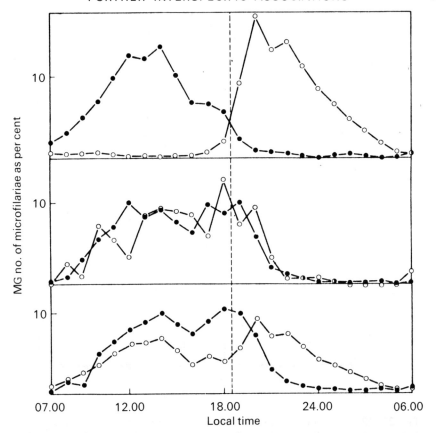

Fig. 23-21. The periodicities of microfilariae of *Loa loa* in *Mandrillus leucophaeus*. ● human strain; ○ simian strain. Note that the human strain retains its natural periodicity in the simian host. From Duke, 1972. Behavioural aspects of parasite transmission. Supplement 1, *Zool. J. Linn. Soc. Lond.*, **51**, 97–107. (E. U. Canning and C. A. Wright, eds.).

and in any case they are alert to catch and eat the insects. The simian vectors could, in theory, bite man, but only in the unlikely event of him exposing himself in the canopy at dusk. Duke points out that the parasite, vector, and host are here linked in a complex of behavioural factors which must have been contributing to the divergent evolution of the two strains. At present, they can still produce fertile crosses, but true specific status may eventually be reached.

One unexpected intrusion of human behaviour into this situation is the lighting of fires, which attract the human vectors down to ground level (apparently because of some chemical attractant in the smoke) and thus facilitate the spread of the parasite. But there are many other ways in which the behaviour patterns of our species, ranging as they do from the strictly utilitarian, through the mildly eccentric, to the frankly disgusting, have helped our parasites to complete their life cycles, and, indeed, still do so. The effect of artificial irrigation in promoting an increased incidence and spread of schistosomiasis through the provision of a good habitat for

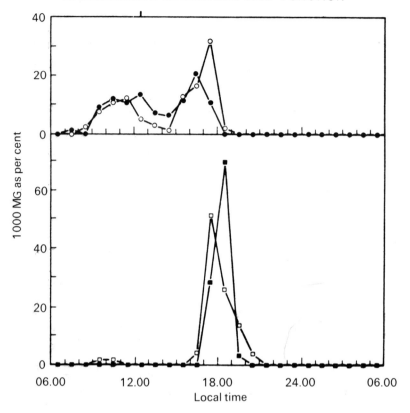

Fig. 23-22. The 24 hr biting cycles, using human bait, of four species of *Chrysops* at canopy level in the rain-forest. The hourly biting densities have been computed from the geometric means (multiplied by 1,000 to give 1,000 M_G) of the catches for each hour, and are expressed as percentages of the total 24 hr biting density. ○ *C. silacea;* ● *C. dimidiate;* □ *C. langi;* ■ *C. centurionis.* From Duke, 1972. *op. cit.*

the snail vector is only one illustration, although an immensely important one. It presents a formidable problem, for it is asking too much of human nature to expect children, living in desert heat, to keep away from the infected water.

Often, however, there is less excuse. Housewives are well known to infect themselves with trichinosis (*Trichinella spiralis*) by nibbling raw sausage meat in their kitchens, and thereby swallowing encapsulated larvae from infected pork sausage meat. But there are other ways of ensuring infection, reminding one only too painfully that 'The world is a bundle of hay, Mankind are the asses who pull'. The eating of exotic meat can be a hazard, as was found by a group of New York gourmets who developed severe trichinosis after enjoying dishes prepared from black-bear meat. And there is a lesson, well worth considering by barbecue addicts, in the experience of a group of teenage Kenyans who, ignoring the traditional wisdom of their tribe, ate the undercooked flesh of bush pig and found themselves with the first trichinosis infection encountered in tropical Africa, and, indeed, with the heaviest infection of it ever recorded in man.

Religious traditions have been helpful in combating parasitism. The avoidance by Jews of pork keeps them free of trichinosis, while Mohammed has reduced the incidence of hydatid disease in Moslems by forbidding close association with dogs, which are the commonest hosts of the tapeworm *Echinococcus granulosus*. Since a single hydatid cyst, which is the larval stage of this parasite, may contain up to 17 litres of liquid, with perhaps 1 million tapeworm scolices per litre, the prophet's guidance is well worth attention. The use of nurse dogs by the Turkana of Kenya, who are said to use them to mop up the vomit and excreta of their babies, takes us to the opposite extreme. Since the dogs will already have infected their muzzles with *Echinococcus* eggs by licking their own anuses, the scene is well set for infecting the babies with hydatid disease.

Nelson, quoting Jenner at the end of his sobering review of this field, remarks that man exposes himself to major parasitological hazard from 'the love of splendour, from the indulgence of luxury and from his fondness of amusement'. Certainly the tourists who set biology at naught by overlooking the possibility that there may be schistosomes in the water, hookworm larvae in the moist sand, and *Ascaris* eggs in the salad, are putting themselves at no small risk. Parasites are not slow to take advantage of these weaknesses.

Once an association has been established between a parasite and a host its further evolution will depend upon the reactions of both organisms. Indeed, the initial establishment depends upon an over-riding of the normal tendency of animals to react to invasions with responses that immobilize or destroy the invader. Such a response is the origin of natural oriental pearls, which usually contain a plerocercoid (a solid larva with an adult scolex) of a tetraphyllidean cestode that parasitizes elasmobranchs. The rarity of pearls suggest that this is a chance infection of the oyster, perhaps a result of it ingesting an infected planktonic crustacean that is acting as the first intermediate host. The life history of the dog tapeworm *Dipylidium caninum* suggests how such a barrier can be overcome. Flea larvae become infected with the onchospheres, but these do not develop further in the larva because they become enclosed by phagocytes; at pupation, however, these cells leave the oncospheres, which can then develop into cystercoid larvae with a scolex which, unlike that of the cysticercus of *Taenia*, is not invaginated. Thus the adult flea is fully infective. Here the reactions of the intermediate host are nicely attuned to the requirements of the parasite, providing a balance that is as necessary for successful parasitism as we have seen it to be for mutualism.

In some instances, however, the achievements of parasites go far beyond this, to secure a success which it is perhaps too easy to take for granted, because it is so familiar. This is their ability to maintain continuing (chronic) infections in their vertebrate hosts, leading, in man, to death or at least to prolonged ill-health. The persistence of the parasite is a considerable success, for it means that in some way it has evaded the host's immune response. This is a complex protective device of vertebrates to ensure the recognition and destruction of foreign organisms, a result achieved in principle by the production of immunoglobulins (antibodies) in response to the presence of invading antigens. How do parasites achieve this evasion? To be able to answer this question would be to make a major advance in the treatment of parasitic infections, but a full understanding lies in the future. Nevertheless, something is becoming known of the devices that parasites employ.

Trypanosomes certainly stimulate antibody production by their hosts at the beginning of their invasion, and they can be destroyed by antibodies. They would seem, therefore, to be ill-adjusted for survival. However, they overcome the difficulty by producing new antigens through the process called antigenic variation. How trypanosomes achieve this is not understood, but the practical result for the host is that following an initial reduction of parasite numbers through antibody action, a new wave of parasites appears with a new antigenic constitution. The process is thought to be aided by some depression of the immune response of the host, and perhaps also by the parasites occupying what are called priviliged sites, in which they are protected from the full effects of the host's antibodies.

Another chronic disease, mentioned earlier, is schistosomiasis, a burden which is believed to affect over 200 million people, producing in them a prolonged debilitation, together with severe liver damage and perhaps eventual death. The condition is due to continued infection by the blood-dwelling digenean trematode, *Schistosoma*, which is remarkable for being dioecious and for living with the female attached to the ventral surface of the male. The eggs (nearly 200 per day from a single pair) pass into the lumen of the intestine and are discharged with the faeces, the life cycle being completed, as we have already noted, in aquatic planorbid snails. The persistence of the parasite within the human host is now thought to depend upon a strange phenomenon. Adult worms evoke a powerful immune response in monkeys, which thereby become resistant to further infection by cercariae. The original adults, however, continue to survive, so that in some way they must have protected themselves against the immune response that they have themselves evoked. It is thought that this protection is due to the presence over their surface of antigens similar to those of the host, which is thus unable to recognize them as invaders. One view is that these host-like antigens are actually synthesized by the host and acquired by the developing parasites from the blood stream. If this interpretation is correct it means that the worms, having donned their disguise and ensured their own survival, can still release schistosome antigens and, by evoking the host's immune response, protect it from new infection.

This concept introduces a degree of subtlety into host–parasite relationships that is not fully encompassed by the conventional definition of parasitism. The adult schistosomes are promoting the survival both of themselves and of their hosts (although doubtless not in the condition which the hosts would prefer), and are doing this by destroying other and younger members of their own species. It has been argued that this pattern of relationship could have still wider implications in natural conditions, for wild mammals may carry multiple infections due to a number of different species of parasites. Perhaps one factor regulating such a parasite–host complex is the establishment of the well-authenticated condition of heterologous immunity, in which one invading species protects the host from invasion by another, or at least ameliorates the infection. Cox points out that this could lead to a parasite actually benefiting its host, and that this might 'make it impossible to produce any overall definition of parasitism'. Which takes us back to the beginning of Chapter 22.

The reactions of the parasite to its host are determined by the fact that the chief benefits that it obtains are protection and a ready source of nutrition; the latter is so richly provided that it permits the impressive fecundity that is such a feature of parasitic life. Thus the reproductive system tends to predominate in the organization of the adult, so much so that the systematic relationships of parasites may become unrecognizable. For example *Entoconcha mirabilis* (Fig. 23–23), a gastropod parasitic in the holothurian *Synapta digitata*, consists of a narrow tube attached at one end by its mouth to the ventral vessel of the host. Within the tube are little more than a blind alimentary canal and a reproductive system. The parasite, in fact, is totally unrecognizable as a gastropod, its true affinities being only revealed when it produces typical veliger larvae. Remarkably enough, when this parasite was first described in 1852 it was interpreted as an example of an alternation of molluscan and echinoderm generations, the assumption being that the holothurian was actually giving rise to veligers.

No less remarkable is the transformation of appearances seen in parasitic crustaceans. Thus the adult female of *Portunion maenadis*, an isopod parasitic within the body cavity of *Carcinus maenas*, is unrecognizable as an isopod, although prior to its final metamorphosis, when it is moving within its host, it is more evidently crustacean. In this group the supreme importance of reproduction in parasitic life is shown by the marked sexual dimorphism. The males, known as dwarf males, are much reduced in size and normally live on the body of the female; often they are found within the brood pouch, in which, as in free-living isopods, the eggs are incubated.

The cirripede Rhizocephala (cf. Fig. 17–7) have equally striking structural modifications. *Sacculina*, the well-known parasite of crabs, passes through typical nauplius and cypris stages, but after settling on a host it becomes a mass of cells which passes into the crab. Eventually it comes to lie externally, as a hermaphroditic, tumour-like structure that absorbs food through branching processes that extend throughout its host's body. This method of nutrition has led in *Thompsonia* (Fig. 23–24) to the development of asexual reproduction. The adults are of simpler structure than

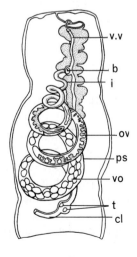

Fig. 23-23. *Entoconcha mirabilis* (after Baur) and its relationships with its host, *Synapta*. *b*, mouth, fixed to the ventral vessel, *v.v*, of the host's intestine; *cl*, ciliated canal; *i*, intestine; *ov*, ovary; *ps*, pseudopallium; *t*, testes; *vo*, masses of developing eggs. From Caullery, 1952. *Parasitism and Symbiosis*. Sidgwick and Jackson, London.

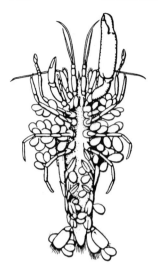

Fig. 23-24. *Thompsonia* sp. on *Synalpheus brucei*. After Potts, from Caullery, 1952. *op. cit.*

Sacculina, and are budded off in large numbers from the foot system used in feeding. They are periodically discarded when the host moults, and a new generation is budded off to take their place, much 'as successive crops of cultivated mushrooms develop from the mycelium', to quote Caullery.

Probably much of the subtlety of host-parasite relationships has yet to be revealed. We have repeatedly emphasized the complexity of nutritional relationships between organisms, and the position of parasites is unlikely to be different in this regard. The ability of the host to provide not only constructional and energy-rich molecules, but also vitamins and trace elements, is thus likely to be an important factor is establishing host-parasite specificity. Relatively little is yet known of this aspect, but, to take one example, the readiness with which trypanosomes can be cultured *in vitro* has made it possible to show that two essential requirements for their growth are haematin and cytochrome oxidase.

In certain instances the physiology of the parasite has become so closely integrated with that of the host that it depends upon the latter for the regulation of its reproduction. An elegant example of this is seen in *Polystoma*, which enters on sexual maturation in spring, when the host frog will be preparing to enter water to effect its own reproduction. As a result, the parasite's eggs are released when the frogs are in water. Moreover, the rate of development of the eggs is so timed that the larvae developing from them are ready to enter the frog tadpoles when these are at the internal gill stage. Initially they are attached to the gills, like the adult stages of more typical Monogenea, but they finally pass down to the bladder when the tadpole metamorphoses. Sometimes this synchrony may be less than perfect, and the larvae become attached to the external gills of the tadpole; should this happen, the larva develops within 20 days into a neotenous form, sexually mature, and capable of giving rise to fertile eggs. This adjustment of adult parasite to the habits of the host is believed to be regulated by the hormones of the latter, for precocious maturity can be brought about in the parasites if pituitary extracts are injected into immature frogs.

This taking-over by entozoic forms of the chemical regulating systems of their hosts may prove to be common. It is found also in *Opalina ranarum* and *Nyctotherus cordiformis*, familiar protozoans of the intestine of frogs; both of these, like *Polystoma*, enter upon their sexual reproductive phase in the early months of the year, in parallel with the reproductive cycle of the host. Another example is seen in the symbiotic flagellates of *Cryptocercus punctulatus*. These reproduce asexually, but some of them enter upon sexual reproduction at the time of the host's ecdysis. Since this sexual phase can be abolished either by transferring the flagellates to a non-moulting host, or by removing the pars intercerebralis from the host, the symbionts apparently are reacting to the host's moulting hormone. We have suggested earlier that these reactions (p. 539) may be to metabolic effects of the hormones, rather than directly to the hormones themselves. Even so, this does not lessen the remarkable intimacy of these host-parasite relationships.

With these relationships, as with all other aspects of invertebrate biology, new and improved techniques of investigation continue to enlarge our appreciation of the elegance and precision of the adaptations by which animals have succeeded in coming to terms with their environment. It cannot be said that these advances have as yet taken us far in our search for the answers to the questions that are posed by the study of animal life. Rather they lead us on to new and more penetrating questions, demanding an increasingly subtle level of analysis. Indeed, at the present stage of biological investigation we should be suspicious of facile answers, which can only too readily emerge as a result of asking the wrong questions. Arthur Koestler's experience ought to be a common one. In his youth he regarded the universe as an open book, but later he came to see it as a text written in invisible ink; a text of which we are sometimes able to decipher small fragments. It is to be hoped that our review of invertebrate structure and function will have given some guidance in this deciphering; but it will have sadly failed in its purpose if it has not also demonstrated how few and small are the fragments that we have so far been able to expose.

Classification

The classification of the animal kingdom is under continuous review, so that the treatment of particular groups is liable to be modified from time to time. The scheme adopted here, which is intended to help the reader to place in an acceptable context the main groups referred to in the text, is largely based upon the classification used in *The Zoological Record* (published annually by the Zoological Society of London). Authorities for the scheme used in this publication are listed in Section 1 of each volume.

Many groups are classified in detail in the multi-volume *Treatise on Invertebrate Palaeontology* (R. C. Moore, ed.), published at Lawrence, Kansas, by the University of Kansas and the Geological Society of America.

Further information may also be obtained from the following:

Hartman, W. D., 1958. A re-examination of Bidder's classification of the Calcarea. *Syst. Zool.* **7**, 97–110.
Honigberg, W. D. *et al.*, 1967. A revised classification of the phylum Protozoa. *J. Protozool.*, **11**, 7–20.
Rothschild, Lord, 1965. *A Classification of Living Animals* (2nd ed.) Longmans, London.
Stephen, A. A. and Edmonds, S. J., 1972. *The phyla Sipuncula and Echiura.* British Museum (Natural History), London.
Vokes, H. E., 1967. Genera of the Bivalvia. *Bull. Ann. Palaeont.* **51**.

Phylum Protozoa

   Subphylum Sarcomastigophora
     Superclass Mastigophora (flagellates)
       Class Phytomastigophorea (phytoflagellates)
       Class Zoomastigophorea (zooflagellates)
     Superclass Opalinata
     Superclass Sarcodina
       Class Rhizopodea
         Subclass Lobosia
           Order Amoebida
           Order Arcellinida

CLASSIFICATION

      Subclass Filosia
      Subclass Granuloreticulosia (including the Order Foraminiferida)
      Subclass Mycetozoa
    Class Actinopodea
      Subclass Radiolaria
      Subclass Heliozoia
      Subclass Acantharia
      Subclass Proteomyxidia
Subphylum Ciliophora
    Class Ciliatea
      Subclass Holotrichia
      Subclass Peritrichia
      Subclass Suctoria
      Subclass Spirotrichia
Subphylum Apicomplexa (= Sporozoa)
    Class Sporozoea
      Subclass Gregarinia
        Order Archigregarinida
        Order Eugregarinida
        Order Neogregarinida
      Subclass Coccidia
        Order Protococcida
        Order Eucoccida
    Class Piroplasmea
Subphylum Myxospora (= Cnidospora in part)
    Class Myxosporea
Subphylum Microspora (= Cnidospora in part)
    Class Microsporea
    Class Haplosporea

Phylum Porifera (= Parazoa)
    Class Calcarea
      Order Homocoela
      Order Heterocoela
    Class Hexactinellida
    Class Demospongia
      Subclass Tetractinellida
      Subclass Monoaxonida
      Subclass Keratosa

Phylum Cnidaria
    Class Hydrozoa
      Order Trachylinida
        Suborder Trachymedusae
        Suborder Narcomedusae
      Order Hydroida
        Suborder Gymnoblastea
        Suborder Calyptoblastea
      Order Milleporina
      Order Stylasterina
      Order Siphonophorida
    Class Scyphozoa
      Subclass Scyphomedusae
      Subclass Conulata
    Class Anthozoa
      Subclass Ceriantipatharia

Order Antipatharia
Order Ceriantharia
Subclass Octocorallia (soft corals)
Order Alcyonacea
Order Gorgonacea
Order Pennatulacea
Subclass Zoantharia
Order Actiniaria (sea anemones)
Order Zoanthiniaria
Order Corallimorphoria (coral-like anemones)
Order Madreporaria (true or stony corals)

Phylum Ctenophora
Class Tentaculata
Order Cydippida
Order Lobata
Order Cestida
Order Platyctenea
Class Nuda
Order Beroida

Phylum Platyhelminthes
Class Turbellaria
Order Acoela
Order Rhabdocoela (includes the *Temnocephalida*)
Order Alloeocoela
Order Tricladida
Order Polycladida
Class Trematoda
Subclass Aspidogastrea
Subclass Digenea
Class Monogenea
Class Cestodaria
Class Cestoda
Subclass Cestodaria
Subclass Eucestoda

Phylum Nemertinea
Subclass Anopla
Order Palaeonemertini
Order Heteronemertini
Subclass Enopla
Order Hoplonemertini
Order Bdellonemertini

Phylum Mesozoa
Order Dicyemida
Order Orthonectida

Phylum Acanthocephala

Phylum Aschelminthes
Class Nematoda (round worms)
Class Nematomorpha (horsehair worms)

Phylum Rotifera

Phylum Gastrotrichia

Phylum Kinorhyncha (Echinodera)

Phylum Entoprocta (=Endoprocta, Polyzoa Endoprocta)

Phylum Polyzoa (=Bryozoa, Ectoprocta, Polyzoa Ectoprocta)
    Class Gymnolaemata
    Class Stenolaemata
    Class Phylactolaemata

Phylum Brachiopoda
    Class Inarticulata
    Class Articulata

Phylum Annelida
    Class Polychaeta
        Order Errantia
        Order Sedentaria
    Class Archiannelida
    Class Myzostomaria (=Myzostomida)
    Class Oligochaeta
    Class Hirudinea
        Order Acanthobdellida
        Order Rhynchobdellida
        Order Gnathobdellida

Phylum Priapulida

Phylum Sipuncula

Phylum Echiura

Phylum Tardigrada

Phylum Pogonophora

Phylum Arthropoda (but see p. 186)
    Class Crustacea
        Subclass Cephalocarida
        Subclass Mystacocarida
        Subclass Branchiura
        Subclass Branchiopoda
        Subclass Ostracoda
        Subclass Copepoda
        Subclass Cirripedia
        Subclass Malacostraca
            Superorder Phyllocarida
            Superorder Hoplocarida
            Superorder Syncarida
            Superorder Peracarida
                Order Mysidacea
                Order Cumacea
                Order Spelaeogriphacea
                Order Tanaidacea
                Order Isopoda
                Order Amphipoda
            Superorder Pancarida
            Superorder Eucarida

Order Decapoda
   Suborder Reptantia
     Infraorder Brachyura
     Infraorder Anomura
     Infraorder Astacura
     Infraorder Palinura
   Suborder Natantia
Order Euphausiacea
Class Trilobita
Class Onychophora
Class Myriapoda
  Subclass Pauropoda
  Subclass Diplopoda
  Subclass Chilopoda
  Subclass Symphyla
Class Insecta (Hexapoda)
  Subclass Apterygota
    Order Diplura
    Order Protura
    Order Collembola
    Order Thysanura
  Subclass Pterygota
      Division Hemimetabola
      Division Holometabola
Class Merostomata
  Subclass Xiphosura
  Subclass Eurypterida
Class Arachnida
    Order Scorpionida
    Order Araneae
    Order Acarina
    Order Pedipalpi
    Order Pseudoscorpiones
    Order Solifugae
    Order Opiliones
    (and other orders)
Class Pycnogonida (=Pantapoda)
Class Pentastomida

Phylum Mollusca
  Class Monoplacophora
  Class Polyplacophora
  Class Aplacophora
  Class Gastropoda
    Subclass Prosobranchia
      Order Archaeogastropoda
      Order Mesogastropoda
      Order Neogastropoda (Stenoglossa)
    Subclass Opisthobranchia
      Order Tectibranchia
      Order Pteropoda
      Order Nudibranchia
    Subclass Pulmonata
      Order Basommatophora
      Order Stylommatophora
  Class Scaphopoda

Class Bivalvia (Lamellibranchia; Pelecypoda)
    Order Protobranchia
    Order Filibranchia
    Order Eulamellibranchia
    Order Septibranchia

A more recent classification of this class is as follows (cf. p. 266):

    Order Protobranchia
    Order Eutaxodonta
    Order Anisomyaria
    Order Schizodonta
    Order Heterodonta
    Order Desmodonta
    Order Septibranchia
Class Cephalopoda
  Subclass Nautiloidea
  Subclass Ammonoidea
  Subclass Coleoidea
    Order Decapoda
    Order Vampyromorpha
    Order Octopoda

Phylum Echinodermata
  Subphylum Homalozoa
  Subphylum Crinozoa
    Class Crinoidea
    (and a number of extinct classes)
  Subphylum Asterozoa
    Class Stelleroidea
      Subclass Somasteroidea
      Subclass Asteroidea
      Subclass Ophiuroidea
  Subphylum Echinozoa
    Class Echinoidea
    Class Holothuroidea
    (and a number of extinct classes)

Phylum Phoronidea

Phylum Chaetognatha

Phylum Hemichordata
  Class Pterobranchia
    Order Rhabdopleurida
    Order Cephalodiscida
  Class Enteropneusta
  Class Planctosphaeroidea

Phylum Chordata
  Subphylum Urochordata (Tunicata)
    Class Ascidiacea
    Class Larvacea
    Class Thaliacea
  Subphylum Cephalochordata
  Subphylum Vertebrata (Craniata)

Selected Bibliography

GENERAL

Multi-volume treatises on Zoology

Grassé, P.-P. (ed.), 1948 onwards. *Traité de Zoologie*. Masson et Cie, Paris.
Harmer, S. F. and A. E. Shipley (eds.), 1895–1909. *The Cambridge Natural History*. Macmillan, London.
Hyman, L. H., 1940 onwards. *The Invertebrates*. McGraw-Hill, New York.
Kükenthal, W. and T. Krumbach (eds.), 1923 onwards. *Handbuch der Zoologie*. de Gruyter, Berlin and Leipzig.
Lankester, E. Ray (ed.), 1900 onwards. *A Treatise on Zoology*. Black, London.

Other general works

There are sometimes reference to elsewhere in a shortened form (e.g. In Rees, 19–50).

Baldwin, E., 1967. *Dynamic aspects of Biochemistry* (5th ed.). Cambridge University Press, London.
Baldwin, E., 1964. *An Introduction to Comparative Biochemistry* (4th ed.). Cambridge University Press, London.
Boolootian, R. A. (ed.), 1966. *Physiology of Echinodermata*. Wiley, New York.
Carter, G. S., 1961. *A General Zoology of the Invertebrata*. Sidgwick and Jackson, London.
Chapman, R. F., 1969. *The Insects*. English Universities Press, London.
Clarke, K. C., 1973. *The Biology of the Arthropoda*. Arnold, London.
Davson, H., 1970. *A Textbook of General Physiology* (4th ed.), 2 vols. Churchill, London.
Dogiel, V. A., 1965. *General Protozoology*. Clarendon Press, Oxford.
Dougherty, E. C. (ed.), 1963. *The Lower Metazoa: Comparative Biology and Phylogeny*. University of California Press, Berkeley and Los Angeles.
Florkin, M. and H. S. Mason (eds.), 1960 onwards. *Comparative Biochemistry*. Academic Press, New York.
Florkin, M. and B. T. Scheer, (eds.), 1967 onwards. *Chemical Zoology*. Academic Press, New York.
Fry, W. G. (ed.), 1970. *The Biology of the Porifera. Symp. zool. Lond.*, **25**.

Hall, R. P., 1953. *Protozoology*. Prentice-Hall, New Jersey.
Harrison, K., 1965. *A Guide-book to Biochemistry* (2nd ed.). Cambridge University Press, London.
Hoar, W. S., 1966. *General and Comparative Physiology*. Prentice-Hall, New Jersey.
Imms, A. D., 1957. *A General Textbook of Entomology* (9th ed., revised O. W. Richards and R. G. Davies). Methuen, London.
Kudo, R. R., 1954. *Protozoology* (4th ed.). Thomas, Springfield, Illinois.
Lehninger, A. L., 1970. *Biochemistry*. Worth, New York.
Lenhoff, H. M. and W. F. Loomis (eds.), 1961. *The Biology of Hydra*. University of Miami Press, Coral Gables, Florida.
MacGinitie, G. E. and N. MacGinitie, 1949. *Natural History of Marine Animals*. McGraw-Hill, New York.
Meglitsch, P. A., 1972. *Invertebrate Zoology* (2nd ed.). Oxford University Press, New York.
Millott, N. (ed.), 1967. *Echinoderm Biology. Symp. zool. Soc. Lond.*, **20**.
Morton, J. E., 1967. *Molluscs* (4th ed.). Hutchinson, London.
Muscatine, L. and H. M. Lenhoff (eds.), 1974. *Coelenterate Biology*. Academic Press, New York.
Nichols, D., 1969. *Echinoderms* (4th ed.). Hutchinson, London.
Nicol, J. A. C., 1960. *The Biology of Marine Animals*. Pitman, London.
Prosser, C. L. and F. A. Brown, 1961. *Comparative Animal Physiology* (2nd ed.). Saunders, Philadelphia and London.
Ramsay, J. A., 1968. *A Physiological Approach to the Lower Animals* (2nd ed.). Cambridge University Press, London.
Rees, W. J. (ed.), 1966. *The Cnidaria and their Evolution. Symp. zool. Soc. Lond.*, **16**.
Rockstein, M., 1964–65. *The Physiology of Insects*, vols. 1–3. Academic Press, New York.
Sleigh, M., 1973. *The Biology of Protozoa*. Arnold, London.
Stephenson, J., 1930. *The Oligochaeta*. Clarendon Press, Oxford.
Waterman, T. H. (ed.), 1960–61. *The Physiology of Crustacea*. vols. 1 and 2. Academic Press, New York.
Wigglesworth, V. B., 1972. *The Principles of Insect Physiology* (7th ed.). Methuen, London.
Wilbur, K. M. and C. M. Yonge (eds.), 1964–66. *Physiology of Mollusca*, vols. 1 and 2. Academic Press, New York.
Wood, D. W., 1974. *Principles of Animal Physiology* (2nd ed.). Arnold, London.

CHAPTER 1

Many citations are relevant to more than one chapter, but they are usually not repeated.

Arber, A., 1954. *The Mind and the Eye* Cambridge University Press, London.
Bernal, J. D., 1954. The origin of life. *New Biol.*, **16**, 28–40.
Bernal, J. D., 1967. *The Origin of Life*. Weidenfeld and Nicolson, London.
Blum, H. F., 1962. *Time's Arrow and Evolution* (2nd ed.). Harper and Row, New York.
Calvin, M., 1962. Evolution of photosynthetic mechanisms. *Perspect. Biol. Med.*, **5**, 147–172.
Calvin, M., 1956. Chemical evolution and origin of life. *Am. Scient.*, **44**, 248–263.
Colinvaux, P., 1973. *Introduction to Ecology*. Wiley, New York.
Crick, F. H. C., 1962. The genetic code. *Scient. Am.*, **207** (4), 66–74.
Echelin, P. and I. Morris, 1965. The relationship between blue-green algae and bacteria. *Biol. Rev.*, **40**, 143–187.

Florkin, M., 1960. *Unity and Diversity in Biochemistry.* Pergamon, Oxford.
Fox. S. W., 1960. How did life begin? *Science, N.Y.*, **132**, 200–208.
Gaffron, H., 1960. The origin of life. In *Evolution after Darwin* (S. Tax, ed.), **1**, 39–84. University of Chicago Press.
Glaessner, M. F., 1962. Pre-Cambrian fossils. *Biol. Rev.*, **37**, 467–494.
Haldane, J. B. S., 1954. The origins of life. *New Biol.*, **16**, 12–27.
Hall, D. O. and K. K. Rao, 1972. *Photosynthesis.* Arnold, London.
Hardy, Sir Alister, 1965. *The Living Stream: Evolution and Man.* Collins, London.
Jevons, F. R., 1964. *The Biochemical Approach to Life.* Allen and Unwin, London.
Keosian, J., 1964. *The Origin of Life.* Chapman and Hall, London.
Kirschner, R. P., 1976. Supernovas in other galaxies. *Scient. Amer.* **235** (6), 89–101.
Lovell, Sir Bernard, 1975. In the centre of immensities. *Advmt. Sci., Lond., New Issue no. 1*, 2–6.
Miller, S. L., 1953. A production of amino acids under possible primitive earth conditions. *Science, N.Y.*, **117**, 528–529.
Needham, A. E., 1959. Origination of life. *Q. Rev. Biol.*, **34**, 189–209.
Needham, A. E., 1965. *The Uniqueness of Biological Materials.* Pergamon, Oxford.
Nirenberg, M. W., 1963. The genetic code, II. *Scient. Am.*, **208** (3), 80–94.
Nursall, J. R., 1962. On the origins of the major groups of animals. *Evolution*, **16**, 118–123.
Oparin, A. I., 1961. *Life: Its Nature, Origin and Development.* Oliver and Boyd, Edinburgh.
Orgel, L. E., 1973. *The Origin of Life: Molecules and Natural Selection.* Chapman and Hall, London.
Owen, D. F., 1974. *What is Ecology?* Oxford University Press, London.
Phillipson, J., 1966. *Ecological Energetics.* Arnold, London.
Poonamperuma, C., R. Mariner, and C. Sagan, 1963. Formation of adenosine by ultra-violet irradiation of a solution of adenine and ribose. *Nature, Lond.*, **198**, 1199–1200.
Popper, K. R., 1965. *The Logic of Scientific Discovery.* Hutchinson, London.
Wald, G., 1963. Phylogeny and ontogeny at the molecular level. In *Evolutionary Biochemistry* (A. I. Oparin, ed.). Pergamon, Oxford.
Whitrow, G. J., 1961. *The Structures and Evolution of the Universe.* Hutchinson, London.

CHAPTER 2

Ambrose, E. J., and D. M. Easty, 1970. *Cell Biology.* Nelson, London.
Anfinsen, C. B., 1959. *The Molecular Basis of Evolution.* Wiley, New York.
Baker, J. R., 1948. The Cell-theory: a restatement, history, and critique. *Q. Jl microsc. Sci.* **89**, 103–125; **90**, 87–108, 331; **93**, 157–190.
Boyden, A., 1957. Are there any 'acellular animals'? *Science, N.Y.*, **125**, 155–156, 990.
Corliss, J. O., 1957. Concerning the 'Cellularity' or Acellularity of the Protozoa. *Science, N.Y.*, **125**, 988.
Davson, H. and J. F. Danielli, 1952. *The Permeability of Natural Membranes* (2nd ed.). Cambridge University Press, London.
Dillon, L. S., 1962. Comparative cytology and the evolution of life. *Evolution*, **16**, 102–117.
Gray, J., 1961. Quoted by Ramsay, J. A. in *The Cell and the Organism* (Ramsay, J. A. and V. B. Wigglesworth, eds.), p. 158. Cambridge University Press, London.
Grimstone, A. V., 1959. Cytology, homology, and phylogeny—a note on 'organic design'. *Am Nat.*, **93**, 273–282.

Grimstone, A. V., 1961. Fine structure and morphogensis in Protozoa. *Biol. Rev.*, **36**, 97–150.
Hutner, S. H. and L. Provasoli, 1957. Concerning the 'Cellularity' or 'Acellularity' of the Protozoa. *Science, N. Y.*, **125**, 989.
Needham, J., 1934. *A History of Embryology.* Cambridge University Press, London.
Pantin, C. F. A., 1951. Organic design. *Advmt. Sci., Lond.*, **8**, 138–150.
Picken, L. E. R., 1960. *The Organization of Cells and other Organisms.* Clarendon Press, Oxford.
Russell, E. S., 1916. *Form and Function.* John Murray, London.
Waterman, T. H., 1961. Comparative physiology. In *The Physiology of Crustacea* (T. H. Waterman, cd.), vol. 2, 521–593.
Woodger, J. H., 1929. *Biological Principles, a Critical Study.* Kegan Paul, London.

CHAPTER 3

Allen, R. D., 1962. Amoeboid movement. In *The Cell* (J. Brachet and A. E. Mirsky, eds.), vol. 2, 135–216. Academic Press, New York.
Allen, R. D., 1962. Amoeboid movement. *Scient. Am.*, **206** (2), 112–122.
Bovee, E. C., 1964. Morphological differences among pseudopodia of various small Amoebae and their functional significance. In *Primitive Motile Systems in Cell Biology* (R. D. Allen and N. Kamiya, eds.), 189–219. Academic Press, New York.
Bradfield, J. R. G., 1955. Fibre patterns in animal flagella and cilia. *Symp. Soc. exp. Biol.*, **9**, 306–334.
Cohen, C., 1975. The protein switch of muscle contraction. *Scient. Am.*, **233** (5), 36–45.
de Bruyn, P. P. H., 1947. Theories of amoeboid movement. *Q. Rev. Biol.*, **22**, 1–24.
Fawcett, D., 1961. Cilia and flagella. In *The Cell* (J. Brachet and A. E. Mirsky, eds.), vol. 2, 212–297. Academic Press, New York.
Gray, J., 1928. *Ciliary Movement.* Cambridge University Press, London.
Hanson, J. and J. Lowy, 1960. Structure and function of the contractile apparatus in the muscles of invertebrates animals. In *The Structure and Function of Muscle* (G. H. Bourne, ed.). Academic Press, New York.
Huxley, H. E., 1958. The contraction of muscle. *Scient. Am.*, **199** (5), 66–86.
Jahn, T. L. and E. C. Bovee, 1965. Protoplasmic movements and locomotion of Protozoa. In *Biochemistry and Physiology of Protozoa* (S. H. Hutner, ed.), vol. 3, 62–129. Academic Press, New York.
Knowles, D., 1962. *The Evolution of Mediaeval Thought.* Longmans, London.
Lowndes, A. G., 1943. The swimming of unicellular flagellate organisms. *Proc. zool. Soc. Lond. A*, **113**, 99–107.
Mast, S. O., 1926. Structure, movement, locomotion, and stimulation in Amoeba. *J. Morph.*, **41**, 347–425.
Mercer, E. H., 1959. An electron microscope study of *Amoeba proteus*. *Proc. R. Soc. Lond. B*, **150**, 216–232.
Noland, L. E., 1957. Protoplasmic streaming: a perennial puzzle. *J. Protozool.* **4**, 1–6.
Pantin, C. F. A., 1956. Comparative physiology of muscle. *Br. med. Bull.*, **12**, 199–202.
Satir, P., 1974. How cilia move. *Scient. Am.*, **231** (4), 44–52.
Sleigh, M. A., 1962. *The Biology of Cilia and Flagella.* Pergamon, Oxford.
Sleigh, M. A., 1964. Flagellar movements of the sessile flagellates *Actinomonas, Codonosiga, Monas*, and *Poteriodendron*. *Q. Jl. microsc. Sci.*, **105**, 405–414.
Wichterman, R., 1953. *The Biology of Paramecium.* McGraw-Hill, New York.

Wilkie, D. R., 1968. *Muscle*. Arnold, London.
Willmer, E. N., 1970. *Cytology and Evolution* (2nd ed.). Academic Press, New York.

CHAPTER 4

Batham, E. J. and C. F. A. Pantin, 1950. Muscular and hydrostatic action in the sea-anemone *Metridium senile* (L.). *J. exp. Biol.*, **27**, 264–288.
Batham, E. J. and C. F. A. Pantin, 1951. The organization of the muscular system of *Metridium senile*. *Q. Jl microsc. Sci.*, **92**, 27–54.
Carlisle, D. B., 1961. Locomotory powers of adult ascidians. *Proc. zool. Soc. Lond.*, **136**, 141–146.
Chapman, D. M., C. F. A. Pantin, and E. A. Robson, 1962. Muscle in coelenterates. *Revue can. Biol.*, **21**, 267–278.
Chapman, G., 1953. Studies of the mesogloea of coelenterates. 1. Histology and chemical properties. *Q. Jl. microsc. Sci*, **94**, 155–176.
Chapman, G., 1958. The hydrostatic skeleton in the invertebrates. *Biol. Rev.*, **33**, 338–371.
Chapman, G., 1975. Versatility of hydraulic systems. *J. exp. Zool.*, **194**, 249–269.
Clark, R. B., 1964. *Dynamics in Metazoan Evolution*. Clarendon Press, Oxford.
Gibson, R., 1972. *Nemerteans*. Hutchinson, London.
Harris, J. E. and H. D. Crofton, 1957. Structure and function in the nematodes: internal pressure and cuticular structure in *Ascaris*. *J. exp. Biol.*, **34**, 116–130.
Lee, D. L., 1965. *The Physiology of Nematodes*. Oliver and Boyd, Edinburgh.
Pantin, C. F. A., 1960. Diploblastic animals. *Proc. Linn. Soc. Lond.*, **171**, 1–14.
Robson, E. A., 1957. The structure and hydromechanics of the musculo-epithelium in *Metridium*. *Q. Jl. microsc. Sci.* **98**, 265–278.
Trueman, E. R., 1975. *The Locomotion of Soft-bodied Animals*. Arnold, London.

CHAPTER 5

Barrington, E. J. W., 1965. *The Biology of Hemichordata and Protochordata*. Oliver and Boyd, Edinburgh.
Currey, J. D., 1976. Further studies on the mechanical properties of mollusc shell material. *J. Zool., Lond.* **180**, 445–453.
Goodrich, E. S., 1945. The study of nephridia and genital ducts since 1895. *Q. Jl. microsc. Sci.*, **86**, 113–392.
Harmer, S. F., 1930. Polyzoa. *Proc. Linn. Soc. London, 141st session*, 68–118.
Lemche, H. and K. G. Wingstrand, 1959. The anatomy of *Neopilina galatheae* Lemche, 1957. *Galathea Rep., vol. 3*. Danish Science Press Ltd., Copenhagen.
Lissman, H. W., 1946. The mechanism of locomotion in gastropod molluscs, II. Kinetics. *J. exp. Biol.*, **22**, 37–50.
Packard, A., 1972. Cephalopods and fish: the limits of convergence. *Biol. Rev.*, **47**, 241–307.
Ryland, J. S., 1967. Bryozoa. *Ocean. Mar. Biol. Ann. Rev.* **5**, 343–366.
Smith, J. E., 1947. The mechanics and innervation of the starfish tube foot-ampulla system. *Phil. Trans. R. Soc. B*, **232**, 279–310.

CHAPTER 6

Chapman, G., 1950. Of the movement of worms. *J. exp. Biol.*, **27**, 29–39.
Edwards, C. R. and J. R. Roffey, 1972. *Biology of Earthworms*. Chapman and Hall, London.
Gray, J., 1939. Studies in animal locomotion, VIII. The kinetics of locomotion of *Nereis diversicolor*. *J. exp. Biol.*, **16**, 9–17.

Gray, J., 1968. *Animal Locomotion*. Prentice-Hall, New Jersey.
Gray, J. and H. W. Lissman, 1938. Studies in animal locomotion, VII. Locomotory reflexes in the earthworm. *J. exp. Biol.*, **15**, 506–517.
Mann, K.H., 1962. *Leeches (Hirudinea)*. Pergamon, Oxford.
Smith, J.E., 1957. The nervous anatomy of the body segments of nereid polychaetes. *Phil. Trans. R. Soc. B*, **240**, 135–196.
Trueman, E.R., and A.D., Ansell, 1969. The mechanism of burrowing into soft-substrates by marine animals. *Oceanogr. Mar. Biol. Ann. Rev.*, **7**, 315–366.
Wells, G. P., 1950. Spontaneous activity cycles in polychaete worms. *Symp. Soc. exp. Biol.*, **4**, 127–142.
Wells, G.P., 1954. The mechanism of proboscis movement in *Arenicola*. *Q. Jl. microsc. Sci.*, **95**, 251–270.
Ubaghs, G., 1971. Diversité et spécialisation des plus anciens Echinodermes que l'on connaisse. *Biol. Rev.*, **46**, 157–200.

CHAPTER 7

Alexander, R. McN., 1968. *Animal Mechanics*. Sidgwick and Jackson, London.
Alexander, R. McN., 1971. *Size and Shape*. Arnold, London.
Boettiger, E. G., 1957. The machinery of insect flight. In *Recent Advances in Invertebrate Physiology* (Scheer, T., ed.), 117–142. University of Oregon, Eugene, Oregon.
Boettiger, E. G., and E. Furshpan, 1952. The mechanics of flight movements in Diptera. *Biol. Bull. mar. biol. Lab., Woods Hole*, **102**, 200–211.
Chadwick, L. E., 1953. The motion of the wings. In *Insect Physiology* (K. D. Roeder, ed.), 577–614. Wiley, New York.
Chadwick, L. E., 1953. Aerodynamics and flight metabolism. *ibid.*, 615–636.
Chadwick, L. E., 1953. The flight muscles and their control. *ibid.*, 637–655.
Cloudsley-Thompson, J. L., 1958. *Spiders, Scorpions, Centipedes, and Mites*. Pergamon, Oxford.
Currey, J. D., 1967. The failure of exoskeletons and endoskeletons. *J. Morph.*, **123**, 1–16.
Currey, J. D., 1970. *Animal Skeletons*. Arnold, London.
Dennell, R., 1960. Integument and Exoskeleton. In *The Physiology of Crustacea* (T. H. Waterman, ed.), vol. 1, 449–472.
Evans, H. E., 1959. Some comments on the evolution of the Arthropoda. *Evolution*, **13**, 147–149.
Gilmour, D., 1961. *The Biochemistry of Insects*. Academic Press, New York.
Manton, S. M., 1950. The evolution of arthropodan locomotor appendages. Part 1: The locomotion of *Peripatus*. *J. Linn. Soc. (Zool.)*, **41**, 529–570.
Manton, S. M., 1952. Part 2: General introduction of the locomotory mechanisms of the Arthropoda. *ibid.*, **42**, 93–117.
Manton, S. M., 1952. Part 3: The locomotion of the Chilopoda and Pauropoda. *ibid.*, **42**, 118–167.
Manton, S. M., 1954. Part 4: The structure, habits and evolution of the Diplopoda. *ibid.*, **42**, 299–368.
Manton, S. M., 1956. Part 5: The structure, habits, and evolution of the Pselaphognatha (Diplopoda). *ibid.*, **43**, 153–187.
Manton, S. M., 1958a. Part 6: Habits and evolution of the Lysiopetaloidea (Diplopoda), some principles of the leg design in Diplopoda and Chilopoda, and limb structure in Diplopoda. *ibid.*, **43**, 487–556.
Manton, S. M., 1958b. Habits of life and evolution of body design in Arthropoda. *ibid.*, **44**, 58–72.
Manton, S. M., 1960. Concerning head development in the arthropods. *Biol. Rev.*, **35**, 265–282.

Manton, S. M., 1961. Part 7: Functional requirements and body design in Colobognatha (Displopoda), together with a comparative account of diplopod burrowing techniques, trunk musculature, and segmentation. *J. Linn. Soc. (Zool.)*, **44**, 383-461.

Manton, S. M., 1964. Mandibular mechanisms and the evolution of arthropods. *Phil. Trans. R. Soc. B*, **247**, 1-183.

Manton, S. M., 1973. Arthropod phylogeny—a modern synthesis. *J. Zool. Lond.*, **171**, 111-130.

Manton, S. M., 1977. *The Arthropoda: Habits, Functional Morphology, and Evolution*. Clarendon Press, Oxford.

Pringle, J. W. S., 1948. The gryoscopic mechanism of the halteres of Diptera. *Phil. Trans. R. Soc. B*, **233**, 347-384.

Pringle, J. W. S., 1957. *Insect Flight*. University Press, Cambridge.

Pryor, M. G. M., 1962. Sclerotization. In *Comparative Biochemistry* (M. Florkin and B. T. Scheer, eds.), Vol. IVB, 371-396.

Richards, A. G., 1951. *The Integument of Arthropods*. University of Minnesota Press, Minneapolis.

Sharov, A. G., 1966, *Basic Arthropodan Stock*. Pergamon, Oxford.

Snodgrass, R. E., 1952. *A Textbook of Arthropod Anatomy*. Cornell University Press, Ithaca, New York.

Tiegs, O. W. and S. M. Manton, 1958. The evolution of the Arthropoda. *Biol. Rev.*, **33**, 255-337.

Weis-Fogh, T., 1972. Energetics of hovering flight in humming birds and in *Drosophila*. *J. exp. Biol.*, **56**, 79-104.

CHAPTER 8

Barrington, E. J. W., 1962. Digestive enzymes. In *Adv. Comp. Physiol. Biochem.* (Lowenstein, O., ed.), **1**, 1-65.

Chen, Y. T., 1950. Investigations of the biology of *Peranema trichophorum* (Euglenineae). *Q. Jl microsc. Sci.*, **91**, 279-308.

Corliss, J. O., 1959. Comments on the phylogeny and systematics of the Protozoa. *Syst. Zool.*, **8** 169-190.

de Duve, C., 1963. The lysosome. *Scient. Am.*, **208** (5), 64-72.

Holter, H., 1959. Pinocytosis. *Int. Rev. Cytol.*, **8**, 481-504.

Jennings, J. B., 1972. *Feeding, Digestion and Assimilation in Animals* (2nd ed.). MacMillian, London.

MacKinnon, D. M. and R. G. J. Hawes, 1961. *An Introduction to the Study of Protozoa*. Clarendon Press, Oxford.

Mast, S. O., 1942. The hydrogen ion concentrations of the contents of the food vacuoles and the cytoplasm in *Amoeba* and other phenomena concerning the food vacuoles. *Biol. Bull. mar. biol. Lab., Woods Hole*, **83**, 173-204.

Mast, S. O., 1947. The food vacuole in *Paramecium. ibid.*, **92**, 31-72.

Mercer, E. H., 1959. An electron microscope study of *Amoeba proteus. Proc. R. Soc. B*, **150**, 216-232.

CHAPTER 9

Bidder, G. P., 1923. The relation of the form of a sponge to its currents. *Q. Jl microsc. Sci.*, **67**, 293-323.

Hardy, Sir Alister, 1958. *The Open Sea: The World of Plankton*. Collins, London.

Jennings, J. B., 1957. Studies on feeding, digestion, and food storage in free-living flatworms. *Biol. Bull. mar. biol. Lab., Woods Hole*, **112**, 63-80.

Jennings, J.B., 1960. Observations on the nutrition of the Rhynchocoelan *Lineus ruber. ibid.*, **119**, 189-196.

Jennings, J. B., 1962. Further studies on feeding and digestion in triclad Turbellaria. *ibid.*, **123**, 571-581.

Pantin, C. F. A., 1942. The excitation of nematocysts. *J. exp. Biol.*, **19**, 294–310.
Picken, L. E. R. and R. J. Skaer, 1966. A review of researches on nematocysts. In Rees, 19–50.
Southward, A. J., 1955. Observations on the ciliary currents of the jelly-fish *Aurelia aurita* L. *J. mar. biol. Ass. U.K.*, **34**, 201—216.
Southward, E. C., 1975. Fine structure and phylogeny of the Pogonophora. *Symp. zool. Soc. Lond.*, **36**, 235–251.
van Weel, P. B., 1949. On the physiology of the tropical fresh-water sponge *Spongilla proliferens* Annand. 1: Ingestion, digestion, and excretion. *Physiologia comp. Oecol.*, **1**, 110–126.
Yonge, C. M., 1937. Evolution and adaptation in the digestive system of the Metazoa. *Biol. Rev.*, **12**, 87–115.

CHAPTER 10

Barrington, E. J. W., 1965. *The Biology of Hemichordata and Protochordata*. Oliver and Boyd, Edinburgh.
Bidder, A., 1950. Digestive mechanisms of European squids. *Q. Jl microsc. Sci.*, **91**, 1–43.
Cannon, H. G., 1933. On the feeding mechanism of the Branchiopoda. *Phil. Trans. R. Soc. B*, **222**, 267–352.
Dales, R. P., 1955. Feeding and digestion in terebellid polychaetes. *J. mar. biol. Ass. U.K.*, **34**, 55–79.
Graham, A., 1949. The molluscan stomach. *Trans. R. Soc. Edinb.*, **61**, 737–778.
Jørgensen, C. B., 1955. Quantitative aspects of filter feeding in invertebrates. *Biol. Rev.*, **30**, 391–454.
Jørgensen, C. B., 1966. *Biology of Suspension Feeding*. Pergamon, Oxford.
MacGinitie, C. E., 1939. The method of feeding of *Chaetopterus*. *Biol. Bull. mar. biol. Lab., Woods Hole*, **77**, 115–118.
Marshall, S. M., 1973. Respiration and feeding in copepods. *Adv. mar. Biol.*, **11**, 57–120.
Marshall, S. M. and A. P. Orr, 1955. *The Biology of a Marine Copepod*. Oliver and Boyd, Edinburgh.
Morton, J. E., 1960. The functions of the gut in ciliary feeders. *Biol. Rev.*, **35**, 92–140.
Nicol, E. A. T., 1930. The feeding mechanism, formation of the tube, and physiology of digestion in *Sabella pavonina*. *Trans. R. Soc. Edinb.*, **56**, 537–598.
Owen, G., 1974. Feeding and digestion in the Bivalvia. *Adv. Comp. Physiol. Biochem.*, **5**, 1–35.
Sutton, M. F., 1957. The feeding mechanism, functional morphology, and histology of the alimentary canal of *Terebella lapidaria*. *Proc. zool. Soc. Lond.*, **129**, 487–523.
Thomas, J. G., 1940. *Pomatoceros, Sabella, and Amphitrite. L. M.B.C. Mem. typ. Br. mar. Pl. Anim.*, **33**.
Yonge, C. M., 1926. Structure and physiology of the organs of feeding and digestion in *Ostrea edulis*. *J. mar. biol. Ass. U.K.*, **14**, 295–386.
Yonge, C. M., 1928. Feeding mechanisms in the invertebrates. *Biol. Rev.*, **3**, 21–76.
Yonge, C. M., 1932. The crystalline style of the Mollusca. *Sci. Prog., Lond.*, **26**, 643–653.
Yonge, C. M., 1937. Evolution and adaptation in the digestive system of the Metazoa. *Biol. Rev.*, **12**, 87–115.
Yonge, C. M., 1939. The protobranchiate Mollusca: a functional interpretation of their structure and evolution. *Phil. Trans. R. Soc. B*, **230**, 79–147.

CHAPTER 11

Barrington, E. J. W., 1968. *The Chemical Basis of the Physiological Regulation.* Scott Foresman, Glenview, Illinois.
Fox, H. M., 1949. Blood pigments. *Endeavour*, **8**, 43–47.
Fox, H. M. and G. Vevers, 1960. *The Nature of Animal Colours.* Sidgwick and Jackson, London.
Gratzer, W. B. and A. C. Allison, 1960. Multiple haemoglobins. *Biol. Rev.*, **35**, 459–506.
Krogh, A., 1941. *Comparative Physiology of Respiratory Mechanisms.* Pennsylvania University Press, Philadelphia.
Manwell, C., 1960. Comparative physiology: blood pigments. *A. Rev. Physiol.*, **22**, 191–244.
Mill, P. J., 1972. *Respiration in the Invertebrates.* Macmillan, London.
Miller, P. L., 1960. Respiration in the desert locusts. 1. The control of ventilation. *J. exp. Biol.*, **37**, 224–236.
Morton, J. E. and C. M. Yonge, 1964. Classification and structure of the Mollusca. In Wilbur and Yonge, **1**, 1–58.
Redfield, A. C., 1934. The haemocyanins. *Biol. Rev.*, **9**, 175–212.
Steen, J. B., 1974. The Comparative Physiology of Respiratory Mechanisms. Academic Press, New York.
Wells, G. P., 1950. Spontaneous activity cycles in polychaete worms. *Symp. Soc. exp. Biol.*, **4**, 127–142.
Wigglesworth, V. B., 1930. A theory of tracheal respiration in insects. *Proc. R. Soc, B*, **106**, 229–250.
Wigglesworth, V. B., 1935. The regulation of respiration in the flea, *Xenopsylla cheopsis Roths.* (Pulicidae). *ibid.*, **118**, 397–419.
Yonge, C. M., 1947. The pallial organs in the aspidobranch Gastropoda and their evolution throughout the Mollusca. *Phil. Trans. R. Soc. B*, **232**, 443–518.

CHAPTER 12

Bahl, K. N., 1947. Excretion in the Oligochaeta. *Biol. Rev.*, **22**, 109–147.
Bairati, A. and F. E. Lehmann, 1956. Structural and chemical properties of the contractile vacuole of *Amoeba proteus. Protoplasma*, **45**, 525–539.
Cohen, P. P. and G. W. Brown, 1960. Ammonia metabolism and urea biosynthesis. In Florkin and Mason, **2**, 161–244.
Delaunay, H., 1931, L'excrétion azotée des invertébrées. *Biol. Rev.*, **6**, 265–301.
Edney, E. B., 1954. Woodlice and the land habitat. *ibid.*, **29**, 185–219.
Edney, E. B., 1957. *The Water Relations of Terrestrial Arthropods.* Cambridge University Press, London.
Edney, E. B., 1960. Terrestrial adaptation. In *The Physiology of Crustacea.* (T. H. Waterman, ed.), vol. 1, 367–393.
Goodrich, E. S., 1945. The study of nephridia and genital ducts since 1895. *Q. Jl microsc. Sci.*, **86**, 113–393.
Jepps, M., 1947. Contribution to the study of the sponges. *Proc. R. Soc. B*, **134**, 408–417.
Kitching, J. A., 1938. Contractile vacuoles. *Biol. Rev.*, **13**, 403–444.
Kitching, J. A., 1952. Contractile vacuoles. *Symp. Soc. exp. Biol.*, **6**, 145–165.
Laverack, M. S., 1963. *The Physiology of Earthworms.* Pergamon, Oxford.
Manton, S. M., 1937. Studies on the Onychophora. II: The feeding, digestion, excretion, and food storage of *Peripatopsis. Phil. Trans. R. Soc. B*, **227**, 411–464.
Martin, A. W., 1957. Recent advances in knowledge of invertebrate renal function. In Scheer, 247–276.
Martin, A. W., 1958. Comparative physiology (Excretion). *A. Rev. Physiol.*, **20**, 225–242.

Moffat, D. B., 1971. *The control of water balance by the kidney.* Oxford Univ. Press, London.
Needham, J., 1938. Contributions of chemical physiology to the problem of reversibility in evolution. *Biol. Rev.,* **13**, 225–251.
Pantin, C. F. A., 1947. The nephridia of *Geonemertes dendyi. Q. Jl microsc. Sci.,* **88**, 15–25.
Parry, G., 1960. Excretion. In *The Physiology of Crustacea.,* (T. H. Waterman, ed.), vol. 1, 341–366.
Pitelka, D., 1963. *Electron-microscopic Structure of Protozoa.* Pergamon, Oxford.
Potts, W. T. W., 1967. Excretion in molluscs. *Biol. Rev.,* **42**, 1–41.
Ramsay, J. A., 1961. The comparative physiology of renal function in invertebrates. In *The Cell and the Organism* (Ramsay, J. A. and V. B. Wigglesworth, eds.), 158–174. Cambridge University Press, London.
Wigglesworth, V. B., 1931. The physiology of excretion in a blood-sucking insect *Rhodnius prolixus* (Hemipter, Reduviidae). *J. exp. Biol.,* **8**, 411–451.
Sutton, S. L., 1972. *Woodlice,* Ginn & Co., London.

CHAPTER 13

A number of the references given in Chapter 12 are also relevant to this chapter.

Beadle, L. C., 1957. Osmotic and ionic regulation in aquatic animals. *A. Rev. Physiol.,* **19**, 329–358.
Kitching, J. A., 1954. Osmoregulation and ionic regulation in animals without kidneys. *Symp. Soc. exp. Biol.,* **8**, 63–75.
Krogh, A., 1939. *Osmotic Regulation in Aquatic Animals.* Cambridge University Press, London.
Lockwood, A. P. M., 1962. The osmoregulation of Crustacea. *Biol. Rev.,* **37**, 257–306.
Lockwood, A. P. M., 1968. *Aspects of the Physiology of Crustacea.* Oliver and Boyd, Edinburgh.
Potts, W. T. W. and G. Parry, 1964. *Osmotic and Ionic Regulation in Animals.* Pergamon, Oxford.
Ramsay, J. A., 1949. The osmotic relations of the earthworm. *J. exp. Biol.,* **26**, 46–56, 65–75.
Ramsay, J. A., 1954. Movements of water and electrolytes in invertebrates. *Symp. Soc. exp. Biol.,* **8**, 1–15.
Robertson, J. D., 1957. Osmotic and ionic regulation in aquatic invertebrates. In Scheer, 229–246.
Robertson, J. D., 1960. Studies of the chemical composition of muscle tissue. *J. exp. Biol.,* **37**, 879–888.
Robertson, J. D., 1960. Osmotic and ionic regulation. In *The Physiology of Crustacea* (T. H. Waterman, ed.), vol. 1, 317–339.
Wigglesworth, V. B., 1933. The effect of salts on the anal gills of the mosquito larva. *J. exp. Biol.,* **10**, 1–15.
Wigglesworth, V. B., 1933. The function of the anal gills of the mosquito larva. *ibid.,* 16–26.
Wigglesworth, V. B., 1933. The adaptation of mosquito larvae to salt water. *ibid.,* 27–37.
Wigglesworth, V. B., 1938. The regulation of osmotic pressure and chloride concentration in the haemolymph of mosquito larvae. *ibid.,* **15**, 235–247.

CHAPTER 14

Amoore, J. E., J. R. Johnstone, and M. Rubin, 1964. A stereochemical theory of odor. *Scient. Am.,* **210** (2), 42–49.
Bishop, G. H., 1956. The natural history of the nerve impulse. *Physiol. Rev.,* **36**, 376–399.

Bodian, D., 1962. The generalized vertebrate neurone. *Science, N.Y.*, **137**, 323–326.
Clarkson, E. N. K. and R. Levi-Setti, 1975. Trilobite eyes and the optics of DesCartes and Huygens. *Nature, Lond.*, **254**, 663–667.
Cohen, M. J. and S. Dijkgraaf, 1961. Mechanoreception. In *The Physiology of Crustacea* (T. H. Waterman, ed.), vol. 2, 65–108.
Davies, J. T., 1962. The mechanism of olfaction. *Symp. Soc. exp. Biol.*, **16**, 170–179.
Dethier, V. G., 1953. Chemoreception. In Roeder, 544–576.
Dethier, V. G., 1955. The physiology and the histology of the contact chemoreceptors of the blowfly. *Q. Rev. Biol.*, **30**, 348–371.
Dethier, V. G., 1962. Chemoreceptor mechanisms in insects. *Symp. Soc. exp. Biol.*, **16**, 180–196.
Dethier, V. G., 1963. *The Physiology of Insect Senses*. Methuen. London.
Hartline, H. K., H. G. Wagner, and E. F. MacNichol, 1952. The peripheral origin of nervous activity in the visual system. *Cold Spring Harb. Symp. quant. Biol.*, **17**, 125–141.
Lowenstein, O., 1962. Frontiers of knowledge in the study of sensory function. *Advmt. Sci., Lond.*, **19**, 222–235.
Waterman, T. H., 1961. Light, sensitivity, and vision. In *The Physiology of Crustacea* (T. H. Waterman, ed.), vol. 2, 1–64.
Wells, M. J., 1961. What the octopus makes of it; our world from another point of view. *Advmt. Sci., Lond.*, **17**, 461–471.
Wigglesworth, V. B., 1961. *The Principles of Insect Physiology* (5th ed.). Methuen, London.
Wolken, J. J., 1971. *Invertebrate Photoreceptors: A Comparative Analysis*. Academic Press, New York and London.
Wulff, V. J., 1956. Physiology of the compound eye. *Physiol. Rev.*, **36**, 145–163.
Young, J. Z., 1960. The statocysts of *Octopus vulgaris*. *Proc. R. Soc. B*, **152**, 3–29.
Young, J. Z., 1961. Learning and discrimination in the octopus. *Biol. Rev.*, **36**, 32–96.

CHAPTER 15

Carthy, J. D., 1958. *An Introduction to the Behaviour of Invertebrates*. Allen and Unwin, London.
Evans, F. G. C., 1951. An analysis of the behaviour of *Lepidochitona cinereus* in response to certain physical features of the environment. *J. Anim. Ecol.*, **20**, 1–10.
Horridge, G. A., 1954. Observations on the nerve fibres of *Aurellia aurita*. *Q. Jl microsc. Sci.*, **95**, 85–92.
Horridge, G. A., 1956. The nervous system of the ephyra larva of *Aurellia aurita*. *ibid.*, **97**, 59–74.
Jones, W. Clifford, 1962. Is there a nervous system in sponges? *Biol. Rev.*, **37**, 1–50.
Knight-Jones, E. W., 1952. On the nervous system of *Saccoglossus cambrensis* (Enteropneusta). *Phil. Trans. R. Soc. B*, **236**, 315–354.
Pantin, C. F. A., 1950. Behaviour patterns in lower invertebrates. *Symp. Soc. exp. Biol.*, **4**, 175–195.
Pantin, C. F. A., 1952. The elementary nervous system. *Proc. R. Soc. B*, **140**, 147–168.
Parker, G. H., 1919. *The Elementary Nervous System*. Lippincott, Philadelphia.
Ross, D. M., 1960. The association between the hermit crab *Eupagurus bernhardus* (L.) and the sea anemone *Calliactis parasitica* (Couch). *Proc. zool. Soc. Lond.*, **134**, 43–57.

Smith, J. E., 1937. On the nervous system of the starfish *Marthasterias glacialis* (L.). *Phil. Trans. R. Soc. B*, **227**, 111–173.

Smith, J. E., 1945. The role of the nervous system in some activities of starfishes. *Biol. Rev.*, **20**, 29–43.

Smith, J. E., 1946. The mechanics and innervation of the starfish tube foot-ampulla system. *Phil. Trans. R. Soc. B*, **232**, 279–310.

Smith, J. E., 1950. Some observations on the nervous mechanisms underlying the behaviour of starfishes. *Symp. Soc. exp. Biol.*, **4**, 196–220.

Usherwood, P. N. R., 1973. *Nervous Systems*. Arnold, London.

CHAPTER 16

Barnes, G. E., 1955. The behaviour of *Anodonta cygnea* L., and its neurophysiological basis. *J. exp. Biol.*, **32**, 158–174.

Boycott, B. B. and J. Z. Young, 1950. The comparative study of learning. *Symp. Soc. exp. Biol.*, **4**, 432–453.

Bullock, T. H. and G. H. Horridge, 1965. *Structure and Function in the Nervous System of Invertebrates*, 2 vols. Freeman, San Francisco.

Bünning, E., 1973. *The Physiological Clock*. Springer-Verlag, New York.

Dethier, V. G., 1964. Microscopic brains. *Science, N.Y.*, **143**, 1138–1145.

Dorsett, D. A., 1964. The sensory and motor innervation of *Nereis. Proc. R. Soc. B*, **159**, 652–667.

Gray, J. and H. W. Lissman, 1938. Studies in animal locomotion. VII: Locomotory reflexes in the earthworm. *J. exp. Biol.*, **15**, 506–517.

Jennings, H. S., 1923. *The Behaviour of Lower Organisms*. Columbia University Press, New York.

Manning, A., 1977. *An Introduction to Animal Behaviour* (3rd ed.). Edward Arnold, London.

Mittelstaedt, H., 1957. Prey capture in mantids. In Scheer, 51–71.

Nicol, J. A. C., 1948. The giant axons of annelids. *Q. Rev. Biol.*, **23**, 291–323.

Smith, J. E., 1957. The nervous anatomy of the body segments of nereid polychaetes. *Phil. Trans. R. Soc. B*, **240**, 135–196.

Thorpe, W. H., 1963. *Learning and Instinct in Animals* (2nd ed.). Methuen, London.

Tinbergen, N., 1951. *The Study of Instinct*. Clarendon Press, Oxford.

Wells, M. J., 1962. *Brain and Behaviour in Cephalopods*. Heinemann, London.

Wells, M. J., 1965. Learning by marine invertebrates. *Adv. mar. Biol.*, **3**, 1–62.

Wiersma, C. A. G., 1961. Reflexes and the central nervous system. In Waterman, **2**, 241–279.

Young, J. Z., 1961. Learning and discrimination in the octopus. *Biol. Rev.*, **36**, 32–96.

Young, J. Z., 1964. *A Model of the Brain*. Clarendon Press, Oxford.

CHAPTER 17

Barrington, E. J. W., 1975. *An Introduction to General and Comparative Endocrinology* (2nd ed.). Clarendon Press, Oxford.

Birch, M. C. (ed.), 1974. *Pheromones*. North-Holland, Amsterdam.

Butler, C. G., 1974. *The World of the Honeybee (rev. ed.)*. Collins, London.

Carlisle, D. B. and F. G. W. Knowles, 1959. *Endocrine Control in Crustacea*. Cambridge University Press, London.

Clark, R. B., 1961. The origin and formation of the heteronereis. *Biol. Rev.*, **36**, 199–236.

Clark, R. B. and P. J. W. Olive, 1973. Recent advances in polychaete endocrinology and reproductive biology. *Ocean. Mar. Biol. Ann. Rev.*, **11**, 176–222.

Finlayson, L. H. and M. P. Osborne, 1975. Secretory activity of neurons and related chemical activity. *Adv. comp. Physiol. Biochem.*, **6**, 165–258.
Galtsoff, P. S., 1961. Physiology of reproduction in molluscs. *Am. Zool.* **1**, 273–289.
Gilbert, L. I., 1963. Hormones controlling reproduction and moulting in invertebrates. In *Comparative Endocrinology* (U.S. von Euler and H. Heller, eds.), vol. 2, 1–46. Academic Press, New York.
Hauenschild, C., 1966. Der hormonale Einfluss des Gehirns auf die sexuelle Entwicklung bei dem Polychaeten *Platynereis dumerilii. Gen. Comp. Endocr.*, **6**, 26–73.
Jacobson, M., 1972. *Insect Sex Pheromones.* Academic Press, New York.
Jacobson, M. and M. Beroza, 1964. Insect attractants. *Scient. Am.*, **211** (2), 20–27.
Kleinholz, L. H., 1961. Pigmentary effectors. In *The Physiology of Crustacea* (T. H. Waterman, ed.), vol. 2, 133–169. Academic Press, New York.
Lucas, C. E., 1949. External metabolites and ecological adaptation. *Symp. Soc. exp. Biol.*, **3**, 336–356.
Wells, M. J., 1960. Optic glands and the ovary of *Octopus. Symp. zool. Soc. Lond.*, **2**, 87–107.
Wells, M. J. and J. Wells, 1959. Hormonal control of sexual maturity in *Octopus. J. exp. Biol.*, **36**, 1–33.
Welsh, J. H., 1961. Neurohumors and neurosecretion. In *The Physiology of Crustacea*, vol. 2, 281–311. *op. cit.*
Whittaker, R. H. and P. P. Feeny, 1971. Allelochemics: chemical interactions between species. *Science*, **171**, 757–770.
Wigglesworth, V. B., 1936. The function of the corpus allatum in the growth and reproduction of *Rhodnius prolixus* (Hemiptera). *Q. Jl. microsc. Sci.*, **79**, 91–121.
Wigglesworth, V. B., 1954. *The Physiology of Insect Metamorphosis.* Cambridge University Press, London.
Wigglesworth, V. B., 1964. The hormonal regulation of growth and reproduction in insects. *Adv. Insect Physiol.*, **2**, 247–336.
Wigglesworth, V. B., 1970. *Insect Hormones.* Oliver and Boyd, Edinburgh.
Williams, C. M., 1961. Insect Metamorphosis: an approach to the study of growth. In *Growth in Living Systems* (Zarrow, M. X., ed.), 313–320. Basic Books, New York.
Williams, C. M., 1952. The physiology of insect diapause. *Biol. Bull. mar. biol. Lab., Woods Hole*, **103**, 120–138.

CHAPTER 18

Anderson, D. T., 1973, *Embryology and Phylogeny of Annelids and Arthropods.* Pergamon Press, Oxford.
Beale, G. H., 1954. *The Genetics of Paramecium aurelia.* Cambridge University Press, London.
Berrill, N. J., 1928. Regeneration in the polychaete *Chaetopterus variopedatus. J. mar. biol. Ass. U.K.*, **15**, 151–158.
Carter, G. S., 1954. On Hadži's interpretations of phylogeny. *Syst. Zool.*, **4**, 163–167, 173.
Dalcq, A. M., 1938. *Form and Causality in Early Development.* Cambridge University Press, London.
Hadži, J., 1963. *The Evolution of the Metazoa.* Pergamon, Oxford.
Hand, Cadet, 1959. On the origin and phylogeny of the Coelenterates. *Syst. Zool.*, **8**, 191–202.
Hanson, E. D. 1958. On the origin of the Eumetazoa. *ibid.*, **7**, 16–47.
Hardy, A. C., 1953. On the origin of the Metazoa. *Q. Jl microsc. Sci.*, **94**, 441–443.

Harvey, L. A., 1961. Speculations on ancestry and evolution. *Sci. Prog., Lond.*, **49**, 11–121.
Jägersten, G., 1955. On the early phylogeny of the Metazoa. The bilaterogastrea theory. *Zool. Bidr. Upps.*, **30**, 321–354.
Jägersten, G., 1959. Further remarks on the early phylogeny of the Metazoa. *ibid.*, **33**, 79–108.
Jefferies, R. P. S., 1975. Fossil evidence concerning the origin of the chordates. *Symp. zool. Soc. Lond.*, **36**, 253–318.
Mackinnon, D. L. and R. S. J. Hawes, 1961. *An Introduction to the Study of Protozoa.* Clarendon Press, Oxford.
Marcus, E., 1958. On the evolution of the animal phyla. *Q. Rev. Biol.*, **33**, 24–58.
Needham, A. E., 1964. *The Growth Process in Animals.* Pitman, London.
Pantin, C. F. A., 1960. Diploblastic animals. *Proc. Linn. Soc. Lond.*, **171**, 1–14.
Rees, W. J., 1966. The evolution of the hydrozoa. *Symp. zool. Soc. Lond.*, **16**, 199–222.
Remane, A., 1963. The evolution of the Metazoa from colonial flagellates *vs.* plasmodial ciliates. In *The Lower Metazoa, Comparative Biology and Phylogeny* (Dougherty, E. C. et al., eds.), 78–90. University of California Press, Berkeley.
Russell, E. S., 1930. *The Interpretation of Development and Heredity: A Study in Biological Method.* Clarendon Press, Oxford.
Sonneborn, T. M., 1957. Breeding systems, reproductive methods, and species problems in Protozoa. In *The Species Problem* (E. Mayer, ed.), 155–324. A.A.A.S. Pub., Washington, D.C.
Wilson, E. B., 1892. Cell lineage of Nereis. *J. Morph.*, **6**, 361–480.
Waddington, C. H., 1956. *Principles of Embryology.* Allen and Unwin, London.
Willmer, E. N., 1975. The possible contribution of the nemertines to the problem of the phylogeny of the protochordates. *Symp. zool. Soc. Lond.*, **36**, 319–345.

CHAPTER 19

Crofts, D. R., 1955. Muscle morphogenesis in primitive molluscs and its relation to torsion. *Proc. zool. Soc. Lond.*, **125**, 711–750.
Fell, H. B., 1948. Echinoderm embryology and the origin of chordates. *Biol. Rev.*, **23**, 81–107.
Garstang, W., 1928. Origin and evolution of larval forms. *Rep. Br. Ass. Advmt. Sci.*, section D, p. 77.
Garstang, W., 1929. The morphology of the Tunicata, and its bearings on the phylogeny of the chordata. *Q. Jl microsc. Sci.*, **72**, 51–187.
Garstang, W., 1951. *Larval Forms, and Other Zoological Verses.* Blackwell, Oxford.
Ghiselin, M. T., 1966. The adaptive significance of gastropod torsion. *Evolution*, **20**, 337–348.
Gurney, R., 1942. *Larvae of Decapod Crustacea.* Ray Society, London.
Hardy, Sir Alister, 1956. *The Open Sea: its Natural History.* **1**: *The World of Plankton.* Collins, London.
Knight-Jones, E. W. and D. J. Crisp, 1953. Gregariousness in barnacles in relation to the fouling of ships and to anti-fouling research. *Nature, Lond.*, **171**, 1109–1110.
Smith, J. E., 1953. The maintenance and spread of sea-shore faunas. *Advmt. Sci., London.*, **10**, 145–156.
Thorsen, G., 1946. *Reproduction and Larval Development of Danish Marine Bottom Invertebrates.* C. A. Reitzels Forlag, Copenhagen.
Thorsen, G., 1950. Reproduction and larval ecology of marine bottom invertebrates. *Biol. Rev.*, **25**, 1–45.

Wilson, D. P., 1932. On the mitraria larva of *Owenia fusiformis* Delle Chiaje. *Phil. Trans. R. Soc. B*, **221**, 231–334.

CHAPTER 20

Carpenter, K. E., 1928. *Life in Inland Waters*. Sidgwick and Jackson, London.
Knight-Jones, E. W., 1953. Laboratory experiments on gregariousness during settling in *Balanus balanoides* and other barnacles. *J. exp. Biol.*, **30**, 584–598.
Knight-Jones, E. W. and J. Moyse, 1961. Intraspecific competition in sedentary marine animals. *Symp. Soc. exp. Biol.*, **15**, 72–95.
Lucas, C. E., 1961. On the significance of external metabolites in ecology. *ibid.*, 190–206.
Mather, K., 1961. Competition and cooperation. *ibid.*, 264–281.
Meadows, P. S. and J. I. Campbell, 1972. Habitat selection by aquatic invertebrates. *Adv. mar. Biol.*, **11**, 1–56.
Needham, J., 1930. On the penetration of marine organisms into fresh water. *Biol. Zbl.*, **50**, 504–509.
Steams, S. C., 1976. Life-history tactics: a review of the ideas. *Q. Rev. Biol.*, **51**, 3–47.
Wilson, D. P., 1948a. The larval development of *Ophelia bicornis* Savigny. *J. mar. biol. Ass. U.K.*, **27**, 540–553.
Wilson, D. P., 1948b. The relation of the substratum to the metamorphosis of *Ophelia* larvae. *ibid.*, **27**, 723–760.
Wilson, D. P., 1952. On the influence of the nature of the substratum on the metamorphosis of the larvae of marine animals, especially the larvae of *Ophelia bicornis* Savigny. *Annls. Inst. océanogr., Monaco*, **27**, 49–156.

CHAPTER 21

Berrill, N. J., 1935. Studies in Tunicate development. Part IV: Asexual reproduction. *Phil. Trans. R. Soc. B*, **225**, 327–379.
Braverman, M. H. and R. G. Schrandt, 1966. Colony development of a polymorphic hydroid as a problem in pattern formation. In Rees, 169–198.
Butler, C. G., (1974, revised). *The World of the Honeybee*. Collins, London.
Butler, C. G., R. K. Callow, and N. C. Johnston, 1962. The isolation and synthesis of queen substance, 9-oxydec-*trans*-2-enoic acid, a honeybee pheromone. *Proc. R. Soc. B*, **155**, 417–432.
Child, C. M., 1941. *Patterns and Problems of Development*. University of Chicago Press.
Harris, W. V., 1961. *Termites: Their Recognition and Control*. Longmans, London.
Lindauer, M., 1961. *Communication among Social Bees*. Harvard University Press, Cambridge, Mass.
Richards, O. W., 1953. *The Social Insects*. Macdonald, London.
Robertson, P. L., 1971. Pheromones involved in aggressive behaviour in the ant, *Myrmecia gulosa*. *J. Ins. Physiol.*, **17**, 691–715.
von Frisch, K., 1950. *Bees—Their Vision, Chemical Senses, and Language*. Cornell University Press, Ithaca, New York.
von Frisch, K., 1954. *The Dancing Bees*. Collins, London.
von Frisch, K., 1962. Dialectics in the language of bees. *Scient. Am.*, **207** (2), 79–87.
von Frisch, K., 1967. *The Dance Language and Orientation of Bees*. Harvard University Press, Cambridge, Mass.
Wilson, E. O. and F. E. Regnier, 1971. The evolution of the alarm-defense system in the formicine ants. *Am. Nat.*, **105**, 279–289.

CHAPTER 22

Baer, J. G., 1951. *Ecology of Animal Parasites*. University of Illinois Press, Urbana.
Baerends, G. P., 1950. Specializations in organs and movements with a releasing function. *Symp. Soc. exp. Biol.*, **4**, 337–360.
Caullery, M., 1952. *Parasitism and Symbiosis*. Sidgwick and Jackson, London.
Dales, R. P., 1957. Commensalism. *Mem. geol. Soc. Am.*, **67**, vol. 1, 391–412.
Davenport, D., 1955. Specificity and behaviour in symbioses. *Q. Rev. Biol.*, **30**, 29–46.
Faurot, L., 1910. Étude sur les associations entre les Pagures et les Actinies: *Eupagurus prideauxi* Heller et *Adamsia palliata* Forbes, *Pagurus striatus* Latreille et *Sagartia parasitica* Gosse. *Arch. Zool. exp. gén.*, Sér. 5, **5**, 421–486.
Faurot, L., 1932. Actinies et Pagures. Études de psychologie animale. *ibid.*, **74**, 139–154.
Gosse, P. H., 1857. On a new form of corynoid polypes. *Trans. Linn. Soc. Lond.*, **22**, 113–116.
Gosse, P. H., 1860. *A History of the British Sea-anemones and Corals*. Van Voorst, London.
Hand, C. and J. R. Hendrickson, 1950. A two-tentacled, commensal hydroid from California (Limnomedusae, Proboscidactyla). *Biol. Bull. mar. biol. Lab., Woods Hole*, **99**, 74–93.
Henry, S. M. (ed.), 1966. *Symbiosis*, vol. 1. Academic Press, New York.
Jennings, D. H. and D. L. Lee (eds.), 1975. *Symbiosis. Symp. Soc. exp. Biol.*, **29**.
Lees, A. D., 1948. The sensory physiology of the sheep tick, *Ixodes ricinus* L. *J. exp. Biol.*, **25**, 145–207.
Ross, D. M., 1967. Behavioural and ecological relationships between sea anemones and other invertebrates. *Ocean. Mar. Biol. Ann. Rev.*, **11**, 176–222.
Ross, D. M. and L. Sutton, 1961. The response of the sea-anemone *Calliactis parasitica* to shells of the hermit crab, *Pagurus bernhardus. Proc. R. Soc. B*, **155**, 266–281.
Starr, M. P., 1975. A generalized scheme for classifying organismic associations. *Symp. Soc. exp. Biol.*, **29**, 1–20.
Welsh, J. H., 1930. Reversal of phototropism in a parasitic water mite. *Biol. Bull. mar. biol. Lab., Woods Hole*, **61**, 165–169.

CHAPTER 23

Barnett, A. J. G. and R. L. Reid, 1961. *Reactions in the Rumen*. Arnold, London.
Cleveland, L. R., 1926. Symbiosis among animals with special reference to termites and their intestinal flagellates. *Q. Rev. Biol.*, **1**, 51–60.
Darwin, C., 1851. *The Structure and Distribution of Coral Reefs* (2nd ed.). Smith Elder, London.
Goodey, T., 1951. *Soil and Fresh-water Nematodes*. Wiley, London.
Hopkins, S. H., 1957. Parasitism. *Mem. geol. Soc. Am.*, **67**, vol. 1, 413–428.
Keeble, F., 1910. *Plant-Animals, A Study in Symbiosis*. University of Cambridge Press, London.
Koestler, A., 1959. *The Sleepwalkers*. Hutchinson, London.
Llewellyn, J., 1965. The evolution of parasitic platyhelminths. In *Evolution of Parasites* (Third Symposium of the British Society for Parasitology), 47–78. Blackwell, Oxford.
Llewellyn, J., 1972. Behaviour of monogeneans. In *Behavioural Aspects of Parasite Transmission* (E. U. Canning and C. A. Wright, eds.). *Symposium 1, Zool. Jl. Linn. Soc. London*, **51**, 19–30.

Nelson, G. S., 1972. Human behaviour in the transmission of parasite diseases. *ibid.*, 102–122.

Noble, E. R. and G. A. Noble, 1964. *Parasitology: The Biology of Animal Parasites* (2nd ed.). Lea and Febiger, Philadelphia.

Smith, D. C., 1973. *Symbiosis of Algae with Invertebrates.* Oxford University Press, London.

Smyth, J. D., 1962. *Introduction to Animal Parasitology.* English Universities Press, London.

Smyth, J. D., 1966. *The Physiology of Trematodes.* Oliver and Boyd, Edinburgh.

Smyth, J. D., 1969. *The Physiology of Cestodes.* Oliver and Boyd, Edinburgh.

Wells, J. W., 1957. Coral reefs. *Mem. geol. Soc. Am.*, 67, vol. 1, 609–631.

Whitfield, P. J., 1977. *The Biology of Parasitism.* Edward Arnold, London.

Yonge, C. M., 1936. Mode of life, feeding, digestion, and symbiosis with xooxanthellae in the Tridacnidae. *Gt. Barrier Reef Exped. Sci. Rep.*, **1**, 283–321.

Yonge, C. M., 1944. Experimental analysis of the association between invertebrates and unicellular algae. *Biol. Rev.*, **19**, 68–80.

Yonge, C. M., 1951. The form of coral reefs. *Endeavour*, **10**, 136–144.

Yonge, C. M., 1958a. Darwin and coral reefs. In *A Century of Darwin* (Barnett, S. A., ed.), 245–266. Heinemann, London.

Yonge, C. M., 1958b. Ecology and physiology of reef-building corals. In *Perspectives in Marine Biology* (Buzzati-Traverso, A.A., ed.), 117–135. University of California Press, Berkeley.

Yonge, C. M., 1963. The biology of coral reefs. *Adv. mar. Biol.*, **1**, 209–260.

Yonge, C. M., 1975. Giant clams. *Scient. Am.*, **232** (4), 96–105.

Index

Acanthaster, 124
Acanthobdella, 170
Acanthocephala, pseudocoel, 108; nephromixium, 336
acetylcholine, 492
Achatina, 361
Acoela, 93, 253
actin, 42
Actinomonas, 56
actinopharynx, 71
Actinophrys, 59
active transport, in mammalian nephron, 356; in invertebrates, 358; in insects, 386; of water, 387
active uptake, in Pogonophora, 230
adaptation, in receptors, 394, 423
adenosine diphosphate, 13, 19, 39, 44
adenosine triphosphate, 13, 19, 39, 44, 47, 54
ADP, see adenosine diphosphate
adrenaline, 493
Aedes, anal papillae of larvae, 386
aerobic life, 15, 39
Aeschna, 211, 327
Alcyonidium, larval settling, 633
algae, 29, 217, 233; symbiotic in sponges and coelenterates, 691; pre-Cambrian, 65
allelochemicals, 539
Allolobophora, 159, 301
alternation of generations, 68; in in protozoans, 545, 709

Agriolimax, movement, 137
Alpheus and *Cryptocentrus*, 674
Amathes, reproductive isolation in, 538
amitosis, in ciliates, 30
Ammophila, learning and orientation, 468
Amoeba, 38, 58, 63; digestion, 226; life history, 545; osmotic relationships, 364; response to light, 402
amoeboid movement, 49, 58
y-aminobutyric acid, 494
ammonotely, 343, 349
amphiblastula larva, 561
amphioxus, digestion, 280; larva, 614; protonephrida, 331
anaerobic life, 14, 33
anal papillae, 387
analogy, 27
Anaspides, 318
androgenic gland, 505
Anemonia, cnidoblasts, 245; zooxanthellae, 704
angle of supply, in sponges, 241
animal clocks, 305, 472, 657
Annelida, metamerism, 153; brain, 454; endocrinology, 516; giant nerve fibres, 474; learning 455; nephridal system, 333; neural control of locomotion, 449; pre-Cambrian, 65
Anodonta, ctenidia, 266; osmoregulation, 371
Anopla, 255

753

ants, social life, 662
Antedon, tube-fact, 128; ciliary feeding, 276
antennal gland, 337, 374
Antherea and red oak, 539
Anthozoa, feeding 252; functional morphology, 69
Aphelenchoides, movement, 105
aphodal sponge, 240
aphrodisiac, 538
Apicomplexa, 224
Apis, wing movement, 212; colonial life, 655; evolution of social habits, 661; sensory capacity, 657; merry song, 659
Aplysia, movement, 140; digestion, 274; osmotic relationships, 369
appetitive behaviour, 472
apposition eye, 404
Arachnida, coxal glands, 337; feeding, 231; haemocyanin, 300; malpighian tubules, 340
archaeocyte, 90
archenteron, 108
archiannelida, 448
Archidoris, giant neurones, 475
Arenicola, burrowing, 166; feeding, 257; gills, 305; haemoglobin, 303; irrigation movements, 305
Architeucthis, 149
Arctonoë, commensal complex, 680
Arenicola, 81, 143, 236; digestion, 264; digging cycles, 159, 166; feeding, 257; gills, 305; haemoglobin, 302; hydrostatic adaptations, 168; irrigation movements, 305; mixonephridia, 335; osmotic regulation, 369; spawning, 525; spontaneous rhythms, 450, 472
arginine phosphate, 44
Argonauta, 147
Arion, respiration, 312
Aristotle's lantern, 119, 236
Artemia, nervous system, 448
arthropodin, 177
arthropodization, 172
Arthropoda, 153; evolution of neurosecretion, 535; nervous system, 460; segmentation of head, 183; success of, 172; types of mandible, 184; phylogeny, 186
Ascaris, hydrostatics, 102
ascidian tadpole, 614
asconoid sponge, 88, 238
Astacus, 207; antennal gland and osmoregulation, 377

Asterias, tube-feet, 124; behaviour, 436; nervous system, 433
Asteroidea, 119
Asterozoa, 119
Astropecten, feeling and movement, 122; chlorocruorohaem, 299
asynchronous flight mechanism, 215
atoll, 698
ATP, see adenosine triphosphate
Atta, efficient farming, 663
Aurelia, 82, feeding, 251; ionic composition, 353; osmotic tolerance, 369
auricularia, 608
autogamy, 549
Autolytus, 552
autotomy, 551
avicularia, 646
axial gradients, 644
axial sinus, 120
axiom, 33
axocoel, 117, 120
axon, 389
axopodia, 48
Aysheaia, 198

bacteria, 16, 21, 29, 51; and mitochondria, 39; and excretion in leeches, 344
bacteroids, 684
barnacles, gregarious habit, 623; larval settling, 629
barrier reef, 698
basal granules, 52
Beroë, feeding response, 86
biogenetic law, 560
bipinnaria, 607
Birgus, terrestrial adaptation, 382
Bivalvia, adductor muscles, 481; feeding 264; foot and movement 141; nervous system, 481
blastocoel, 108
blue-green algae, 29
Bodo, 220
Bohr effect, 301
Boloceroides, 82
Bombus, 45
bombykol, 537
Bombyx, 537
botryoidal tissue, 345
brachiolaria, 607
Brachiopoda, 236, 307; haemerythrin in, 300
brackish water, decapods, 373
brain, in annelids, 454, 459; in crustaceans, 462; in octopus, 483

Branchiobdellidae, 170
Branchionecta, 194
Branchiopoda, filter feeding, 281; larval stages, 637
Bryozoa, 114
buoyancy, 354
byssus, 182

Calanus, feeding, 288
Calcarea, 90, 237, 561
Calliactis, responses, 423; and *Buccinum*, 425, 429; and *Pagurus*, 676
Calliphora, 510; photoreceptors, 408; puparium, 180
Cambarus, giant nerve fibres, 475
camouflage, 678
campaniform sensilla, 213, 397
Campodea, 207
canal systems of sponges, 92
Cancer, 376
Carchesium, myonemes, 47; colonial life, 640
Carcinus, 207; osmoregulation, 373
carridoid facies, 195, 292
Carinaria, 140
carotenoids, visual, 401
cell theory, history of, 31
cellulose, digestion of, 685; in termites, 686; in mammals, 698
central nervous system, 444
cephalization, 183, 444, 481
Cephalochordata, 109
Cephalopoda, digestion, 274; foot, 147; colour change, 145, 497; giant nerve fibres, 476; parallelism with fish, 146; nervous system, 482; postures, 145; respiratory mechanishm, 315
Cephalothrix, 304, 586
Ceratium, 220
Cerebratulus, 97, 99
Cestoda, 93; life cycles, 718; origin of, 717
Cestodaria, 93
Cestum (*Cestus*), 87
Chaetopterus, autotomy, 551; feeding, 263
Chaos, 60
Chelicerata, relationships, 186; head, 190; locomotion, 210
chemical transmitters, 492
chemoreceptor, 395, 399
Chilodonella, 222
chilopods, sclerotization, 178; locomotion, 202

Chininiaster, 130
Chironex, 245
Chironomus, anal papillae, 387; haemoglobin, 297, 299
Chitin, 176
chitons, gills, 311; shell, 134; nervous system, 480
chloragogenous tissue, 344
chlorophyll, 14
chlorocruorin, 299
choanocytes, 89, 238, 242
cholinesterase, 493
Chordates, 177
chordotonal sensilla, 397
Chorthippus, songs of, 398
Chromactivating hormone, 500
chromaffin reaction, 493
chromatophore, 497
chromatosomes, 497
chromosomes, 29
Chthalamus, 633
Chydoridae, 285
ciliary feeding, 236, 257
ciliates, 30; cellulose digestion by, 689; colonial, 640
Ciliophora, 221
cilium, movement, 48, 53; structure, 51
Ciona, light sense, 402
cirri, in bivalves, 267; in *Euplotes*, 58
citric acid cycle, see tricarboxylic acid cycle
Cladocera, life cycles, 637
cleaner shrimps, 674
cleavage, determinate, 86
cleidoic egg, 346
click (snap) mechanism, 216
Clione, swimming, 140
Cnidaria, 68, 84; chromosomes, 30; colonies, 643; excretion, 330; hermatypic forms, 698; life histories, 566; nervous system, 421
cnidoblasts, 243, 413
cnidotrichocysts, 223
coccidians, 546, 709
Codonosiga, collar, 56
coelenterate, original connotation, 84 See Cnidaria
coelom, 108; *Arenicola*, 166; functions of, 111; leeches, 170; molluscs, 133; oligochaetes, 156; origin of, 108; polychaetes, 160; reduction of, 137, 174
coelomoduct, 328; arthropods, 337; molluscs, 342

Coeloplana, 87
collencytes, 89
colonial life, 68, 639
colour change, 497
colour vision, 408, 657
combs, of ctenophores, 86
commensalism, 670, 682, in vorticellids, 671; in hydroids, 671
compound eye, 500; retinal pigment migration in, 500
conditioned reflex, 419
conjugation, in ciliates, 547
consummatory act, 472
contractile vacuole, in Protozoa, 364; in sponges, 367
contraction, 42
Convoluta, digestion, 257; symbiosis, 693
coral reefs, origin of, 699
corals, 698; and zooxanthellae, 701; calcium metabolism, 705
Corophium, adaptive behaviour, 416
corpus allatum, 512
corpus cardiacum, 509
coxal glands, 337
Cordylophora, feeding reaction, 248
Crassostraea, ventilation, 315
creatine phosphate, 44
Crepidula, crystalline style, 273; endostyle, 279
Crinozoa, 118
Cristatella, 114
Crustacea, relationships, 186; cephalization, 186; colour change, 497; larvae, 597; limbs, 192; moulting, 502; parasitic forms, 720; retinal pigment migration, 500; sinus gland, 498; terrestrial adaptation, 382
crustecdysone, see 20-hydroxyecdysone
Cryptocercus, ecdysone and symbionts, 539, 729; cellulose digestion, 687
crystalline style, 271
Ctenoplana, 87
Culex, anal papillae of larvae, 387
ctenidium, 236, 264, 308; evolution of, in bivalves, 315
Ctenolepisma, 685
cuticle, arthropods, 174
cuttlebone, and buoyancy, 148, 355
Cyanea, 83
cycles of nutrients, 16; in symbiosis, 691

cyclic AMP, as second messenger, 535; in aggregation of *Dictyostelium*, 536
Cycloporus, digestion, 254
cydippid larva, 86
cystid, 114
cytochromes, 20, 297
cytogamy, 549

Dalyellia, symbiosis, 693
Danaus, courtship behaviour, 539
dances of bees, 660
Daphnia, feeding, 285; capability of eye, 488
Darwin, view of life, 8; and struggle for existence, 21; and *Triatoma*, 713
death, a condition of life, 16
Decapoda, statocyst, 395; calcification, 178
Demospongia, 91
dendritic field, 389
deoxyhaemoglobin, 297
deoxyribose nucleic acid, 10, 29
dermal light sense, 402
Dermocystidium, and oysters, 708
Desor's larva, 594
detritus feeding, 23, 235
Deuterostomis, 109, 117, 237, 439, 584; feeding and digestion, 277; nervous system, 439
diapause, 514, 528
Diaptomus, filter feeding, 288
diatoms, aquatic pasturage, 233
Dictyostelium, aggregation of, 536
Didinium, feeding 222
diffusion, passive pore, 37; facilitated, 37
Digenea, 715; life cycles and host ecology, 717, 719, 722
digestion, modes of, 249; anthozoans, 252; bivalves, 271; coelenterates, 249; platyhelminths, 253
Dinoflagellida, 219, 233; zooxanthellae, 702
Dinophilus, protonephridia, 331
dipleurula, 584, 603
diploblastic structure, 70, 85, 93
diplodal sponge,
diplopods, calcification, 178; locomotion, 202
Dipylidium, 725
direct development, 587
ditaxy, 136
Dixippus, 386, 463

DNA, see deoxyribose nucleic acid
Donax, 113, 142; burrowing, 144
dorsal ciliated organ, 334
dorsal sac, 120
Dufour's gland, 664
dynein, 55

earthworms, action potentials, 450; chemoreceptors, 399; diapause, 528; hydrostatics, 158; neurosecretion, 529; water relationships, 344
β-ecdysone, see 20-hydroxyecdysone
Echinocardium, 13
Echinococcus, 725
Echinodermata, coelom, 109, 117; ciliary feeding, 276; larvae, 603; nervous system, 431; regulation of spawning, 534; skeleton, 119; phylogeny, 117, 611
Echinoidea, 119
Echinozoa, 119
echinopluteus, 605
Echinus, function of madreporite, 126
echiuroids, 236
ecosystems, 20, 621, 669; chemical communication in, 541
ectoderm, 70, 93
ectomesoderm, 93
ectoneural system, echinoderms, 435
eelgrass, see *Zostera*
Eimeria, 224, 711
electromagnetic receptors, 395
elementary granules, in neurosecretion, 496
Eleutherozoa, 118
Elliobiophyra, 671
Elminius, 629
Elphidum, (*Polystomella*), 49, 545
Emerita, terrestrial adaptations, 382
Enchytraeus, metanephridia, 336
endocrine gland, 496
endocuticle, 174
endoderm, 70, 93
endomesoderm, 93
endoplasmic reticulum, 40
energy flow, 22
endostyle, protochordates, 277; *Crepidula*, 279
enteronephridia, 380
Enteropneusta, behaviour, 440; feeding, 277; giant neurones, 442; nervous system, 439

Entoprocta, 108, 236, 307; protonephridia, 330
Enopla, 255
enterocoel, 109
Enteromyxa, 233
Entoconcha, 727
enzymes, digestive, 226
ephippium, 637
Ephydatia, contractile vacuole, 367; gemmules, 565
ephyra larva, 566; nervous system and behaviour, 427
epicuticle, 174
epidermis, 70, 86
epigenesis theory, 555
Epistylis, absence of spasmonemes, 47; colonial life, 640
epitheliomuscular cells, 70
epitoky, 519, 553
Eriocheir, 374; osmoregulation, 374, 376
ergostaplasm, 40
errant polychaetes, 165, 257
Escherichia coli, 29
eserine, 493
eucaryotes, 29
Euchlora, nematocysts, 85
Euglena, pellicle, 38; movement, 53; contractile vacuole, 365; nutrition, 217
Euglenida, and evolution of heterotrophy, 219
Eulalia, neurosecretory system, 517; regeneration, 522
Euplotes, motorium, 58
euryhaline animals, 369
Eurypterida, 319
eurypylous sponge, 240
Euscorpius, haemocyanin, 300
Euspongia, form and feeding, 241
evolution, 2; chemical, 7, 28, 217; comparative biochemistry and, 347
excitatory post-synaptic potential (EPSP), 418
excretion, nature of, 328
exoskeleton, arthropods, 172; molluscs, 134
experimental design, problems of, 421, 457
exteroceptor, 395, 450
eyes, simple, 402; compound, 403

facile answers, suspect nature of, 729
facilitation, 418

Fasciola, egg capsules, 182
fermentation, 18
fibrillar muscles, insects, 214
Filograna, 623
filter feeding, 235; curstaceans, 192; 281 deuterostomia, 275; molluscs, 264; polychaetes, 256; sponges, 237
filtration/resorption mechanism, 363
flagellates, cellulose digestion, 687; colonial, 639
flagellum, movement, 48, 53; structure, 51
flight, 210; neural influence in insects, 463
fluid feeding, 231
food chain, 21, 233
food resources, 229
food webs, 22
Foraminiferida, 58
Forficula, 207
fringing reef, 698
Fungia, 704

Gammarus, antennal gland, 377
gastrodermis, 70, 86
Gastrodes, 86
Gastropoda, feeding, 273; foot, 136; torsion, 311. 596; nitrogen excretion, 346
Gastrotricha, pseudocoel, 108
gastrovascular canals, in ctenophores, 86
Gecarcinus, terrestrial adaptations, 382
gemmules, in sponges, 564
genetic code, 10
geodesic fibre systems, 97, 102
Geonemertes, movement, 97; form, 99; protonephridia, 333
Geophilus, movement and hydrostatics, 208
giant clam, see *Tridacna*.
giant nerve fibres, 149; *Archidoris*, 475; cephalopods, 476; distribution of, 474; function, 474
gills, crustaceans, 318; molluscs, 308; pallial, 313; polychaetes, 305; spiracular, 326; tracheal, 327
gill books, 318
Glossiphonia, 345
glutamic acid, 494
glutathione, in feeding response of *Hydra*, 247
glycolysis, 18

Golgi bodies, 40
Gonactinea, movements, 81
Graptolebris, feeding habits, 286
gregarines, 709; life cycles, 546
guanine, excretion by arachnids, 350
Gymnolaemata, 114
Gyrodinium, feeding, 219

habitat, 25; selection by larvae, 625
habituation, 419; annelids, 455; arthropods, 464
haemerythrin, 300
haemocoel, 174
haemocyanin, 300, 316
haemoglobin, 296, 316; distribution, 297; oxygen dissociation curves, 301
Haemogregarina, 711
Haemonchus, 103, 105
hair sensilla, 397
Halammohydra, statocysts in, 570
Haliotis, 274, 342, 361
halteres, 213, 396
Harmothoë. commensal relationships, 681
heads, 182
Heliozoia, 48, 59
Helix, movement, 138; digestion, 274
Hemichordata, 109, 117; development, 612
hemimetabolous life cycle, 618
Hemimysis, feeding, 290
Heterodera, 105, chitin in cysts, 182
heteronereid, 519
heterotropy, primary, 14; secondary, 15
Hexactinellida, 91, 235, 242
higher and lower animals, 3
Hirudinea, 170; endocrinology, 527; excretion, 344; movements, 451; locomotion, 170
Hirudo, excretion, 345
holistic approach, 420
holometabolous life cycle, 618
Holothuria, light sense, 402
homology, 27; latent, 28
Homalozoa, 118
honey bee, see *Apis*
hormones, 419, 496; chromactivating, 500; diuresis, 515; growth regulation, polychaetes, 517; mode of action, 535; moulting in crustaceans, 502; moulting in

insects, 509; reproduction in polychaetes, 519; retinal pigment migration, 500
Hutchinsoniella, 193
Hyalophora (Platysamia), 617, dispause, 514
Hyas, an osmo-conformer, 368
Hydra, strictive, 70; relationships, 569; digestion, 250; feeding 246; freshwater life, 367; nematocycts, 243; neurosecretion, 534, 645; symbiosis, 691
Hydractinia, 643; and *Pagurus*, 674
Hydrobia, 347
hydrocoel, 117, 120
9-hydroxydecenoic acid, 656
20-hydroxyecdysone, 503, 510; and symbiosis, 539
5-hydroxytryptamine (5-HT), 494
Hydrozoa, 68
Hymenoptera, social life, 661
hyponeural system, echinoderms, 435
hypothesis, 33; testing to destruction, 5

imprisoned phytoplankton, 703
independent effector, 413
indirect development, 587
information theory, inadequacy of, 420
inquilinism, 670
insecticides, 493
insects, aquatic respiration, 326; diuretic hormone, 515; flight, 210; fluid feeding. 232; life histories, 617; locomotion, 208; moulting and metamorphosis, 508; mycetocytes, 684; neural control of movements, 463; social life, 654; tracheae, 320; and terrestrial life, 384
insight learning, 419
instinctive behaviour, 414; in arthropods, 471
integration in behaviour, 416
interneurones, 444
interoceptor, 395
inulin and kidney activity, 357
ionic regulation, 352, 358, 368
isopods, excretion of ammonia gas, 349; terrestrial adaptations, 383
Isoptera, social life, 665; digestion of cellulose, 686
isosmotic intracellular regulation, 369

jet propulsion, cephalopods, 146; giant nerve fibres and, 476; in scallops, 146
juvenile hormone, 512; in adults, 516

keratin, 179
kinetodesmata, 47, 57
Kinorhyncha, pseudocoel, 108
Kronbergia, parasitism, 714

labial palps, 264, 268
lamellar muscles, 214
lamellibranchs, see Bivalvia
Lar, see *Proboscidactyla*
larva, in insects, 618; types and significance of, 587
Larvacea, 233
larvae and freshwater life, 635
larval behaviour, 599, 625
lazy tongs, 97
learning, 419; types of, 419; in arthropods, 464; in headless cockroach, 464; in intact cockroach, 466; in honey bee, 657
lecithotrophic larva, 587
leeches, see Hirudinea
leghaemoglobin, 298
Lepidocaris, 193, 281
Lepidochiton, adaptive behaviour, 414
Leptosynapta, 120
leuconid sponge, 240
light compass orientation, 657
Ligia, 207; movement, 197; osmotic relationships, 383
Limax, mucus and movement, 139
Limnodrilus, 25
Limnoria, 685
Limulus, feeding, 191; eyes, 404; gill books, 318; haemocyanin, 300
Lineus, 96; form and movement, 99; digestion, 255; life history, 586
Lingula, 300
Lithobius, movement, 204
Littorina, 587; uric acid excretion, 347
Loa loa, periodicity and transmission, 722
lobopodia, 199
Locusts, ootheca, 180
Loligo, 148, 149; giant nerve fibres, 476; Magnesium level, 353

Lophomonas, in cockroach, 220
lophophore, 114, 300, 307
Loxosoma, 115
Lucilia, 348
lunar periodicty, 413, 526, 536
Luidia, 299
Lumbricus, movement, 159; regeneration, 527; oxygen dissociation curves, 301
See also earthworms
lung, pulmonates, 312; decapods, 319
lung books, 320
Lymnaea, lung, 313; uric acid, 346; as intermediate host, 717, 719; neurosecretory cells, 531
lysosomes, 40

Macoma, excretion of ammonia, 370
macrophagy, 236
Macropodia, 207
madreporic vesicle, 120
madreporite, 121, 124, 127
Macrostella, escape to the plankton, 290
Magelona, 300
malaria parasite, see *Plasmodium*
Malpighian tubules, 202, 339, 385
nephron, mammalian, compared with invertebrate organs, 363
mankind, regrettable habits, 723
Mantis, copulatory behaviour, 472
mantle, molluscs, 308
Mastigophorea, 49, 217
mechanoreceptors, 395; insects, 397
medusa, 68; movement, 82; history of, 567; reduction of, 569
megalopa larva, 601
meiosis, in life cycles, 542
Melia, and anemones, 678
membranes of cell, 34
Membranipora, 114, 369
mesentery, in Anthozoa, 72
Mesozoa, 574
metanephridium, 329, 333; and terrestrial life, 379
Metazoa, origin of, 225, 571
Metridium, functional morphology, 71; through-conduction tracts, 424
mesoderm, 93
mesogloea, 70, 86, 90
metagenesis, 68
metamerism, annelids, 152; arthropods, 173; characteristics of, 153; origin of, 154

metamorphosis, echinoderms, 605; insects, 618; crustaceans, 598; polychaetes, 589; sponges, 563; ascidians, 614
metaneophromixium, 335
methaemoglobin, 297
microphagy, 114, 117, 235, 270, 275
Microspora, 224
mind, 420
miroestrol, 113
mitochondria, 39, 45
mitosis, 30
mitraria larva, 591
mixonephridium, 335
Monogenea, 714; host location, 721
monotaxy, 137
mosaic development, 86
mosaic egg, 556, 575
motorium, 57
moulting, 175; regulation of, in crustaceans, 502, in insects, 508
Mollusca, coelomoducts, 340; ctenidia, 264; evolution of nervous system, 480 exoskeleton, 134; foot, 136; ground plan, 133, 308; neurosecretory cells, 530; stomach, 271
Monas, feeding, 56
Monogenea, 93
mucus, in locomotion, 25, 95, 97
Müller's larva, 592
multiple-hormone hypothesis, 499
Murex, 274
mutation, 2, 17
mutualism, 425, 670
Mya, use of syphons, 144; salinity relationships, 369
mycetocytes, 684
Myrianida, stolonization, 552
myocytes, 89
myofibrils, 42, 45
myofilaments, 42
myosin, 42
Myriapoda, relationships, 202; head, 189; locomotion, 202; Malpighian tubules, 345
Mysidacea, 395
Mysis, 188
Mytilus, 266; ventilation rate, 315; osmotic stress, 370
Myxospora, 224

Naegleria, phase change, 49
natural selection 2, 622
nauplius, 597

Nautilus, locomotion, 148; respiration, 316
Nebalia, feeding, 290
nectochaetous larva, 589
nematocysts, 243; action of, 245
Nematoda, 48, 108; hydrostatics, 101; one model? 106; feeding, 231
Nematomorpha, pseudocoel, 108
Nemertinea, 96; proboscis, 96; development, 592; protonephridia, 330
Neopilina, 134, 308, 340
nephridium, 328, 344; distribution, 336
nephrostome, 329
Nephtys, burrowing, 165; feeding, 256; haemoglobin, 304; gills, 305; infracerebral gland, 517
Neptune's cup, 241
nereids, nervous system and locomotion, 448; neurosecretory regulation, 516.
Nereis, propulsion, 56; larva, 589; endocrinology, 517; feeding, 256; locomotion, 160; osmotic relationships, 371
nerve net, 70
nerve impluse, 389
nerve net, *Hydra*, 70; coelenterates, 421; *Aurelia*, 427; enteropneusts, 439; reduction of, 444
neurocrine activity, 496
neurohaemal organ, 391, 497
neurohormone, 389, 494
neurohumours, 389, 492
neuroid transmission, 389
neurone (neuron), 389, 493
neurosecretion, 494; in lower invertebrates, 534
neurosecretory cell, 494
Neurospora, mutant strains, 17
niche, 25
nitrogen excretion, 342; flexibility of, 348
Noctiluca, 220, 543
noradrenaline, 493
nucleolus, 29
nucleus, 29
Nucula, 315; use of foot, 141
Nudibranchia, 312
nymph, 509, 618

Obelia, 79, 82
Ockham's razor, 62
Octopus, 148; benthic life, 151; ultra filtration, 361; digestion, 274; eye, 409; learning, 483; reproduction, 529
oligochaetes, locomotion and hydrostatics, 157; endocrinology, 527
olynthus, 88
ommatidium, 404, 411
Onchidella, uric acid excretion, 347
oncomiracidium, host location, 721
ophiopluteus, 605
Onychophora, cuticle, 174; development, 580; excretion, 345; eye, 404; mandibles, 184. See also *Peripatus*
Opalina, ciliary beat, 58; relationships, 221
open systems, 1
Ophabinia, greeted with laughter, 200
Ophiuroidea, 119
Opisthobranchia, pelagic life, 140; detorsion, 312
optic gland, *Octopus*, 529
Orchestia, female sex hormone, 505
Orconectes, antennal gland functioning, 377
organism, definitions of, 34
organs, 93
orthokinesis, 414
osmo-conformers, 368, 371
osmoregulation, 371
osmotrophy, 16, 217, 231
Ostrea, feeding, 267
Otohydra, statocysts, 570
Owenia, cataclysmic metamorphosis, 592
oxygen, availability, 293; diffusion, 294; utilization in molluscs, 315
oxygen dissociation (equilibrium) curve, 301
oxyhaemoglobin, 297
9-oxodecenoic acid, 656

Pagurus, and *Hydractinia*, 674
Palaemon 207; statocyst, 396
Palaemonetes, eye pigment movements, 500
pallial complex, molluscs, 311
Palolo worms, 526
Pandalus, synthesis of hormone, 500; protandry, 508
Paramecium, surface, 38; motorium, 57; behaviour, 420; conjugation, autogamy and cytogamy, 547; digestion, 226; trichocysts, 222; food requirements, 224

parapodium, polychaetes, 160; opisthobranchs, 140; in respiration of polychaetes, 305
parasitic castration, 507
parasitism, 670; ecological aspects, 706; evolution in Protozoa, 709; evolution in Platyhelminths, 714; specificity, 706; periodicity and transmission, 722; physiological integration with host, 725; difficulty of definition, 726; copepods, 290
Parazoa, 85, 88
parenchyma, 93, 96
pars intercerebralis, 509
parthenogenesis, Cladocera, 637
particulate feeding, 235
Patella, feeding, 236, 274; trochophore larva, 595; movement, 136
Peachia, burrowing, 79
Pecten, feeding, 267
pedicellariae, 119
Pediculus, mycetome, 685
perihaemal system, 120
Pelagia, 83
Pelmatozoa, 118
Pelomyxa, weighted down, 61
Pennaria, axial gradients, 645
Pennella, parasitism on whale, 290
Pentastomida, 186
Peripatus, reduction of coelom, 134, 136; haemocoel, 174; coxal glands, 337; habits, 197; locomotion, 200; relationships, 198; tracheae, 320; city financier and, 4
Peranema, feeding, 219, 231
Perinereis, an osmo-conformer, 371
Periplaneta, mechanoreceptors, 397; diuretic hormone, 515; learning, 464, 466; neural control of walking, 463. See also cockroach
peristomium, *Nereis*, 183
Peritrichia, colonies, 640
peritrophic membrane, 345
Phakellia, water flow, 241
Phagocyta, cannibalism of Root Spring, 25
pharyngotremy, 277
Phascolosoma, haemerythrin, 300
Pheretima, urine, 344, 380
pheromones, 536; ants, 664; bees, 656; termites, 668
Philoscia, cellulose digestion by bacteria, 685
Pholas, 142
Phormia, action and myosin, 45; chemoreceptors, 399; feeding, 400, 416, 419; wing movement, 211
Phoronis, 300; lophophore, 307; pseudo-colony, 623
phosphagens, 44
photoreceptors, 401
phototrophs, 14, 232
Phylactolaemata, 144
phyllopodium, 192
phyllosoma larva, 603
physa, 79
Physalia, predation, 245; polymorphism, 571
Phytomastigophorea, 217
Phytomonadida, colonies, 639
phytoplankton, 233
pilidium larva, 592
pinacocytes, 89
pinocytosis, 37, 230
planktotrophic larva, 587
Planorbis, 297; respiration and haemoglobin, 314
Platasterias, 119; tube-feet, 130
platform reef, 698
Platyhelminthes, 93; development, 592; nervous system, 445; parasitism, 714; protonephridia, 330; shape 295; feeding and digestion, 253
Plasmodium, life cycle, 710
plastron respiration, 326
Pleurobrachia, effect on *Beroë*, 87
Plumatella, 114
Pogonophora, organization, 229; feeding, 230
poikilosmotic animals, 368
polarized light, discrimination of, 408, 657
Polian vesicles, 127
Polycelis, digestion, 253
polychaetes, growth and regeneration, 517; locomotion and hydrostatics, 160. See also *Nereis*, nereids, etc
polymorphism, coelenterate colonies, 68, 645; polyzoan colonies, 646
polypide, 114
Polyplacophora, 134
Polystoma, 714
Polystomella, see *Elphidium*
polyp, 68
Polyzoa (Ectoprocta), 236; hydrostatics, 114; Pophophore, 307; organization of colony, 646
Pomatoceros, feeding, 262

porocyte, 90, 237
porphyrins, 296
Portunion, dwarf males, 727
Potamon, freshwater adaptation, 376
potential, action, 391, 450; generator, 394; receptor, 391; resting, 352, 389
preformation theory, 555
prototrochophore larva, 589
Priapulida, 108; haemerythrin, 300
primitive characters, 4
Proboscidactyla, 671
proboscis, nemertines, 96
Procambarus, neural basis of limb movements, 460
procaryotes, 29
proprioceptors, 395, 450; *Octopus*, 487
Prorodon, cnidotrichocysts, 223
prosobranchs, 140
prostomium, *Nereis*, 183
Protochordata, pharyngotremy, 277; larvae, 614
Protolaelaps, 539; allelochemicals and, protonephridium, 90, 96, 329, 380, 382; distribution, 336; moluscan larvae, 340
protonephromixium, 334
Protostomia, 109, 133
Protozoa, nuclear division, 30; interpretation of, 34; mitochondria, 40; colonial life, 639; contractile vacuole, 364; phylogeny, 224; surface of, 37
protozoea larva, 599
pseudocoel, 102
pseudo-colonies, 623
pseudometamerism, 154
Pseudospora, feeding, 233
pseudotracheae, 319
psoas muscle, 42
Pterobranchia, nervous system, 442; feeding, 277
Pulmonata, lung, 312
pupa, 618
pupariation, in *Calliphora*, 510
puparium, 180
pupation, 510
Pycnogonida, 186

Queen butterfly, see *Danaus*
queen substance, 656

recapitulation theory, 560
radula, sclerotization, 182; use in feeding, 236, 264

receptors, stretch, 391
rectal glands, 386
reductionist approach, 420, 432
reflex arc, 413
regeneration, 551
regulation egg, 556
releaser, 414, 416, 472
replication, 2, 10
resilin, 177, 211
respiration, external, 293; gaseous exchange, 304
respiratory pigments, 296; storage and transport functions, 301
rhagon, 240
Rhodnius, cuticle, 175; moulting, 509; diuretic hormone, 516; and *Actinomyces*, 685
rhynchocoel, 96
Rhynchodemus, 95
rhythms, endogenous basis of, 473
ribonucleic acid, 11, 29
RNA, see ribonucleic acid
Root Spring, ecosystem, 24
Rotifera, pseudocoel, 108; protonephrida, 331

Sabella, feeding, 257
Sabellaria, pseudocolony, 623; habitat selection, 628
Sacculina, 727
sarcolemma, 42
Sarcodina, 221
Sarcomastigophora, 217
Sarcophaga, flight mechanism, 215
sarcoplasm, 42
sarcosomes, 45
Sarsia, budding, 570
Scala Naturae, 2
Scaphopoda, 307
Schellackia, life cycle, 711
Schistocerca, click mechanism, 216
Schistosoma, life cycle, 717; immune response, 726
schizocoel, 109
scleroblast, 90
sclerotin, 179
sclerotization, 178
Scolopendra, location, 204
scorpion, 207
Scrobicularia, siphons, 145
Scutigera, fleetest centipede, 205
scyphistoma, 426, 566
Scyphozoa, 69, 82
second messengers, 535
secretion, 41
sedentary polychaetes, 165, 257

segment, characteristics of, 153
sense cell, primary, 394; in enteropneusts, 440; secondary, 394
Sepia, 342 buoyancy, 148, 354; digestion, 274
Septibranchia, feeding, 273
sex attractants, 537
sexual differentiation, origin of, 543
sexual exploitation by orchids, 683
Sialis, larva and freshwater life, 387
Silver Springs, ecosystem, 22
sinus gland, 498
siphonoglyph, 71, 78
Sipunculus, burrowing, 112; haemerythrin, 300
size, 295, 313, 320, 323
skeleton, hydrostatic, 66; arthropods, 174; echinoderms, 120; molluscs, 136; coelenterates, 70; nematodes, 102; platyhelminths, 92
slaughter by ants, 665
sleeping sickness, see trypanosomiasis
social insects, bees, 419; ants, 662; termites, 665; wasps, 662
social life, evolution of hymenopteran, 661
sodium pump, 352
solenocyte, 331, 334
Somasteroidea, 119, 130
somatocoel, 117, 120
spasmoneme, 47
specialized characters, 4
Sphaeronellopsis, parasitism, 290
Sphinx, flight, 211
spicules, sponges, 90; echinoderms, 120
spiracles, 320; closing mechanisms, 324
spiral cleavage, 575, 592
spirocysts, 245
Spirorbis, respiratory pigments, 299; larval settling, 629
sponges, 46, 88, 225; colonies, 642; contractile vacuole, 367; digestion, 242; hydraulics, 238; reproduction, 560
Spongillidae, contractile vacuole, 367; and algae, 691; gemmules, 564
spongin, 90
spontaneous movement, *Metridium*, 73
spontaneous neural rhythms, annelids, 450

statocyst, crustaceans, 395; *Octopus*, 412, 486; ascidian larva, 614
stenohaline animals, 369
stenopodium, 192
Stentor, myonemes, 47; kinetodesma, 57; habituation, 420
stolonization, in polychaetes, 552; in urochordates, 651
stomoblastula, 561
Stomphia, swimming, 81, 431
stone canal, 121, 124, 127
strobilation, *Aurelia*, 426
struggle for existence, 620
suboesophageal ganglion, influence in annelids, 455
Subulamoeba, 58
success, 5
summation, 418
suspension feeding, 235, 257
superposition eye, 404
swans, not all necessarily white, 5
syconoid sponge, 238
Syllis, budding, 552; reproduction, 523, 526
symbiosis, 232, 670; chemical communication in, 680
symmetry, anthozoan, 71, 78; bilateral, 182
Symphyla, 190
synapse, 391, 418
synaptic vesicles, 492
syngamy, in life cycles, 543

tagmosis, 182
tanning, 179, 510
Tardigrada, 186
taxis, 416
Tellina, burrowing, 142
Temnocephalida, 673
Tenebrio, conservation of water, 387
Terebella, feeding, 257
Teredo, 142; use of cellulose, 685
termites, see Isoptera
terrestrial adaptations, crustaceans, 382
Tetrahymena, food requirements, 223
Thalassicola, central capsule, 182
Thermobia, active transport of water, 387
Theromyzon, 345
theory, 32
Thompsonia, budding, 728
thoracotropic hormone, 509
tornaria larva, 612
torsion, in gastropods, 311, 596

tracheae, 320
tracheal gills, 331, 387
transduction, 394
transport, active, 37; carrier-facilitated, 37
Trematoda, 93, 714
trial and error learning, 419
Triatoma, and Darwin, 713
tricarboxylic acid cycle, 17, 19, 39
trichocysts, 222
Trichonympha, 39; absence of mitochondria, 40
Tridacna, symbiosis, 695
Trilobita, relationships, 186
triploblastic structure, 93
trochophore larva, polychaetes, 333; Polygordius, 588; molluscs, 595; *Ophelia*, 625
tropomyosin, 42, 45
trypanosomiasis, 713, 726
Trypanomonads, nutrition, 220; polymorphism, 712
tube-feet, *Antedon*, 127; *Echinocardium*, 131; starfish, 121
Tubifex, metabolism, 303
Tubularia, anatomy, 69; actinula, 566, 569; axial gradients, 644
Turbatrix, locomotion, 105
Turbellaria, 93
tritocerebral commissure, 500

Uca, colour change, 498; moulting, 502
ultrafiltration, in antennal gland, 359; in mammalian nephron, 355; in molluscs, 359
undulatory propulsion, 103
Unionicola, and *Anodonta*, 680
Uniramia, 186
ureotely, 314
uricotely, 343, 345; in gastropods, 346
Urochordata, colonies, 647

Vampyrella, feeding, 249
veliger larva, 595
Venus, acetylcholine, 493
vertebrates, origin of, 363, 615
vibraculum, 646
vitamins, 16, 218, 223
vitellaria doliolaria, 608
Viviparus, uric acid, 347; ultrafiltration, 361
Vorticella, myonemes, 47

water-vascular system, 120

X-organ, 498
Xenopsylla, spiracles, 324

Y-organ, 502
Yungia, decerebration, 445

Zeitgeber, 473
Zoantharia, 245
zoea larva, 599
Zooecium, 114
Zoomastigophorea, 217
zooplankton, 233
Zoothamnium, myonemes, 47
zooxanthellae, 691; in corals, 701
Zoostera, and *Labyrinthula*, 709